D1072665

Handbook of Optical
Constants of Solids

Academic Press Handbook Series

EDWARD D. PALIK (ed.). Handbook of Optical Constants of Solids, 1985

In preparation
N. P. BANSAL AND R. H. DOREMUS. Handbook of Glass Properties
NORMAN G. EINSPRUCH (ed.). VLSI Handbook
K. NARAHARI RAO (ed.). Handbook of Infrared Standards
AKIRA YAMAGUCHI. NMR Spectra and Data Handbook

Handbook of Optical Constants of Solids

Edited by

EDWARD D. PALIK

Naval Research Laboratory
Washington, D.C.

1985

ACADEMIC PRESS, INC.

(Harcourt Brace Jovanovich, Publishers)

Orlando San Diego New York London
Toronto Montreal Sydney Tokyo

ACADEMIC PRESS, INC.
Orlando, Florida 32887

United Kingdom Edition published by
ACADEMIC PRESS INC. (LONDON) LTD.
24–28 Oval Road, London NW1 7DX

Library of Congress Cataloging in Publication Data
Main entry under title:

Handbook of optical constants of solids.

1. Solids––Optical properties––Handbooks, manuals,
etc. I. Palik, Edward D.
QC176.8.06H36 1985   530.4'1   84–15870
ISBN 0–12–544420–6 (alk. paper)

PRINTED IN THE UNITED STATES OF AMERICA

85 86 87 88   9 8 7 6 5 4 3 2 1

Before using the tables and figures of the critiques in Part II of this handbook, the reader should refer to the editor's introduction to these materials (Chapter 1, pp. 6–7).

Contents

List of Contributors   xv
Preface   xvii

Part I DETERMINATION OF OPTICAL CONSTANTS

Chapter 1 Introductory Remarks   3
EDWARD D. PALIK

    I.   Introduction   3
    II.   The Chapters   4
    III. The Critiques   5
    IV. The Tables   6
    V.   The Figures of the Tables   7
    VI. General Remarks   8
        References   9

Chapter 2 Basic Parameters for Measuring Optical Properties   11
ROY F. POTTER

    I.   Introduction   11
    II.   Intrinsic Material Parameters in Terms of Optical Constants   16
    III. Reflectance, Transmittance, and Absorptance
        of Layered Structures   18
    IV. The General Lamelliform — Phase Coherency Throughout   19
    V.   The General Lamelliform — Phase Incoherency in Substrate   21
    VI. Summary   24
Appendix A. Basic Formulas for Fresnel Coefficients   24
Appendix B. General Formulas for the Case of a Parallel-Sided Slab   25
Appendix C. Reflectance, R_{jk} at $j-k$ Interface   26
Appendix D. Reflectance of Single Layer on Each Side of a Slab and Single
        Layer on Either Side of a Slab   26
Appendix E. Critical Angle of Incidence   30
        Definition of Terms   33
        References   34

Chapter 3 Dispersion Theory, Sum Rules, and Their Application
to the Analysis of Optical Data   35
D. Y. SMITH

    I.   Introduction   36
    II.   Optical Sum Rules and Their Physical Interpretation   36

III.   Finite-Energy Sum Rules   45
IV.   Sum Rules for Reflection Spectroscopy   51
V.   Analysis of Optical Data and Sum-Rule Applications   55
VI.   Summary   64
    References   64

Chapter 4   Measurement of Optical Constants in the Vacuum Ultraviolet Spectral Region   69

W. R. HUNTER

I.   Introduction   69
II.   General Discussion of Reflectance Methods   70
III.   Reflectance Method for Two Media   85
    References   87

Chapter 5   The Accurate Determination of Optical Properties by Ellipsometry   89

D. E. ASPNES

I.   Reflection Techniques; Background and Overview   89
II.   Measurement Configurations   92
III.   Accurate Determination of Optical Properties: Overlayer Effects   96
IV.   Living with Overlayers   99
V.   Eliminating Overlayers   102
VI.   Bulk and Thin-Film Effects; Effective-Medium Theory   104
VII.   Conclusion   108
    References   110

Chapter 6   Interferometric Methods for the Determination of Thin-Film Parameters   113

JOSEPH SHAMIR

I.   Introduction   113
II.   Basic Principles   114
III.   Nonlaser Interferometers   117
IV.   Kösters-Prism Interferometers   123
V.   A Self-Calibrating Method   126
VI.   Surface Effects   131
VII.   Conclusions   132
    References   133

Chapter 7   Thin-Film Absorptance Measurements Using Laser Calorimetry   135

P. A. TEMPLE

I.   Introduction   135
II.   Single-Layer Films   138
III.   Wedged-Film Laser Calorimetry   139
IV.   Electric-Field Considerations in Laser Calorimetry   143
V.   Entrance versus Exit Surface Films   147

VI. Experimental Determination of α_f, a_{af}, and a_{fs} 149
     References   153

Chapter 8   Complex Index of Refraction Measurements
             at Near-Millimeter Wavelengths   155
             GEORGE J. SIMONIS

I.   Introduction   155
II.   Fourier Transform Spectroscopy   156
III.   Free-Space Resonant Cavity   161
IV.   Mach–Zehnder Interferometer   163
V.   Direct Birefringence Measurement   164
VI.   Overmoded Nonresonant Cavity   165
VII.   Crystal Quartz as Index Reference   165
VIII.   Conclusion   167
        References   167

Chapter 9   The Quantum Extension of the Drude–Zener Theory
             in Polar Semiconductors   169
             B. JENSEN

I.   Introduction   169
II.   Quantum Theory of Free-Carrier Absorption   172
III.   Theoretical Results   174
IV.   Comparison with Experimental Data   176
        Appendix   187
        References   188

Chapter 10   Interband Absorption — Mechanisms and Interpretation   189
             DAVID W. LYNCH

I.   Introduction   189
II.   One-Electron Model   190
III.   Electron–Hole Interaction, Excitons   198
IV.   Local Field Effects   203
V.   Examples   204
        References   210
        General References   211

Chapter 11   Optical Properties of Nonmetallic Solids
             for Photon Energies below the Fundamental Band Gap   213
             SHASHANKA S. MITRA

I.   Introduction   213
II.   Infrared Dispersion by Polar Crystals   215
III.   Kramers–Kronig Dispersion Relations   227
IV.   Determination of Absorption Coefficient
        in the Intermediate Region   229
V.   Absorption Coefficient in the Transparent Regime   230
VI.   Multiphonon Absorption   232

VII.   Infrared Absorption by Defects and Disorders   254
VIII. Infrared Dispersion by Plasmons   263
       References   267

Part II   CRITIQUES

Subpart 1   Metals

Comments on the Optical Constants of Metals and an Introduction to the Data for Several Metals   275
DAVID W. LYNCH AND W. R. HUNTER

I.   Introduction   275
II.   Anomalous Skin Effect   277
       References   279
III.   Copper (Cu)   280
       References   280
IV.   Gold (Au)   286
       References   287
V.   Iridium (Ir)   296
       References   296
VI.   Molybdenum (Mo)   303
       References   304
VII.   Nickel (Ni)   313
       References   314
VIII. Osmium (Os)   324
       References   324
IX.   Platinum (Pt)   333
       References   334
X.   Rhodium (Rh)   342
       References   342
XI.   Silver (Ag)   350
       References   351
XII.   Tungsten (W)   357
       References   358

The Optical Properties of Metallic Aluminum   369
D. Y. SMITH, E. SHILES, AND MITIO INOKUTI

I.   General Features   369
II.   Optical Measurements and Sample Conditions   372
III.   Tabulated Data   377
       References   383

Subpart 2   Semiconductors

Cadmium Telluride (CdTe)   409
EDWARD D. PALIK

References   413

Gallium Arsenide (GaAs)   429
EDWARD D. PALIK

References   432

Gallium Phosphide (GaP)   445
A. BORGHESI AND G. GUIZZETTI

References   449

Germanium (Ge)   465
ROY F. POTTER

References   469

Indium Arsenide (InAs)   479
EDWARD D. PALIK AND R. T. HOLM

References   481

Indium Antimonide (InSb)   491
R. T. HOLM

References   494

Indium Phosphide (InP)   503
O. J. GLEMBOCKI AND H. PILLER

References   506

Lead Selenide (PbSe)   517
G. BAUER AND H. KRENN

References   518

Lead Sulfide (PbS)   525
G. GUIZZETTI AND A. BORGHESI

References   528

Lead Telluride (PbTe)   535
G. BAUER AND H. KRENN

References   538

Silicon (Si)   547
DAVID F. EDWARDS

References   552

Silicon (Amorphous) (a-Si)   571
H. PILLER

References   573

Silicon Carbide (SiC)   587
W. J. CHOYKE AND EDWARD D. PALIK

References   589

Zinc Sulfide (ZnS)   597
EDWARD D. PALIK AND A. ADDAMIANO

References   602

Subpart 3   Insulators

Arsenic Selenide (As₂Se₃)   623
D. J. TREACY

References   625

Arsenic Sulfide (As₂S₃)   641
D. J. TREACY

References   644

Cubic Carbon (Diamond)   665
DAVID F. EDWARDS AND H. R. PHILIPP

References   668

Lithium Fluoride (LiF)   675
EDWARD D. PALIK AND W. R. HUNTER

References   678

Lithium Niobate (LiNbO₃)   695
EDWARD D. PALIK

References   697

Potassium Chloride (KCl)   703
EDWARD D. PALIK

References   706

Silicon Dioxide (SiO$_2$), Type α (Crystalline)   719
H. R. PHILIPP

References   721

Silicon Dioxide (SiO$_2$) (Glass)   749
H. R. PHILIPP

References   752

Silicon Monoxide (SiO) (Noncrystalline)   765
H. R. PHILIPP

References   766

Silicon Nitride (Si$_3$N$_4$) (Noncrystalline)   771
H. R. PHILIPP

References   772

Sodium Chloride (NaCl)   775
J. E. ELDRIDGE AND EDWARD D. PALIK

References   779

Titanium Dioxide (TiO$_2$) (Rutile)   795
M. W. RIBARSKY

References   798

List of Contributors

Numbers in parentheses indicate the pages on which the authors' contributions begin.

A. ADDAMIANO (597), Naval Research Laboratory, Washington, D.C. 20375

D. E. ASPNES (89), Bell Communications Research, Inc., Murray Hill, New Jersey 07974

G. BAUER (517, 535), Institut für Physik, Montanuniversität Leoben, A-8700 Leoben, Austria

A. BORGHESI (445, 525), Dipartimento di Fisica "A. Volta" and Gruppo Nazionale di Struttura della Materia del CNR, Università di Pavia, Pavia, Italy

W. J. CHOYKE (587), Westinghouse Research and Development Center, Pittsburgh, Pennsylvania 15235, and University of Pittsburgh, Pittsburgh, Pennsylvania 15260

DAVID F. EDWARDS* (547, 665), University of California, Los Alamos National Laboratory, Los Alamos, New Mexico 87545

J. E. ELDRIDGE (775), Department of Physics, University of British Columbia, Vancouver V6T 1W5, British Columbia, Canada

O. J. GLEMBOCKI (503), Naval Research Laboratory, Washington, D.C. 20375

G. GUIZZETTI (445, 525), Dipartimento di Fisica "A. Volta" and Gruppo Nazionale di Struttura della Materia del CNR, Università di Pavia, Pavia, Italy

R. T. HOLM (479, 491), Naval Research Laboratory, Washington, D.C. 20375

W. R. HUNTER† (69, 275, 675), Naval Research Laboratory, Washington, D.C. 20375

MITIO INOKUTI (369), Argonne National Laboratory, Argonne, Illinois 60439

B. JENSEN (169), Department of Physics, Boston University, Boston, Massachusetts 02215

H. KRENN (517, 535), Institut für Physik, Montanuniversität Leoben, A-8700 Leoben, Austria

* Present address: Lawrence Livermore National Laboratory, Livermore, California 94550.
† Present address: Sachs/Freeman Associates, Inc., Bowie, Maryland 20715.

DAVID W. LYNCH (189, 275), Department of Physics and Ames Laboratory, U.S. Department of Energy, Iowa State University, Ames, Iowa 50011

SHASHANKA S. MITRA (213), Department of Electrical Engineering, University of Rhode Island, Kingston, Rhode Island 02881

EDWARD D. PALIK (3, 409, 429, 479, 587, 597, 675, 695, 703, 775), Naval Research Laboratory, Washington, D.C. 20375

H. R. PHILIPP (665, 719, 749, 765, 771), General Electric Research and Development Center, Schenectady, New York 12301

H. PILLER (503, 571), Department of Physics and Astronomy, Louisiana State University, Baton Rouge, Louisiana 70803

ROY F. POTTER (11, 465), Department of Physics and Astronomy, Western Washington University, Bellingham, Washington 98225

M. W. RIBARSKY (795), School of Physics, Georgia Institute of Technology, Atlanta, Georgia 30332

JOSEPH SHAMIR (113), Faculty of Electrical Engineering, Technion — Israel Institute of Technology, Haifa 32000, Israel

E. SHILES* (369), Argonne National Laboratory, Argonne, Illinois 60439

GEORGE J. SIMONIS (155), Electronics Research and Development Command, Harry Diamond Laboratories, Adelphi, Maryland 20783

D. Y. SMITH (35, 369), Argonne National Laboratory, Argonne, Illinois 60439, and Max-Planck-Institut für Festkörperforschung, 7 Stuttgart 80, Federal Republic of Germany

P. A. TEMPLE (135), Michelson Laboratory, Physics Division, Naval Weapons Center, China Lake, California 93555

D. J. TREACY (623, 641), Physics Department, U.S. Naval Academy, Annapolis, Maryland 21402

* Present address: Gulf Research and Development Company, Houston, Texas 37048.

Preface

Those of us who do optical experiments have on many occasions needed the optical constants of a specific solid in a specific wavelength region. These numbers are often not readily available, and this situation requires a search in AIP, OSA, and CRC handbooks, as well as in reports never published in archival journals, or a visit to the library to sift through computer printouts of papers selected by key words. Then, once the papers are found, the spectral range presented is often narrow (not the one in which we are interested). I know of only two materials, namely, Si and NaCl, that have been treated in great detail (by H. H. Li), and these have served as models for the present work to some extent.

This handbook was roughed out by an *ad hoc* committee during the May 1980 meeting on the basic optical properties of materials at the National Bureau of Standards. Several colleagues, some of whom later became contributors to this volume, made up the committee, including David W. Lynch, H. E. Bennett, Roy F. Potter, M. Hass, and W. R. Hunter, as well as E. V. Cohen of Academic Press. The intent was to present as many materials as feasible over the widest possible spectral range with some discrimination, so that only a single set of numbers would be presented. Also, brief mention was to be made of the experimental technique, sample characterization, and accuracy and precision of the data (if available).

As completed, this handbook enlists the aid of 11 chapter contributors who summarize methods for the determination of the index of refraction n and the extinction coefficient k in various spectral regions (Chapters 1–9) and the absorption mechanisms in solids (Chapters 10 and 11). Significant time was spent on the film-on-a-substrate type of sample, which is so prevalent today in the areas of antireflection and reflection coatings, metals and insulators on semiconductors, multiple semiconductor layers, and deposited and epitaxial films on substrates in general. In addition, 21 contributors have read the published papers on the optical properties of 37 solids of scientific and technological interest and provided critiques. They have extracted from these papers what they feel is the best set of single values of n and k in the spectral range from x rays to millimeter waves and have produced detailed numerical tables and coarse graphs of these data. The original references are also provided should further detail be needed.

These data will be of interest to (1) those optical researchers who spend their careers working with one material or a class of materials (metals, semiconductors, insulators), (2) those who are interested in optical properties in general, (3) those who are interested in the film on a substrate or multilayer thin films, and (4) those college professors who teach optics. For the teacher, the bird's-eye view of optical properties over the wide spectral range afforded by the graphs and data serves as a good introduction to understanding a wide variety of absorption and dispersion processes and, consequently, optical properties.

As editor, I have had the usual problems of getting results from 27 different scientists in a timely manner. I thank them for their efforts, from which I have learned a lot of optical physics. In addition, I thank Donna D. Strasburg and Dinah W. Smith for their secretarial help.

November 1984   EDWARD D. PALIK

Part **I**

Determination of Optical Constants

Chapter 1

Introductory Remarks

EDWARD D. PALIK

Naval Research Laboratory
Washington, D.C.

I.	Introduction	3
II.	The Chapters	4
III.	The Critiques	5
IV.	The Tables	6
V.	The Figures of the Tables	7
VI.	General Remarks	8
	References	9

INTRODUCTION I

During a professional career covering molecular spectroscopy, magneto-optical properties of semiconductors, internal-reflection spectroscopy, and general optical properties of electronic materials, I rarely measured refractive index n or extinction coefficient k to better than two significant figures and then only when the experiment really demanded it. However, there was always a need for n and k values for estimating an optical effect, internal reflection in a prism with multilayers, band-edge transmission of a thin semiconductor film on a substrate, or free-carrier reflection effects, for example. Such numbers were not very handy, especially for a variety of metals into the infrared to the far infrared and for semiconductors with various free-carrier concentrations.

The present handbook is an attempt to collect references and tabulate n and k for a reasonably large number of materials of technological and physics interest, so that a person having the need could easily find the numbers over the widest spectral range, ideally x-ray to microwave, in a grid fine enough to yield the number by inspection or by a simple interpolation, and provide the pertinent references if more details were needed. For each material a critique is provided to indicate how the numbers were obtained from experiment and how good they are.

3

HANDBOOK OF OPTICAL CONSTANTS OF SOLIDS

ISBN 0-12-544420-6

II THE CHAPTERS

To preface and supplement the critiques for each material, there are 11 chapters (including this one) that discuss various aspects of optical constants with some emphasis on experimental techniques and sample preparation. While n and k are oblivious to the units used in Maxwell's equations, deriving these numbers both theoretically and experimentally can be done in various units that the individual contributors have indicated when necessary (primarily cgs and mks).

In Chapter 2 Roy F. Potter develops R, T, and A for a general sample of multiple films on both sides of a thick substrate. This simplifies to the single slab and the single film on a slab substrate, the two most prevalent kinds of samples. In Chapter 3 D. Y. Smith reviews the world of sum rules over infinite and finite frequency ranges and stresses how such analysis can spot experimental errors.

In Chapter 4 W. R. Hunter discusses polarized nonnormal incidence reflection techniques especially as they are practiced in the ultraviolet–visible (UV–vis) spectral region, with emphasis on how errors in R are propagated to errors in n and k and the effects of oxide layers. In Chapter 5 D. E. Aspnes outlines wavelength ellipsometry with emphasis on the effects of surface preparation and the properties of a deposited or grown film on a substrate. In Chapter 6 Joseph Shamir reviews standard interferometric techniques for transparent samples and then discusses his more recent laser interferometric technique for measuring both n and k for a film on a substrate. In Chapter 7 P. A. Temple deals with calorimetry for bulk material and for a film on a substrate to determine small absorption constants. In Chapter 8 George J. Simonis discusses the more-specialized techniques used in the far-IR and submillimeter-wave region to determine optical constants. These are of considerable interest now with the advent of gyratron sources. In Chapter 9 B. Jensen summarizes the quantum free-carrier-absorption theory for semiconductors and gives numerical results for the contributions to n and k for GaAs, InP, and InAs for various free-carrier densities. Thus, the reader can calculate from scratch (or use an abbreviated table) to determine the free-carrier contribution to ε_1 and ε_2 and thus n and k. The regular critiques for GaAs, InP, and InAs were done for "pure" material, and these values of n and k can be combined with the free-carrier contributions to estimate the optical constants for any sample of arbitrary free-carrier concentration.

Finally in Chapter 10 David W. Lynch discusses the absorption mechanisms above the fundamental band gap with emphasis on the variety of interband, plasma, core-level, exciton, and Urbach-tail absorption processes. In Chapter 11 Shashanka S. Mitra describes the absorption mechanisms below the fundamental band gap, such as single- and multiphonon processes, free-carrier absorption, and impurity absorption.

Special attention has been given to aluminum by D. Y. Smith in a long critique since it is the most-studied metal and its optical constants have been determined over a wide spectral range with the aid of dispersion relations. It would be hoped that the data given for some of the other materials might spur further such analysis.

We have stressed the film-on-a-substrate sample because this is becoming more and more the sample confronting us in materials studies since the upsurge in (1) epitaxial film growth, (2) use of insulator films on semiconductors, (3) surface studies of adsorbed layers under ultrahigh-vacuum (UHV) conditions, and (4) reflection and antireflection coatings.

THE CRITIQUES III

Thirty-seven materials have been examined with some crystalline and amorphous forms included for a few materials. The critiques have been done by 21 people. Thus, the styles vary somewhat. We wanted a common thread of discussion of experimental techniques, reasons for choices of numbers used in the table, the most important literature references, and some idea of accuracy and precision. This last requirement has been met head-on by some of the contributors (and original workers) who quote plus/minus variations in experimental scatter for a Sellmeier-type fit to express the data, who discuss limits of systematic and random errors, or who (more vaguely) discuss which decimal place is becoming unreliable or uncertain. Because of the wide spectral range along with the number of different experimental techniques used for a given material, this becomes a task impossible to do in detail, and we can only refer the reader back to the original references. In most tables we have overlapped n and k values from different experiments to give the reader an idea of the mismatch. Sometimes, we can figure out the reasons for this mismatch and suggest which is the better set of data. For example, Kramers–Kronig (KK) analysis deteriorates when k becomes smaller than n and at the edges of a particular spectral region; also, the purity and structure of the material can vary among laboratories. Unfortunately, the variations in n and k are such that for nontransmission experiments generally, only one or two figures should be used in quoting absolute values (accuracy), while more are often carried along to show spectral structure as a function of wavelength (precision). Only in the case of measurement of refractive index by minimum deviation or fringes are we able to set down values to four decimal places with some confidence. While we can give examples of ellipsometric measurements and KK analysis of reflectivity in an overlap region (1.5–6.0 eV) that agree to two figures, we also have examples in which two laboratories measuring R and doing KK analysis hardly agree to two figures among themselves.

Much of the work cited was done by physicists interested in absorption peaks so as to be able to identify phonon absorption and interband absorption transitions in detail. Their precision is generally high in order to distinguish peaks and shoulders, but their accuracy is not necessarily high, as a comparison of two experiments measuring multiphonon absorption in silicon or gallium arsenide will show, for example.

We do think that a single set of values of n and k is desirable in contrast to a collection of all the work ever done since the dawn of time with no selectivity or hints as to what is what. To some extent, we feel this handbook accomplishes this goal.

There have been a number of previous collections of optical data that have been helpful. The most useful is "The Refractive Index of Alkali Halides," by Li [1], in which Sellmeier-type fits have been given to many materials. The work of Haelbich et al. [2], "Optical Properties of Some Insulators in the Vacuum Ultraviolet Region," was especially useful because it catalogued all the x-ray to UV work of a number of insultors and proved to be a good starting point. Another older collection was published by Moses [3], "Refractive Index of Optical Materials in the Infrared Region." My copy of Ballard et al. [4], "Optical Materials for Infrared Instrumentation," is dogeared from use. Another more recent starting point was Wolfe's [5] "Properties of Optical Materials." The table of Hass and Hadley [6], "Optical Properties of Metals," was useful, as was the more recent compilation of Weaver et al. [7] based on synchrotron data for metals.

IV THE TABLES

A few general comments about the format of the tables are in order. We have used photon energy in electron volts, wave number in reciprocal centimeters, and wavelength in micrometers for each material to aid the reader. It is natural to think in electron volts and angstroms in the x-ray–UV region, while in the infrared reciprocal centimeters and micrometers are more comfortable. We will never get used to a number like $110,000$ cm^{-1}, although this usage appears in older atomic spectroscopy books [8]. Therefore, we introduce reciprocal centimeters usually near $50,000$ cm^{-1} (~ 6 eV), keeping them to all lower energies.

It is usually obvious what the units were for the original data because of the systematic trends of the simple numbers given to one, two, or three figures. The other two units are usually given to four figures just for consistency and to help identify the original units. To convert electron volts to reciprocal centimeters we use 8065.48 cm^{-1}/eV rounded off from the most recently determined [9] fundamental constants e/hc.

The real and imaginary parts of the dielectric function ε are given by $\varepsilon = \varepsilon_1 \pm i\varepsilon_2 = (n \pm ik)^2$. One obtains $n \pm ik$ by choosing the form of the plane-wave solution to Maxwell's equations as $\exp[i(\mp \omega t \mp \mathbf{q} \cdot \mathbf{r})]$. This choice also determines signs in the Lorentz-oscillator formula for lattice vibrations. We have elected to tabulate n and k because it is then an easy matter to obtain $\varepsilon_1 = n^2 - k^2$ and $\varepsilon_2 = 2nk$ as well as reflectivity $R = [(n - 1)^2 + k^2]/[(n + 1)^2 + k^2]$.

The extinction coefficient is often given in powers of 10. This power is understood to be the same down the column until this power repeats itself just before a new power of 10 appears or the number becomes a pure decimal at 10^{-1}.

In most instances, the values for n and k have been obtained from tables and graphs in the original references. In some cases, the original workers have kindly supplied detailed tables to the critiquers. The tables are acknowledged as private communications. To obtain n and k from a graph produces all sorts of eyestrain problems. Many authors do not provide fiduciary marks on all four sides of the graph or omit the marks for 2–9 on log plots. Tables that are too small to read must be expanded. For some reason, artists and/or experimenters cannot draw a linear scale with evenly spaced fiduciary marks. Therefore, some of these numbers probably have an uncertainty larger than the experimental uncertainty quoted in the original paper.

When absorption is due to extrinsic mechanisms such as impurities and surface quality, the critiquer has mentioned this and either omitted the data or noted its extrinsic character.

The references are given in brackets next to the beginning n and k and are understood to apply down the column until a new reference appears. This may get a little confusing when a significant overlap of two sets of data occurs.

THE FIGURES OF THE TABLES   V

The tables are plotted for each critique, usually as Fig. 1, on a log–log scale, so that the reader can see at a glance what is going on. Admittedly, the details are lost. The solid line is for n; the dashed line is for k. After plotting over 30 such tables, I find the optical constants of solids to be a boring subject with the same old reststrahlen bands, the same old multiphonon bands, the same old free-carrier absorption, the same fundamental band gap and higher-lying band gaps and the same K, L, M atomic absorption edges.

VI   GENERAL REMARKS

A few interesting observations have been made while I was doing several of the critiques and reading all of them. Room temperature was anything from 292 to 300 K, and it was surprising how many authors neglected to report this number at all. This is no problem except for the refractive-index variation typically in the fourth decimal place for a 1-K change in temperature. Some critiques do quote temperature dependences.

Description of the sample surface preparation was often neglected, as was the free-carrier concentration in semiconductors. Mechanical (nonchemical) polishing of the surface should have been outlawed 20 years ago, although whether it affects an optical measurement often depends on the wavelength-to-damage size and the kind of material—metal, semiconductor, or insulator. We sometimes run into trouble with surface roughness at reststrahlen minima at which reflection is low ($R \approx 0.02$) and at maxima at which surface-optical-phonon features weakly occur because roughness can couple light to the surface-phonon modes. Deposited films were often undercharacterized as to state of crystallinity.

The great majority of far-IR determinations of n and k are indirect via the Lorentz-oscillator-model and Drude-model fits to R. The former is notoriously poor in the wings, but it is all we have in many cases, and many of the materials in the handbook are analyzed in the IR by this technique. In a few cases, KK analysis for R was also done, with some agreement with the results of an oscillator-model fit. Fortunately, asymmetric Fourier transform spectroscopy can now give both the real and imaginary parts of R, ε_1, and ε_2, and n and k directly, thus avoiding the oscillator model and KK analysis. The other, more direct, technique of polarized nonnormal incidence reflection is not so widespread in the IR yet.

The nomenclature for optics gets a little mixed up on occasion in the original references and in the present chapters and critiques. While reflection and transmission are general processes, the measurement one makes on a laboratory sample yields numbers, the reflectance and transmittance. This, of course, includes multiple-reflection effects. As the sample becomes opaque or as the sample is wedged, the reflectance approaches the single-surface reflectivity, and this is the quantity that is used in KK analysis.

Since $\alpha = 4\pi k/\lambda$ defines the absorption constant, most people measure α directly and do not convert to k. In converting some α data to k data over a considerable range of λ, I noticed that detailed structure was often washed out somewhat in the final plot of k in contrast to the plot of α. Also, a peak in α did not necessarily fall at the same wavelength as a peak in k. Since power absorption is proportional to α, measurement of α is a better approach to locating resonances.

Very few of the measurements were done to determine n and k for their own sake, except when the material was new, such as in the case of InSb in 1953. The interest in laser-window materials spurred such measurements in selected transparent materials such as NaCl and KCl. The availability of synchrotron-radiation sources is now producing physics interests in material structure and will yield better values.

Some general reviews of how to calculate and measure optical constants are also listed in the references [10–14].

REFERENCES

1. H. H. Li, *J. Phys. Chem. Ref. Data* **5**, 329 (1976).
2. R.-P. Haelbich, M. Iwan, and E. E. Koch, "Physik Daten, Physics Data," Fach-information zentrum, Karlsruhe, 1977.
3. A. J. Moses, Refractive Index of Optical Materials in the Infrared Region, Report on Contract F33615-68-C-1225, Air Force Systems Command, Hughes Aircraft Company, Culver City, California, 1970.
4. S. S. Ballard, K. A. McCarthy, and W. L. Wolfe, Optical Materials for Infrared Instrumentation, Report from Infrared Information and Analysis Center, University of Michigan, Willow Run Laboratories, Ann Arbor, Michigan, 1959.
5. W. L. Wolfe, *in* "Handbook of Optics OSA," Section 7, McGraw-Hill, New York, 1978.
6. G. Hass and L. Hadley, *in* "AIP Handbook," Section 6, pp. 118–169, McGraw-Hill, New York, 1972.
7. J. H. Weaver, C. Krafka, D. W. Lynch, and E. E. Koch, "Physik Daten, Physics Data," "Optical Properties of Metals," Vol. 18-1, Fach-information zentrum, Karlsruhe, 1981.
8. H. E. White, "Introduction to Atomic Spectra," McGraw-Hill, New York, 1934.
9. M. Aguilar-Benitez, R. L. Crawford, F. Frosch, G. P. Gopal, R. E. Hendrick, R. L. Kelly, M. J. Losty, L. Montanet, F. C. Porter, A. Rittenberg, M. Roos, L. D. Roper, T. Shimada, R. E. Shrock, T. G. Trippe, C. Walek, C. G. Wohl, and G. P. Yost, *Phys. Lett.* **111B**, 1 (1982).
10. E. E. Bell, *in* "Handbuch der Physik," Vol. 25, Part 2a, p. 1 (S. Flugge, ed.), Springer-Verlag, Berlin, 1967.
11. F. Abeles, *in* "Progress in Optics," Vol. 11, p. 251 (E. Wolf, ed.), North-Holland Publ., Amsterdam, 1963.
12. H. E. Bennett and J. M. Bennett, *in* "Physics of Thin Films," Vol. 4, p. 1 (G. Hass and R. E. Thun, eds.), Academic Press, New York, 1967.
13. O. S. Heavens, "Optical Properties of Thin Solid Films," Dover, New York, 1965.
14. P. H. Berning, *in* "Physics of Thin Films," Vol. 1, p. 69 (G. Hass, ed.), Academic Press, New York, 1963.

Chapter **2**

Basic Parameters
for Measuring
Optical Properties

ROY F. POTTER

Department of Physics and Astronomy
Western Washington University
Bellingham, Washington

I. Introduction	11
II. Intrinsic Material Parameters in Terms of Optical Constants	16
III. Reflectance, Transmittance, and Absorptance of Layered Structures	18
IV. The General Lamelliform—Phase Coherency Throughout	19
V. The General Lamelliform—Phase Incoherency in Substrate	21
A. Absorptance of a General Lamelliform	22
B. Reflectance of Single Film on Nontransmitting Substrate	23
VI. Summary	24
Appendix A. Basic Formulas for Fresnel Coefficients	24
Appendix B. General Formulas for the Case of a Parallel-Sided Slab	25
Appendic C. Reflectance R_{jk} at $j–k$ Interface	26
Appendix D. Reflectance of Single Layer on Each Side of a Slab and Single Layer on Either Side of a Slab	26
Appendix E. Critical Angle of Incidence	30
Definitions of Terms	33
References	34

INTRODUCTION **I**

The interaction of electromagnetic (em) radiation with matter has provided over the years an effective and powerful tool for understanding and probing the electronic and vibrational states of condensed materials. Maxwell's equations describing the propagation of em waves are so well established and accepted that they serve as the underlying bases for optical studies. Based on Faraday's law of induction and Maxwell's extension of Ampere's

11

law, they are given, in partial differential form, as

$$\mathbf{V} \times \mathbf{E} = -\partial\mathbf{B}/\partial t, \qquad \mathbf{V} \times \mathbf{H} = \mathbf{J} + \partial\mathbf{D}/\partial t, \tag{1}$$

along with the divergence relations for induction and displacement

$$\mathbf{V} \cdot \mathbf{B} = 0, \qquad \mathbf{V} \cdot \mathbf{D} = 0$$

(assuming that no free charge exists), where $\mathbf{E}$ is the electric field, $\mathbf{H}$ the magnetic field, $\mathbf{B}$ the magnetic induction, $\mathbf{D}$ the electric displacement, and $\mathbf{J}$ the electric current flux [1]. (Expressions are in mks units.)

However, these expressions are not sufficient in themselves to completely describe em propagation through a medium; i.e., no "optical constants" are contained in Maxwell's equations. A full description must include the trio of "material equations"

$$\mathbf{D} = \varepsilon\varepsilon_0\mathbf{E}, \qquad \mathbf{B} = \mu\mu_0\mathbf{H}, \qquad \mathbf{J} = \sigma\mathbf{E} \quad \text{(Ohm's law)}, \tag{2}$$

where the coefficients $\varepsilon\varepsilon_0$, $\mu\mu_0$, and σ are restricted for present purposes to be time- and field- independent scalar quantities. (See Born and Wolf [2] for more general tensor treatments.) These restrictions follow the assumption of isotropic media and linear responses. An additional important assumption is that these quantities are continuous in the *immediate* vicinity of the points where the fields are considered or measured. The coefficients $\varepsilon\varepsilon_0$, $\mu\mu_0$, and σ are known as the electric permittivity, the magnetic permeability, and the electrical conductivity, respectively; while ε_0 and μ_0 are the free-space permittivity and permeability, respectively. The so-called optical constants are contained in the coefficients ε, μ, and σ. Used for describing the macroscopic optical properties of matter, these coefficients can be related to polarizabilities on the atomic scale and provide details on the physics of the condensed state.

The expressions used in this section are based on the rationalized mks system of units. A consequence is that the propagation velocity in a given medium is given by $(\mu\varepsilon\mu_0\varepsilon_0)^{-1/2} = c(\mu\varepsilon)^{-1/2}$, where $c = (\mu_0\varepsilon_0)^{-1/2}$ is the free-space velocity of light. However, when Gaussian units are used, the propagation velocity is stated explicitly in Maxwell's equations, and the free-space velocity defined by $(\mu_0\varepsilon_0)^{-1/2}$. At the risk of belaboring the point, the refractive index for a given medium c/v is defined by the dimensionless quantity $(\mu\hat{\varepsilon})^{1/2}$. The Fresnel coefficients, reflectivity and transmittivity, are also defined by dimensionless ratios, namely, of the field strengths. Thus the complex refractive indices (as well as the corresponding complex dielectric functions $\hat{\varepsilon}$) carry no reference to any system of units. When they are used to determine field strength values relative to given input values, the fields must be stated in a consistent set of units (see Table I).

An additional point to keep in mind is that the dielectric susceptibility term in the polarization vector $\mathbf{P}$ in the rationalized system of units is 4π times the susceptibility term in an unrationalized system.

TABLE I

Conversion of Units

Quantity	mks	Practical cgs	Gaussian
Permittivity of free space ε_0	$\dfrac{10^{-9}\ \text{F}}{4\pi a^2\ \text{m}}$	$\dfrac{10^{-11}\ \text{F}}{4\pi a^2\ \text{cm}}$	$\dfrac{1}{4\pi}$
Permeability of free space μ_0	$4\pi \times 10^{-7}\ \dfrac{\text{H}}{\text{m}}$	$4\pi \times 10^{-9}\ \dfrac{\text{H}}{\text{cm}}$	$\dfrac{4\pi}{c^2}$
Speed of light $c = (\mu_0\varepsilon_0)^{-1/2}$	$a \times 10^8\ \dfrac{\text{m}}{\text{sec}}\ (a = 2.9979)$	$a \times 10^{10}\ \dfrac{\text{cm}}{\text{sec}}$	$a \times 10^{10}\ \dfrac{\text{cm}}{\text{sec}}$
Impedance of free space $Z_0 = (\mu_0/\varepsilon_0)^{1/2}$	$40\pi a\ \Omega$	$40\pi a\ \Omega$	$4\pi/c$
Electric intensity E	$\dfrac{\text{V}}{\text{m}}$	$\dfrac{\text{V}}{\text{cm}}$	$\dfrac{1}{300}\dfrac{\text{V}}{\text{cm}}$
Electric displacement D	$\dfrac{\text{C}}{\text{m}^2}$	$4\pi\ \dfrac{\text{C}}{\text{cm}^2}$	$12\pi \times 10^9\ \dfrac{\text{C}}{\text{cm}^2}$
Magnetic intensity H	$\dfrac{\text{ampere-turn}}{\text{m}}$	Oe	Oe
Magnetic induction B	$\dfrac{\text{Wb}}{\text{m}^2}\ (10^4\ \text{G})$	G	G

Table I lists several parameters in terms of three systems of units in common use.

Introducing the material equations permits Maxwell's equations to be expressed as a modified set of partial differential equations involving only two of the four field parameters, namely, **E** and **H**. Assuming that the fields have time dependence proportional to $e^{i\omega t}$, the modified Maxwell's equations become

$$\mathbf{V} \times \mathbf{E} = -\mu\mu_0\dot{\mathbf{H}} = -i\mu\mu_0\omega\mathbf{H} = -i\frac{2\pi\mu}{\lambda}Z_0\mathbf{H},$$

$$\mathbf{V} \times \mathbf{H} = \mathbf{J} + \dot{\mathbf{D}} = \mathbf{J} + \varepsilon\varepsilon_0\mathbf{E}\omega = i\left(\varepsilon - \frac{i\sigma}{\varepsilon_0\omega}\right)\varepsilon_0\omega\mathbf{E} = \frac{2\pi}{\lambda}\frac{\hat{\varepsilon}}{Z_0}\mathbf{E},$$

(3)

where

$$\hat{\varepsilon} = \varepsilon - (i\sigma/\varepsilon_0\omega) = \varepsilon - i\sigma\lambda Z_0/2\pi = \varepsilon' - i\varepsilon''$$

and

$$c = (\mu_0\varepsilon_0)^{-1/2}, \qquad Z_0 = (\mu_0/\varepsilon_0)^{1/2}, \qquad \lambda = 2\pi c/\omega.$$

The complex function $\hat{\varepsilon}$ has a real term equal to the ratio of the medium permittivity to that of free space, i.e., ε, and an imaginary term proportional to the frequency-dependent conductivity σ. The wave velocity, impedance, and wavelength for free space are given by c, Z_0, and λ, respectively.

For the infrared, visible, and ultraviolet spectral regions, the em radiation is considered to be propagating as plane waves. When such waves are incident upon a specimen composed of one or more homogeneous phases separated by specular surfaces, the amount and degree of transmission, reflection, and refraction are direct measures of the material parameters. Many of the usual and useful experimental arrangements have the em waves incident upon a set of parallel-sided slabs, with zero separation between the slabs (lamelliforms). The two field components E and H are orthogonal to each other and are contained in a plane orthogonal to the propagation direction. One needs to consider two cases only: (1) the electric field E normal to the plane of incidence (POI) (TE mode) and (2) the magnetic field component H normal to the POI (TM mode). (The POI is defined as the plane containing the propagation vector and the normal to the surface upon which the wave is incident.) Consider that the normal to the incident surfaces lies parallel to the z axis and that the x axis is normal to the POI. The y axis lies in the POI and is parallel to the surface.

Both the TE (also known as s or perpendicular polarization) and TM (p or parallel polarization) modes are treated in similar manners by using separation of variables techniques. The following are taken as solutions for each polarization:

	TE polarization	TM polarization	
	$E_x = U(z)X(t, y)$	$H_x = U(z)X(t, y)$	(4)
	$H_y = V(z)X(t, y)$	$E_y = V(z)X(t, y)$	
	$H_z = W(z)X(t, y)$	$E_z = W(z)X(t, y)$	

where

$$X = \exp i(\omega t - \kappa S y),$$

in which

$$\kappa = \omega/c = 2\pi/\lambda = (\mu_0\varepsilon_0)^{1/2}\omega, \qquad S = (\mu\hat{\varepsilon})^{1/2}\sin\phi,$$

and ϕ is the angle of incidence in the medium. Operating on these with the modified Maxwell's equations,

$$\mathbf{V} \times \mathbf{E} = -\mu\mu_0 \, \partial\mathbf{H}/\partial t, \qquad \mathbf{V} \times \mathbf{H} = \hat{\varepsilon}\varepsilon_0 \, \partial\mathbf{E}/\partial t,$$

yields for either polarization the expressions

$$U' = -i\hat{m}V, \qquad V' = -i\hat{\gamma}^2/\hat{m}U, \qquad SU = -\hat{m}W, \qquad (5)$$

where $\hat{\gamma} = \alpha - i\beta$, $\hat{\gamma}^2 = \mu\hat{\varepsilon} - S^2$, and $\mu\hat{\varepsilon} = \hat{n}^2 = (n - ik)^2$. The $\hat{m}$ is defined for each mode as

$$\hat{m} = \begin{cases} \mu Z_0 = \hat{n}\hat{Z} & \text{(TE)}, \\ -\varepsilon/Z_0 = -\hat{n}/\hat{Z} & \text{(TM)}. \end{cases} \qquad (6)$$

It should be noted that in this section the prime denotes the use of the operator $(1/\kappa)\,\partial/\partial z$. Eliminating W yields two second-order ordinary differential equations in U and V:

$$U'' + \hat{\gamma}^2(z)U - f'(z)U' = 0,$$
$$V'' + \hat{\gamma}^2(z)V + g'(z)V' = 0, \tag{7}$$

where

$$f = \ln(\hat{m}) = \ln(\zeta/\hat{\gamma}),$$
$$g = \ln(\hat{m}/\hat{\gamma}^2) = \ln(\zeta/\hat{\gamma}),$$

and where the material parameters m and ζ are still to be considered to vary along z; also:

$$\zeta = \frac{\hat{m}}{\hat{\gamma}} = \begin{cases} \hat{Z}\hat{n}/\hat{\gamma} = Z_0\mu/\hat{\gamma} & \text{(TE)}, \\ -\hat{n}/\hat{Z}\hat{\gamma} = -\hat{\epsilon}/\hat{\gamma}Z_0 & \text{(TM)}. \end{cases} \tag{8}$$

The pair of differential equations (7) can be solved analytically for only a few special cases. (See Jacobsson [3] for a comprehensive discussion of this problem.) Two cases are considered here, the first of which is that for an abrupt interface; i.e., $\hat{n}'$ is infinite. This is expressed in matrix form as

$$\begin{bmatrix} U_0 \\ V_0 \end{bmatrix} = \begin{bmatrix} 1 & 0 \\ 0 & 1 \end{bmatrix} \begin{bmatrix} U_f \\ V_f \end{bmatrix}, \tag{9}$$

where the subscripts refer to the initial side (0) and the opposite side (f) of the slab or interface. This matrix expression is the statement that tangential fields are equal across the interface.

The second case is that for a medium characterized as having a uniform refractive index n, i.e., f' and g' vanish. This results in the usual homogeneous wave equation whose solutions are expressed in the "characteristic matrix"

$$\begin{bmatrix} U_0 \\ V_0 \end{bmatrix} = \begin{bmatrix} \cos(\hat{\gamma}z) & i\zeta\sin(\hat{\gamma}z) \\ [i\sin(\hat{\gamma}z)]/\zeta & \cos(\hat{\gamma}z) \end{bmatrix} \begin{bmatrix} U_f \\ V_f \end{bmatrix} = \hat{M} \begin{bmatrix} U_f \\ V_f \end{bmatrix}, \tag{10}$$

where U, V, and ζ are appropriately defined for either polarization mode. Note that z is the dimensionless product of the distance between points 0 and f with $\kappa = \omega/c$. The determinants for both of these 2×2 matrices are constants and the particular solutions have been chosen to make both equal to unity.

The characteristic matrix for a uniform layer is the basis for calculating the resultant matrix M_R for a series of parallel layers. Applying matrix multiplication to the matrices for the individual layers gives

$$\hat{M}_0 \times \hat{M}_1 \times \cdots \times \hat{M}_f = \hat{M}_R = \begin{bmatrix} \hat{A}_R & \hat{B}_R \\ \hat{C}_R & \hat{D}_R \end{bmatrix} \tag{11}$$

(see Macleod [4]).

An alternative approach is to apply Eqs. (5) and express the matrix for each interface and layer in terms of the forward and reverse field components:

$$U = U^+ + U^- = (V^+ - V^-)\zeta, \qquad V = V^+ + V^- = (U^+ - U^-)\zeta^{-1} \quad (12)$$

(note that $\zeta V^\pm = \pm U^\pm$). The superscripts $+$ and $-$ refer to forward and reverse components, respectively.

For the ith abrupt interface, the matrix equations are

$$\begin{bmatrix} U_0^+ \\ U_0^- \end{bmatrix} = \frac{1}{\hat{t}_{0f}} \begin{bmatrix} 1 & \hat{r}_{0f} \\ \hat{r}_{0f} & 1 \end{bmatrix} \begin{bmatrix} U_f^+ \\ U_f^- \end{bmatrix} = \hat{C}_s(U) \begin{bmatrix} U_f^+ \\ U_f^- \end{bmatrix},$$

$$\hat{r}_{0f} = \frac{\zeta_f - \zeta_0}{\zeta_f + \zeta_0}, \qquad \hat{t}_{0f} = 1 + \hat{r}_{0f} = \frac{2\zeta_f}{\zeta_f + \zeta_0},$$

$$\hat{C}_s(V) = \frac{1}{\hat{t}_{0f}} \begin{bmatrix} 1 & -\hat{r}_{0f} \\ -\hat{r}_{0f} & 1 \end{bmatrix}, \qquad \hat{C}_s(W) = \frac{\hat{m}_f}{\hat{m}_0} \hat{C}_s(U),$$

$$\hat{t}'_{0f} = 1 - \hat{r}_{0f} = (\zeta_0/\zeta_f)t_{0f},$$

while that for the medium with a uniform index is

$$\begin{bmatrix} U_0^+ \\ U_0^- \end{bmatrix} = \begin{bmatrix} e^{i\hat{\gamma}d} & 0 \\ 0 & e^{-i\hat{\gamma}d} \end{bmatrix} \begin{bmatrix} U_f^+ \\ U_f^- \end{bmatrix} = \hat{C}_l \begin{bmatrix} U_f^+ \\ U_f^- \end{bmatrix} \quad (14)$$

with $d = z_f - z_0$.

Each interface and intervening homogeneous layer can be described by its appropriate matrix. Matrix multiplication must be applied in the proper sequence, beginning with the initial interface matrix (s0) and ending with that for the final interface (sf) and alternating uniform layer–interface–uniform layer, etc., matrices. The result, C_R, corresponding to Eq. (11) is

$$\hat{C}_R = \hat{C}_{s0} \times \hat{C}_{l1} \times \hat{C}_{s1} \times \hat{C}_{l2} \times \hat{C}_{s2} \times \cdots \times \hat{C}_{lf} \times \hat{C}_{sf}. \quad (15)$$

The matrices expressed in Eqs. (13) and (14) are given in terms of the Fresnel terms $\hat{r}$ and $\hat{t}$ as well as the phase factor $\hat{\gamma}$ and are the bases for ellipsometry. The reader is referred to Heavens for a discussion of this approach [5].

II INTRINSIC MATERIAL PARAMETERS IN TERMS OF THE OPTICAL CONSTANTS

This section outlines the connections between the macroscopic optical constants and the electric-dipole excitations on the atomic scale. Thus, the accurate determination of the optical parameters of a given medium provide intrinsic information about that medium. At optical frequencies these parameters exhibit rich and varied features unique to the particular composition under study. Even the presence of a small amount of an impurity can significantly alter portions of the spectral behavior. In addition, these parameters can be modified by temperature variations and by additional spectral features introduced by the application of external fields such as electric, magnetic, and stress fields, which can bring about symmetry changes, injection of charges, and so on.

It should be understood that the expanded Ampere's law is a generalized statement, that the motion of charges in an oscillating electric field has two components: (1) 90° out of phase with the field with no absorption of power and (2) in phase with energy from the field being dissipated in the material, corresponding to the usual Ohm's law. The optical conductivity is not to be confused with the dc conductivity σ_0, however.

From Ohm's law [Eq. (2)] the mean power P dissipated per unit volume is given by

$$P = \sigma \langle E^2 \rangle. \tag{16}$$

Let $p(t)$ be the probability that a photon of energy $\hbar\omega$ is absorbed in time t per unit volume; then

$$P = \hbar\omega p(t)/t. \tag{17}$$

This is a consequence of the observation that on the atomic scale, radiation must be treated as if it is absorbed or emitted in quanta or photons with magnitude $\hbar\omega$.

Equating the macroscopic term with that of the atomic gives

$$P = \varepsilon'' \omega \varepsilon_0 \langle E^2 \rangle = \hbar\omega p(t)/t \quad \text{or} \quad \varepsilon'' = \hbar p(t)/\varepsilon_0 \langle E^2 \rangle t. \tag{18}$$

These expressions relate the macroscopic parameter ε'' to that of $p(t)$ at the atomic level for the interaction of electromagnetic radiation while propagating through matter. If the physics of condensed matter gives a value or estimate for $p(t)$, it can be checked and validated by measurements of the optical constants [6].

For example, the semiclassical quantum-mechanical treatment of a bound electron shows that ε'' is proportional to the product of the matrix element for this transition and the joint density of states for the initial and final states. The matrix element provides the base for determining selection rules based on lattice symmetry as well as indicating the strength of the transition. The joint density of states, on the other hand, gives a measure of the availability of sites available for transitions between states. These topics are handled in further detail elsewhere in this handbook.

The classical-oscillator case for an electron is an example of absorption and dispersion due to free carriers. Since there are no restoring forces, the real and imaginary parts of the dielectric function are written as

$$\varepsilon' - \varepsilon'_\infty = \frac{(e^2/\varepsilon_0)(N/m^*)}{\omega^2 + g^2}, \qquad \varepsilon'' = \frac{ge^2}{\varepsilon_0 \omega} \frac{N/m^*}{(\omega^2 + g^2)}, \tag{19}$$

where ε_∞ is the dielectric function at frequencies sufficiently high that free-carrier effects are negligible, N is the free-carrier concentration, g is a damping term, and m^* is the isotropic free-carrier effective mass. These expressions describe the optical properties of many semiconductors in the spectral region between the reststrahlen frequencies and those for the interband electronic transitions (see Moss [6]).

Other absorption mechanisms can exist which contribute to the optical properties of the condensed phase. The classical treatment already sketched illustrates the type of microscopic information on the atomic scale to be inferred and derived from the macroscopic measurements of the so-called optical constants.

III REFLECTANCE, TRANSMITTANCE, AND ABSORPTANCE OF LAYERED STRUCTURES

A common type of specimen structure for use in the experimental determination of the optical parameters is that of the lamelliform, which is a structure consisting of parallel-sided lamellae in intimate contact and with each interface a specular surface. Such structures include the plain slab as well as a slab with a single film.

Ellipsometry is a general class of experimental techniques for studying such specimens. Matrices (13) and (14) are the bases for describing the Fresnel coefficients and determining the ratio of the amplitudes of the reflected TM and TE modes as well as the phase difference between them (see Chapter 5).

The other principal class of experiments consists of those that measure the intensity ratios: reflectance R, transmittance T, and absorptance A. For a given experimental setup, these quantities are related by

$$A + R + T = 1. \tag{20}$$

Most measurements of one or more of these quantities have been made at zero, or near zero, angles of incidence. This section will develop expressions for them that are generalized for either polarization as well as for *all* angles of incidence. (See Chapter 4 for more details of nonnormal-incidence reflectance measurements.)

The source of em radiation is considered to be on one side of the lamelliform only, i.e., the incident side (subscript 0), and no backward wave exists in the medium opposite the source (subscript f), i.e., $U_f^- = 0$. Taking the z axis along the normal to the lamelliform, the Poynting vector components along that axis for the three field amplitudes are proportional to $U_0^+ V_0^+{}^*$, $U_0^- V_0^-{}^*$, and $U_f^+ V_f^+{}^*$ for the incident, the reflected, and the transmitted waves, respectively. In terms of the matrix C_R [Eq. (15)], the field amplitudes are related as

$$\begin{bmatrix} U_0^+ \\ U_0^- \end{bmatrix} = \frac{1}{t_{of}} \begin{bmatrix} a & b \\ c & d \end{bmatrix} \begin{bmatrix} U_f^+ \\ 0 \end{bmatrix} = C_R \begin{bmatrix} U_f^+ \\ 0 \end{bmatrix}, \tag{21}$$

with $t_{of} = t_0 \cdot t_1 \cdot t_2 \cdots t_f$.

The reflectance R and the transmittance T are defined by

$$R_{of} = \frac{U_0^- U_0^{-*}}{U_0^+ U_0^{+*}} = \frac{\hat{c}\hat{c}^*}{\hat{a}\hat{a}^*} = \hat{r}_{of}\hat{r}_{of}^*,$$

$$T_{of} = \frac{\mathrm{Re}[U_f^{+*}\sqrt{(V_f^+)^2 + (W_f^+)^2}\cos\phi_f]}{\mathrm{Re}[U_0^{+*}\sqrt{(V_0^+)^2 + (W_0^+)^2}\cos\phi_f]}$$

$$= \frac{U_f^+ U_f^{+*}\,\mathrm{Re}(1/\zeta_f)}{U_0^+ U_0^{+*}\,\mathrm{Re}(1/\zeta_0)} = \frac{\hat{t}_{of}\hat{t}_{of}^*\,\mathrm{Re}(1/\zeta_f)}{\hat{a}\hat{a}^*\,\mathrm{Re}(1/\zeta_0)}, \tag{22}$$

$$\frac{\mathrm{Re}(1/\zeta_f)}{\mathrm{Re}(1/\zeta_0)} = \zeta_0\,\mathrm{Re}\left(\frac{1}{\zeta_f}\right) = \frac{1 - \hat{r}_{of}\hat{r}_{of}^*}{1 + \hat{r}_{of}\hat{r}_{of}^* - 2\,\mathrm{Re}(\hat{r}_{of})}$$

$$= \frac{1 - R_{of}}{1 + \hat{R}_{of} - 2\,\mathrm{Re}(\hat{r}_{of})}.$$

The resultant matrix C_R can also be expressed in terms of the characteristic matrix M_R of Eq. (11). Although the characteristic matrix Eq. (11) is often preferred for calculations because each submatrix is defined for its own particular lammella, the elements of M_R must be directly related to those of Eq. (21), thus the components a and c of Eq. (21), in terms of Eq. (11), become

$$\hat{a} = \tfrac{1}{2}[\hat{A}_R + \zeta_0\hat{C}_R + (\hat{B}_R/\zeta_f) + (\zeta_0/\zeta_f)\hat{D}_R],$$
$$\hat{c} = \tfrac{1}{2}[\hat{A}_R - \zeta_0\hat{C}_R + (\hat{B}_R/\zeta_f) - (\zeta_0/\zeta_f)\hat{D}_R]. \tag{23}$$

In most instances, the reflectance and the transmittance are measured in the same medium, i.e., $\zeta_0 = \zeta_f$, but experimental conditions exist for which the ratio $\mathrm{Re}(1/\zeta_f)/\mathrm{Re}(1/\zeta_0)$ must be included for T in Eq. (22).

THE GENERAL LAMELLIFORM— IV
PHASE COHERENCY THROUGHOUT

A lamelliform of general interest is that of a parallel-sided dielectric or semiconductor slab, coated in either side with one or more lamellae, each sufficiently thin so as to transmit radiation of a given frequency (or wavelength). The approach here is to use Eqs. (13) and (14) to form $\hat{C}_R$ as the product of three matrices:

(1) $\hat{C}_f$ for the n lamellae on the front or source side,
(2) $\hat{C}_s$ for the substrate slab itself, and
(3) $\hat{C}_b$ for the m lamellae on the back side, opposite the source.

The resulting matrices are represented as

$$C_f = \frac{1}{t_f}\begin{bmatrix} a_f & b_f \\ c_f & d_f \end{bmatrix} = \frac{a_f}{t_f}\begin{bmatrix} 1 & r'_f \\ r_f & D_f \end{bmatrix}, \tag{24a}$$

where $t_f = t_{f1} \cdot t_{f2} \cdots t_{fn}$ (n lamellae), $r_f = c_f/a_f$, $r'_f = b_f/a_f$, and $D_f = d_f/a_f$;

$$C_s = a_s\begin{bmatrix} 1 & 0 \\ 0 & D_s \end{bmatrix}, \tag{24b}$$

where $a_s = \exp i\hat{\gamma}_s d_s$, $D_s = a^{-2} = E\hat{e}$, $E = \exp(-2\beta_s d_s)$, and $\hat{e} = \exp(-2i\alpha_s d_s)$;

$$C_b = \frac{a_b}{t_b}\begin{bmatrix} 1 & r'_b \\ r_b & D_b \end{bmatrix}, \tag{24c}$$

where $t_b = t_{b1} \cdot t_{b2} \cdots t_{bm}$ (m lamellae), $r_b = c_b/a_b$, $r'_b = b_b/a_b$, and $D_b = d_b/a_b$.
Multiplying these matrices in proper sequence gives

$$C_R = \frac{a_f a_s a_b(1 + r'_f r_b D_s)}{t_f t_b}\begin{bmatrix} 1 & r' \\ r & \dfrac{r_f r'_b + D_f D_s D_b}{1 + r'_f r_b D_s} \end{bmatrix},$$

$$r = \frac{r_f + r_b D_s D_f}{1 + r'_f r_b D_s}, \qquad r' = \frac{r'_b + r'_f D_s D_b}{1 + r'_f r_b D_s}. \tag{25}$$

From Eq. (22) the reflectance for either polarization is

$$R = rr^* = \frac{R_f + R_b D_f D_f^* E^2 + 2E \, \mathrm{Re}(r_f^* r_b D_f \hat{e})}{1 + R'_f R_b E^2 + 2E \, \mathrm{Re}(r'_f r_b \hat{e})}, \tag{26}$$

where $R_f = r_f r_f^*$, $R'_f = r'_f r_f^*$, $R_b = r_b r_b^*$, and R_f is the reflectance of the front lamellae when the reflection from the substrate back surface R_b is zero. Here R'_f is the back reflectance from the front lamellae as though the source were located beyond the slab; i.e., the product $R'_f R_b$ is invariant when the slab is reversed 180° with respect to the source. (See Fig. 1 in Appendix D.)
The transmittance for the system is given by

$$T = \frac{T_f T_b T_s}{1 + R'_f R_b E^2 + 2E \, \mathrm{Re}(r'_f r_b \hat{e})},$$

$$T_f = \frac{t_f t_f^*}{a_f a_f^*} \frac{\mathrm{Re}(1/\zeta_s)}{\mathrm{Re}(1/\zeta_{f0})}; \qquad T_b = \frac{t_b t_b^*}{a_b a_b^*} \frac{\mathrm{Re}(1/\zeta_{b0})}{\mathrm{Re}(1/\zeta_s)}, \tag{27}$$

$$T_s = \sqrt{D_s D_s^*} = (a_s a_s^*)^{-1} = \exp(-2\beta_s d_s) = E.$$

Here T_f is the transmittance into the slab (medium s) through the lamellae on the source side, and the incident medium is characterized by $\hat{\zeta}_{f0}$. T_b is the transmittance from the slab through the lamellae on the back side into the b medium characterized by $\hat{\zeta}_{b0}$. $T_s = E$, analogous to the Lambert–Beer law, i.e., the transmittance through the substrate interior characterized by $\hat{\zeta}_s$.

THE GENERAL LAMELLIFORM—PHASE INCOHERENCY V IN SUBSTRATE

A case of particular interest is that for the substrate whose thickness-to-wavelength ratio d_s becomes so large that most experimental setups have insufficient resolution to sense the phase coherence of the transmitted and reflected beams through the substrate. In this case it becomes necessary to average the real part of the phase change across the substrate (assuming $k_s/n_s \ll 1$) [7].

Rearranging the terms of Eq. (26) gives

$$R = \frac{R_f + R_b D_f D_f^* E^2 + 2\sqrt{R_f R_b D_f^* D_f}\, E(\cos\theta\cos Y + \sin\theta\sin Y)}{(1 + R_f' R_b E^2)\left[1 + 2\sqrt{R_f' R_b}\, E\cos Y/(1 + R_f' R_b E^2)\right]} \quad (28)$$

where

$$\cos\theta = \frac{\mathrm{Re}(r_f' r_f D_f^*)}{(R_f' R_f D_f D_f^*)^{1/2}} \quad \text{and} \quad Y = 2\alpha_s d_s + \text{const.}$$

Averaging the quantity Y over $(2m + 1)\pi \geq Y \geq 0$, where the integer m is very large, gives the average reflectance $\bar{R}$ and transmittance $\bar{T}$ as

$$\bar{R} = \frac{1}{(2m + 1)\pi}\int_0^{(2m+1)\pi} R\, dY = \frac{R_f + R_b G_f E^2}{1 - R_f' R_b E^2}, \quad \bar{T} = \frac{T_f T_b E}{1 - R_f' R_b E^2}, \quad (29)$$

where

$$G_f = D_f D_f^* - 2\,\mathrm{Re}(r_f' r_f D_f^*) = (D_f - r_f' r_f)(D_f - r_f' r_f)^* - R_f R_f'.$$

Inductive arguments show that [using Eqs. (13), (14), and (21)]

$$D_f - r_f r_f' = \prod_{j=0}^{f} \det(m_j) = \left[\prod_{j=0}^{f} e^{-2\gamma_j d_j}(1 - \hat{r}_j^2)\right]\bigg/ a_T^2. \quad (30)$$

Therefore,

$$G = T_f T_f' - R_f R_f',$$

where

$$T_f' = \frac{\hat{t}_f' \hat{t}_f\, \mathrm{Re}(1/\hat{\zeta}_{f0})}{\hat{a}_f \hat{a}_f^*\, \mathrm{Re}(1/\hat{\zeta}_s)}, \quad \hat{t}_f' = 1 - \hat{r}_f,$$

giving the result that

$$\bar{R} = \frac{R_{\mathrm{f}} + (T_{\mathrm{f}}T_{\mathrm{f}}' - R_{\mathrm{f}}R_{\mathrm{f}}')R_{\mathrm{b}}E^2}{1 - R_{\mathrm{f}}'R_{\mathrm{b}}E^2} = R_{\mathrm{f}} + \frac{T_{\mathrm{f}}T_{\mathrm{f}}'R_{\mathrm{b}}E^2}{1 - R_{\mathrm{f}}'R_{\mathrm{b}}E^2}, \tag{31}$$

where T_{f}' is the transmittance in the front lamellae, as though the radiation were to propagate from the substrate to the front surface. Gabriel and Nedoluha [8] give similar expressions using a different approach.

A   Absorptance of a General Lamelliform

Most experiments measure either or both of the pair, transmittance and reflectance. The third quantity, absorptance, A (also known as emittance, through Kirchhoff's law), is the fraction of the incident radiation that has been absorbed by the lamelliform. From Eqs. (20), (29), and (30) it follows that

$$\bar{A} = 1 - R_{\mathrm{f}} - E\frac{T_{\mathrm{f}}(T_{\mathrm{f}}'R_{\mathrm{b}}E + T_{\mathrm{b}})}{1 - R_{\mathrm{f}}'R_{\mathrm{b}}E^2} = A_{\mathrm{f}} + T_{\mathrm{f}}\left[1 - E\frac{(T_{\mathrm{b}} + R_{\mathrm{b}}T_{\mathrm{f}}'E)}{1 - R_{\mathrm{f}}'R_{\mathrm{b}}E^2}\right]. \tag{32}$$

These expressions for $\bar{A}$, $\bar{R}$, and $\bar{T}$ bring out an interesting point. When the lamelliform is not symmetric about its center plane, the reflectance and absorptance would differ if the lamelliform were to be inverted 180° with respect to the source. Since both of the products $T_{\mathrm{f}}T_{\mathrm{b}}$ and $R_{\mathrm{f}}'R_{\mathrm{b}}$ are invariant for the 180° reversal, the transmittance is the same. Palik et al. [9] measured the reflectance for a series of asymmetric Si slabs and explicitly showed these results.

When the various lamellae are weakly absorbing, formulas such as these provide insight into the optical properties of an otherwise complex system. For example, if the lamellae on either side of the slab have negligible absorption, then

$$\bar{A} = \frac{(1 - R_{\mathrm{f}})(1 + R_{\mathrm{b}}E)(1 - E)}{1 - R_{\mathrm{f}}'R_{\mathrm{b}}E^2} \simeq \frac{2\beta_{\mathrm{s}}d_{\mathrm{d}}(1 - R_{\mathrm{f}})(1 + R_{\mathrm{b}})}{1 - R_{\mathrm{f}}'R_{\mathrm{b}}}. \tag{33}$$

Whenever the optical properties of the lamellae are symmetrical with respect to the slab, i.e., $R_{\mathrm{b}} = R_{\mathrm{f}}'$, the total absorptance is approximated by

$$\bar{A} = \frac{(1 - E)(1 - R_{\mathrm{f}})}{1 - R_{\mathrm{f}}'E} \simeq 2\beta_{\mathrm{s}}d_{\mathrm{s}} \tag{34}$$

(which is also true for the bare-substrate case). The approximations in Eqs. (33) and (34) hold for the weakly absorbing substrate case. Measurements of these types can be very useful in determining the optical properties in the infrared spectral region. Stierwalt and Potter [10] studied the infrared properties of semiconductors by using this approach.

Reflectance of Single Film on Nontransmitting Substrate B

When the substrate is opaque ($E = 0$) or infinite in depth (or sufficiently nonparallel, i.e., $R_b = 0$), a reflectance experiment measures the front-surface reflectance R_f only. The use of the matrix multiplication for a single layer will demonstrate the approach for determining R_f for Eqs. (29)–(34). The resulting formulas have been found to be useful in optical measurements of metals, dielectrics, and semiconductors.

Using matrices in the form given in Eqs. (13) and (14) [Eqs. (10) and (23) can be used but Eqs. (13) and (14) have the Fresnel relations already given], the front surface resultant matrix m_f is

$$\hat{m}_f = \frac{1}{\hat{t}_{01}} \begin{bmatrix} 1 & \hat{r}_{01} \\ \hat{r}_{01} & 1 \end{bmatrix} \begin{bmatrix} \exp i\hat{\gamma}d & 0 \\ 0 & \exp -i\hat{\gamma}d \end{bmatrix} \frac{1}{\hat{t}_{12}} \begin{bmatrix} 1 & \hat{r}_{12} \\ \hat{r}_{12} & 1 \end{bmatrix}$$

$$= \frac{1}{(Ww)^{1/2}} \begin{bmatrix} 1 + \hat{r}_{01}\hat{r}_{12}W\hat{w} & \hat{r}_{12} + \hat{r}_{01}W\hat{w} \\ \hat{r}_{01} + \hat{r}_{12}W\hat{w} & Ww + \hat{r}_{01}\hat{r}_{12} \end{bmatrix} \frac{1}{\hat{t}_{01}\hat{t}_{12}}, \qquad (35)$$

where $W\hat{w} = \exp(-2i\hat{\gamma}d)$, $W = \exp(-2\beta d)$, $\hat{w} = \exp(-2i\delta)$, $\delta = \alpha d$, and $\hat{\gamma} = \alpha - i\beta$ for the layer.

The reflectivity is written

$$\hat{r}_f = \frac{\hat{r}_{01} + \hat{r}_{12} - \hat{r}_{12}(1 - W\hat{w})}{1 + \hat{r}_{01}\hat{r}_{12} - \hat{r}_{01}\hat{r}_{12}(1 - W\hat{w})}$$

$$= \frac{\hat{r}_{02} - \hat{r}_{12}(1 - W\hat{w})/(1 + \hat{r}_{01}\hat{r}_{12})}{1 - \hat{r}_{01}\hat{r}_{12}(1 - W\hat{w})/(1 + \hat{r}_{01}\hat{r}_{12})},$$

giving, in turn, the reflectance

$$R_f = \hat{r}_f\hat{r}_f^*$$

$$= \frac{R_{02} - \{R_{12}(1 - W^2) + 2(1 - W)\,\mathrm{Re}(\hat{r}_{01}\hat{r}_{12}^*) + 4W \sin \delta\,[\sin \delta\,\mathrm{Re}(\hat{r}_{01}\hat{r}_{12}^*) + \cos \delta\,\mathrm{Im}(\hat{r}_{01}\hat{r}_{12}^*)]\}/D}{1 - \{R_{01}R_{12}(1 - W^2) + 2(1 - W)\,\mathrm{Re}(\hat{r}_{01}\hat{r}_{12}^*) + 4W \sin \delta\,[\sin \delta\,\mathrm{Re}(\hat{r}_{01}\hat{r}_{12}^*) - \cos \delta\,\mathrm{Im}(\hat{r}_{01}\hat{r}_{12}^*)]\}/D} \qquad (36)$$

with

$$D = 1 + R_{01}R_{12} + 2\,\mathrm{Re}(\hat{r}_{01}\hat{r}_{12}^*), \qquad R_{02} = [R_{01} + R_{12} + 2\,\mathrm{Re}(\hat{r}_{01}\hat{r}_{12}^*)]/D$$

(the front surface reflectance of bare substrate).

As W approaches zero, i.e., the film becomes opaque, R_f approaches the value for R_{01}. For the weakly absorbing layer, i.e., $W \cong 1 - 2\beta d$, $\mathrm{Im}(r_{01}) \cong 0$,

$$R_f = \frac{R_{02} - 4[\beta d(R_{12} + r_{01}g) + r_{01} \sin \delta\,(g \sin \delta - h \cos \delta)]/D}{1 - 4[\beta d(R_{01}R_{12} + r_{01}g) + r_{01} \sin \delta\,(g \sin \delta - h \cos \delta)]/D}, \qquad (37)$$

where

$$D = (1 + r_{01}g)^2 + (r_{01}h)^2, \qquad g = \text{Re}(\hat{r}_{12}), \qquad h = \text{Im}(\hat{r}_{12}).$$

The expression for the completely nonabsorbing case ($\beta = 0$) has been given elsewhere [11]. When the layer is so thin that d_s^2 can be ignored compared to unity, Eq. (36) becomes

$$R_f = \frac{R_{02} - 4\{\beta d[R_{12} + \text{Re}(\hat{r}_{01}\hat{r}_{12}^*)] + \delta\,\text{Im}(\hat{r}_{01}\hat{r}_{12}^*)\}/D}{1 - 4\{\beta d[R_{01}R_{12} + \text{Re}(\hat{r}_{01}\hat{r}_{12}^*)] - \delta\,\text{Im}(\hat{r}_{01}\hat{r}_{12}^*)\}/D}. \tag{38}$$

This expression is equivalent to that given by McIntyre and Aspnes [12].

VI SUMMARY

This brief chapter is a summary of the relations between the optical constants and the macroscopic equations (Maxwell's equations for propagation through the material), as well as the constitutive (material) equations and the intrinsic microscopic parameters of the physics of the condensed state. A series of formulas has been developed on the basis of those relationships that are useful to the experimenter in interpreting his optical measurements. The application of either the characteristic or Fresnel matrices permits descriptions of systems of varying complexity. A small sample of the possible lamellar systems has been given here.

APPENDIX A   BASIC FORMULAS FOR FRESNEL COEFFICIENTS

The Fresnel coefficients $\hat{r}_{ij}$ and $\hat{t}_{ij}$ are functions of the real and imaginary parts of the complex dielectric functions $\hat{\varepsilon}_i$, and $\hat{\varepsilon}_j$, the incident dielectric function n_0^2 and the angle of incidence ϕ. Following is a compact set of formulas defining the coefficients in terms of these parameters.

The incident medium is considered to be completely transparent and to have a dielectric function n_0^2. Snell's law defines the quantity $S = n_0 \sin \phi$, which is a real quantity throughout the system; i.e., $\hat{n}_j \sin \phi_j = S$ is real in the jth medium. The phase term $\gamma_0 = n_0 \cos \phi$ is also real but $\hat{\gamma}_j = \hat{n}_j \cos \phi_j$ is, in general, a complex quantity.

In the jth layer one has

$$\hat{\varepsilon}_j \mu_j = \hat{n}_j^2 = \varepsilon_j' - i\varepsilon_j'', \qquad \hat{n}_j = \hat{n}_j - ik_j, \qquad \hat{\gamma}_j = \alpha_j - i\beta_j,$$

where

$$\hat{\gamma}_j \hat{\gamma}_j^* = P_j^2 = \alpha_j^2 + \beta_j^2 = \sqrt{(\varepsilon_j' - S^2)^2 + \varepsilon_j''^2},$$
$$2\alpha_j^2 = P_j^2 + \varepsilon_j' - S^2, \qquad 2\beta_j^2 = P_j^2 - \varepsilon_j' + S^2.$$

The function $\hat{\zeta}_j$ is defined for either polarization as

$$\hat{\zeta}_j = \begin{cases} \mu Z_0/\hat{\gamma}_j & \text{(TE, s)}, \\ -\hat{\varepsilon}_j/(\hat{\gamma}_j Z_0) & \text{(TM, p)}. \end{cases} \tag{8'}$$

The Fresnel coefficients can be expressed for either polarization in terms of the functions $\hat{\zeta}_j$ and $\hat{\zeta}_k$. The positive direction is taken as going from the jth medium to the kth medium:

$$\hat{r}_{jk} = (\hat{\zeta}_k - \hat{\zeta}_j)/(\hat{\zeta}_j + \hat{\zeta}_k) = g_{jk} + ih_{jk},$$

$$\hat{t}_{jk} = 1 + \hat{r}_{jk} = 2\hat{\zeta}_k/(\hat{\zeta}_j + \hat{\zeta}_k).$$

Note. An alternative convention for defining $\hat{n}_j$ and $\hat{n}_j^2$ is in common use. The solutions for Eq. (7) can be expressed as the inverse of the matrix in Eq. (10) as well as Eq. (14). This requires that $\hat{n}_j$ and $\hat{n}_j^2$ be expressed as $\hat{n}_j = n_j + ik_j$ and $\hat{n}_j^2 = \varepsilon_j' + i\varepsilon_j''$. The formulas, as developed in this chapter, are compatible with this convention because the product $h_{jk} \sin \alpha_j d_j$ is invariant.

APPENDIX B   GENERAL FORMULAS FOR THE CASE OF A PARALLEL-SIDED SLAB

a. Asymmetrically Layered Slab (Absorbing Layers)

$$\bar{R} = R_f + \frac{T_f T_f' R_b E^2}{1 - R_f' R_b E^2}, \tag{31}$$

$$\bar{T} = \frac{T_f T_b E}{1 - R_f' R_b E^2}, \tag{29}$$

$$\bar{A} = A_f + T_f \left[1 - \frac{E(T_b + R_b T_f' E)}{1 - R_f' R_b E^2} \right]. \tag{32}$$

Here $E = \exp(-2\beta_s d_s)$ and f, b refer to the front and back sides of the slab, respectively. It is assumed that d_s is sufficiently large that the phase coherency in $\exp(i\alpha_s d_s)$ is averaged out and that $k_s/n_s \ll 1$, where s refers to the substrate.

b. Symmetrically Layered Slab (Absorbing Layers).   For $R_f' = R_b$ and $T_f' = T_b$

$$\bar{R} = R_f + \frac{T_f T_f' R_f' E^2}{1 - R_f'^2 E^2},$$

$$\bar{T} = \frac{T_f T_f' E}{1 - R_f'^2 E^2},$$

$$\bar{A} = (1 - R_f) - \frac{E T_f T_f'}{1 - R_f' E}$$

$$= \frac{(1 - R_f)(1 - E)}{1 - R_f' E} + E \left[\frac{A_f(1 - R_f') + A_f'(1 - R_f) - A_f A_f'}{1 - R_f' E} \right].$$

c. Nonabsorbing Layers with Symmetry. For $R_f = R'_f$ and $T_f \cdot T'_f = (1 - R_f)^2$

$$\bar{R} = R_f \left[\frac{1 + E^2(1 - 2R_f)}{1 - R_f^2 E^2} \right] \simeq \frac{2R_f}{1 + R_f} \left(1 - \frac{2\beta_s d_s}{1 + R_f} \right),$$

$$\bar{T} = \frac{(1 - R_f)^2 E}{1 - R_f^2 E^2} \simeq \frac{1 - R_f}{1 + R_f} \left(1 - 2\beta_s d_s \frac{1 + R_f^2}{1 - R_f^2} \right),$$

$$\bar{A} = \frac{1 - R_f}{1 - R_f E} (1 - E_\infty) \simeq 2\beta_s d_s (1 - 2\beta_s d_s R_f).$$

Approximations hold for $E \simeq 1 - 2\beta_s d_s$. (See Appendix D for single layer and bare substrate.)

APPENDIX C   REFLECTANCE, R_{jk} AT j–k INTERFACE

$$R_{jk} = \hat{r}_{jk} \hat{r}_{jk}^* = \begin{cases} \dfrac{P_j^2 + P_k^2 - 2(\alpha_j \alpha_k + \beta_j \beta_k)}{P_j^2 + P_k^2 + 2(\alpha_j \alpha_k + \beta_j \beta_k)} = R_{jks} & \text{(TE, s)} \\[2ex] R_{jks} \dfrac{S^4 + P_j^2 P_k^2 - 2S^2(\alpha_j \alpha_k - \beta_j \beta_k)}{S^4 + P_j^2 P_k^2 + 2S^2(\alpha_j \alpha_k - \beta_j \beta_k)} = R_{jkp} & \text{(TM, p).} \end{cases}$$

Whenever $j = 0$ or $k = 0$, i.e., whenever the interface is between a nonabsorbing ambient and another medium, terms in S^2 and γ_0 can be consolidated to give

$$R_{0ks} = \frac{\gamma_0^2 + P_k^2 - 2\gamma_0 \alpha_k}{\gamma_0^2 + P_k^2 + 2\gamma_0 \alpha_k}$$

and the ratio

$$\mathcal{R}_{0k} = \frac{R_{0kp}}{R_{0ks}} = \frac{X_0^2 + P_k^2 - 2X_0 \alpha_k}{X_0^2 + P_k^2 + 2X_0 \alpha_k}, \quad \text{where} \quad X_0 = \frac{S^2}{\gamma_0}.$$

APPENDIX D   REFLECTANCE OF SINGLE LAYER ON EACH SIDE OF A SLAB AND SINGLE LAYER ON EITHER SIDE OF A SLAB

In Eqs. (29), (31), and (32), the transmittance $\bar{T}$, reflectance $\bar{R}$, and absorptance $\bar{A}$ are given for a transparent, paralled-sided slab, with the phase coherency averaged out in terms of the front and back-surface reflectances and transmittances. These terms will be evaluated for the case of a single layer on each side of the slab. If each layer has identical properties and thickness, the system is said to be symmetrical. For that case $T_b = T'_f$ and $R'_f = R_b$; otherwise, they must be evaluated separately in the same manner as that illustrated below for the front surface.

The following expressions for the front-surface reflectance R_f for a single layer on a substrate are given in terms of parameters that are explicit functions of the real and imaginary parts of the dielectric functions $\hat{\varepsilon}_j$, the angle of incidence ϕ, and the front-layer-thickness-to-wavelength ratio d. [For example, 0 labels the incident ambient, 1 the film, 2 the substrate, and 0 the back ambient (film on front surface); for film on back surface, 0 labels the incident ambient, 2 the substrate, 1 the film, and 0 the back ambient.]

Expressions for single layer on transparent substrate with interfaces 01 and 12 (Eq. 36) for use in Eqs. (31), (29), and (32):

$$R_f = \{R_{02} - \{R_{12}(1 - W^2) + 2(1 - W)(g_{01}g_{12} + h_{01}h_{12})$$
$$+ 4W[(g_{01}g_{12} + h_{01}h_{12})\sin^2\delta$$
$$- (g_{01}h_{12} - g_{12}h_{01})\sin\delta\cos\delta]\}/D\}/D_f,$$

$$D_f = 1 - \{R_{01}R_{12}(1 - W^2) + 2(1 - W)(g_{01}g_{12} - h_{01}h_{12})$$
$$+ 4W[(g_{01}g_{12} - h_{01}h_{12})\sin^2\delta$$
$$- (g_{01}h_{12} + g_{12}h_{01})\sin\delta\cos\delta]\}/D; \qquad (36')$$

$$R_{02} = [R_{01} + R_{12} + 2(g_{01}g_{12} + h_{01}h_{12})]/D$$
$$= g_{02}^2 + h_{02}^2 \qquad \text{(see Appendix C)},$$

$$D = 1 + R_{01}R_{12} + 2(g_{01}g_{12} - h_{01}h_{12}),$$

$$R_{01} = g_{01}^2 + h_{01}^2 \qquad \text{(see Appendix C)},$$

$$R_{12} = g_{12}^2 + h_{12}^2 \qquad \text{(see Appendix C)},$$

$$W = \exp(-2\beta_1 d_1), \qquad \delta = \alpha_1 d_1.$$

TE, s polarization $\hat{\zeta}_j = \mu Z_0/\hat{\gamma}_j$:

$$g_{jk} = (P_j^2 - P_k^2)/D_{jk}, \qquad h_{jk} = 2(\alpha_j\beta_k - \alpha_k\beta_j)/D_{jk},$$
$$D_{jk} = P_j^2 + P_k^2 + 2(\alpha_j\alpha_k + \beta_j\beta_k).$$

TM, p polarization $\hat{\zeta}_j = -\hat{\varepsilon}_j\mu/\hat{\gamma}_j$:

$$g_{jk} = [(P_j^2 P_k^2 - S^2)(P_k^2 - P_j^2) - 4S^2(\alpha_j^2\beta_k^2 - \alpha_k^2\beta_j^2)]/D_{jk},$$
$$h_{jk} = 2[(\varepsilon_j'\varepsilon_k' - \varepsilon_j''\varepsilon_k'')(\alpha_j\beta_k - \alpha_k\beta_j)$$
$$+ (\varepsilon_j''\varepsilon_k' - \varepsilon_k''\varepsilon_j')(\alpha_j\alpha_k + \beta_j\beta_k)]/D_{jk},$$
$$D_{jk} = [P_j^2 + P_k^2 + 2(\alpha_j\alpha_k + \beta_j\beta_k)][(P_j^2 P_k^2 + S^4 + 2S^2(\alpha_j\alpha_k - \beta_j\beta_k)].$$

R_f' is evaluated from the expression for R_f by interchanging the subscripts 01 and 12. Figure 1 shows the configurations for a film on the front surface of a slab substrate and a second configuration for a film on the back surface. The film is labeled 1, the substrate 2, and the front and back ambients 0 in each case.

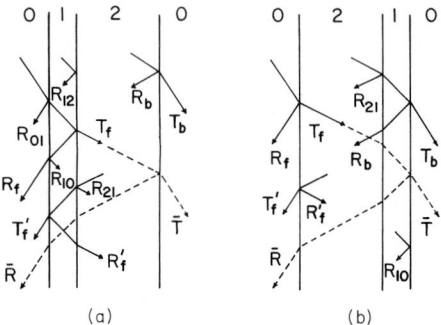

Fig. 1. Single layer on either side of parallel-sided slab. The number zero denotes ambient; one, single layer; and two, slab. (a) Layer on front side, (b) layer on back side.

For the first configuration Ia, R_f and R'_f are as given above;

$$R_b = R_{02}, \qquad T_f = (1 + R_{02} + 2g_{02})\frac{\zeta_2}{\zeta_0}\frac{W}{D_f},$$

$$T'_f = (1 + R_{02} - 2g_{02})\frac{\zeta_2}{\zeta_0}\frac{W}{D_f};$$

$$T_b(1 - R_{02})\frac{\zeta_2}{\zeta_0} = T_{02}, \qquad g_{jk} = -g_{kj}, \qquad h_{jk} = -h_{kj}, \qquad R_{jk} = R_{kj}.$$

For the second configuration Ib,

$$R_f = R_{02} = R'_f, \qquad R_b = R'_f \qquad \text{(configuration Ia)},$$

$$T_f = (1 - R_{02})\frac{\zeta_0}{\zeta_2}, \qquad T'_f = (1 - R_{02})\frac{\zeta_2}{\zeta_0},$$

$$T_b = T'_f \qquad \text{(configuration Ia)}.$$

For a symmetrically layered slab, $R_b = R'_f$ and $T_b = T'_f$. Also,

$$T_f = (1 + R_{02} + 2g_{02})\frac{\zeta_0}{\zeta_2}\frac{W}{D_f}, \qquad T'_f = (1 + R_{02} - 2g_{02})W\frac{\zeta_2}{\zeta_0},$$

$$T_f T'_f = \frac{[(1 - R_{02})^2 + 4h_{02}^2]W^2}{D_f^2}, \qquad T_f T'_f = T_f T_b \frac{W_f}{D_f}.$$

(N.B.: h_{02} vanishes for the nonopaque slab because $k_2/n_2 = k_s/n_s \ll 1$, so that $\beta_0 = \beta_2 = 0$.)

Here R_f is expressed in a manner that emphasizes the presence of the substrate reflectance R_{02}. Whenever the absorption of the film is small, this substrate reflectance will be apparent. However, when the layer is opaque, the expression for R_f becomes that for R_{01}, the front-surface reflectance of the layer itself.

For the symmetrical case of an identical film on each side, Eqs. (31), (29), and (32) can be expressed as

$$\bar{R} = R_f + \frac{[(1 - R_{02})^2 + 4h_{02}^2]E^2W^2R_f'}{(1 - R_f'^2E^2)D_f^2},\tag{31'}$$

$$\bar{T} = \frac{[(1 - R_{02})^2 + 4h_{02}^2]E^2W^2}{(1 - R_f'^2E^2)D_f^2},\tag{29'}$$

$$\bar{A} = 1 - R_f - \frac{E[(1 - R_{02})^2 + 4h_{02}^2]W2}{D_f^2(1 - R_f'E)}.\tag{32'}$$

These are general expressions for the three measurable parameters $\bar{A}$, $\bar{R}$ and $\bar{T}$, but two special cases are often encountered.

(1)   For a transparent film for which β_1, h_{02}, h_{12} vanish, $W = 1$, $D_f = 1 - C$, $R_f = R_f' = (R_{02} - C)/(1 - C)$, and $C = 4g_{01}g_{12}\sin^2\delta$, these expressions become

$$\bar{R} = \frac{R_f[1 + E^2(1 - 2R_f)]}{1 - R_f^2E^2} = \frac{(R_{02} - C)[1 + E^2(1 - 2R_{02})] - C(1 - E)]}{1 - R_{02}^2E^2 - 2C[1 - 2R_{02}E^2 + C^2(1 - E^2)]},\tag{31''}$$

$$\bar{T} = \frac{(1 - R_f)^2E}{1 - R_f^2E^2} = \frac{(1 - R_{02})^2E}{1 - R_{02}^2E^2 - 2C[1 - 2R_{02}E^2 + C^2(1 - E^2)]},\tag{29''}$$

$$\bar{A} = \frac{(1 - R_f)(1 - E)}{1 - R_fE} = \frac{(1 - R_{02})(1 - E)}{1 - R_{02}E - C(1 - E)}.\tag{32''}$$

It is evident why the reflectance $\bar{R}$ exhibits the strongest interference effects due to the presence of a film while the absorbtance $\bar{A}$ shows the least. These same expressions hold for the bare-substrate case since $C = 0$ and $R_f = R_{02}$.

(2)   The other special case is that for nonabsorbing films on a weakly absorbing substrate, i.e., $E \simeq 1 - 2\beta_s d_s$:

$$\bar{R} \simeq \frac{2R_f}{1 + R_f}\left[\frac{1 - \dfrac{2\beta_s d_s(1 - 2R_f)}{1 - R_f}}{1 + \dfrac{4\beta_s d_s R_f^2}{1 - R_f^2}}\right] \simeq \frac{2R_f}{1 + R_f}\left[1 - \frac{2\beta_s d_s}{1 + R_f}\right],\tag{31'''}$$

$$\bar{T} \simeq \left[\frac{1 - R_f}{1 + R_f}\right]\frac{[1 - 2\beta_s d_s]}{\left[1 + \dfrac{4R_f^2\beta_s d_s}{1 - R_f^2}\right]} \simeq \frac{1 - R_f}{1 + R_f}\left[1 - \frac{2\beta_s d_s(1 + R_f^2)}{1 - R_f^2}\right],\tag{29'''}$$

$$\bar{A} \simeq 2\beta_s d_s\left[1 - \frac{2R_f\beta_s d_s}{1 - R_f}\right].\tag{32'''}$$

Again, as above, these hold for the bare-substrate case in which $R_f = R_{02}$.

At normal incidence ($\phi = 0$),

$$R_{02} = \left(\frac{n_2 - n_0}{n_2 + n_0}\right)^2, \qquad g_{01}g_{12} = \frac{n_1 - n_0}{n_1 + n_0}\frac{n_2 - n_1}{n_1 + n_2},$$

$$\beta_s = k_2 \qquad \text{and} \qquad \delta = n_2 d_s.$$

APPENDIX E   CRITICAL ANGLE OF INCIDENCE

Whenever a layer in a lamelliform is characterized with the real part of its dielectric function ε_j' less than that of the incident medium n_0^2, a special condition exists for the reflectivity from that jth layer at those angles at which $S^2 > \varepsilon_j'$. The transition angle is commonly referred to as the "critical angle of incidence."

In the convention used here $\gamma_j = \alpha_j - i\beta_j$ is complex with α_j and β_j both real and positive quantities, where

$$\alpha_j^2 + \beta_j^2 = P_j^2 = [(\varepsilon_j' - S^2)^2 + \varepsilon_j'']^{1/2},$$

$$2\alpha_j^2 = P_j^2 + \varepsilon_j' - S^2, \qquad 2\beta_j^2 = P_j^2 + S^2 - \varepsilon_j'.$$

If $n_0^2 > \varepsilon_j'$, there exists a critical angle of incidence ϕ_c at which

$$\alpha_j(\phi_c) = \beta_j(\phi_c) = \varepsilon_j''/2^{1/2},$$

where

$$\phi_c = \arcsin(\varepsilon_j'/n_0^2)^{1/2}.$$

(N.B.: Many texts define ϕ_c as $\arcsin(n_j/n_0)$, but this is the special case for a nonabsorbing medium with $\alpha_j(\phi_c) = \beta_j(\phi_c) = 0$.) Although the Fresnel relations for $\phi > \phi_c$ can differ markedly from those for $\phi < \phi_c$, the formalism developed in this chapter is directly applicable.

Two examples for the three-medium expression of Eq. (36) will illustrate the point.

Example 1. Frustrated Total Reflectance.   Consider that the third medium is optically identical to the incident medium and that the intervening medium is completely nonabsorbing, i.e., $n_2 = n_0$, $\varepsilon_1'' < 0$, and $n_1 < n_0$. The term R_{02} in Eq. (36) is zero for all angles of incidence. Also note that $R_{01} = R_{12} = g_{01}^2 + h_{01}^2$, $g_{01} = -g_{12}$, and $h_{01} = -h_{12}$. Also $\delta_1 = \alpha_1 d_1$ and $W = \exp(-2\beta_1 d_1)$.

Rearranging Eq. (36) with $R_{02} = 0$ gives

$$R_f = \frac{R_{01}[4W \sin^2 \delta_1 + (1 - W)^2]}{(1 - R_{01}W)^2 + 4W[h_{01}^2 + (g_{01}^2 - h_{01}^2)\sin^2 \delta_1 - 2g_{01}h_{01}\sin \delta_1 \cos \delta_1]}. \tag{36''}$$

It is evident that below the critical angle the reflectance varies between zero and $4R_{01}/(1 + R_{01})^2$ as a function of $\alpha_1 d_1$, while above the critical angle, $R_{01} = 1$, $\alpha_1 = 0$ and R_f has values between zero (zero thickness) and unity (i.e. thickness d_1 is sufficiently large that $\beta_1 d_1$ is such a large number that $W \to 0$).

What is termed *frustrated total reflection* occurs for intermediate thicknesses. For small values of $2(S^2 - n_1^2)^{1/2} d_1 = \beta_1 d_1$,

$$R_f \simeq \frac{(S^2 - n_1^2)d_1^2}{h_{01}^2 + (S^2 - n_1)d_1^2}$$

vanishes as d_1^2 and becomes zero, as expected. On the other hand, as the thickness-to-wavelength ratio d_1 is increased, W decreases exponentially; and as it becomes small in value, R_f can be approximated as

$$R_f \simeq 1 - [4Wh_{01}^2/(1 - W)^2].$$

Although the front-surface reflectance R_{01} is unity, it is modified by the factor W. This is the result of penetration of the fields into the rarer medium. The W term is a measure of the damping of these fields, and if the two surfaces are not separated by a suitable amount, the evanescent waves couple into the third medium—frustrating the totality of the reflectance. Equations (13), (14), and (22) contain the information necessary for studying the amplitudes and intensities of the fields present in the rarer medium.

The following list of equations gives the parameters needed for the frustrated total-reflectance case:

$$R_f = \frac{R_{01}[4W \sin^2 \delta + (1 - W)^2]}{(1 - R_{01})^2 + 4W[h_{01}^2 + (g_{01}^2 - h_{01}^2) \sin^2 \delta - 2g_{01}h_{01} \sin \delta \cos \delta]},$$

$$\gamma_1 = \alpha_1 - i\beta_1, \quad P_1^2 = \alpha_1^2 + \beta_1^2, \quad \gamma_1^2 = \alpha_1^2 - \beta_1^2 - 2i\alpha_1\beta_1 = n_1^2 - S^2,$$

$$\gamma_0 = n_0 \cos \phi, \quad X_0 = S^2/\gamma_0, \quad S^2 = n_0^2 \sin^2 \phi.$$

	$\phi < \phi_c$	$\phi > \phi_c$
α_1:	$(n_1^2 - S^2)^{1/2}$	0
β_1:	0	$(S^2 - n_1^2)^{1/2}$
γ_1:	α_1	$-i\beta_1$
P_1:	$\alpha_1^2 = \gamma_1^2$	$\beta_1^2 = -\gamma_1^2$
W:	1	$\exp(-2\beta_1 d_1)$
δ:	$\gamma_1 d_1$	0

$$g_{01}: \begin{cases} \dfrac{\gamma_0 - \gamma_1}{\gamma_0 + \gamma_1} & \text{(TE, s)} \quad \dfrac{\gamma_0^2 + \gamma_1^2}{\gamma_0^2 - \gamma_1^2} = \dfrac{n_0^2 + n_1^2 - 2S^2}{n_0^2 - n_1^2} \\[3mm] \dfrac{(\gamma_1 - X_0)(\gamma_1 - \gamma_0)}{(\gamma_1 + X_0)(\gamma_1 + \gamma_0)} & \text{(TM, p)} \quad \dfrac{\gamma_1^2(X_0 + \gamma_0)^2 + (\gamma_1^2 + X_0\gamma_0)^2}{(X_0^2 - \gamma_1^2)(\gamma_0^2 - \gamma_1^2)} \\[3mm] & \qquad\qquad = \dfrac{(\gamma_1^2/\gamma_0)n_0^4 + n_1^4}{(n_0^2 - n_1^2)[(n_0^2/\gamma_0)X_0 - n_1^2]} \end{cases}$$

(continued on next page)

(continued from preceding page)

$$h_{01}: \begin{cases} 0 & \text{(TE, s)} & \dfrac{2i\gamma_1\gamma_0}{\gamma_0^2 - \gamma_1^2} = \dfrac{2\{(S^2 - n_1^2)(n_0^2 - S^2)\}^{1/2}}{n_0^2 - n_1^2} \\[4mm] 0 & \text{(TM, p)} & \dfrac{2i\gamma_1(\gamma_0 + X_0)(\gamma_1^2 + S^2)}{(X_0^2 - \gamma_1^2)(\gamma_0^2 - \gamma_1^2)} \\[4mm] & & = \dfrac{2(S^2 - n_1^2)^{1/2}(n_0^2/\gamma_0)n_1^2}{[(X_0/\gamma_0)n_0^2 - n_1^2](n_0^2 - n_1^2)} \end{cases}$$

$$R_{01}: \begin{cases} \left(\dfrac{\gamma_0 - \gamma_1}{\gamma_0 + \gamma_1}\right)^2 & \text{(TE, s)} & \qquad\qquad 1 \\[5mm] \left[\dfrac{(\gamma_1 - X_0)(\gamma_1 - \gamma_0)}{(\gamma_1 + X_0)(\gamma_1 + \gamma_0)}\right]^2 & \text{(TM, p)} & \qquad\qquad 1 \end{cases}$$

Example 2. Attenuated Total Reflectance.   A more interesting experimental situation exists when either medium 1 or 2 absorbs a fraction of the radiation; i.e., $\varepsilon_j'' \neq 0$. In this case, the radiation is also attenuated by the factors $\exp(-\beta_j d_j)$, hence the name *attenuated total reflectance* (ATR). Two cases are considered: (1) the layer is the rarer medium but slightly absorbing and (2) the third medium is the rarer one, slightly absorbing and sufficiently thick that $\exp(-\beta_2 d_2)$ is zero. The weak absorption is determined by $\varepsilon_j''^2/(\varepsilon_j' - S^2)^2 \ll 1$. Above the critical angle the following approximations can be made:

$$P_j^2 = [(S^2 - \varepsilon_j')^2 + \varepsilon_j''^2]^{1/2} \simeq (S^2 - \varepsilon_j')\left[1 + \frac{\varepsilon_j''^2}{2(S^2 - \varepsilon_j')^2}\right] \simeq (S^2 - \varepsilon_j'),$$

$$2\alpha_j^2 \simeq \frac{\varepsilon_j''^2}{2(S^2 - \varepsilon_j')}, \qquad 2\beta_j^2 \simeq 2(S^2 - \varepsilon_j')\left[1 + \frac{\varepsilon_j''^2}{2(S^2 - \varepsilon_j')}\right] \simeq 2(S^2 - \varepsilon_j'),$$

$$\alpha_j \simeq \frac{\varepsilon_j''}{2(S^2 - \varepsilon_j')^{1/2}}, \qquad \beta_j \simeq (S^2 - \varepsilon_j')^{1/2}.$$

Case 1.   Medium 1 is rarer than the incident medium; $\varepsilon_1' < n_0^2$. An example is a low-index thick film deposited on a high-index internal-reflection prism. For angles above the critical angle where $(S^2 - \varepsilon_1')^{1/2}d$ is sufficiently large that $W = \exp(-2\beta_1 d_1) \ll 1$ and $W\sin\delta$ also becomes vanishingly small, Eq. (36) reduces to $R_f = R_{01}$, and the reflectance measurement probes the properties of the layer.

For TE, s polarization

$$R_f = \frac{\gamma_0^2 + P_1^2 - 2\alpha_1\gamma_0}{\gamma_0^2 + P_1^2 + 2\alpha_1\gamma_0} \simeq 1 - \frac{4\alpha_1\gamma_0}{\gamma_0^2 + P_1^2},$$

$$R_f \simeq 1 - \frac{2\varepsilon_1''\gamma_0}{(S^2 - \varepsilon_1')^{1/2}(n_0^2 - \varepsilon_1')}.$$

Similarly for TM, p polarization

$$R_f \simeq 1 - \frac{4\alpha(P_1^2 + X_0\gamma_0)(X_0 + \gamma_0)}{(P^2 + X_0^2)(P^2 + \gamma_0^2) + 4\alpha^2 X_0\gamma_0}$$

$$\simeq 1 - \frac{2\varepsilon_1''(2S^2 - \varepsilon_1')\gamma_0 n_0^2}{(S^2 - \varepsilon_1')^{1/2}(n_0^2 - \varepsilon_1')(n_0^2 S^2 - \varepsilon_1'\gamma_0^2)}.$$

It should be noted that the ratio $\mathcal{R} = R_s/R_p$ can also be used as a probe in ATR:

$$\mathcal{R} = 1 - \frac{2\varepsilon_1'' S^2 \gamma_0}{(S^2 - \varepsilon_1')^{1/2}(S^2 n_0^2 - \varepsilon_1'\gamma_0^2)}.$$

Whenever W does not vanish, terms involving it must be included in Eq. (36).

Case 2. Medium 2 is considered to be rarer, i.e., $n_0^2 > \varepsilon_2'$, as well as weakly absorbing. An example is a coupling prism spaced by air from a sample surface. The intervening layer, medium 1, is nonabsorbing with a refractive index such that $\varepsilon_1' > S^2$ for all angles of interest so that $\beta_1 \simeq 0$ and $W = 1$, so that

$$R_f = \frac{R_{02} - C_{s,p}}{1 - C_{s,p}}, \qquad C_{s,p} = 4g_{01}\sin\delta\,(g_{12}\sin\delta - h_{12}\cos\delta).$$

Evaluating R_{02} in the same manner as that for R_{01} for case 1 gives

$$R_{02} \simeq 1 - \frac{2\varepsilon_2''\gamma_0}{(S^2 - \varepsilon_2')^{1/2}(n_0^2 - \varepsilon_2')}$$

$$R_f \simeq 1 - \frac{2\varepsilon_2''\gamma_0}{(S^2 - \varepsilon_2')^{1/2}(n_0^2 - \varepsilon_2')(1 - C_s)} \qquad \text{(TE, s)},$$

$$R_{02} \simeq 1 - \frac{2\varepsilon_2''\gamma_0 n_0^2(2S^2 - \varepsilon_2')}{(S^2 - \varepsilon_2')^{1/2}(n_0^2 - \varepsilon_2')(n_0^2 S^2 - \varepsilon_2'\gamma_0^2)}$$

$$R_f \simeq 1 - \frac{2\varepsilon_2''(\gamma_0 n_0^2)(2S^2 - \varepsilon_2')}{(S^2 - \varepsilon_2')^{1/2}(n_0^2 - \varepsilon_2')(n_0^2 S^2 - \varepsilon_2'\gamma_0^2)(1 - C_p)} \qquad \text{(TM, p)}.$$

Depending on the thickness of the intervening layer, the ATR for case 2 is modulated by the factor $(1 - C)^{-1}$. In both cases, the reflectance becomes total as the absorption vanishes. Although the ratio $\mathcal{R} = R_p/R_s$ can be useful as a probe, care must be taken in case 2 with terms $C_{s,p}$.

DEFINITIONS OF TERMS

$\mu\mu_0$ Magnetic permeability of a medium
μ_0 Magnetic permeability of free space
$\varepsilon\varepsilon_0$ Permittivity of a medium

ε_0 Permittivity of free space

c Speed of light in free space: $c = 1/\sqrt{\mu_0\varepsilon_0}$

$\hat{n}$ Complex refractive index of medium: $\hat{n} = \sqrt{\mu\hat{\varepsilon}} = n - ik$

$\hat{n}^2$ Complex dielectric function: $\hat{n}^2 = (\varepsilon' - i\varepsilon'') = n^2 - k^2 - 2ink$

S Snell's law, S is constant: $S = \hat{n} \sin \phi = n_0 \sin \phi_0 = n_2 \sin \phi_2$

$\hat{\gamma}$ Complex phase term: $\hat{\gamma} = \hat{n} \cos \phi = \sqrt{\hat{n}^2 - S^2} = \alpha - i\beta$

Z_0 Impedance of free space: $Z_0 = \sqrt{\mu_0/\varepsilon_0}$

Z Complex impedance of medium: $\hat{Z} = \sqrt{\mu/\hat{\varepsilon}}\, Z_0$

λ Wavelength of light in free space: $\lambda = c/\nu$

$\hat{r}_{if}$ Fresnel coefficient for reflectivity: $\hat{r}_{if} = (\hat{\zeta}_f - \hat{\zeta}_i)/(\hat{\zeta}_f + \hat{\zeta}_i)$

$\hat{t}_{if}$ Fresnel coefficient for transmittivity across interface in forward direction:
$$\hat{t}_{if} = 1 + \hat{r}_{if} = 2\hat{\zeta}_f/(\hat{\zeta}_f + \hat{\zeta}_i)$$

$\hat{t}'_{if}$ Fresnel coefficient for transmittivity across interface in backward direction:
$$\hat{t}'_{if} = 1 - \hat{r}_{if} = 2\hat{\zeta}_i/(\hat{\zeta}_f + \hat{\zeta}_i)$$

κ Propagation vector in free space: $\kappa = 2\pi/\lambda$

z Dimensionless parameter for propagation vector $\times$ distance along z axis:
$z = \kappa \times$ physical distance

d Dimensionless parameter for fixed distance: $d = (z_f - z_i)$

REFERENCES

1. J. C. Slater and N. I. T. Frank, "Electromagnetism," McGraw-Hill, New York, 1947, or Dover, New York, 1969.
2. M. Born and E. Wolf, "Principles of Optics," Pergamon, New York, 1975.
3. R. Jacobsson, "Progress in Optics," Vol. 5 (E. Wolf, ed.), North-Holland Publ., New York, 1966.
4. I. A. Macleod, "Thin Film Optical Filter," American Elsevier, New York, 1969.
5. O. S. Heavens, "Optical Properties of Thin Films," Butterworth, London, 1955, or Dover, New York, 1965.
6. T. Moss, "Optical Properties of Semi-Conductors," Butterworth, London, 1959.
7. R. F. Potter, *SPIE Proceedings* **276**, 204 (1981).
8. C. Gabriel and A. Nedoluha, *Opt. Acta* **18**, 415 (1971).
9. Edward D. Palik, N. Ginsburg, H. B. Rosenstock, and R. T. Holm, *Appl. Opt.* **17**, 3345 (1978).
10. D. L. Stierwalt and R. F. Potter, "Semiconductor and Semimetals," Vol. 3 (R. K. Willardson and A. C. Beer, eds.), p. 71, Academic Press, New York, 1967.
11. Roy F. Potter, "Optical Properties" (S. Nudelman and S. S. Mitra, eds.), Plenum, New York, 1969.
12. J. D. McIntyre and D. E. Aspnes, *Surf. Sci.* **24**, 417 (1971).

Chapter **3**

Dispersion Theory, Sum Rules, and Their Application to the Analysis of Optical Data*

`D. Y. Smith

Argonne National Laboratory
Argonne, Illinois
and
Max-Planck-Institut für Festkörperforschung
Stuttgart, Federal Republic of Germany

I.	Introduction	36
II.	Optical Sum Rules and Their Physical Interpretation	36
	A. Superconvergence Sum Rules for $\varepsilon(\omega)$	38
	B. Superconvergence Sum Rules for $N(\omega)$	41
	C. Superconvergence Sum Rules for Functions of N and ε	42
	D. Superconvergence Relations Employing Weighting Functions	44
	E. Static-Limit Sum Rules	44
III.	Finite-Energy Sum Rules	45
	A. Defect Absorptions and Smakula's Equation	47
	B. Finite Oscillator Strength Sums and $n_{\text{eff}}(\omega)$	49
IV.	Sum Rules for Reflection Spectroscopy	51
	A. Normal-Incidence Spectroscopy	51
	B. Reflectance between Media at Nonnormal Incidence, Ellipsometry, and Transmission	55
V.	Analysis of Optical Data and Sum-Rule Applications	55
	A. Construction of Composite Data Sets	56
	B. Sum-Rule Tests	60
VI.	Summary	64
	References	64

* Work supported by the U.S. Department of Energy and the Max-Planck Gesellschaft.

35

I INTRODUCTION

The basic optical properties of solids [1, 2], despite their great diversity, are rigorously limited by nature. These limitations take the form of sum rules and dispersion relations that reflect the physical laws governing the dynamics of matter and its interaction with light. Many of these limitations were enumerated in the first half of this century, but in the early 1970s, the discovery of a large number of new sum rules [3–15] deepened our understanding of these constraints and their application. Thus, it is now known that in addition to the celebrated f sum rule for absorption processes, there are companion sum rules for dispersive processes. For example, the real part of the complex refractive index $N(\omega) = n(\omega) + ik(\omega)$ satisfies [3, 4]

$$\int_0^\infty [n(\omega) - 1] \, d\omega = 0. \tag{1}$$

That is, the refractive index averaged over all frequencies must be unity. Moreover, it may be shown that this restriction arises in part from the inertial property of matter. Related rules—with a similar physical interpretation—have been shown to apply to the complex dielectric function and its inverse.

Formally rules such as these are a consequence [4] of the asymptotic behavior of the optical functions and the Kramers–Kronig dispersion relations [16–18]. They are the optical analog of the superconvergence relations [19–21] of high-energy physics. However, physically they arise simply from causality and the dynamical laws of motion [7] and may be viewed as a restatement of these laws in frequency space.

Besides giving insight into the mathematical structure of the optical functions, the various sum rules provide a means of relating different physical properties without model fits to spectra and are valuable as self-consistency tests. This chapter is intended as a brief introduction to this topic[†] with emphasis on the underlying physics and applications. Throughout, the cgs system of units will be used since this is most common in practice. In addition, isotropic media and scalar dielectric functions are assumed. In the case of anisotropic media or symmetry-breaking external fields, the individual dielectric tensor elements must be treated separately. This is reviewed for magneto-optics [11–13] by Smith [23a] and for natural optical activity by Thomaz and Nussenzveig [23b].

II OPTICAL SUM RULES AND THEIR PHYSICAL INTERPRETATION

The best-known optical sum rule is the f sum rule [1, 24], which may be written in a variety of forms useful in optical analysis [1, 14]:

$$\int_0^\infty \omega \varepsilon_2(\omega) \, d\omega = 2 \int_0^\infty \omega n(\omega) k(\omega) \, d\omega = \frac{\pi}{2} \omega_{\mathrm{p}}^2, \tag{2}$$

[†] For some representative references for a broader treatment of linear response theory and sum rules, see Gross, Macdonald *et al.*, Pines *et al.*, Marten *et al.*, and Kubo *et al.* [22].

$$\int_0^\infty \omega k(\omega)\, d\omega = \frac{\pi}{4}\, \omega_p^2, \tag{3}$$

and

$$\int_0^\infty \omega\, \text{Im}[\varepsilon^{-1}(\omega)]\, d\omega = -\frac{\pi}{2}\, \omega_p^2, \tag{4}$$

where $\varepsilon(\omega) = \varepsilon_1(\omega) + i\varepsilon_2(\omega)$ is the complex dielectric function. In the case[†] of nonrelativistic electrons in a solid under consideration here, ω_p denotes the plasma frequency $[4\pi \mathcal{N} e^2/m]^{1/2}$; with $\mathcal{N}$ the total electron density and e and m the electronic charge $[e < 0]$ and mass, respectively. As will become apparent, the factor of $\frac{1}{2}$ difference between the right-hand sides of Eqs. (2) and (3) arises because the complex refractive index is the square root of the dielectric function [cf. Eqs. (8) and (22)].

These rules are occasionally rewritten in terms of the optical oscillator strength density [18]

$$f(\omega)|_\varepsilon = (m/2\pi^2 e^2)\, \omega \varepsilon_2(\omega), \tag{5}$$

or the absorption coefficient,

$$\alpha(\omega) = 2\omega k(\omega)/c, \tag{6}$$

where c is the speed of light *in vacuo*. Note that Eq. (5) is the traditional optical definition of oscillator strength density and arises from the classical notion of the density of Lorentz oscillators. In other applications—such as charged particle energy loss—ε^{-1} plays the fundamental role, and the oscillator strength density for energy-loss processes is taken as [30]

$$f(\omega)|_{\varepsilon^{-1}} = -(m/2\pi^2 e^2)\omega\, \text{Im}[\varepsilon^{-1}(\omega)]. \tag{7}$$

The f sum rule was first proved by Thomas and Reiche [31] and by Kuhn [32] for nonrealistic systems in 1925. It has played a crucial part in the development of quantum mechanics [33], atomic [24, 34] and nuclear [35][‡] spectroscopy, and it is fundamental to determining the concentrations of defects and dopants in condensed phases via Smakula's equation [36–38].[§] In

---

[†] The f sum rules given here are exact for a system of nonrelativistic electrons and infinitely heavy nuclei interacting via a local, velocity-independent potential. Finite nuclear mass may be taken into account via an effective mass. The formalism required for arbitrary nonrelativistic particles of arbitrary mass and a generalization to all electric multipole orders has been given by Sachs and Austern [25]. However, for practical purposes this effective mass correction is generally negligible in treating the optical properties of solids. Relativistic corrections are important for the inner shells of heavy elements. In first order these corrections arise from the mass–velocity term in the approximate relativistic wave equation describing the electrons. These appear to have been first described by Levinger *et al.* [26] and later found independently by Dogliani and Bailey [27]. Relativistic oscillator strengths for individual transitions have been given by Jacobsohn [28] and by Payne and Levinger [29].

[‡] Chapter 1 of Levinger [35] contains an insightful introduction to these rules.

[§] For a modern treatment see Dexter [37].

addition to the f sum rule, several less familiar sum rules for moments of the absorption spectrum have been given by Vinti [39] and others [40, 41]. These have been widely applied to atomic spectra [34], nuclear photodis-integration [35] and quantum chemistry [40], but only limited applications have been made to the spectroscopy of solids [41–43]. Values of the various moments of the absorption spectrum involved in several sum rules have been shown to be interrelated by the uncertainty principle [35, 44].

The existence of numerous sum rules for the absorptive process suggests analogous rules hold for the dispersive process. However, no hint of them is given by the commonly employed [24] proofs of the absorption rules that involve combining the time-dependent Schrödinger equation and the fer-mion commutator. Here, an alternative approach [4], employing the high-frequency (or short-time) behavior of matter, will be used. This has the advantages of allowing the systematic construction of a large class of sum rules and of emphasizing the underlying physics.

A Superconvergence Sum Rules for $\varepsilon(\omega)$

1 Formal Development

At frequencies higher than any characteristic absorption of the system, the excitation frequency dominates the expression for the polarizability, and the complex dielectric function has the asymptotic form [1, 45]

$$\lim_{\omega \to \infty} \varepsilon(\omega) = 1 - (\omega_p^2/\omega^2) + \cdots. \tag{8}$$

This series expansion is universal: The term in ω^{-1} is missing; the term in ω^{-2} involves only the electron density through the plasma frequency. These first terms are independent of interactions within the system. Physically, the reason is that at frequencies well above the highest characteristic absorption, inertial effects, not restoring or dissipative forces, dominate the dynamics. The latter forces enter into succeeding terms. [For example, the next term in the expansion of $\varepsilon(\omega)$ for a Lorentz oscillator with damping constant γ is $i\omega_p^2\gamma/\omega^3$.]

Formally, the physics and mathematics may be combined to get the sum rules by comparing the limiting behavior of $\varepsilon(\omega)$, Eq. (8), with the Kramers–Kronig relations [1, 16–18]:

$$\varepsilon_1(\omega) - 1 = \frac{2}{\pi} P \int_0^\infty \frac{\omega' \varepsilon_2(\omega')}{\omega'^2 - \omega^2} \, d\omega', \tag{9}$$

$$\varepsilon_2(\omega) - 4\pi \frac{\sigma(0)}{\omega} = -\frac{2}{\pi} \omega P \int_0^\infty \frac{\varepsilon_1(\omega') - 1}{\omega'^2 - \omega^2} \, d\omega'. \tag{10}$$

Here P denotes the principal value integral, and $\sigma(0)$ is the dc value of the conductivity. The conductivity term is, of course, not present for insulators. These relations hold for linear, causal systems, and their use guarantees the functions involved are analytic, square-integrable, causal response functions [46, 47].

To proceed, note that Eq. (8) actually makes two statements, one for the real part of $\varepsilon(\omega)$,

$$\lim_{\omega \to \infty} \varepsilon_1(\omega) = 1 - (\omega_p^2/\omega^2), \tag{11}$$

and a second for the imaginary part,

$$\lim_{\omega \to \infty} \varepsilon_2(\omega) \text{ falls off faster than } \omega^{-2}. \tag{12}$$

The first of these may be compared with the high-frequency limit of the first Kramers–Kronig relation

$$\lim_{\omega \to \infty} \varepsilon_1(\omega) - 1 = -\frac{2}{\pi} \frac{1}{\omega^2} \int_0^\infty \omega' \varepsilon_2(\omega') \, d\omega' + \cdots . \tag{13}$$

Equating powers of ω^{-2} in Eqs. (11) and (13) yields the f sum rule[†]

$$\int_0^\infty \omega \varepsilon_2(\omega) \, d\omega = \frac{\pi}{2} \omega_p^2. \tag{14}$$

Similarly, the high-frequency limit of the second Kramers–Kronig relation is

$$\lim_{\omega \to \infty} \varepsilon_2(\omega) = \frac{4\pi\sigma(0)}{\omega} + \frac{2}{\pi} \frac{1}{\omega} \int_0^\infty [\varepsilon_1(\omega') - 1] \, d\omega' + \cdots . \tag{15}$$

Since by Eq. (12) there is no term in ω^{-1} in ε_2, equating powers of ω^{-1} in Eqs. (12) and (15) yields [4].[‡,§]

$$\int_0^\infty [\varepsilon_1(\omega) - 1] \, d\omega + 2\pi^2\sigma(0) = 0. \tag{16}$$

[†] The first proof of the f sum rule employing this method appears to have been given by R. de Laer Kronig [17] in his original work on the Kramers–Kronig relations.

[‡] An alternative proof [8] of Eq. (16) can be made by simply integrating $\varepsilon(\omega)$ over the contour consisting of the real axis and the semicircle at infinity in the upper half-plane. In the case of metals $\varepsilon(\omega)$ has a pole at the origin $i4\pi\sigma(0)/\omega$; this contributes the term in $\sigma(0)$ to Eq. (16), and the remainder of the real axis gives the integral over $\varepsilon(\omega)$, while the semicircle at infinity gives a vanishing contribution. At first this proof seems simpler, but the vanishing of the contribution of the semicircle at infinity does not follow directly from the asymptotic behavior along the real axis [Eq. (8)]. Rather, an argument based on the Phragmén–Lindelöf theorem is required. For details on this point see Titchmarsh [48]; Sec. 5.61. See also Nussenzveig [19], Theorems 2.5.4 and 7.5.1.

[§] $\sigma(\omega)$ has units sec^{-1} in cgs electrostatic units. To convert σ in $\text{ohm}^{-1} \text{cm}^{-1}$ to σ in sec^{-1} multiply by 8.9876×10^{11} ohm cm sec^{-1}. Generally, sum-rule integrals are done over an energy scale resulting in values of $\hbar\sigma$. The corresponding multipliers for electron volt and cm^{-1} energy scales are 5.9158×10^{-4} ohm cm eV and 4.7713 ohm cm cm^{-1}, respectively.

This is the analog for $\varepsilon_1(\omega)$ of Eq. (1). Note that this sum rule provides independent information on the parameters describing free-carrier absorption. Traditionally, these are found by making Drude-model fits to $\varepsilon(\omega)$. However, the integral of $[\varepsilon_1(\omega) - 1]$ provides an additional constraint since $\sigma(0) = ne^2\tau/m_0$, where n is the carrier concentration, τ their lifetime, and m_0 their optical mass (see Stern [1] for details).

In taking the high-frequency limit of the dispersion relations, we have relied on the intuitive notion that at large ω the denominator of the integrals is dominated by ω^2 [or, more precisely, that $(\omega^2 - \omega'^2)^{-1}$ may be expanded in a power series in inverse powers of ω^2]. This will be true provided that the numerator decreases fast enough as $\omega' \to \infty$. The necessary restrictions on the numerator are embodied in the superconvergence theorem [49–50].[†] For the present, it is sufficient to note that these limits hold provided that $\varepsilon_2(\omega)$ falls off at least as fast as $\omega^{-2} \ln^{-\alpha} \omega$, for $\alpha > 1$, as $\omega \to \infty$. This is a mild restriction on $\varepsilon_2(\omega)$ since it decreases faster than this in actual systems. For example, the contribution to $\varepsilon_2(\omega)$ from K-shell electrons that dominate the high-frequency absorption falls off as $\omega^{-4.5}$ (neglecting retardation and relativistic effects) [51]. (For more information see Bethe et al. [24] and Rau et al. [51].)

In discussing the optical properties of metals and superconductors, it is common to rewrite the rules in terms of the complex frequency-dependent conductivity as

$$\sigma(\omega) = -(i\omega/4\pi)\varepsilon(\omega). \tag{17}$$

The f sum rule then takes the form

$$\int_0^\infty \text{Re } \sigma(\omega)\, d\omega = \frac{\pi}{2}\frac{\mathcal{N}e^2}{m}. \tag{18}$$

This rule and its generalizations [9, 52] as well as related rules for the surface impedance [53, 54] have been used extensively to relate the difference in conductivity between the superconducting and the normal states in the region of the superconducting gap to the δ-function conductivity at zero frequency in superconductors [53, 55–57].

2   Physical Picture

The physical basis of the f sum rule [Eq. (14)] and the dc conductivity rule [Eq. (16)] lies in the dynamical laws of motion. This is most easily seen using the linear-response-theory [7] result that the polarization $P_\delta(\tau)$ induced by a δ-function pulse electric field as a function of the time τ following the

---

† The term *superconvergent* was introduced into particle physics to describe the convergence properties of dispersion relation integrals; the corresponding sum rules are often referred to as superconvergence relations. (For an introduction to this theorem see the appendix of Altarelli et al. [4]. Details are given in de Alfaro et al. [49] and Frey et al. [50].)

pulse is just the Fourier transform of the electric susceptibility $[\varepsilon(\omega) - 1]/4\pi$. That is,

$$P_\delta(\tau) = \frac{1}{2\pi} \int_{-\infty}^{\infty} \left(\frac{\varepsilon(\omega) - 1}{4\pi} \right) e^{-i\omega\tau} \, d\omega. \tag{19a}$$

Taking account of the pole in $\varepsilon(\omega)$ at the origin for conductors and by using the fact that $\varepsilon_1(\omega)$ is an even function of ω while $\varepsilon_2(\omega)$ is odd leads to

$$P_\delta(\tau) = \frac{1}{4\pi^2} \left\{ \int_0^{\infty} \text{Re}[\varepsilon(\omega) - 1] \cos(\omega\tau) \, d\omega, \right.$$

$$\left. + \int_0^{\infty} \text{Im } \varepsilon(\omega) \sin(\omega\tau) \, d\omega + 2\pi^2\sigma(0) \right\}. \tag{19b}$$

The sum rules then follow [7] by using elementary arguments to determine $P_\delta(\tau)$ at small τ.

For example, inertia requires that the response cannot be instantaneous; $P_\delta(\tau)$ for $\tau \geq 0$ must join onto $P_\delta(\tau)$ for $\tau < 0$ without a discontinuity. By causality the latter values are all zero (no response before excitation) so that $P_\delta(\tau = 0) = 0$. Setting $\tau = 0$ in Eq. (19) then leads to the dc conductivity rule [Eq. (16)]. This rule and, by analogy, the corresponding relation for $n(\omega)$ [Eq. (1)] are therefore direct consequences of causality and the law of inertia; they are thus often referred to as inertial sum rules.

The f sum rule may be similarly derived by calculating the initial rate of change of polarization $P_\delta(\tau)$ in the impulse approximation and equating this to the time derivative of Eq. (19) at $\tau = 0$. In this case the determining factor is the dynamical law of motion—Newton's law in a classical picture. The details are given by Altarelli and Smith [7]. In the commonly quoted quantum-mechanical derivation employing commutation relations [24], the dynamics enters through the time-dependent Schrödinger equation.

Superconvergence Sum Rules for N(ω)   B

A parallel development [7] may be given for the refractive index. The only difference lies in that for metals there is no pole in the index at the origin arising from the conductivity term. Rather, there is a square-root singularity

$$[i4\pi\sigma(0)/\omega]^{1/2} = (1 + i)\sqrt{2\pi\sigma(0)} \, \omega^{-1/2},$$

at the origin, and the Kramers–Kronig dispersion relations take the form[†]

$$n(\omega) - 1 = \frac{2}{\pi} P \int_0^{\infty} \frac{\omega' k(\omega')}{\omega'^2 - \omega^2} \, d\omega', \tag{20}$$

[†] For a derivation of these relations for insulator see Section 1.9 in Nussenzveig [19] and for conductors see Smith [58].

$$k(\omega) = -\frac{2\omega}{\pi} P \int_0^\infty \frac{n(\omega') - 1}{\omega'^2 - \omega^2} \, d\omega'. \tag{21}$$

Comparing the high-frequency limits of these with the limit for $N(\omega)$ obtained from the square root of Eq. (8),

$$\lim_{\omega \to \infty} N(\omega) = 1 - \tfrac{1}{2}(\omega_p^2/\omega^2) + \cdots, \tag{22}$$

yields the f sum rule for $k(\omega)$ and the companion inertial sum rule for $n(\omega)$

$$\int_0^\infty \omega k(\omega) \, d\omega = \frac{\pi}{4} \omega_p^2, \tag{23}$$

and

$$\int_0^\infty [n(\omega) - 1] \, d\omega = 0. \tag{24}$$

A physical interpretation [7] of these rules in terms of a response function is similar to that for $\varepsilon(\omega)$.

In addition to sum rules involving integral equations, a variety of integral inequalities hold for the dielectric function, the refractive index, and the reflectance. These have received some formal treatment recently [6, 59] but have not yet been applied to experimental problems.

C   Superconvergence Sum Rules for Functions of *N* and *ε*

Since an analytic function of an analytic function is also analytic, it is possible to construct analytic functions of $N(\omega)$ and $\varepsilon(\omega)$ and develop sum rules and dispersion relations for them. For example, starting with $N(\omega)$, $N^2(\omega)$ would yield the sum rules for $\varepsilon(\omega)$. The number of such rules is limited only by one's imagination and a large number of generalized rules involving powers of ω and the real and imaginary parts of $[N(\omega) - 1]^m$, and so on, have been given in the literature [6–8]. In general the higher the power m, the faster the convergence of the sum rule, so that a given rule is more sensitive to a particular spectral range than to others. A few preliminary applications [7] of these rules have been made to test optical data. Since they are based on the same asymptotic behavior as the rules for $N(\omega)$ and $\varepsilon(\omega)$, they must contain the same physics, but with emphasis on different spectral regions.

A number of examples [7] of these rules for an insulating system are

$$\int_0^\infty \mathrm{Re}\{[N(\omega') - 1]^m\} \, d\omega' = 0, \tag{25}$$

and

$$\int_0^\infty \omega' \, \mathrm{Im}\{[N(\omega') - 1]^m\} \, d\omega' = \begin{cases} \dfrac{1}{4} \pi \omega_p^2 & (m = 1), \\ 0 & (m > 1). \end{cases} \tag{26}$$

For $m = 1$ these yield the familiar inertial and f sum rules, respectively. For higher powers Eq. (25) yields

$$m = 2: \quad \int_0^\infty \{[n(\omega') - 1]^2 - k^2(\omega')\} \, d\omega' = 0, \tag{27}$$

$$m = 3: \quad \int_0^\infty [n(\omega') - 1]\{[n(\omega') - 1]^2 - 3k^2(\omega')\} \, d\omega' = 0. \tag{28}$$

Similarly, Eq. (26) yields the rules

$$m = 2: \quad \int_0^\infty \omega' k(\omega')[n(\omega') - 1] \, d\omega' = 0, \tag{29}$$

$$m = 3: \quad \int_0^\infty \omega' k(\omega')\{3[n(\omega') - 1]^2 - k^2(\omega')\} \, d\omega' = 0. \tag{30}$$

Equation (29) has been previously noted by Stern [1]. Observe that this rule, together with the inertial sum rule for $n(\omega)$ [Eq. (24)] states that $[n(\omega) - 1]$ averages to zero *either alone or when weighted by* $\omega k(\omega)$. Analogous rules may be derived [7] for conductors by taking $\omega[N(\omega) - 1]$ as the starting point. The reader is referred to the literature [6–8] for details and related rules involving $(N - 1)^m$ multipled by powers of ω.

Equation (27) deserves further comment. It can be rewritten in the more suggestive form

$$\int_0^\infty [n(\omega') - 1]^2 \, d\omega' = \int_0^\infty k^2(\omega') \, d\omega'. \tag{31}$$

As such it provides a potentially useful check on Kramers–Kronig inversions. It is a particular case of the general theorem [60, Theorems 90 and 91] that the norm of a function is preserved in a Hilbert transformation. Using the identity $\varepsilon_1(\omega) = n^2(\omega) - k^2(\omega)$ and the inertial sum rule for $n(\omega)$, it follows from Eq. (16) that the generalization of Eq. (31) for conductors is

$$\int_0^\infty \{[n(\omega) - 1]^2 - k^2(\omega)\} \, d\omega + 2\pi^2 \sigma(0) = 0. \tag{32}$$

The extra term in $\sigma(0)$ arises from the pole at the origin. (Note: Separately $[n(\omega) - 1]^2$ and $k^2(\omega)$ have poles at the origin but their difference does not; hence, they have been written together under the same integral to avoid undefined expressions.)

Since $N(\omega)$ has at most a $\omega^{-1/2}$ singularity for conductors, $\omega N(\omega)$ is square integrable for both insulators and conductors. It then follows that its real and imaginary parts are related by Hilbert transforms. Thus, in analogy with Eq. (31), the norm of the transform is preserved [6],

$$\int_0^\infty \omega^2 [n(\omega) - 1]^2 \, d\omega = \int_0^\infty \omega^2 k^2(\omega) \, d\omega. \tag{33}$$

The fact that both Eqs. (33) and (31) or, in the case of a conductor Eq. (32), must hold is another remarkable example of the constraints imposed on the optical functions.

Before leaving this section, it might be noted that a generalization of the f sum rule for higher powers of ω_p has been found by Villani and Zimerman [5], who considered powers of $\omega^2[N(\omega) - 1]$ and $\omega^2[\varepsilon(\omega) - 1]$. In the case of the refractive index their result is

$$\int_{-\infty}^{\infty} \omega^{2m-1}[N(\omega) - 1]^m \, d\omega = (-1)^{m+1}2^{-m}i\pi\omega_p^{2m}. \tag{34}$$

D Superconvergence Relations Employing Weighting Functions

Superconvergence techniques can be applied to quantities consisting of the optical functions multiplied by an appropriate analytic weighting function. Weighting with powers of ω mentioned in the preceding section is perhaps the simplest example of this. The technique has been applied by using algebraic, exponential, and Gaussian weighting functions by Villani and Zimerman [6], King [8], and Kimel [61]; a general discussion has been given by Fischer *et al.* [62]. A major result of Villani and Zimmerman based on the Lin–Okubo [63] weighting function is

$$\int_0^a \frac{[n(\omega) - 1] \, d\omega}{(a^2 - \omega^2)^{1/2}} = \int_a^\infty \frac{k(\omega) \, d\omega}{(\omega^2 - a^2)^{1/2}}. \tag{35}$$

This is particularly interesting since it relates the index in the interval $[0, a]$ to the extinction coefficient in the remainder of the spectrum. While this relation does not seem to have been used in practical data analysis to date, it should be a helpful guide to constructing composite sets of data. For example, in semiconductors and insulators $n(\omega)$ is often available with high accuracy for ω less than the band gap, ω_g, while above ω_g, $k(\omega)$ may be imprecisely or only partially known. By using Eq. (35) with $a \leq \omega_g$ and requiring that the f sum rule be also satisfied, it should be possible to sharpen the limits on the values of $k(\omega)$ for $\omega > \omega_g$.

E Static-Limit Sum Rules

The sum rules discussed thus far all derive from the high-frequency behavior. The static limit also yields several well-known rules [18] that follow directly from the dispersion relations. For insulators both $\varepsilon(\omega)$ and $N(\omega)$ are well behaved at the origin, and the Kramers–Kronig relations yield the dc dielectric constant and index as

$$\varepsilon_1(0) = 1 + (2/\pi) \int_0^\infty \omega^{-1}\varepsilon_2(\omega) \, d\omega, \tag{36}$$

and

$$n(0) = 1 + (2/\pi) \int_0^\infty \omega^{-1} k(\omega) \, d\omega. \tag{37}$$

The other dispersion relations give the trivial results $\varepsilon_2(0) = 0$ and $k(0) = 0$.

In the case of conductors $\varepsilon(\omega)$ has a pole at the origin arising from the term $i4\pi\sigma(0)/\omega$ in the dielectric function. Along the real axis this yields an ω^{-1} divergence in $\varepsilon_2(\omega)$, but $\varepsilon_1(\omega)$ remains finite. The latter finite value of $\varepsilon_1(0)$ can be recovered by considering dispersion relations for the function $\varepsilon(\omega) - [i4\pi\sigma(0)/\omega] - 1$. The result is

$$\varepsilon_1(0) = 1 + \frac{2}{\pi} \int_0^\infty \frac{\omega' \varepsilon_2(\omega') - 4\pi\sigma(0)}{\omega'^2} \, d\omega'. \tag{38}$$

Note that this limit holds only as ω approaches zero along the real axis. Because of the singularity, the limit of $\mathrm{Re}\,\varepsilon(\omega)$ depends on the direction in the complex ω plane along which the limit is taken.

In the case of the refractive index of a conductor, both $n(\omega)$ and $k(\omega)$ have $\omega^{-1/2}$ singularities and diverge in the static limit.

FINITE-ENERGY SUM RULES   III

The sum rules discussed to this point involve the infinite frequency interval $[0, \infty]$ and include all absorptive processes. However, a typical spectrum consists of different regions corresponding primarily to a single class of absorptions such as excitation of phonons, conduction, valence, or core electron. It is tempting to seek a means of treating these individual processes. In those favorable cases in which a particular absorptive process is not overlapped by other absorptions, finite-energy sum rules and dispersion relations may be developed [14, 64].

The key requirement [14] is that the absorption in question be sufficiently isolated that it can be viewed to a good approximation as taking place in a transparent medium with real background dielectric function $\varepsilon_b(\omega)$ arising from the dispersion of absorptive processes in all other spectral ranges.

The situation is illustrated in Fig. 1. Examples include phonon absorptions in polar insulators and free-carrier absorption in wide-gap semiconductors. In both instances, the absorptions in question are well separated from interband transitions that provide the polarizable background. On a wider spectral range, the valence electrons of a solid may be regarded as moving in a dielectric "medium" consisting of the polarizable ion cores in those materials for which valence and core absorptions do not overlap.

To proceed, consider an energy $\hat{\omega}$ large compared with that of the isolated absorption—or group of absorptions—but well below the energy of the

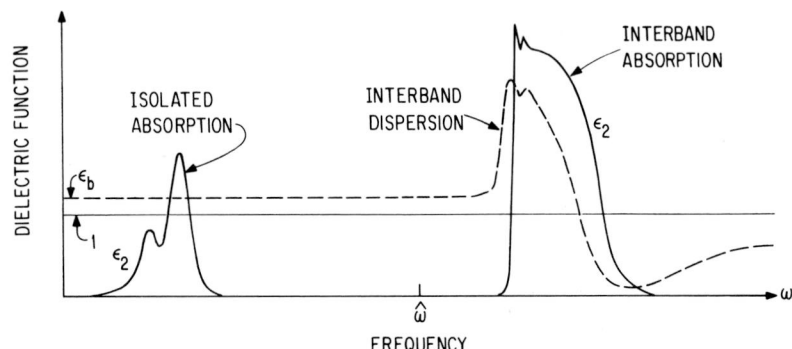

Fig. 1. Absorption spectra for nonoverlapping bands. (Solid curve, absorption; dashed curve, interband dispersion.)

transitions responsible for the dispersive dielectric background. Such an energy is indicated in Fig. 1. The dielectric function may then be expanded in the neighborhood of $\hat{\omega}$ as [14]

$$\lim_{\omega \to \hat{\omega}} \varepsilon(\omega) = \varepsilon_b(\omega) - \omega_{pi}^2/\omega^2 + \cdots, \tag{39}$$

where ω_{pi} is the plasma frequency associated with the isolated absorption. In the limit of large separation between the isolated absorption and the other absorptions of the system, $\varepsilon_b(\omega)$ may be taken as a constant so that

$$\lim_{\omega \to \hat{\omega}} \varepsilon(\omega) = \varepsilon_b - \omega_{pi}^2/\omega^2 + \cdots, \tag{40}$$

$$\lim_{\omega \to \hat{\omega}} N(\omega) = \varepsilon_b^{1/2} - \omega_{pi}^2/2\,\varepsilon_b^{1/2}\omega^2 + \cdots, \tag{41}$$

$$\lim_{\omega \to \hat{\omega}} \varepsilon^{-1}(\omega) = \varepsilon_b^{-1} + \omega_{pi}^2/\varepsilon_b^2\omega^2 + \cdots. \tag{42}$$

In taking the square root and the inverse of Eq. (40) to obtain Eqs. (41) and (42), it has been assumed that $\omega \gg \omega_{pi}$. Notice particularly that the dielectric constant ε_b appears in the coefficient of the ω^{-2} term in the limiting values of $N(\omega)$ and $\varepsilon^{-1}(\omega)$.

Finite-energy sum rules for the frequency interval containing the isolated absorption then follow either by applying superconvergence methods to the idealized problem of an isolated absorption in a medium with a constant ε_b at all frequencies or by employing [21] a Cauchy-theorem integration over a semicircular path of radius $\hat{\omega}$ in the upper half-plane.

The resulting finite-energy f sum rules are

$$\int_0^{\hat{\omega}} \omega' \varepsilon_2(\omega')\,d\omega' \approx \frac{\pi}{2}\,\omega_{pi}^2, \tag{43}$$

$$\int_0^{\hat{\omega}} \omega' k(\omega')\,d\omega' \approx \frac{\pi}{4}\,\frac{\omega_{pi}^2}{\varepsilon_b^{1/2}}, \tag{44}$$

$$\int_0^{\hat{\omega}} \omega' \operatorname{Im}[\varepsilon^{-1}(\omega')]\, d\omega' \approx -\frac{\pi}{2}\frac{\omega_{\text{pi}}^2}{\varepsilon_{\text{b}}^2}. \tag{45}$$

Similarly, the inertial sum rules are

$$\int_0^{\hat{\omega}} [\varepsilon_1(\omega') - \varepsilon_b]\, d\omega' + 2\pi^2\sigma(0) \approx 0, \tag{46}$$

$$\int_0^{\hat{\omega}} [n(\omega') - \varepsilon_b^{1/2}]\, d\omega' \approx 0, \tag{47}$$

$$\int_0^{\hat{\omega}} \{\operatorname{Re}[\varepsilon(\omega')^{-1}] - \varepsilon_b^{-1}\}\, d\omega' \approx 0. \tag{48}$$

These rules are exact in the limit of $\hat{\omega} \to \infty$. For finite $\hat{\omega}$ the f sums are accurate to within terms of the order of $\omega_{\text{pi}}^2(\gamma/\hat{\omega})$, and the inertial rules hold to terms of order $\omega_{\text{pi}}^2/\hat{\omega}$, where γ is the effective damping constant for the low-energy absorption.

The inertial sum rules are what one would expect intuitively: The dispersion associated with the isolated absorption is superimposed—though not linearly in the case of $N(\omega)$ or $\varepsilon^{-1}(\omega)$—on the background ε_b, $\varepsilon_b^{1/2}$, or ε_b^{-1} and averages to zero (or $-2\pi^2$ times the dc conductivity) with respect to that background.

The finite-energy f sum rules are more noteworthy in that the background dielectric constant appears in the right-hand side of the rules for $k(\omega)$ and $\operatorname{Im}[\varepsilon^{-1}(\omega)]$. That is, the integral of $\omega\varepsilon_2(\omega)$ gives ω_{pi} for an isolated absorption directly. However, the areas under $\omega k(\omega)$ or $\omega \operatorname{Im}[\varepsilon^{-1}(\omega)]$ do not; in these f sums *it is necessary to include the effect of virtual processes from all other transitions* via ε_b. This is manifest in the factors $\varepsilon_b^{1/2}$ and ε_b^2 in Eqs. (44) and (45), respectively, which may be thought of as correcting for the dielectric shielding of the isolated transition by the remainder of the system. The greater the background dielectric function, the weaker the apparent strength of an absorption as measured by $k(\omega)$ or $\operatorname{Im}[\varepsilon^{-1}(\omega)]$ for a fixed ω_{pi} (i.e., for a fixed oscillator strength of the isolated absorption).

Defect Absorptions and Smakula's Equation A

A particularly striking example [64] of the dependence of the amplitude of an isolated absorption on the remainder of the system occurs for defects or impurities that introduce absorption bands in the region of transparency of insulators. This is illustrated in Fig. 2, which shows the modeled absorption spectra for a series of four different defects. Each exhibits a single absorption band; they all have the same strength as measured by ω_{pi}^2 and the same half-width but have different absorption frequencies ω_d. In the figure the bands are labeled by $\nu = \omega_d/\omega_{\text{host}}$, where ω_{host} is the frequency of the host's fundamental absorption edge. As the defect absorption frequency approaches

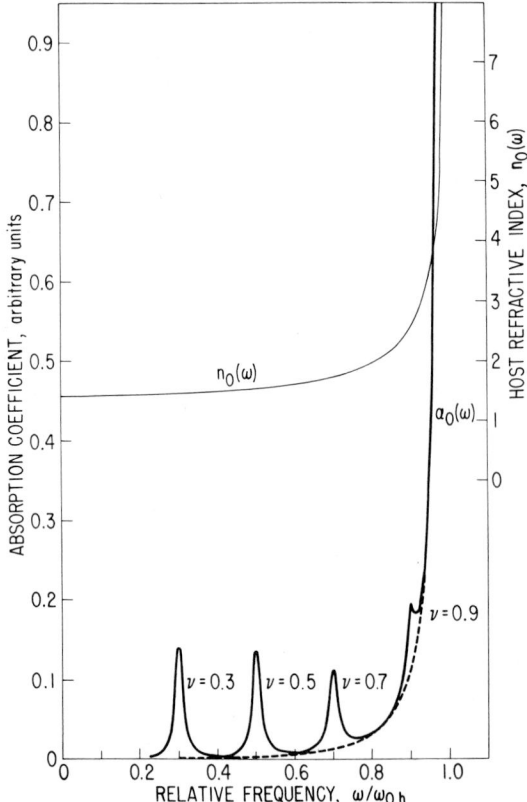

Fig. 2. Modeled spectra of a weak defect absorption on the low-energy side of the fundamental crystal absorption. Four possible defect absorptions are shown for energies $\omega_d = \nu\omega_{0,h}$, where $\omega_{0,h}$ is host absorption energy. Although all four absorptions have the same oscillator strengths, the absorption coefficient for bands at higher energies are reduced as a result of the larger values of the host refractive index, $n_0(\omega)$. (After Smith and Graham [64].)

that of the host's fundamental absorption edge, $\varepsilon_b(\omega)$ increases sharply with frequency and the apparent strength of the defect absorption as measured by the absorption coefficient $\alpha(\omega) = 2\omega k(\omega)/c$ decreases markedly.

In this case, it is necessary to treat the case in which the background dielectric function varies over the band. The result is a generalization of Smakula's equation [36–38] for the integrated oscillator strength $f = (m/4\pi e^2)\omega_{pi}^2$ of a defect or impurity in a transparent medium [64]

$$\rho f = \frac{m}{\pi^2 e^2} \int \omega n_0(\omega)\,\Delta k(\omega)\,d\omega. \tag{49}$$

Here ρ is the density of defects that is taken to be much less than the density of host atoms and effective-field effects have been omitted for simplicity. The refractive index of the host is denoted by $n_0(\omega) = \varepsilon_b^{1/2}(\omega)$ (the change in index induced by the defect is negligible), and $\Delta k(\omega)$ is the absorption induced by the defect.

Finite Oscillator Strength Sums and $n_{\mathrm{eff}}(\omega)$ B

In analogy with the f sum rule it has become common practice to define the effective number density of electrons contributing to the optical properties up to an energy ω by the partial f sums [14, 45]

$$n_{\mathrm{eff}}(\omega)\big|_\varepsilon = \frac{m}{2\pi^2 e^2} \int_0^\omega \omega' \varepsilon_2(\omega')\, d\omega', \qquad (50)$$

$$n_{\mathrm{eff}}(\omega)\big|_k = \frac{m}{\pi^2 e^2} \int_0^\omega \omega' k(\omega')\, d\omega, \qquad (51)$$

$$n_{\mathrm{eff}}(\omega)\big|_{\varepsilon^{-1}} = -\frac{m}{2\pi^2 e^2} \int_0^\omega \omega'\, \mathrm{Im}[\varepsilon^{-1}(\omega')]\, d\omega'. \qquad (52)$$

Here the subscripts ε, k, and ε^{-1} are used to distinguish the partial f sums involving $\varepsilon_2(\omega)$, $k(\omega)$, and $\mathrm{Im}[\varepsilon^{-1}(\omega)]$, respectively. The f sum rule guarantees that in the limit $\omega \to \infty$ all three values of $n_{\mathrm{eff}}(\omega)$ approach the number of electrons $\mathcal{N}$ in the system.

For intermediate values of ω the three values generally do not coincide. This is illustrated in Fig. 3a for the simple classical Drude-model of a metal in which core-electron x-ray transitions are neglected. Starting at zero for $\omega = 0$, the three values increase monotonically, but on different paths until in the vicinity of the plasma frequency they draw together and asymptotically approach $\mathcal{N}$ at infinity.

In discussing the spectra of materials with valence or conduction electron absorptions that are well separated from those of core transitions, it is common to calculate $n_{\mathrm{eff}}(\hat{\omega})$ for the valence or conduction electrons. Then $\hat{\omega}$ is chosen to exhaust the valence- or conduction-electron oscillator strength but to lie below the lowest-energy core transitions. A comparison of this situation with the finite-energy f sum rules [Eqs. (43)–(45)] shows that the various $n_{\mathrm{eff}}(\hat{\omega})$ values for conduction or valence electrons should not be equal when the polarizability of the core electrons is accounted for. Rather [14],

$$n_{\mathrm{eff}}(\hat{\omega})\big|_\varepsilon = \varepsilon_b^{1/2} n_{\mathrm{eff}}(\hat{\omega})\big|_k = \varepsilon_b^2 n_{\mathrm{eff}}(\hat{\omega})\big|_{\varepsilon-1}, \qquad (53)$$

where ε_b is the background dielectric constant of the fictitious medium consisting of the polarizable atomic cores alone.

This is illustrated in Fig. 3b. Again a Drude model has been employed, but a background dielectric constant $\varepsilon_b = 1.05$ has been included to take

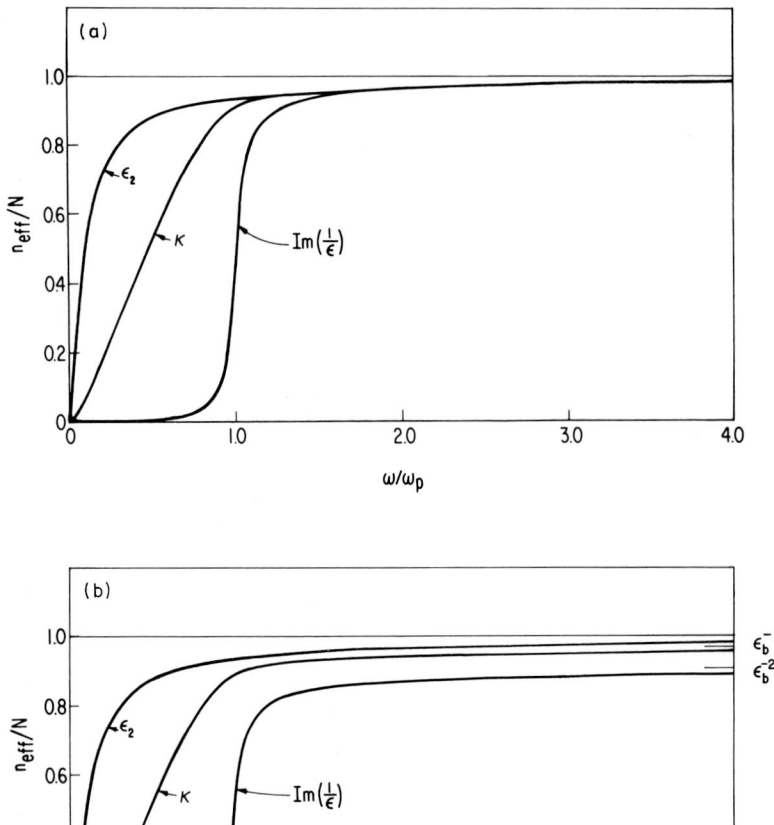

Fig. 3. The functions $n_{eff}(\omega)$ for a Drude metal (a) with unpolarizable cores, i.e., $\varepsilon_b = 1$, and (b) embedded in a dielectric medium with $\varepsilon_b = 1.05$ corresponding to the core polarizability for a light element such as aluminum. (From Smith and Shiles [14].)

account of the core transitions which lie at higher energies. This value is typical for a light element such as aluminum. A plot of the actual $n_{eff}(\omega)$ for both conduction and core transitions of aluminum is discussed in Section V.

It should be further noted that the asymptotic limit of $n_{eff}(\hat{\omega})|_\varepsilon$ is generally *not* the actual number of valence or conduction electrons. It is their oscillator strength. To start with the distinction between valence and conduction electrons is only approximate. The separation is never complete because of Coulomb and exchange interactions, and in the one-electron approximation this leads to an "exchange" of oscillator strength between core and valence electrons [2, 38, 65]: The major contribution to this effect can be understood by observing that the Pauli principle prohibits transitions of core electrons to the occupied valence or conduction states, so that the core-electron oscillator strength is less than their number. Similarly, electrons in higher-lying states cannot make transitions to the occupied core states, and they consequently exhibit an oscillator strength greater than their number (Note: Emissive processes have negative oscillator strength.)

Typically this Pauli principle effect increases the valence-electron oscillator strength by a few percentages for light elements and from 10 to 20% for heavy elements.[†] In addition a small correction arises from the exchange part of the potential that occurs when electrons are treated in the one-electron approximation. Briefly, the f sum rule for dipole matrix elements applies strictly only for systems in which the interaction operator and spatial coordinates commute [24]. This does not hold for the (nonlocal) exchange part of one-electron Hamiltonian that leads to a small correction in the one-electron model oscillator strength. This has been treated by Fock [66], who found the effect to be less than 20% of the Pauli principle correction. However, note that what is a prohibited absorption for one particular electron is a prohibited emission for another, so that the f sum taken over transitions to all levels still totals to the electronic number of the whole system in accord with the f sum rule.

SUM RULES FOR REFLECTION SPECTROSCOPY   IV

Reflection spectroscopy has become a widely used method for determining optical constants of highly absorbing materials, since it was shown that reflectance spectra could be analyzed for the unknown phase by dispersion methods [67, 68].

Normal-Incidence Spectroscopy   A

The simplest situation occurs for normal incidence from vacuum or, to a good approximation, air. In this case the complex reflectivity (for the magnetic field vector) is given by [1]

$$\tilde{r}(\omega) = r(\omega)e^{i\theta(\omega)} = [N(\omega) - 1]/[N(\omega) + 1]. \tag{54}$$

[†] See, for example, Table 1 of Smith and Dexter [38].

Since N is analytic, $\tilde{r}(\omega)$ is also analytic, and its real and imaginary parts satisfy sum rules and dispersion relations similar to those for $N(\omega)$ [69–71]. For example,

$$\int_0^\infty r^m(\omega) \cos[m\theta(\omega)]\, d\omega = 0 \qquad (m = 1, 2, 3, \ldots) \tag{55}$$

and

$$\int_0^\infty \omega r^m(\omega) \sin[m\theta(\omega)]\, d\omega = \begin{cases} \dfrac{\pi}{8}\, \omega_p^2 & (m = 1), \\[2mm] 0 & (m \geq 2). \end{cases} \tag{56}$$

These and similar superconvergence relations for powers of $\tilde{r}(\omega)$, and so on, all involve a mixture of reflectivity amplitude and phase. They are, therefore, currently of little practical value since the phase is generally not measured.

The phase and amplitude may be separated by considering the analytic function

$$\ln \tilde{r}(\omega) = \ln r(\omega) + i\theta(\omega). \tag{57}$$

There is, however, a major difference between this and the dielectric function, refractive index, and so on. The latter approach unity as $\omega \to \infty$, whereas $\ln \tilde{r}(\omega)$ diverges logarithmically since $\lim_{\omega \to \infty} r(\omega) = 0$. Thus, the dispersion relations and sum rules for $\ln \tilde{r}(\omega)$ are not of the same form as those for $\varepsilon(\omega)$ or $N(\omega)$.

The dispersion relations for $\ln \tilde{r}(\omega)$ may be obtained by a number of methods[†] that yield [68, 72, 73].[‡]

$$\theta(\omega) = -\frac{2\omega}{\pi} P \int_0^\infty \frac{\ln r(\omega')}{\omega'^2 - \omega^2}\, d\omega' \tag{58}$$

and its inverse [76, 77]

$$\ln r(\omega) - \ln r(\omega_0) = \frac{2}{\pi} P \int_0^\infty \omega'\theta(\omega') \left(\frac{1}{\omega'^2 - \omega^2} - \frac{1}{\omega'^2 - \omega_0^2} \right) d\omega', \tag{59}$$

where it is necessary to assume that the reflectance is known at some frequency ω_0. The equation for $\theta(\omega)$ [Eq. (58)] has the usual Kramers–Kronig form, while the second is a subtracted dispersion relation. Note that, while the second appears to simply involve the difference between two Kramers-Kronig integrals—one for $r(\omega)$ and one for $r(\omega_0)$—such an interpretation is invalid. The reason is that $\theta(\omega) \to \pi$ as $\omega \to \infty$, leading to integrals that are divergent when taken separately.

Equation (59) displays a further point. The phase does not determine the

---

[†] For a detailed discussion of the phase retrieval problem see Toll [46], Roman and Marathay [74], and Burge *el al.* [75]; a brief, but physical, discussion is given by Stern [1].

[‡] The sign convention used here assumes a time dependence $\exp[-i\omega t]$. If $\exp[i\omega t]$ is used, the index must be written as $N = n - ik$ and Eq. (58) for $\theta(\omega)$ changes sign.

reflectance uniquely [77]. One must know the reflectivity amplitude at one point $\omega = \omega_0$, to fix the ratio $r(\omega)/r(\omega_0)$. In other words, two systems with reflectances differing by a constant multiplier have the same phase. [This is clear from Eq. (58) since multiplying the reflectance by a constant c contributes $\ln c \int_0^\infty (\omega'^2 - \omega^2)^{-1} \, d\omega'$ to the phase which is identically zero.[†]]

Sum rules are similarly limited [71]. Several methods have been proposed for circumventing this. They are (1) use of convergence factors [13, 70], (2) consideration of the normalized function [71] $r(\omega)/r(0)$ or of the ratio of the reflectances of two materials [13], and (3) finite-energy sum rules [71]. In general, the reflectance sum rules are too new to have been widely studied or utilized, but a number of them appear promising and will be mentioned briefly.

For insulators a sum rule for the derivative of the phase in the static limit that involves the ratio $\ln|r(\omega)/r(0)|$ has been developed [71].

$$
\int_0^\infty \omega^{-1} \left[\frac{r'(\omega)}{r(\omega)} \right] d\omega = \int_0^\infty \omega^{-2} \ln \left| \frac{r(\omega)}{r(0)} \right| d\omega
$$

$$
= -\frac{\pi}{2} \frac{d\theta}{d\omega}\bigg|_{\omega=0} = -\frac{\pi}{n^2(0) - 1} \frac{dk}{d\omega}\bigg|_{\omega=0}, \qquad (60)
$$

where in the first integral $r'(\omega)$ denotes the derivative. The two integrals are related by integration by parts, and $d\theta/d\omega|_{\omega=0}$ has been written in terms of $dk/d\omega|_{\omega=0}$ by employing Eq. (54).

In most instances, $dk/d\omega|_{\omega=0}$ is small so that there is a large degree of cancellation within the integrals of Eq. (60). This has been demonstrated in an application [71] to the reststrahl absorption of NaCl. The negative contribution to $\omega^{-1}[r'(\omega)/\omega(r)]$ exceeds the positive contribution by roughly 10%, the remainder giving $dk/d\omega|_{\omega=0}$ to within experimental accuracy.

The derivative form of the normalized reflectance sum rule is of potential interest in connection with modulation spectroscopy [78–80], where the measured quantity is the relative change $\Delta R/R$ induced in the intensity reflection coefficient $R = |\tilde{r}|^2$ by an oscillatory external perturbation or wavelength modulation.

Finite-energy sum rules are particularly useful in reflection spectroscopy of systems with well-isolated absorptions such as polar insulators or wide-band-gap semiconductors in which the infrared vibrational modes or free-carrier absorption is separated from band-to-band electronic transitions by a wide region of transparency. In this region of transparency the reflectance arising from the band-to-band transition r_b is, to a good approximation, constant except as the fundamental edge is approached, and the infrared reflectance spectra appears superimposed on this constant "optical" reflectance as a background. This superposition is complicated and nonlinear, but one can

---

[†] Note that $P \int_0^\infty [dx/(x^2 - a^2)] = \frac{1}{2}aP \int_{-\infty}^\infty [dx/(x - a)] = \frac{1}{2}aP \int_{-\infty}^\infty (dy/y) \equiv 0$.

proceed in analogy with the development of finite-energy rules for $N(\omega)$, etc., in Section III.

Finite-energy dispersion relations and sum rules are then found for the quantity $\tilde{r}(\omega)/r_b$ over the finite energy internal $[0, \hat{\omega}]$, where, as in Section III , $\hat{\omega}$ is a frequency in the region of transparency below the point at which dispersion in r_b becomes significant but well above the highest of the isolated absorption bands in the infrared. The two sum rules that appear to have the greatest utility are [71]

(1)   A reflectance conservation rule for isolated reflection bands

$$\int_0^{\hat{\omega}} \omega[r'(\omega)/r(\omega)]\, d\omega = \int_0^{\hat{\omega}} \ln|r(\omega)/r_b|\, d\omega \approx 0. \tag{61}$$

Here $r_b = (\varepsilon_b^{1/2} - 1)/(\varepsilon_b^{1/2} + 1)$ is the background reflectance with ε_b the "optical" dielectric constant. The integrals are related by integration by parts as in Eq. (60).

(2)   A phase f sum rule

$$\int_0^{\hat{\omega}} \omega\theta(\omega)\, d\omega \approx \frac{\pi\omega_{pi}^2}{2\varepsilon_b^{1/2}(\varepsilon_b - 1)}, \tag{62}$$

where ω_{pi}^2 is the plasma frequency of the infrared absorption.

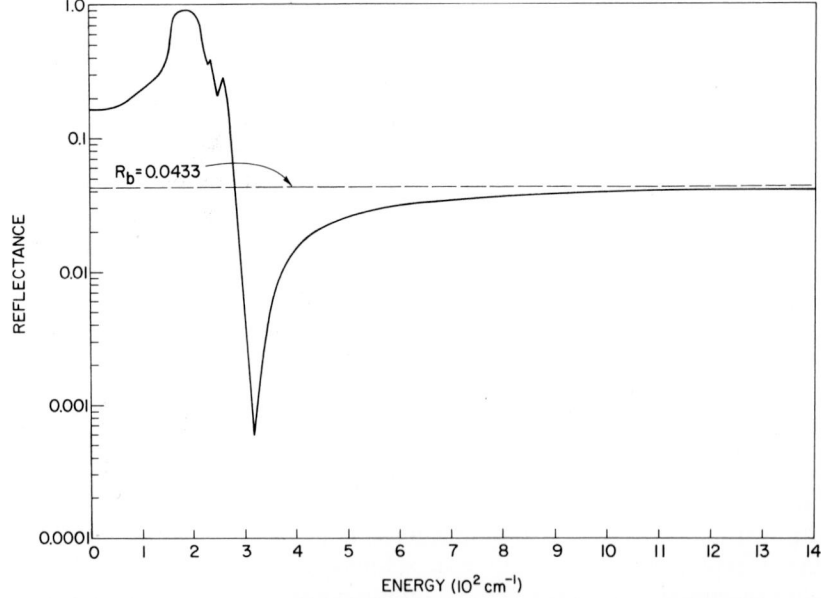

Fig. 4. The reflectance of crystalline NaCl in the infrared plotted on a logarithmic scale versus energy. Note the equality of areas between the curve above and below the background reflectance line at $R_b = 0.0433$, the reflectance in the visible. (From Smith and Manogue [71].)

The content of the reflectance conservation rule [Eq. (61)] is that, regardless of the processes involved, the logarithm of the reflectance of the low-energy process averages to the logarithm of the background reflectance. This is illustrated [71] for the NaCl reststrahl band in Fig. 4. Notice the equality of the areas above and below the dashed line, which denotes the reflectance in the visible or background reflectance $R_b = r_b^2$.

The second rule [Eq. (62)] allows the direct calculation of the oscillator strength of the low-energy absorption without the usual intermediate calculation of $\varepsilon_2(\omega)$ from the phase and reflectance. It must be stressed that this only applies for a finite frequency interval.

Reflectance between Media at Nonnormal B
Incidence, Ellipsometry, and Transmission

The case of normal incidence already discussed leads to particularly simple dispersion relations and sum rules. In the case of reflection in a medium other than vacuum [81–84] or at nonnormal incidence [81, 84–86], the dispersion relations must be modified to account for more complex terms in the phase shift [46]. The interested reader is referred to the literature [8–86] for details. While this more complex situation has led to discussions of the validity of dispersion analysis [87, 88], sum rules have not been investigated for these more general cases except in the case of ellipsometry. In this procedure the ratio of the reflectivities for light polarized parallel and perpendicular to the plane of incidence is considered. Dispersion relations [89] and sum rules [90] for the real and imaginary parts of the ellipsometric function have been developed. Their use has been limited by the small wavelength range over which ellipsometry is now feasible. However, applications to highly localized absorptions in a finite energy interval is an attractive possibility.

Dispersion relations have also been developed for the analysis of transmission measurements [91–93].

ANALYSIS OF OPTICAL DATA AND SUM-RULE APPLICATIONS V

Wide-spectral-range composites of optical data have been constructed with increasing frequency as sufficient measurements become available. Some examples of this work are given in Ehrenreich et al. [94], Hagemann et al. [95], Inagaki et al. [96], Shiles et al. [97], and Cardona [98]. The various sum rules provide a means of testing such composites against both theoretical constraints and independently measured quantities. We briefly consider several aspects of dispersion analysis and the use of sum rules to eliminate systematic errors.

A   Construction of Composite Data Sets

In principle, exact knowledge of both the real and imaginary parts of a single-valued analytic function over a finite interval is sufficient to determine the function for all values of the argument by analytic continuation. However, in practice optical data are subject to experimental errors, so any such continuation outside the measured interval has virtually no meaning. It is thus necessary to have optical measurements at all significant frequencies to determine optical properties over the complete spectral range. While this strict requirement is generally not satisfied, good results over a limited range can be had in favorable cases. We consider this situation first.

1   Dispersion Analysis over a Limited Range

For argument, suppose a single optical function is known (or can be estimated) over all frequencies. Then the others are given by a Kramers–Kronig analysis. As a specific example, if the reflectance for normal incidence from vacuum is known, the phase can be found by using the Šimon–Robinson–Price method [67, 68] by the phase dispersion relation Eq. (58):

$$\theta(\omega) = -\frac{\omega}{\pi} P \int_0^\infty \frac{\ln R(\omega')}{\omega'^2 - \omega^2} \, d\omega', \tag{63}$$

where the expression has been rewritten in terms of the intensity reflection coefficient $R(\omega) = |\tilde{r}(\omega)|^2$, the quantity usually measured. Jahoda's proof [72] of this dispersion relation for the optical case was completed by Velický [73] using Toll's general theory [46]. The other optical functions then follow from Fresnel's equations [1]. In practice dispersion integrals are usually integrated numerically after eliminating [1] the principle value by subtracting an integral that is identically zero[†] to get

$$\theta(\omega) = -\frac{\omega}{\pi} \int_0^\infty \frac{\ln R(\omega') - \ln R(\omega)}{\omega'^2 - \omega^2} \, d\omega'. \tag{63'}$$

Graphical [68, 99] and, more recently, Fourier-integral [100] and Fourier-series [101–103][‡] methods have also been developed to evaluate Eq. (63). Usually only reflection measurements over a limited frequency interval $\omega_1 \leq \omega' \leq \omega_2$ are available and without further information $\theta(\omega)$ cannot be

[†] See footnote to the discussion of Eq. (59) in Section IV.
[‡] Similar methods have been applied to phase determination in electron microscopy [104, 105].

determined since the contributions to Eq. (63) from $0 \leq \omega' < \omega_1$ and $\omega_2 < \omega' < \infty$ are unknown. However, "physically reasonable" estimates and extrapolations of $R(\omega')$ can often be made and, for ω in the measured interval, considerable information about the structure of $\theta(\omega)$—though not its absolute value—can be found.

The important point is that the denominator of Eq. (63) tends to emphasize values of $R(\omega')$ near ω more than those far away. This is particularly evident if Eq. (63) is integrated by parts [72, 106, 107]

$$\theta(\omega) = -\frac{1}{2\pi} \int_0^\infty \frac{d \ln|R(\omega')|}{d\omega'} \ln\left|\frac{\omega' + \omega}{\omega' - \omega}\right| d\omega'. \tag{64}$$

The factor $\ln|(\omega' + \omega)/(\omega' - \omega)|$ is sharply peaked at ω and weights $R(\omega')$ near ω strongly. Thus, for ω within the measured range, the portion of the dispersion integral from ω_1 to ω_2 contributes the structure to $\theta(\omega)$. The values of $R(\omega')$ outside this measured interval typically supply a slowly varying "background" contribution $\Delta\theta(\omega)^\dagger$. Only under special conditions is this background negligible [111, 112], and tests indicate that errors may be significant particularly in regions of small k [113].

The presence of a slowly varying background component in $\theta(\omega)$ is crucial to estimating $\theta(\omega)$ since if $\theta(\omega)$ is known even at a single point, the transform Eq. (63) or (64) can be "anchored" there. The more such points, the better the estimate of $\Delta\theta(\omega)$. The additional information required can be obtained by ellipsometry in reflecting regions and by absorption or index measurements in transparent regions.

Traditionally, such additional data have been used either (1) to establish an extrapolation for $R(\omega)$ or (2) to determine the slowly varying component $\Delta\theta(\omega)$ of the phase directly. The two methods are equivalent but have been compared in only a few cases [114].

In the first approach methods used include trial-and-error searches [115] for an extrapolation of $R(\omega)$ that reproduces the additional data, and the determination of free parameters in a simple reflectance extrapolation function. Some common examples of high-energy parametric extrapolations are the power law [116, 117]‡ $R(\omega) = R(\omega_2)(\omega_2/\omega)^p$, the exponential [119] $R(\omega) = R(\omega_2) \exp[B(\omega_2 - \omega)]$, and a closely related power-law development [120] of $\ln R(\omega)$. These are useful when reflectance measurements extend to energies high enough so that the strength of conduction or valence electrons is largely exhausted. The parameters p and B are generally chosen to make $\theta(\omega)$ zero below the fundamental edge in nonmetals.

In infrared studies of molecular and lattice modes, constant reflectance extrapolations $R(\omega') = R_L$, $0 \leq \omega' < \omega_1$, and $R(\omega') = R_H$, $\omega_2 < \omega' < \infty$, for the reflectance above and below the measured interval are common [76, 121]. The two constants R_L and R_H are chosen to reproduce the auxiliary data.

---

† Explicit examples of this are given in Roessler [108], Velický [109], and MacRae et al. [110].

‡ For correction to Cardona and Greenaway [117] see Scouler [118].

A closely related method employs a mean-value approximation to the dispersion integrals in the unmeasured regions [107, 108]. More detailed schemes [122] using parameters to define harmonic-oscillator extrapolations have been developed for cases in which sufficient data is at hand. Arkatova *et al.* [123] have critically reviewed several of these extrapolation methods.

The second approach to the finite measurement interval problem involves finding the slowly varying component of the phase. Methods include calculating $\Delta\theta(\omega)$ at specific points at which additional data are available and then fitting the result with a simple function [109, 124] or determining a power-series expansion of $\Delta\theta(\omega)$ [125–127]. The latter approach is based on Velický's observation [109] that since $\ln \tilde{r}(\omega)$ is analytic, the difference between the actual phase $\theta(\omega)$ and the phase calculated from Eq. (63) by using measured data plus a "smooth" extrapolation can be expressed in a power series. The coefficients of the series are determined from the auxiliary measurements of $\theta(\omega)$. [Since $\theta(\omega)$ is an odd function, only odd powers of ω are present.]

The remarkable success of these methods in treating reflectance measurements over a finite interval plus one or more auxiliary data points to anchor the phase can be best understood in terms of subtracted dispersion relations, a technique widely used in high-energy physics [19, 20]. If $\theta(\omega)$ is known at a frequency $\omega = a$, then by Eq. (63)

$$\theta(a) = -\frac{a}{\pi} P \int_0^\infty \frac{\ln R(\omega')}{\omega'^2 - a^2} \, d\omega'. \tag{65}$$

Subtracting this from Eq. (63) yields [128]

$$\theta(\omega) = \left(\frac{\omega}{a}\right)\theta(a) - \frac{\omega}{\pi}(\omega^2 - a^2)P \int_0^\infty \frac{\ln R(\omega')}{(\omega'^2 - \omega^2)(\omega'^2 - a^2)} \, d\omega'. \tag{66}$$

This new expression manifestly gives the correct value of $\theta(\omega)$ at $\omega = a$ and is odd in ω as required. By way of qualitative comparison with the Velický [109] series expansion of $\theta(\omega)$, Eq. (66) has the form of a modified dispersion relation plus a term linear in ω. Most important, the modified dispersion integral converges more rapidly than Eq. (63); at large ω' the denominator of the integrand behaves as ω'^4 compared with ω'^2 for Eq. (63). Thus, the knowledge of additional data in the form of the phase at $\omega = a$ has anchored $\theta(\omega)$ at the point $\theta(a)$ and has significantly reduced the dependence of the dispersion integral on the unknown details of $R(\omega)$ far from the measured interval.

The procedure may be carried as far as one has extra data points. For example, with two points the doubly subtracted dispersion relation becomes

$$\theta(\omega) = \frac{\omega}{a}\frac{\omega^2 - b^2}{a^2 - b^2}\theta(a) + \frac{\omega}{b}\frac{\omega^2 - a^2}{b^2 - a^2}\theta(b)$$

$$- \frac{\omega}{\pi}(\omega^2 - a^2)(\omega^2 - b^2)P \int_0^\infty \frac{\ln R(\omega') \, d\omega'}{(\omega'^2 - \omega^2)(\omega'^2 - a^2)(\omega'^2 - b^2)}. \tag{67}$$

Now $\theta(\omega)$ is anchored at two points a and b; moreover, the dispersion integral is even less dependent on the high-energy extrapolation.

The method of subtracted dispersion relations has had only limited applications [128, 129] in optics, but it should provide a systematic scheme utilizing all available data and offering significant savings in effort particularly over trial-and-error extrapolation.

Dispersion Analysis over a Wide Spectral Range   2

For an increasing number of substances some optical data are available in all significant spectral ranges. However, the same type of data are not available at all frequencies. Reflectance or ellipsometric measurements are practical in opaque spectral regions, while the absorption coefficient or refractive index can be measured in regions of near transparency. As a consequence of this mix of data, the Kramers–Kronig method cannot be applied directly since it requires knowledge of a single optical function over all frequencies. However, the combination of Kramers–Kronig and Fresnel equations can be solved as coupled equations.

In practice, one employs a method of successive approximations in which a trial value of one of the optical functions is estimated as best as can be from the available data. A Kramers–Kronig analysis is performed on this trial function. A new trial function is then constructed (generally by substitution of measured values in place of the calculated ones) and the procedure repeated until measurements and calculations agree [97, 132].

In principle, any optical function could be chosen as the starting point for an iteration. For materials with large regions of transparency—such as insulators—the extinction coefficient has been found to be convenient in practice [96]. However, transmission data can also be treated directly by using a dispersion relation for the phase shift in terms of the transmittance [91–93] that is analogous to Eq. (63). In the case of metals [97], the reflectance is a particularly favorable starting point since in opaque regions reflectance data are the most common, while the reflectance can often be approximated from Fresnel's equations in transparent regions. In the latter, both absorption coefficient and refractive index data can be measured and, even if only absorption data is at hand in the region of transparency, simple extrapolations or model fits give reasonable starting values for the index.

This procedure, or at least one cycle of the procedure, has been carried out in a large number of studies by using the reflectance (for example, Ehrenreich et al. [94]) or a combination of reflectance and transmittance data (for example, Hagemann et al. [95]). In principle, there should be no difference in the final result whatever starting function is chosen, provided that the calculation is iterated to self-consistency, although this is not often done.

A full self-consistent iterative analysis has been applied to polystyrene by Inagaki et al. [96] and to aluminum by Shiles et al. [97] and preliminary results have been reported for silicon [130] and gallium arsenide [131]. Aluminum is particularly favorable for this procedure since one form or another of optical data are now available from approximately 0.04 to 10,000 eV. In addition, in the region near the plasma frequency where optical experiments are difficult, there are accurate electron energy-loss measurements that give the shape of $\text{Im}[\varepsilon^{-1}(\omega)]$ (though not its absolute magnitude). Specific results are discussed in the section of this handbook devoted to the optical properties of aluminum.

B Sum-Rule Tests

Sum rules provide significant guidance in the construction of composite sets of data, particularly in selecting the most probable values from the available optical measurements and in pinpointing systematic errors. In the case of metallic aluminum mentioned earlier [97], it has proven possible to generate a comprehensive set of optical functions that satisfies all principal sum rules and is compatible with electron-energy-loss and stopping-power data.

1 The Inertial Sum Rule for n(ω)

The rule

$$\int_0^\infty [n(\omega) - 1]\, d\omega = 0 \tag{68}$$

must be satisfied for any $n(\omega)$ corresponding to a physically acceptable absorption spectrum. Provided that $k(\omega)$ has an acceptable high-frequency behavior (no ω^{-1} component), the rule tests the acceptability of $n(\omega)$ as part of an analytic function. Hence, it tests the accuracy of the Kramer–Kronig transform procedure and any high- or low-frequency extrapolation of $n(\omega)$ that may have been used.

In practice, it is convenient to define a verification parameter [7]

$$\xi = \int_0^\infty [n(\omega) - 1]\, d\omega \Big/ \int_0^\infty |n(\omega) - 1|\, d\omega \tag{69}$$

to assess the extent to which a spectrum satisfies the rule. Shiles et al. [97] found that values of the order of 2×10^{-3} were obtained for self-consistent solutions with ordinary numerical procedures.

Application of the rule to several composite data sets in the literature disclosed large values of ξ (some as large as 0.2) [7, 97]. These unacceptable values appear to arise primarily from nonanalytic modifications of the results of a Kramers–Kronig transformation. Such modifications include fitting inappropriate extrapolations or otherwise forcing $n(\omega)$ to pass through data points inconsistent with the absorption spectrum.

The DC-Conductivity Sum Rule 2

A second consequence of causality and inertia is the expression for the dc conductivity

$$\int_0^\infty [\varepsilon_1(\omega) - 1]\, d\omega = -2\pi^2 \sigma(0). \tag{70}$$

In the case of insulators, there are only interband transitions, and the right-hand side is zero. In metals both inter- and intraband transitions occur. To a good approximation the two processes contribute to $\varepsilon(\omega)$ by linear super-position, and the nonzero term on the right-hand side of Eq. (70) arises primarily from the intraband part. Thus, the dc-conductivity rule measures the conduction-electron contribution to $\varepsilon_1(\omega)$. Generally, this dispersive contribution extends over a wider and experimentally more accessible range than the corresponding absorptive contribution to $\varepsilon_2(\omega)$. Hence, Eq. (70) can be expected to give a more reliable optically derived value of $\sigma(0)$ for comparison with the measured dc conductivity than a Drude-model fit to $\varepsilon_2(\omega)$ in the far infrared.

In comparing values of $\sigma(0)$ derived from optical measurements with the bulk dc conductivity, it must be borne in mind that optical measurements are sensitive to the state of the sample at the surface. Moreover, many measurements are made on evaporated films that may have properties differing significantly from those of the bulk. In applying this rule to aluminum Shiles *et al.* [97] found that the optical and electrical values were equal within experimental error for unoxidized evaporated samples prepared in ultrahigh vacuum. This would generally not have been the case for contaminated films.

The f Sum Rule 3

The various f sum rules test the conformity of optical data with the known electron density. Further, to the extent that absorptions can be attributed to a particular group of levels, the finite-energy rules test the distribution of oscillator strength within broad spectral regions.

The concept of the effective number of electrons contributing to processes up to energy ω [14, 45]

$$n_{\text{eff}}(\omega)\big|_\Phi = \frac{m}{2\pi^2 e^2} \int_0^\infty \omega' \Phi(\omega')\, d\omega', \tag{71}$$

where $\Phi(\omega')$ may be $\varepsilon_2(\omega')$, $2k(\omega')$, or $-\text{Im}[\varepsilon^{-1}(\omega')]$, is particularly useful in applying the f sum rule. A plot of these three quantities for aluminum is

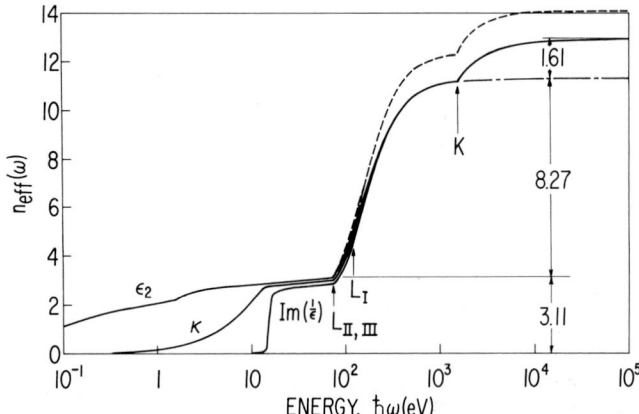

Fig. 5. The functions $n_{\text{eff}}(\omega)$ for metallic aluminum. The dashed curve shows the results for experimental data exhibiting a total oscillator strength over 14 rather than 13 e/at. The solid curve shows $n_{\text{eff}}(\omega)$ after correction of the magnitude of the L-shell data. (From Shiles *et al.* [97].)

shown in Fig. 5. Notice that the three quantities differ significantly as a function of energy. They describe different processes: The $\varepsilon_2(\omega)$ function concerns the dissipation of energy from an electromagnetic wave, the $k(\omega)$ function the attenuation of the wave's amplitude, and Im $\varepsilon^{-1}(\omega)$ the energy loss of a charged particle.

The requirement that the three forms of $n_{\text{eff}}(\omega)$ be equal at $\omega = \infty$, while trivial, is a test of the arithmatic consistency of ε, N, and ε^{-1}. The further requirement that $n_{\text{eff}}(\infty)$ is the observed electron density has more substance and serves as a test for net systematic error. In applying it, one must remember that reflection data measures the state of matter near the surface and that transmission measurements on thin films may not be representative of the bulk. In both instances, surface contamination and impurities introduced in preparation, or differences in density between thin-film and bulk samples may lead to spurious effects.

An example of an apparent systematic error can be seen in the aluminum data of Fig. 5. The dashed curve shows $n_{\text{eff}}(\omega)$ for what initially appeared to be the best available experimental data. However, the total f sum approaches a limit of approximately 14.1 electrons/atom (e/at.) rather than the actual density of 13 e/at. This indicates a significant systematic error in an important spectral region.

The partial f sum rules serve as a guide to the location of such errors. Aluminum is an especially favorable case for this since the conduction, L-shell, and K-shell absorptions are well separated and the oscillator strength of one absorption is virtually exhausted before the onset of the next. The partial f sums for the conduction electrons show a plateau region from the plasmon frequency (~ 15.0 eV) to the $L_{\text{II,III}}$ edge at ~ 72.7 eV. As explained in Section

TABLE I

The "Effective" Number of Electrons per Atom for the Various Energy Levels in Aluminum[a]

Electronic shell	Atomic parentage	Occupation	Raw data	$n_{\text{eff}}(\omega)\big|_{\varepsilon}$ modified data	HFS theory
Conduction	$(3s^2, 3p)$	3	3.1_1	3.1_1	3.12
L shell	$(2s^2, 2p^6)$	8	9.3_5	8.2_7	8.33
K shell	$(1s^2)$	2	1.6_1	1.6_1	1.55
Total		13	14.0_8	12.9_9	13.00

[a] After Shiles *et al.* [97].

III, the three forms of $n_{\text{eff}}(\omega)$ differ in this asymptotic region because of virtual processes in the polarizable cores. A fit of Eq. (53) to the plateau yields the core background dielectric constant $\varepsilon_b \approx 1.03_5$. The total conduction electron oscillator strength is given by the limiting value of $n_{\text{eff}}(\omega)\big|_{\varepsilon}$; extrapolating the plateau gives a total oscillator strength for the three conduction electrons of 3.1_1 e/at. Above the $L_{II,III}$ edge the effects of shielding decrease sharply, and the three forms of $n_{\text{eff}}(\omega)$ draw together.

The oscillator strengths for the various energy levels in metallic aluminum as derived from a self-consistent composite of the reported data are listed in Table I under raw data. Also given is the oscillator strength calculated for the corresponding *atomic* energy levels in a one-electron Hartree–Fock–Slater (HFS) model that takes into account the Pauli principle redistribution of oscillator strength. In going from the atom to the solid, the valence-electron energies and wave functions are vastly altered, leading to a completely different absorption spectrum of the outermost electrons. However, the core levels are not changed significantly, and the net oscillator strength of each group of levels in the metal and the atom should be qualitatively comparable. Making this comparison in Table I, one sees that the L-shell strength of the reported data is too strong by approximately 1 e/at. corresponding to a systematic overestimate of 12–13% averaged over the whole L absorption.

In the high-absorption region between the L edge and 500 eV, virtually all measurements have been made on thin evaporated films prepared in conventional vacuum and subsequently exposed to the atmosphere. It is known that such films are highly reactive and form surface layers. Moreover, their absorption shows a surface as well as a bulk component [133, 134].

It was found by trial and error that an *ad hoc* reduction of the reported thin-film absorption data by 14% from the L edge to 300 eV followed by a smooth interpolation to the rolled-foil data above 500 eV gave satisfactory agreement with the partial *f* sum rules. Independent experiments by Balzarotti *et al.* [134] provide additional evidence for this correction. Using films of various thickness, these authors separated surface and bulk effects and found absorption coefficients some 15% below those generally reported in this region. Unfortunately, their measurements extend only 10 eV above the L edge, but as far as they go, they are in agreement with the values inferred by demanding the f sum rule be satisfied.

VI SUMMARY

The optical constants obey a large number of integral constraints in the form of sum rules. We have shown that some of the simpler of these rules arise directly from the requirements of causality, inertia, and the dynamical laws. These optical rules may therefore be viewed as the ω-space equivalent of the dynamical laws of motion in time space. The sum rules provide a useful means of testing optical measurements, particularly wide-range composite data, both against theoretical constraints and independently measured quantities.

ACKNOWLEDGMENTS

The author wishes to express his thanks to Professor Dr. H. Bilz for his hospitality and encouragement and to the Max-Planck-Institut für Festkörperforschung for support during the early stages of preparation of this review. Thanks are also due to Dr. M. Inokuti for stimulating discussions and Mrs. Janice Grant for her aid in preparing the manuscript.

REFERENCES

1. F. Stern, *in* "Solid State Physics" (F. Seitz and D. Turnbull eds.), Vol. 15, p. 299, Academic Press, New York, 1963.
2. F. Wooten, "Optical Properties of Solids," Academic Press, New York, 1972.
3. W. M. Saslow, *Phys. Lett.* **33A**, 157 (1970).
4. M. Altarelli, D. L. Dexter, H. M. Nussenzveig, and D. Y. Smith, *Phys. Rev. B* **6**, 4502 (1972).
5. A. Villani and A. H. Zimerman, *Phys. Lett.* **44A**, 295 (1973).
6. A. Villani and A. H. Zimerman, *Phys. Rev. B* **8**, 3914 (1973).
7. M. Altarelli and D. Y. Smith, *Phys. Rev. B* **9**, 1290 (1974); M. Altarelli and D. Y. Smith, *Phys. Rev. B* **12**, 3511 (1975).
8. F. W. King, *J. Math. Phys.* **17**, 1509 (1976).
9. K. Furuya, A. H. Zimerman, and A. Villani, *Phys. Rev. B* **13**, 1357 (1976).
10. D. Y. Smith, *J. Opt. Soc. Am.* **66**, 547 (1976).
11. D. Y. Smith, *Phys. Rev. B* **13**, 5303 (1976).
12. K. Furuya, A. H. Zimerman, and A. Villani, *J. Phys. C* **9**, 4329 (1976).
13. K. Furuya, A. Villani, and A. H. Zimerman, *J. Phys. C* **10**, 3189 (1977).
14. D. Y. Smith and E. Shiles, *Phys. Rev. B* **17**, 4689 (1978).
15. C. K. Mukhtarov, *Dok. Akad. Nauk SSSR* **249**, 851 (1979); C. K. Mukhtarov, *Sov. Phys. Dokl.* **24**, 991 (1979).
16. H. A. Kramers, *Nature* **117**, 775 (1926) [abstract]; H. A. Kramers, *Atti Congr. Intern. Fisici, Como-Pavia-Roma* (N. Zanichelli, Bologna, 1928) **2**, 545–557 [reprinted in H. A.

Kramers, "Collected Scientific Papers," North-Holland Publ., Amsterdam, 1956]; H. A. Kramers, *Phys. Z.* **30**, 522 (1929).

17. R. de L. Kronig, *J. Opt. Soc. Am.* **12**, 547 (1926).

18. L. D. Landau and E. M. Lifshitz, "Electrodynamics of Continuous Media," Pergamon, Oxford, 1960.

19. H. M. Nussenzveig, "Causality and Dispersion Relations," Academic Press, New York, 1972.

20. H. Burkhardt, "Dispersion Relation Dynamics," North Holland Publ., Amsterdam, 1969.

21. C. Ferro Fontán, N. M. Queen, and G. Violini, *Riv. Nuovo Cimento* **2**, 357 (1972).

22. B. Gross, *Nuovo Cimento Suppl.* **3**, 235 (1956); J. R. Macdonald and M. K. Brachman, *Rev. Mod. Phys.* **28**, 393 (1956); D. Pines and P. Nozières, "The Theory of Quantum Liquids," W. A. Benjamin, New York, 1966; P. C. Martin, *Phys. Rev.* **161**, 143 (1967); J. des Cloizeaux, *in* "Theory of Condensed Matter," p. 325 Int. At. Energy Ag., Vienna, 1968; R. Kubo and M. Ichimura, *J. Math. Phys.* **13**, 1454 (1972).

23a. D. Y. Smith, *in* "Theoretical Aspects and New Developments in Magnetooptics" (J. T. Devreese ed.), p. 133, Plenum, New York, 1980.

23b. M. T. Thomaz and H. M. Nussenzveig, *Ann. Phys. (N.Y.)* **139**, 14 (1982).

24. H. A. Bethe and E. E. Salpeter, "Quantum Mechanics of One- and Two-Electron Atoms," Sec. 61 and 62, pp. 357 and 358, Springer-Verlag, Berlin, 1957.

25. R. G. Sachs and N. Austern, *Phys. Rev.* **81**, 705 (1951).

26. J. S. Levinger, M. L. Rustgi, and K. Okamoto, *Phys. Rev.* **106**, 1191 (1957).

27. H. O. Dogliani and W. F. Bailey, *J. Quant. Spectrosc. Radiat. Transfer* **9**, 1643 (1969).

28. B. Jacobsohn, Ph.D. dissertation, University of Chicago, Chicago, 1947 (unpublished). [Available in part through DOE Technical Information Center, Oak Ridge, Tennessee, as ANL-HDY-423 (Del).]

29. W. B. Payne and J. S. Levinger, *Phys. Rev.* **101**, 1020 (1956).

30. U. Fano, *in* "Annual Review of Nuclear Science" (E. Segrè, G. Friedlander, and H. P. Noyes eds.) Vol. 13, p. 1, Annual Reviews, Palo Alto, California, 1963.

31. W. Thomas, *Naturwiss.* **28**, 627 (1925); and F. Reiche and W. Thomas, *Z. Phys.* **34**, 510 (1925).

32. W. Kuhn, *Z. Phys.* **33**, 408 (1925).

33. B. L. van der Waerden, "Sources of Quantum Mechanics," North-Holland Publ., Amsterdam, 1967.

34. U. Fano and J. W. Cooper, *Rev. Mod. Phys.* **40**, 441 (1968); U. Fano and J. W. Cooper, *Rev. Mod. Phys.* **41**, 724 (1969).

35. J. S. Levinger, "Nuclear Photo-Disintegration," Oxford Univ. Press, London, 1960.

36. A. Smakula, *Z. Phys.* **59**, 603 (1930).

37. D. L. Dexter, *in* "Solid State Physics" (F. Seitz and D. Turnbull eds.), Vol. 6, p. 353, Academic, New York, 1958.

38. D. Y. Smith and D. L. Dexter, *in* "Progress in Optics" (E. Wolf, ed.), Vol. 10, Sec. 3.1, North-Holland Publ., Amsterdam, 1972.

39. J. P. Vinti, *Phys. Rev.* **41**, 432 (1932).

40. J. O. Hirschfelder, W. B. Brown, and S. T. Epstein, *in* "Advances in Quantum Chemistry" (P.-O. Löwdin, ed.), Vol. 1, Sec. 10. Academic Press, New York, 1964.

41. J. J. Hopfield, *Phys. Rev. B* **2**, 973 (1970).

42. M. Brauwers, R. Evrard, and E. Kartheuser, *Phys. Rev. B* **12**, 5864 (1975).

43. C.-H. Wu, G. Mahler, and J. L. Birman, *Phys. Rev. B* **18**, 4221 (1978).

44. J. Gardavský, *Coll. Phen.* **1**, 193 (1974).

45. H. Ehrenreich, *in* "The Optical Properties of Solids" (J. Tauc, ed.), p. 106, Academic Press, New York, 1966.

46. J. S. Toll, Ph.D. thesis Princeton University, Princeton, New Jersey, 1952; J. S. Toll, *Phys. Rev.* **104**, 1760 (1956).

47. N. G. van Kampen, *J. Phys, Rad.* **22**, 179 (1961).

48. E. C. Titchmarsh, "The Theory of Functions," Oxford Univ. Press, London, 1968.
49. V. de Alfaro, S. Fubini, G. Rossetti, and G. Furlan, *Phys. Lett.* **21**, 576 (1966).
50. G. Frey and R. L. Warnock, *Phys. Rev.* **130**, 478 (1963).
51. A. R. P. Rau and U. Fano, *Phys. Rev.* **162**, 68 (1967).
52. R. A. Ferrell and R. E. Glover, *Phys. Rev.* **109**, 1398 (1958).
53. G. Brändli, *Phys. Rev. Lett.* **28**, 159 (1972); G. Brändli and A. J. Sievers, *Helv. Phys. Acta* **45**, 847 (1972); G. Brandli and A. J. Sievers, *Phys. Rev. B* **5**, 3550 (1972).
54. G. Brändli, *Phys. Rev. B* **9**, 342 (1974).
55. M. Tinkham and R. A. Ferrell, *Phys. Rev. Lett.* **2**, 331 (1959).
56. D. M. Ginsberg and M. Tinkham, *Phys. Rev.* **118**, 990 (1960).
57. L. H. Palmer and M. Tinkham, *Phys. Rev.* **165**, 588 (1968).
58. D. Y. Smith, *J. Opt. Soc. Am.* **66**, 454 (1976).
59. F. W. King, *J. Math. Phys.* **22**, 1321 (1981); F. W. King, *Phys. Rev. B* **25**, 1381 (1982).
60. E. C. Titchmarsh, "Introduction to the Theory of Fourier Integrals," Oxford Univ. Press, London, 1962.
61. I. Kimel, *Phys. Rev. B* **25**, 6561 (1982).
62. J. Fischer, J. Pišút, P. Prešnajder, and J. Šebesta, *Czech. J. Phys. B* **19**, 1486 (1969).
63. Y. C. Liu and S. Okubo, *Phys. Rev. Lett.* **19**, 190 (1967).
64. D. Y. Smith and G. Graham, *J. Phys. (Paris)* **41**, Colloque C6, C6-80 (1980).
65. R. de L. Kronig and H. A. Kramers, *Z. Phys.* **48**, 174 (1928).
66. V. Fock, *Z. Phys.* **89**, 744 (1934).
67. I. Šimon, *J. Opt. Soc. Am.* **41**, 336 (1951).
68. T. S. Robinson, *Proc. Phys. Soc. (London) B* **65**, 910 (1952); T. S. Robinson and W. C. Price, *Proc. Phys. Soc. (London) B* **66**, 969 (1953); T. S. Robinson and W. C. Price, *in* "Molecular Spectroscopy" (G. Sell ed.), p. 211, Inst. Petroleum, London, 1955; J. K. Wilmshurst and S. Senderoff, *J. Chem. Phys.* **35**, 1078 (1961).
69. T. Inagaki, A. Ueta, and H. Kuwata, *Phys. Lett.* **66A**, 329 (1978).
70. F. W. King, *J. Chem. Phys.* **71**, 4726 (1979); I. Kimel, *Phys. Lett.* **88A**, 62 (1982).
71. D. Y. Smith and C. A. Manogue, *J. Opt. Soc. Am.* **71**, 935 (1981).
72. F. C. Jahoda, Ph.D. thesis, Cornell University, Ithaca, New York, 1957; F. C. Jahoda, *Phys. Rev.* **107**, 1261 (1957).
73. B. Velický, *Czech. J. Phys. B* **11**, 541 (1961).
74. P. Roman and A. S. Marathay, *Nuovo Cimento* **30**, 1452 (1963).
75. R. E. Burge, M. A. Fiddy, A. H. Greenaway, and G. Ross, *J. Phys. D* **7**, L65 (1974).
76. M. Gottlieb, Ph.D. thesis, University of Pennsylvania, Phildelphia, Pennsylvania 1959; M. Gottlieb, *J. Opt. Soc. Am.* **50**, 343 (1960); M. Gottlieb, *J. Opt. Soc. Am.* **50**, 350 (1960).
77. D. Y. Smith, *J. Opt. Soc. Am.* **67**, 570 (1977).
78. G. Bonfiglioli and P. Brovetto, *Phys. Lett.* **5**, 248 (1963); G. Bonfiglioli and P. Brovetto, *Appl. Opt.* **3**, 1417 (1964).
79. B. O. Seraphin, *in* "Optical Properties of Solids" (F. Abelès, ed.), p. 163, North-Holland Publ. Amsterdam, 1972.
80. A Balzarotti, E. Colavita, S. Gentile, and R. Rosei, *Appl. Opt.* **14**, 2412 (1975).
81. J. S. Plaskett and P. N. Schatz, *J. Chem. Phys.* **38**, 612 (1963).
82. E. A. Lupashko, V. K. Miloslavskii, and I. N. Shklyarevskii, *Opt. Spectrosk.* **29**, 789 (1970); E. A. Lupashko, V. K. Miloslavskii, and I. N. Shklyarevskii, *Opt. Spectrosc.* **29**, 419 (1970).
83. E. A. Lupashko, V. K. Miloslavskii, and I. N. Shklyarevskii, *Opt. Spectrosk.* **24**, 257 (1968); Lupashko, Miloslavskii, and Shklyarevskii, *Opt. Spectrosc.* **24**, 132 (1968).
84. A. A. Clifford, M. I. Duckels, and B. Walker, *J. Chem. Soc. (London), Faraday Transactions II*, **68**, 407 (1972).
85. D. M. Roessler, *Brit. J. Appl. Phys.* **16**, 1359 (1965).
86. D. W. Berreman, *Appl. Opt.* **6**, 1519 (1967).
87. W. G. Chambers, *Infrared Phys.* **15**, 139 (1975).
88. R. H. Young, *J. Opt. Soc. Am.* **67**, 520 (1977).

89. G. M. Hale, W. E. Holland, and M. R. Querry, *Appl. Opt.* **12**, 48 (1973); M. R. Querry and W. E. Holland *Appl. Opt.*, **13**, 595 (1974).
90. T. Inagaki, H. Kuwata, and A. Ueda, *Phys. Rev. B* **19**, 2400 (1979); T. Inagaki, H. Kuwata, and A. Ueda, *Surf. Sci.* **96**, 54 (1980).
91. S. Maeda, G. Thyagarajan, and P. N. Schatz, *J. Chem. Phys.* **39**, 3474 (1963); K. Kozima, W. Suëtaka, and P. N. Schatz, *J. Opt. Soc. Am.* **56**, 181 (1966).
92. P.-O. Nilsson, *Appl. Opt.* **7**, 435 (1968).
93. J. D. Neufeld and G. Andermann, *J. Opt. Soc. Am.* **62**, 1156 (1972).
94. H. Ehrenreich, H. R. Philipp, and B. Segall, *Phys. Rev.* **132**, 1918 (1963); H. R. Philipp and H. Ehrenreich, *J. Appl. Phys.* **35**, 1416 (1964).
95. H.-J. Hagemann, W. Gudat, and C. Kunz, *J. Opt. Soc. Am.* **65**, 742 (1975); H.-J. Hagemann, W. Gudat, and C. Kunz, DESY Report SR 74/7, "Optical Constants from the Far Infrared to the X-Ray Region," Deutsches Elektronen-Synchrotron, Hamburg, 1974.
96. T. Inagaki, E. T. Arakawa, R. N. Hamm, and M. W. Williams, *Phys. Rev. B* **15**, 3243 (1977).
97. E. Shiles, T. Sasaki, M. Inokuti, and D. Y. Smith, *Phys. Rev. B* **22**, 1612 (1980).
98. M. Cardona, *in* "Optical Properties of Solids," (S. Nudelman and S. S. Mitra eds.), p. 137 Plenum, New York, 1969.
99. L. D. Kislovskii, *Opt. Spektrosk.* **1**, 672 (1956); L. D. Kislovskii, *Opt. Spektrosk.* **2**, 186 (1957); L. D. Kislovskii, *Opt. Spektrosk.* **5**, 66 (1958); L. D. Kislovskii, *Opt. Spektrosk.* **6**, 810 (1959); L. D. Kislovskii, *Opt. Spectrosc.* **6**, 529 (1959); L. D. Kislovskii, *Opt. Spektrosk.* **7**, 311 (1959); L. D. Kislovskii, *Opt. Spectrosc.* **7**, 201 (1959).
100. C. W. Peterson and B. W. Knight, *J. Opt. Soc. Am.* **63**, 1238 (1973).
101. D. W. Johnson, *J. Phys. A* **8**, 490 (1975).
102. F. W. King, *J. Phys. C* **10**, 3199 (1977).
103. F. W. King, *J. Opt. Soc. Am.* **68**, 994 (1978).
104. W. O. Saxton, *J. Phys. D* **7**, L63 (1974).
105. D. L. Misell, *J. Phys. D* **7**, L69 (1974).
106. H. W. Bode, "Network Analysis and Feedback Amplifier Design," D. Van Nostrand, New York, 1945.
107. D. M. Roessler, *Brit. J. Appl. Phys.* **16**, 1119 (1965); D. M. Roessler, *Brit. J. Appl. Phys.* **16**, 1777 (1965).
108. D. M. Roessler, *Brit. J. Appl. Phys.* **17**, 1313 (1966).
109. B. Velický, *Czech. J. Phys. B* **11**, 787 (1961).
110. R. A. MacRae, E. T. Arakawa, and M. W. Williams, *Phys. Rev.* **162**, 615 (1967).
111. G. R. Anderson and W. B. Person, *J. Chem. Phys.* **36**, 62 (1962).
112. H. J. Bowlden and J. K. Wilmshurst, *J. Opt. Soc. Am.* **53**, 1073 (1963).
113. W. G. Spitzer and D. A. Kleinman, *Phys. Rev.* **121**, 1324 (1961).
114. O. Castaño, M. de Dios, and R. Pérez, *comunicaciones (Havana)* **17**, 65 (1975); O. D. Castaño, M. de Dios Leyva, and R. Pérez Alvarez, *Phys. Stat. Solidi B* **71**, 111 (1975).
115. B. W. Veal *in* "The Actinides: Electronic Structure and Related Properties, Vol. 2, pp. 87–88, Academic Press, New York, 1974.
116. H. R. Philipp and E. A. Taft, *Phys. Rev.* **113**, 1002 (1959); H. R. Philipp and E. A. Taft, *Phys. Rev.* **136**, A1445 (1964).
117. M. Cardona and D. L. Greenaway, *Phys. Rev.* **133**, A1685 (1964).
118. W. J. Scouler, *Phys. Rev.* **178**, 1353 (1969).
119. M. P. Rimmer and D. L. Dexter, *J. Appl. Phys.* **31**, 775 (1960); M. G. Doane, M.S. thesis, University of Rochester, Rochester, New York, 1961.
120. V. B. Tulvinskii and N. I. Terentev, *Opt. Spektrosk.* **28**, 894 (1970); V. B. Tulvinskii and N. I. Terentev, *Opt. Spectrosc.* **28**, 484 (1970).
121. P. N. Schatz, S. Maeda, and K. Kozima, *J. Chem. Phys.* **38**, 2658 (1963); D. W. Barnes and P. N. Schatz, *J. Chem. Phys.* **38**, 2662 (1963).
122. G. Andermann, A. Caron, and D. A. Dows, *J. Opt. Soc. Am.* **55**, 1210 (1965); G. Andermann

and D. A. Dows, *J. Phys. Chem. Sol.* **28**, 1307 (1967); C.–K. Wu and G. Andermann, *J. Opt. Soc. Am.* **58**, 519 (1968).

123. T. G. Arkatova, N. M. Gopshtein, E. G. Makarova, and B. A. Mikhailov, *Opt. Mekh. Promst.* **48**, 44 (1981); T. G. Arkatova, N. M. Gopshtein, E. G. Makatova, and B. A. Mikhailov, *Sov. J. Opt. Technol.* **48**, 552 (1981).
124. D. G. Thomas and J. J. Hopfield, *Phys. Rev.* **116**, 573 (1959).
125. V. K. Miloslavskii, *Opt. Spektrosk.* **21**, 343 (1966); V. K. Miloslavskii, *Opt. Spectrosc.* **21**, 193 (1966).
126. P.-O. Nilsson and L. Munkby, *Phys. Kondens. Materie* **10**, 290 (1969).
127. V. K. Zaitsev and M. I. Fedorov, *Opt. Spektrosk.* **44**, 1186 (1978); V. K. Zaitsev and M. I. Fedorov, *Opt. Spectrosc.* **44**, 691 (1978).
128. R. K. Ahrenkiel, *J. Opt. Soc. Am.* **61**, 1651 (1971); R. K. Ahrenkiel, *J. Opt. Soc. Am.* **62**, 1009 (1972).
129. R. Z. Bachrach and F. C. Brown, *Phys. Rev. B* **1**, 818 (1970); N. J. Carrera and F. C. Brown, *Phys. Rev. B* **4**, 3651 (1971).
130. E. Shiles and D. Y. Smith, *Bull. Am. Phys. Soc.* **23**, 226 (1978).
131. E. Shiles and D. Y. Smith, *Bull. Am. Phys. Soc.* **24**, 335 (1979).
132. G. Leveque, *J. Phys. C* **10**, 4877 (1977).
133. D. H. Tomboulian and E. M. Pell, *Phys. Rev.* **83**, 1196 (1951).
134. A. Balzarotti, A. Bianconi, and E. Burattini, *Phys. Rev. B* **9**, 5003 (1974).

Chapter **4**

Measurement of Optical Constants in the Vacuum Ultraviolet Spectral Region[*]

W. R. HUNTER[†]
Naval Research Laboratory
Washington, D.C.

I. Introduction	69
II. General Discussion of Reflectance Methods	70
A. Reflection at Normal Incidence—Measurement of Index of Refraction	70
B. Reflection at Oblique Incidence	71
C. Effect of Layer Thickness	78
D. Critical-Angle Method	79
E. Measurement of Extinction Coefficients	81
III. Reflectance Method for Two Media	85
References	87

INTRODUCTION   I

The optical constants of an isotropic material are the index of refraction n and the extinction coefficient k. They are the real and imaginary components of the complex index of refraction $N = n - ik$. They can be measured at a given wavelength by direct methods or inferred from photometric or polarimetric measurements. For wavelengths less than 2000 Å, the vacuum ultraviolet (VUV) spectral region, polarimetric methods are seldom useful because the phase-discriminating components of the system are strongly absorbing, consequently, reflectance methods are almost exclusively used for obtaining

---

[*] This chapter is a condensed version of a paper, "Measurement of Optical Properties of Materials in the Vacuum Ultraviolet Spectral Region," published in *Applied Optics* **21**, 2103 (1982).

[†] Present address: Sachs/Freeman Associates, Inc., Bowie, Maryland.

n and k. A number of methods exist for extracting n and k from specular-reflectance measurements at both normal and oblique incidence and for semi-infinite media as well as layers on substrates.

This chapter will discuss the most useful methods for obtaining n and k in the VUV and the problems encountered in applying these methods. The Kramers–Kronig method will not be discussed here (see Chapter 3) because it cannot be used to deduce n and k from measurements at a single wavelength.

II   GENERAL DISCUSSION OF REFLECTANCE METHODS

The specular reflectance of a substance at an angle of incidence ϕ, measured from the surface normal, is related to the complex index of refraction by the generalized Fresnel reflection coefficients,

$$R_s = [(a - \cos \phi)^2 + b^2]/[(a + \cos \phi)^2 + b^2]$$
$$R_p = R_s[(a - \sin \phi \tan \phi)^2 + b^2]/[(a + \sin \phi \tan \phi)^2 + b^2]$$
$$R_a = \tfrac{1}{2}[R_p(1 + p) + R_s(1 - p)],$$

where R_s is the component of specular reflection perpendicular to the plane of incidence, R_p the parallel component, and R_a the reflectance for non-polarized radiation. The polarization of the incident radiation is given by

$$p = (I_p - I_s)/(I_p + I_s),$$

and

$$a^2 = \tfrac{1}{2}\{[(n^2 - k^2 - \sin^2 \phi) + 4n^2k^2]^{1/2} + (n^2 - k^2 - \sin^2 \phi)\},$$
$$b^2 = \tfrac{1}{2}\{[(n^2 - k^2 - \sin^2 \phi) + 4n^2k^2]^{1/2} - (n^2 - k^2 - \sin^2 \phi)\}.$$

These equations cannot be solved explicitly for n and k; hence one must resort to graphical or numerical means to solve them [1–11]. Since the equations have two unknowns, two independent measurements are necessary if a solution is to be obtained.

A   Reflection at Normal Incidence—Measurement of Index of Refraction

At normal incidence the two components of reflection are indistinguishable and, for nonabsorbing media, the equations may be reduced to

$$R = (n - 1)^2/(n + 1)^2.$$

Unlike the generalized coefficients, this equation can be solved for n as

$$n = (1 + R^{0.5})/(1 - R^{0.5}).$$

The values of n of transparent media in the VUV usually range from 1.3 to 2.0. Over this range the absolute error in determining n from reflectance measurements is approximately 10 times the absolute error in measuring R. Thus, n is sensitive to small errors in R and the accuracy of the method makes it suitable only for determining provisional values to two significant figures at best.

This simple method may be used for any material that is transparent in the VUV and can be polished or cleaved to give a specularly reflecting surface. Precautions must be taken to measure reflectance from the first surface only. Reflections from the second surface can be eliminated by making the sample wedge shaped or can be suppressed by grinding the second surface and blackening it.

Reflection at Oblique Incidence   B

Humphreys–Owen [6] lists nine methods by which n and k can be deduced from reflectance measurements at oblique incidence. These methods are divided into two classes: (1) two reflectance measurements at one angle of incidence or one reflectance measurement at each of two angles of incidence and (2) one reflectance measurement at any angle of incidence and measurement of a special angle of incidence capable of supplying the necessary second measurement. There are only two special angles: (1) the principal angle of incidence, which must be determined by polarimetric methods and so it will not be discussed, and (2) the Brewster, or pseudo-Brewster (pB), angle. The Brewster angle is defined only in terms of dielectric media and is given by $n = \tan \phi_B$. At this angle $R_p = 0$. As k increases, R_p always has a minimum value, but it is not zero. The angle at which this minimum occurs in the pB angle and, if $k > 0$, $\phi_{pB} > \phi_B$. Humphreys–Owen's list has been rearranged as follows:

Class 1

Method 1   Reflectance at two angles of incidence using natural or polarized radiation; sometimes referred to as *the* reflectance-versus-angle of incidence method

Method 2   The ratio R_p/R_s at two angles of incidence

Method 3 R_s and R_p at one angle of incidence

Class 2

Method 4   Pseudo-Brewster angle and R_s or R_p at that angle

Method 5   Pseudo-Brewster angle and R_p/R_s at that angle

Method 6   Pseudo-Brewster angle and R_s, R_p, or R_p/R_s at any other angle
             of incidence

In principle, only two measurements are required, but, because of possible errors in the measurements, a redundancy of measurements is more useful and will give an indication of the errors involved in determining n and k.

Although these methods can be used at any wavelength, their sensitivities to errors in measurements of reflectance, of the angles involved, or of the state of polarization are functions of both n and k, as well as the angle of incidence and the state of polarization. Thus, there may be parts of the n, k plane, angles of incidence, and states of polarization, for which the lack of sensitivity reduces their accuracy to unacceptable values.

Since the sensitivities of the methods are dependent on n and k, hence on the shape of the R-versus-ϕ curves, a collection of these curves is shown in Fig. 1 [11] for reference during the ensuing discussion. The large numbers give the values of n and k used in calculating a particular set of curves. These values cover the range of n and k most likely to be encountered in the VUV. The small numbers along the abscissa and ordinate show the angle of incidence and percent reflectance, respectively. The upper curve is always R_s, the lower curve R_p, and the curve between R_a. Generally, for a given n, as k increases

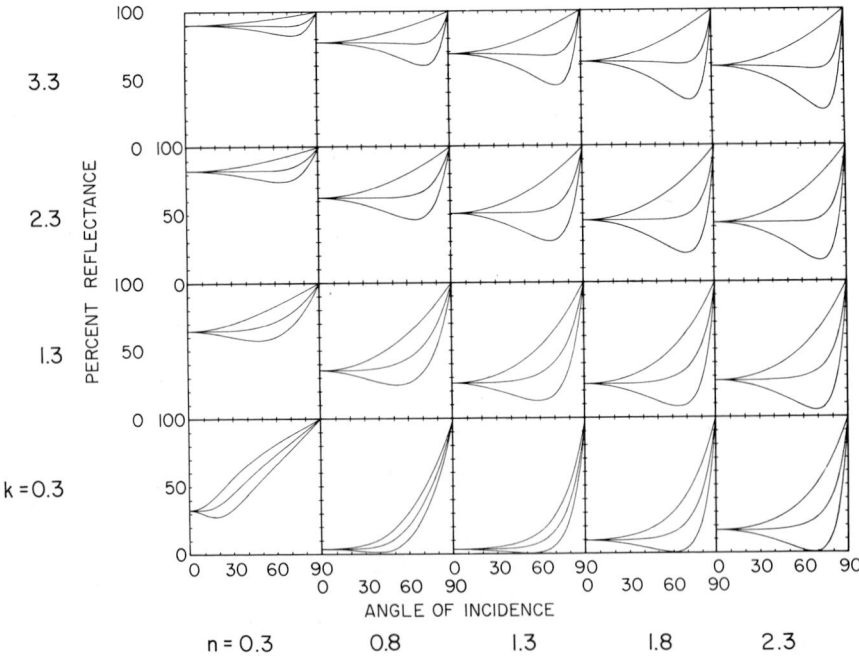

Fig. 1. R-versus-ϕ curves calculated using the optical constants shown by the large numbers. The small numbers shown on the abscissa and ordinate are the values of ϕ and R, respectively.

the reflectance at normal incidence increases, the change in R with ϕ becomes small, especially for R_a, except at the pB angle, and the pB angle becomes larger. For a given k, as n increases the reflectance at normal incidence decreases. the change in R with ϕ becomes more pronounced, especially for R_p, and again the pB angle becomes larger. The value of R_p at the pB angle appears to be zero for $n \geq 1.3$ and $k = 0.3$. This is not so, but the values are extremely small.

Method 1   1

The sensitivity of this method is most easily investigated by means of isoreflectance curves. An isoreflectance curve is the locus of points in the n, k plane corresponding to a given value of R for a specific value of ϕ. If isoreflectance curves are plotted for perfect data, i.e., reflectance values calculated using the Fresnel formula, an insight into the sensitivity of the method can be obtained. Figure 2 [11] shows such a set of curves for the parallel component covering that part of the n, k plane $0.3 \leq n \leq 2.3$ and $0.3 \leq k \leq 3.3$.

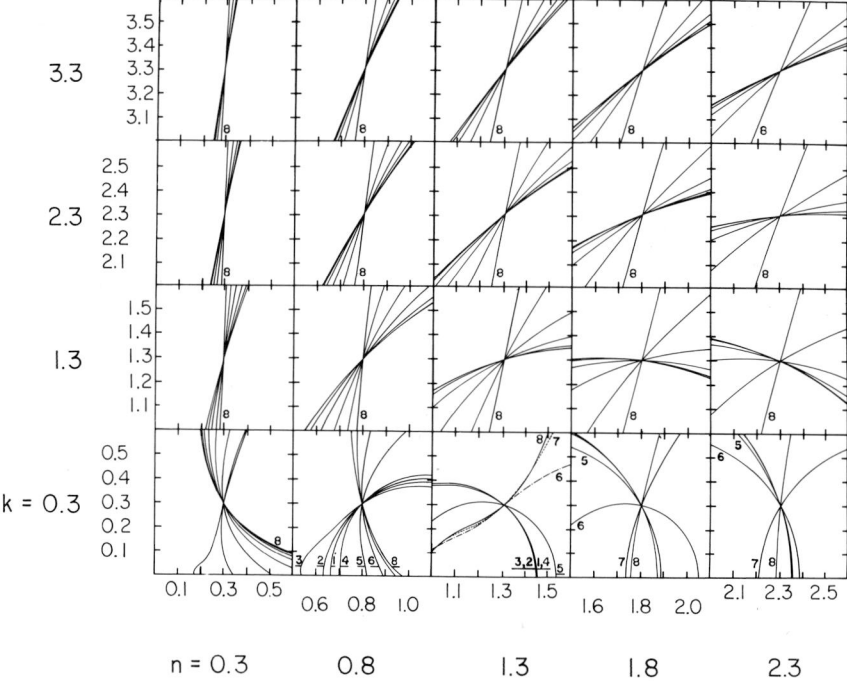

Fig. 2. Isoreflectance curves for R_p calculated using the optical constants shown by the larger numbers. The small numbers used for the abscissa and ordinate show the scale in the n, k plane. The numeral 8 designates the isoreflectance curve corresponding to 80°. The other curves are for angles of incidence 70, 60, through 10°. The curves usually occur in descending order; when they do not, each curve is labeled with a single digit to avoid confusion (method 1).

An indication of the sensitivity is the angle of intersection of the isoreflectance curves. If there is an error in the reflectance, the isoreflectance curve will be shifted parallel to itself by an amount depending on the magnitude of the error and in a direction depending on its sign. Thus, if two curves intersect at a small angle, a slight displacement of one with respect to the other may shift the intersection point by a large amount. Such a displacement could be due either to an error in measuring R or an error in measuring ϕ. The figure indicates that for $n = 0.3$, the sensitivity with respect to k decreases as k increases. Thus, for small n and $k > n$, the R-versus-ϕ method can provide reasonable values for n but may result in large errors for k. Metals with large reflectance values, e.g., Al, Mg, etc., usually have small n and $k > n$. For example, at 1216 Å, Al has $n \approx 0.06$ and $k \approx 1.0$.

For $n > 0.3$, the isoreflectance curves for small angles of incidence are more nearly parallel than those for the larger angles. Thus, the sensitivity is greatest if measurements are made at the larger angles of incidence. For $n = 0.3$, however, the maximum sensitivity for small k is obtained using small angles and as n increases, maximum sensitivity is shifted only slowly to larger angles. Generally, the angles with which maximum sensitivity is obtained are those angles at which the curvature in R is maximum, which is not necessarily the angles at which R changes most rapidly with ϕ. Similar sets of isoreflectance curves have been calculated for R_s and R_a [11].

Although the curves in Fig. 2 furnish an idea of the sensitivity of the method, the accuracy of the method can only be found by using false data. By adding positive and negative errors to perfect reflectance data, the magnitude of the displacement of the isoreflectance curves can be obtained and the accuracy of the method determined. Hunter [11] has investigated the accuracy of method 1 using angles of incidence of 20 and 70°. His results are shown in Fig. 3 for both R_p and R_a. The blackened parallelopipeds represent the error in determining n and k for $\pm 1\%$ errors in measuring the reflectance. The parallelopipeds correspond to the values of n and k shown in large numbers. Lines connect the extremities of these parallelopipeds to indicate the magnitude of the errors at intermediate points. The magnitude of the errors is in keeping with the sensitivity as discussed in connection with Fig. 2. For example, for small n (0.3), the error in n as k increases grows comparatively slowly, but the corresponding error in k becomes large quite rapidly because the parallelopipeds are almost parallel to the k axis. As n increases and for large k, the long axis of the parallelopipeds rotates in a clockwise direction so that the error in determining n increases while that for k decreases. It is evident that R_p is more tolerant of errors than R_a.

In using the R-versus-ϕ method it is not necessary to know the actual R values [11]. The fact that reflectance is defined as the ratio of the reflected intensity at angle ϕ to the incident intensity is equivalent to normalizing the reflected intensity at angle ϕ to the reflected intensity at 90° angle of incidence, which is defined as unity. For use in the R-versus-ϕ method, normalization

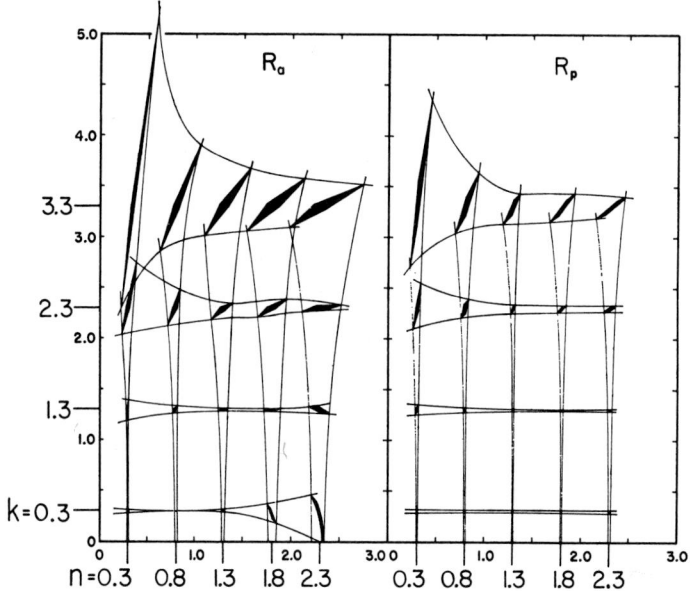

Fig. 3. The effect of $\pm 1\%$ error in the measurement of reflectance on the determination of n and k using method 1. The angles of incidence are 20 and 70°.

can be done at any angle of incidence and is referred to as oblique normalization. Field and Murphy [12] have published an analysis of the R-versus-ϕ method for oblique normalization. For oblique normalization the angle between the isoreflectance curves may not be a good indicator of sensitivity, and one should plot known errors in the obliquely normalized reflectance values to get a correct indication of sensitivity. Oblique normalization has no advantages over the usual mode of normalization and is only used when the physical arrangement of the reflectometer makes measurement of the incident intensity impractical.

Method 2   2

Figure 4 shows two sets of curves for the ratio R_p/R_s. To the left is a set for small angles of incidence and to the right a set for large angles. According to Humphreys–Owen, an indication of the sensitivity is the spacing between contours—if the spacing is large, the sensitivity is good and vice versa. Assuming that his assertion is correct, the large-angle contours are obviously not useful if n and k are large but might be used to obtain good values if n and k are small. The small-angle contours are only useful if n and k are small. Humphreys–Owen showed a limited set of contours of R_p/R_s for 80 and 60°, and a more extended, calculated set indicates that these two angles may be more useful than those shown here for larger values of n and k.

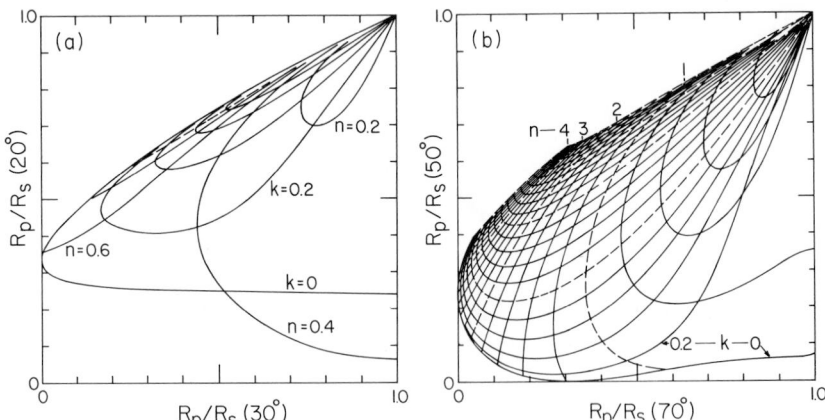

Fig. 4. Curves of constant n and k as a function of R_p/R_s for (a) small (30–20°) and (b) large (70–50°) angles of incidence. Δn and Δk are each 0.2. Curves for integer n and k values are dashed (method 2).

3   Method 3

Figure 5 shows two sets of curves for R_s and R_p at both 20 and 70° angles of incidence. Judging from sets of curves obtained for other angles of incidence, these two angles are the most useful. For the 20° angle of incidence, the curves of constant k are spaced 0.01 apart from $k = 0$ to $k = 0.1$. The next k contour is for $k = 0.2$ and thereafter $\Delta k = 0.2$. The n contours have $\Delta n = 0.1$.

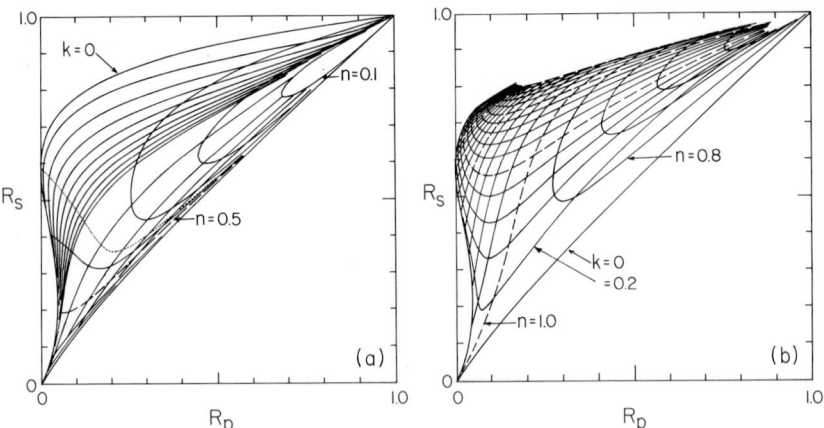

Fig. 5. Curves of constant n and k as a function of R_p and R_s for (a) small (20°) and (b) large (70°) angles of incidence. For (a) Δk is 0.01 for $0 \le k \le 0.1$; otherwise Δk is 0.2. Δn is 0.1 and the curve for $n = 0.5$ is dashed. The dotted line corresponds to the value of n for the Brewster angle, $n = 0.364$. For (b) $\Delta k = \Delta n = 0.2$. Curves for integer n and k values are dashed (method 3).

The dotted line is the contour of $n = 0.364 = \tan 20°$. This set of curves might be useful for $n < 0.5$ and $k < 0.4$.

The set of curves of $70°$ is less useful for small n and k but more useful for larger values. For this set $R_p = 0$ when $n = 2.747 = \tan 70°$. Both Δn and Δk are 0.2 for this set.

The envelope of the $20°$ contours consists of the curves $k = 0$ and, ultimately, the curve for large n values. The $70°$ contours are enveloped by the two curves for $k = 0$, which coincide at $R_s = R_p = 0$. At $45°$, where $R_p = R_s^2$ for any n and k, the envelope is compressed to a line that has the form of a parabola.

Methods 4 and 5   4

Figure 6 shows sets of curves for R_p/R_s (method 4) and R_s (method 5) versus the pB angle. The R_p/R_s set favors small values of n and k and the R_s set favors somewhat larger n and k. Note that the R_s curves can give double values. For example, if the pB angle is $15°$ and $R_s = 0.56$, k would be 0.3 but n could be 0.2 or 0.3. Thus, the region with double values is not useful. Contours for R_p-versus-pB angle are similar to those for R_p/R_s.

According to Humphreys–Owen, method 6 gives no advantages over the other methods.

In order to obtain quantitative data on the accuracy of methods 2–5, it is necessary to plot the contours for $R \pm \Delta R$, or pB $\pm \Delta$pB, which has not been done. The existence of large spacings between the contours suggests but does not guarantee small displacements of the contours when errors are introduced Humphreys–Owen points out that R_p, at the pB angle, can be quite

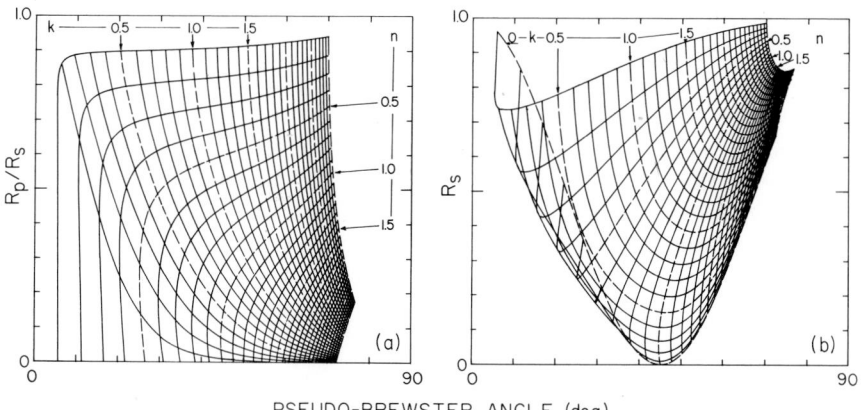

PSEUDO-BREWSTER ANGLE (deg.)

Fig. 6. Curves of constant n and k as a function (a) of R_p/R_s and the pseudo-Brewster (pB) angle and (b) of R_s and the pB angle (methods 4 and 5, respectively). $\Delta n = \Delta k = 0.1$ and curves for integer and half-integer; n and k values are dashed.

small; thus, there may be a large error in measuring it. Furthermore, the minimum in R_p at the pB angle is usually broad, so accurate location of the angle may be difficult.

If method 2 or 5 is to be used, normalization to get actual reflectance values is unnecessary.

The equation for R_p can be manipulated to obtain the pB angle in terms of n and k [6]. It is given below,

$$2(p^2 + q)v^3 + p^2(p^2 - 3)v^2 - 2p^4v + p^4 = 0,$$

where

$$p = n^2 + k^2, \qquad q = n^2 - k^2, \qquad \text{and} \qquad v = \sin(pB).$$

The experimental values for n and k obtained from the curves can be substituted in the formula to help verify the correctness of the results.

C Effect of Layer Thickness

Preparing a material for reflectance measurements from which n and k are to be obtained is often done by vacuum evaporation of the material onto a highly polished substrate. Since methods 1–5 are intended for use with semi-infinite media, the layer must be thick enough to qualify in this respect. Generally, as the thickness of a vacuum-deposited layer increases, so does its surface roughness. If the layer is not too thick, the error introduced by scattering due to the surface roughness will be small compared to errors from other causes. If, however, the thickness is too small, radiation will penetrate to the substrate/layer boundary, be reflected back to the layer/vacuum boundary, and interfere with the radiation reflected at the layer/vacuum boundary. If this interference is significant, the measured reflectances will not be representative of the reflectances of a semi-infinite medium and the reflectance methods 1–5 will not be useful. Hunter and Hass [13] have investigated the effects of interference on method 1 and found that the thickness required to reduced errors in n and k to approximately 1% depends on the n, k values of the layer. Their Table I shows that for $n = k = 0.3$, the geometrical thickness of the layer must be approximately one wavelength, and as n and k increase, the required thickness decreases. At $n = 2.3$ and $k = 3.3$, for example, a geometrical thickness of approximately 0.1 wavelength is required.

The effect of layer thickness on methods 2–5 has not been investigated but is expected to be generally the same as for method 1.

The general behavior of R-versus-ϕ and isoreflectance curves indicates that for large k/n, the method is not very accurate. Since metals with large reflectance values in the visible and infrared usually have large k/n, reflectance methods appear not to be suitable for obtaining n and k of metals in these spectral regions.

There are wavelength ranges in the VUV in which the n of some substances is less than unity and $k < 0.1$. This occurs for Al, Mg, Si, for example [14], and the alkali metals [15] at wavelengths less than their critical wavelength. For such values of n and k, the R-versus-ϕ curves exhibit the behavior shown in Fig. 7 [16]. This figure shows curves for a substance with $n = 0.707$, corresponding to a critical angle ϕ_c of 45°, and for different values of k. For $k = 0$, ϕ_c is well defined and $\sin \phi_c = n$. As k increases, however, ϕ_c is no longer well defined because the R-versus-ϕ curves become rounded at angles larger than ϕ_c. The point of maximum slope of the curve, however, occurs quite close to ϕ_c and can be used as a good approximation to ϕ_c. Since only the angle of incidence at which the maximum slope occurs is required, the reflectance values need not be known. To get correct values of n, k must be known and

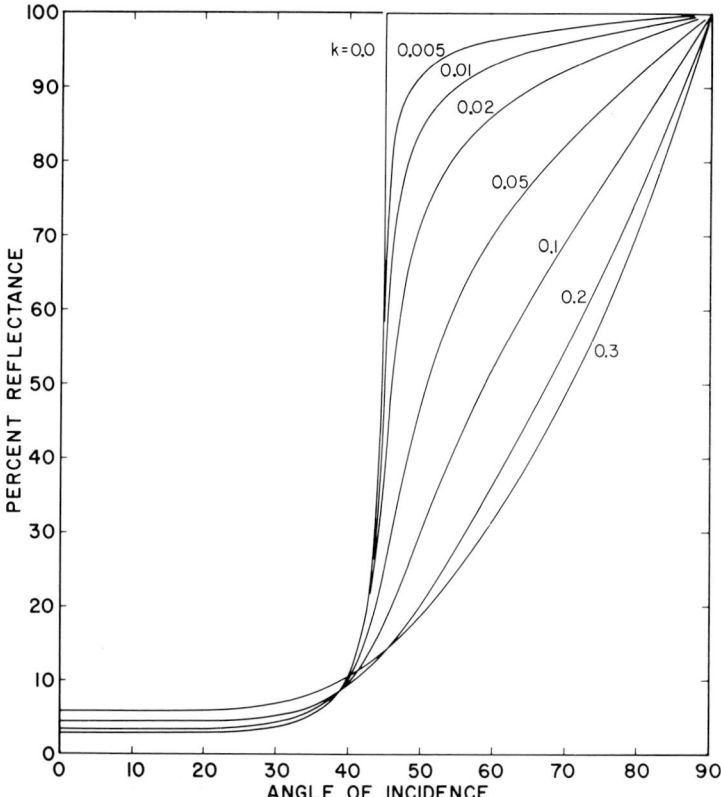

Fig. 7. Calculated R-versus-ϕ curves for $n = 0.707$, $\phi_c = 45°$, and different values of k showing the behavior of the reflectance for nonpolarized light in the vicinity of the critical angle.

a correction made. The figure shows that as k increases, the correction required to obtain the true n also increases.

Corrections must also be made for the state of polarization of the incident radiation. These corrections decrease with increasing n, becoming small for $n = 0.7$.

Hunter [16] has used this technique to measure n of evaporated layers of quasi-free electron metals such as Al, Mg, Si, and others at wavelengths less than their critical wavelengths. In order to correct the measured angle of maximum slope for the effect of k, he used extinction coefficients obtained from transmittance measurements through thin, unbacked films of the metal under investigation. The presence of the oxide layer that forms readily on the chemically active metals can shift the position of maximum slope. Hunter found that the natural oxide thickness for Al (30–40 Å) tends to shift the

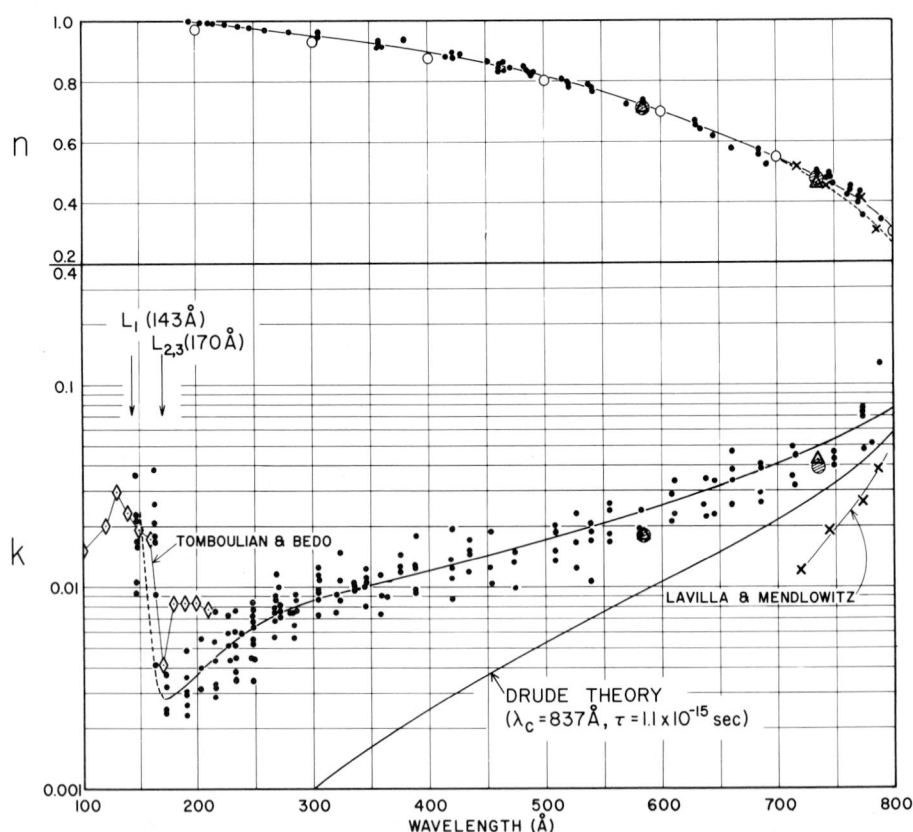

Fig. 8. The curve at the top shows n values for Al obtained using the critical angle method. The curve at the bottom shows the extinction coefficient of Al deduced from transmittance measurements. The scatter of points indicates that the critical angle method gives fairly accurate values for n, but that the transmittance method for finding k can give quite variable results.

maximum slope to larger angles, especially for small values of n, but that the magnitude of the shift is small.

For this method the layer thickness is also important because interference effects can shift the position of maximum slope. Usually, if the thickness exceeds 2000 Å, interference effects are negligible. This method is also advantageous because of its relative insensitiveness to surface roughness frequently encountered for layer thicknesses in excess of 2000–3000 Å.

Whang et al. [17] have extended the method to obtain k from the maximum slope of the R-versus-ϕ curve. Because the value of the slope is required when using their method, actual reflectance values must be measured. In addition, the polarization of the incident radiation must be known. They estimate that the critical-angle method, in spectral regions in which it can be used, gives more accurate results than using method 1. Hunter [18] found that the scatter of data points is quite small when using this method, as illustrated by the top part of Fig. 8.

Measurement of Extinction Coefficients   E

The extinction coefficient can be determined by measuring the transmittance of a known thickness of material and calculating k using Lambert's law after correcting the transmittance measurements for reflection losses at the surfaces. If the light is not coherent, these corrections involve only reflected intensities, and if two different thicknesses of the material are used, the corrections usually can be eliminated.

Lambert's law relates transmittance T, thickness t, and k through the equation

$$T = I/I_0 = e^{-4\pi kt/\lambda}.$$

For two thicknesses, k is obtained from

$$k = \lambda \ln(T_1/T_2)/[4\pi(t_2 - t_1)],$$

where λ is the wavelength in the same units as t.

If the film has surface layers, such as oxide layers, they are frequently much more absorbing than the material to be measured. If the oxide layers are always the same thickness, however, their absorptance does not affect the ratio T_1/T_2 and their reflectance values are independent of the test-material film thickness for all practical film thicknesses. Therefore, transmittance values obtained using two different test-material thicknesses should give reasonable values for k.

This method for determining k can be applied to thin, unbacked films or thin films on a transparent substrate if the value of k is small enough so that the transmitted signal has a reasonable signal/noise ratio. Hass et al. [19]

obtained k values for unbacked films of Al at 736 and 584 Å, and Hunter [18, 20] has made extensive measurements on unbacked films of Al and other materials.

Despite the apparent simplicity of the method, there are two possible sources of error: (1) anomalous dependence of transmittance on thickness and wavelength caused by interference effects and (2) methods of preparation of the sample.

For most materials in the VUV, the absorptance is so large that transmittances can only be obtained through thin films a few thousand angstroms thick. The coherence length of the light is usually larger than the film thickness, so interference effects will be present, and the simple intensity correction for reflection losses may not be valid.

Figure 9 shows the calculated transmittance of an Al film as a function of the Al thickness with a 40-Å-thick oxide coating on either surface. The k value used for Al is 0.0233. As the Al film thickness increases, the interference effects decrease in magnitude. At a thickness of 2000 Å the amplitude of the interference effect is about 10% of the transmittance, and at 3000 Å about 4%. By choosing thicknesses corresponding to interference maxima and minima, k values as large as 0.0698 can be obtained, and even negative k values

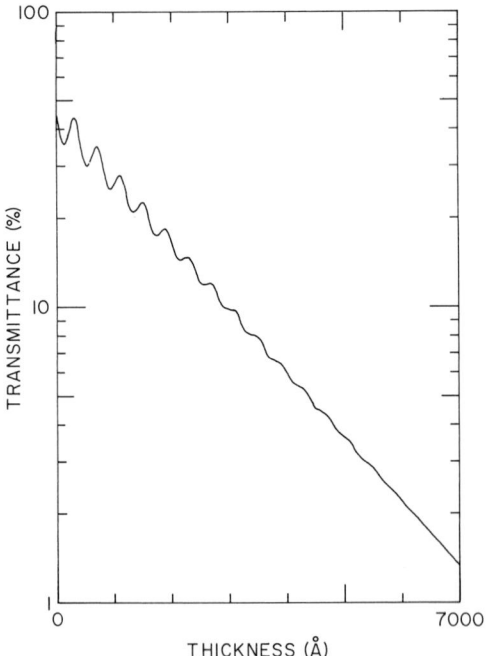

Fig. 9. Calculated transmittance versus thickness at 584 Å, and at normal incidence, for an unbacked Al film with 40 Å of oxide on both surfaces. The curve shows how interference effects alter the transmittance as the Al thickness changes.

can be calculated. Calculated from transmittance values at thicknesses of 6000 and 7000 Å, k is 0.0216, about a 7% error, and for thicknesses of 9000 and 10,000 Å, not shown in Fig. 9, a value of 0.0299 was obtained, an error of 1.7%. Thus, in principle, more accurate values of k are obtained as the thickness is increased. In practice, the transmittance values become too small for accurate measurement, so thinner films must be used.

Although interference effects are always present in the transmittance values, the experimental error in measuring the transmittance can mask the error caused by interference. For example, if the experimental error is $\pm 1\%$ of full scale, a transmittance of 16.5% ($t = 2000$ Å) can be in error by $\pm 6\%$, but the error introduced by interference would be 1.6% (10% of the transmittance value). Thus, for this example, the error in transmittance caused by interference would contribute little to the error in determining k for film thicknesses in excess of 2000 Å.

The transmittance of a thin film can also show interference effects as a function of wavelength. An example is given in Fig. 10 [21] which shows the measured transmittance of an unbacked film of Al 800 Å thick as a function of wavelength. Also shown for comparison are the calculated transmittances of an Al film of the same thickness with and without oxide layers 40 Å thick on either surface. The oxide is much more strongly absorbing than the Al, which decreases the transmittance of the oxidized film over that of the nonoxidized film, and its presence accentuates the interference effects. If interference-modulated transmittance spectra are used to provide transmittance values for the equation for k, k will also have a wavelength-dependent modulation. Reduction of this modulation is achieved by using thick films.

The transmittance method for finding k using thin films is perhaps more susceptible to errors arising from the method of preparation of the sample than from oscillatory terms in the transmittance equation or interference-modulated transmittance spectra. Unbacked metal films are made most simply by evaporating the metal onto a water-soluble layer [22]. Both the act of removal of the evaporated film from the substrate and the evaporation conditions under which it is made can strongly affect its optical properties. After the layer is dissolved, the films are allowed to float free on the water surface until picked up on a suitable frame for transmittance measurements. No two exposures to water will be alike; therefore, there may be considerable variability in the thicknesses of the oxide layers on either side of the metal film and also in k deduced from transmittance measurements made on such films. Figure 8 [18] shows measurements of k for Al. The scatter of points is quite large, primarily because of differences in the thicknesses of the oxide layers on either side of a metal film and because of differences in thicknesses of the oxide films on the different samples. Some of the scatter, however, is caused by interference effects of the type shown in Figs. 9 and 10.

Evaporation conditions are most apt to affect chemically active metals, such as Al, which adsorb oxygen readily. The amount of residual oxygen

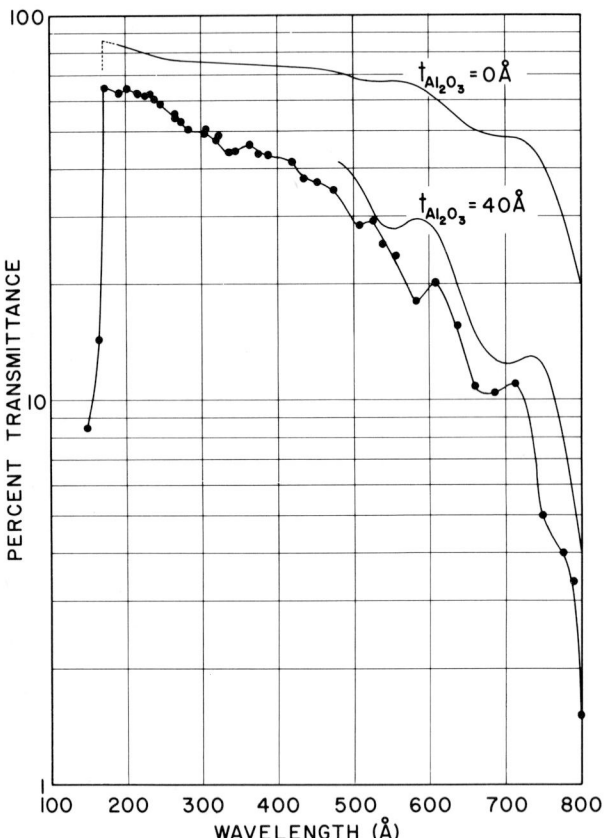

Fig. 10. The measured transmittance of an unbacked Al film 800 Å thick as a function of wavelength. Included for comparison are calculated curves for Al with and without 40-Å oxide layers on both surfaces. Interference effects cause the fluctuations in transmittance that are accentuated by the presence of the oxide layers.

captured in the film during evaporation must be reduced as much as possible. This can be done by evaporating in an ultrahigh-vacuum evaporator, or by doing a very fast evaporation in a conventional evaporator. If the residual oxygen is not removed or reduced sufficiently, Al films tend to have transmittances almost independent of thickness.[†]

A more specialized technique exists for measuring k of a material, if it is small ($k < 0.1$) and if n is known. At a given wavelength the reflectance at near-normal incidence of a substrate/layer combination, wherein the material to be measured forms the layer, is determined as a function of layer thickness.

[†] D. N. Steele, Luxel Corporation, 515 Tucker Avenue, Friday Harbor, Washington 98250, and C. Kunz, DESY HASYLAB, Notkestieg 1, Hamburg 52, Federal Republic of Germany; private communications.

The reflectance of the combination is then calculated as a function of layer thickness, with different k values, until a best fit to the experimental points is found. For this technique n and k of the substrate must be known. Canfield et al. [23] have used this method to determine k of MgF_2 layers on Al at 1216 Å, and Cox et al. [24] have used it to determine k of LiF layers on Al at 1026 Å.

REFLECTANCE METHOD FOR TWO MEDIA   III

In principle, reflectance methods can be used to measure n and k of either component of a layer–substrate combination. One must know n and k of either layer or substrate, the layer thickness, and the layer–substrate transition region must be small compared to the wavelength used to make the measurement. Using such measurements, one could determine the optical constants of the natural oxide that forms on aluminum or of iron or cobalt films that cannot be made opaque because of their large internal strains.

Juencker [10] has discussed using method 1 for substrate–layers but did not discuss the conditions under which the method is valid. It is difficult to generalize for two media because of the large number of parameters involved, but there are some avenues of approach as well as some problems that will be discussed in the following paragraphs.

In contrast with single-medium reflectance methods, the two-medium reflectance method requires interference between the light reflected from the vacuum–layer and layer–substrate interfaces. Therefore, the layer must be thin enough for this interference to take place. The amount of light available for interference depends on the optical constants of both the layer and substrate, on the layer thickness, and on the angle of incidence. An example will be given to illustrate the method.

Assume a substrate with $N = 0.4 + i0.1$ and a layer with $N = 2.0 + i0.1$. The upper left panel of Fig. 11 shows how the reflectance of the substrate–layer varies with layer thickness. Choosing the first half-wave thickness, $0.13t/\lambda$, which may not be the optimum thickness, one calculates the R-versus-ϕ curves shown in 11b. Then, isoreflectance curves are plotted. Studies showed that for this particular layer–substrate combination, the parallel component gave the best isoreflectance curves for both layer and substrate. They are shown in 11c and d, respectively. Note that if only the large angles of incidence are used, there appear to be two solutions. This ambiguity can be resolved by using some reflectance measurements at smaller angles of incidence or by using a second, different thickness. In the latter case the correct intersection remains unmoved, but the spurious intersection shifts to a new location in the n, k plane.

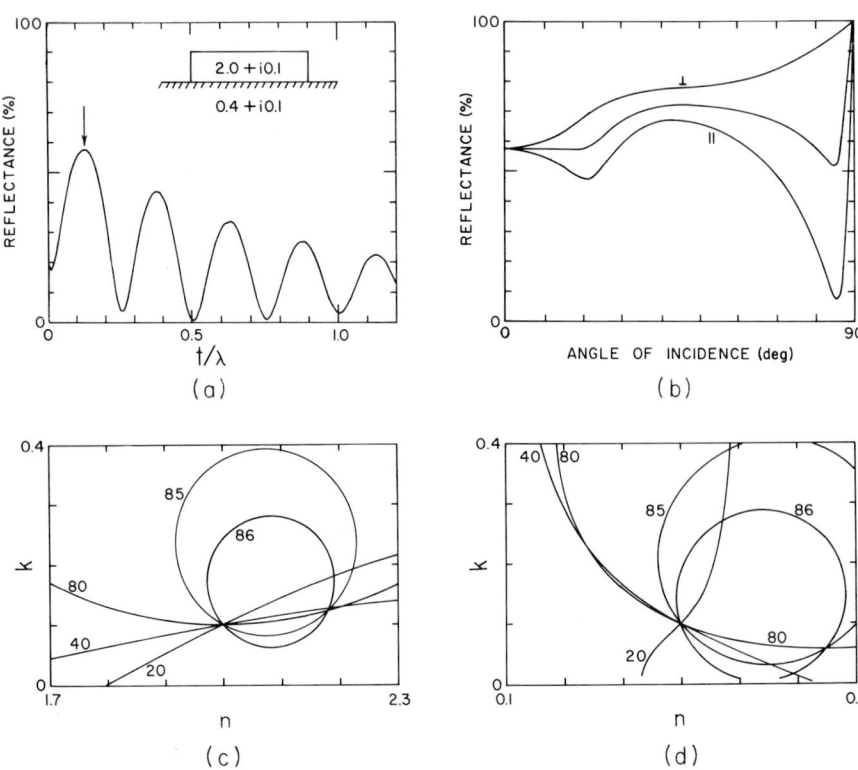

Fig. 11. Illustration of the procedure tor obtaining isoreflectance curves for determining n and k using two media. (a) Reflectance versus thickness is shown at normal incidence; choosing $t/\lambda = 0.13$ (arrow), (b) R-versus-ϕ curves are calculated; isoreflectance curves are obtained (c) for the layer assuming that the substrate is known and (d) for the substrate assuming the layer is known.

There are a number of problems associated with the two-medium reflectance method. First, there is the possibility of interdiffusion between substrate and layer, giving rise to a transition-region dimension comparable to, or greater than, the wavelength. If glass or fused silica is used for the substrate, the probability of interdiffusion is remote. There may be situations, however, when a substrate with a large reflectance adds to the contrast of the interference effects; for example, the effective substrate could be an opaque layer of Al on glass. Unoxidized Al interdiffuses readily with some other metals, notably Au [25]. Thus, some advance research must be done to ascertain the compatibility of the substrate and layer vis-à-vis interdiffusion.

Another problem is nonuniformity of the layer. The two-medium reflectance method tacitly assumes that the layer has a constant n and k as a function of distance into the layer. Such an assumption is not necessarily true. Reflectance measurements made on aged Al + MgF$_2$ mirrors at 1216 Å with polarized radiation indicate that n or k or both may not be constant throughout the layer. Magnesium fluoride is known to adsorb water vapor which can modify the optical constants of the MgF_2 layer.

Fabre *et al.* [26] have devised a method for measuring n and k of slightly absorbing films in the VUV and have applied it to a study of MgF_2 films [27]. The film is vacuum deposited onto a suitable substrate (glass) such that its thickness varies along the long dimension of the coated area. After the actual thickness and its variation have been determined, the reflectance is measured as a function of thickness by moving the substrate/film uniformly past the radiation beam. The resultant reflectance curve resembles the curve shown in Fig. 11a. Using the enveloping curves for the reflectance maxima and minima, and the position of these extremes, Fabre *et al.* were able to calculate values for n and k.

Daude *et al.* [28] have used a technique similar to that employed by Canfield *et al.* and Cox *et al.* to determine both n and k of MgF_2 films on Al but that they consider applicable to any thin film with small absorptance. They measure the reflectance of the substrate/film combination as a function of film thickness and fit a curve to the measurements by calculating the reflectance as a function of layer thickness while varying n and k to achieve best fit.

REFERENCES

1. C. Boeckner, *J. Opt. Soc. Am.* **19**, 7 (1919).
2. R. Tousey, *J. Opt. Soc. Am.* **29**, 235 (1939).
3. J. R. Collins and R. O. Bock, *Rev. Sci. Inst.* **14**, 135 (1943).
4. I. Simon, *J. Opt. Soc. Am.* **41**, 336 (1951).
5. D. G. Avery, *Proc. Phys. Soc. (London)* **65**, 425 (1952).
6. S. P. F. Humphreys–Owen, *Proc. Phys. Soc. (London)* **77**, 949 (1961).
7. T. Sasaki and K. Ishiguro, *Japan J. Appl. Phys.* **2**, 289 (1963).
8. A. P. Prishivalko, "Otrazhenie Sveta ot Poglishchaoiuchikh Sred" ["Reflection of Light from Absorbing Media"], Izdate'lstvo, Akademii Nauk, Beloruskoi SSR Minsk, 1963.
9. A Vasicek, "Tables of Determination of Optical Constants from the Intensities of Reflected Light," Nakladatelstvi Ceskoslovenske Akademie Vec, Prague, 1964.
10. D. W. Juenker, *J. Opt. Soc. Am.* **55**, 295 (1965).
11. W. R. Hunter, *J. Opt. Soc. Am.* **55**, 1197 (1965).
12. G. R. Field and E. Murphy, *Appl. Opt.* **10**, 1402 (1971).
13. W. R. Hunter and G. Hass, *J. Opt. Soc. Am.* **64**, 429 (1974).
14. D. Pines, *Rev. Mod. Phys.* **28**, 184 (1956).
15. C. Zener, *Nature*, **132**, 968 (1933).
16. W. R. Hunter, *J. Opt. Soc. Am.* **54**, 15 (1964).
17. U. S. Whang, R. N. Hamm, E. T. Arakawa, and M. W. Williams, *J. Opt. Soc. Am.* **63**, 305 (1973).
18. W. R. Hunter, *J. de Physique* **25**, 154 (1964).
19. G. Hass, W. R. Hunter, and R. Tousey, *J. Opt. Soc. Am.* **47**, 120A (1957).
20. W. R. Hunter, "Proceedings of the International Colloquium on Optical Properties and Electronic Structure of Metals and Alloys" (F. Abélès, ed.), p. 136, North-Holland Publ., Amsterdam, 1966.

21. W. R. Hunter and R. Tousey, *J. de Physique* **25**, 148 (1964).
22. W. R. Hunter, *in* "Physics of Thin Films" (G. Hass, M. Francombe, and R. W. Hoffman, eds.), p. 43, Academic Press, New York, 1973.
23. L. R. Canfield, G. Hass, and J. E. Waylonis, *Appl. Opt.* **5**, 45 (1966).
24. J. T. Cox, G. Hass, and J. E. Waylonis, *Appl Opt.* **7**, 1535 (1968).
25. W. R. Hunter, T. L. Mikes, and G. Hass, *Appl. Opt.* **11**, 1594 (1972).
26. D. Fabre, J. Romand, and B. Vodar, *J. Phys. Radium* **21**, 263 (1960).
27. D. Fabre and J. Romand, *J. Phys. Radium* **22**, 324 (1961).
28. A. Daude, A. Savary, A. Seignac, and S. Robin, *Opt. Acta* **20**, 353 (1973).

Chapter **5**

The Accurate Determination of Optical Properties by Ellipsometry

D. E. ASPNES

Bell Communications Research, Inc.
Murray Hill, New Jersey

I.	Reflection Techniques; Background and Overview	89
II.	Measurement Configurations	92
III.	Accurate Determination of Optical Properties: Overlayer Effects	96
IV.	Living with Overlayers	99
V.	Eliminating Overlayers	102
VI.	Bulk and Thin-Film Effects; Effective-Medium Theory	104
VII.	Conclusion	108
	References	110

REFLECTION TECHNIQUES; BACKGROUND AND OVERVIEW I

In this section, we give some background in electromagnetic theory to aid in understanding the different techniques used to determine optical properties of materials. An incident plane wave propagating in the z-direction of a local orthogonal coordinate system can be represented as [1, 2]

$$\mathbf{E}_i(\mathbf{r}, t) = \mathrm{Re}[(\hat{x}E_x + \hat{y}E_y)e^{ikz - i\omega t}], \tag{1}$$

where E_x and E_y are the *complex field coefficients* describing the amplitude and phase dependences of the projections of $\mathbf{E}_i(\mathbf{r}, t)$ along the x and y axes. If the electromagnetic wave is reflected by a smooth surface, the outgoing wave can be represented in the absence of anisotropic effects in another local coordinate system as

$$\mathbf{E}_r(\mathbf{r}, t) = \mathrm{Re}[(\hat{x}r_pE_x + \hat{y}r_sE_y)e^{ikz' - i\omega t}], \tag{2}$$

89

where in both local coordinate systems the x axes are in the plane of incidence, the y axis is perpendicular to the plane of incidence, and the z axes define the plane of incidence.

In this approximation the effect of the surface is described by two coefficients r_p and r_s. These *complex reflectances* describe the action of the sample on the field components parallel (p) and perpendicular (s = German senkrecht) to the plane of incidence. Thus, four parameters—two amplitudes and two phases—are necessary to describe completely the incoming wave and four more to describe the sample. The properties of the sample therefore are obtainable if the properties of both incident and reflected waves are known. *Reflectometry* and *ellipsometry* are two techniques for obtaining this information. Neither extracts all the available data, but each is an incomplete form of a still more general technique, *polarimetry* [3].

To appreciate the different types of information obtained in reflectometry, ellipsometry, and polarimetry, it is useful to recast the four parameters describing the plane waves of Eqs. (1) and (2) into four new parameters by means of the concept of the *polarization ellipse*. The polarization ellipse is the locus traced out in time by the real (observable) electric field vector $\mathbf{E}(\mathbf{r}, t)$ at any fixed plane $z = z_i$. The polarization ellipse has two attributes that can be termed *size* and *shape*. The size $E = (|E_x|^2 + |E_y|^2)^{1/2}$ is a scalar whose square is proportional to the power flow or *intensity I*. The intensity is the only quantity of interest in a reflectance or transmittance measurement and one of the quantities of interest in a polarimetric measurement.

The shape is an intensity-independent quantity that clearly requires two parameters to specify, such as the minor/major axis ratio and the azimuth angle of the major axis of the polarization ellipse. One convenient representation of the shape is the *polarization state* $\chi = E_x/E_y$ [2]. Since E_x and E_y are complex, χ is also complex and therefore contains the required two parameters. For example, if E_x and E_y are in phase, then χ is real, both field components are strictly proportional at all times, and the light is linearly polarized with an azimuth angle $\psi = \tan^{-1} \chi$. If E_x and E_y are 90° out of phase, then χ is purely imaginary and one field component is at a maximum while the other is at zero and vice versa; if, in addition, $|E_x| = |E_y|$, then $\chi = \pm i$, and the light is circularly polarized. The polarization state is the only quantity of interest in ellipsometry and another of the quantities of interest in polarimetry. In addition, polarimetry is also concerned with depolarized or unpolarized light, the superposition of electromagnetic waves in uncorrelated states of polarization.

Given either intensities or polarization states for the incident and reflected beams, the sample properties can be calculated by taking appropriate ratios. In reflectometry, we distinguish between p- and s-polarized light. By Eqs. (1) and (2), the intensity ratios are

$$(I_r/I_i)_p = |r_p E_x|^2/|E_x|^2 = |r_p|^2 = R_p, \tag{3a}$$

$$(I_r/I_i)_s = |r_s E_y|^2 / |E_y|^2 = |r_s|^2 = R_s. \tag{3b}$$

The *reflectances* R_p and R_s are the absolute squares of the respective complex reflectance coefficients. At normal incidence both polarizations are equivalent and $R_p = R_s$. In ellipsometry, the ratio of polarization states yields a somewhat different perspective of the sample. By Eqs. (1) and (2)

$$\chi_r/\chi_i = (r_p E_x/r_s E_y)/(E_x/E_y) = r_p/r_s \tag{4a}$$

$$= \rho = (\tan \psi) e^{i\Delta}. \tag{4b}$$

The *complex reflectance ratio* ρ is often expressed as an amplitude $(\tan \psi)$ and a phase Δ. Depolarization or cross-polarization effects determined by polarimetry are important in more complex sample configurations and in the presence of scattering [3], topics that we shall not discuss here.

We can now make some general remarks about the relative advantages and disadvantages of reflectometry, ellipsometry, and polarimetry. For isotropic samples ellipsometry is strictly a nonnormal-incidence technique, while reflectometry and polarimetry can be performed at either normal or nonnormal incidence. Reflectometry deals with intensities and is therefore a power measurement, while ellipsometry deals with intensity-independent complex quantities and is therefore more nearly analogous to an impedance measurement. Because more complicated quantities are involved, an ellipsometer is necessarily more complicated than a reflectometer and a polarimeter is more complicated than an ellipsometer. The most serious drawback of ellipsometers and polarimeters is that transmission through optical elements (e.g., polarizers) is generally required. Therefore, such measurments are limited to those wavelength ranges in which good-quality transmitting elements are available.

Leaving polarimetry as a topic not yet sufficiently developed for detailed discussion, we now consider the relative advantages and disadvantages of reflectometry and ellipsometry. Ellipsometry is unquestionably the more powerful for a number of reasons. First, two parameters instead of one are independently determined in any single-measurement operation. Consequently, both real and imaginary parts of the complex dielectric function ε of a homogeneous material can be obtained directly on a wavelength-by-wavelength basis without having to resort to multiple measurements or to Kramers–Kronig analysis. Two independent parameters also place tighter constraints on models representing more complicated, e.g., laminar microstructures. While two independent parameters R_s and R_p are also available in nonnormal-incidence reflectance measurements, these parameters can be obtained separately only after additional adjustments of system components. Second, ellipsometric measurements are relatively insensitive to intensity fluctuations of the source, temperature drifts of electronic components, and macroscopic roughness. Macroscopic roughness causes light loss by scattering the incident radiation out of the field of the instrument, which can be a serious problem in reflectometry but not in ellipsometry, for which absolute intensity measurements are not required.

Third, accurate reflectometric measurements are difficult, in general requiring double-beam methods. In contrast, ellipsometry is intrinsically a double-beam method in which one polarization component serves as amplitude and phase reference for the other. Finally, ρ explicitly contains phase information that makes ellipsometry generally more sensitive to surface conditions. Insensitivity to surface conditions is often considered to be an advantage of reflectometry, but this is not correct if the objective is to obtain accurate values of the intrinsic dielectric responses of bulk materials. Because small differences in R can result in large differences in ε, reflectances in general must be measured more accurately. Surface artifacts affect ellipsometric results at the measurement level at which they can be identified and often corrected on the spot, whereas they affect reflectometric results at the data-reduction level at which it may be too late to do anything about them.

II   MEASUREMENT CONFIGURATIONS

Despite the intrinsic advantages of ellipsometry, reflectometry historically has been a far more significant factor in advancing our general knowledge of the optical properties of materials. Reflectometers are well suited for spectral measurements, while the classical null ellipsometer is by nature a single-wavelength instrument. The development of photometric ellipsometers has now changed the picture, although spectroscopic instruments must still be built by the user. For this reason, we review in this section some configurations of interest (as of July 1982). More detailed discussions have been given by Azzam and Bashara [2], Hauge [3], Muller [4], and Rzhanov and Svitashev [5].

A schematic diagram of the general optical configuration for reflectometry, ellipsometry, or polarimetry is shown in Fig. 1. The source provides collimated

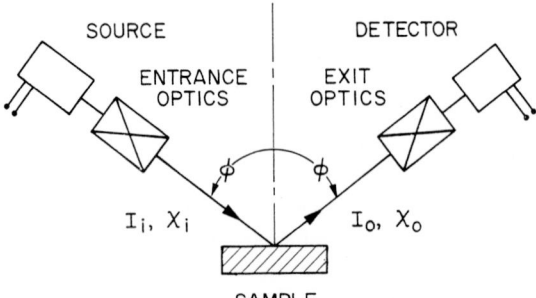

Fig. 1. General configuration for determining optical properties by reflectance techniques.

monochromatic light. The entrance and exit optics establish and analyze, respectively, the intensities, and/or polarization states. The detector may determine absolute or relative intensities or their time dependences.

The simplest configuration is that of a normal-incidence reflectometer. Here, there are no entrance or exit optics. The detector may be moved to intercept either entrance or exit beam or a rotating light pipe may be used. The movable-detector approach is particularly well suited for synchrotron radiation sources because they provide light in spectral ranges in which transmission elements are not available.

Because reflectometry is discussed in detail elsewhere in this volume [6], we now restrict the discussion to ellipsometry. As already implied, the most important instrument until recently was the classical null ellipsometer. Here, the entrance optics consist of a polarizer and compensator, or quarter-wave plate, and the exit optics consist of a second polarizer, or analyzer. The polarizer–compensator combination functions as a general *elliptical* polarizer. In normal operation the ellipticity is adjusted so that it is exactly cancelled by reflection. The reflected beam can then be blocked by properly adjusting the azimuth of the analyzer. The null ellipsometer is the optical analog of the ac impedance bridge.

The twin advantages of the null ellipsometer are mechanical simplicity and high accuracy. If the light source is sufficiently intense, the accuracy limit is determined solely by the mechanical components and their alignment and is not affected by detector nonlinearity. However, these advantages are relatively minor compared to its disadvantages. The null condition cannot be achieved by independent adjustments because the polarizer and analyzer azimuth settings interact. The intensity near null depends quadratically on both azimuths, so the null is not well defined. While Faraday-cell modulation of the plane of polarization can be used to improve null sensitively [7, 8], the detector still operates at minimum light levels so the rate of information flow is small and dark-current and shot noise may be significant. The traditional approach toward dealing with these problems has been to increase the source intensity, which means that in practice most null ellipsometric measurements have been performed at the Hg green line $\lambda = 5461$ Å or more recently at the HeNe laser line $\lambda = 6328$ Å. Consequently, the use of a compensator, a component with a relatively narrow useful wavelength range, has not proved to be the significant handicap.

The modern era in ellipsometry can essentially be traced back to the automatic photometric ellipsometer described by Cahan and Spanier in 1969 [9]. Photometric instruments operate on entirely different principles than do null ellipsometers. One of the components is a modulator that is used to produce a time dependence on the transmitted intensity that is later decoded for the properties of interest. Because most of the light is transmitted, photometric systems are compatible with the relatively weak continuum sources suitable for infrared–visible–near-UV spectroscopy. In the Cahan–Spanier design

[9], the modulator consisted simply of the final analyzer prism, which was rotated mechanically to produce a sinusoidal variation of the transmitted intensity (see the next paragraph). In addition, the data were obtained automatically, with operator feedback not required. A number of configurations [2–5] have also been developed involving photoelastic modulators [10–13] and Pockels cells [14] as well as mechanically rotating elements [15–20] and featuring also digital encoding and analysis of the transmitted intensity. The more elaborate designs [3, 17, 18] also can distinguish between polarized and unpolarized light. Photometric ellipsometers now totally dominate the field as far as research and spectroscopic applications are concerned, with null ellipsometers now mainly relegated to routine measurements on simple systems, such as determining oxide thicknesses on semiconductor wafers.

Only a few of the many photometric designs have been considered sufficiently practical to warrant the time and effort of actual development. The simplest of these are the rotating-analyzer ellipsometer (RAE) [15, 16] and its complement, the rotating-polarizer ellipsometer (RPE) [20]. In these configurations, the entrance and exit optics consist of single polarizing elements, the polarizer and analyzer, respectively. As the names imply, one of these is rotated mechanically while the other is held fixed. The advantages are simplicity and the absence of a compensator; the only wavelength-dependent element is the sample itself. A second advantage is that the transmitted intensity has a very simple Fourier spectrum consisting of a single ac component on a dc background. Disadvantages include the requirement of either a rigorously unpolarized source for an RPE or a rigorously polarization-insensitive detector for an RAE. In addition, as with all photometric systems, the detector must be rigorously linear. Finally, these configurations cannot distinguish between circularly polarized or unpolarized light or the handedness of circularly polarized light, and a significant loss of accuracy occurs for ε_2 if $\varepsilon_2 \cong 0$.

Because photodiodes and end-on photomultipliers are relatively polarization insensitive while equivalently unpolarized sources do not exist, the RAE is preferred over the RPE unless the samples or their environments are themselves sources of significant amounts of radiation. Stray light suppression is much better in an RPE because the monochromator can be placed between sample and detector. With respect to the other limitations, depolarization is not usually a factor unless sample quality is poor. Handedness can be determined from spectral dependences. The inability to measure ε_2 accurately for $\varepsilon_2 \cong 0$ can be circumvented by using a compensator, but this introduces another wavelength-dependent element into the system. A combination of a rotating analyzer followed by a fixed analyzer has been found useful in the infrared [19], where the principal objective is to determine threshold energies of line structures rather than to determine accurate values of dielectric functions.

If a compensator must be used, then the rotating-compensator ellipsometer (RCE) [18] is a more practical alternative. In this configuration, the entrance optics consist of a fixed polarizer and the exit optics of a rotating compensator followed by a fixed analyzer. The fixed polarizers at either end completely eliminate polarization sensitivity artifacts due to source or detector. Moreover, the RCE is capable of measuring handedness and can also distinguish between circularly polarized and unpolarized light [18]. The only reason that RCEs have not completely replaced RAEs and RPEs is that suitable achromatic compensators do not yet exist. With currently available elements, the usable wavelength range of an RCE is about a factor of 2. For fixed-wavelength operation the RCE is superior to an RAE, an RPE, or a null ellipsometer.

Photometric systems that rely on mechanically rotating components are relatively slow, with complete optical cycles requiring about 10 msec. Much higher speeds can be obtained with photoelastic modulators or with Pockels cells. Ellipsometers using these elements are identical to the null ellipsometer except that the compensator is replaced with the modulating element. Modulation frequencies are of the order of 100 kHz for photoelastic units [10–13] and upwards of 1 MHz for Pockels cells [14]. Disadvantages of birefringent-modulation systems include the introduction of another degree of freedom, the modulation amplitude, which in principle should also be measured on a wavelength-by-wavelength basis and should also be correlated with the wavelength for optimum sensitivity. In principle, the modulation amplitude can be obtained from an analysis of the higher-frequency harmonics. Because the operational frequency is much higher, digital analysis may be impractical and several lock-in amplifiers may be required. This introduces the usual calibration and gain uncertainties associated with several independent demodulators. Higher speeds also require higher source intensities.

Other configurations have also been developed, but those already discussed have accounted for most published work and appear entirely adequate when used within their limitations. Future developments will take advantage of the capabilities of photometric configurations to measure absolute intensities as well as polarization states and therefore to determine reflectances as well as complex reflectance ratios. At present, this capability has only been exploited differentially in thin-film measurements in which changes in the reflectance, rather than the reflectance itself, are measured [21]. The development of a truly achromatic compensator would be significant because an RCE using this element could perform spectroscopic measurements over a wide energy range without the limitations of an RAE or RPE.

But for most purposes, at least in the near IR–visible–near UV, instrumentation is a solved problem. The most significant factor responsible for differences among ε spectra is now sample preparation and maintenance, not measurement method or instrumentation. We discuss sample effects in the next section.

III   ACCURATE DETERMINATION OF OPTICAL PROPERTIES: OVERLAYER EFFECTS

The accurate determination of the intrinsic optical properties of a material in its pure bulk form is a double challenge, requiring not only accurate instrumentation but also well-characterized samples. Indeed, all reflection techniques for determining ε are model dependent, and the key to success is to prepare and maintain the sample in such a way that it can be described by fewer variables than independent experimental quantities. This is not a trivial matter: visible–near-UV optical properties are so strongly affected by microstructure and surface conditions that optical measurements are a convenient means of characterizing both microstructure [22] and surface quality [23]. As a particular case, the optical properties of evaporated metal films may not be accurately described by intrinsic bulk properties of the constituent material because the samples generally contain a significant volume fraction of voids [24]. We shall return to bulk problems after first discussing difficulties associated with imperfect surfaces.

It is useful to begin with the ideal case, in which the substrate is homogeneous and isotropic and the surface is mathematically sharp. Then ε and the normal incidence reflectance R are given by the two-phase (substrate–ambient) model [2]:

$$\varepsilon/\varepsilon_a = \sin^2 \phi + \sin^2 \phi \tan^2 \phi \left[(1 - \rho)/(1 + \rho)\right]^2, \tag{5a}$$

$$R = |(n - n_a)/(n + n_a)|^2, \tag{5b}$$

where ϕ is the angle of incidence and $n = \varepsilon^{1/2}$ and $n_a = \varepsilon_a^{1/2}$ are the complex and ordinary indices of refraction of substrate and ambient, respectively. (We assume that the ambient is transparent.) Although ε and ρ may be transformed interchangeably by Eq. (5a), there are no corresponding expressions for ε and R, R_s, and R_p. The analogous equations for these quantities must be solved numerically for ε [25].

For real samples, the assumption of a mathematically sharp interface leads to immediate practical difficulties. Transition-region widths are finite even if the lattice termination is atomically perfect. Under typical laboratory conditions, surfaces are covered with oxides or adsorbed contaminants. Surfaces are usually microscopically rough, often as a result of cleaning processes used to remove the other overlayers. If such effects are not taken into account, the optical parameters calculated from Eqs. (5) may be considerably in error [26]. However, in the analysis of ellipsometric data it is often useful to assume perfection anyway and to convert the measured quantity ρ into a derived quantity, the *pseudodielectric function* $\langle \varepsilon \rangle$, by Eq. (5a). The pseudodielectric function necessarily is an average of the dielectric responses of substrate and overlayer, and the accuracy by which it actually represents ε depends on the accuracy by which the sample configuration approximates that of the ideal two-phase model. The pseudodielectric function is useful simply because it expresses ρ in a form related to the fundamental quantity of interest. The complex reflectance ratio can always be recovered from $\langle \varepsilon \rangle$ by inverting Eq. (5a).

The importance of surface conditions can be appreciated by noting that penetration depths of light in opaque materials are typically 100–500 Å in the near UV, whence a 1-Å surface film can affect ρ, $\langle \varepsilon \rangle$, or R as much as several percentage points. This can be shown quantitatively by expanding ρ, $\langle \varepsilon \rangle$, and R to first order in d/λ, where d is the film thickness. We have [26, 27]

$$\rho \cong \rho_0 \left[1 + \frac{4\pi i d n_a \cos \phi}{\lambda} \frac{\varepsilon(\varepsilon - \varepsilon_o)(\varepsilon_o - \varepsilon_a)}{\varepsilon_o(\varepsilon - \varepsilon_a)(\varepsilon \cot^2 \phi - \varepsilon_a)} \right]; \qquad (6a)$$

$$\langle \varepsilon \rangle \cong \varepsilon + \frac{4\pi i d}{\lambda} \frac{\varepsilon(\varepsilon - \varepsilon_o)(\varepsilon_o - \varepsilon_a)}{\varepsilon_o(\varepsilon - \varepsilon_a)} (\varepsilon - \varepsilon_a \sin^2 \phi)^{1/2}; \qquad (6b)$$

$$\cong \varepsilon + (4\pi i d/\lambda)\varepsilon^{3/2}; \qquad (6c)$$

$$R \cong R_0 \left[1 + \frac{8\pi n_a d}{\lambda} \text{Im}\left(\frac{\varepsilon_o - \varepsilon_a}{\varepsilon - \varepsilon_a} \right) \right]. \qquad (6d)$$

In Eqs. (6), ε_o is the effective dielectric function of the overlayer and ρ_0 and R_0 are the corresponding values for the ideal system. The quantity ε_o may be anisotropic because ε_o can be defined in terms of spatial averages that are different for s- and p-polarized light [28]. However, effects due to anisotropy are usually small, and we ignore them here. Equation (6c) follows from (6b) in the limit that $|\varepsilon| \gg |\varepsilon_o| \gg |\varepsilon_a|$. Note that ε_o has dropped out of Eq. (6c) completely, so that relative changes in d can be investigated through $\langle \varepsilon \rangle$ without having to identify the nature of the overlayer materials [29].

Effects of oxide overlayers and microscopic roughness on $\langle \varepsilon \rangle$ and R are shown in Figs. 2 and 3 for a typical semiconductor, GaAs. These calculations were done by using the exact three-phase (substrate–overlayer–ambient) model [2] instead of Eqs. (6). The dielectric function of the oxide was assumed to be that of an electrochemically grown anodic film [30]. Microscopic roughness was modeled in the Bruggeman effective medium approximation [31] as a layer consisting of 60% substrate material and 40% voids, values typical of microscopically rough surface on semiconductors [29, 32]. The effect on $\langle \varepsilon \rangle$ of a microscopically rough layer 14 Å thick is identical to that of a 10-Å oxide on the scale of Fig. 2. Figures 2 and 3 show that at higher energies $\langle \varepsilon \rangle$ is about three times more sensitive than R to the presence of overlayers; the quantitative ratio of the relative sensitivities of R and $\langle \varepsilon \rangle$ is seen by Eqs. (6c) and (6b) to be

$$\left| \frac{\delta R/R}{\delta \langle \varepsilon \rangle / \langle \varepsilon \rangle} \right| \cong \frac{2n_a}{|\varepsilon|^{3/2}} \text{Im}\left(\frac{\varepsilon_o - \varepsilon_a}{\varepsilon - \varepsilon_a} \right), \qquad (7)$$

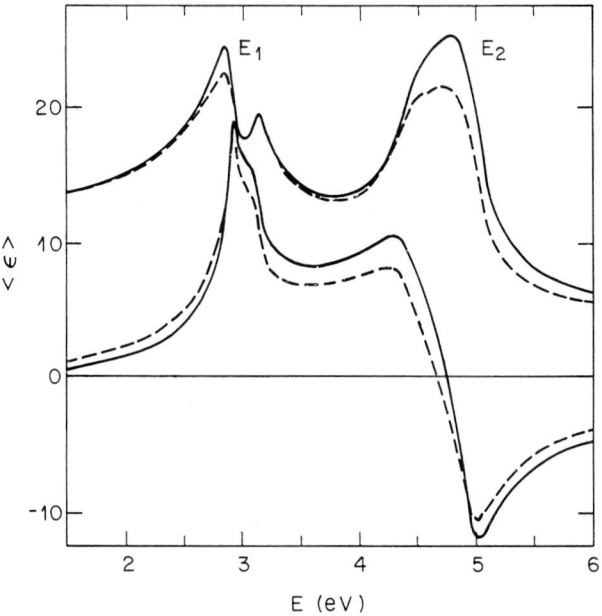

Fig. 2. Effect of a 10-Å oxide on $\langle \varepsilon \rangle$ for GaAs. The results for a 14-Å-thick microscopically rough surface are identical to this scale. [——, abrupt; ————, oxidized ($d = 10$ Å).]

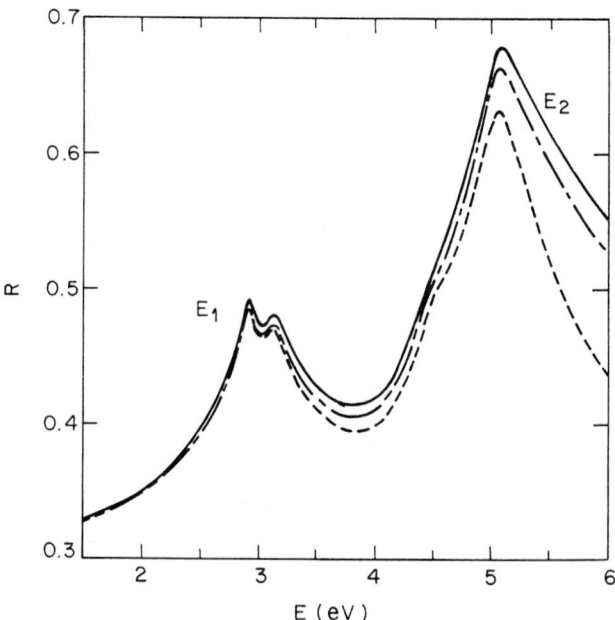

Fig. 3. As in Fig. 2 but for R. Note that the film thicknesses are larger than those in Fig. 2. [——, abrupt; ————, rough ($d = 30$ Å); ————, oxidized ($d = 30$ Å).]

about $\frac{1}{3}$ at the 4.78-eV E_2 peak in ε_2 for an oxide on GaAs. At low energies where ε and ε_o are real, the first-order dependence of R on d vanishes. Figure 3 shows that the effect of microscopic roughness on R, as with $\langle\varepsilon\rangle$, is also qualitatively similar to that of an oxide overlayer.

The spectroscopic dependence of the effect of an overlayer on $\langle\varepsilon\rangle$ can be understood from Eq. (6c). At low energies $\varepsilon^{3/2}$ is real, so overlayers affect only $\langle\varepsilon_2\rangle$. Or, equivalently, Eq. (6a) shows that $\Delta = \arg(\rho)$ is affected, but not $\psi = \tan^{-1}|\rho|$. The importance of Δ in providing information about surface films has long been known [26, 33–42]. As the energy increases, $\varepsilon^{3/2}$ becomes purely imaginary and the effect of the overlayer shifts entirely over to $\langle\varepsilon_1\rangle$. At still higher energies, $\varepsilon^{3/2}$ becomes completely real and the effect again shifts back to $\langle\varepsilon_2\rangle$.

LIVING WITH OVERLAYERS IV

Basically, there are two courses of action to follow in dealing with overlayers. First, overlayers can be accepted as a necessary evil and their effects identified by proper data acquisition and reduction. In this approach the sample is represented by a more complicated three-phase (substrate–overlayer–ambient) model and additional measurements are made to determine the extra parameters. For example, auxiliary measurements can be used to independently determine the overlayer thickness. Alternatively, the data can be effectively overdetermined and the inadequacy of the two-phase model made apparent by varying the angle of incidence or the refractive index of the ambient medium. To use this approach successfully, all parameters of the three-phase model (itself an approximation) must be obtained independently. Second, overlayers may be removed as far as possible and the sample maintained in a state that most nearly approximates the ideal two-phase model. Both approaches have their advantages and limitations.

We consider the laissez-faire approach first. Most work of this type has been done with null ellipsometers at single wavelengths on samples for which overlayers are nominally transparent oxides. Overlayer transparency already represents a considerable simplification because if $\text{Im}(\varepsilon_o) = 0$, then only the thickness and ordinary refractive index of the overlayer need be determined. If the substrate is also transparent or nearly transparent, as is the situation for Si at the 5461- and 6328-Å wavelengths commonly used with null ellipsometers, then ψ and Δ become essentially independent, with ψ being only a function of ε_1 and Δ of ε_2 and the overlayer thickness in the form of the product $(n_o d)$. Thus, ε_1 can be obtained directly with essentially no interference from the overlayer.

Even with these simplications, considerably more effort is required to obtain ε_2, and a number of approaches have been tried. The most direct is simply to measure d by an alternative method and then to assume that the thickness so determined is also appropriate for the optical measurement. As an example, Claussen and Flower [35] used reflection electron scattering to determine the thickness of the residual SiO_2 layer on hydrogen-annealed Si wafers and obtained a complex refractive index of $n = 4.05 + i0.03$ at 5461 Å for Si, in rather good agreement with the current best value [43] $n = (4.086 \pm 0.003) + i(0.031 \pm 0.0015)$. However, the validity of the assumption of equivalent optical and SiO_2 thicknesses is questionable in view of the recent work by Chang and Boulin [44], who showed that ellipsometrically determined "SiO_2" thicknesses for overlayers with $d \lesssim 120$ Å were consistently larger than those determined by Auger spectroscopy. The discrepancy was attributed to a Si-rich interface region.

A second approach is to use transmission measurements to determine a small substrate extinction coefficient and to use this value in the data-reduction process. Where applicable, transmission measurements are invariably much more accurate than reflection measurements simply because more material interacts with the electromagnetic radiation. The combined transmission–ellipsometric method works well for Si at 5461 and 6328 Å and was used by Taft [43] to obtain the accurate refractive index value quoted in the preceding paragraph. This approach has been used numerous times in the analysis of optical properties and oxide thicknesses on Si [39, 40, 43].

"Self-contained" approaches involving only ellipsometry must rely on multiple measurements at the same wavelength to determine all parameters of the system. Most methods that have been developed are useful only when the overlayer is transparent. In fact, most of these have only been tested on Si, a material whose technological importance and ready availability in high-quality, single-crystal form has made it a favorite for this purpose. Thus, measurements of a sample in air can be compared with measurements of the same sample immersed in liquids of various indices of refraction [37, 45, 46]. The assumption is that immersion does not affect the properties of the sample. Another approach is to make measurements at different angles of incidence [47–50] or to modulate directly the angle of incidence [51–53]. The latter approach was used by Bermudez in a spectroscopic ellipsometric study of F-center generation by UV irradiation of KCl [53]. Hunderi [50] has shown that to terms linear in d/λ the quantity $\bar{N} = (\langle \varepsilon \rangle - \varepsilon_a \sin^2 \phi)^{1/2}$ [as represented by using parameters defined in Eq. (5a) and the ensuing discussion] differs from $N = (\varepsilon - \varepsilon_a \sin^2 \phi)^{1/2}$ by a quantity independent of ϕ. Therefore, by examining the rate of change of $\bar{N}$ with ϕ, one can determine ε regardless of the values of ε_o or d. Several workers have also proposed to use the approximate invariance of R to provide a third condition and thereby to place another constraint on the system [54, 55].

The capabilities and limitations of these techniques with respect to very thin films can be understood from Eq. (6b). In this expression, the leading terms $\langle \varepsilon \rangle$ and ε are already independent of d, ε_a, and ϕ, so the possibility of determining the correction term from its functional dependence on these variables can be examined directly. The dependences on d and λ are not very useful because d is not generally under direct experimental control (or we would simply make it zero), and the functional dependence of ε on λ is the quantity being sought. In principle, the choice $\varepsilon_a = \varepsilon_o$ eliminates the correction term entirely; a good example [46] is the use of CCl_4 with Si because CCl_4 index-matches fused silica almost exactly. However, the method does not work if ε_o is complex, and in fact the approach can never provide unique information unless ε_a is also allowed to take on complex values. The same limitation is encountered by varying ϕ. Moreover, the variation of the correction term with ϕ is already very weak, especially for large values of ε, as seen directly in Eq. (6b). The same comments apply to Hunderi's approach [50], since his expression is also equivalent to Eq. (6b). The condition provided by the invariance of R likewise cannot provide unique information because only a single-phase projection is involved. In short, nothing works well if $d/\lambda \ll 1$ and if ε_o takes on arbitrary values.

However, variations in d can be useful if the film is transparent and $d \sim \lambda$. The method of Lukes et al. [45] involves measurements at various thicknesses as well as immersion in various ambients, although later refinements [49] incorporated measurements at multiple angles of incidence in addition to circumventing the immersion requirement. Taft and Cordes [56] used controlled etching to systematically reduce the thickness of a 7000-Å thermally grown oxide on a Si wafer; ellipsometric measurements at 5461 Å at various stages of etching yielded not only the ordinary refractive index of the oxide but also the small strain birefringence of the film and showed in addition the existence of a 6-Å interfacial layer of refractive index 2.8. Similar single-wavelength ellipsometric measurements were performed on anodized GaAs samples by Dinges [57]. It is noteworthy that oxide-removal techniques were the first to determine by optical means the existence of fairly wide ($\sim 5-10$-Å) interface regions between oxides and substrates, a consequence of the sensitivity of optical measurements to interface conditions at wavelengths and angles of incidence corresponding to antireflection [58]. In this regard, we mention the reflectance technique developed by Jungk [59] in which half the sample is stripped of its oxide and the angle of incidence adjusted so there is no difference in R_p between the two halves. Then ϕ is just equal to the Brewster's angle of the oxide, and its index can be readily calculated.

Spectral measurements offer other alternatives for overlayer correction, and they also illustrate shortcomings not evident from single-wavelength measurements. In principle, a thickness determined at a wavelength where the overlayer is known to be transparent should be valid at all wavelengths. However, attempts [60] to determine substrate optical properties for GaSb by solving simultaneously the exact three-phase equations for both ε and ε_o from data for two samples with 30- and 2000-Å oxides showed about a 10%

overcorrection in ε_2 in the near UV when compared with more recent data [61]. Daunois and Aspnes [62] assumed an SiO_2 layer on a Si wafer and established the thickness by requiring that $\varepsilon_2 \to 0$ as $E \to 1.5$ eV; the resultant height of the E_2 peak in $\langle\varepsilon_2\rangle$ went from 40.46 at 4.22 eV to 52.35 at 4.25 eV compared to the current "best" value 46.8 at 4.25 eV [61]. Consequently, the assumption of uniform overlayers with the optical properties of ideal oxides cannot always be justified, although it continues to be used [12, 63].

In fact, there is mounting evidence [30, 56, 57, 64, 65] that interfaces do not have the optical properties of either oxide or substrate. Philipp [65] noted that the characteristic SiO_2 structure did not appear as expected at 10.2 eV in normal-incidence reflectance data for Si [66] and so corrected these data for 15 Å of SiO instead of SiO_2. In single-wavelength null-ellipsometric work on the Si–thermally oxidized SiO_2 system, Taft and Cordes [56] found a 6-Å interface of refractive index 2.8 as already mentioned, while in spectroscopic ellipsometric measurements on similar oxides Aspnes and Theeten [64] found 7 ± 2 Å of $SiO_{0.4\pm0.2}$. Some $\langle\varepsilon\rangle$ data have been corrected by assuming that the residual interface layer after chemical stripping is that of the oxide interface itself [30, 64]. While the variations in the "best" values of ε obtained by *post hoc* correction procedures are getting smaller, this approach obviously still leaves something to be desired. The subject has been discussed recently by Jellison and Modine [67].

V   ELIMINATING OVERLAYERS

The preparation of high-quality samples with minimal overlayer thickness is an approach that actually has seen surprisingly little use. The importance of sample preparation and of maintaining surfaces in inert ambients such as dry N_2 was stressed as early as 1958 by Archer [68] and re-emphasized by Donovan *et al.* in 1963 [69]. It is no accident that the discrete-wavelength ellipsometric data of Archer [68] and the reflectance spectra of Donovan *et al.* [69], both on Ge, agree within several percentage points with recent work [61] in a field where discrepancies of 10–20% are not uncommon.

To successfully eliminate overlayers requires an unambiguous means of assessing overlayer thickness in real time. Null ellipsometers are simply too slow, and moreover the long-wavelength sensitivity of Δ to the presence of overlayers is not useful because it is not unique. For example, the refractive index of amorphous silicon at long wavelengths is about 10% greater than that of crystalline material, so changes in Δ can have either sign according to whether the overlayer is an oxide or is simply disordered material. Some techniques, such as vacuum ultraviolet reflectometry, are incompatible with real-time assessment, and also involve environments in which samples undergo some oxidation during preparation (loading and pumpdown). The early reflectance data of Philipp and co-workers [66, 70, 71] show evidence of such artifacts [61].

Photometric instrumentation has solved the speed problem because measurements can be made within seconds of processing if the sample is optically prealigned [72]. The uniqueness problem can also be solved by measuring at the energy corresponding to the E_2 peak in $\langle \varepsilon \rangle$ [29]. The argument is as follows. The magnitude of the peak and the energy at which it occurs are the combined result of the type of atoms constituting the crystal and the long-range order. If either is modified, e.g., by oxidation, disorder, or microscopic roughness, then the dielectric response of the surface region will drop below its bulk value. This tends to impedance-match ambient to substrate, thereby lowering R and $\langle \varepsilon \rangle$. The effect can be summarized qualitatively as "biggest is best" and is illustrated in Fig. 4, which shows $\langle \varepsilon_2 \rangle$ spectra of a single InSb sample after various surface treatments [29, 72]. The similarity to Fig. 2 is apparent. The spectrum that most nearly approximates the true bulk dielectric response is that with the highest value of $\langle \varepsilon_2 \rangle$. Other advantages of measurements at the E_2 peak are that the penetration depth of light is near its minimum value so that surface effects are most pronounced. Also, the wavelength is shorter so the expansion term d/λ is larger for a given d. A similar connection between sample quality and the height of the E_2 peak also occurs in reflectance [73, 74].

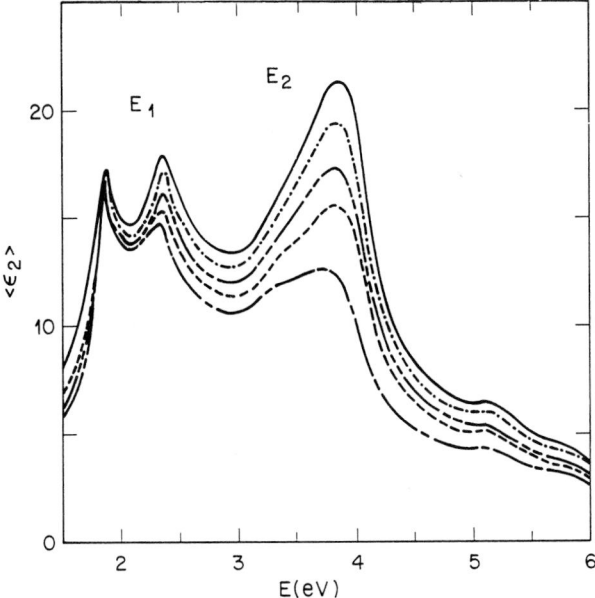

Fig. 4. $\langle \varepsilon_2 \rangle$ spectra for InSb for various surface conditions: (a–c) $\langle 100 \rangle$ surface after chemomechanical polishing and after 6- and 27-h exposures to 1:1 HCl:methanol; (d and e) $\langle 110 \rangle$ surface after 1:1 HCl:methanol and after the procedure described by Aspnes and Studna [61]. (——, e; —·—·, d; —··—, c; ————, b; ———, a.)

Aspnes and Studna [72, 75] investigated various procedures for obtaining surfaces with narrow transition regions on semiconductors. The best surfaces on Si and Ge were obtained by chemical etching. Atomically clean surfaces prepared in ultrahigh vacuum by ion bombardment and annealing were poorer [76], probably as the result of subsurface damage and reconstruction that occurs in the direction normal to the surface plane. Cleavage was also observed to give poor results: Transition regions were broader, and the surfaces oxidized at a much faster rate than chemically etched surfaces, consistent with previous work [37, 39] and indicative of a large number of active sites from cracks or damage. The best chemical procedures involved a chemomechanical polish with Syton or a weak ($\sim$0.05-vol. %) solution of bromine in methanol, followed by crystal- and orientation-specific procedures for removing residual overlayers [72, 75]. Standard chemical polishing solutions such as CP-4 ($HNO_3:HF:CH_3COOH$ 5:3:3) left the surfaces microscopically rough. However, microscopically smooth $\langle 111 \rangle$ surfaces could be obtained on Si with an etch consisting of $HNO_3:HF$ 10:1. Other surfaces were not investigated in detail.

In retrospect, the superior performance of chemical processing probably follows because the correct combination of reagents should produce an etch that only attacks edge and corner atoms and leaves the surface atomically flat. The process is the inverse of habit plane formation in crystal growth. While such surfaces are not atomically clean, the chemisorbed monolayer(s) provides a more natural termination of the electronic wave functions and appears to preserve bulk dielectric function behavior closer to the surface.

Even greater control should be obtainable with the sample immersed in an electrolyte. Here, various oxidation potentials could be established with proper reactants, and interface potentials could also be controlled externally. Ellipsometric measurements on such surfaces have been reported for InSb [77] and appear to be promising for preparing high-quality surfaces on other materials.

While attractive for low-index faces of single crystals, the chemical preparation approach may not be useful for composite materials or for highly reactive materials such as Al or Ti. In the former case, sharp interfaces probably cannot be obtained owing to the microstructure of the material, while in the latter case, the only appropriate environment is ultrahigh vacuum.

VI BULK AND THIN-FILM EFFECTS; EFFECTIVE-MEDIUM THEORY

The preceding discussion on overlayers presumes homogeneous bulk material. It is often necessary to determine the optical properties of heterogeneous or composite systems that are more accurately described as mixtures of separate regions of two or more materials, each of which retains its own dielectric identity. This is the objective of effective-medium theory. Examples of composite materials include metal films, which can be described as heterogeneous mixtures of material and voids owing to the inability of forming grain boundaries in close-packed systems without some loss of material [24]. Other examples include polycrystalline films, amorphous materials, and glasses; in the latter two cases, the inhomogeneity is essentially on the atomic scale and concerns the number of polarizable elements (bonds) per unit volume. A microscopically rough surface can also be considered a heterogeneous medium, being a mixture of bulk and ambient on a microscopic scale. In this discussion, we assume that the characteristic dimensions of the microstructure are large enough ($\gtrsim 10$–20 Å) so that the individual regions retain essentially their bulk dielectric responses, but small ($\lesssim 0.1$–0.2λ) compared to the wavelength of light. Then, the macroscopic **E** and **H** fields of Maxwell's equations will not vary appreciably over any single region, and quasistatic theory can be used. This avoids complications due to scattering and retardation effects that are dominant in macroscopically inhomogeneous systems [78].

The optical properties of heterogeneous systems are calculated in the same way that macroscopic quantities or observables are calculated in thermodynamics or quantum mechanics. First, the microscopic problem is solved exactly; then the microscopic quantities are averaged to obtain their macroscopic counterparts. The dielectric function is obtained from the macroscopic average electric field **E** and polarization **P** according to

$$\mathbf{D} = \varepsilon\mathbf{E} = \mathbf{E} + 4\pi\mathbf{P}, \tag{7a}$$

$$\mathbf{P} = \frac{1}{V}\sum q_i\,\Delta\mathbf{x}_i, \tag{7b}$$

where $\Delta\mathbf{x}_i$ is the displacement of the charge q_i under the action of the local field at q_i. It is the appearance of the volume normalizing factor in Eq. (7b) that is responsible for the sensitivity of ε to density.

A qualitative understanding of effective-medium effects can be obtained by considering the configuration of a metal sphere in an insulating dielectric, as shown in Fig. 5 [79]. Owing to its large polarizability, screening charge collects on the surface of the sphere and prevents the electric field from penetrating. The distortions in the microscopic field and polarization that result are what give rise to "local-field effects" and dielectric functions that differ from simple averages of those of the constituents. As indicated in Fig. 5, the more-polarizable species in any composite provides the screening charge and tends to contribute less than expected from simple volume arguments. Consequently, the less-polarizable species (e.g., voids) tends to dominate the dielectric response of composite materials.

If the electrostatic solution for the dielectric sphere in a dielectric host is volume averaged, then the Maxwell–Garnett effective-medium expression is

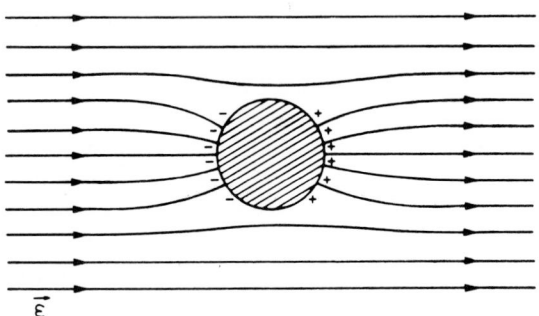

Fig. 5. Effect of screening on the local field near a metal sphere in an insulating matrix.

obtained. A general representation of two-phase effective-medium theories appropriate to three-dimensional isotropic systems is [32]

$$\frac{\varepsilon - \varepsilon_h}{\varepsilon + 2\varepsilon_h} = f_a \frac{\varepsilon_a - \varepsilon_h}{\varepsilon_a + 2\varepsilon_h} + f_b \frac{\varepsilon_b - \varepsilon_h}{\varepsilon_b + 2\varepsilon_h}, \tag{8}$$

where ε_a, ε_b and f_a, f_b are the dielectric functions and volume fractions of the phases a and b. The quantity ε_h is a host dielectric function that is assigned different values according to the model. To obtain the Maxwell–Garnett expressions [80], a or b is considered the host medium, so $\varepsilon_h = \varepsilon_a$ or ε_b, and one of the two terms on the right-hand side of Eq. (8) vanishes. The Bruggeman effective-medium approximation (EMA) [31] is obtained by making the self-consistent choice $\varepsilon_h = \varepsilon$, in which case the left-hand side vanishes. The Lorentz–Lorenz expression [81, 82] is obtained by choosing $\varepsilon_h = 1$, i.e., empty space.

The various effective-medium theories actually differ only in the choice of host material, but this choice does imply different microstructures. Thus, the Maxwell–Garnett theories, with one medium completely surrounded by another, are appropriate to the cermet or coated sphere microstructures, while the Bruggeman theory is appropriate to random or aggregate configurations [83]. The aggregate microstructure can also be considered as an average over grain shapes, and probably more accurately describes most thin films.

Effective-medium theory can be used to understand differences in optical properties of metal films, as shown in Fig. 6. Here, representative ε spectra from among the 30 or so available in the literature are presented for Au. The data of Thèye [84] were obtained by combined transmittance–reflectance measurements of thin (~ 200-Å) films deposited and annealed in ultrahigh vacuum. The spectroscopic ellipsometric data of Aspnes et al. [85] fall into three categories: films deposited by electron-beam evaporation under conditions similar to those of Thèye but not annealed and films prepared at moderate ($\sim 10^{-6}$-torr) pressures by filament evaporation onto room-temperature

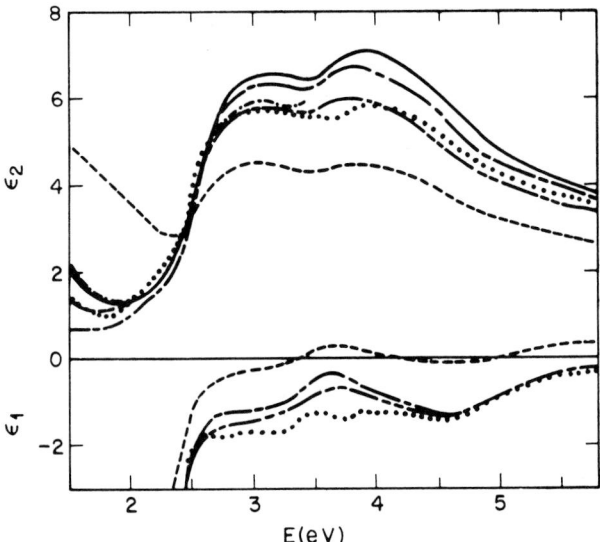

Fig. 6. Representative ε spectra for Au as described in the text. Thèye [84] (———); Johnson and Christy [86] (·······); Winsemius [87] (—·—·—·—); Aspnes *et al.* [85]: electron-beam deposition (———), evaporation onto room-temperature substrate (—··—), and evaporation onto liquid-nitrogen-cooled substrate (————).

and liquid-nitrogen-cooled substrates. The data of Johnson and Christy [86] were obtained by combined transmittance–reflectance measurements on films deposited by filament evaporation on room-temperature substrates. The agreement between ε_2 spectra for the two films deposited under basically identical conditions indicates again that sample preparation is more important than measurement method. The data of Winsemius [87] were obtained ellipsometrically on bulk samples cleaned and annealed at 700 K for several hours in ultrahigh vacuum.

All data clearly show the plasma edge at 2.5 eV and higher-energy structures, but there are significant differences in amplitude. These can be understood by supposing that the densities of the various films are different. If $\varepsilon_b = 1$ represents the void phase and ε_a the metal, with $f_b \ll 1$, then to first order in f_b either the Bruggeman or Maxwell–Garnett forms of Eq. (8) yields

$$\varepsilon = \varepsilon_a\{1 - 3f_b[(\varepsilon_a - 1)/(2\varepsilon_a + 1)]\}, \tag{9a}$$

$$\varepsilon \cong \varepsilon_a(1 - \tfrac{3}{2}f_b), \tag{9b}$$

where Eq. (9b) follows if $|\varepsilon_a| \gg 1$. For increasing concentrations of voids, the dielectric response simply scales to lower values approaching $\varepsilon_b = 1 + i0$. The decrease in ε is $\tfrac{3}{2}$ times the actual density deficit, a result of the metal screening itself more effectively as discussed in connection with Fig. 5. The void fraction of 0.21 calculated by comparing the electron-beam-evaporated and liquid-nitrogen-cooled film data of Fig. 6 is consistent with TEM micrographs, which show 12% clear aperture in the 500-Å film deposited on the cooled substrate. The remaining 9% missing material can be accounted as voids in the film network. Below 2.5 eV a pronounced Drude tail appears in ε_2 for the porous film, indicating increased broadening from surface collisions and therefore a smaller grain size. Surface scattering becomes important when the grain size drops significantly below the intrinsic mean free path, which for Au is 400 Å at room temperature.

A careful examination of Fig. 6 shows that both ε_1 and ε_2 spectra of the films evaporated on room-temperature substrates have been displaced downward with respect to the corresponding spectra for the electron-beam-evaporated film, while Eqs. (9) would predict that the ε_1 spectra should be displaced upward. This effect is due to surface roughness, which is an increasingly important factor for thicker films. The separation of surface roughness and density effects can only be done if both ε_1 and ε_2 spectra are available.

From the preceding analysis, we can conclude that the ε_2 spectrum most representative of the intrinsic dielectric response of Au is that of Thèye, while the spectrum most representative of the dielectric response of films deposited e.g., under the pressure and substrate temperature conditions used by Johnson and Christy, are those of Johnson and Christy.

The effective-medium arguments apply in general to optical data on metal films and allow differences among conflicting handbook data for a particular metal to be understood. For example, the "American Institute of Physics Handbook" [88] gives three sets of data for the complex refractive index of Ag at $\lambda = 10\ \mu m$, from which we calculate $\varepsilon = -2700 + i1400$ [89], $-3570 + i1310$ [90], and $-4650 + i1480$ [91]. From this discussion, it follows that both real and imaginary parts of ε are related to film density but that ε_2 depends in addition on grain size. It follows that the first datum [89] corresponds to the least dense material with the smallest grain size and that the last datum [91] most accurately represents the true bulk dielectric function of Ag.

VII CONCLUSION

We have discussed a number of methods developed to determine accurate values for bulk dielectric properties of materials. Which is best? The following comments, applicable primarily but not exclusively to spectral work in the visible–near UV, can be made.

Kramers–Kronig and related analyses of reflectance data are probably least accurate because they require reflectance data over the entire energy range or complete ε_1, ε_2 data at certain wavelenghts or ranges of wavelengths [92, 93] or accurate extrapolations into experimentally inaccessible spectral regions. Unfortunately, these procedures are also sensitive both to small errors in reflectances and to extrapolations, and reflectances are generally difficult to measure accurately. For example, reflectance measurements on Si have been performed and analyzed from 0 to 25 eV by Philipp and coworkers [65, 66] and from 0.5 to 10 eV by Verleur [94]. The values obtained for the E_2 peak in R near 4.6 eV were 0.73 and 0.82, respectively. While one would expect Verleur's data to yield the higher value for the corresponding E_2 peak in ε_2 at 4.25 eV, the actual values determined in the analysis of the reflectance data were 41.2 and 27.8, respectively. The difference was due essentially to extrapolation procedures, even though data were available in both analyses to 10 eV, seemingly sufficiently far from the evaluated point.

In a more extreme case, Morrison [73] calculated a "peak" value of 5 at the 4 eV E_2 peak of ε_2 for InAs from reflectance results determined experimentally from 0 to 6 eV and extrapolated via a power-law relationship to higher energy. Here, the error is due almost entirely to the extrapolation procedure; in a Kramers–Kronig analysis of their 0–27-eV reflectance data, Philipp and Ehrenreich [70] obtained 21 at 4.45 eV, in quite reasonable agreement with the value of 22.81 at the same energy obtained via an optimized-surface approach [61].

Extrapolation errors can be reduced in principle by making use of fixed-phase points established by independently measured values of ε_1 and ε_2 [92, 93]. However this procedure can also lead to considerable error. For example, Grasso et al. [95] obtained an E_2 peak of 31 for ε_2 at 4.8 eV for GaAs from an R peak of 0.59, compared to ellipsometric values [61] of 25.22 and 0.677, respectively. The low value of R observed by Grasso et al. indicates that the actual sample was covered with a significant overlayer. Nevertheless, the calculated ε_2 value was larger, which indicates a probable error in the Kramers–Kronig analysis of the order of 50%. The error arises because even though the integrand in the Ahrenkel algorithm [92] used in the calculation converges as E^{-4}, the prefactor is proportional to E^3. Thus, extrapolation is still important. The message is clear: A Kramers–Kronig analysis should be attempted only if accurate reflectance data are available over a very, very wide energy range.

The second approach is to overdetermine the data enough to be able to determine the substrate parameters and those describing the overlayer independently. This approach requires some assumptions about the nature of the surface and substrate, but it may be the only alternative for reactive materials for which clean surfaces are difficult to obtain or for composite materials for which smooth surfaces may be difficult to obtain. This approach has been used widely for Si at 5461 Å and has given fairly consistent results, although in some respects the transparent oxide and nearly transparent substrate characteristic of this system make it particularly favorable for this type of analysis. The results at higher energies are less convincing: at 3655 Å,

Ibrahim and Bashara [48] and So and Vedam [49] obtained $36.93 + i36.69$ and $(39.43 \pm 0.44) + i(29.57 \pm 0.6)$, respectively, a substantial difference in view of the expected accuracy of the technique. Nevertheless, over-determination seems preferable to the Kramers–Kronig approach if for no other reason than that it offers some self-consistency checks.

The third approach is to concentrate on preparing high-quality samples and eliminating surface overlayers as far as possible. The most consistent results for a given material, regardless of measurement technique, have been obtained in this way. The observed consistency instills some confidence that the values obtained are representative of the intrinsic dielectric response. Although arbitrarily accurate values can never be obtained in this manner, it appears that errors introduced by the neglect of finite transition region widths are less than those resulting from overdetermination, correction, or Kramers–Kronig analysis procedures.

We have not discussed explicitly the determination of intrinsic optical properties of glasses, polycrystalline semiconductors, polymers, anisotropic materials, or other important classes of solids and thin films. The same general remarks apply, but each material seems to give rise to its own special problems. The problem of the accurate determination of intrinsic dielectric properties has not yet been solved in general.

REFERENCES

1. J. D. Jackson, "Classical Electrodynamics," 5th ed., Wiley, New York, 1975.
2. R. M. A. Azzam and N. M. Bashara, "Ellipsometry and Polarized Light," North-Holland Publ., Amsterdam, 1977.
3. P. S. Hauge, *Surf. Sci.* **96**, 108 (1979).
4. R. H. Muller, *Surf. Sci.* **56**, 19 (1976).
5. A. V. Rzhanov and K. K. Svitashev, *Adv. Electron. Electron Phys.* **49**, 1 (1979).
6. See Chapter 4, this volume.
7. A. B. Winterbottom, *in* "Ellipsometry in the Measurement of Surfaces and Thin Films" (E. Passaglia, R. R. Stromberg, and J. Kruger eds.), p. 97, National Bureau of Standards Special Publication 256, 1963.
8. H. J. Mathieu, D. E. McClure, and R. H. Muller, *Rev. Sci. Instrum.* **45**, 798 (1974).
9. B. D. Cahan and R. F. Spanier, *Surf. Sci.* **16**, 166 (1969).
10. S. N. Jasperson and S. E. Schnatterly, *Rev. Sci. Instrum.* **40**, 761 (1969).
11. V. M. Bermudez and V. H. Ritz, *Appl. Opt.* **17**, 542 (1978).
12. G. E. Jellison, Jr., and F. A. Modine, *J. Appl. Phys.* **53**, 3745 (1982).
13. B. Drevillon, J. Perrin, R. Marbot, A. Violet, and J. L. Dalby, *Rev. Sci. Instrum.* **53**, 104 (1982).
14. B. H. Billings, *J. Opt. Soc. Am.* **42**, 12 (1952).
15. P. S. Hauge and F. H. Dill, *IBM J. Res. Dev.* **17**, 472 (1973).
16. D. E. Aspnes and A. A. Studna, *Appl. Opt.* **14**, 220 (1975).
17. D. E. Aspnes, *J. Opt. Soc. Am.* **65**, 1274 (1975).
18. P. S. Hauge and F. H. Dill, *Opt. Commun.* **14**, 431 (1975).
19. R. W. Stobie, B. Rao, and M. J. Dignam, *Appl. Opt.* **14**, 999 (1975).
20. J. B. Theeten, R. P. H. Chang, D. E. Aspnes, and T. E. Adams, *J. Electrochem. Soc.* **127**, 379 (1980).
21. B. D. Cahan, *Surf. Sci.* **56**, 354 (1976).

22. D. E. Aspnes, *SPIE Proc.* **276**, 188 (1981).
23. D. E. Aspnes and A. A. Studna, *SPIE Proc.* **276**, 227 (1981).
24. P. Rouard and A. Meessen, *in* "Progress in Optics" (E. Wolf ed.), Vol. 15, p. 77, North-Holland, Publ., Amsterdam, 1977.
25. S. P. F. Humphreys–Owen, *Proc. Roy. Soc. (London)* **77**, 949 (1961).
26. D. K. Burge and H. E. Bennett, *J. Opt. Soc. Am.* **54**, 1428 (1964).
27. D. E. Aspnes, in "Optical Properties of Solids: New Developments" (B. O. Seraphin, ed.), p. 799, North-Holland Publ., Amsterdam, 1976.
28. W. J. Plieth and K. Naegele, *Surf. Sci.* **64**, 84 (1977).
29. D. E. Aspnes, *J. Vac. Sci. Technol.* **17**, 1057 (1980).
30. D. E. Aspnes, G. P. Schwartz, A. A. Studna, G. J. Gualtieri, and B. Schwartz, *J. Electrochem. Soc.* **128**, 590 (1981).
31. D. A. G. Bruggeman, *Ann. Phys. (Leipzig)* **24**, 636 (1935).
32. D. E. Aspnes, J. B. Theeten, and F. Hottier, *Phys. Rev. B* **20**, 3292 (1979).
33. A. B. Winterbottom, *K. Nor. Vidensk. Selsk. Sk.* **1**, 61 (1955).
34. R. J. Archer, *J. Electrochem. Soc.* **104**, 619 (1957).
35. B. H. Claussen and M. Flower, *J. Electrochem. Soc.* **110**, 983 (1963).
36. A. N. Saxena, *Appl. Phys. Lett.* **7**, 113 (1965).
37. F. Lukes, *Optik* **31**, 83 (1970).
38. K. Vedam and S. S. So, *Surf. Sci.* **29**, 379 (1972).
39. S. S. So and K. Vedam, *J. Opt. Soc. Am.* **62**, 596 (1972).
40. Y. J. van der Meulen, *J. Electrochem. Soc.* **119**, 530 (1972).
41. K. Vedam, *Surf. Sci.* **56**, 221 (1976).
42. B.-L. Twu, *J. Electrochem. Soc.* **126**, 1589 (1979).
43. E. A. Taft, *J. Electrochem. Soc.* **125**, 968 (1978).
44. C. C. Chang and D. M. Boulin, *Surf. Sci.* **69**, 385 (1977).
45. F. Lukes, W. H. Knausenberger, and K. Vedam. *Surf. Sci.* **16**, 112 (1969).
46. G. A. Egorova, N. S. Ivanova, E. V. Potapov, and A. V. Rakov, *Opt. Spektrosk.* **36**, 773 (1974); G. A. Egorova, N. S. Ivanova, E. V. Potapov, and A. V. Rakov, *Opt. Spectrosc.* **36**, 449 (1974).
47. J. Shewchun and E. C. Rowe, *J. Appl. Phys.* **41**, 4125 (1970).
48. M. M. Ibrahim and N. M. Bashara, *J. Opt. Soc.* **61**, 1622 (1971).
49. S. S. So and K. Vedam, *J. Opt. Soc. Am.* **62**, 16 (1972).
50. O. Hunderi, *Surf. Sci.* **61**, 515 (1976).
51. O. Hunderi, *Appl. Opt.* **11**, 1572 (1972).
52. P. Picozzi, S. Santucci, and A. Balzarotti, *Surf. Sci.* **45**, 227 (1974).
53. V. M. Bermudez, *Surf. Sci.* **94**, 29 (1980).
54. K. Vedam, W. Knausenberger, and F. Lukes, *J. Opt. Soc. Am.* **59**, 64 (1969).
55. R. C. O'Handley, *Surf. Sci.* **46**, 24 (1974).
56. E. A. Taft and L. Cordes, *J. Electrochem. Soc.* **126**, 131 (1979).
57. H. W. Dinges, *Thin Solid Films* **50**, L17 (1978).
58. J. B. Theeten and D. E. Aspnes, *Thin Solid Films* **60**, 183 (1979).
59. G. Jungk, *Phys. Status Solidi A* **34**, 69 (1976).
60. D. E. Aspnes, *Surf. Sci.* **56**, 322 (1976).
61. D. E. Aspnes and A. A. Studna, *Phys. Rev. B* **27**, 985 (1983).
62. A. Daunois and D. E. Aspnes, *Phys. Rev. B* **18**, 1824 (1978).
63. H. Burkhard, H. W. Dinges, and E. Kuphal, *J. Appl. Phys.* **53**, 655 (1982).
64. D. E. Aspnes and J. B. Theeten, *J. Electrochem. Soc.* **127**, 1359 (1980).
65. H. R. Philipp, *J. Appl. Phys.* **43**, 2835 (1972).
66. H. R. Philipp and H. Ehrenreich, *Phys. Rev.* **129**, 1550 (1963).
67. G. E. Jellison and F. A. Modine, *J. Opt. Soc. Am.* **72**, 1253 (1982).
68. R. J. Archer, *Phys. Rev.* **110**, 354 (1958).
69. T. M. Donovan, E. J. Ashley, and H. E. Bennett, *J. Opt. Soc. Am.* **53**, 1403 (1963).

70. H. R. Philipp and E. A. Taft, *Phys. Rev.* **120**, 37 (1960).
71. H. R. Philipp and H. Ehrenreich, in "Optical Properties of Solids: New Developments" (R. K. Willardson and A. C. Beer eds.), p. 93, Academic Press, New York, 1967.
72. D. E. Aspnes and A. A. Studna, *SPIE Proc.* **276**, 227 (1981).
73. R. E. Morrison, *Phys. Rev.* **124**, 1314 (1961).
74. M. Cardona, *J. Appl. Phys.* **32**, 958 (1961).
75. D. E. Aspnes and A. A. Studna, *Appl. Phys. Lett.* **39**, 316 (1981).
76. D. E. Aspnes and A. A. Studna, *Surf. Sci.* **96**, 294 (1980).
77. N. P. Syoseva and B. M. Ayupov, *Opt. Spectrosk.* **47**, 300 (1979).
78. P. Beckmann, *The Depolarization of Electromagnetic Waves*, Golem, Boulder, Colorado, 1968.
79. D. E. Aspnes, *Am. J. Phys.* **50**, 704 (1982).
80. J. C. M. Garnett, *Philos. Trans. Roy. Soc. (London)* **203**, 385 (1904); J. C. M. Garnett, *Philos. Trans. Roy. Soc.* (London). *A* **205**, 237 (1906).
81. L. Lorenz, *Ann. Phys. Chem. (Leipzig)* **11**, 70 (1880).
82. H. A. Lorentz, "Theory of Electrons," 2nd ed., Teubner, Leipzig, 1916.
83. G. A. Niklasson, C. G. Granquist, and O. Hunderi, *Appl. Opt.* **20**, 26 (1981).
84. M.-L. Thèye, *Phys. Rev. B* **2**, 3060 (1970).
85. D. E. Aspnes, E. Kinsbron, and D. D. Bacon, *Phys. Rev. B* **21**, 3290 (1980).
86. P. B. Johnson and R. W. Christy, *Phys. Rev. B* **6**, 4370 (1972).
87. M. Guerrisi, R. Rosei, and P. Winsemius, *Phys. Rev. B* **12**, 557 (1975).
88. D. E. Gray (ed.), pp. 6–150, "American Institute of Physics Handbook," McGraw-Hill, New York, 1972.
89. B. Dold and R. Mecke, *Optik* **22**, 435 (1965).
90. V. G. Padalka and I. N. Shklyarevskii, *Opt. Spectrosk.* **11**, 527 (1961); V. G. Padalka and I. N. Shklyarevskii, *Opt. Spectrosc.* **11**, 285 (1961).
91. J. R. Beattie, *Physica* **23**, 898 (1957).
92. R. K. Ahrenkiel, *J. Opt. Soc. Am.* **61**, 1651 (1971).
93. R. Hulthén, *J. Opt. Soc. Am.* **72**, 794 (1982).
94. H. W. Verleur, *J. Opt. Soc. Am.* **58**, 1356 (1968).
95. V. Grasso, G. Mondio, G. Saitta, S. U. Campisano, G. Foti, and E. Rimini, *Appl. Phys. Lett.* **33**, 632 (1978).

Chapter **6**

Interferometric Methods for the Determination of Thin-Film Parameters

JOSEPH SHAMIR

Faculty of Electrical Engineering
Technion—Israel Institute of Technology
Haifa, Israel

I.	Introduction	113
II.	Basic Principles	114
III.	Nonlaser Interferometers	117
	A. The Fizeau Method	117
	B. The FECO Method	118
	C. Shearing Interferometers	119
	D. Discussion	121
IV.	Kösters-Prism Interferometers	123
	A. Transmission Phase Measurement	123
	B. Reflection Phase Measurement	124
	C. Ellipsometric Interferometry	124
	D. Measuring Thermal Expansion and Index Variation	125
	E. Response Linearization	126
V.	A Self-Calibrating Method	126
VI.	Surface Effects	131
VII.	Conclusions	132
	References	133

INTRODUCTION I

The optical parameters of thin films frequently deviate from those determined for bulk material. Therefore, special methods are required for the measurement of optical parameters of thin films in their final form and, as far as possible, with no special preparation for the measurement. While some of the methods discussed can also be applied for measuring the optical properties of bulk material, emphasis will be on thin films deposited on known bulk slabs.

113

Among the optical measuring techniques, interferometry is capable of the highest accuracy. However, before the invention of the laser, its usefulness was quite limited, and other methods had been developed to a high level of sophistication. The highly advanced methods for thin-film analysis, such as the ellipsometric methods discussed in Chapter 5, made interferometric techniques a second choice at best. Nevertheless, recent advances in laser interferometry should lead to a renewal of interest in this field and result in the synthesis of novel methods with extremely high accuracies and simplified operating procedures.

The first part of this chapter will deal with the description of traditional interferometric methods for the measurement of thin-film parameter [1–5], while the rest will be devoted to the description of novel laser interferometric techniques.

II   BASIC PRINCIPLES

For typical experimental situations the investigated sample will be assumed to consist of a uniform slab of thickness h and complex refractive index

$$n_2 = n - ik, \tag{1}$$

situated between two media of respective refractive indices n_1 and n_3. In most cases $n_1 \simeq 1$ (air) and n_3 is a known, transparent substrate (Fig. 1); however, opaque substrates may be used in reflection measurements. For high-sensitivity measurements the simplest practical situation is a bare substrate (n_3) with a thin, equivalent layer representing the combined effect of surface roughness and adsorbed matter, to be discussed in Section VI.

The sample is probed by light beams incident on either or both sides. Each polarization component of a light beam will be represented by its analytic signal

$$V = E \exp(i\omega t). \tag{2}$$

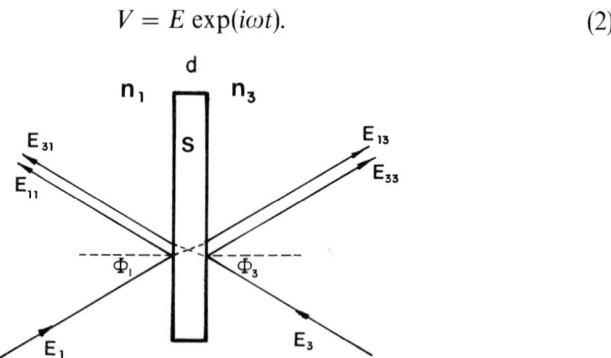

Fig. 1. Illumination configurations. S is the sample film imbedded between two media of respective refractive indices n_1 and n_3. Two possible illuminating beams are E_1 and E_3 that produce the respective two pairs of outgoing beams E_{11}, E_{13} and E_{33}, E_{31}.

In the figure, E_1 represents one polarization component of the complex amplitude of the light incident from medium I. A fraction

$$E_{11} = r_1 E_1 \tag{3}$$

of the beam is reflected into medium I, while the fraction

$$E_{13} = t_1 E_1 \tag{4}$$

is transmitted into medium III. In a similar way, the sample may be interrogated by a beam E_3 from the other side to produce the two outgoing beams

$$E_{33} = r_3 E_3 \tag{3a}$$

and

$$E_{31} = t_3 E_3. \tag{4a}$$

The complex amplitude reflectances and transmittances are defined by

$$r_j = |r_j| \exp(i\rho_j), \tag{5a}$$
$$(j = 1, 3)$$
$$t_j = |t_j| \exp(i\theta_j), \tag{5b}$$

where ρ_j and θ_j are the phase shifts upon reflection and transmission, respectively.

It should be noted here that usually the light beams inside the substrate medium n_3 are inaccessible for detection, and measurements made outside the substrate should take into account the effects of the back surface and possibly internal losses as well. In particular, back-surface reflections may interfere with all detected beams and require special arrangements for their elimination. Such arrangements may be based on spatial filtering configurations and index matching.

Most noninterferometric methods involve the measurement of intensity reflectances

$$R_j = |r_j|^2 \tag{6}$$

and transmittances

$$T_1 = (p_3/p_1)|t_1|^2; \qquad T_3 = (p_1/p_3)|t_3|^2, \tag{7}$$

where

$$p_j = n_j \cos\phi_j, \tag{8}$$

with ϕ the incidence angle.

At first sight, there appear to be eight real measurable parameters defined in Eqs. (5). However, these parameters are not independent and some basic relations interconnect them. Apart from the two simple general relations [1, 6]

$$T_1 = T_2 \equiv T, \tag{9}$$

and

$$\theta_1 = \theta_2 \equiv \theta, \tag{10}$$

there are a number of other, more complicated ones. If a single, homogeneous film on a known substrate is considered, these relations may be used to reduce the number of independent measurables to three, adequate in principle for the determination of the three film parameters n, k, and h. Unfortunately, the transcendental and multivalued nature of these relations [2, 3, 7] prohibits, in general, the unique determination of film parameters from only three measurements. Therefore, in most cases, some *a priori* information about the film is required, or additional measurements should be performed.

The phase shift under transmission θ and the two phase shifts due to reflection ρ_1 and ρ_3 are the main subjects for interferometric measurements. The major problem in interferometry is that these phases cannot usually be determined absolutely in an independent way. In fact, most methods are only able to measure the variations of these phases in space and time. The parameters of Eqs. (5) are functions of wavelength, angle of incidence, and polarization. Thus, by varying these factors, the required phase shift changes may be obtained. Sometimes, external parameters such as pressure or temperature may be utilized as well; however, the common procedure in interferometry is the utilization of an actual reference surface.

The attainable accuracies in interferometric setups are limited by the quality of the optical components, stability of the system, and stability and coherence of the light source. The restricted coherence of nonlaser sources was the main cause for the neglect of interferometry as a major technological tool in the field of thin-film analysis. Nevertheless, a number of interferometric principles were developed for application with conventional thermal sources. Some of the interferometers representative of these principles will be discussed in Section III.

While nonlaser interferometers should involve very small optical path differences, this restriction can be greatly relaxed when lasers are used. However, even laser interferometers are strongly affected by environmental noise and special designs are required for high-accuracy measurements. A compact and relatively immune laser interferometer will be introduced in Section IV, and its various applications for optical parameter determination will be described.

Most interferometric methods require a reference for a comparative measurement. Such a reference may be an optical flat with a very high-quality surface like in the Fizeau and fringes of equal chromatic order (FECO) interferometers, [1, 4] to be described in Section III, a mirror in a Michelson interferometer, or a portion of uncoated substrate as used in some of the Kösters-prism configurations described in Section IV. A self-calibrating method that dispenses with a reference surface will be described in Section V.

At this point one should mention also a new concept in interferometry that is based on phase conjugation [8]. This new method is very promising because information can be evaluated from the comparison of a wave front with its complex conjugate providing a measure of absolute phase value. However, since they are still in a research stage, phase conjugating methods will not be considered here. This chapter will be concluded by a discussion of the effects of surface quality and some comparisons among the various interferometric methods.

NONLASER INTERFEROMETERS   III

The term *nonlaser interferometer* refers to systems designed primarily for application with conventional light sources. Naturally, this does not imply that these interferometers cannot be used also with laser light, and in most cases the converse is true—laser illumination will improve the performance of the system.

A number of common characteristics for the nonlaser interferometers can be pointed out. First of all, the fact that they have to utilize very small optical path differences between the interfering beams results also in a relatively high stability. Usually, these interferometers are easy to operate for low-sensitivity measurement and can be applied in quite hostile environments; however, when accuracy requirements become more stringent, laser interferometers may become preferable. In this section, we shall discuss three methods: (1) the Fizeau method that uses multiple reflection of monochromatic light, (2) the FECO method that uses multiple reflection of white light, and (3) shearing methods.

The Fizeau Method   A

Probably the most popular method for the optical measurement of the thickness of thin films is the Fizeau method [1, 4]. The advantages here are simple construction and ease of use, while the main drawback is the special sample preparation that is required for high-sensitivity measurements. The sample film is deposited onto a high-quality optical surface, usually with a channel left uncoated (Fig. 2a), and then the whole surface is overcoated by a high-reflectance metal coating. Fizeau fringes are obtained by multiple reflection of diffused monochromatic light between the sample and a second optical flat (overplate) placed above it at a slight inclination.

The Fizeau method has been extensively studied in the cited literature; thus, here we shall only mention its main characteristics and applicability. Fizeau fringes originate from the interference among the multiple reflected beams and are described by the Airy formula that gives the intensity distribution

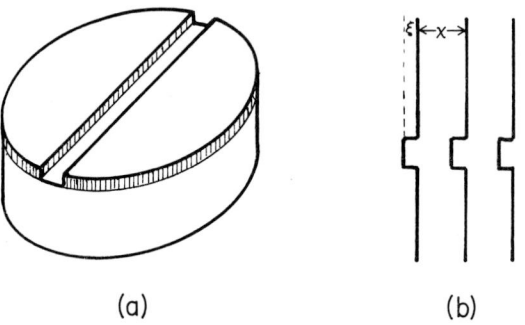

Fig. 2. (a) Thin-film sample on optical flat for Fizeau interferometry and (b) fringe pattern.

over the surface [1]:

$$I = \frac{I_{\max}}{1 + F \sin^2 \pi[(2nt \cos \phi)/\lambda]}, \tag{11}$$

where $I_{\max}$ is the maximum intensity at fringe center, nt the equivalent optical distance (including phase shift on reflection) between the two reflecting surfaces, ϕ the illumination angle, λ the observation wavelength, and F the finesse of the system defined by

$$F = 4R/(1 - R)^2, \tag{12}$$

with $R = |r_1||r_2|$ being the average intensity reflectance of the two surfaces. With monochromatic illumination, fringes are observed over the whole sample surface with lateral spacing x determined by the overplate inclination. With normal-incidence illumination, the vertical optical path difference between adjacent fringes is $\lambda/2$. Film thickness is determined by the ratio between fringe shift ε (Fig. 2b) and x.

The Fizeau method is very useful for high-accuracy (up to 0.001λ) thickness measurement of properly prepared samples and also for quick, low-sensitivity estimation of step height in any reflective sample. In principle, this method is also applicable for the determination of ρ_j; however, the interpretation of the result may be ambiguous, since it is difficult to separate the phase shift due to reflection from that induced by optical path length.

Improvements in the sensitivity of the Fizeau method are possible by multiple-pass arrangements [9, 10].

B   The FECO Method

The FECO method [1, 4, 11] differs from the Fizeau method by two main features. First, here, too, reference is made to Eq. (11); however, instead of varying nt, in the FECO method the wavelength is varied. Second, instead of making observation over a large portion of the sample, only a narrow line is observed at a time. The second difference may be considered a drawback, but at the same time it is an advantage in that it greatly reduces the quality requirements of the sample and absolute determination of *nt* is, in principle, possible at each point on its surface. The basic constituent of the FECO method is a multiple-beam interferometer similar to the one used for the Fizeau method; however, here the illumination is by white light, and the interferometer surface is imaged onto the entrance slit of a spectrometer. On the output plane of the spectrometer, one obtains a strong spectral line for each wavelength that, according to Eq. (11), produces a maximum for the portion of the interferometer imaged on the slit.

For the purpose of the determination of general optical parameters, the FECO method appears to be superior to the Fizeau method because it is capable of the absolute determination of the optical thickness of transparent films (reflectively coated on two sides) through the whole visible region. As a matter of fact, there is no reason not to use the method on both sides of the visible spectrum as well.

In summary, one can state that the Fizeau and FECO methods are based on a quite stable interferometric setup; thus, they are applicable, in relatively hostile environment, for the determination of optical thickness and surface topography with moderate accuracy. High accuracies are also attainable, but with highly sophisticated sample preparation and very careful experimental procedures.

A third way of using Eq. (11) was also proposed by Pliskin and Conrad [12]: In this method the illumination angle ϕ was varied during measurement also facilitating the determination of the thickness and refractive index of transparent films [12].

Shearing Interferometers C

The requirement of a reflective overcoat on the sample limits the possible applications of the Fizeau and FECO methods to the determination of film thickness only, with some exceptions when refractive index can be determined as well. Furthermore, the apparently simple method may become quite expensive and inconvenient due to the requirement of a well-aligned, high-quality optical reference flat. All these difficulties are resolved by using the shearing interferometric method. Instead of using a physical reference, in the shearing techniques the wave front, transmitted or reflected from the object, is compared with itself after a transversal shift has been imparted to it. Any nonuniformity in the wave front will result in the production of interference fringes determined by the local phase difference δ (Fig. 3), provided that the shear Δ does not exceed the spatial coherence of the source. In most systems,

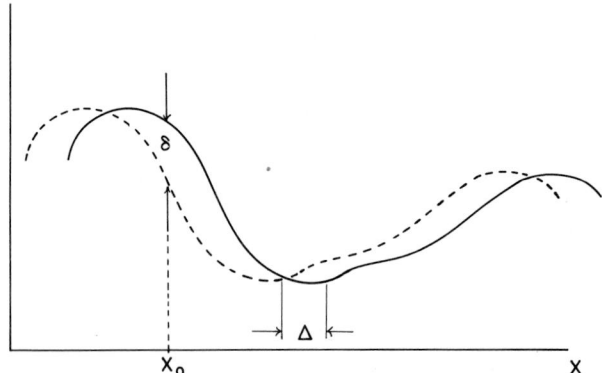

Fig. 3. One-dimensional representation of a sheared wave front. Δ is the shear and δ is the measured phase shift at point x_0.

the shear is very small, which leads to the determination of phase gradients rather than phase magnitudes.

For the purpose of analyzing thin films, usually the gradient measurement is adequate for the determination of the optical thickness, i.e., *nh* for transmission measurement and *h* for reflection measurement of an overcoated sample. For shearing interferometric measurement, the sample should be prepared in the same way as for the Fizeau method or just with a film edge on a useful part of the substrate. For the measurement of such a sample a mere shear is inadequate as demonstrated by the single "infinite" broad fringe in Fig. 4a. To produce a useful, "finite" fringe pattern, a slight inclination is imparted to the sheared wave front as in Fig. 4b. Similar to the Fizeau fringes, here too the phase shift δ can be determined from the ratio of the fringe shift ε to the fringe spacing.

Shearing interferometers differ from each other mainly by the optical arrangement used for imparting the required shear. For the analysis of small objects, as is the case in thin-film measurements, the best choice will be a single-component shearing configuration, which makes the interferometer into a "common path" interferometer in which both interfering waves traverse effectively the same path. Common-path interferometers are very stable and can be easily used in noisy and unfavorable environments.

Probably the most popular component to produce the required shear, together with the inclination mentioned earlier, is the Wollaston prism. The Wollaston prism is made of a doubly refracting crystal that splits a light beam into its two polarization components separating them by a small angle. A number of useful configurations, with a Wollaston prism as their central component, have been described for various applications by Nomarski [13] and Francon and Mallick [5]. Another component worth mentioning here is the Kösters prism [14, 15], which will be discussed in Section IV from a different point of view.

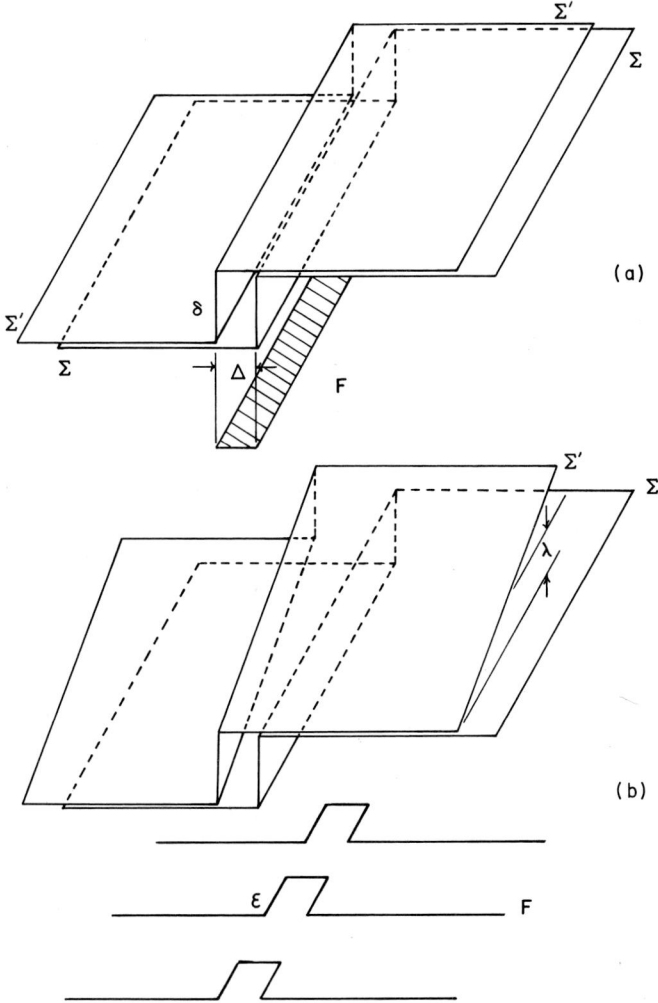

Fig. 4. Sheared wave fronts from a sample with a step of height δ with lateral shear Δ; Σ is the original wave front, Σ' is the sheared one, and F represents the shape of the fringe pattern. (a) Lateral shear only and (b) lateral shear with inclination.

Discussion   D

Three approaches have been described for nonlaser interferometric analysis of thin films. The Fizeau and shearing methods are used to visualize a whole region of the sample, while the FECO method investigates only a line at a time. The FECO method is capable of measuring absolute optical thickness;

this usually is not possible with the two other methods unless measurements are performed at a number of wavelengths.

Owing to the high finesse of multiple-beam interference fringes, one would expect much higher accuracy for the Fizeau and FECO methods than for the shearing methods, which are essentially double-beam interferometers. Actually, however, the monochromatic shearing interferometers may be used with laser light that produces very high-luminosity fringes. These fringes may be analyzed by simple electronic means [16] to yield high-sensitivity measurements. At first sight, it appears that the Fizeau method may also be improved by laser illumination, but for high-accuracy measurements the benefits of coherent light may be upset by the coherent noise (i.e., due to speckle, dust particles, and irregularities in the optical components of the whole system) that will effectively reduce the finesse of the fringes.

The rest of this chapter is devoted to laser-illuminated interferometers all of them being essentially double-beam interferometers where high luminosity and electronic detection play a cardinal role.

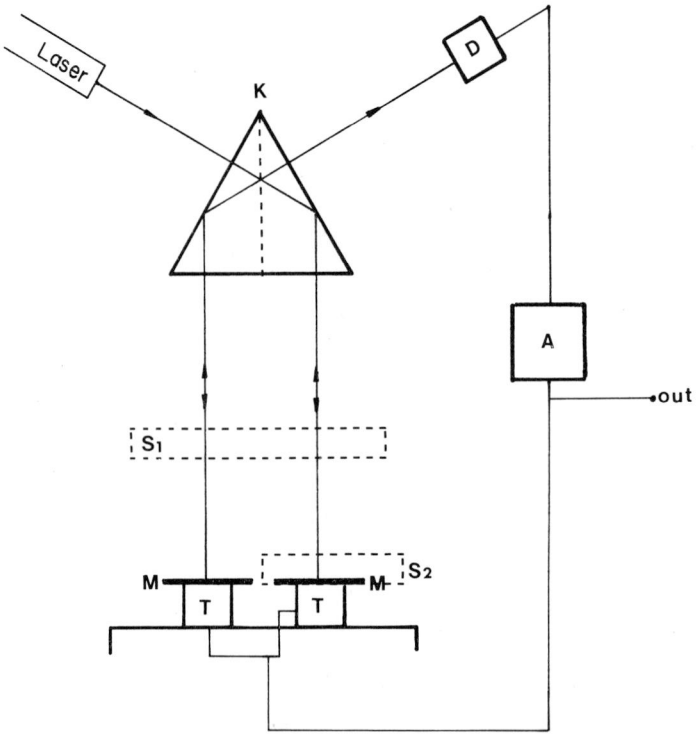

Fig. 5. Kösters-prism interferometer. K, Kösters prism; M, mirrors; T, piezoelectric transducers; D, fringe position detector; A, integrating amplifier; S, possible sample positions.

The main advantage of the Kösters-prism interferometer as represented schematically in Fig. 5 is its high immunity to environmental noise due to the close vicinity of the two interfering beams. This immunity makes it an ideal tool for high-precision measurements [15–19]. Here, the basic principle will be discussed, while the description of actual measuring systems can be found in the literature cited.

The central constituent of the interferometer is the Kösters-prism beam splitter [14] K, which produces two parallel coherent beams (see Fig. 5). Transparent samples may be inserted in region S_1, while reflective samples will replace one or both reflecting mirrors M at position S_2. The two reflected beams recombine in K, and the interference is observed in detector D. The rest of the system constitutes a feedback arrangement to be discussed later.

The principle of measurement is demonstrated in Fig. 6; the thin film F is deposited on part of the substrate S. The two parallel beams from the Kösters prism E_1 and E_2 impinge on the sample as shown, part of them is reflected and part transmitted.

Transmission Phase Measurement   A

For the measurement of the transmission phase θ [Eq. (10)], the sample is inserted at S_1. The phase shift is determined by detecting the interference fringe motion while translating the sample in such a way that the beams cross

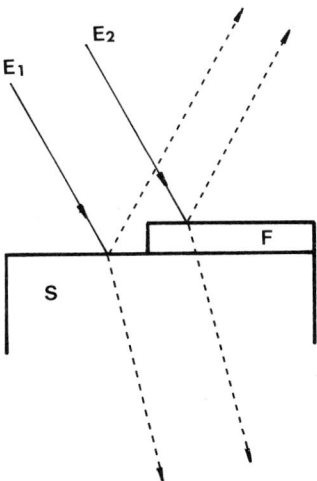

Fig. 6. Thin film (F) on substrate (S) configuration for measurement with two parallel beams E_1 and E_2 in Kösters-prism interferometer.

the film border in succession. This system is capable of measuring θ with a sensitivity of 10^{-4} rad [18]. For films with small absorption, this phase shift is independent of k; therefore, the method is suitable for the accurate determination of either n or d, provided that the other is known *a priori*. When neither is known, additional measurements may be performed at different angles or different wavelengths.

B Reflection Phase Measurement

The second mode of operation of the interferometer is in reflection for measuring ρ_1 or ρ_2. With the sample inserted at S_2, the phase shifts can be measured by sample translation as before. In contrast to transmission phase shift θ, the reflection phase shift is strongly dependent on all three optical parameters. Thus, in principle, more information is attainable by reflection measurements than by transmission measurements; however, the transmission mode of operation is more convenient for transparent samples since it is much less sensitive to sample vibration and misalignment. Furthermore, reflection measurements on transparent samples may be greatly disturbed by spurious reflection, especially from the substrate's second surface.

To summarize these two operating modes, we note that for transparent samples one should use transmission measurements if possible. Usually, these will be adequate for the determination of n and h. Reflection measurements may be used as well, but they are more difficult. Nevertheless, for reflecting samples, especially overcoated ones, reflection measurements are ideal for thickness measurements. Accuracies better than 0.5 Å are attainable without too much effort.

C Ellipsometric Interferometry

The main drawback of the procedures described in the preceding subsections is the requirement of a physical, relative translation between the scanning beams and the sample. This translation induces some stringent quality requirements on the sample and mechanical (or electro-optic) translation systems. To overcome this difficulty, in the third mode of operation the sample is stationary and the *polarization* of the incident illumination is varied. In principle, this converts the system into a kind of ellipsometer [19] in which the ellipsometric phase shift is directly measured. The ellipsometric phase shifts are the changes of θ and ρ upon $90°$ rotation of the polarization. Alternative interferometric ellipsometers have also been suggested [20].

Measuring Thermal Expansion and Index Variation   **D**

This subsection describes a system that is proposed for the accurate determination of the thermal expansion together with thermal index variation of
optical materials. For the measuring process, a sample with two flat surfaces
(not necessarily parallel) should be prepared and inserted in the system of
Fig. 5. However, this time the sample transmits only one of the beams, and
it serves also as a second interferometer (Fig. 7). Part of the beam is reflected
by the two surfaces of the sample, and the two reflected beams interfere to
form fringes in the detector D_2. The position of the interference fringes is a
function of the optical path $2nt$ in the sample, where n is its refractive index
and t the geometrical path length in the sample. A temperature change dT of
the sample introduces a change in its optical thickness $l = nt$:

$$\frac{dl}{dT} = n\frac{dt}{dT} + t\frac{dn}{dT}. \tag{13}$$

The change in the optical path difference of the interfering beams in D_2 is

$$dl_2 = 2dl. \tag{14}$$

However, for the second pair of interfering beams we obtain a change

$$dl_1 = 2dl - 2\frac{dt}{dT}dT, \tag{15}$$

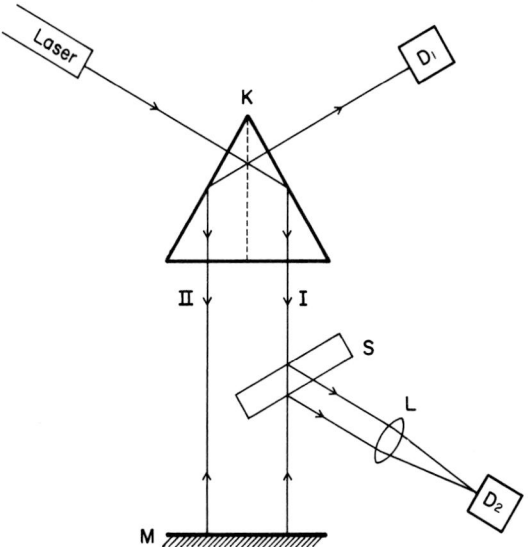

Fig. 7. Modified interferometer for measuring thermal effects.

since the expanding sample occupies a portion of the original path in air. From Eqs. (13–15) one may determine dt/dT and dn/dT separately.

The expected sensitivity of this method can easily be estimated. Assuming a conservative fringe shift measuring sensitivity of 10^{-3} fringe and a measuring wavelength of 600 nm, an expansion of 0.3 nm is measurable. Thus, for a sample 3 mm thick, a thermal expansion coefficient of 10^{-7} is measurable with a temperature change of 1°C. The thermal variation of n may be determined to the same order of magnitude. With a stable system, thicker sample and larger temperature change, the measuring sensitivity could be increased about two orders of magnitude, but then the effects of air refraction and possible system deformation should be taken into account.

E Response Linearization

One drawback of double-beam interferometers is their sinusoidal response. In classical interferometry this fact limited measuring capability to about $1/10\lambda$. Although this limitation does not exist with laser interferometers [16], still the nonlinearities may be troublesome. The solution of this problem involves the electronic part of the system in Fig. 5; the signal from the detector is amplified, integrated, and fed into piezoelectric (PZT) transducers T that impart a relative translation to the two reflecting mirrors. The result is a cancellation of the phase shift induced by the sample. Thus, the fringe pattern is locked into position, and the measure of the phase shift is the voltage on the transducer, the quality of which determines the linearity.

An additional benefit of the feedback arrangement is its directional dependence that facilitates phase measurements accompanied by their algebraic sign.

V A SELF-CALIBRATING METHOD

As mentioned in the introduction, most interferometric methods require a reference for comparison: The Fizeau method uses a reference flat and so does the FECO method unless the investigated film itself can be reflectively coated on its two sides. All interferometers based on the Michelson interferometer, including the Kösters prism, configurations utilize a reference surface or a comparison between two parts of the sample itself. Even shearing interferometers are based on a comparison between pairs of points on the surface of the sample.

The need for a reference constitutes a drawback of the interferometric methods as compared with other techniques such as ellipsometry or photometry. In this section we describe an interferometric method that performs the measurement at a single point and, being self-calibrating, does not require any reference. The penalties paid for these advantages are some inconvenient alignment problems and the fact that the method is applicable only for transparent samples.

The basic principle, theory, and experimental verification have already been discussed in the literature [21, 22, 6]. Nevertheless, considering the new concepts involved, the method will be treated here in some length, though in a simplified manner.

In Fig. 1, the sample S may be any stratified layer situated between two media of respective indices of refraction n_1 and n_3. In the measuring system, the sample serves as a beam splitter in an interferometric arrangement in which two coherent beams of complex amplitudes E_1 and E_3 are incident on its two sides. The alignment is such that the reflected fraction of each beam is superposed on the transmitted fraction of the other beam. Assuming a linear medium, one will observe two emerging beams with respective intensities given by

$$I_1 = a_1|E_{11} + E_{31}|^2, \tag{16}$$

and

$$I_3 = a_3|E_{33} + E_{13}|^2, \tag{17}$$

where a_1 and a_3 are constants depending on the environment and angle of incidence. The components of each beam are given by Eqs. (3) and (4). Naturally, if n_3 is a substrate, measurements are performed outside, and a corresponding modification has to be made in the relevant expressions. Furthermore, reflections from the back surface of the substrate should be removed by spatial filtering or index matching.

The definition of a complex ratio between the two incident complex amplitudes

$$E_3 = \eta E_1 \exp(j\phi), \tag{18}$$

leads to

$$I_1 = a_1|E_1|^2[|r_1|^2 + \eta^2|t_3|^2 + 2|r_1||t_3|\eta \cos(\theta - \rho_1 + \phi)], \tag{19a}$$

$$I_3 = a_3|E_1|^2[\eta^2|r_3|^2 + |t_1|^2 + 2|r_2||t_1|\eta \cos(\rho_3 - \theta + \phi)]. \tag{19b}$$

If one records the two emerging intensities and plots them one against the other on an xy recorder in such a way that

$$x \propto I_1; \qquad y \propto I_3, \tag{20}$$

one may write,

$$x = x_0 + A_x \cos(\phi - \rho_1 + \theta), \tag{21a}$$

$$y = y_0 + A_y \cos(\phi + \rho_2 - \theta), \tag{21b}$$

where

$$x_0 = F_x(|r_1|^2 + \eta^2|t_3|^2), \tag{22a}$$

$$y_0 = F_y(\eta^2|r_3|^2 + |t_1|^2), \tag{22b}$$

$$A_x = 2F_x|r_1||t_3|\eta, \tag{22c}$$

and

$$A_y = 2F_y|r_3||t_1|\eta, \tag{22d}$$

where F_x and F_y are some scaling factors. Equations (21) may be considered as a parametric representation of an ellipse, with ϕ as parameter. Thus, if y is plotted against x as ϕ is continuously varied, an ellipse is produced (Fig. 8). An ellipse is determined by three independent parameters; therefore, in principle, measurements made on the ellipse should be adequate for the determination of three independent sample parameters. The scaling factors F_x and F_y depend on the calibration of the electronic measuring instruments, the quality of the optical components, and the back surface of the sample

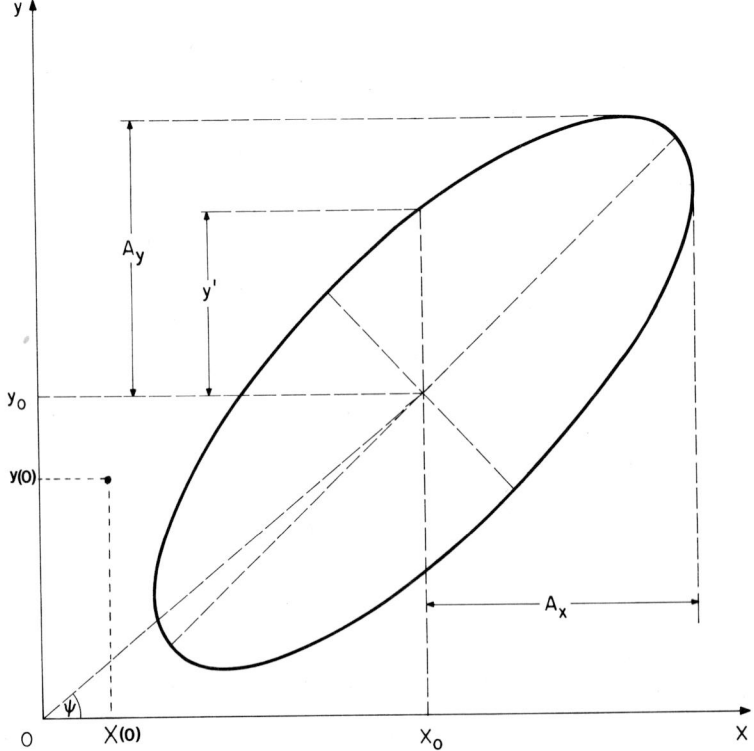

Fig. 8. Definition of parameters on interferometer xy output recording.

substrate. One of the advantages of this method is that it is possible to define measurable parameters that are independent of these uncertain factors. One such quantity is the phase difference between x and y that can be determined by putting $x = x_0$, which implies

$$\phi - \rho_1 + \theta = \pi/2.$$

Therefore, at this point,

$$y - y_0 \equiv y' = A_y \sin \Delta, \tag{23}$$

where y' is defined in the figure and

$$\Delta = \rho_1 + \rho_3 - 2\theta \tag{24}$$

is a dimensionless quantity that may be determined from the measured ratio between y' and A_y. Thus, Δ is our first observable that has already been noted as an indicator of absorption [2]. The angle ψ, which is a function of the intensity ratio of the incident beams η, defines the direction toward the center of the ellipse. One may define a function

$$f(\eta) \equiv \tan \psi = y_0/x_0$$

or, according to Eqs. (22),

$$f(\eta) = \frac{F_y}{F_x} \frac{\eta^2 |r_3|^2 + |t_1|^2}{|r_1|^2 + \eta^2 |t_3|^2}, \tag{25}$$

which is also a dimensionless observable, but dependent on F_y/F_x. Another dimensionless quantity also depending on F_y/F_x is the ratio between the respective variation amplitudes (or modulation depth) of the two intensities

$$A_y/A_x = \frac{F_y}{F_x} \frac{|r_3||t_1|}{|r_1||t_3|}. \tag{26}$$

The uncertain scaling factor F_y/F_x may be cancelled between Eqs. (25) and (26), but this would reduce the number of independent measurables. Instead, an additional measurement is possible by recording one point with one of the incident beams blocked. Blocking the beam E_3 is equivalent to a substitution $\eta = 0$. The point recorded has the coordinates

$$x(0) = F_x |r_1|^2; \qquad y(0) = F_y |t_1|^2, \tag{27}$$

from which one has

$$f(0) = \frac{F_y}{F_x} \frac{|t_1|^2}{|r_1|^2}. \tag{28}$$

A new measurable is obtained by dividing Eq. (26) by (28)

$$A \equiv \frac{A_y/A_x}{f(0)} = \frac{|r_1||r_3|}{|t_1||t_3|}, \tag{29}$$

which can be evaluated from measurements performed solely on the recorded intensities.

Two additional measurables can be defined by dividing Eqs. (22a) and (22b) by the respective Eqs. (27). A rearrangement of terms yields the two new measurables

$$B \equiv [x_0/x(0) - 1]\eta^2 = |t_3|^2/|r_1|^2,$$
$$C \equiv [y_0/y(0) - 1]\eta^2 = |r_3|^2/|t_1|^2,$$

(30)

which are also independent of the factors F_x and F_y but depend on the intensity ratio of the two incident beams that have to be measured separately. The two quantities B and C are not independent since they are related through Eq. (29). Other observables that are derivable by blocking the other beam are also simply related to the previous ones, and they may be used for cross-checking.

The schematic diagram of Fig. 9 shows a possible setup for the application of this method; beam splitter BS splits a laser beam to produce the two coherent measuring beams. The transmitted fraction is directed into the sample via mirrors M_1 and M_2 as beam E_1. The reflected fraction is incident on mirror M_3 mounted on a piezoelectric transducer T and returned through the beam splitter as E_3. The transducer is used to vary ϕ during measurement

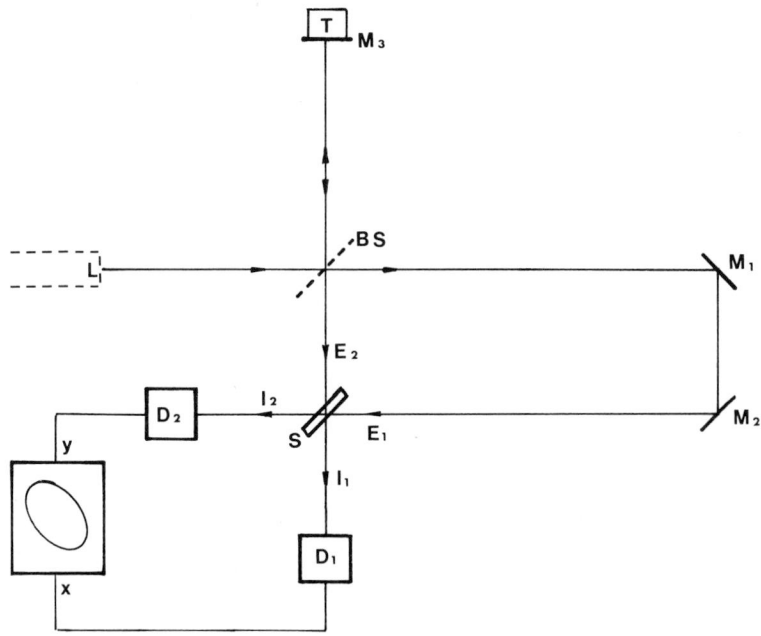

Fig. 9. Possible configuration for self-calibrating interferometer. L, laser; BS, beam splitter; S, sample; D, detectors; M, mirrors; T, piezoelectric transducer; xy recorder.

while the two emerging intensities are recorded on the xy recorder. An ellipse is recorded when ϕ is varied through 2π and the point $[x(0); y(0)]$ is obtained by blocking E_3. Equations (24), (29), and (30) are now utilized in a computer program to find the film parameters h, n, and k that correspond to the measured quantities Δ, A, B, and C. The four measured quantities are required to determine three parameters since the scaling factor F_y/F_x is also unknown.

This scheme of measuring procedure is one in a number of possibilities, leading to accuracies indicated by the results cited in Section VI. More sophisticated electronics and high-quality optics provide at least one order of magnitude improvement. Such an improvement is very helpful for extremely thin layers where ambiguities may arise [23, 24].

The phase shift Δ [Eq. (24)] is very sensitive to the extinction coefficient k, and its measurement is the most accurate. Therefore, this method is most useful for the direct determination of k in very weakly absorbing films. An important advantage is also the fact that the method, unlike many other techniques, can be applied for measurements in a wide spectral range [25] and also for anisotropic samples [23].

SURFACE EFFECTS   VI

For high-accuracy measurements surface roughness and material adsorption may be quite troublesome. The effects of surface roughness have been extensively discussed in Bennett [26] and Elson et al. [27], but not much has been done with regard to optical effects of adsorption.

A comprehensive treatment of these problems is outside the scope of this chapter, and here we shall briefly discuss only their influence on the measuring methods described.

Surface effects are important for very thin films that are, in most cases, coherent with the substrate (i.e., the film surface is nearly identical in form with the substrate surface). For these cases, the effects of roughness may be cancelled by measurements of the type described in Subsections IV.A and IV.B if the size of the measuring spot is large enough to perform an averaging process. The form of fringes in the Fizeau and FECO methods may be used as a direct measure for the roughness; however, irregular fringes greatly reduce measuring accuracy.

Most sensitive to surface effects are single-point measurements demonstrated by some results obtained by using the method of Section V; measurements at 6328 Å performed on bare fused quartz flats [6, 22] yielded equivalent surface layers with $h = 10$–20 nm, $n \simeq 1.3$, and $k \simeq 10^{-4}$. This result corresponds to expectations and is due to a combined effect of surface roughness and adsorbed water and hydrocarbons. A more striking effect has been observed in measurements performed with a CO_2 laser at 10.6 μm by using the same method [25]: (1) The surface layer on an NaCl window (that was completely transparent for visible light) was found to have the parameters $h = 1.7 \ \mu m$, $n = 1.8$, and $k = 0.03$. Integral absorption calculated by using these values was in full agreement with transmission measurements at the same wavelength. The measured effect should be attributed to water absorbed in the outer salt layer. (2) An amorphous layer of As_2Se_3 deposited on an NaCl substrate was also investigated, and the derived parameters were $h = 3.67 \pm 0.04 \ \mu m$, $n = 2.80 \pm 0.03$, and $k = 0.02$. Here, too, the results were in correspondence with transmission measurements.

The large influence of surface effects was conclusively demonstrated by independent measurements performed on the As_2Se_3 film by using a method that is very insensitive to surface effects when they are restricted to a layer that is much less than the wavelength and the total film thickness. This method is based on the wave-guiding properties of thin dielectric films [28]: The electromagnetic wave is coupled into the thin film by using a prism or grating. The coupling angle and guiding properties are strong functions of film thickness, absorption, and refractive index; thus, they can be used for accurate determination of these parameters. Measurements performed on this layer yielded results that are identical with the interferometric measurements for film thickness and refractive index. However, the guiding properties revealed an absorption that should be at least one order of magnitude less than the mentioned result. The definite conclusion is that the waveguiding was not affected by a thin absorbing layer that is responsible for absorption of light transversally incident on the sample.

VII CONCLUSIONS

The interferometric methods described in this chapter are, almost exclusively, adapted to the optical analysis of thin films. General properties of bulk material have not been considered since conventional, noninterferometric techniques are usually better suited for these purposes. The advantages of interferometry manifest themselves when the quantities to be measured are small. The best method for each application should be selected by considering a number of factors. First of all, one should specify the parameters to be measured and the parameters already known. Among the other factors, one should mention sample preparation, required accuracy, environment, available equipment, and personnel. For example, if only thickness is the required parameter, the sample may be prepared in any form and there are no interferometric experts around, one should choose the Fizeau method. On the other hand, if for the same purpose the sample is given with a film edge, the Kösters-prism interferometer may be the best choice. Actually, this would probably be the best choice also for thickness and index determination, especially if the substrate and film are of relatively low optical quality. If all three optical parameters are required for a weakly absorbing film or just the absorption of that film, one should choose the method of Section V, provided that the film surface can be properly cleaned and at least one good technician is available.

As a final remark, one should remember that interferometric parameters are usually multivalued and, in many cases, measured parameters will be in correspondence with a number of possible parameter sets. This is especially so if all three film parameters are not known in advance. Therefore, it is very useful to obtain additional information on the sample, even if it is of relatively low accuracy.

REFERENCES

1. S. Tolansky, "Multibeam Interferometry of Surfaces and Films" Clarendon Press, Oxford, 1948.
2. F. Abélès, *in* "Progress in Optics" (E. Wolf, ed.), Vol. 2, p. 251, North-Holland Publ. Amsterdam, 1963.
3. O. S. Heavens, *in* "Physics of Thin Films" (G. Hass and R. E. Thun, eds.), Vol. 2, p. 193, Academic Press, New York, 1964.
4. H. E. Bennett and J. M. Bennett, *in* "Physics of Thin Films" (G. Hass and R. E. Thun, eds.), p. 1, Academic Press, New York, 1967.
5. M. Francon and S. Mallick, "Polarization Interferometers," Wiley, New York, 1971.
6. Y. Demner and J. Shamir, *Appl. Opt.* **17**, 3738 (1978).
7. M. Born and E. Wolf, "Principles of Optics," Pergamon, New York, 1959.
8. Y. Fainman, E. Lenz, and J. Shamir, *Appl. Opt.* **20**, 158 (1981).
9. H. D. Polster, *Appl. Opt.* **8**, 522 (1969).
10. P. Langenbeck, *in* "Optical Instrument and Techniques" (J. H. Dickson, ed.), p. 276, Oriel Press, England, 1970.
11. J. N. Israelachvili, *J. Colloid Interface Sci.* **44**, 259 (1973).
12. W. A. Pliskin and E. E. Conrad, *IBM J. Res. Devel.* **8**, 43 (1964).
13. M. G. Nomarski, *J. Phys. Rad.* **16**, 9 (S) (1955).
14. W. Kösters, "Interferenzdopplprisma für Messwecke," German patent 595211, 1934.
15. J. B. Saunders, *J. Res. Nat. Bur. Stand. Sec. C*, **73**, 1 (1969); J. B. Saunders, *J. Opt. Soc. Am.* **62**, 6 (1972)
16. J. Shamir, R. Fox, and S. G. Lipson, *Appl. Opt.* **8**, 103 (1969).
17. Th. Kwaaitaal, "A New Interferometric Method for Measuring Small Magnetostrictive Effects," private communication, 1976.
18. J. Shamir, *J. Phys. E* **9**, 499 (1976).
19. H. Rosen and J. Shamir, *J. Phys. E* **11**, 905 (1978).
20. H. F. Hazebroek and A. A. Holscher, *J. Phys. E* **6**, 822 (1973).
21. J. Shamir and P. Graff, *Appl. Opt.* **14**, 3053 (1975).
22. J. Shamir, *Appl. Opt.* **15**, 120 (1976).
23. R. A. Soli and H. H. Soonpaa, *Appl. Opt.* **18**, 3367 (1979).
24. J. Shamir and Y. Demner, *Appl. Opt.* **19**, 2658 (1980).
25. D. Apter and J. Shamir, *Appl. Opt.* **21**, 1512 (1982).
26. J. M. Bennett, *Appl. Opt.* **15**, 2705 (1976).
27. J. M. Elson, H. E. Bennett, and J. M. Bennett, *in* "Applied Optics and Optical Engineering" (R. R. Shannon and J. C. Wyant, eds.), Vol. 7, p. 191, Academic Press, New York, 1979.
28. R. Th. Kersten, *Opt. Acta* **22**, 503 (1975); R. Th. Kersten *Opt. Comm.* **13**, 327 (1975).

Chapter **7**

Thin-Film Absorptance Measurements Using Laser Calorimetry

P. A. TEMPLE

Michelson Laboratory, Physics Division
Naval Weapons Center
China Lake, California

I.	Introduction	135
II.	Single-Layer Films	138
III.	Wedged-Film Laser Calorimetry	139
IV.	Electric-Field Considerations in Laser Calorimetry	143
V.	Entrance versus Exit Surface Films	147
VI.	Experimental Determination of α_f, a_{af}, and a_{fs}	149
	References	153

INTRODUCTION I

Laser calorimetry has been used for several years as a method for measuring the absorptive loss in materials in both bulk and thin-film form. Two review articles on bulk absorption coefficient measurements in low-absorption, transparent materials were written by Skolnik [1] and Hordvik [2]. Skolnik includes a table showing bulk absorption coefficients as low as 10^{-6} cm^{-1}. The technique is not limited to low-absorption materials or to those wavelengths at which intense laser lines are available. In the work by Bos and Lynch [3], calorimetry was used to measure the absorption of chromium and two chromium alloys over the spectral range from 0.24 to 14.0 μm at liquid-helium temperatures.

Calorimetry has the advantage of simplicity and high sensitivity. The fundamental quantity measured is the absorptance A, where

$$A = \text{absorbed energy/incident energy.} \tag{1}$$

135

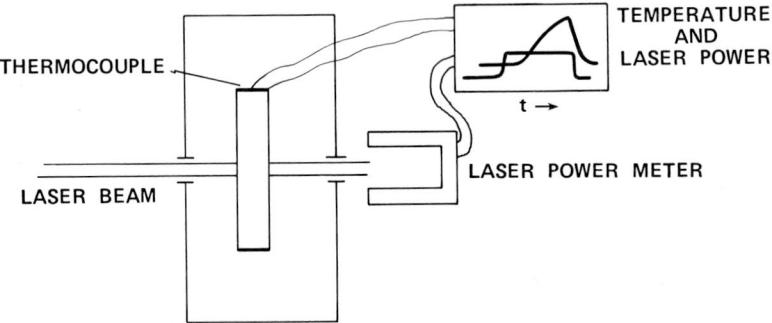

Fig. 1. Typical laser rate calorimeter.

With optical materials, laser calorimetry is generally used to measure samples with absorptances of 1×10^{-2} and less. Figure 1 is a schematic of a typical laser calorimeter. In this apparatus, the sample is heated by a laser beam of constant intensity for a period of time sufficient to raise the temperature of the sample above ambient. The laser is then turned off, and the sample slowly returns to thermal equilibrium with its surroundings. Both the laser power and sample temperature are recorded on a strip-chart recorder. Several analysis techniques have been described in the literature. Figure 2 shows schematically five sample temperature-versus-time curves. In technique (a) Pinnow and Rich [4] placed the sample in a laser cavity to take advantage of high intercavity power, allowed it to come to thermal equilibrium, and abruptly turned the laser off and observed the cooling curve. In technique (b) Allen *et al.* [5] irradiated the sample for 4 min and, using computer modeling, determined the absorptance by measuring the temperature

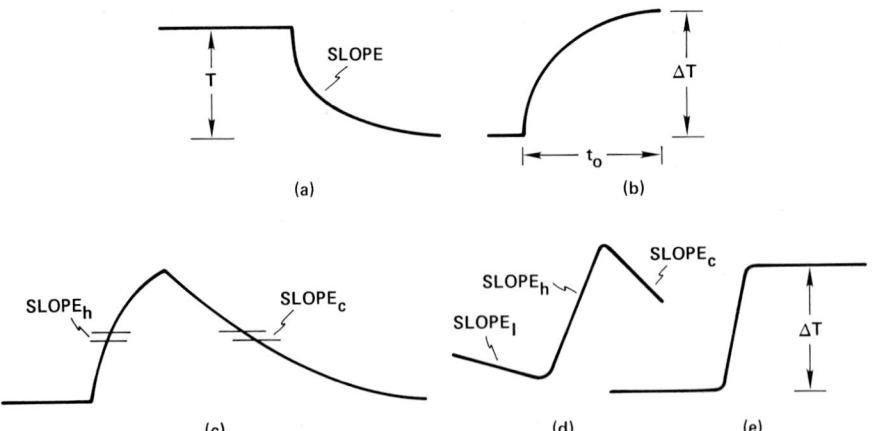

Fig. 2. Sample temperature as a function of time for five temperature-versus-time analysis schemes.

rise at the 4-min point. Technique (c), called the rate method, has been used by various workers and is probably the most common analysis technique [1, 2]. The three-slope method, shown in (d), is not unlike the rate technique shown in (c), but it has the advantage of not requiring the sample to be initially in thermal equilibrium with its surroundings [6]. Finally, technique (e), described by Decker and Temple [7], is a truly adiabatic technique in which the sample enclosure is temperature controlled and is caused to follow the temperature of the sample as it is heated by the laser. This technique overcomes some of the temperature-gradient problems of the other methods and is easily calibrated electrically.

There are a number of instrumental considerations involved in accurately determining the absorptance [8]. Probably the most serious of these is that of temperature nonuniformity in the sample during irradiation. In the usual analysis, it is assumed that the rate of temperature increase is uniform throughout the sample. As discussed by Bernal G. [9], this is correct only if the sample's thermal diffusivity is large and the heat losses are small. Only techniques (b), in which a theoretical correction is made, and (e), in which the sample is allowed to come to thermal equilibrium with itself, are free of this problem.

A second experimental consideration that is often neglected for both bare and coated samples is that of substrate interference effects. In any calorimetric technique using transparent substrates, the sample will have a rear surface from which the beam must emerge and from which the beam will be partially reflected. Because of the long coherence length of laser radiation, the substrate may act as an added thin film of some unknown thickness. This will cause a substantial uncertainty in the transmittance of the substrate–film system and in the electric-field distribution within both the substrate and any films present. In order to avoid this undesirable situation, we are forced to defeat laser coherence effects in the substrate. Figure 3 shows three ways of

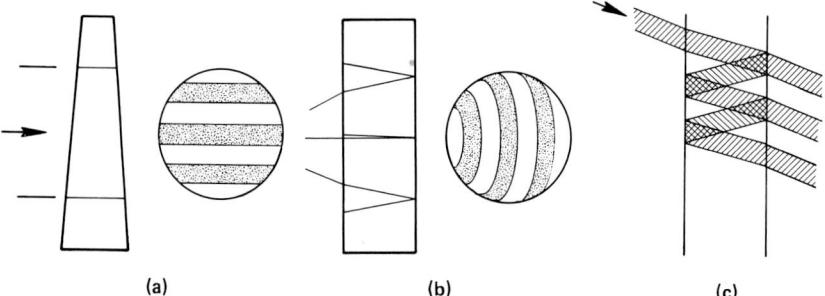

(a)   (b)   (c)

Fig. 3. Three techniques for defeating laser coherence effects in a substrate. (a) Wedged substrate and a large parallel beam, (b) plane-parallel substrate and divergent beam, tilted sample shown, and (c) plane-parallel sample using small beam at nonnormal angle of incidence to avoid overlap.

accomplishing this. All three methods result in the same average electric-field intensity distribution in the various regions of the sample. In the first case, the substrate is slightly wedged and the beam is quite large. The wedge causes the substrate to go through several cycles in optical thickness. The second case uses a plane-parallel substrate but a large diverging beam. The optical path length is different for various angles of incidence within the beam, causing a mixing similar to the preceding case. In this case, tilting the sample slightly enhances the mixing. In the third case, a small-diameter beam is incident on a plane-parallel sample at slightly off-normal incidence. Nonoverlap of beams on the entrance (or exit) surface assures us that no interference will exist within the substrate. This is essentially the classical incoherent-beam, multiple-bounce picture, and this model is used to calculate the effect of a real substrate in data given later in this chapter.

When these experimental considerations are taken into account, any of the calorimetric analysis techniques described can be used to measure the absorptance of either uncoated or coated substrates. In this chapter, we are concerned with measuring the absorptive properties of a single-layer, thin film deposited on a highly transparent substrate.

II   SINGLE-LAYER FILMS

The most common way of determining the absorption properties of a material in thin-film form is to first measure the absorptance A_1 of a suitable substrate. A film of known thickness l, usually a half-wave in optical thickness (OT), is then deposited on the substrate, and the absorptance A_2 is again measured. The film absorption coefficient β, where

$$\beta = (A_2 - A_1)/l, \tag{2}$$

is then calculated. The variation in the electric field both throughout the film and from material to material is not accounted for in Eq. (2). Because of this, one typically finds that β measured when the sample is illuminated from the filmed side will be different from that measured when illuminated from the uncoated side. It is generally recognized that the increased absorption of the filmed substrate is due not only to the bulk of the film, but also to the newly introduced film–substrate interface and the film–air interface, where there is a change in material type with possible stoichiometry problems, as well as residual cleaning compound, etc., that may lead to higher absorption. In addition, the work of MacLeod [10] suggests that water migrates into the film system through pinholes and spreads out laterally into the interface regions. Donovan [11] has shown, by a nuclear resonance technique, that water is present in the film–substrate interface for discontinuous silicon films but is absent when the film is continuous. His work suggests that the interface layer in this system is at most a few hundred angstroms thick. In spite of the presence of interface absorption, nearly all thin-film, absorption-coefficient data reported in the literature have been calculated by using Eq. (2) since it is a simple and convenient scheme for quickly characterizing coating materials.

A more complete analysis is possible by using a larger set of samples or, more elegantly, a single sample with a wedge-shaped film deposited on the incident face [12]. In this technique a simple model of a film is assumed, where the bulk absorption coefficient of the film is constant throughout the thickness of the film and all the interface absorption occurs in infinitesimally thin sheets at the two film boundaries. Then, a minimum of two measurements, using $\lambda/4$ and $3\lambda/4$ OT (or $\lambda/2$ and 1λ OT) films, will give the bulk absorption coefficient of the film without interface absorption present. Two more measurements, one on the bare substrate and one on a $\lambda/2$ (or $3\lambda/4$ OT) coating, can be used to determine the contribution from the film–substrate interface and the air–film interface. This technique will now be described in detail.

WEDGED-FILM LASER CALORIMETRY   III

Figure 4 shows a sample with a beam normally incident from the left. The time average of the square of the electric field of the beam far from the sample is $\langle E_0^2 \rangle$. Reflection of the beam from the various film and substrate interfaces causes a redistribution of electric field within the sample. The time average of the square of this field is $\langle E^2(z) \rangle$, a quantity that varies sinusoidally in the vicinity of sample interfaces. For a sample of total thickness L including any films, Lambert's law can be written to give the absorptance as

$$A = \int_0^L \frac{n(z)\langle E^2(z)\rangle}{n_0 \langle E_0^2 \rangle} \alpha(z)\, dz, \tag{3a}$$

$$A = \int_0^L p(z)\alpha(z)\, dz, \tag{3b}$$

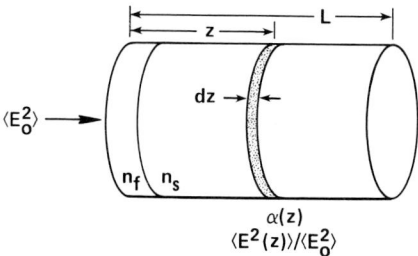

Fig. 4. Configuration for calculating the absorptance of a sample as per Eq. (3).

where $\alpha(z)$ and $n(z)$ are the absorption coefficient and the real part of the index of refraction, respectively, of the sample at the position z and n_0 is the index of the material outside the sample, which is normally air. Here $\alpha(z)$ is related to the extinction coefficient by $\alpha(z) = 4\pi k(z)/\lambda$.

The relative energy density, which contains the incident beam normalization factor, is $p(z)$. The purpose of wedged-film laser calorimetry is to determine $\alpha(z)$ for the film and the film interfaces by using Eq. (3b). This equation relates the measured absorptance A and the calculable quantity $p(z)$ to the material property $\alpha(z)$. Equation (3a) makes it clear that the contribution that any region in the sample makes to sample heating is proportional to $\alpha(z)$ in that region times the electric-field strength in that same region.

In order to make use of Eq. (3b), one assumes that the bulk absorption coefficient of the film α_f is constant throughout the film and is independent of film thickness. We also assume that all of the air–film interface absorption takes place in a thin sheet of material of absorption coefficient α_{af} and thickness Δ_{af}, and the film–substrate interface absorption takes place in a thin sheet of thickness Δ_{fs}, with absorption coefficient α_{fs}. Then the air–film and film–substrate interface absorption are characterized by the specific absorptances $a_{af} = \alpha_{af}\Delta_{af}$ and $a_{fs} = \alpha_{fs}\Delta_{fs}$. Finally, in calculating $p(z)$, which will be described in Section IV, we assume no absorptive attenuation of the illuminating beam in passing through the sample. In samples where $A < 10^{-2}$, this approximation is certainly acceptable. Without this simplification, knowledge of $\alpha(z)$ is necessary to calculate $p(z)$, and the entire problem becomes more difficult.

Under these approximations, the absorptance of a single-layer film on a transparent substrate can be written as[†]

$$A = p_{af}a_{af} + \bar{p}_f\alpha_f l_f + p_{fs}a_{fs} + \bar{p}_s A_0, \tag{4}$$

where $\bar{p}_f$ and $\bar{p}_s$ are the spatially averaged relative energy densities within the film and substrate and A_0 is the absorptance measured on the uncoated substrate. Here p_{af} and p_{fs} are the relative energy density at the air–film and film–substrate interfaces, respectively. As the film thickness is varied, a_{af}, $\alpha_f a_{fs}$, and A_0 remain constant and A, l_f, p_{af}, $\bar{p}_f$, p_{fs}, and $\bar{p}_s$ change. Of these A, A_0, and l_f are the experimentally measured absorptances and physical film thickness. And p_{af}, $\bar{p}_f$, p_{fs}, and $\bar{p}_s$ are calculated quantities. Values of p_{af}, $\bar{p}_f$, p_{fs}, and $\bar{p}_s$ for $\lambda/4$ and $\lambda/2$ thickness films are given in graphical form at the end of this chapter. By measuring the absorptances of an appropriate set of film thicknesses, we obtain a set of linearly independent equations that can be solved for α_f, a_{af}, and a_{fs}, the three coefficients that characterize a single film on a transparent substrate.

In order to use Eq. (4) to determine the bulk absorption of the film α_f, one measures the absorptance of a pair of film thicknesses that have the same

---

[†] The last term in this expression has a further approximation, which is discussed in Section V.

relative power densities at the interfaces, in the bulk of the film, and in the substrate. Such a pair consists of any two films that differ by $\lambda/2$ in optical thickness. Differentiating Eq. (4) and using absorptances from this pair of film thicknesses, we obtain

$$\alpha_f = (A_{t+\lambda/2} - A_t)/\bar{p}_f l_{\lambda/2}, \tag{5}$$

where $\bar{p}_f$ is calculated for film thickness t and $l_{\lambda/2}$, the physical thickness of a film of $\lambda/2$ OT. Like Eq. (2), two measurements are required; but in contrast to Eq. (2), this expression accounts for interface absorption and electric-field strength in the film and gives the bulk absorption coefficient of the thin-film coating.

A convenient sample configuration for this measurement is a sample one-half of which has been masked for part of the deposition. In this way, one side of a diameter has a coating of OT t while the other side has a coating of OT $t + \lambda/2$. While any thickness t will suffice, values of $\bar{p}_f$ for the $\lambda/2$, λ pair (or the $\lambda/4$, $3\lambda/4$ pair) are given here. When measuring the absorptances of these two coatings, the beam is aligned slightly off-center so that it passes through one or the other of the coating halves. The use of one sample eliminates substrate-to-substrate and deposition-to-deposition variations that are present when two substrates and two depositions are needed to obtain the t and $t + \lambda/2$ samples.

With additional measurements on the bare substrate and on a $\lambda/4$ OT film, we can determine a_{af} and a_{fs}. The relative power densities at both interfaces, within the film, and within the substrate all change in going from a $\lambda/4$ OT film to a $\lambda/2$ OT film. This fact is utilized in the determination of a_{af} and a_{fs}. By rewriting Eq. (4) for these two cases (unprimed for $\lambda/2$ OT and primed for $\lambda/4$ OT) with all measured and/or calculated quantities on the right, including α_f, which is determined by using Eq. (5), we have

$$a_{af}p_{af} + a_{fs}p_{fs} = [A - \bar{p}_f\alpha_f l_f - \bar{p}_s A_0] = [F], \tag{6a}$$

$$a_{af}p'_{af} + a_{fs}p'_{fs} = [A' - \bar{p}'_f\alpha_f l'_f - \bar{p}'_s A_0] = [F'], \tag{6b}$$

which have the solutions

$$a_{af} = \frac{[F]p'_{fs} - [F']p_{fs}}{p_{af}p'_{fs} - p'_{af}p_{fs}}, \tag{7a}$$

$$a_{fs} = \frac{[F']p_{af} - [F]p'_{af}}{p_{af}p'_{fs} - p'_{af}p_{fs}}. \tag{7b}$$

Figure 5 shows a convenient sample configuration for making this type of measurement. The maximum film thickness is $\sim 1\lambda$ OT, allowing measurement at the four regions required for obtaining A_0, $A_{1/4}$, $A_{1/2}$, and A_1. A small-diameter laser beam is used with the sample tilted slightly. The required thickness is found by moving the sample until the transmittance minimum

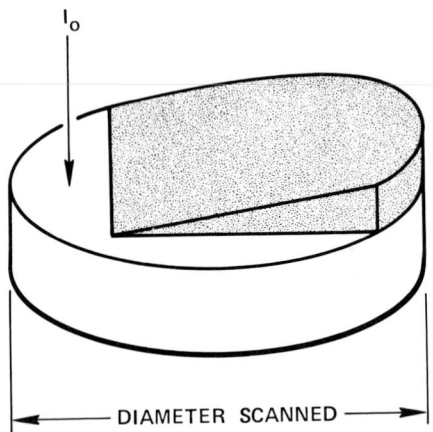

I_0

DIAMETER SCANNED

Fig. 5. Wedged-film sample used to obtain film bulk and interface absorption for single-layer films. The film varies from zero to ~1λ OT. (From Temple [12].)

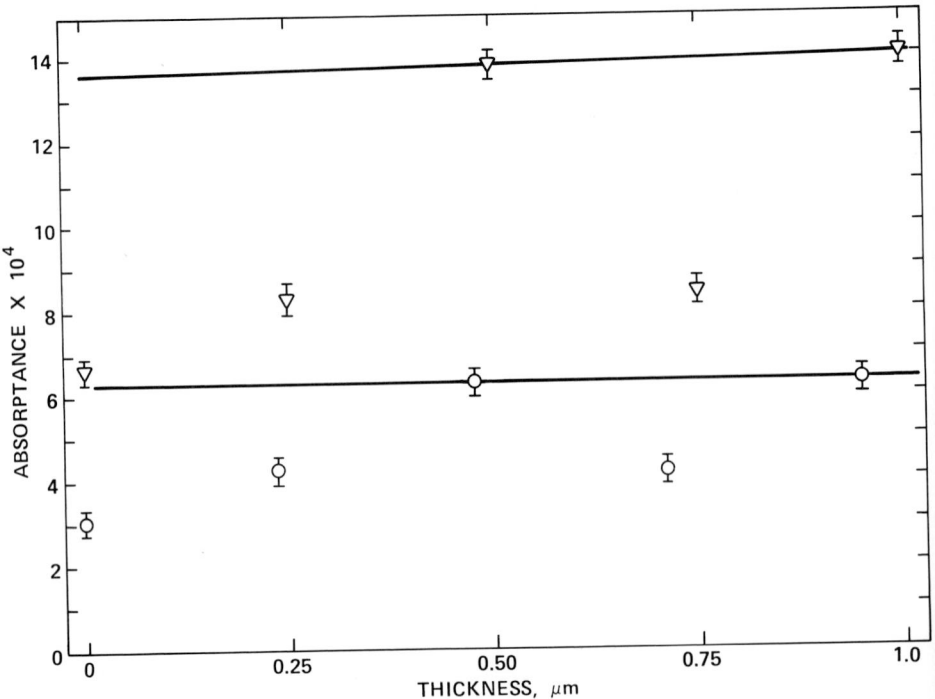

Fig. 6. The absorptance of an entrance surface film on As_2Se_3 on a CaF_2 substrate measured at two wavelengths and at λ/4, λ/2, 3λ/4, and 1λ OT. The two data points at 0.0-μm thickness are bare substrate absorptances A_0. The two lines have been drawn through λ/2 and 1λ OT data points. The λ/4 and 3λ/4 data points have lower absorptances because of reduced p_{fs} for these thicknesses. (∇, 2.87 μm; ○, 2.72 μm; for the upper curve $\alpha = 0.4$ cm^{-1}, $a_{af} \cong 0.0$, and $a_{fs} = 4.2 \times 10^{-4}$; for the lower curve, $\alpha = 0.0$ cm^{-1}, $a_{af} \cong 0.0$, and $a_{fs} = 2.3 \times 10^{-4}$.) (From Temple [12].)

or maximum is found. This type of sample is particularly useful when absorptances are to be measured at a variety of wavelengths, since precise $\lambda/4$, $\lambda/2$, and 1λ OTs can be found for any wavelength. As in the case in which two film thicknesses are present on the same substrate, this type of sample eliminates substrate-to-substrate and deposition-to-deposition variation present when three separate depositions are made to obtain the $\lambda/4$, $\lambda/2$, and λ OT films.

Figure 6, which is an example of this type of measurement [12, 13], shows the absorptance measured on the bare CaF_2 substrate (shown at 0.0 film thickness) and at $\lambda/4$, $\lambda/2$, $3\lambda/4$, and λ OT positions of an As_2Se_3 wedged film. Measurements were made at 2.72- and 2.87-μm wavelength by using a line-tuned HF cw gas laser. The substrate index is 1.42 and the film index is 2.6, which causes a $\lambda/2$ OT film to be ~ 0.5 μm thick. The values of α_f were calculated from the $\lambda/2$ and λ OT data (solid lines through data). The presence of water in the filmed substrate is evidenced by the higher absorptance as measured at 2.87 rather than at 2.72 μm. However, the bulk absorption coefficient of the film is quite low. The analysis of interface absorption indicates that nearly all the absorption is taking place at the film–substrate interface. Analysis of the bare substrate and $\lambda/2$ OT data and Eq. (2) gives a β of 14.60 cm^{-1}, which is ~ 35 times greater than α_f calculated by the wedged-film technique.

ELECTRIC-FIELD CONSIDERATIONS IN LASER CALORIMETRY   IV

The technique and considerations in calculating $p(z)$ for use in Eqs. (7a) and (7b) will now be described. The accurate determination of $p(z)$ must take into consideration (1) the coherent interaction of a single beam with a dielectric interface and (2) the effect of the multiple (but incoherent) bounces the beam undergoes between the two faces of the sample.

A convenient method of making the first calculation was first given by Leurgans [14] and was more recently detailed by DeBell [15]. As originally described, the technique uses the Smith chart, but it is easily adapted to a computer or even a hand calculator. Leurgans's method, which we will briefly describe, is restricted to normal-incidence plane waves and lossless films. While only single films are of concern here, the general multilayer description will be given.

Figure 7 shows the layer-labeling system. The substrate is of index n_0, and the incident medium is of index n_k. The layers are labeled 1 through $k-1$ and are of physical thickness l_1 to l_{k-1}.

A right-traveling, normally incident plane wave of wavelength λ is represented by $\hat{E}^+ = E^+ e^{i(\omega t - \delta)}$, where $\delta = 2\pi n z/\lambda$. At a dielectric boundary, the

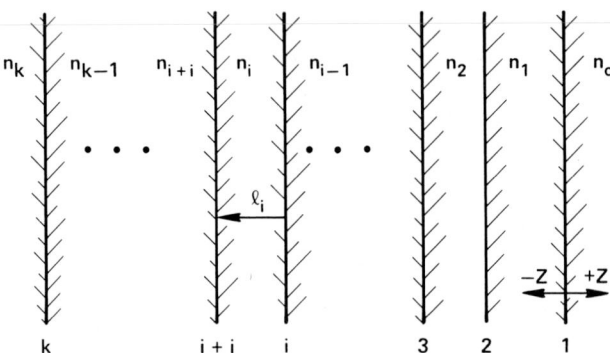

Fig. 7. The layer-labeling system for a thin-film stack.

partially reflected left-going field is given by

$$\hat{E}^- = E^+ \hat{\rho} e^{i(\omega t + \delta)},$$

where the reflection coefficient $\hat{\rho}$ is complex. Now, in general,

$$\hat{\rho}(z) = |\hat{\rho}(z)| e^{i(2\delta + \phi)}$$

within any given layer. In the lossless case, $|\hat{\rho}(z)|$ is constant within a layer and $\hat{\rho}$ only changes in phase through the layer. To calculate $\hat{\rho}$ across a boundary, we use the boundary condition requiring continuity of E_{tan} and H_{tan}. This is done by using the reduced admittance $\hat{y}(z)$:

$$\hat{y}(z) = \hat{H}(z)/n\hat{E}(z).$$

The boundary condition then requires that

$$\hat{y}_{i+1}(z_b)n_{i+1} = \hat{y}_i(z_b)n_i,$$

where z_b is the position of the boundary between the ith and $(i + 1)$th layers. Finally, $\hat{y}(z)$ and $\hat{\rho}(z)$ are related by

$$\hat{\rho}_i(z) = \frac{1 - \hat{y}_i(z)}{1 + \hat{y}_i(z)}; \qquad \hat{y}_i(z) = \frac{1 - \hat{\rho}_i(z)}{1 + \hat{\rho}_i(z)}.$$

These expressions, along with the condition that there is no beam incident on the exit surface from the right, are sufficient to uniquely determine $\hat{\rho}(z)$.

The calculation begins within the substrate and proceeds from right to left. The procedure is the following:

(1) Realize that $\hat{\rho}_0 = 0$ and $\hat{y}_0 = 1$;
(2) use $\hat{y}_1(0) = \hat{y}_0 n_0/n_1$ to cross the substrate–film (1) interface;
(3) use $\hat{\rho}_1(0) = [1 - \hat{y}_1(0)]/[1 + \hat{y}_1(0)]$ to find $\hat{\rho}_1(0)$;
(4) use $\hat{\rho}_1(z) = |\hat{\rho}_1(0)| e^{i2\delta}$ to find $\hat{\rho}_1(z)$ in the first film;
(5) use $\hat{y}_1(z_1) = [1 - \hat{\rho}_1(z_1)]/[1 + \hat{\rho}_1(z_1)]$ and
(6) $\hat{y}_2(z_1) = \hat{y}_1(z_1)n_1/n_2$ to find $\hat{y}_2(z_1)$;
(7) now proceed as at step (3) above to find $\hat{\rho}_2(z)$, and so on.

In the work by Leurgans [14] and DeBell [15], the calculation was used to determine the reflectance R of the stack, which is given by $R = \hat{\rho}_k \hat{\rho}_k^*$.

Having found $\hat{\rho}(z)$ throughout the system by the method of Leurgans [14], we will now calculate the electric-field strengths throughout the layered system. At any point z in the ith layer, the net field is

$$\hat{E}_i(z) = \hat{E}_i^+(z) + \hat{E}_i^-(z) = [1 + \hat{\rho}_i(z)]\hat{E}^+(z).$$

The time average of the square of the field is given by

$$\langle E^2(z)\rangle = \frac{1}{T}\int_0^T \frac{[\hat{E}_i(z) + \hat{E}_i^*(z)]^2}{2}\, dt = \frac{E_{imax}^{+2}}{2}[1 + \hat{\rho}_i(z)][1 + \hat{\rho}_i^*(z)],$$

or, finally

$$\langle E_i^2(z)\rangle = \langle E_i^{+2}\rangle[1 + \hat{\rho}_i(z)][1 + \hat{\rho}_i^*(z)]. \tag{8}$$

While $\langle E^2(z)\rangle$ in general varies throughout a layer, $\langle E_i^{+2}\rangle$ is constant in any given layer. Using the continuity of E_{tan} across a boundary, we derive the recursion relation

$$\langle E_i^{+2}\rangle = \langle E_{i+1}^{+2}\rangle\frac{[1 + \hat{\rho}_{i+1}(z_{i+1})][1 + \hat{\rho}_{i+1}^*(z_{i+1})]}{[1 + \hat{\rho}_i(z_{i+1})][1 + \hat{\rho}_i^*(z_{i+1})]}, \tag{9}$$

where the numerator is evaluated just to the left of the $(i + 1)$th boundary and the denominator just to the right of it. Knowing ρ and using the recursion relation, we can now determine the electric-field distribution throughout the system.

The starting point in calculating the set of $\langle E_i^{+2}\rangle$ is within the incident medium where the right-traveling field is $\langle E_k^{+2}\rangle$, a known quantity. Equation (9) is used to find $\langle E_i^{+2}\rangle$ for successive layers to the right of the incident medium. Equation (8) is used to determine $\langle E_i^2(z)\rangle$. Finally, the relative energy density is

$$p_i(z) = n_i\langle E_i^2(z)\rangle/n_k\langle E_k^{+2}\rangle.$$

This value does not take into account the multiple bounces that the beam undergoes between the two faces of the sample; this will be done next.

Figure 8 shows the right- and left-going beams within and to the right of the substrate. These are expressed in terms of the incident field $\langle E_{inc}^2\rangle$ and the first and second surface reflectance and transmittance of the coated sample R_1, R_2 and T_1, T_2 (calculable from Leurgans's method); n_s is the substrate index.

When using Eq. (3), we must include the contribution of the field of each of the individual beams in Fig. 8. Since there is no beam overlap and therefore no interference, this can be done by adding all of the right-going beams to form an effective right-going beam and all the left-going beams to form

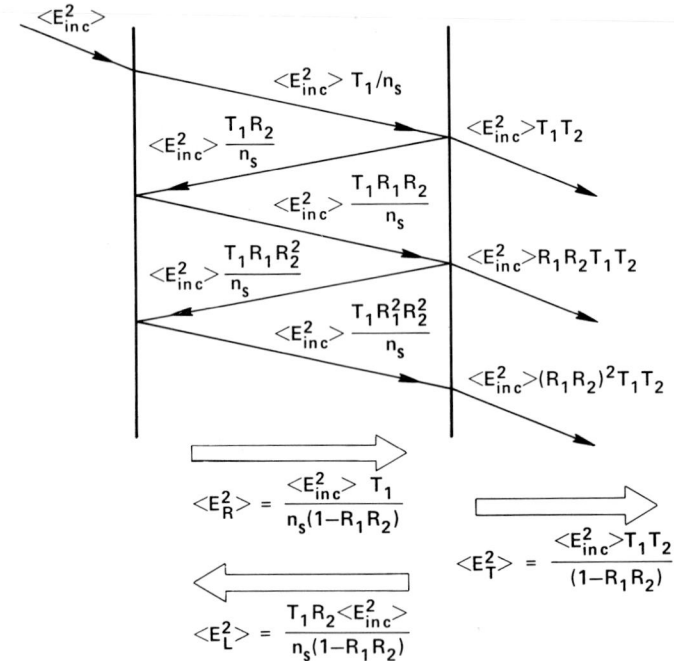

Fig. 8. The electric-field intensities within a substrate for multiple bounces of nonoverlapping beams. The entrance surface has reflectance R_1 and the exit surface R_2. Assumed are $T_1 = 1 - R_1$ and $T_2 = 1 - R_2$.

an effective left-going beam. These values are

$$\langle E_R^2 \rangle = \frac{\langle E_{\text{inc}}^2 \rangle T_1}{n_s(1 - R_1 R_2)}, \tag{10a}$$

$$\langle E_L^2 \rangle = \frac{\langle E_{\text{inc}}^2 \rangle T_1 R_2}{n_s(1 - R_1 R_2)}. \tag{10b}$$

To calculate $p(z)$ in the region near the exit surface of the sample, the right-going field $\langle E_R^2 \rangle$ is used as the starting value in the recursion relation Eq. (9). This and Eq. (8) are used to calculate $\langle E^2(z) \rangle$. The average value of $\langle E^2 \rangle$ within the substrate is the sum of Eqs. (10a) and (10b)

$$\langle E_{\text{subst}}^2 \rangle = \frac{\langle E_{\text{inc}}^2 \rangle T_1(1 + R_1)}{n_s(1 - R_1 R_2)}. \tag{11}$$

To calculate $p(z)$ in the region near the entrance surface, we add the field resulting from the incident beam $\langle E_{\text{inc}}^2 \rangle$ to the field resulting from the left-going beam $\langle E_L^2 \rangle$. This is done since, while the transmittance of a lossless filmed surface is the same when illuminated from either side, the electric-field distribution is not. This is illustrated in Fig. 9, in which the values of $\langle E(z)^2 \rangle / \langle E_{\text{inc}}^2 \rangle$ (solid lines) and $p(z)$ (dashed lines) are shown for a single

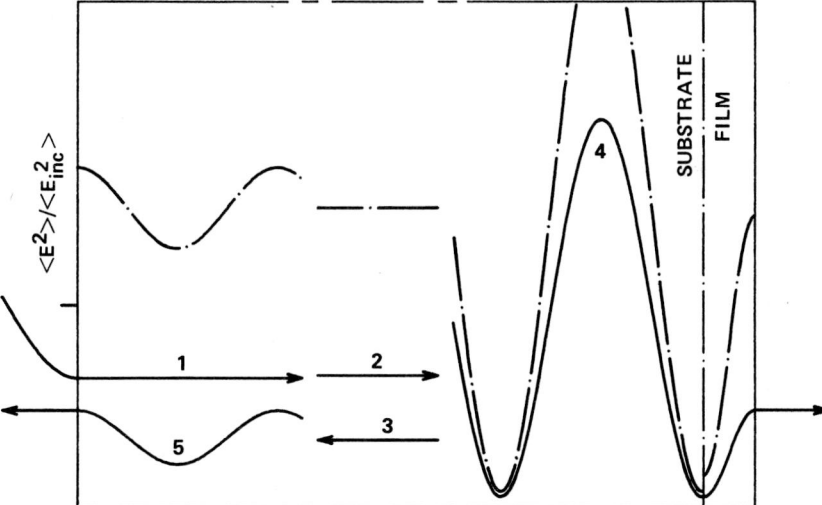

Fig. 9. The relative electric-field intensity (solid curve) and the relative energy density (dashed curve) for a $\lambda/4$ exit surface film. Line 1 is the incident beam, $\langle E_{inc}^2 \rangle T_1/n_s$, while lines 2 and 3 are $\langle E_R^2 \rangle$ and $\langle E_L^2 \rangle$. Line 4 is the coherent combination of 2 and 3 near the exit surface where the beams overlap; line 5 is the coherent combination of 2 and 3 *less* 1 near the entrance surface. The dashed lines are n_s/n_0 times 1 plus 5, 2 plus 3, and 4 at the entrance surface, in the central region, and near the exit surface, respectively. (Substrate index = 1.50; film index = 3.00; film thickness = 0.250; transmittance = 0.480.)

layer of material of index 3.0 and $\lambda/4$ OT deposited on the exit surface of a substrate of index 1.5. A film of this thickness and index has a high reflectance and accentuates the various effects just described. The line labeled 1 is the value of $\langle E^2(z) \rangle / \langle E_{inc}^2 \rangle$ resulting from the incident beam encountering the entrance surface of the substrate. Lines 2 and 3 are the net right- and left-traveling beams $\langle E_R^2 \rangle / \langle E_{inc}^2 \rangle$ and $\langle E_L^2 \rangle / \langle E_{inc}^2 \rangle$, respectively. Line 4 is the value of $\langle E^2(z) \rangle / \langle E_{inc}^2 \rangle$ in the vicinity of the exit surface of the substrate. Line 5 is the value of $\langle E^2(z) \rangle / \langle E_{inc}^2 \rangle$ near the entrance surface due to the left-traveling beam. The total relative field intensity near the entrance surface is the sum of lines 1 and 5, and the total within the bulk is the sum of lines 2 and 3. The values of $p(z)$ are then these sums multiplied by n_s/n_0. It is this value of $p(z)$, which includes all substrate multibounce effects, that is to be used in Eq. (3).

ENTRANCE VERSUS EXIT SURFACE FILMS   V

Figure 10 shows the value of $\langle E^2 \rangle / \langle E_{inc}^2 \rangle$ at the entrance surface of the substrate, within the bulk of the substrate, at the film–substrate interface, the average value within the film, and at the air–film interface for an exit

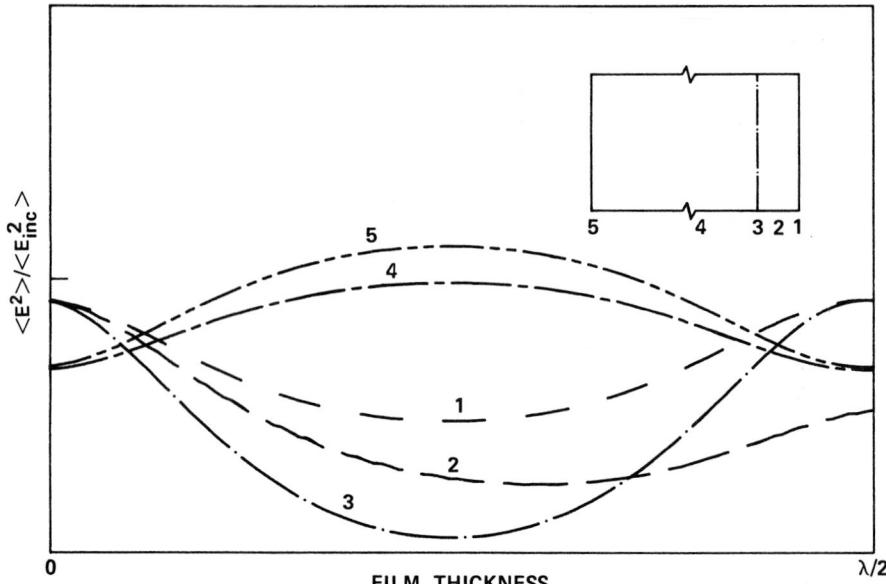

Fig. 10. The value of $\langle E^2 \rangle \langle E^2_{\text{inc}} \rangle$ for five regions in a single-layer exit surface film. The film thickness varies from zero to $\lambda/2$ OT. (Substrate index = 1.50; film index = 3.00.)

surface film, all as functions of film thickness. This graph shows how the fields gradually change with changing film thickness, causing each region in the system to contribute a different proportion to the measured absorptance.

The three significant regions of the substrate are the central or bulk region and the two interfaces, shown as regions 4, 5, and 3, respectively, in Fig. 10. Also in Fig. 10 we see that these three quantities change relative to each other as film thickness changes. For example, a $\lambda/2$ OT film causes nearly equal weighting, while a $\lambda/4$ OT film accentuates the incident surface and the bulk over the exit surface. Without knowing the values of $\alpha(z)$ throughout the substrate and in particular at the interfaces, it is not possible to properly account for the substrate contribution for an exit surface film.

Figure 11 is similar to Fig. 10, except here the film is on the entrance surface. In this case, $\langle E^2 \rangle / \langle E^2_{\text{inc}} \rangle$ for the two interfaces and in the bulk of the substrate (i.e., at 3, 5, and 4) maintain a nearly constant ratio to each other for all film thicknesses. A careful inspection shows that the ratio between the bulk and the exit surface values of p is identically constant and that the ratio between the bulk and the film–substrate interface value of p is constant to within ~2% over the thickness range shown. This is an important advantage in analysis, since now the substrate can be treated as a single entity whose contribution is directly proportional to the transmittance of the entrance surface of the sample. We do not need prior knowledge of $\alpha(z)$ for the

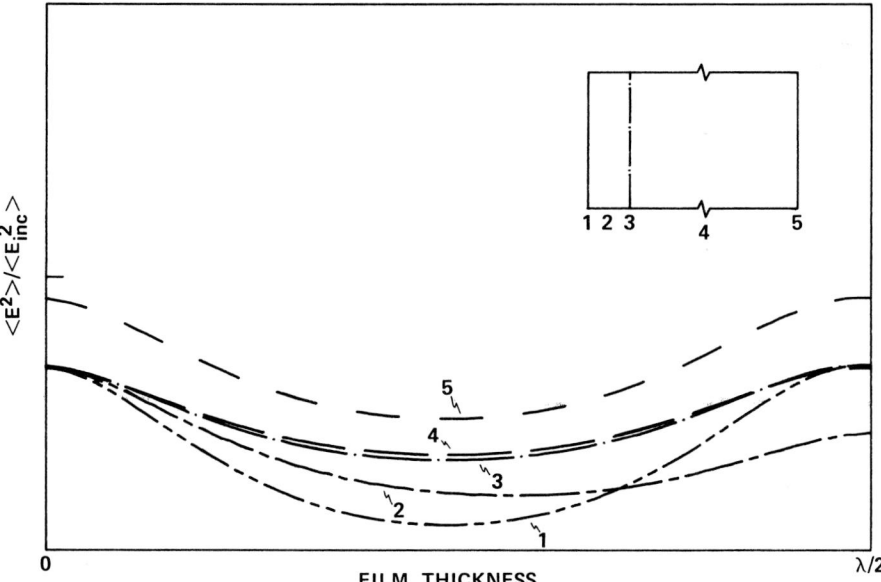

Fig. 11. The value of $\langle E^2 \rangle / \langle E_{inc}^2 \rangle$ for five regions in a single-layer entrance surface film. The film thickness varies from zero to $\lambda/2$ OT. (Substrate index = 1.50; film index = 3.00.)

substrate as we would in the exit surface analysis. It is this property of entrance surface films that permits us to account for the substrate contribution with the single term $\bar{p}_s A_0$ in Eq. (4).

EXPERIMENTAL DETERMINATION OF α_f, a_{af}, AND a_{fs} VI

Several values of relative energy density p are needed in the analysis of two-thickness films using Eq. (5) or of wedge-shaped films using Eqs. (7a) and (7b). Figures 12–17 show plots of relative energy densities for substrates illuminated at normal incidence from the film side for $\lambda/4$ and $\lambda/2$ OT films. These calculations include substrate multibounce effects. All the figures are families of curves for substrates ranging in index from 1.0 to 4.0 by steps of 0.2. The film index also ranges from 1.0 to 4.0 and is given as the abscissa. The corresponding relative energy density is shown as the ordinate.

Figure 12 uses Eq. (11) and shows $\bar{p}_s$, the average substrate relative power density for a $\lambda/4$ OT film. The value of $\bar{p}_s$ for $3\lambda/4$, $5\lambda/4$, ... OT films are identical to that for the $\lambda/4$ OT film. The value of $\bar{p}_s$ for a bare substrate and for $\lambda/2$, λ, ... OT films are identically 1.0 for all film and substrate indices and are not shown graphically.

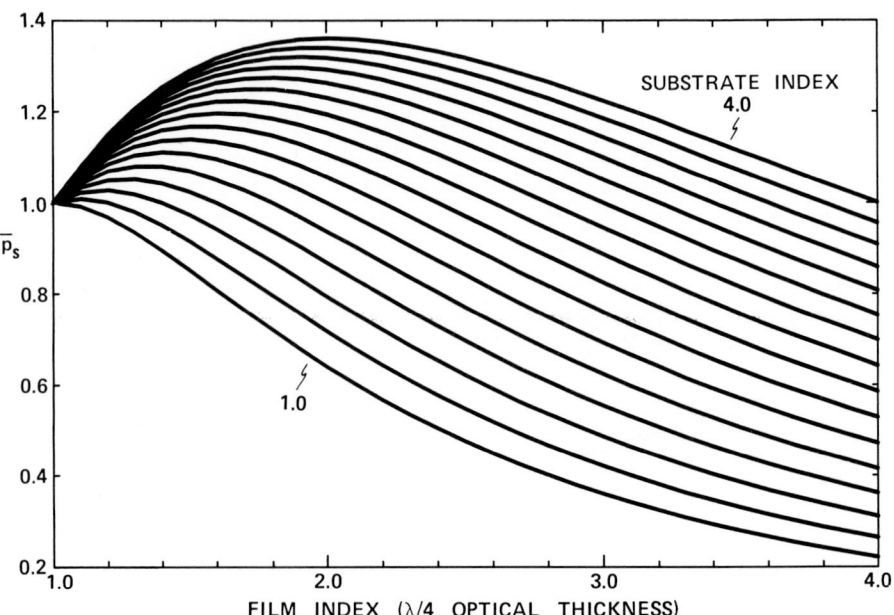

Fig. 12. The average relative energy density $\bar{p}_s$ within the bulk of a substrate with a $\lambda/4$, $3\lambda/4, \ldots$ film on the entrance surface.

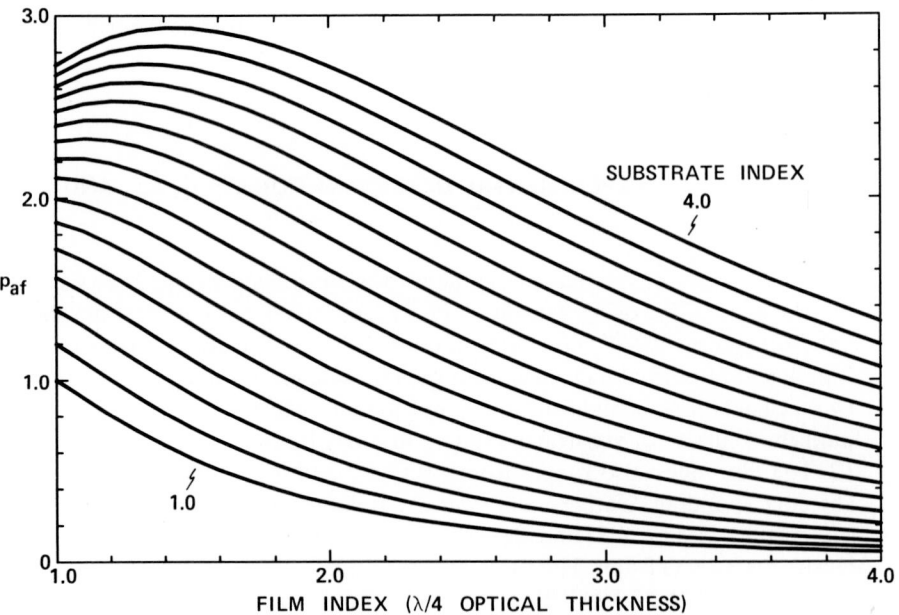

Fig. 13. The relative energy density p_{af} at the air–film interface for an entrance surface film of $\lambda/4, 3\lambda/4, \ldots$ OT.

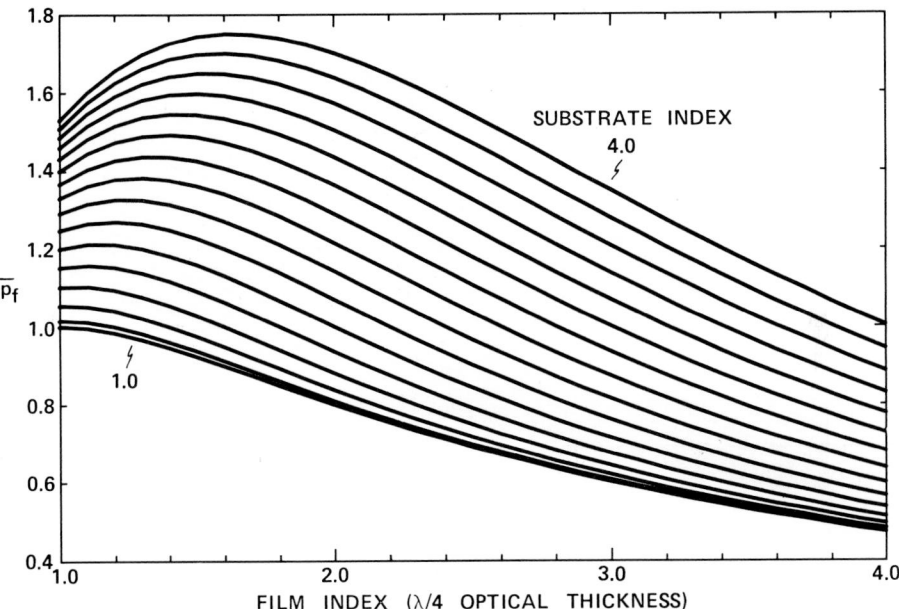

Fig. 14. The average relative energy density $\bar{p}_f$ within the bulk of an entrance surface film of $\lambda/4, 3\lambda/4, \ldots$ OT.

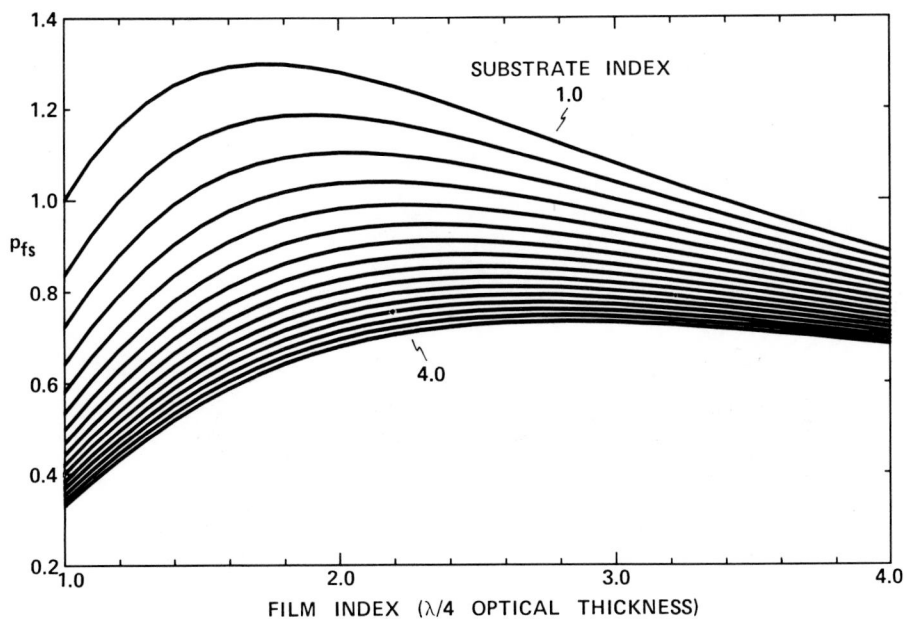

Fig. 15. The relative energy density p_{fs} at the film–substrate interface for an entrance surface film of $\lambda/4, 3\lambda/4, \ldots$ OT.

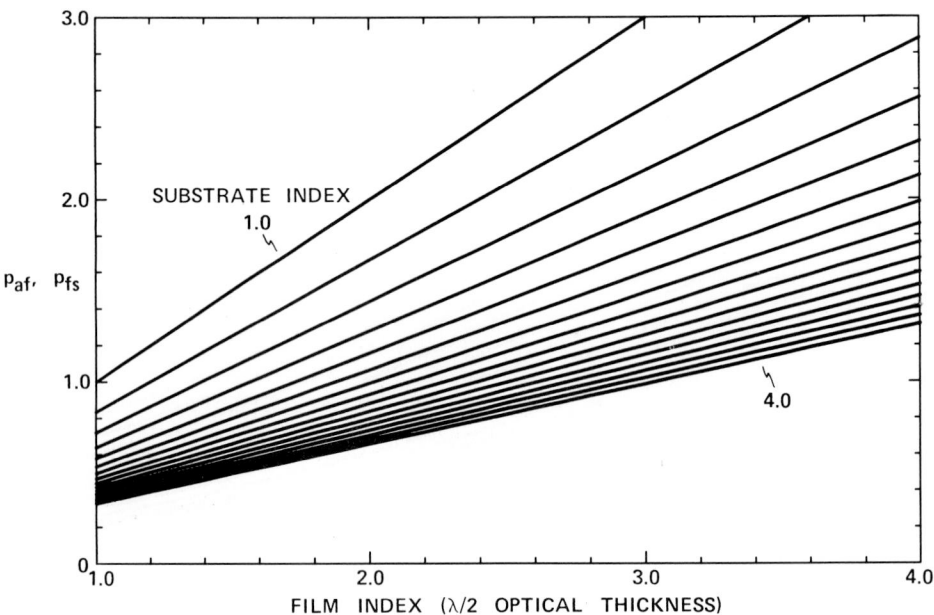

Fig. 16. The relative energy density at the air–film p_{af} and the film–substrate p_{fs} interface for an entrance surface film of $\lambda/2, \lambda, \ldots$ OT.

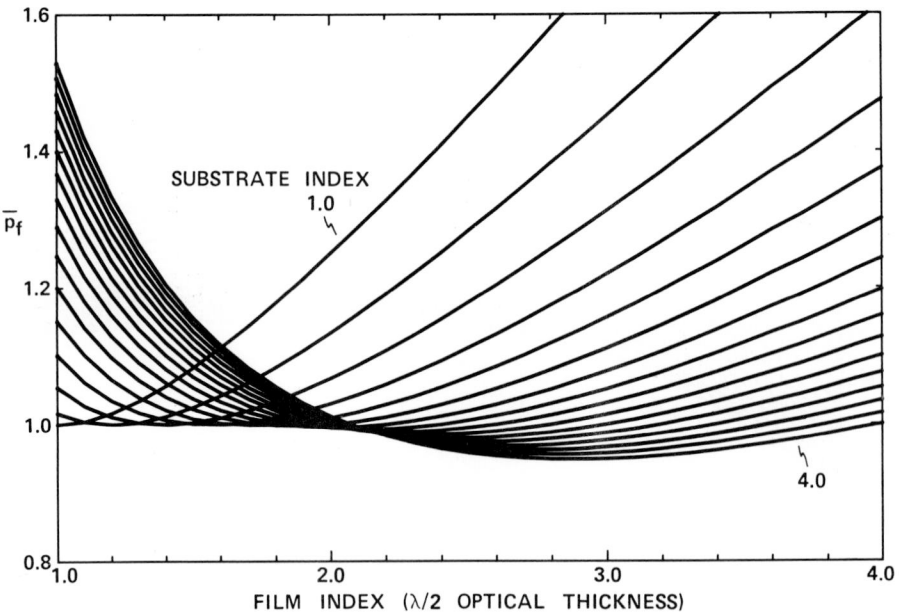

Fig. 17. The average relative energy density $\bar{p}_f$ within the bulk of an entrance surface film of $\lambda/2, \lambda, \ldots$ OT.

Figures 13–15 show p_{af}, $\bar{p}_f$, and p_{fs}, the air-film, average in the film, and film–substrate relative power density for a $\lambda/4$ OT film. As in the preceding section, these same values apply to a $3\lambda/4$, $5\lambda/4$, ... OT film. Figure 16 shows p_{af} and p_{fs} for films of $\lambda/2$, λ, ... thickness. Figure 17 shows $\bar{p}_f$ for $\lambda/2$, λ, ... OT films.

In evaluating a sample, the values of p_{af}, $\bar{p}_f$, $\bar{p}_{fs}$, and $\bar{p}_s$ for the applicable substrate and film index are extracted from Figs. 12–17. If only the bulk absorption coefficient of the film is to be determined, then the absorptance for $\lambda/2$ and 1λ films are measured calorimetrically, and Eq. (5) is used, along with the value of $\bar{p}_f$ from Fig. 17, to find α_f.

If the interface absorptances are to be determined, then the absorptance of the bare substrate and a $\lambda/4$ OT film must be measured, in addition to the $\lambda/2$ and 1λ OT films. The values of p_{af}, $\bar{p}_f$, p_{fs}, and $\bar{p}_s$ for $\lambda/4$, $\lambda/2$, and 1λ OT films are taken from Figs. 12–17. Then, Eqs. (7a) and (7b) are used to determine a_{af} and a_{fs}.

REFERENCES

1. L. Skolnik, *in* "Optical Properties of Highly Transparent Solids" (S. S. Mitra and B. Bendow, eds.) pp. 405–533, Plenum, New York, 1975.
2. A. Hordvik, *Appl. Opt.* **16**, 2827 (1977).
3. L. W. Bos and D. W. Lynch, *Phys. Rev. B* **2**, 4567 (1970).
4. D. A. Pinnow and T. C. Rich, *Appl. Opt.* **12**, 984 (1973).
5. T. H. Allen, J. H. Apfel, and C. K. Carniglia, *in* "Laser Induced Damage in Optical Materials: 1978" (A. J. Glass and A. H. Guenther, eds.), pp. 33–36, National Bureau of Standards Spec. Publ. 541, Washington, D.C., 1978.
6. M. J. Hass, J. W. Davisson, P. H. Klein, and L. L. Boyer, *J. Appl. Phys.* **45**, 3959 (1974).
7. D. L. Decker and P. A. Temple, *in* "Laser Induced Damage in Optical Materials: 1977" (A. J. Glass and A. H. Guenther, eds.), pp. 281–285, National Bureau of Standards, Spec. Publ. 509, Washington, D.C., 1977.
8. P. A. Temple, *SPIE Proc.* **325**, 156 (1982).
9. E. Bernal G., *Appl. Opt.* **14**, 314 (1975).
10. H. A. MacLeod and D. Richmond, *Thin Solid Films* **37**, 163 (1976).
11. T. M. Donovan, P. A. Temple, Shiu-Chin Wu, and T. A. Tombrello, *in* "Laser Induced Damage in Optical Materials: 1979" (H. E. Bennett, A. J. Glass, A. H. Guenther, and B. E. Newnam eds.) pp. 237–246, National Bureau of Standards, Spec. Publ. 568, Washington, D.C., 1980.
12. P. A. Temple, *Appl. Phys. Lett.* **34**, 677 (1979).
13. P. A. Temple, D. L. Decker, T. M. Donovan, and J. W. Bethke, *in* "Laser Induced Damage in Optical Materials: 1978" (A. J. Glass and A. H. Guenther, eds.), pp. 37–42. National Bureau of Standards, Washington, D.C., 1978.
14. P. J. Leurgans, *J. Opt. Soc. Am.* **41**, 714 (1951).
15. G. W. DeBell, *SPIE Proc.* **140**, 2 (1978).

Chapter **8**

Complex Index of Refraction Measurements at Near-Millimeter Wavelengths

GEORGE J. SIMONIS
Electronics Research and Development Command
Harry Diamond Laboratories
Adelphi, Maryland

I.	Introduction	155
II.	Fourier Transform Spectroscopy	156
III.	Free-Space Resonant Cavity	161
IV.	Mach–Zehnder Interferometer	163
V.	Direct Birefringence Measurement	164
VI.	Overmoded Nonresonant Cavity	165
VII.	Crystal Quartz as Index Reference	165
VIII.	Conclusion	167
	References	167

INTRODUCTION I

The measurement of the complex index of refraction $\hat{n}$ at wavelengths in the range from hundreds of micrometers to several millimeters, or "near-millimeter" wavelengths (roughly 100 to 1000 GHz), can draw on a diversity of techniques. These wavelengths fall between the far-infrared wavelengths, at which the *optical* techniques are predominant, and the conventional millimeter wavelengths, at which *microwave* techniques are predominant. The wavelengths are short enough that quasi-optical apparatus can have clear apertures of many wavelengths for acceptable apparatus dimensions in the laboratory and long enough that single-mode microwave apparatus can have dimensions and tolerances accessible with established fabrication techniques. However, if the measurement accuracy requirements are high,

155

ISBN 0-12-544420-6

then many optical techniques can be a serious challenge because of diffraction effects, as can microwave techniques due to dimension tolerances. Many different approaches have been used in dealing with these challenges, but adaptations of optical techniques seem to predominate for this frequency range.

An extensive review of available data on the near-millimeter wave (NMMW) properties of materials is available [1]. Some of the different approaches used to measure these properties will be introduced and discussed briefly in the following sections. Space limitations will not allow a comprehensive discussion of this topic. The precise measurement of n and α at *fixed* NMMW frequencies, at which $\hat{n} = n - \hat{i}(\alpha\lambda/4\pi)$ and α is the exponential power absorption coefficient, is very useful because there are few strong narrow material resonances compared to those in the infrared range. The NMMW index dispersion is generally lower, and measurement at a fixed frequency can generally be extrapolated to neighboring frequencies to more significant figure accuracy than would be possible in the infrared. Debye relaxation effects are most often more pronounced at lower frequencies. The losses for "low-loss" materials typically get progressively higher as the NMMW frequency increases and acquire a nonlinear frequency dependence more typical of the far-infrared with broad features as the wings of infrared absorption bands are approached.

II FOURIER TRANSFORM SPECTROSCOPY

The vast majority of the interferometers in use for far-infrared Fourier transform spectroscopy (FTS) are of a Michelson interferometer configuration with Mylar beam splitters as portrayed in Fig. 1. Many good discussions of FTS techniques are available [2–5]. Fourier transform spectroscopy techniques have the advantage over grating spectrometer techniques of being able to collect data simultaneously at many frequencies and over a large solid angle. This becomes progressively more important, when blackbody sources are used, as measurements are sought at lower frequencies at which the signal level decreases as the frequency squared. Fourier transform spectroscopy techniques have the attractive feature of being able to provide measurements of n and α continuously as a function of frequency over large frequency ranges. This is in contrast to some of the fixed-frequency techniques.

Fourier transform spectroscopy techniques also have several unattractive features for NMMW studies. Even with their signal-to-noise advantage, it is most often very difficult to obtain useful data below 8 cm^{-1} (240 GHz) unless the instrumentation is specially adapted for low-frequency operation. The instrumentation and data processing are conceptually simple, but the implementation can become quite complicated. Also, at low frequencies the beam splitters must be very thick in order to get constructive interference from the

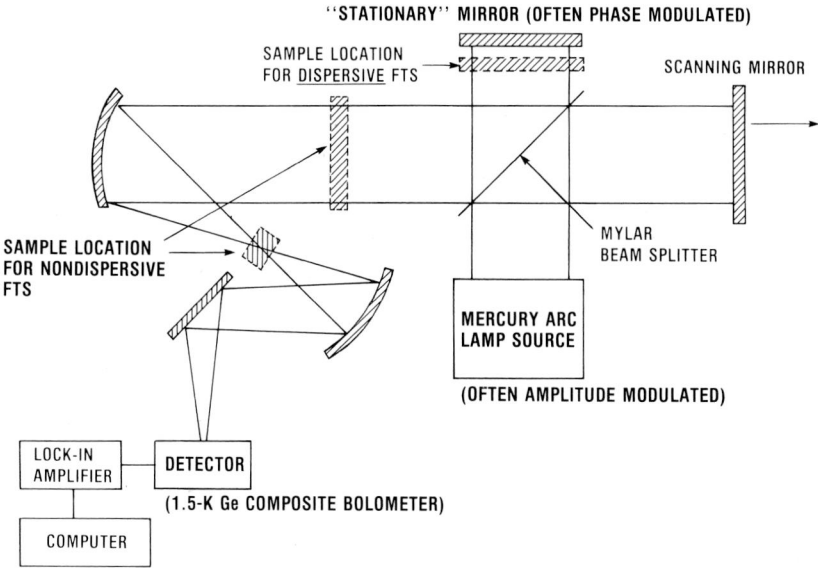

Fig. 1. Fourier transform spectroscopy configuration with a Mylar beam splitter. Dispersive or nondispersive data can be obtained depending on sample position.

two splitter surface reflections, and even then there is a low-frequency roll-off to a null reflected signal level. This thickness introduces significant beam splitter absorption losses at Mylar absorption frequencies and periodic null signal levels due to coherent interference effects from the two surfaces of the splitter. The low NMMW signal levels lead to the use of detectors cooled to 1.5 K, extensive filtering that limits the frequency coverage, and averaging of many scans. Enhanced diffraction effects at NMMW frequencies lead to a need for larger beam and sample diameters. Even under optimized conditions, it is difficult to get accurate data on lossy samples or accurate loss information on low-loss samples. In spite of these difficulties, remarkable results have been achieved at frequencies as low as 100 GHz with specially adapted FTS systems [6]. Polarizing interferometer configurations using wire grating beam splitters alleviate some of the beam splitter problems [6–8]. Such beam splitters do not have material resonance absorptions or periodic reflection nulls. They also have a flat, uniform, high efficiency for all wavelengths considerably greater than the grating wire spacing. The use of phase modulation [5] along with careful detector choice, filtering, source regulation, and cooling can make dramatic improvements in the NMMW signal-to-noise ratio.

Fourier transform spectroscopy measurements of $\hat{n}$ can be divided up into two basic approaches—dispersive and nondispersive FTS. The two are distinguished by the different placement of the sample in the interferometer (Fig. 1). In dispersive FTS, the sample is placed in one arm of the interferometer, while in nondispersive FTS, the sample is placed in the output where the two beams are already recombined. In the nondispersive FTS configuration, the output beam is often focused onto the sample, allowing measurements to be made on relatively small samples. This is also an attractive feature for use with temperature- and pressure-controlled cells, in which small apertures are needed. However, if thick samples are needed for measurement accuracy, then errors can be introduced in focused-beam configurations due to beam defocusing by the sample. The real-index component is determined from multiple-reflection interference effects and the known spacing of the two surfaces of the sample. The imaginary component is obtained from throughout signal attenuation. The nondispersive approach must deal with a progressively smaller signal for real n determination as the sample index gets smaller or as α gets bigger. The multiple-surface reflection signal gets smaller for decreasing n until, finally, with materials such as styrofoam ($n = 1.017$) there are no observable multiple reflections, and the real index cannot be measured.

The dispersive FTS approach can be contrasted with the nondispersive approach inasmuch as the transmitted signal unreflected in the sample carries

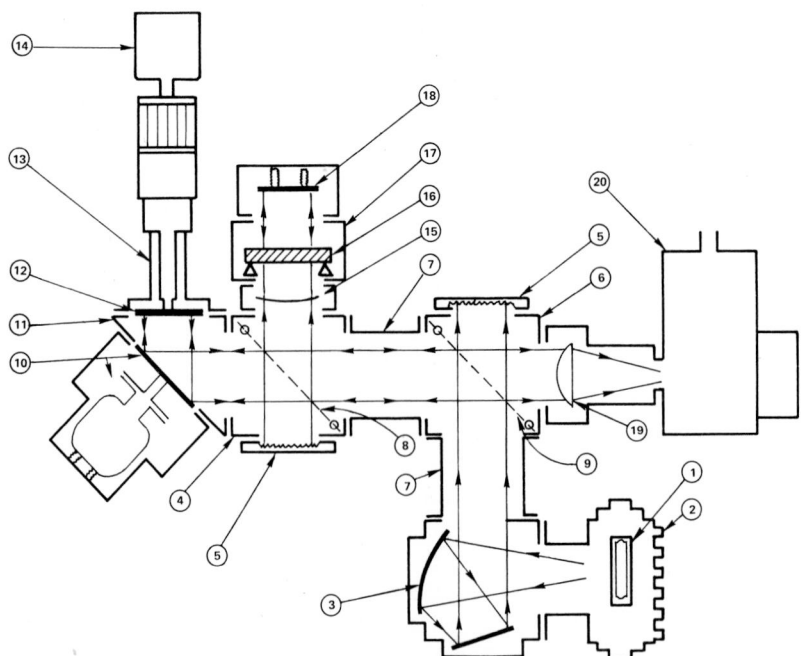

Fig. 2. Polarizing interferometer using wire-grating beam splitters and phase modulation to conduct dispersive Fourier transform spectroscopy. (1) Mercury lamp, (2) water-cooled lamp housing, (3) two-mirror collimator, (4) central cube, (5) radiation dump, (6) cube 2, (7) connectors (spacers), (8) wire-grid beam splitter, (9) wire-grid polarizer/analyzer, (10) 45° mirror and phase modulation assembly, (11) half-cube, (12) scanning mirror, (13) micrometer, (14) stepping motor, (15) Mylor window, (16) solid specimen, (17) specimen holder, (18) adjustable fixed mirror, (19) focusing lens, and (20) InSb Rollin detector. (From Afsar and Button [6].)

the real-index information for dispersive FTS. Thus, a relatively good signal-to-noise ratio exists for low index or lossy materials. The dispersive FTS index information is obtained from the shifting of the primary feature of the interferogram rather than in the appearance of secondary features, an important distinction when signal-to-noise ratio is such a serious problem. An extensive review of the dispersive FTS approach is available [9]. The sample is placed in the collimated beam of one arm that, combined with a requirement for small diffraction effects, leads to a need for NMMW samples with clear apertures of several inches. A dispersive FTS interferometer [6] employing wire-grating beam splitters, phase modulation, and a cooled-InSb Rollin detector is presented in Fig. 2 [6]. One polarization is reflected into

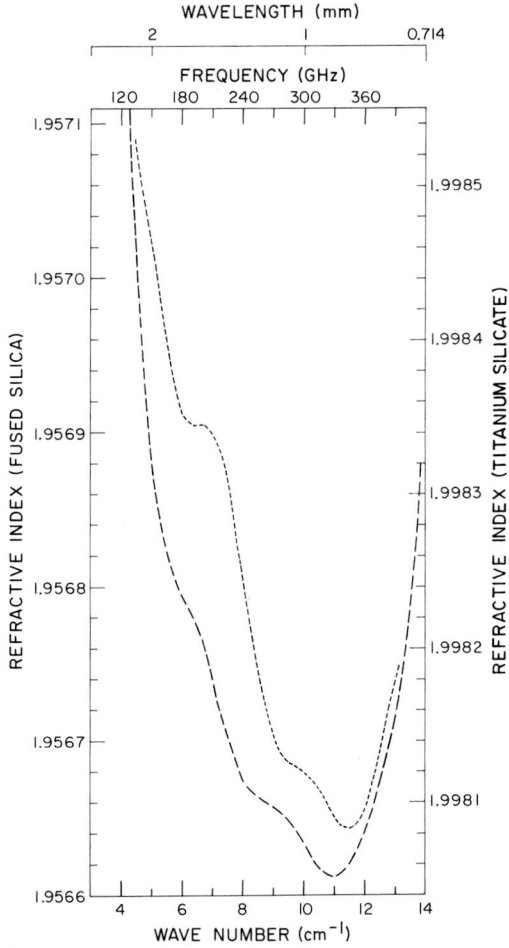

Fig. 3. Refractive index data obtained at 25°C using the instrumentation of Fig. 2. [······, Corning 7940 (fused silica); ————, Corning 7971 (titanium silicate).] (From Afsar and Button [6].)

the interferometer from the source with the other being disposed of. The reflected linear polarization is split into two orthogonal linear components that are separately phase shifted in different amounts, recombined, and sent on to the detector through the first grating, which also functions as a polarization filter for the detector. When sufficiently large flat plane-parallel samples are available and the technique is carried to extreme, remarkable results can be achieved with this approach. Figure 3 shows recent NMMW index of refraction data reported to six significant figures [6]. It is expected that NMMW features observed in such data for the first time will be relatable to physical sample characteristics. Figure 4 shows associated material-loss data reported here in terms of the imaginary component of the dielectric constant [6].

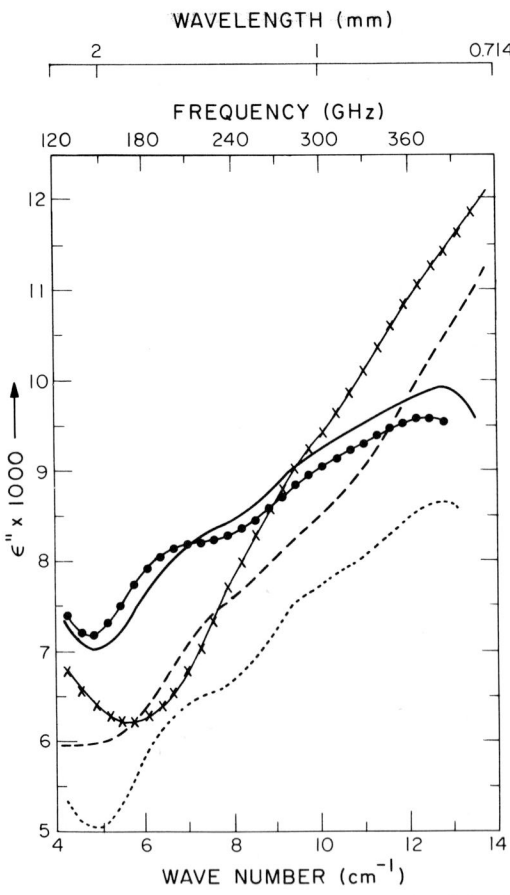

Fig. 4. Measurements of the imaginary component of the relative dielectric constant at 25°C using the instrumentation of Fig. 2. [×, alumma 995; ●, BeO; ——, Corning 9606 (glass ceramic); ······, Corning 7940 (fused silica); and ————, Corning 7971 (titanium silicate).] (From Afsar and Button [6].)

Thus, it can be seen that, despite the difficulties of the FTS approach, very useful results can be achieved across the NMMW range with appropriate samples and experimental procedure.

FREE-SPACE RESONANT CAVITY   III

The NMMW free-space resonant cavity techniques (usually Fabry–Perot cavities; see Figs. 5 and 6) are somewhat analogous to the popular microwave closed-resonant-cavity techniques in that they also rely on the perturbation of the cavity Q and resonant frequencies (or, alternatively, resonant-cavity length at fixed frequency) for the determination of the complex index. These cavities are driven by narrow-band sources providing a very good signal-to-noise ratio. The closed-microwave-resonant-cavity technique is difficult

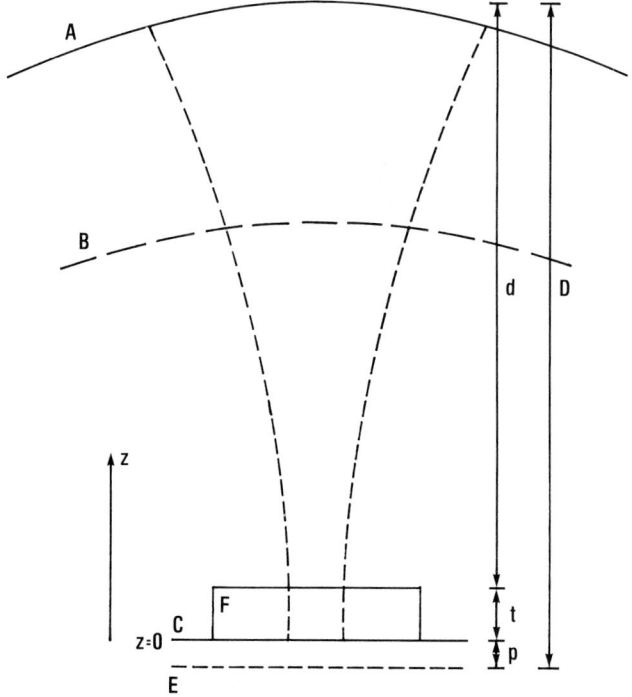

Fig. 5. Semiconfocal cavity configuration used for the measurement of the complex index of refraction. A, concave mirror; B, phase front; C, plane mirror at resonant position with sample; D, resonance-cavity length without sample; E, plane mirror at resonant position without sample; F, sample; t, sample thickness; $t + d$, resonance-cavity length with sample; and p, shift in resonance-cavity length due to sample. (From Jones [10]. Copyright 1976, The Institute of Physics.)

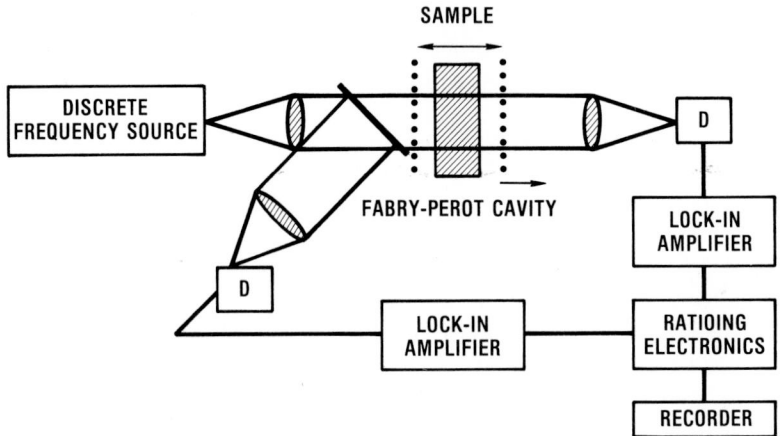

Fig. 6. Plane-parallel mesh-cavity configuration used for the measurement of the index of refraction. Detectors are labeled D.

to apply in the NMMW range because the dimensions and tolerances on fundamental-mode cavities are difficult to work with and larger overmoded closed cavities have too many competing modes to deal with. Also, it is difficult to insert and remove the sample without perturbing the tiny cavity. The free-space resonant cavities have dimensions large relative to the wavelength, require close tolerances only on two quasi-optical end reflectors (easily achievable by using optical fabrication techniques), allow the simple insertion and removal of samples, and have only a few low-loss modes if properly apertured. Also, they can achieve very high Q values because no sidewall losses are involved. The high Q values can be thought of as providing the equivalent of many passes through the sample, making the technique a sensitive method of measuring the material losses of very-low-loss samples. The sharp cavity resonance also allows accurate index determinations to be made.

Figure 5 presents schematically a semiconfocal configuration (used by Jones [10] at 35 GHz) that several groups including ourselves are extending to the NMMW range. Insertion of the sample, labeled F, causes the resonance-cavity length to be shortened by an amount p. Errors in the determination of the real dielectric constant on the order of 0.1% were reported. References to several similar efforts in resonant cavities can be found in Cullon and Yu [11], and Simonis and Felock [12]. Various plane-parallel, confocal, and semiconfocal Fabry–Perot cavities have been used. Simonis and Felock [12] used a plane-parallel cavity having metal-mesh reflectors selected for use with the 245-GHz output of an optically pumped $C^{13}H_3F$ laser (Fig. 6). Such crossed-wire meshes can be produced to high accuracy by using photo-lithographic techniques and are commercially available. Single or multiple meshes can serve the purpose in the NMMW range that single or multiple dielectric layers often serve in the infrared for achieving desired reflectivities.

A number of publications cited in the preceding references also discuss the use of changing open-cavity Q with sample insertion for the determination of the sample loss tangent, power absorption coefficient, or imaginary component of the dielectric constant. Some careful examinations of the sources of error of this approach are provided. The resonant-cavity approach provides a high sensitivity to small sample losses. However, samples with large losses degrade the cavity Q so badly that this technique becomes very difficult to employ with lossy samples.

The resonant-cavity technique requires a highly stabilized and narrow-band-source frequency so that small shifts in resonance frequency or, equivalently, resonance-cavity length can be accurately measured. The use of optically pumped NMMW lasers alleviates some of these problems. Another difficulty with resonant-cavity techniques is that there are multiple roots to the index equations, providing multiple possible answers. The correct index must be distinguished by working with multiple sample thicknesses or by knowing what the approximate value should be.

MACH–ZEHNDER INTERFEROMETER   IV

The Mach–Zehnder interferometer is the optical analog of the balanced-bridge techniques used at microwave and lower frequencies. The sample is placed in one arm and balanced against a calibrated variable phase shift and attenuation in the other arm. This configuration has the advantage of requiring only one pass through the sample. Thus, higher-loss samples can be dealt with.

The results are more straightforward to interpret and source-frequency drift is less of a problem than with resonant-cavity techniques. The signal-to-noise ratio is generally very good. However, one does need calibrated variable phase shifters and attenuators. The phase shift is readily obtainable as a variable optical path length [13] or as a variable sample thickness in the case of liquids [14]. The attenuator can be based on total internal reflection in a closely spaced pair of prisms [15, 16] or on crossed wire gratings [17].

The beam splitters generally are sheet material that has pronounced frequency-dependent properties. This frequency dependence can be removed by using wire-grating beam splitters as discussed in Section II. Such a polarizing Mach–Zehnder interferometer is shown in Fig. 7. Instead of having an attenuator in one arm, the relative signal level in the two arms is changed by rotating the initial polarizer [18]. The Mach–Zehnder approach also yields several possible index values that must be distinguished as with the resonant-cavity method.

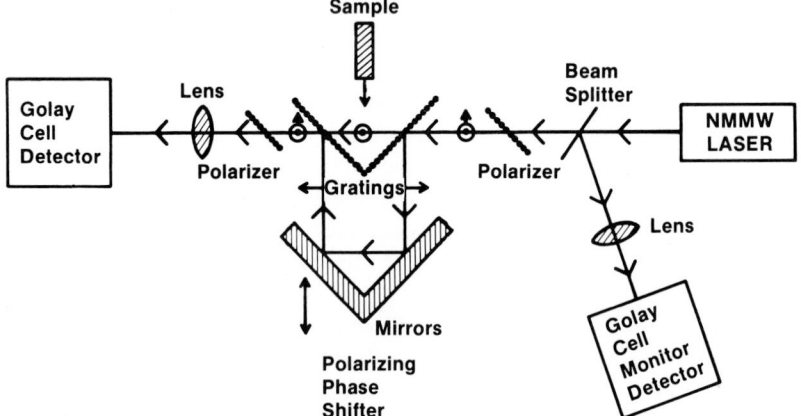

Fig. 7. Polarizing Mach–Zehnder configuration for the measurement of the index of refraction. Dotted beam splitters are wire gratings.

V   DIRECT BIREFRINGENCE MEASUREMENT

The birefringence of a material can be measured *directly* without first measuring the individual indices by canceling its birefringence with a tunably birefringent wire-grating configuration as shown in Fig. 8. The sample and tunably birefringent grating assembly are placed in series between crossed polarizers and adjusted until a null signal level is achieved [19]. Effects due

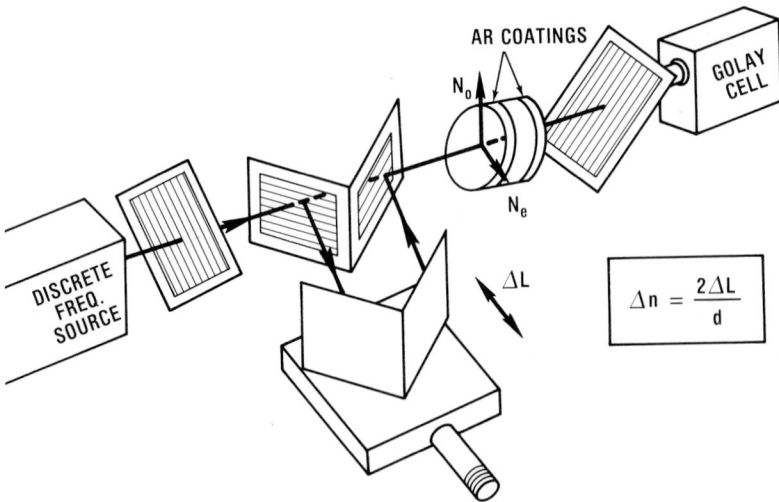

$$\Delta n = \frac{2\Delta L}{d}$$

Fig. 8. A configuration similar to that of Fig. 7 but used to measure material birefringence directly. (From Simonis [19].)

to multiple reflections from the sample surfaces can be minimized by antireflection coating the sample or can be allowed for in the analytical treatment of the data. Measurements on crystal quartz yielded a value of $\Delta n = 0.047$ at 245 GHz [19], in good agreement with other reported measurements at nearby frequencies.

OVERMODED NONRESONANT CAVITY VI

Overmoded nonresonant cavities have been in use for many years for microwave dielectric studies. Their most common utilization has been for gas spectroscopy. Energy from a blackbody or narrow-band source is coupled into a greatly overmoded closed metal cavity through a small hole and sensed by detectors in the cavity or through a small output coupling hole. Mode stirring devices are often rotated inside the cavity (much as in microwave ovens) or the source frequency is slightly dithered in order to average out standing-wave patterns. There are no critical cavity dimensions. The use of such cavities for the study of the NMMW properties of solid materials has been explored [20–23]. The technique allows the study of irregularly shaped samples and powders since the reflected radiation is just returned to the cavity and eventually returns to the sample again. The technique can facilitate the NMMW study of the dielectric properties of biological material (including those in *in vivo*), a topic of current interest. This technique could also provide good NMMW metal reflectivity data. However, the analysis can get complicated for lossy materials. The technique shows promise for some special applications but is not likely to displace the techniques previously described for routine measurements.

CRYSTAL QUARTZ AS INDEX REFERENCE VII

As mentioned in Section I, many techniques are being employed for the determination of the NMMW properties of materials. It would be useful to have materials with well-known indices of refraction that reproduced well from one sample to another. Such materials could be used to prove out and compare apparatuses and techniques. Many materials have not been carefully measured or vary significantly in their NMMW properties from sample to sample [1]. These NMMW variations may be due, for example, to variations in crystallinity, density, chemical impurities, or available free carriers.

Single-crystal quartz appears to be a useful NMMW reference material for index of refraction measurements. Its NMMW index has been measured

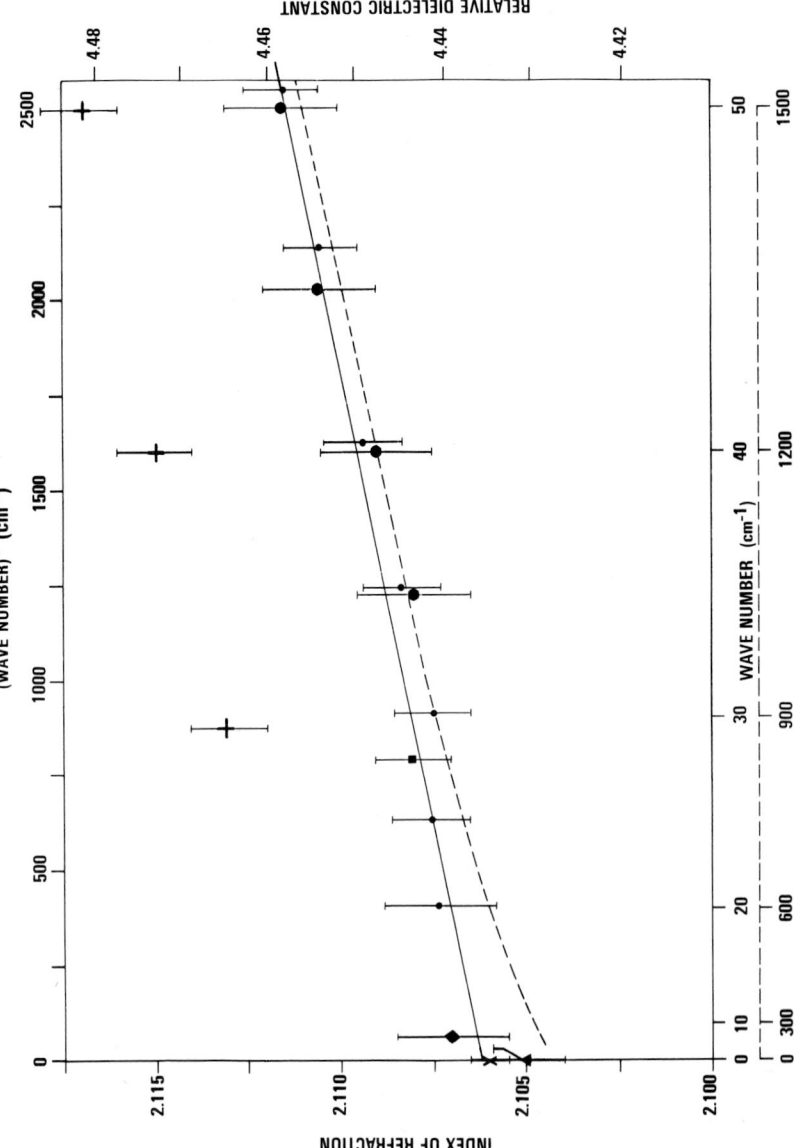

Fig. 9. The ordinary index of refraction of crystalline quartz as measured by many researchers using many techniques. The solid line is a frequency-square dependence fit to the zero-frequency and 50-cm^{-1} values of Russell and Bell [25]. ●, Russell and Bell [25]; ▲, Jones [10]; ◆, Simonis and Ferlock [12]; ×, Fontanella [24]; ●, Passchier et al. [26]; +, Loewenstien et al. [27]; ■, Charlemagne and Hadni [28]; ——, Afsar [29]; and ——Russell and Bell [25].

accurately by a number of researchers using various techniques. The absorption losses of crystal quartz are extremely small and therefore present no problems. Its ordinary index is well behaved in the NMMW range, and there is good agreement between different measurements. Figure 9 [12] presents the semiconfocal resonant-cavity measurements of Jones [10] at 35 GHz, the bridge-substitution measurements of Fontanella [24] at 1 kHz, the mesh Fabry–Perot measurements of Simonis and Felock [12] at 245 GHz, the dispersive FTS measurements of Russell and Bell [25], the dispersive FTS measurements of Passchier *et al.* [26], the nondispersive FTS measurements of Loewenstein *et al.* [27], the laser measurement of Charlemagne and Hadni [28] at 890 GHz, and the dispersive FTS measurements of Afsar [29]. The solid line is a fit of a frequency-squared dependence to the "zero frequency" and 50-cm^{-1} values of Russell and Bell [25] when dispersion from an infrared resonance is assumed. The discrepancies between the data of Loewenstein *et al.* [27] and those of others remain unexplained (Loewenstein *et al.* [27] were aware of the discrepancy).

One concern about the use of crystalline quartz as a reference is that it is birefringent. A misoriented Z-cut sample would show a mix of the two indices, yielding erroneously large ordinary-index results.

CONCLUSION   VIII

Adaptations of several quasi-optical techniques of complex index measurement can provide NMMW values of sufficient accuracy for most purposes. Several examples have been given. Many materials do not have sufficiently reproducible NMMW properties to be reliable confirmations of techniques being tested by different laboratories, but some materials, such as Z-cut crystalline quartz, can be very helpful in this regard. Equipment designed and employed for NMMW characterization of materials is readily available if such data are needed.

REFERENCES

1. G. J. Simonis, *Int. J. Infrared Millimeter Waves* **3**, 439 (1982).
2. R. J. Bell, "Introductory Fourier Transform Spectroscopy," Academic Press, New York, 1972.
3. G. W. Chantry, "Submillimetre Spectroscopy," Academic Press, New York, 1971.
4. P. R. Griffiths, "Transform Techniques in Chemistry," Plenum, New York, 1978.
5. J. E. Chamberlain, "The Principles of Interferometric Spectroscopy," Wiley, New York, 1979.
6. M. N. Afsar and K. J. Button, *Int. J. Infrared Millimeter Waves* **2**, 1029 (1981).
7. D. H. Martin and E. Puplett, *Infrared Phys.* **10**, 105 (1969).
8. T. J. Parker, D. A. Ledsham, and W. G. Chambers, *Infrared Phys.* **18**, 179 (1978).

9. J. R. Birch and T. J. Parker, *in* "Infrared and Millimeter Waves" (K. J. Button, ed.), Vol. 2, p. 138, Academic Press, New York, 1979.
10. R. G. Jones, *J. Phys. D* **9**, 819 (1976).
11. A. L. Cullen and P. K. Yu, *Proc. R. Soc. London Ser. A.* **325**, 493 (1971).
12. G. J. Simonis and R. D. Felock, *Appl. Opt.* **22**, 194 (1983).
13. J. E. Allnutt and J. A. Staniforth, *J. Phys. E* **4**, 730 (1971).
14. J. Chamberlain, J. Haigh, and M. J. Hine, *Infrared Phys.* **11**, 75 (1971).
15. E. R. Schineller, *Proc. Symp. Quasi-Optics* **14**, 517 (1964).
16. J. J. Taub, H. J. Hindin, O. F. Hinchselmann, and M. L. Wright, *IEEE Trans. Microwave Theory Tech.* **MTT-11**, 338 (1963).
17. J. A. Staniforth and J. E. Allnutt, *Proc. IEE* **118**, 343 (1971).
18. G. J. Simonis and B. C. Redman (to be published).
19. G. J. Simonis, *IEEE Trans. Microwave Theory Tech.* **MTT-31**, 356 (1983).
20. D. T. Llewellyn-Jones, R. J. Knight, P. H. Moffat, and H. A. Gebbie, *Proc. IEE A*, **127**, 535 (1980)
21. H. A. Gebbie and D. T. Llewellyn-Jones, *Int. J. Infrared Millimeter Waves* **2**, 197 (1981).
22. J. R. Izatt and F. Kremer, *Appl. Opt.* **20**, 2555 (1981).
23. F. Kremer and J. R. Izatt, *Int. J. Infrared Millimeter Waves* **2**, 675 (1981).
24. J. Fontanella, *J. Appl. Phys.* **45**, 2852 (1974).
25. E. E. Russell and E. E. Bell, *J. Opt. Soc. Am.* **57**, 341 (1967).
26. W. F. Passchier, D. D. Konijk, M. Mandel, and M. N. Afsar, *J. Phys. D* **10**, 509 (1977).
27. E. V. Loewenstien, D. R. Smith, and R. L. Morgan, *Appl. Opt.* **12**, 398 (1973).
28. D. Charlemagne and A. Hadni, *Opt. Acta* **16**, 53 (1969).
29. M. N. Afsar, Ph.D. thesis, University of London, London, England, 1978.

Chapter **9**

The Quantum Extension
of the Drude–Zener Theory
in Polar Semiconductors*

B. JENSEN

Department of Physics
Boston University
Boston, Massachusetts

I. Introduction	169
II. Quantum Theory of Free-Carrier Absorption	172
III. Theoretical Results	174
IV. Comparison with Experimental Data	176
Appendix	187
References	188

INTRODUCTION **I**

The classical Drude model for the complex dielectric constant of a semi-conductor can be used to extract the mobility and the free-carrier density n_e from an analysis of the reflectivity and transmittance data in the far infrared [1–4]. The dielectric constant ε is the square of the complex refractive index, which determines the optical properties of a given material. One has

$$\varepsilon = \varepsilon_1 - i\varepsilon_2 = N^2, \tag{1}$$

where the real and imaginary parts of the complex dielectric constant ε_1 and ε_2 are functions of the complex refractive index N as

$$N = n - ik, \tag{2}$$

* This research was supported by the Department of Energy under contract No. DE-AC02-79ER10444.A000

$$\varepsilon_1 = n^2 - k^2, \tag{3}$$

$$\varepsilon_2 = 2nk = 4\pi\sigma/\omega. \tag{4}$$

The choice of $n - ik$ rather than $n + ik$ is determined by the original use of $\exp i(\omega t - \mathbf{q} \cdot \mathbf{r})$ in the assumed plane-wave solution of Maxwell's equations.

In Eqs. (1)–(4), n is the real part of the complex refractive index, k the imaginary part or extinction coefficient, and σ the optical conductivity. The absorption coefficient α is proportional to σ, to ε_2, and to k:

$$n\alpha = 4\pi\sigma/c = (\omega/c)\varepsilon_2, \tag{5}$$

$$\alpha/2 = (\omega/c)k = 1/\delta. \tag{6}$$

The extinction coefficient k is essentially the ratio of the free-space wavelength of light of frequency ω to the skin depth δ.

The Drude theory gives the free-carrier contribution to ε_1 and ε_2 in terms of the plasma frequency $\bar{\omega}_p$ and the electron scattering time τ as

$$\varepsilon_1 = \varepsilon_\infty(1 - \bar{\omega}_p^2/\omega^2\eta), \tag{7}$$

$$\varepsilon_2 = (\bar{\omega}_p^2/\omega^2\eta)1/\omega\tau, \tag{8}$$

where

$$\omega_p^2 = 4\pi n_e e^2/m_n, \qquad\qquad \bar{\omega}_p^2 = \omega_p^2/\varepsilon_\infty, \tag{9}$$

$$\eta = 1 + (1/\omega^2\tau^2) \to 1, \qquad \omega\tau \gg 1, \tag{10}$$

and ε_∞ is the high-frequency lattice dielectric constant. The real and imaginary parts of the complex refractive index are obtained from ε_1 and ε_2. One has

$$\varepsilon = (\varepsilon_1^2 + \varepsilon_2^2)^{1/2} = n^2 + k^2, \tag{11}$$

$$n = [(\varepsilon + \varepsilon_1)/2]^{1/2}, \tag{12}$$

$$k = \varepsilon_2/2n = [(\varepsilon - \varepsilon_1)/2]^{1/2}. \tag{13}$$

Experimentally, n and k are found from measurements of the reflectivity R of a bulk, opaque sample and the transmittance T of a slab, which are given in terms of n and k as

$$R = \frac{(n - 1)^2 + k^2}{(n + 1)^2 + k^2} \tag{14}$$

and

$$T = \frac{(1 - R)^2 e^{-2\omega kd/c}}{1 - R^2 e^{-4\omega kd/c}}, \tag{15}$$

where d is the sample thickness. For the slab multiple-reflection effects are averaged, so that interface fringes are not resolved.

For the III–V and II–VI semiconducting compounds GaAs, InP, InAs, CdTe, and ZnSe, it is known that free-carrier absorption at frequencies just below the fundamental absorption edge, which lie in the near-infrared region of the spectrum, requires a quantum-mechanical description. The experimentally observed λ^3 wavelength dependence [5–9] of the optical conductivity in this region can be accounted for by a quantum-mechanical treatment based on time-dependent perturbation theory [10–12] that utilizes the band structure of the Kane theory [13]. In the far infrared, for photon energies small compared with $k_0 T$ (k_0 is the Boltzmann constant) and with the energy $\hbar\omega_Q$ of the phonon involved in the scattering, the quantum result reduces to the λ^2 dependence given by the Drude theory, and the quasi-classical Boltzmann transport equation [1–3]. The departures from the Drude theory at high frequencies are associated mainly with k rather than n, and hence, the transmission is affected more than the reflectivity. The latter depends on k in the region of the reflectivity minimum, where $n \simeq 1$, but is determined essentially by n over the region of the absorption spectrum for which $n > k$, which is the region where departures from the Drude theory would occur.

The quantum theory of free-carrier absorption has been treated by various investigators [10–12]. The problem is of theoretical as well as practical interest, since it is related to the question of the limits of validity of the quasi-classical Boltzmann transport equation, which describes many transport phenomena in solids. It is expected that deviations from the quasi-classical Boltzmann equation, or Drude theory, which gives the same result, should occur when photon energies become large compared with electron energies. The latter are of the order of $k_0 T$ or E_f for nondegenerate or degenerate materials, respectively, where E_f is the Fermi energy measured relative to the conduction band edge. The quantum result, which predicts a λ^3 wavelength dependence of the absorption coefficient for polar-optical-mode scattering and a $\lambda^{3.5}$ to λ^4 dependence for ionized-impurity scattering, can be derived from the equation of motion of the quantum density matrix [14]. The response of electrons to a driving field may be followed from the quasi-classical limit of small ω to the quantum limit that occurs when $\hbar\omega$ is no longer small compared with characteristic energies of the system. In this case, a generalized Boltzmann equation is obtained that reduces to the quasi-classical Boltzmann transport equation when the electron wave vector q tends to zero and ω is small [14–17]. When ω is appreciable, one obtains, under certain conditions, a solution of the Boltzmann equation in terms of a frequency-dependent relaxation time. This relaxation rate, which has been tabulated as a function of frequency and carrier concentration for various materials [18–20], can be used in the usual expressions of the classical Drude theory to obtain the quantum result. In particular, the low-frequency $\hbar\omega \simeq k_0 T$ limit gives a good estimate for the dc mobility as a function of carrier concentration. At high frequencies, in lightly doped materials in which polar scattering dominates, $n\alpha$ is proportional to λ^3 and ε_2 and k are proportional to λ^4 rather than λ^3. The real part of the dielectric constant is given approximately by the Drude-theory expression and $n \simeq (\varepsilon_\infty)^{1/2}$ for $\bar{\omega}_p \ll \omega \ll G/\hbar$, where $G/\hbar$ is the frequency of the fundamental absorption edge and G is the direct-band-gap energy of the semiconductor. As ω approaches $G/\hbar$ there is a small quantum-mechanical correction to ε_1 and hence to n. A summary of the results of the quantum theory is given in Section II.

II   QUANTUM THEORY OF FREE-CARRIER ABSORPTION

Free-carrier absorption in semiconductors refers to the process of optical absorption in which the electron makes a transition from an initial state to a final state in the same band. The absorption or emission of a photon is then accompanied by scattering by optical- or acoustical-mode phonons or by charged impurities. The scattering mechanism is necessary for the conservation of crystal momentum, since the interaction with the photon provides energy but negligible momentum transfer.

To formulate the problem in terms of the quantum density-matrix equation of motion, we consider a system of dynamically independent electrons in interaction with a scattering system and an equilibrium radiation field. A perturbing electromagnetic field is then applied. The total system is now described by a quantum density matrix R with the equation of motion

$$R = R^0 + p, \qquad i\hbar\dot{p} = [H, R], \tag{16}$$

where R^0 is the equilibrium density matrix and p the deviation from equilibrium. Here H is the total Hamiltonian of the system and is given by

$$H = H^0 + H' + H^{er}, \tag{17}$$

where H^0 is the sum of the Hamiltonians of the noninteracting subsystems consisting of electrons, phonons, and photons, respectively. The scattering interaction of the electrons with the phonon field is denoted by H' and the interaction of the electrons with the quantum radiation field by H^{er}

$$H^{er} = -(e/mc)\mathbf{A} \cdot \mathbf{P}, \tag{18}$$

where $\mathbf{A}$ is the vector potential of the radiation field, $\mathbf{P}$ the momentum of the electron system, and the Coulomb gauge is used. This implies that $\mathbf{A}$ and $\mathbf{E}$ are transverse, where

$$\mathbf{E} = -(1/c)\dot{\mathbf{A}}. \tag{19}$$

The form of H' depends on the scattering mechanism. Various cases are considered [12, 14]. The complete derivation is given in Jensen [14].

We work in the number representation and take matrix elements of p between the eigenstates of $|N\rangle$ and $|N'\rangle$ of H^0. We define*

$$|N\rangle = |n\rangle|v\rangle|\gamma\rangle, \tag{20}$$

where $|n\rangle$, $|v\rangle$, and $|\gamma\rangle$ are the wave functions of the subsystems consisting of electrons, phonons, and photons.

A generalized Boltzmann equation can be derived for the matrix element of p. The deviation from equilibrium of the Fourier component of the electron distribution function g_k^q is obtained by taking the trace over $|n\rangle|v\rangle$ of

$$p_{nv}^{(2)} = (nv|p^{(2)}|nv) \tag{21}$$

and the electron number operator $(n|\hat{N}_k|n) = n_k$ as follows

$$\sum_{nv} n_k p_{nv}^{(2)} = R(\gamma)g_k^q. \tag{22}$$

The superscript on p_{nv} in Eqs. (21) and (22) denotes the result of an expansion to second order of the density-matrix equation of motion in powers of $(1/\omega)$, which is permissible when $\omega\tau \gg 1$, and $R(\gamma)$ is the photon density matrix. The electron current $\mathbf{j}^q$ is given by the expression

$$\mathbf{j}^q = \sum_k e\mathbf{v}_k g_k^q, \tag{23}$$

where $\mathbf{v}_k$ is the matrix element of the electron momentum operator between the Bloch functions of the Kane theory divided by the electron mass.

The optical conductivity is obtained by taking the trace over the remaining photon subspace of

$$\langle \mathbf{j} \cdot \mathbf{E} \rangle_q = \sigma \langle E^2 \rangle_q. \tag{24}$$

The coefficient of $\langle E^2 \rangle_q$ in Eq. (24) is identified as the optical conductivity and is related to the absorption coefficient α obtained by using second-order perturbation theory by Eq. (5). That is, one obtains the same result when using either method. The advantage of the more cumbersome density-matrix approach is that it allows insight into the extension of the Boltzmann equation at high frequencies and permits an identification of the frequency-dependent relaxation rate that reduces to a constant in the far infrared and predicts the dc mobility. In the derivation of $n\alpha$ or ε_2 using second-order, time-dependent perturbation theory, one finds

$$\alpha = n_e\Sigma, \tag{25}$$

where Σ is the scattering cross section per electron, which at high frequencies may be written as an analytical function of $\hbar\omega/G$ [12–21], where G is the

---

* No confusion should arise from the double definition of N and n.

band-gap energy in electron volts. One does not address the question of an electron relaxation time in this formulation of the problem. While the cross section is, in fact, proportional to the generalized relaxation rate, this relaxation rate does not reduce to that obtained from transport theory unless scattering is elastic and the photon energy is small compared with the average electron energy. In the case of an inelastic scattering mechanism, such as interaction with optical mode phonons (polar scattering), a modified scattering rate is found that, however, does reduce to a constant in the far infrared so that the λ^2 dependence of the Drude theory is obtained for $\omega\tau \gg 1$.

III THEORETICAL RESULTS

The results obtained for the conductivity σ by using the above method can be written in terms of a generalized relaxation rate $1/\tau$, where τ reduces to the usual relaxation time under conditions previously discussed. When H' is the interaction between electrons and polar-optical-mode phonons, the conductivities $\sigma_\pm$ are found as follows, where the subscripts $+$ and $-$ correspond to photon absorption accompanied by phonon absorption and phonon emission, respectively, and the total optical conductivity for polar scattering is the sum of the two terms:

$$\sigma_\pm = (\omega_p^2/4\pi\omega^2)\langle 1/\tau \rangle_\pm, \qquad \omega\tau \gg 1. \qquad (26)$$

The generalized relaxation rate is defined as follows:

$$\langle 1/\tau \rangle_\pm = (1/\tau_p^0)[F_\pm(\omega, \omega_Q)/X](I_\pm/12n_eV_c). \qquad (27)$$

The conductivity and generalized relaxation times in Eqs. (26) and (27) are given in terms of the following parameters, which are obtained experimentally:

$$1/\tau_p^0 = g_p/\pi\hbar^2\alpha_0 c, \qquad\qquad \alpha_0 = (G/2m_nc^2)^{1/2}, \qquad (28)$$

$$g_p = 2\pi e^2\hbar\omega_Q[(1/e_\infty) - (1/e_0)], \qquad X = \hbar\omega/G, \qquad (29)$$

$$1/V_c = 2/\pi^2\lambda_c^3, \qquad\qquad \lambda_c = \hbar/m_n\alpha_0 c. \qquad (30)$$

In Eq. (27), $F_\pm(\omega, \omega_Q)$ is a function of temperature and frequency that involves the optical-phonon occupation probability, G is the band-gap energy at $k = 0$, m_n is the effective mass at the band edge, $\hbar\omega_Q$ is the optical-phonon energy, and ε_0 and ε_∞ are the static and high-frequency lattice dielectric constants, respectively. The term V_c is a characteristic volume element associated with the electron density of states, and it has been written in terms of the wavelength $\lambda_c \to \hbar/mc$, $m_n \to m$, $\alpha_0 \to 1$ to utilize the analogy between the Kane theory and the Dirac theory [12, 13]. The quantity $I_+/12n_e V_c X$, where I_+ is an integral over initial electron energies, is essentially frequency independent at low frequencies and becomes independent of carrier concentration and a function only of frequency at high frequencies. In this limit, the absorption coefficient becomes proportional to the carrier concentration times a frequency-dependent cross section $\Sigma_\pm$, which is related to the scattering rate as follows for $\omega\tau \gg 1$:

$$n\Sigma_\pm = (1/n_e)(\omega_p^2/\omega^2 c)\langle 1/\tau \rangle_\pm = n\Sigma_0 \frac{F_\pm(\omega, \omega_Q)}{X^3} \frac{I_\pm}{n_e V_c}, \tag{31}$$

$$n\Sigma_0 = (1/12n_e)(\omega_p^2/\omega_g^2 c)1/\tau_p^0, \tag{32}$$

$$w_g = G/\hbar. \tag{33}$$

For the case of impurity scattering, the optical conductivity σ^I assumes the form

$$\sigma^I = (\omega_p^2/4\pi\omega^2)\langle 1/\tau^I \rangle, \tag{34}$$

$$\langle 1/\tau^I \rangle = (1/\tau_I^0)I_+^I/6n_e V_c X, \tag{35}$$

$$1/\tau_I^0 = g^I/\pi\hbar^2\alpha_0 c, \tag{36}$$

$$g^I = 8N_i V_c(e^2/e_0)^2/\lambda_c, \tag{37}$$

where N_i is the impurity concentration and I_+^I is again an integral over initial electron energies of a function appropriate to impurity scattering. The optical conductivity can be expressed in terms of a cross section per electron Σ^I calculated from perturbation theory as

$$4\pi\sigma^I/c = n\alpha^I = nn_e\Sigma^I, \tag{38}$$

$$n\Sigma^I = (n\Sigma_0^I/X^3)I_+^I/2n_e V_c, \tag{39}$$

$$n\Sigma_0^I = (1/12n_e)(\omega_p^2/\omega_g^2 c)1/\tau_I^0. \tag{40}$$

The optical conductivity due to impurity scattering thus assumes the form of the usual Drude expression if written in terms of an effective relaxation rate $\langle 1/\tau^I \rangle$ that is related to the cross section per electron as

$$n\Sigma^I = (1/n_e)(\omega_p^2/\omega^2 c)\langle 1/\tau^I \rangle. \tag{41}$$

The total conductivity σ, which is the sum of the conductivities due to polar optical mode and impurity scattering, is proportional to the total relaxation rate and the total absorption coefficient α, where

$$4\pi\sigma = 4\pi(\sigma_+ + \sigma_- + \sigma^I) = (\omega_p^2/\omega^2)1/\tau = cn\alpha, \tag{42}$$

$$1/\tau = \langle 1/\tau \rangle_+ + \langle 1/\tau \rangle_- + \langle 1/\tau^I \rangle. \tag{43}$$

Equation (42) gives the high-frequency limit in which $\omega\tau \gg 1$. The transition to the low-frequency regime $\omega\tau \leq 1$ is given by

$$\sigma = \sigma_0/\omega^2\tau^2\eta = \sigma_0/(1 + \omega^2\tau^2) \rightarrow \begin{cases} \sigma_0, & \omega\tau \ll 1 \\ \sigma_0/\omega^2\tau^2, & \omega\tau \gg 1, \end{cases} \tag{44}$$

where

$$\sigma_0 = n_e e^2\tau/m_n = en_e\mu, \qquad \mu = e\tau/m_n. \tag{45}$$

For degenerate materials, m_n in Eq. (45) should be replaced by

$$m^* = m_n(1 + 2E_f/G), \tag{46}$$

which is the effective mass at the Fermi surface from the Kane theory. Detailed calculations of the scattering rates are given by Jensen [12, 14, 21]. Numerical results for various compounds are also given by Jensen [18–20] for $1/\tau$ as a function of n_e and ω, from which the complex dielectric constant in Eq. (8) can be immediately calculated. The remaining quantities, including the complex refractive index, absorption coefficient, optical conductivity, and skin depth, then follow as discussed in Section I.

Values of n, k, and R versus wavelength, calculated by using the method discussed for GaAs, InP, and InAs, are tabulated in Section IV.

IV   COMPARISON WITH EXPERIMENTAL DATA

The theoretical results are calculated by using experimental values for the various quantities given in Table I. For polar-optical mode scattering, no adjustable parameters are involved. Theoretical results are obtained for the scattering rate $1/\tau$ in Eq. (43). This enables one to calculate ε_1 and ε_2 in Eqs. (7) and (8), from which n and k are obtained by using Eqs. (11–13). The reflectivity R is found as a function of n and k from Eq. (14). This is the quantity measured experimentally.

A calculation of ε_1 appropriate to electrons in polar semiconductors with the band structure of the Kane theory and based on the quantum density-matrix equation of motion yields a high-frequency modification to Eq. (7). For $\omega\tau \gg 1$ and $X < 0.1$, one obtains [18, 22]

$$\begin{aligned} \varepsilon_1 &= [\varepsilon_\infty/(1 - X)][1 - (X/\varepsilon_\infty) - \bar{\omega}_p^2/\omega^2] \\ &\cong [\varepsilon_\infty/(1 - X)](1 - \bar{\omega}_p^2/\omega^2) \\ &\rightarrow \varepsilon_\infty(1 - \bar{\omega}_p^2/\omega^2), \qquad X \ll 1. \end{aligned} \tag{47}$$

TABLE I

Parameters Used to Calculate Optical Constants

	InP	GaAs	InAs
G (eV)	1.35	1.43	0.36
m_n/m	0.073	0.071	0.024
$\hbar\omega_Q$ (eV)	0.0428	0.0349	0.0299
ε_∞	9.61	11.1	11.6
$1/\varepsilon_\infty - 1/\varepsilon_0$	0.02767	0.0135	0.01619
α_0	$\frac{1}{234}$	$\frac{1}{224}$	$\frac{1}{266}$
g (eV^2 cm)	1.13×10^{-9}	4.25×10^{-10}	4.39×10^{-10}
$1/\tau_p$ sec^{-1}	6.46×10^{12}	2.33×10^{12}	2.86×10^{12}
$e\tau_p/m_n$ (cm^2)/V sec[a]	3.37×10^3	1.06×10^4	2.59×10^4
μ (cm^2)/V sec[b]	4.53×10^3	8.3×10^3	
μ (cm^2)/V sec[c]	4.5×10^3	8.5×10^3	3×10^4

[a] $T = 300$ K.
[b] Theoretical value.
[c] Experimental value.

We note that $1/\varepsilon_\infty \lesssim 0.1$ and hence $X/\varepsilon_\infty \ll 1$ for compounds we consider, and this term can be neglected. In the limit $X \ll 1$, the quasi-classical high-frequency Drude result is recovered, as required. For $X \sim 0.1$, there is a high-frequency correction given by Eq. (47), which is used to calculate the numerical values of $\varepsilon_1 = n^2 - k^2$. The major modification of the classical result is dispersion in n as one approaches the fundamental absorption edge [22].

Figure 1 shows the impurity scattering rate calculated for various samples of InP described in Table II. As the free-carrier density increases from sample 1 to sample 8, the scattering rate increases. In Fig. 2 the relaxation rate for polar scattering $1/\tau_p = \langle 1/\tau_+ \rangle + \langle 1/\tau_- \rangle$ is shown for the same samples. In the quantum limit, this quantity varies as λ and is roughly independent of carrier concentration. At low frequencies, such that $\hbar\omega \sim k_0 T$, the Drude limit is recovered and the polar scattering rate converges to a concentration-dependent constant. If the scattering rate for impurity scattering is now included, the total relaxation time τ in the Drude limit can be used to predict the dc mobility. The total scattering rate is $1/\tau$ plotted for the samples of InP in Fig. 3. The calculated optical constants n and k for some of the samples of Table II are listed in Table III.

Figure 4 shows the total scattering rate for various samples of GaAs (Table IV) versus $X = \hbar\omega/G$ at $T = 300$ K. The intersection of the curve $1/\tau$ with ω gives the point at which $\omega\tau = 1$ for a given sample. Sample a shows the contribution of polar scattering only while all other samples include the contributions of polar plus impurity scattering for various values of the

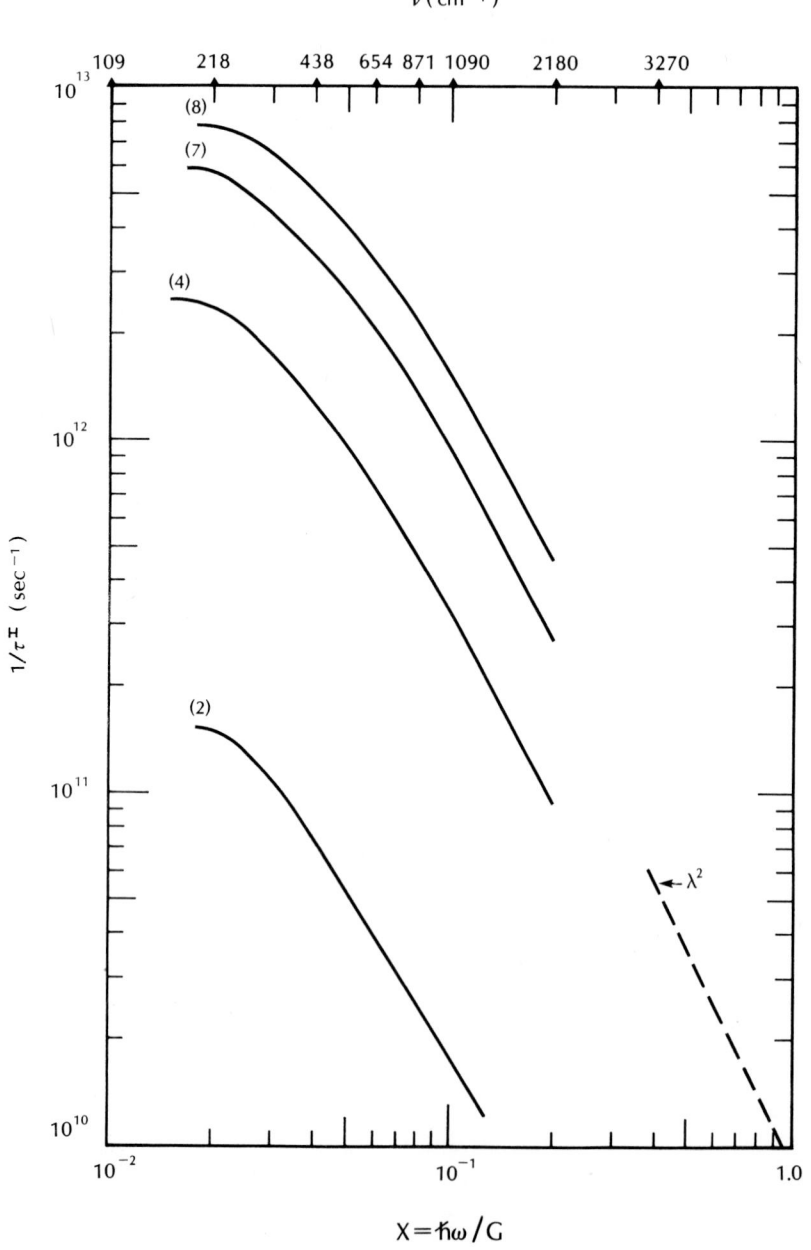

Fig. 1. The relaxation rate for impurity scattering calculated for the samples of InP at 300 K as described in Table II. (Sample numbers in parentheses.)

TABLE II

InP Samples of Varying Free-Carrier Concentration[a]

Sample	Carrier concentration n_e (cm^{-3})
1	4.69×10^{15}
2	2.25×10^{16}
3	6.27×10^{16}
4	4.13×10^{17}
6	6.17×10^{17}
7	1.23×10^{18}
8	2.13×10^{18}

[a] Data tabulated in Table III and/or plotted in Figs. 1–3.

carrier concentration. Uncompensated samples such that $N_i = n_e$ are considered. Polar scattering dominates at high frequencies for all but the most heavily doped samples shown, while in the Drude limit one obtains a frequency-independent scattering rate that is a function of carrier concentration. In Fig. 5, the Drude limit is used to calculate the dc mobility at $T = 300$ K as a function of free-carrier density for GaAs and comparison is made with various experimental results. In Figs. 6 and 7, the scattering rate at $T = 300$ K has been used to calculate the real and imaginary parts of the complex index of refraction of GaAs samples that are listed in Table IV.

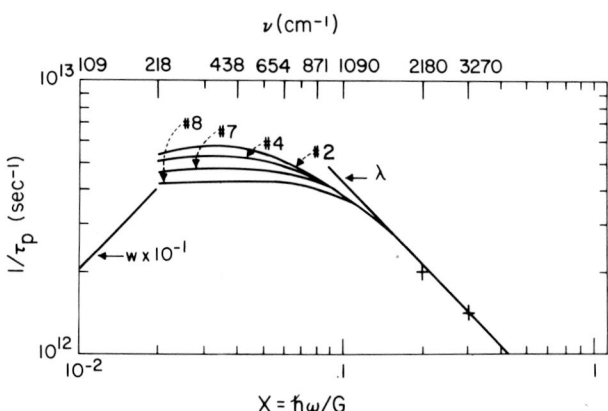

Fig. 2. The relaxation rate for polar scattering calculated for the samples of InP described in Table II. The high-frequency quantum limit is indicated by crossmarks.

$\nu \, (\mathrm{cm}^{-1})$

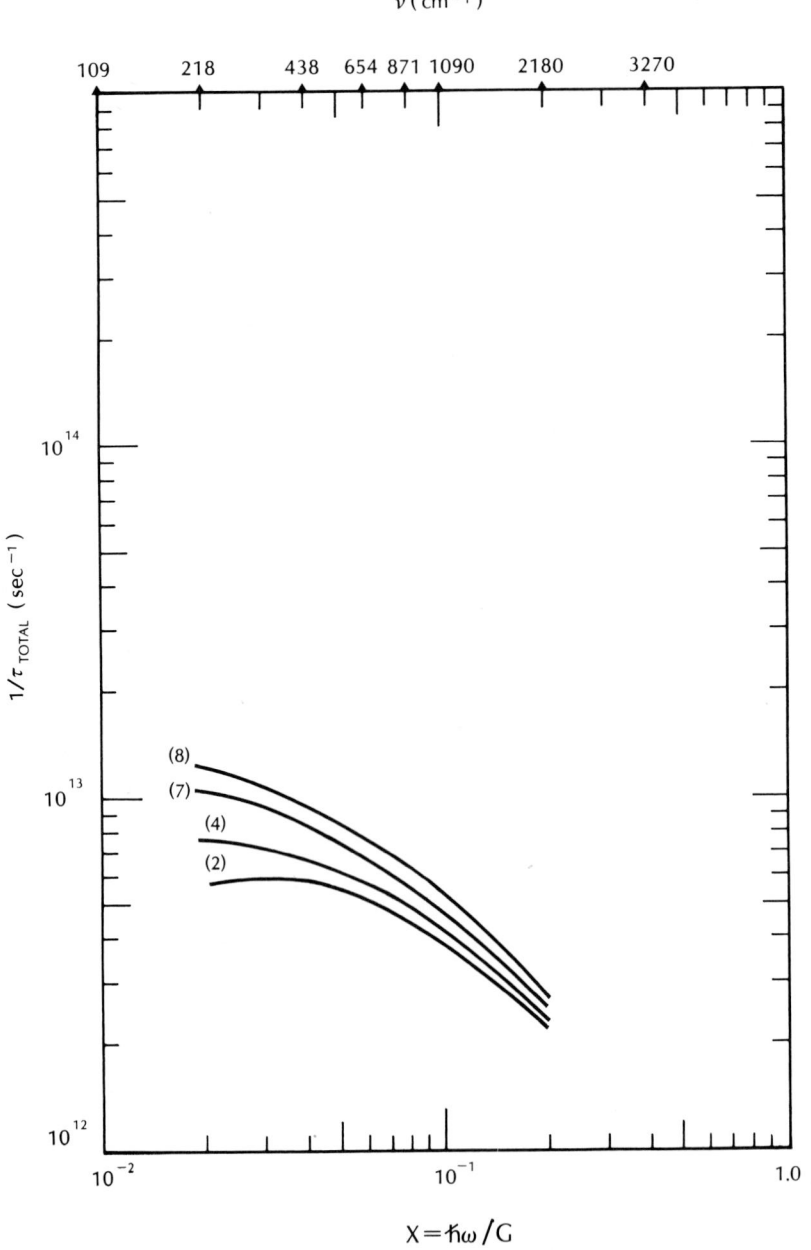

Fig. 3. The total relaxation rate for samples of InP at 300 K, given by the sum of the relaxation rates for polar and for impurity scattering. (Sample numbers in parentheses.)

TABLE III

Optical Constants of InP at 300 K as a Function of Free-Carrier Density

Sample			1		3		6		8	
eV	cm⁻¹	μm	n	k	n	k	n	k	n	k
0.1351	1089	9.18	3.25	1.12×10^{-5}	3.24	1.53×10^{-4}	3.14	1.88×10^{-3}	2.85	1.01×10^{-2}
0.1078	869.6	11.5	3.22	2.53×10^{-5}	3.20	3.47×10^{-4}	3.05	4.51×10^{-3}	2.57	2.66×10^{-2}
0.08103	653.6	15.3	3.19	7.02×10^{-5}	3.16	9.82×10^{-4}	2.88	1.37×10^{-2}	1.92	0.106
0.06199	500.0	20	3.16	1.74×10^{-4}	3.115	2.46×10^{-3}	2.62	3.92×10^{-2}	0.516	1.04
0.05391	434.8	23	3.15	2.72×10^{-4}	3.09	3.90×10^{-3}	2.42	6.83×10^{-2}	0.421	2.08
0.04959	400.0	25	3.15	3.52×10^{-4}	3.07	5.08×10^{-3}	2.25	9.98×10^{-2}	0.458	2.62
0.04133	333.3	30	3.14	6.02×10^{-4}	3.03	8.87×10^{-3}	1.73	0.241	0.597	3.72
0.03542	285.7	35	3.13	9.40×10^{-4}	2.98	1.43×10^{-2}	0.999	0.702	0.795	4.68
0.03100	250	40	3.12	1.36×10^{-3}	2.93	2.15×10^{-2}	0.662	1.61	1.02	5.52
0.02701	217.9	45.9	3.11	2.03×10^{-3}	2.86	3.35×10^{-2}	0.674	2.49	1.33	6.49
0.01351	108.9	91.8	3.04	1.61×10^{-2}	1.91	0.384	1.72	6.88	4.14	12.5

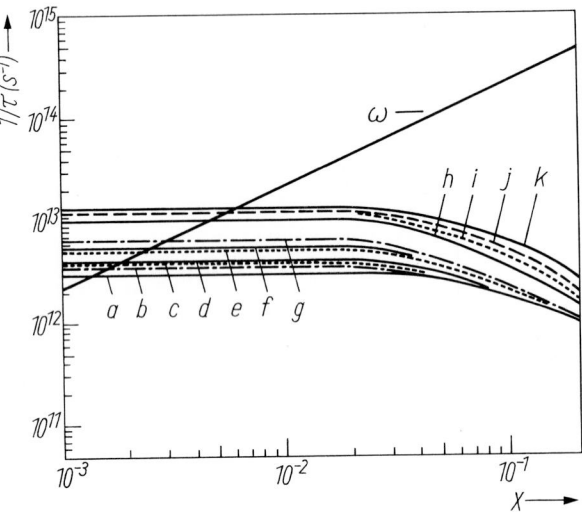

Fig. 4. The total frequency-dependent relaxation rate versus $x = \hbar\omega/G$ for samples of GaAs of varying carrier concentration listed in Table IV. The intersection of the curve ω with the curve $1/\tau$ for a given sample specifies the point at which $\omega\tau = 1$.

TABLE IV

GaAs Samples of Varying Free-Carrier (Electron) Concentration[a]

Sample	Carrier concentration n_e (cm^{-3})
a	3.38×10^{15}
b	2.75×10^{16}
c	6.33×10^{16}
d	1.41×10^{17}
e	3.37×10^{17}
f	4.27×10^{17}
g	5.84×10^{17}
h	2.16×10^{18}
i	3.38×10^{18}
j	4.85×10^{18}
k	8.73×10^{18}

[a] Data tabulated in Table V and/or Figs. 4–8.

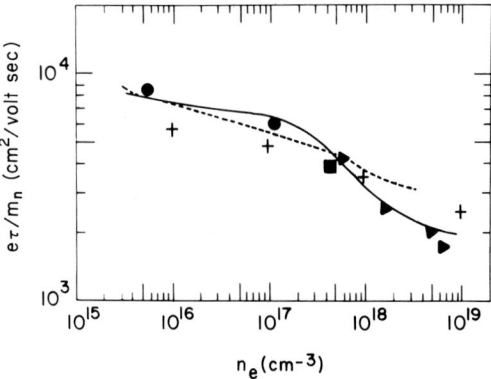

Fig. 5. Theoretical results (for GaAs) for mobility $e\tau/m_n$, $m_n = 0.071$ m, are given by the solid curve. The effective mass at the Fermi surface should be used to calculate the mobility at high carrier concentrations $\gtrsim 10^{18}$ cm^{-3}. The ratio of the effective mass at the Fermi surface to m_n is 1.10, 1.14, 1.18, and 1.26 for samples h, i, j, and k, described in Table IV, respectively. Prior results are: theoretical: –––, Madelung [31]; +, Perhowitz [3]; ■, Vakulenko [23]; experimental: ▲, ●, Madelung [31].

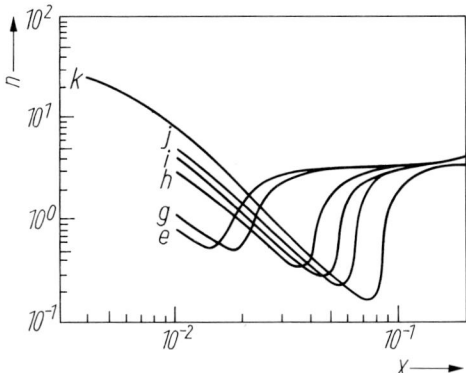

Fig. 6. The real part of the refractive index versus the photon energy in units of G for samples of GaAs described in Table IV.

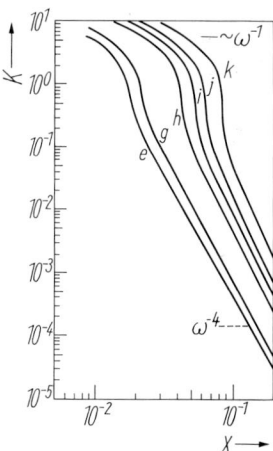

Fig. 7. The imaginary part of the refractive index versus the photon energy in units of G for samples of GaAs described in Table IV.

TABLE V

Optical Constants of GaAs at 300 K as a Function of Free-Carrier Density

Sample			a		b		g		h		j		k	
eV	cm^{-1}	μm	n	k	n	k	n	k	n	k	n	k	n	k
0.1430	1153	8.67	3.51	3.57×10^{-6}	3.51	2.89×10^{-5}	3.42	8.14×10^{-4}	3.17	4.90×10^{-3}	2.69	1.89×10^{-2}	1.78	6.37×10^{-2}
0.1287	1038	9.63	3.49	8.19×10^{-6}	3.49	4.34×10^{-5}	3.38	1.24×10^{-3}	3.07	7.89×10^{-3}	2.44	3.15×10^{-2}	0.972	0.171
0.1148	925.9	10.8	3.47	8.29×10^{-6}	3.47	6.72×10^{-5}	3.33	1.99×10^{-3}	2.93	1.32×10^{-2}	2.07	5.89×10^{-2}	0.186	1.40
0.09999	806.5	12.4	3.45	1.37×10^{-5}	3.45	1.11×10^{-4}	3.27	3.41×10^{-3}	2.73	2.41×10^{-2}	1.37	0.144	0.169	2.48
0.09537	769.2	13									0.920	0.259		
0.08856	714.3	14									0.306	1.03		
0.08551	689.7	14.5	3.44	2.38×10^{-5}	3.43	1.93×10^{-4}	3.19	6.20×10^{-3}	2.40	4.90×10^{-2}	0.262	1.34	0.205	3.53
0.07167	578.0	17.3	3.42	4.49×10^{-5}	3.40	3.65×10^{-4}	3.06	1.27×10^{-2}	1.76	0.132	0.244	2.76	0.293	4.78
0.06888	555.6	18							1.55	0.173				
0.06526	526.3	19							1.19	0.280				
0.06199	500.0	20							0.732	0.542				
0.05714	460.8	21.7	3.40	9.48×10^{-5}	3.38	8.00×10^{-4}	2.82	3.12×10^{-2}	0.394	1.35	0.345	4.26	0.453	6.45
0.04290	346.0	28.9	3.38	2.38×10^{-4}	3.34	2.05×10^{-3}	2.27	0.108	0.425	3.44	0.632	6.35	0.835	9.00
0.03875	312.5	32	3.37	3.30×10^{-4}	3.32	2.90×10^{-3}	1.94	0.187	0.540	4.10	0.778	7.17	1.12	10.0
0.03647	294.1	34	3.36	4.00×10^{-4}	3.31	3.49×10^{-3}	1.67	0.263	0.599	4.51	0.892	7.69	1.28	10.7
0.03444	277.8	36	3.36	4.75×10^{-4}	3.30	4.17×10^{-3}	1.36	0.390	0.664	4.90	1.01	8.20	1.45	11.3
0.03263	263.2	38	3.36	5.59×10^{-4}	3.29	4.93×10^{-3}	1.01	0.627	0.742	5.27	1.12	8.69	1.61	12.0
0.02857	230.4	43.4	3.35	8.47×10^{-4}	3.27	7.49×10^{-3}	0.575	1.70	0.895	6.25	1.48	9.99	2.07	13.6
0.02254	181.8	55	3.34	1.70×10^{-3}	3.20	1.57×10^{-2}	0.598	3.31	1.38	8.21	2.27	12.7	3.26	17.1
0.01879	151.5	66					0.765	4.53	1.86	9.97	3.15	15.1	4.45	20.3
0.01746	140.8	71	3.32	3.66×10^{-3}	3.08	3.47×10^{-2}	0.833	5.00	2.17	10.7	3.58	16.1	5.12	21.6
0.01430	115.3	86.7	3.30	6.57×10^{-3}	2.94	6.54×10^{-2}	1.16	6.46	3.00	13.0	5.06	19.2	8.08	24.4
0.01240	100.0	100	3.28	1.02×10^{-2}	2.79	0.105	1.46	7.59	3.90	14.7	6.40	21.6	9.05	28.5
0.007948	64.10	156					3.11	11.8	7.72	21.1	12.4	29.9	17.1	39.1
0.004959	40.00	250	2.98	0.155	1.34	2.98	6.50	17.4	14.4	28.1	21.9	39.4	29.3	50.4
0.003723	30.03	333					9.62	21.1	19.4	33.2	28.8	45.3	38.2	58.3

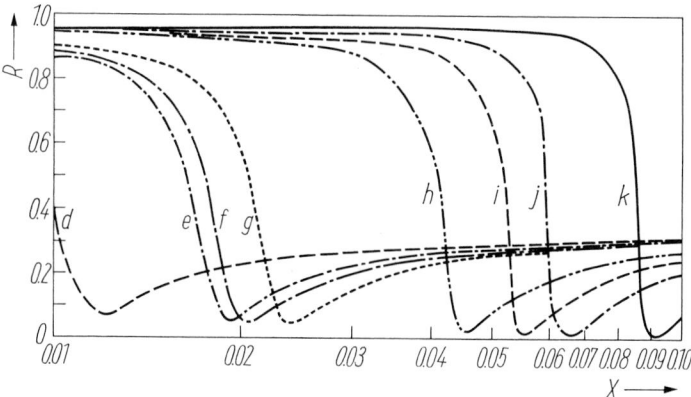

Fig. 8. Reflectivity curves for samples of GaAs of varying carrier concentration calculated from values of n and k shown in Figs. 6 and 7.

The ω^{-4} dependence of k at high frequencies, which corresponds to a λ^3 rather than a λ^2 dependence of the absorption coefficient, has been reported by many authors [23–30]. The absorption coefficient is calculated and compared with experimental results by Jensen [12]. The reflectivity calculated from the n and k values is shown in Fig. 8 for GaAs. Note that the reststrahlen features near $X = 0.024$ are not present. The addition of these features is discussed in the appendix.

For InAs samples listed in Table VI, the optical constants are given in Table VII. Enough points are tabulated to allow the reader to plot the data and interpolate for other wavelengths.

In summary, the Drude theory of the optical constants of III–V and II–VI polar semiconductors can be extended into the near infrared in terms of a frequency-dependent electron scattering rate formulated to give the quantum result when used in the classical Drude theory. In the far infrared, this scattering rate reduces to a concentration-dependent but frequency-independent constant that predicts the dc mobility as a function of carrier concentration for the various samples.

TABLE VI

InAs Samples of Varying Free-Carrier (Electron) Concentration[a]

Sample	Carrier concentration n_e (cm^{-3})
E	5.07×10^{16}
H	1.12×10^{17}
J	2.16×10^{17}

[a] Data tabulated in Table VII.

TABLE VII

Optical Constants of InAs at 300 K as a Function of Free-Carrier Density

Sample			E		H		J	
eV	cm⁻¹	μm	n	k	n	k	n	k
0.07208	581.4	17.2	3.68	7.91×10^{-4}	3.60	1.75×10^{-3}	3.42	3.90×10^{-3}
0.03604	290.7	34.4	3.21	1.04×10^{-2}	2.84	2.59×10^{-2}	1.93	8.25×10^{-2}
0.03444	277.8	36	3.17	1.22×10^{-2}	2.60	3.72×10^{-2}	1.12	0.194
0.03263	263.2	38	3.11	1.50×10^{-2}	2.46	4.77×10^{-2}	0.473	0.561
0.03100	250.0	40	3.05	1.80×10^{-2}	2.31	6.06×10^{-2}	0.256	1.23
0.02877	232.0	43.1	2.95	2.30×10^{-2}	2.04	8.85×10^{-2}	0.210	1.91
0.02480	200.0	50	2.71	4.14×10^{-2}	1.14	0.259	0.221	3.03
0.02160	174.2	57.4	2.38	7.10×10^{-2}	0.572	0.712	0.265	3.43
0.01440	116.1	86.1	0.357	1.61	0.349	3.92	0.478	6.37
0.01078	86.95	115	0.367	3.68	0.536	6.00	0.798	8.94
0.007208	58.14	172	0.676	6.66	1.10	9.64	1.70	13.7
0.003604	29.07	344	2.37	14.1	3.93	19.2	6.08	26.4
0.002877	23.20	431	3.55	17.4	5.81	23.4	8.89	31.8
0.002160	17.42	574	5.86	22.4	9.33	29.6	14.0	39.6
0.001440	11.61	861	11.2	30.8	16.8	39.3	24.3	51.6
0.0007200	5.807	1772	26.5	46.9	36.1	57.2	49.1	73.4
0.0003600	2.904	3444	48.1	64.7	61.6	78.0	81.3	99.7
0.0002880	2.323	4305	56.2	71.4	71.1	86.0	93.4	110
0.0002160	1.742	5740	67.7	81.0	84.8	97.7	111	125
0.00007200	0.5807	17220	126	134	155	163	201	210
0.00003600	0.2904	34440	181	187	223	228	288	294

APPENDIX

The optical constants n and k have been calculated in this chapter for the free carriers in the limit that the lattice dielectric constant can be approximated by a constant $\varepsilon_l = \varepsilon_\infty$. This approximation is not valid in the frequency range in which the reststrahlen contribution of the lattice is important, i.e., for $\omega_T \lesssim \omega \lesssim \omega_L$, where ω_L and ω_T are the angular frequencies of the longitudinal and transverse optical phonons, respectively. This covers a range from approximately 268 to 292 cm^{-1} for GaAs, from 307 to 351 cm^{-1} for InP, and from 219 to 243 cm^{-1} for InAs [31]. The purpose of this appendix is to consider how the results given here can be combined with the results in the critiques of InP, GaAs, and InAs to obtain the total n_t and k_t in the reststrahlen region and elsewhere.

The results in Tables III, V, and VII of this chapter are n_f and k_f for free carriers such that

$$(n_f - ik_f)^2 = \varepsilon_f = \varepsilon_{f1} - i\varepsilon_{f2}, \tag{A.1}$$

where ε_{f1} and ε_{f2} are as given by Eqs. (7) and (8).

The results in the critiques for InP, GaAs, and InAs, n_l and k_l may be considered as having been obtained from the Lorentz oscillator model with

$$(n_l - ik_l)^2 = \varepsilon_l = \varepsilon_{l1} - i\varepsilon_{l2} = \varepsilon_\infty\left(1 + \frac{\omega_L^2 - \omega_T^2}{\omega_T^2 - \omega^2 + i\Gamma\omega}\right). \tag{A.2}$$

Generally, we can assume that dielectric functions are additive (while n and k are not), so that the total dielectric function is

$$\varepsilon_t = \varepsilon_f + \varepsilon_l - \varepsilon_\infty. \tag{A.3}$$

It is necessary to subtract off one ε_∞ either from ε_f or ε_l since each separate consideration of ε_f and ε_l contains an ε_∞. Then it is straightforward, for example, to construct $\varepsilon_{l1} = n_l^2 - k_l^2$ and $\varepsilon_{l2} = 2n_lk_l$ from Table I in the GaAs critique as well as $\varepsilon_{f1} = n_f^2 - k_f^2$ and $\varepsilon_{f2} = 2n_fk_f$ from Table V of this Chapter. We have

$$(n_t - ik_t)^2 = \varepsilon_{t1} - i\varepsilon_{t2} = \varepsilon_{f1} - i\varepsilon_{f2} + \varepsilon_{l1} - i\varepsilon_{l2} - \varepsilon_\infty$$
$$= (\varepsilon_{f1} + \varepsilon_{l1} - \varepsilon_\infty) - i(\varepsilon_{f2} + \varepsilon_{l2}). \tag{A.4}$$

Then we find that

$$\varepsilon_{t1} = n_t^2 - k_t^2 = \varepsilon_{f1} + \varepsilon_{l1} - \varepsilon_\infty, \qquad \varepsilon_{t2} = 2n_tK_t = \varepsilon_{f2} + \varepsilon_{l2}. \tag{A.5}$$

We then extract n_t and k_t from

$$2n_t^2 = \varepsilon_{t1} \pm \sqrt{\varepsilon_{t1}^2 + \varepsilon_{t2}^2}, \qquad 2k_t^2 = -\varepsilon_{t1} \pm \sqrt{\varepsilon_{t1}^2 + \varepsilon_{t2}^2}. \tag{A.6}$$

Since the ε_∞ used to construct Table V is slightly different from the ε_∞ used to obtain Table I in the GaAs critique and approximations are made in the calculation of ε_f, this procedure is not accurate for short wavelengths when $\lambda < 10~\mu$m.

Well below the reststrahlen region (say $\omega < \omega_T/2$), the lattice contribution to ε_t approaches ε_0, the static dielectric constant. Then the determination of ε_t becomes especially simple, only requiring the subtraction of ε_∞ from $\varepsilon_{f1} = n_f^2 - k_f^2$ and the addition of ε_0. Nothing need be done to $\varepsilon_{f2} = 2n_f k_f$.

REFERENCES

1. E. D. Palik and R. T. Holm., *in* "Nondestructive Evaluation of Semiconductor Materials and Devices" (J. N. Zemel, ed.), Chap. 7, Plenum, New York, 1979.
2. R. T. Holm, J. W. Gibson, and E. D. Palik, *J. Appl. Phys.* **48**, 212 (1977).
3. S. Perkowitz and R. H. Thorland, *Phys. Rev. B* **9**, 545 (1974); S. Perkowitz, *J. Phys. Chem. Solids* **32**, 2267 (1971).
4. H. Y. Fan, *in* "Semiconductors and Semimetals" (R. K. Willardson and A. C. Beer eds.), Vol. 3, Academic Press, New York, 1967.
5. J. K. Kung and W. G. Spitzer, *J. Electrochem. Soc.* **121**, 1482 (1974).
6. W. G. Spitzer and J. W. Whelan, *Phys. Rev.* **114**, 59 (1959).
7. W. P. Dumke, M. R. Lorenz, and G. D. Pettit, *Phys. Rev. B* **1**, 4668 (1970).
8. B. V. Dutt, O. Kim, and W. G. Spitzer, *J. Appl. Phys.* **48**, 2110 (1977).
9. B. V. Dutt, M. Al-Delaimi, and W. G. Spitzer, *J. Appl. Phys.* **47**, 565 (1976).
10. W. P. Dumke, *Phys. Rev.* **124**, 1813 (1961).
11. E. Haga and H. J. Kimura, *J. Phys. Soc. Jpn.* **18**, 777 (1963); E. Haga and H. J. Kimura *J. Phys. Soc. Jpn.* **19**, 471, 658, 1596 (1964).
12. B. Jensen, *Ann. Phys.* **80**, 284 (1973).
13. E. O. Kane, *J. Phys. Chem. Solids* **1**, 249 (1957).
14. B. Jensen, *Ann. Phys.* **95**, 229 (1975).
15. P. J. Price, *IBM J. Res. Dev.* **10**, 395 (1966).
16. P. N. Argyres, *J. Phys. Chem. Solids* **19**, 66 (1961).
17. W. Kohn and J. M. Luttinger, *Phys. Rev.* **108**, 590 (1957); W. Kohn and J. M. Luttinger, *Phys. Rev.* **109**, 1892 (1958).
18. B. Jensen, *Phys. Status Solidi* **86**, 291 (1978); B. Jensen, *Solid State Commun.* **24**, 853 (1977).
19. B. Jensen, *J. Appl. Phys.* **50**, 5800 (1979).
20. B. Jensen, *in* "Laser Induced Damage in Optical Materials; 1980" (H. E. Bennett, A. J. Glass, A. H. Guenther, and B. D. Newman, eds.), p. 416, National Bureau of Standards Special Publication 620, Boulder, Colorado, 1981.
21. B. Jensen, *in* "Infrared and Submillimeter Waves" (K. J. Button, ed.), Vol. 8, Academic Press, New York, 1983.
22. B. Jensen, *IEEE J. Quantum Electron.* **QE-18**, 1361 (1982); B. Jensen and A. Torabi, *IEEE J. Quantum Electron.* **QE-19**, 448–457, 877–882, 1362–1365 (1983); B. Jensen and A. Torabi, *J. Appl. Phys.* **54**, 2030–2035, 3623–3625, 5945–5949 (1983).
23. O. V. Vakulenko and M. P. Lisitsa, *Sov. Phys, Solid State* **9**, 769 (1967).
24. M. G. Mil'vidskii, V. B. Osvenskii, E. P. Rashevskaya, and T. G. Yugova, *Sov. Phys. Solid State* **7**, 2784 (1966).
25. E. P. Rashevskaya and V. I. Fistul', *Sov. Phys. Solid State* **9**, 2849 (1968).
26. J. Dixon, *Proc. Int. Conf. Phys. Semicond., Prague, 1960*, p. 366, Publishing House Czech. Acad. Sci., Prague, 1961.
27. R. M. Culpepper and J. R. Dixon, *J. Opt. Soc. Am.* **58**, 96 (1968).
28. A. J. Strauss and G. W. Isler, *Bull. Am. Phys. Soc.* **17**, 326 (1972).
29. R. Newman, *Phys. Rev.* **111**, 1518 (1958).
30. A. Kahan, AFCRL, Physical Sciences Research Paper. 537, AFCRL-TR-73-0122 (1973).
31. O. Madelung, "Physics of the III–V Compounds," Wiley, New York, 1964.

Chapter **10**

Interband Absorption
—Mechanisms and Interpretation

DAVID W. LYNCH

Department of Physics and Ames Laboratory, U.S. Department of Energy
Iowa State University
Ames, Iowa

I.	Introduction	189
II.	One-Electron Model	190
III.	Electron–Hole Interaction, Excitons	198
IV.	Local Field Effects	203
V.	Examples	204
	References	210
	General References	211

INTRODUCTION   I

In this chapter we discuss the origin or interpretation of the optical properties of solids at and above the band gap. We thus discuss interband transitions but not intraband transitions (Drude or free-carrier absorption) or
optical absorption arising from the excitation of phonons or electrons on
impurities. Plasmons, collective excitations of an electron gas, can be excited
optically in special geometries, but we omit discussing them. At rather high
energies, e.g., above 10 or 20 eV, electrons can be excited from core levels
that are localized in space and in energy. These have some characteristics of
interband transitions, but space does not permit a description of the phenomena contributing to the edge shapes and structures above the edges that
are observed in some spectra above about 10 eV. These, too, are optical
properties, but they are studied mainly by transmission through thin films or
by photoemission. This chapter will be without great depth, and we refer the
reader to many references that cover more extensively the material to be
presented in this chapter.

189

We shall begin with the one-electron picture and go as far as we can with it. Specific many-body effects will be introduced in this chapter. The one-electron picture can be applied quantitatively to many solids, although if the number of atoms per unit cell becomes large, it is impractical to do so. Many-body effects are much more difficult to incorporate in an arbitrary material. As expected, most progress has been made on a model solid, the free-electron gas, with extensions to a free-electron gas with a band gap, which may resemble adequately a semiconductor or insulator. However, to calculate realistically many-body effects in an arbitrary material is not yet feasible, and we can make only qualitative interpretations based on calculations on model systems. We conclude this chapter with a few examples of spectra of typical solids and attempt to point out what we do and do not know at this time.

II ONE-ELECTRON MODEL

The calculation of the one-electron optical properties of solids arising from interband transitions can be carried out in a number of ways, all leading to essentially the same result. These calculations often are analogous to the calculation of the optical absorption spectrum of an atom. In the outline that follows, the electric field of the incident electromagnetic wave is treated as a time-dependent perturbation that induces transitions from filled eigenstates to empty eigenstates of the crystal Hamiltonian or an approximation to it. The same method is applicable to amorphous and crystalline solids, and to alloys as well, although only in the case of crystalline solids are the other quantum numbers known, simplifying the problem considerably.

The solution to the Schrödinger equation for a single electron in the crystal is assumed to be known. The eigenfunctions are Bloch functions characterized by an energy $E_n(\mathbf{k})$, a wave vector $\mathbf{k}$ in the first Brillouin zone, a band index n, and a spin index. The relationship $E_n(\mathbf{k})$ is the band structure of the crystal. We suppress the spin index and write the wave function as $\psi_{n,\mathbf{k}}(\mathbf{r})$.

In the sample there is an electromagnetic wave with electric field

$$\mathbf{E}(\mathbf{r}, t) = \text{Re}[\mathbf{E}_0 \exp(i\mathbf{q} \cdot \mathbf{r} - i\omega t)]$$

and a corresponding magnetic field. The latter plays a negligible role, as can be seen by examining the classical expression for the force on an electron,

$$\mathbf{F} = (-e)(\mathbf{E} + (\mathbf{v}/c) \times \mathbf{B}).$$

In free space $|\mathbf{B}| = |\mathbf{E}|$ in Gaussian units, and since $v/c \ll 1$, the magnetic force is negligible when dealing with transitions as strong as those to be considered in Section IV. The effect of the magnetic field on the electron spin is also negligible. The electric and magnetic fields can be derived from a vector potential $\mathbf{A}$ as

$$\mathbf{B} = \nabla \times \mathbf{A}, \qquad \mathbf{E} = (1/c)(\partial \mathbf{A}/\partial t),$$

with

$$\mathbf{A}(\mathbf{r}, t) = \text{Re}[-(ic/\omega)\mathbf{E}_0 \exp(i\mathbf{q} \cdot \mathbf{r} - i\omega t)].$$

In the presence of the electromagnetic field the Hamiltonian for an electron in the crystal is modified by replacing the momentum operator $\mathbf{p}$ by $(\mathbf{p} + e\mathbf{A}/c)$, which introduces several new terms into the Hamiltonian

$$(1/2m)[(e/c)(\mathbf{p} \cdot \mathbf{A} + \mathbf{A} \cdot \mathbf{p}) + (e^2/c^2)A^2].$$

The last term is negligible when considering only linear optical absorption, so we drop it. The other two terms are treated as a perturbation. The vector potential is assumed to be known, but, in fact, it may not be known. We assume for now that it is the average $\mathbf{A}$ in the material, averaged over a volume large compared with the wavelength of the radiation, that is itself larger than unit-cell dimensions. We return to this problem in Section IV.

We assume the eigenfunctions and eigenvalues of the unperturbed Hamiltonian are known and apply time-dependent perturbation theory to obtain the transition rate between an initial state i and a final state f as

$$W_{if} = (2\pi/\hbar)|\langle f|M|i\rangle|^2 f_{\text{FD}}(E_i)[1 - f_{\text{FD}}(E_f)] \, \delta(E_f - E_i - \omega),$$

where

$$M = (e/2mc)(\mathbf{p} \cdot \mathbf{A} + \mathbf{A} \cdot \mathbf{p})$$

and f_{FD} is the Fermi–Dirac distribution function and the factors containing it ensure that the initial state is occupied and the final state empty before the transition. The radiation also stimulates transitions from f to i, but, except for the infrared, the distribution functions make this process ignorable for weak exciting fields. By expanding the exponential in the vector potential as

$$A_0 e^{i(\mathbf{q} \cdot \mathbf{r} - \omega t)} = A_0 e^{-i\omega t}[1 - i\mathbf{q} \cdot \mathbf{r} + \cdots]$$

and keeping only the first term, we obtain the electric-dipole approximation, which is adequate for all observable interband transitions at and above the band gap, except at energies high enough that q^{-1} approaches the interatomic distance. Then $\mathbf{p} \cdot \mathbf{A} + \mathbf{A} \cdot \mathbf{p} = 2\mathbf{A} \cdot \mathbf{p}$, so that

$$|\langle f|M|i\rangle|^2 = (e/mc)^2 |\mathbf{A}_0 \cdot \mathbf{p}_{if}|^2,$$

with $\langle f|\mathbf{p}|i\rangle = \mathbf{p}_{if}$ the momentum matrix element.

Next, one sums over initial and final states. For a crystalline solid this can be done by integration over the Brillouin zone, after letting the wave vector be the summation index. Setting $\mathbf{A}_0 = \hat{e}A_0$, we get

$$W(\hbar\omega) = \frac{\pi e^2}{2m^2} \frac{A_0^2}{c^2} \int \frac{d^3k}{4\pi^3} |\hat{e} \cdot \mathbf{p}_{if}|^2 f_{FD}(E_i)[1 - f_{FD}(E_f)] \, \delta(E_f - E_i - \omega).$$

This is the probability per unit time and volume that the electromagnetic wave loses energy $\hbar\omega$ to the electronic system by exciting an electron.

The power absorbed per unit volume is $\hbar\omega W(\hbar\omega)$. The incident flux is

$$\langle S \rangle = (\omega^2 A^2 / 2\pi c)n\hbar k,$$

with n the refractive index (real part). The absorption coefficient μ is defined as

$$\mu = (1/\langle S \rangle)(d/dr)\langle S \rangle = (1/\langle S \rangle)\hbar\omega W.$$

Since $\mu = (\omega\varepsilon_2/cn)$, we end with

$$\varepsilon_2(\omega) = \frac{e^2}{\pi m^2 \omega^2} \int d^3k |\hat{e} \cdot \mathbf{p}_{if}|^2 f_{FD}(E_i)[1 - f_{FD}(E_f)] \, \delta(E_f - E_i - \omega).$$

The real part of the dielectric function can be obtained from the imaginary part by an integral transform, the Kramers–Kronig integral. Both parts of $\tilde{\varepsilon}$ arise from similar transitions, except that they are real in the case of ε_2 but may be virtual in the case of ε_1.

Other functions frequently are used to describe the optical properties of absorbing media. The complex refractive index is $N = n + i\kappa$, where n is the refractive index so useful for transparent ($\kappa = 0$) media. Since $\tilde{\varepsilon} = \varepsilon_1 + \varepsilon_2 = \tilde{N}^2$, we have

$$\varepsilon_1 = n^2 - \kappa^2, \qquad\qquad\qquad \varepsilon_2 = 2n\kappa,$$

or

$$n = [+\varepsilon_1 + \sqrt{\varepsilon_1^2 + \varepsilon_2^2}]^{1/2}/\sqrt{2}, \qquad \kappa = [-\varepsilon_1 + \sqrt{\varepsilon_1^2 + \varepsilon_2^2}]^{1/2}/\sqrt{2}.$$

The absorption coefficient is

$$\mu = (\varepsilon_2 \omega / nc) = (4\pi\kappa/\lambda).$$

Finally, for metals the optical conductivity σ is useful. In Gaussian units it is

$$\sigma = \sigma_1 + i\sigma_2 = -i(\omega/4\pi)(\tilde{\varepsilon} - 1).$$

The foregoing expressions are valid for cubic crystals or randomly oriented polycrystalline samples. For noncubic crystals, $\tilde{\varepsilon}$ and $\tilde{\sigma}$ become tensors and $\tilde{N}$ and μ become anisotropic as well. In the following, we shall drop the distribution–function factors and deal with fully occupied initial states and unoccupied final states.

Shortly we shall neglect the dipole matrix element, but this can be justified only if one is not interested in a complete interpretation of the optical properties. Upon neglecting the matrix element, the optical properties become independent of the orientation of the electric vector with respect to the crystallographic axes of a single crystal; all optical anisotropy is lost. Even for unpolarized radiation some effects of anisotropy are expected. Unpolarized radiation incident normally on a surface parallel to the basal plane of a hexagonal crystal will give a different reflectance spectrum than the same radiation normally incident on a plane face containing the c axis, but an explanation of the anisotropy requires consideration of the matrix elements.

One can show that $\mathbf{p}_{if} = im\omega_{if}\mathbf{r}_{if}$, where $e\mathbf{r}_{if}$ is the transition dipole moment and $\hbar\omega_{if} = E_f - E_i$. This is often used, but is it correct only when one is dealing with exact eigenfunctions. For approximate eigenfunctions the use of each side of the preceding expression may not give the same results, and it is not clear, *a priori*, which will be better. The right-hand side emphasizes the regions of large r in the integrand, while the left-hand side emphasizes the regions of large gradient, closer to the ion cores than the regions of large r. The evaluation of $\mathbf{p}_{if}$ and $\mathbf{r}_{if}$ is very sensitive to small changes in the wave functions used.

When Bloch states are used, the evaluation of the dipole matrix element leads to the selection rule that the wave vectors of the initial and final states are the same. Physically, this corresponds to conservation of momentum $\mathbf{k}$ because the absorbed photon has a wave vector that is very small on the scale of electron wave vectors in the crystal. Thus, in an energy-band diagram allowed optical transitions are vertical.

Group theory is useful for deciding when the dipole matrix elements vanish, i.e., for determining the other selection rules. The symmetry of the Bloch states limits the number of transitions for states with wave vectors lying in symmetry planes, along symmetry lines, and at symmetry points in the Brillouin zone. However, for those transitions for which the matrix element does not vanish, the magnitude of the matrix element cannot be obtained from group theory; the integral must be evaluated. The same application of group theory also gives polarization selection rules, describing some of the anisotropy in noncubic crystals. Often such calculations are carried out on nonrelativistic energy bands that lack spin-orbit effects. Upon including these effects, a similar calculation leads to new selection rules that usually are less "selective." This is the crystal analog of the spin-orbit term in an atomic Hamiltonian that makes spin-forbidden or orbitally forbidden transitions allowed by mixing spin and orbital angular momentum. Again, however, it is difficult to assess whether a transition that is allowed only in the presence of the spin-orbit interaction is strong enough to be detected in an actual spectrum. Tabulations of selection rules for simple crystal structures can be found in Eberhardt and Himpsel [1], Borstel *et al.* [2], and Benbow [3].

The evaluation of the dipole matrix elements can be carried out during a band calculation by "keeping" the wave functions used in the calculation, but the accuracy of the results may not be very high. Such calculations have not been done very often in the past, and, except for semiconductors and a few simple metals, the agreement with experiment has not been very gratifying. Smith [4] and Lässer et al. [5] have used an interpolation scheme, often used for energy bands alone, to obtain dipole matrix elements rather easily and accurately for several noble metals and for a series of alloys. It requires a basis set larger than that normally used for interpolation of the bands, but one that is manageable.

To proceed with the simplification of the expression for ε_2, we make a common approximation that the dipole matrix element may be replaced by an average value. Often this is not evaluated, and only the shape of the ε_2 spectrum, slightly distorted by the energy dependence of the actual matrix element, is considered. This gives

$$\varepsilon_2(\omega) = \frac{e^2}{\pi m^2 \omega^2} \overline{|\hat{e} \cdot \mathbf{p}_{if}|^2} \int d^3k \; \delta(E_f - E_i - \omega),$$

which can be placed in the form

$$\varepsilon_2(\omega) = \frac{e^2}{\pi m^2 \omega^2} \overline{|\hat{e} \cdot \mathbf{p}_{if}|^2} \int \frac{dS}{\nabla_k |E_{if}(\mathbf{k})|}.$$

Often the integrand denominator causes the structure in the spectrum. The points in k space $\mathbf{k}_0$ at which the denominator vanishes are called critical points. At symmetry critical points $\nabla_k E_i(\mathbf{k}) = \nabla_k E_f(\mathbf{k}) = 0$, while general critical points have $\nabla_k E_i(\mathbf{k}) = \nabla_k E_f(\mathbf{k}) \neq 0$. If we expand the interband energy in a series about $\mathbf{k}_0$ we obtain

$$E_f(\mathbf{k}) - E_i(\mathbf{k}) = E_0 + 0 + \sum_{i=1}^{3} \frac{\hbar^2}{2m_i^*} [(\mathbf{k} - \mathbf{k}_0)_i]^2 + \dots,$$

in which local principal axes have been assumed. The critical points are classified by the number of negative values of the principal interband effective masses m_i^* and by the dimensionality. (If the bands are very flat, one or two effective masses become very large and the system is effectively two- or one-dimensional when a particular critical point is being discussed.) In all cases the integral for ε_2 can be evaluated for one critical point at a time, and characteristic shapes in the ε_2 spectrum result. These can be written in compact form as

$$\varepsilon(E) = \frac{C_n}{\omega^2} i^{r-n} \int_0^{E - E_0 + i\Gamma} t^{\left(\frac{n-4}{2}\right)} dt = \varepsilon_1 + i\varepsilon_2,$$

Fig. 1. Characteristic shapes of spectra of $E^2\varepsilon_1$ and $E^2\varepsilon_2$ from interband critical points in one, two, and three dimensions, with schematic energy bands for each type of transition. (a) One-dimensional critical points, (b) two-dimensional critical points, and (c) three-dimensional critical points. The terms m^* are interband effective-mass components in a local system of principal axes.

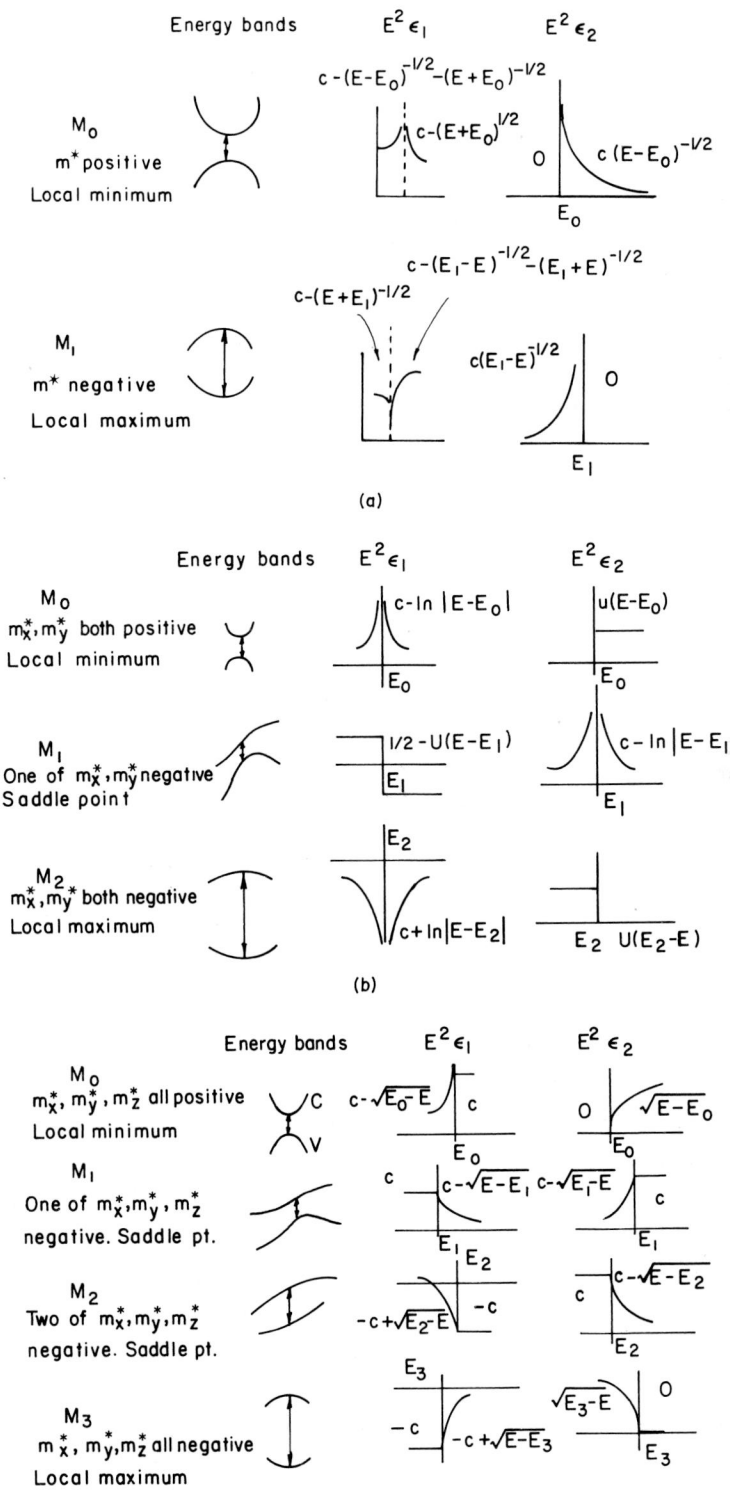

(a)

(b)

(c)

in which n is the dimensionality of the system and r the type of critical point ($r \leq n$), t is a dummy variable, and Γ takes account of broadening processes. The corresponding shapes of ε_1 and ε_2 for all possible n and r are displayed in Fig. 1 with $\Gamma = 0$. In three dimensions the singularity in the integrand for ε_2 leads only to a singularity in the first energy derivative of ε_2. The three-dimensional critical points are better studied in derivative measurements (modulation spectroscopy) unless the energy bands involved are nearly two-dimensional, or one is at the fundamental absorption edge, always an M_0. Interesting illustrations of how critical points evolve as the dimensionality changes are included in Nakao [6], Sasaki *et al.* [7], and Okuyana *et al.* [8]. Diamond- and zincblende-structured semiconductors have a peaked structure in their ε_2 spectra due to a nearly two-dimensional critical point in these cubic materials.

A feature resembling a critical point structure in $\tilde{\varepsilon}$ can occur in metals if there is a very flat band, e.g., a d band, if it is flat enough. Transitions between this band and the others at the Fermi energy "turn on" very suddenly as the photon energy is increased. This provides a sudden increase in ε_2 roughly like that provided by an M_0 critical point, but one whose width is determined by the width of the Fermi–Dirac distribution function edge and by the flatness of the "flat" band.

Critical points are expected to provide features in the ε spectra of all crystals, but they may be difficult to observe. All materials with a band gap will have an M_0 interband critical point at the fundamental absorption edge if the gap is a direct one. However, it may not be found with the expected shape for reasons soon to be given. Parallel bands often occur in simple polyatomic metals. Such a band structure and the resultant optical properties can be treated in closed form by using a simple but realistic model with the result that in the absence of broadening ε_2 displays a singularity of the form

$$\varepsilon_2 \sim (E - E_0)^{-1/2} E \geq E_0, \qquad \varepsilon \sim \text{const } E \langle E_0.$$

This is like an M_0 critical point in one dimension. By and large, however, it is difficult to see critical point effects in metals because they are not large and they lie on a large background. They are best seen in semiconductors when exciton effects are disturbing but not of obliterating strength. Insulators tend to have such strong exciton effects that they dominate the above critical point line shapes.

The foregoing discussion assumed that the electronic excitations were perfectly sharp. The radiative lifetimes of the excited states usually are long enough so that radiative lifetime broadening is negligible. The transitions are broadened by electron–electron scattering of the excited electron and hole and by phonon absorption and emission. Both of these are usually incorporated phenomenologically in the expression for ε_2 by introducing an interband lifetime or energy width, as Γ above, to account for the electron–electron scattering, and convolving the resultant spectrum with a phonon density of states, weighted with a suitable coupling coefficient. If such a convolution is omitted, the interband lifetimes become temperature dependent, representing all broadening processes.

The fundamental band gap is an exception, for here the electron–phonon interaction plays a more obvious role. If the band gap is direct, i.e., if the valence-band maximum and the lowest conduction-band minimum (there may be several of each, all degenerate) lie at the same point in reciprocal space, then one has only an M_0 critical point whose phonon broadening is easy to detect on the low-energy side of the critical point transition. If the two band edges are at different points in k space, the selection rule on k prohibits transitions across the indirect gap. Such transitions can occur only if a second interaction occurs to allow conservation of wave vector, and the most common is the simultaneous absorption or emission of a phonon when the photon is absorbed. This process requires two interactions, so it occurs in second-order perturbation theory only, with consequently weak absorption. Thus, indirect transitions are observed only at the fundamental absorption edge (Fig. 2). The electron at the valence-band maximum may be photoexcited, producing a virtual intermediate state "after" which the electron may be scattered either by absorption or emission of a phonon of the appropriate wave vector. An electron with the wave vector of the conduction-band maximum instead may be photoexcited with the hole and "then" scattered by phonon absorption or emission. Both processes may occur and their matrix elements must be added before squaring, leading to the possibility of interference effects. However, usually one or the other of the routes shown in Fig. 2 dominates, and the resultant expression for the absorption edge

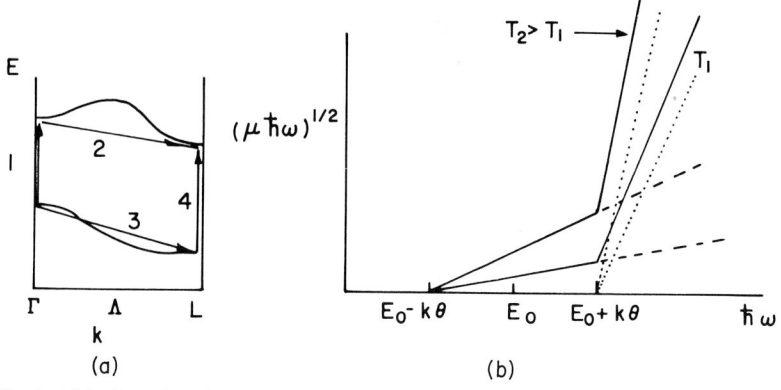

Fig. 2. (a) Indirect interband transitions between a valence band maximum at $\Gamma(k = 0, 0, 0)$ and a conduction-band minimum at $L(k = (\pi/a_0)(1, 1, 1))$, separated by wave vector $k = (\pi a_0)$ $(1, 1, 1)$. Transitions $1 + 2$ and $3 + 4$ are both possible and may interfere. Transitions 1 and 3 are photon absorption processes, while 2 and 4 are electron and hole scattering processes, respectively, caused by phonon absorption or emission. (b) Indirect absorption edge from processes $1 + 2$ or $3 + 4$ at two temperatures. The dashed line represents phonon absorption; the dotted line, phonon emission; and the solid line, the sum. It is assumed (incorrectly) that E_0 does not depend on temperature.

contains two terms, one for phonon absorption and one for phonon emission:

$$\mu = \frac{\text{const}}{(\hbar\omega)^2} \left[\frac{(\hbar\omega - E_0 + k\theta)^2}{e^{\theta/T} - 1} + \frac{(\hbar\omega - E_0 - k\theta)^2}{1 - e^{-\theta/T}} \right],$$

$$\hbar\omega > E_0 - k\theta, \qquad \hbar\omega > E_0 + k\theta.$$

The phonon has energy $k\theta$ and wave vector q, the separation of the band edges in reciprocal space. In general, there are several phonons, one from each branch, which may scatter, so the absorption edge is then a sum of terms like the equations above with different values of θ and different scattering matrix elements (hidden in the constant prefactor). Symmetry considerations sometimes lead to a vanishing scattering probability for some phonons. The resultant absorption coefficients have characteristic shapes, as shown in Fig. 2. (The foregoing assumed the electron excitation by the photon was allowed by the electric-dipole selection rules. If not, then the wave functions are expanded in a series in k. The term in k^1 gives the first nonvanishing absorption. The exponent 2 in the preceding equation then becomes 3, for these indirect forbidden transitions.)

In small-band-gap semiconductors and in heavily doped semiconductors, the foregoing absorption-edge description is incorrect because the Fermi level may lie inside the conduction or valence band. Then one gets a sharp turn-on of absorption with a shape dominated by the Fermi–Dirac distribution function. In this case, the position of the Fermi level rises with increasing doping for n-type material, giving a shift in the edge energy with doping level, the Burstein–Moss shift.

This is as far as we shall go with the one-electron picture. It provides a qualitatively good picture for metals; indeed, often a quantitatively good one. (One must first subtract the intraband contribution to ε_2 from the measured ε_2 spectrum.) For semiconductors the electron–hole interaction, discussed in Section III, alters the expected spectrum appreciably near threshold but apparently not strongly in regions far above threshold. As the band gap increases, such effects become stronger and, in some cases, the band picture may not be the best starting point.

III   ELECTRON–HOLE INTERACTION, EXCITONS

The excited electron and the hole are created at the same lattice site, so their interaction cannot be neglected. This leads to excitons, and to other effects as well. We first consider only the lowest energy M_0 critical point transition with gap energy E_0. There are two extreme models for the electron–hole interaction. The Wannier–Mott model assumes that the electron–hole interaction is weak. The effective-mass model is employed and the hydrogen atom Hamiltonian results, but with the factors e^2 and m replaced by e^2/ε and $\mu^* = m_e^* m_h^* / (m_e^* + m_h^*)$, respectively. The static dielectric constant is ε, and the m^* are the electron and hole effective masses, here assumed to be isotropic. This Hamiltonian leads to discrete levels and a continuum. The former, nonconducting excited states, have energies of $E_0 - R_{ex}/n^2$ with $R_{ex} = 13.6$ eV $(\mu^*/m)\varepsilon^{-2}$ and n an integer. The radii of the exciton "orbits" are $r_n = 0.53$ Å $\{\varepsilon m n^2/\mu^*)$, and if the model is to be self-consistent, the smallest radius must be at least several interatomic spacings in order to make the picture of dielectric screening meaningful. This is usually the case for small-band-gap materials that usually have small effective masses and high dielectric constants, but it may be less valid for the lowest energy exciton than for the others. In such cases a central-cell correction is needed to account for the departure of the potential from the simple $-e^2/\varepsilon r$ assumed when the electron and hole are nearby. This may be done with a nonlocal dielectric function $\varepsilon(r)$.

The other extreme model is the Frenkel exciton, basically an excited atom or ion in the crystal. The electron and hole are localized on one site or a small number of sites. The solid is polarized about this site, so free-atom or free-ion energy levels are not expected. Such a model was formerly believed to be the appropriate one for all large-band-gap insulators, e.g., the solid rare gases and the alkali halides, but now it seems that the Wannier model is quite suitable for all of them, except for the $n = 1$ exciton, and perhaps for a few higher levels in solid Ne. The Frenkel model is very useful for describing excitons in organic crystals.

An electron–hole interaction of intermediate strength requires the use of the Frenkel model with the excitation "localized" on a large number of sites or the Wannier model with Bloch functions employed with many wave vectors, not just those near the critical point wave vector. Thus, as the electron–hole interaction increases in strength, more and more of the band wave functions are used in the Wannier model, and the exciton is no longer associated with a particular critical point; possibly more than one conduction band may have to be used. In such a case, the band picture may not be the best starting point for an understanding of all the optical properties of insulators.

In both of these models the excitation is not permanently localized. It propagates through the crystal with a wave vector $\mathbf{K}$ and, in the case of the Wannier exciton, with a kinetic energy $\hbar^2 K^2/2(m_e^* + m_h^*)$, giving rise to a band. The selection rule on wave vector, however, limits the excitons that can be directly photoexcited to those with $\mathbf{K} = 0$. For indirect transitions, one can excite part of the exciton band because the phonon participating in the transition supplies the proper wave vector to allow this to happen.

The hydrogen-atom problem also has unbound continuum states that differ from free-electron states, especially near threshold. In the crystal, the

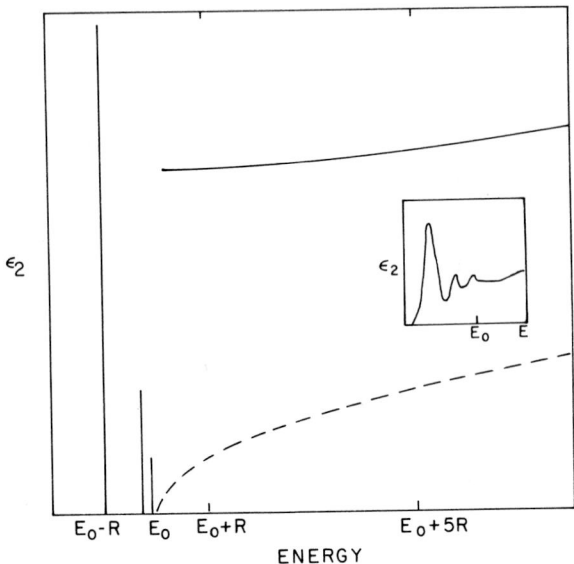

Fig. 3. An M_0 absorption-edge spectrum (dashed line) and how it is modified by the electron–hole interaction (solid line). Only the $n = 1$, 2, and 3 excitons are shown, and these are shown unbroadened as delta functions, with heights proportional to their oscillator strengths. An observed spectrum resembles the spectrum in inset.

electron–hole interaction modifies the absorption above the critical point as well. Elliott [9] has given a description of the effects of the electron–hole interaction for an M_0 critical point for a direct-band-gap material and for an indirect-band-gap absorption threshold. As shown schematically in Fig. 3, the series of excitons merges into a continuum. The enhanced absorption of the Coulomb states above E_0 meets them, the result being a series of lines merging into a continuum with no apparent structure at E_0. If there is any broadening present, as there must be, the continuum begins further below E_0, possibly leaving only the $n = 1$ exciton resolved. For M_0 critical points in two-dimensional crystals, the exciton spectrum is different, with $E = E_0 - R_{ex}/(2n - 1)^2$, and the dependence of absorption strength on n is different from the three-dimensional case [10, 11]. For an indirect gap, the edge is also distorted, and its analysis into two terms according to Fig. 2 is then invalid over an energy range from about R_{ex} below to several times R_{ex} above E_0. For direct-band-gap materials with large values of R_{ex}, the entire valence-to-conduction-band transitions can be distorted by the electron–hole interaction [12].

The electron–hole interaction is present for electron–hole pairs produced by exciting any electron to any excited state, not just at M_0 thresholds. At an M_1 critical point any discrete bound electron–hole pair we try to make will be degenerate with other excited states into which it may decay, for the

M_1 critical point transitions overlap other transitions. Such an excited state is called a hyperbolic exciton, after the nature of the $E_{if}(\mathbf{k})$ surface, and if the decay rate is not too great, it may be observable. Such excitons are believed to have been observed in several semiconductors, although the subject is controversial [13–15]. The electron–hole interaction decreases the absorption strength below that expected for the one-electron picture for M_2 and M_3 critical points, making them *less* prominent in spectra than expected by the one-electron model [16].

At energies above the band-gap energy, it is possible to think of other types of discrete excited states overlapping the continuum of excitations above the fundamental edge. Such discrete excitations could arise from the excitation of excitons below another M_0 critical point involving a second, higher, conduction band or a deeper valence band, e.g., the spin-orbit "partner" of the highest one. Core-electron excitation may also give rise to such a situation. In such cases, the two excitations may not be independent, and the absorption spectrum is not the sum of the two individual spectra. The final excited state of the system is a superposition of both the discrete and continuum excited states. When this superposition is used to evaluate the electric-dipole matrix element, interference effects occur upon squaring, and the strength of the coupling of two types of excitations has a strong effect on the appearance of the resultant absorption spectrum. There can be shifts of the absorption peak, which now is broadened, and a dip may appear on either the high- or low-energy side. Fano [17–19] first described such effects, and the resultant line shapes are called Fano line shapes. (Hyperbolic excitons give rise to Fano line shapes.)

The electron–hole interaction is not just the Coulomb attraction. This is also an exchange term that is important, and the net interaction may be attractive or repulsive. Toyozawa *et al.* [20] have considered the electron–hole interaction at each type of critical point and have summarized the results in a figure, reproduced here as Fig. 4. It shows how the interaction mixes the line shape expected for a given critical point with the line shape for an "adjacent" critical point, the mixing increasing with increasing strength of the electron–hole interaction. The result is that one cannot identify properly a critical point in the observed spectrum of an insulator without consideration of the effect of the electron–hole interaction. The inclusion of exchange also can alter considerably the apparent spin-orbit features in the spectrum [21].

Electrons and holes each interact with the vibrating lattice, and despite the apparent electrical neutrality of an exciton, the exciton–phonon interaction may be rather strong. The strength is described by the ratio of the average coupling energy to the bandwidth of the excitons. The coupling is usually to the longitudinal optical phonons or to zone-boundary acoustic phonons [22], and the bandwidth is a measure of exciton mobility. The exciton line shape is an asymmetric Lorentzian for weak coupling and

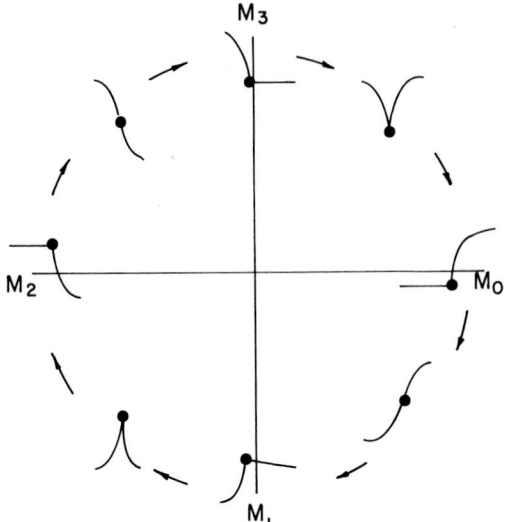

Fig. 4. Metamorphosis of critical points by an attractive electron–hole interaction. A three-dimensional critical point of type M_j ($j = 0, 1, 2, 3$) is as shown with no electron–hole interaction. As this interaction grows, the shape distorts according to the arrow until at very large interaction the M_j has become M_{j+1} ($j = 4 \rightarrow j = 0$). A repulsive interaction acts similarly but with the arrows pointing counterclockwise. (From Toyozawa *et al.* [20].)

Gaussian for strong coupling [23, 24]; both shapes are temperature dependent. As a result of the phonon coupling, there is a shift in the energy of the peak in the exciton from that calculated for the electronic system alone [23, 24].

On the low-energy side of the first exciton absorption peak, the absorption coefficient is found to be exponential with energy over a range of up to four decades for a variety of insulating and semiconducting materials. This is the "Urbach rule" [25], according to which

$$\mu(E) = \mu_0 \exp[-b(E_0 - E)].$$

In some materials, e.g., highly doped semiconductors, b is a constant, while for many other materials, e.g., alkali halides,

$$b = b_0(2kT/\hbar\omega_0) \tanh(\hbar\omega_0/2kT)$$

with $\hbar\omega_0$ an effective phonon energy. An Urbach tail is found in amorphous semiconductors, as well. At low temperatures in some crystals, e.g., CdTe [26], the exponential edge shape changes, and phonon sidebands become resolved. There are two models that give the observed exponential absorption edge. One is based on the electric fields arising from impurities, defects, and phonons [27]. There is convincing evidence that this model applies to heavily doped semiconductors [28] and disordered materials [29]. The other model is the standard exciton–phonon coupling model [23, 24], which can yield an exponential edge [30–34] or phonon sidebands [32], depending on the coupling strength to the phonons. There is convincing evidence that this mechanism is operative in GeS, for the phonons involved have been identified as phonons with no accompanying electric field [37]. In many other materials with exponential absorption edges, the two models cannot be so clearly distinguished. A phenomenological model [35, 36] also has been used for a number of materials.

LOCAL FIELD EFFECTS IV

We have just described what appears to be a two-body effect, but, in fact, all the valence electrons participate in screening the electron–hole interaction. Their roles are subsumed in $\varepsilon(r)$. There are other many-body effects that play a real role. We consider only the local field effect, an old classical problem. The electric field acting on an electron, in our case causing it to become excited, is the local field caused by the incident electromagnetic wave and by the polarization of the medium around the electron in response to that wave and to the polarization in the medium. A self-consistent local field must be found [38–45]. The resultant local field depends on the degree of localization of the electron under consideration. For a spatially delocalized electron, we can use just the average field in the medium. For a highly localized electron, one confined to a unit cell, the local field "correction" becomes the largest. Classically it is given by

$$\mathbf{E}_{loc}(r) = (\varepsilon + 2/3)\mathbf{E}(r)$$

for a cubic or isotropic crystal, and $\varepsilon \to \tilde{\varepsilon}$ in regions of absorption. As the electron becomes more delocalized, the size the correction decreases to unity. Also, at high energies $\tilde{\varepsilon} \to 1$, and the correction is negligible, no matter how localized the electron may be. The use of a local field correction depending on ε alters the shape of the ε_2 spectrum obtained from experiment [46, 47]. In Fig. 5 we show some data for an alkali halide. The dielectric function has been derived from a reflectance measurement by Kramers–Kronig analysis, with two different assumptions about the local field, the most extreme assumptions. The differences in the spectra are very large, and it is not at all clear which may be the more correct in a given spectral region, for we do not really know how much local field correction should be applied to a given transition. This is clearly an area in which more understanding is needed.

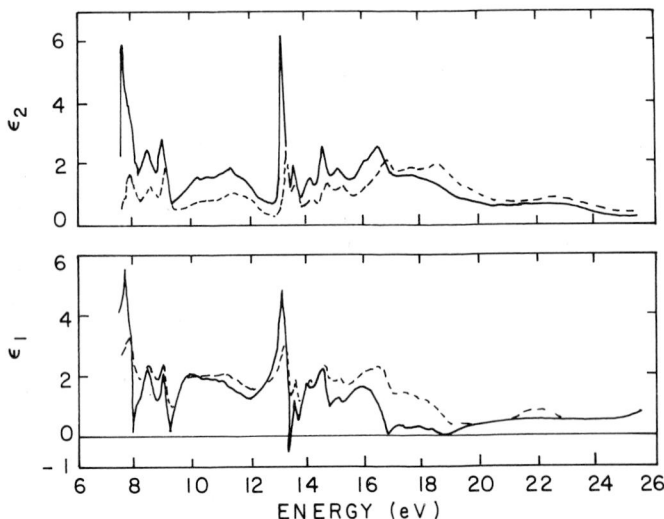

Fig. 5. Dielectric function spectra for CsCl obtained by Kramers–Kronig analysis of reflectance spectra; solid line represents no local field correction used and dashed line full
Lorentz–Lorenz local field correction used. (From Bergstresser and Rubloff [46].)

V   EXAMPLES

We now give some examples of interband spectra for some simple materials. Figure 6 shows the interband optical conductivity of an Al single crystal
at 4.2 K, obtained by a Kramers–Kronig analysis of absorptance and reflectance [48]. The theoretical result presented is a fit to the parallel-band model
for a polyvalent, nearly free-electron metal [49]. The model is based on the
pseudopotential formalism, and in the case of Fig. 6, four parameters have
been used to fit both the shape and the magnitude, with two of the parameters consonant with Fermi-surface data on Al. Moreover, a more elaborate
pseudopotential band calculation has been carried out, and the optical properties calculated from it [50]. The results are in good agreement with the
measured 1.5-eV peak and in reasonably good agreement for the 0.5-eV peak,
but the calculated results are uncertain at low energy because they result from
small differences in nearly equal energy eigenvalues. Reasonably good fits
have been achieved for other similar metals—In, Ga, Pb, and Tl—although
in Pb and Tl, spin-orbit splitting of certain bands contributes features in the
experimental spectra not considered in the model. Other polyvalent
metals—Be, Zn, Cd—and the alkaline earths do not conform to this simple
(local) pseudopotential model.

Figure 7 shows the measured [51] ε_2 for Cu and the results of an empirical,
nonlocal pseudopotential calculation of the band structure and the dielectric

TABLE I

Important Interband Transitions in Cu

Energy (eV)	Bands involved	Location in Brillouin zone	Type of transition
2.1	$4, 5 \rightarrow 6$	$\Delta_5 \rightarrow \Delta_1(E_F)$	Fermi edge
3.2	$4, 5 \rightarrow 6$	Near X	Volume
3.7	$3 \rightarrow 6$	$\Sigma_1 \rightarrow \Sigma_1(E_F)$	Fermi edge
3.7	$3, 4, 5 \rightarrow 6$	Near X	Volume
3.96	$4, 5 \rightarrow 6$	$X_5 \rightarrow X'_4$	M_1
4.25	$6 \rightarrow 7'$	$L'_2 \rightarrow L^u_1$	M_0
4.33	$1 \rightarrow 6$	$\Sigma_1 \rightarrow \Sigma_1(E_F)$	Fermi edge
4.5	$3, 4, 5 \rightarrow 6$		Volume
5.0	$1, 2 \rightarrow 6$		Volume

function [52]. This calculation also used optical and photoemission data to fix a total of eight parameters in the band calculation. Although onsets of structure may be attributed to transitions to the Fermi level from the relatively flat 3d bands and to a few critical points, much of the large background giving the magnitude of ε_2 comes from transitions occurring in large volumes of the Brillouin zone. The contributing transitions are shown in Table I.

The optical properties of Cu have also been calculated quite accurately from first principles. Janak *et al.* [53] have calculated the (self-consistent) energy bands, dielectric function, and photoemission spectra from a potential for copper adjusted only to give the correct Fermi surface. The only other adjustable parameter characterized the electron–electron interaction, and it was adjusted to improve agreement with ε_2. The agreement is about the same as shown in Fig. 7, but only one parameter was adjusted to produce the fit. Finally, Lässer *et al.* [5] have calculated the dielectric function for the noble metals by using an interpolation scheme, evaluating the dipole matrix elements from $E(k)$ by simple differentiation [5]. The fit, although not quite as good as that shown in Fig. 7, extends to an energy of 24 eV, and the departures from the experimental spectra seem understandable on the basis of the model.

Figure 8 shows the measured interband ε_2 for Mo obtained by Kramers–Kronig analysis of reflectance spectra [54, 55], and one calculated ε_2 [56]. The calculation began with energy bands but ignored electric-dipole matrix elements. Thus, the calculated ε_2 is in arbitrary units. By using only specific regions of the Brillouin zone, the origin of structures or large featureless contributions to ε_2 could be identified. However, another group [57] used a different set of energy bands as a starting point and a different technique for locating the origin of structures by using energy, rather than momentum, windows, and obtained comparably good agreement with experiment (again without dipole matrix elements) but different assignments for a number of the

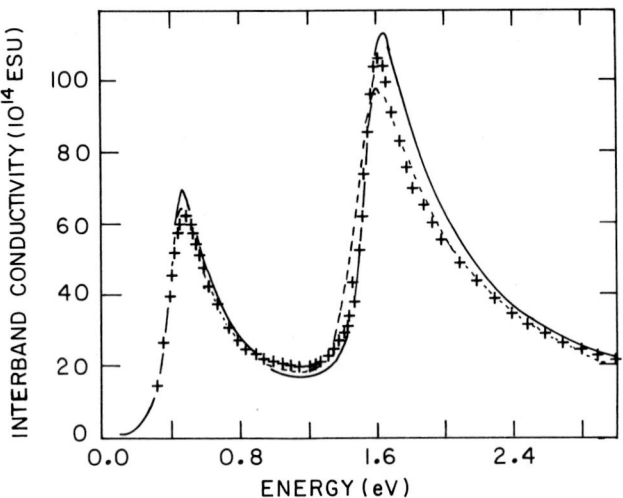

Fig. 6. Interband contribution to the optical conductivity of Al (solid). The crosses and dashed curve represent model calculations of the interband conductivity. (From Benbow and Lynch [48].)

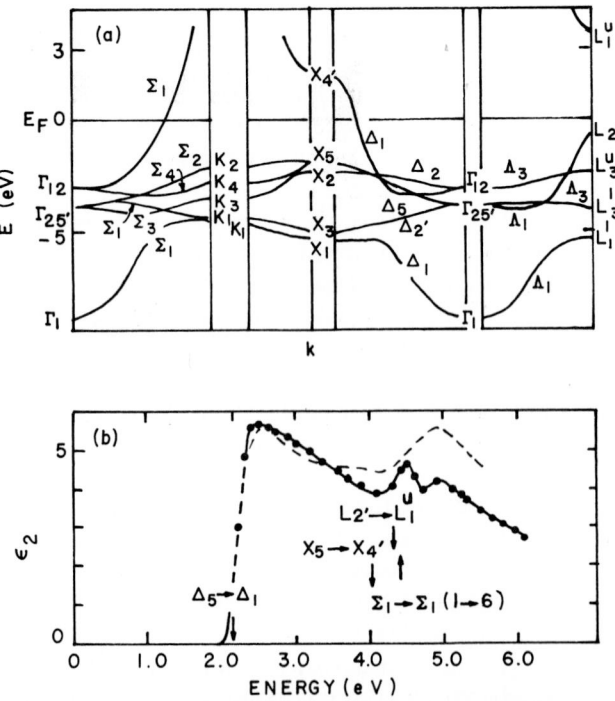

Fig. 7. (a) Band structure of Cu calculated by an empirical pseudopotential method, (b) interband contribution to ε_2 for Cu (dashed) ([51]) and an empirical pseudopotential calculation of ε_2 (solid). Some important transitions are labeled. (See also Table I and Fig. 7a.) (From Fong *et al.* [52].)

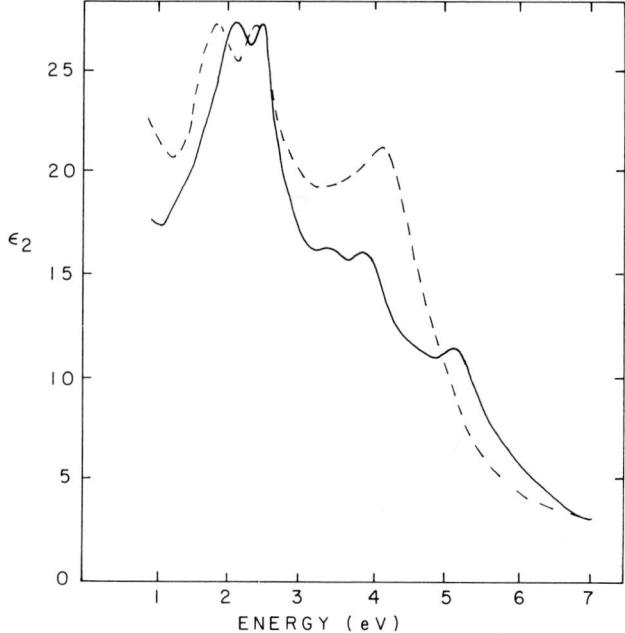

Fig. 8. ε_2 of Mo. Solid line represents values measured by Weaver *et al.* [54]; dashed line values calculated by Pickett and Allen [56].

transitions. It would be most useful to evaluate ε_2 for a transition metal by using dipole matrix elements to see if ε_2 can be calculated well for a material with unfilled d bands.

In both the transition metals and the noble metals, it now is clear that some of the structure and much of the magnitude in the ε_2 spectra arise from transitions throughout large volumes of the Brillouin zone, and such transitions are not easily accounted for in any way other than a numerical integration over the Brillouin zone. Simpler models, such as are appropriate for simple metals, do not suffice. Calcium is an example of a simple metal in which the Fermi surface and critical points play no role in the optical properties [58].

Figure 9 shows the ε_2 spectrum for Si, a covalently bonded semiconductor [59]. Although the one-electron calculation of ε_2 was quite successful for Ge, a similar material it is clear that the peak at 3.5 eV in Si appears as only a shoulder at 3.4 eV in the one-electron calculation, whose integrated area is too small by about 30% [42].

A number of critical points have been identified in derivative spectra, so the problem seems not to be grossly inadequate energy bands in the calculation. The bonding electrons, those being excited in this spectrum, are highly localized in Si, and several calculations of local field effects have shown that they have a large effect, although some of the early calculations did not lead

David W. Lynch

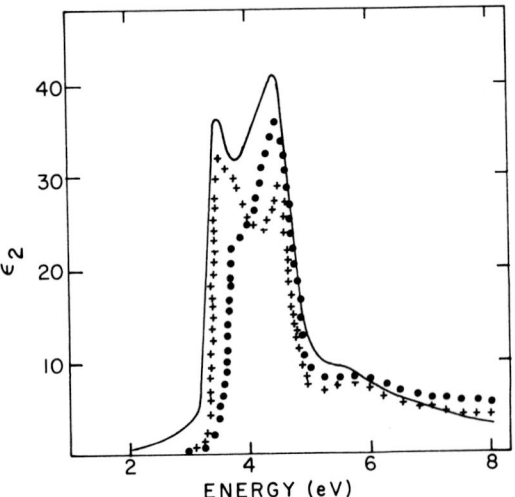

Fig. 9. ε_2 of Si. Solid line represents experiment by Philipp and Ehrenreich [59]; filled circles, values calculated by a one-electron empirical pseudopotential method (Louie *et al.* [41]; crosses) calculated by using a local field correction. (From Hanke and Sham [45].)

to an effect that produced better agreement with experiment. The recent calculations [45] show that proper treatment of the local field and other many-electron effects is a *sine qua non* to the quantitative understanding of the optical spectrum of silicon and of diamond. Such calculations have not been carried out on other solids for which local field effects are believed to be important.

The absorption-edge region is shown in Fig. 10 [60, 61]. These data were taken with high resolution at a number of temperatures. There is considerable resemblance to the spectrum shown schematically in Fig. 2, indicating the occurrence of indirect transitions with the participation of a phonon. Detailed analysis of all the spectra shows that the band-gap energy is temperature dependent, that the slopes near threshold are not as expected because of the electron-hole interaction, and that there are several overlapping structures at higher temperatures because phonons with four different energies (i.e., from four different branches) participate. A second indirect edge, at which the electrons are excited into a higher-lying set of conduction-band minima, has been observed [62] with great difficulty, for it is weak and lies at 1.65 eV, where the absorption shown in Fig. 10 is very large.

As an example of an insulator we can use CsCl, shown in Fig. 5. The sharp structures near 8 and 13 eV are clearly excitons, those near 8 eV arising from the valence-to-conduction-band transitions, and those at 13.5 eV from the excitation of Cs 5p core levels. Considerable analysis has been carried out on the valence and core excitons in alkali halides and solid rare gases, but far less is known about the other structures. (Many weaker exciton features,

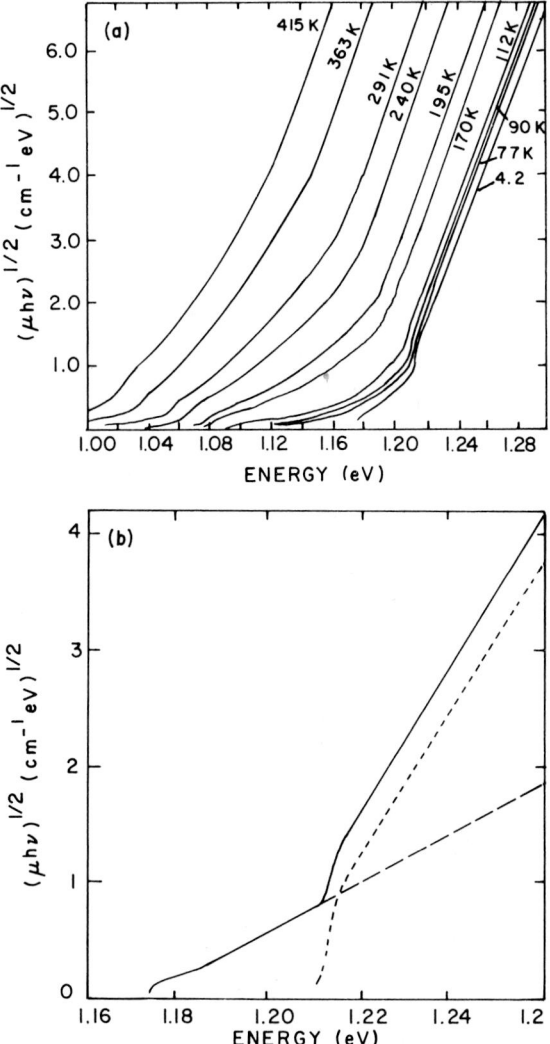

Fig. 10. (a) Absorption-coefficient spectra for Si at the fundamental (indirect) edge at several temperatures (from Macfarlane *et al.* [60]) and (b) decomposition of the 4.2 K spectrum into two components, phonon emission (dotted line beginning at higher energy) and phonon absorption (dashed line lying under the solid line, then leaving it). The solid line is the spectrum from both components together.

not apparent in Fig. 2, are known in many alkali halides.) These have been associated with structures expected in the joint density of states for interband transitions, but the large electron–hole interaction leads one not to expect a one-to-one correspondence between the one-electron spectrum and the experimental spectrum.

ACKNOWLEDGMENT

The Ames Laboratory is operated for the U.S. Department of Energy by Iowa State University under contract No. W-7405-Eng-82. This research was supported by the Director for Energy Research, Office of Basic Energy Sciences.

REFERENCES

1. W. Eberhardt and F. D. Himpsel, *Phys. Rev. B* **21**, 5572 (1980).
2. G. Borstel, M. Neumann and M. Wohlecke, *Phys. Rev. B* **23**, 3121 (1981).
3. R. L. Benbow, *Phys. Rev. B* **22**, 3225 (1980).
4. N. V. Smith, *Phys. Rev. B* **19**, 5019 (1979).
5. R. Lässer, N. V. Smith, and R. L. Benbow, *Phys. Rev. B* **24**, 1895 (1981).
6. K. Nakao, *J. Phys. Soc. Jpn.* **25**, 1343 (1968).
7. Y. Sasaki, C. Hamaguchi, A. Moritani, and J. Nakai, *J. Phys. Soc. Jpn.* **36**, 177 (1974).
8. M. T. Okuyana, T. Nishino, and Y. Hamakawa, *J. Phys. Soc. Jpn.* **37**, 431 (1974).
9. R. J. Elliott, *Phys. Rev.* **108**, 1384 (1957).
10. H. I. Ralph, *Solid State Commun.* **3**, 303 (1965).
11. M. Shinada and S. Sugano, *J. Phys. Soc. Jpn.* **20**, 1274 (1965).
12. U. Rössler and O. Schütz, *Phys. Stat. Solidi B* **56**, 483 (1973).
13. C. B. Duke and B. Segall, *Phys. Rev. Lett.* **17**, 19 (1966).
14. J. Hermanson, *Phys. Rev. Lett.* **18**, 170 (1967).
15. J. E. Rowe, F. H. Pollak, and M. Cardona, *Phys. Rev. Lett.* **22**, 933 (1969).
16. B. Velický and J. Sak, *Phys. Stat. Solidi* **16**, 147 (1966).
17. U. Fano, *Phys. Rev.* **124**, 1866 (1961).
18. K. P. Jain, *Phys. Rev.* **139**, A544 (1965).
19. Y. Onodera, *Phys. Rev. B* **4**, 2751 (1971).
20. Y. Toyozawa, M. Inoue, T. Inui, and E. Hanamura, *J. Phys. Soc. Jpn.* **22**, 1337, 1349 (1967).
21. Y. Onodera and Y. Toyozawa, *J. Phys. Soc. Jpn.* **22**, 833 (1967); Y. Onodera and Y. Toyozawa, *J. Phys. Soc. Jpn.* **22**, 1337, 1349 (1967).
22. Y. Toyozawa and M. Inoue, *J. Phys. Soc. Jpn.* **21**, 1663 (1966).
23. Y. Toyozawa, *Prog. Theor. Phys.* **20**, 53 (1958).
24. H. Sumi, *J. Phys. Soc. Jpn.* **32**, 616 (1972).
25. F. Urbach, *Phys. Rev.* **92**, 1325 (1953).
26. D. T. F. Marple, *Phys. Rev.* **150**, 728 (1966).
27. J. D. Dow and D. Redfield, *Phys. Rev. B* **5**, 594 (1972).
28. D. Redfield and M. A. Aframovitz, *Appl. Phys. Lett.* **11**, 138 (1967).
29. J. Szczyrbowski, *Phys. Stat. Solidi B* **105**, 515 (1981).
30. K. Cho and Y. Toyozawa, *J. Phys. Soc. Jpn.* **30**, 1555 (1971).
31. H. Sumi and Y. Toyozawa, *J. Phys. Soc. Jpn.* **31**, 342 (1971).
32. H. Miyazaki and E. Hanamura, *J. Phys. Soc. Jpn.* **50**, 1310 (1981).
33. S. Schmitt-Rink, H. Haug, and E. Mohler, *Phys. Rev. B* **24**, 6043 (1981).
34. M. Schreiber and Y. Toyozawa, *J. Phys. Soc. Jpn.* **51**, 1544 (1982).
35. T. Skettrup, *Phys. Rev. B* **18**, 2622 (1978).
36. D. J. Dunstan, *J. Phys. C* **30**, L419 (1982).
37. J. D. Wiley, E. Schonherr, and A. Breitschwerdt, *Solid State Commun.* **34**, 891 (1980).
38. S. L. Adler, *Phys. Rev.* **126**, 413 (1962).
39. N. Wiser, *Phys. Rev.* **129**, 62 (1963).
40. Y. Onodera, *Prog. Theor. Phys.* **49**, 37 (1973).
41. S. G. Louie, J. R. Chelikowski, and M. L. Cohen, *Phys. Rev. Lett.* **34**, 155 (1975).
42. W. Hanke and L. J. Sham, *Phys. Rev. B* **12**, 4501 (1975).
43. R. D. Turner and J. C. Inkson, *J. Phys. C* **9**, 3585 (1976).

44. R. Bonneville, *Phys. Rev. B* **21**, 368 (1980).
45. W. Hanke and L. J. Sham, *Phys. Rev. B* **21**, 4656 (1980).
46. T. K. Bergstresser and G. W. Rubloff, *Phys. Rev. Lett.* **30**, 794 (1973).
47. S. R. Nagel and T. A. Witten, *Phys. Rev. B* **11**, 1623 (1975).
48. R. L. Benbow and D. W. Lynch, *Phys. Rev. B* **12**, 5615 (1975).
49. N. W. Ashcroft and K. Sturm, *Phys. Rev. B* **3**, 1898 (1971).
50. D. Brust, *Phys. Rev. B* **2**, 818 (1970).
51. U. Gerhardt, *Phys. Rev.* **172**, 651 (1968).
52. C. Y. Fong, M. L. Cohen, R. R. L. Zucca, J. Stokes, and Y. R. Shen, *Phys. Rev. Lett.* **25**, 1486 (1970).
53. J. F. Janak, A. R. Williams, and V. L. Moruzzi, *Phys. Rev. B* **11**, 1522 (1971).
54. J. H. Weaver, D. W. Lynch, and C. G. Olson, *Phys. Rev. B* **10**, 501 (1974).
55. B. W. Veal and A. P. Paulikas, *Phys. Rev. B* **10**, 1280 (1974).
56. W. E. Pickett and P. B. Allen, *Phys. Rev. B* **11**, 3599 (1975).
57. D. D. Koelling, F. M. Mueller, and B. W. Veal, *Phys. Rev. B* **10**, 1290 (1974).
58. P. O. Nilsson and G. Forssell, *J. Phys. F* **5**, L159 (1975).
59. H. R. Philipp and H. Ehrenreich, *Phys. Rev.* **129**, 1550 (1963).
60. G. G. Macfarlane, T. P. McLean, J. E. Quarrington, and V. Roberts, *Phys. Rev.* **111**, 1245 (1958).
61. G. G. Macfarlane, T. P. McLean, J. E. Quarrington, and V. Roberts, *J. Phys. Chem. Solids* **8**, 388 (1959).
62. R. A. Forman, W. R. Thurber, and D. E. Aspnes, *Sol. State Commun.* **14**, 1007 (1974).

GENERAL REFERENCES

Abélès, F., *in* "Optical Properties of Solids" (F. Abélès, ed.), p. 93, North-Holland Publ., Amsterdam, 1972.
Altarelli, M., *J. de Phys.* **39**, C4–95 (1978).
Aspnes, D. E., *in* "Handbook of Semiconductors" (M. Balkanski, ed.), Vol. 2, p. 109, North-Holland Publ., Amsterdam, 1980.
Bassani, F., *Appl. Opt.* **19**, 4093 (1980).
Bassani, F., and Pastori Parravicini, G., "Electronic States and Optical Transitions in Solids," Pergamon, Oxford, 1975.
Brown, F. C., *in* "Synchrotron Radiation Research" (H. Winick and S. Doniach, eds.), p. 61, Plenum, New York, 1980.
Brown, F. C., *Solid State Phys.* **29**, 1 (1974).
Brown, G. S., and Doniach, S., *in* "Synchrotron Radiation Research" (H. Winick and S. Doniach, eds.), p. 353, Plenum, New York, 1980.
Cardona, M., *in* "Semiconductors and Semimetals" (R. K. Willardson and A. C. Beer, eds.), Vol. 3, p. 125, Academic Press, New York, 1967.
Cardona, M., "Modulation Spectroscopy," *Solid State Phys.* **Supp. 11** (1969).
Cohen, M. H., *in* "Optical Properties of Solids" (J. Tauc, ed.), p. 1, Academic Press, New York, 1966.
Dexter, D. L., and Knox, R. S., "Excitons," Wiley, New York, 1963.
Dow, J. D., *in* "Optical Properties of Solids—New Developments" (B. O. Seraphin, ed.), p. 3, North-Holland Publ., Amsterdam, 1976.
Ehrenreich, H., *in* "Optical Properties of Solids" (J. Tauc, ed.), p. 106, Academic Press, New York, 1966.
Evans, B. L., *in* "Optical and Electrical Properties" (P. A. Lee, ed), p. 1, Reidel, Dordrecht, 1976.
Greenaway, D. L., and Harbeke, G., "Optical Properties and Band Structure of Semiconductors," Pergamon, New York, 1968.

Hamakawa, Y., and Nishino, T., *in* "Optical Properties of Solids—New Developments" (B. O. Seraphin, ed.), p. 255, North-Holland Publ., Amsterdam, 1976.

Hanke, W., *in* "Festkorper Probleme XIX—Advances in Solid State Physics" (J. Treusch, ed.), p. 43, Vieweg, Braunschweig, 1979.

Harbeke, G., *in* "Optical Properties of Solids" (F. Abélès, ed), p. 21, North-Holland Publ., Amsterdam, 1972.

Johnson, E. J., *in* "Semiconductors and Semimetals" (R. K. Willardson and A. C. Beer, eds.), Vol. 3, p. 154, Academic Press, New York, 1967.

Koch, E. E., Kunz, C., and Sonntag, B., *Phys. Rep.* **28C**, 154 (1977).

Kotani, A., and Toyozawa, Y., *in* "Synchrotron Radiation Techniques and Applications" (C. Kunz, ed.), p. 169, Springer-Verlag, Heidelberg, 1979.

Knox, R. S., "Theory of Excitons," *Solid State Phys.* **Supp. 5** (A63).

Kunz, C., *J. de Phys.* **39**, C4–112 (1978).

Kunz, C., *in* "Optical Properties of Solids—New Developments" (B. O. Seraphin, ed.), p. 423, North-Holland Publ., Amsterdam, 1976.

Lynch, D. W., *in* "Synchrotron Radiation, Techniques and Applications" (C. Kunz, ed.), p. 357, Springer-Verlag, Heidelberg, 1979.

Moss, T. S., "Optical Properties of Semiconductors," Butterworth, London, 1959.

Nilsson, P. O., *Solid State Phys.* **29**, 139 (1974).

Pankove, J. I., "Optical Processes in Semiconductors," Prentice-Hall, Englewood Cliffs, New Jersey, 1971.

Petroff, Y., *in* "Handbook of Semiconductors" (M. Balkanski, ed.), Vol. 2, p. 1, North-Holland Publ., Amsterdam, 1980.

Philipp, H. R., and Ehrenreich, H., *in* "Semiconductors and Semimetals" (R. K. Willardson and A. C. Beer, eds.), Vol. 3, p. 93, Academic Press, New York, 1967.

Phillips, J. C., *in* "Optical Properties of Solids" (J. Tauc, ed.), p. 155, Academic Press, New York, 1966.

Phillips, J. C. *Solid State Phys.* **18**, 55 (1966).

Reynolds, D. C., and Collins, T. C., "Excitons, Their Properties and Uses," Academic Press, New York, 1981.

Schwentner, N., *Appl. Opt.* **19**, 4104 (1980).

Skibowski, M., Sprussel, G., and Saile, V., *Appl. Opt.* **19**, 3978 (1980).

Tauc, J., *in* "Optical Properties of Solids" (J. Tauc, ed.), p. 277, Academic Press, New York, 1966.

Toyozawa, Y., *Appl. Opt.* **19**, 4101 (1980).

Wooten, F., "Optical Properties of Solids," Academic Press, New York, 1972.

Zavetova, M., and Velicky, B., *in* "Optical Properties of Solids—New Developments" (B. O. Seraphin, ed.), p. 379, North-Holland, Publ., Amsterdam, 1976.

Chapter **11**

Optical Properties of Nonmetallic Solids for Photon Energies below the Fundamental Band Gap

SHASHANKA S. MITRA
Department of Electrical Engineering
University of Rhode Island
Kingston, Rhode Island

I.	Introduction	213
II.	Infrared Dispersion by Polar Crystals	215
III.	Kramers–Kronig Dispersion Relations	227
IV.	Determination of Absorption Coefficient in the Intermediate Region	229
V.	Absorption Coefficient in the Transparent Regime	230
VI.	Multiphonon Absorption	232
	A. General Remarks	232
	B. Absorption in the Region $\omega > \omega_{LO}$	236
	C. Absorption in the Region $\omega < \omega_{TO}$	249
VII.	Infrared Absorption by Defects and Disorders	254
	A. General Remarks	254
	B. Crystals with Substitutional Impurities	255
	C. Amorphous and Glassy Systems	259
VIII.	Infrared Dispersion by Plasmons	263
	References	267

INTRODUCTION I

There are two chief sources of absorption in a pure solid, viz., lattice vibrations and electronic transitions. For most materials, a sufficiently wide spectral window exists between these two limits where the material is transparent (see Fig. 1). However, this transparent regime displays residual absorption

213

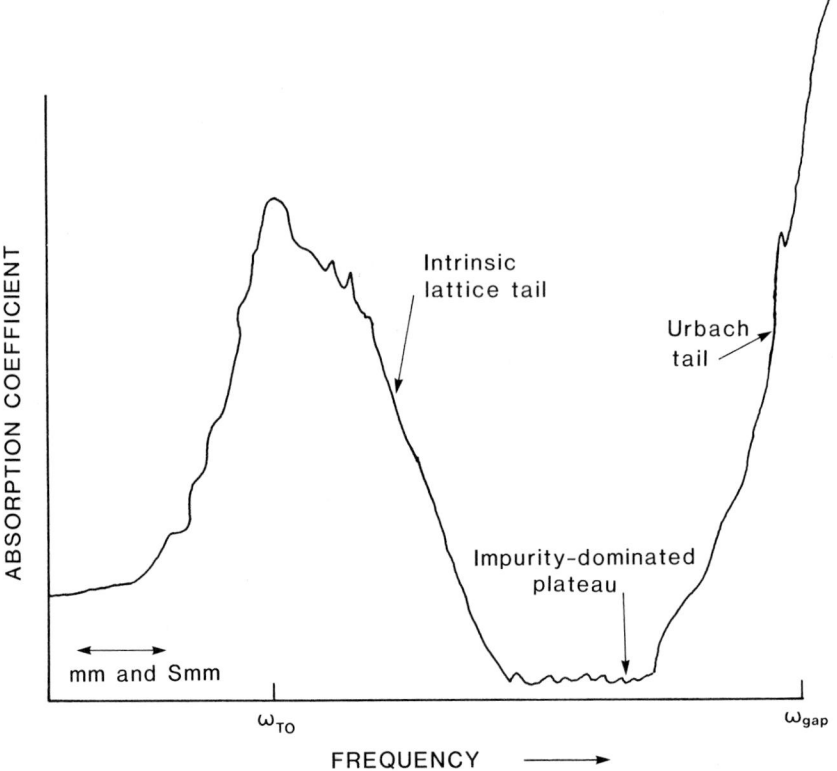

Fig. 1. The absorption gap between the fundamental electronic absorption edge and the one-phonon adsorption process in a polar crystal (schematic).

due to (a) multiphonon processes, (b) defects and impurities, (c) phonon-assisted electronic transitions in the long-wavelength tail of the fundamental absorption edge, and/or (d) multiphoton electronic transitions in the case of high-photon flux. As is evident from Fig. 1, for most solids, the mid- and near-infrared regions fall at energies substantially higher than the reststrahlen region and yet far below the electronic absorption edge. Thus, the primary source of absorption in this region is multiphonon summation processes. The millimeter and submillimeter spectral regions, on the other hand, occur on the low-energy side of the one-phonon reststrahlen region. The primary source of absorption in the region is thus two-phonon difference processes, although there will be some contribution from the sum processes involving low-energy acoustical phonons, and one-phonon absorption processes in the case of structurally disordered solids, and solids containing imperfections and impurities.

In this chapter we discuss optical properties of solids in the infrared spectral region that are caused by interaction of electromagnetic radiation with lattice vibrations. This chapter is organized in the following manner: We start with optical interactions in crystals resulting from one-phonon processes; next we treat multiphonon interactions both at energies higher and lower than the fundamental one-phonon processes; this is followed by a description of phonon processes induced by defects and impurities and their interaction with infrared radiation. Finally, photon–phonon interaction in crystals that have substantial amount of charge carriers (plasma effects) is considered.

We regard a crystal as a mechanical system of nN particles, where n is the number of particles per unit cell and N the number of unit cells. Such a crystal shall have $3nN$ degrees of freedom, of which $3nN - 3$ are the linearly independent normal modes and three are pure translations. The very large number ($\sim 10^{24}$) of modes of a macroscopic crystalline sample necessitates the description of the frequency spectrum in terms of a frequency distribution function. The frequency spectrum of the ionic motions in a solid can be obtained by constructing the classical equations of motion for the lattice points and solving for the normal modes as plane waves of the form $\exp i(\omega_i t - \mathbf{k}_i \cdot \mathbf{r})$, where ω_i and $\mathbf{k}_i$ are, respectively, the frequency and the wave vector of the ith mode. The vibration frequencies occur as the $3n$ roots of the secular equation, involving the wave vector, which may take N values. The $3nN$ modes are thus distributed on $3n$ branches, $3n - 3$ of which are termed optical that have nonvanishing mode frequencies for all values of k and three are acoustic branches that have vanishing frequencies at $k = 0$.

The energy of each of the $3nN$ modes is quantized, and the term *phonon* is used to describe a quantized lattice vibration. Throughout the rest of this chapter we have considered lattice vibrations as phonons or quantized excitations, rather than as waves of vibrating ions; and thus in their interactions with photons, other phonons, electrons, or neutrons, the requirement of the conservation of energy and momentum are tacitly implied. The formulas developed use cgs units.

INFRARED DISPERSION BY POLAR CRYSTALS   II

In the long-wave limit ($k \cong 0$), the optical branch vibrations of a diatomic lattice corresponds to the motion of one type of atoms, all in phase, relative to the other kind. In polar crystals, such a motion is associated with strong electric moments and hence can directly interact with the electric field of proper polarization of the incident electromagnetic radiation, provided that energy and wave vectors are conserved. The wave vectors associated with an infrared photon are essentially zero (10^4–10^5 cm^{-1}) compared with those of most optical phonons (10^9 cm^{-1}) of the same energy. For example, in rock salt, a diatomic crystal, there are three optic branches. At $k = 0$ one corresponds to the long-wavelength longitudinal optical mode and the other two to the doubly degenerate transverse optical mode. The electromagnetic radiation being transverse cannot directly interact with longitudinal phonons in an infinite crystal. The $k \cong 0$ transverse optical mode, on the other hand, will interact and can be observed as a resonant frequency by measuring the transmittance of infrared radiation by thin films. In the vicinity of the resonance frequency, one thus expects drastic changes in the optical properties of such a crystal. In the case of strong photon–phonon coupling encountered in ionic solids, it is thus impossible to distinguish between purely photon or phononlike modes. On the other hand, in a homopolar crystal such as diamond, germanium, or silicon, one does not expect any such resonant interaction.

We will briefly discuss the dispersion of infrared radiation by cubic diatomic ionic crystals with optical isotropy. Huang [1] (also see Born and Huang [2]) was the first to give a phenomenological theory of infrared dispersion by long-wavelength optical phonons of ionic crystals. In terms of a reduced displacement vector $\mathbf{w} = \mathbf{u}(\mu/v)^{1/2}$, where $\mathbf{u}$ represents the displacement of the positive ions relative to the negative ions and μ/v is the reduced ionic mass per Bravais cell. The polar motions and the dielectric polarization may then be described by

$$\ddot{\mathbf{w}} = b_1\mathbf{w} + b_2\mathbf{E} \tag{1}$$

and

$$\mathbf{P} = b_2\mathbf{w} + b_3\mathbf{E} = [(\varepsilon - 1)/4\pi]\mathbf{E}, \tag{2}$$

where $\mathbf{E}$ is the electric field and $\varepsilon(\omega)$ the dielectric response function. The b's are constants characteristic of the solid; b_1 is related to the near-neighbor force constant, b_2 to a transverse effective ionic charge, and b_3 to polarizability. Considering the periodic solutions

$$(\mathbf{w}, \mathbf{E}, \mathbf{P}) = (\mathbf{w}_0, \mathbf{E}_0, \mathbf{P}_0)e^{-i\omega t}, \tag{3}$$

one obtains by manipulation of Eqs. (1)–(3) the expression for ε

$$\varepsilon(\omega) = \varepsilon_\infty + \frac{(\varepsilon_0 - \varepsilon_\infty)\omega_0^2}{\omega_0^2 - \omega^2}, \tag{4}$$

where ω_0 is the dispersion frequency and ε_0 and ε_∞ are static and high-frequency dielectric constants, respectively. The coefficients of Eqs. (1) and (2) may be identified as

$$b_1 = -\omega_0^2, \tag{5}$$

$$b_2 = \omega_0[(\varepsilon_0 - \varepsilon_\infty)/4\pi]^{1/2}, \tag{6}$$

and

$$b_3 = (\varepsilon_\infty - 1)/4\pi. \tag{7}$$

Equations (1) and (2) describe the electrostatic interaction between the ions. To include propagation of the excitations with finite velocity, one must solve these equations along with Maxwell's equations. Now considering $\mathbf{w}$, $\mathbf{P}$, $\mathbf{E}$ and the magnetic field $\mathbf{H}$ all varying as $\exp[i(\mathbf{q} \cdot \mathbf{r} - \omega t)]$, one obtains

$$\mathbf{q} \cdot (\mathbf{E} + 4\pi\mathbf{P}) = 0, \tag{8}$$

$$\mathbf{q} \cdot \mathbf{H} = 0, \tag{9}$$

$$\mathbf{q} \times \mathbf{E} = (\omega/c)\mathbf{H}, \tag{10}$$

$$\mathbf{q} \times \mathbf{H} = -(\omega/c)(\mathbf{E} + 4\pi\mathbf{P}). \tag{11}$$

Equations (2) and (8) yield

$$\varepsilon(\omega)(\mathbf{q} \cdot \mathbf{E}) = 0. \tag{12}$$

Equation (12) is satisfied by either

$$\mathbf{q} \cdot \mathbf{E} = 0 \tag{13}$$

or

$$\varepsilon(\omega) = 0. \tag{14}$$

Let us consider Eq. (13) first, which signifies that $\mathbf{E}$ is perpendicular to $\mathbf{q}$. Considering Eq. (10) also, we find that $\mathbf{E}$, $\mathbf{H}$, and $\mathbf{q}$ are all perpendicular to each other. From Eqs. (1) and (2) it follows that $\mathbf{w}$ and $\mathbf{P}$ are both parallel to $\mathbf{E}$. Thus Eq. (13) describes transverse waves. Elimination of $\mathbf{H}$ between Eqs. (10) and (11) yields

$$\varepsilon(\omega) = (qc/\omega)^2, \tag{15}$$

signifying propagating transverse waves that are simultaneously electromagnetic and lattice waves known as *polaritons*.

To understand the nature of the dielectric response function (4) we shall consider the electrostatic approximation in which the transverse modes are purely mechanical. Considering $\mathbf{E} = 0$, Eq. (1) becomes

$$\ddot{\mathbf{w}}_t = b_1\mathbf{w}. \tag{16}$$

If ω_{TO} is considered the natural frequency of the transverse oscillator, one has

$$\omega_{TO}^2 = -b_1 = \omega_0^2. \tag{17}$$

Thus, the dispersion frequency in Eq. (4) is identical with the long-wavelength transverse optical-mode frequency.

Returning to the second case [Eq. (14)], one can rewrite Eq. (2) as

$$\mathbf{E} = -4\pi\mathbf{P}. \tag{18}$$

Equation (18) along with Eqs. (9) and (11) gives $\mathbf{H} = 0$. Because of Eq. (10) it follows that $\mathbf{q}$ and $\mathbf{E}$ are parallel to each other and also parallel to $\mathbf{w}$ and

P [see Eqs. (1) and (2)]. Thus, Eq. (14) signifies longitudinal waves. Denoting the natural frequency of the longitudinal oscillator as ω_{LO} and solving Eq. (14) with the aid of Eq. (4) one obtains

$$\varepsilon_\infty + \frac{(\varepsilon_0 - \varepsilon_\infty)\omega_0^2}{\omega_0^2 - \omega_{LO}^2} = 0.$$

Solving for ω_{LO} one has

$$\omega_{LO} = (\varepsilon_0/\varepsilon_\infty)^{1/2}\omega_0 = (\varepsilon_0/\varepsilon_\infty)^{1/2}\omega_{TO}. \tag{19}$$

This relation between the long-wavelength longitudinal and transverse optical-mode frequencies is known as the Lyddane–Sachs–Teller (LST) relation [3].

In a homopolar crystal such as Si or Ge, in the absence of polar interactions the atomic motions are determined only by the local elastic restoring forces, i.e., $b_2 = 0$ in Eq. (1). From Eqs. (6) and (19) it follows that for such crystals $\varepsilon_0 = \varepsilon_\infty$ and $\omega_{LO} = \omega_{TO}$.

Electromagnetic radiation with frequencies in the vicinity of the dispersion frequency ω_0, will strongly interact with an ionic solid. Unfortunately, relation (4) is inadequate to describe this interaction realistically. The dielectric function given by Eq. (4) is a real quantity and thus is incapable of describing selective absorption of radiation. However, reflection of radiation could be

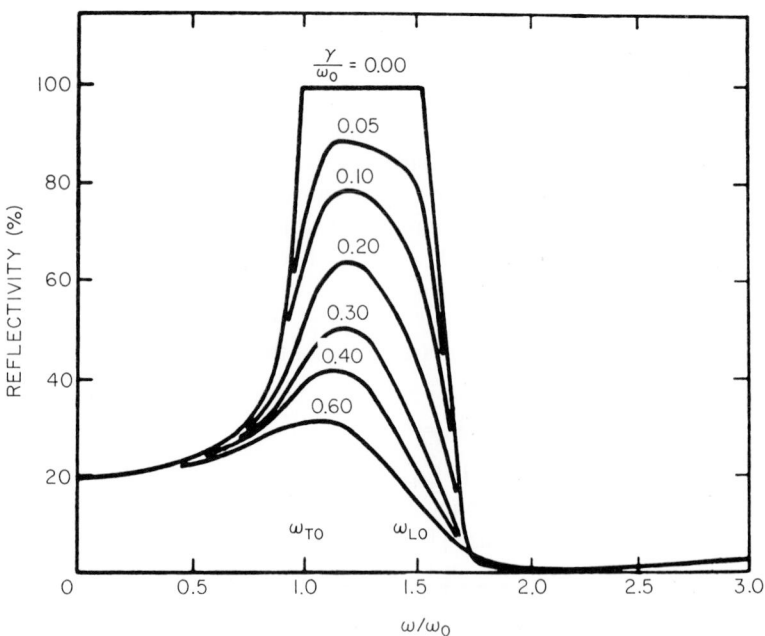

Fig. 2. Reflection spectra of a damped oscillator for various values of the damping factor.

understood by Eq. (4) along with the Fresnel formula

$$R = [(n - 1)/(n + 1)]^2, \qquad (20)$$

where $n = \sqrt{\varepsilon(\omega)}$ is the refractive index. The selective reflection of infrared radiation in the vicinity of optical lattice-mode frequencies is known as the *reststrahlen phenomenon*. The reststrahlen band for an ionic crystal in the harmonic approximation [Eq. (1)] is shown in Fig. 2 (top curve). We notice that such a crystal becomes totally reflecting between the frequencies ω_{TO} and ω_{LO}.

In any real ionic or partially ionic diatomic crystal, the observed reflectivity shows a characteristic reststrahlen band with high reflectivity between the frequencies ω_{TO} and ω_{LO}. However, in shape or intensity the observed reflectivity does not agree quantitatively with the ideal case represented by Eqs. (4) and (20). This is evident from Fig. 3, in which the reflection spectrum of AlSb is shown [4]. In real crystals, the strong reflection in the reststrahlen region is also associated with strong absorption, whereas Eq. (4), being a real quantity for all values of ω, predicts no such selective absorption. Equation

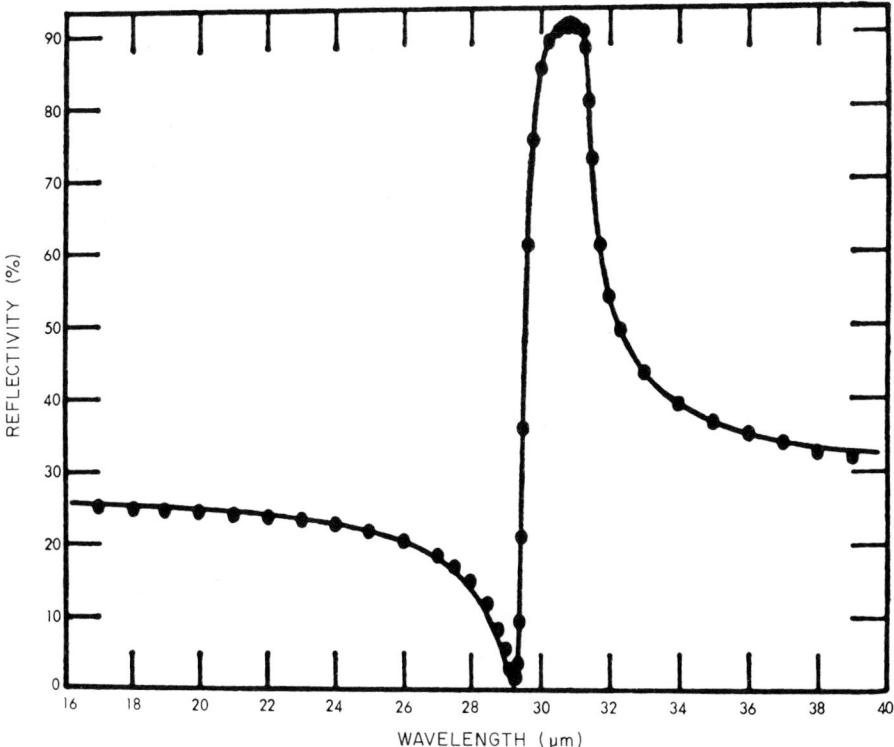

Fig. 3. Reststrahlen spectrum of AlSb. (From Turner and Reese [4].)

(4) was obtained from the phenomenological equations of motion [Eqs. (1) and (2)], which neglected all but linear terms giving rise to mutually independent lattice waves of infinite lifetimes. In real crystals, the lattice waves are coupled by anharmonic and higher-order electric-moment terms, which play an important role in the dissipation of energy, through lattice waves of finite lifetimes.

A more realistic expression for the dielectric response function may be obtained by the phenomenological introduction of a damping term that represents a force always opposed to the ionic motions and proportional to the velocity. The damping factor then represents the phonon half-width or its reciprocal, the phonon lifetime for ω_{TO}. This procedure thus provides a crude method for including the effect of energy dissipation in the neighborhood of ω_0. Equation (1) now takes the form

$$\ddot{\mathbf{w}} = b_1 \mathbf{w} - \gamma \dot{\mathbf{w}} + b_2 \mathbf{E}. \tag{21}$$

Because of the periodic nature of $\mathbf{w}$, the introduction of the additional term amounts to replacing the coefficient b_1 with $(b_1 + i\omega\gamma)$. Therefore, the dielectric response function [Eq. (4)] including damping now becomes

$$\varepsilon(\omega) = \varepsilon_\infty + \frac{\varepsilon_0 - \varepsilon_\infty}{1 - (\omega/\omega_0)^2 - i\gamma(\omega/\omega_0^2)}, \tag{22}$$

which is a complex quantity, as expected of a lossy medium.

A plane electromagnetic wave of phase velocity $c/\sqrt{\varepsilon} = \omega/k$ and frequency ω may be represented by

$$\mathbf{E} = \mathbf{E}_0 \exp i\omega(\mathbf{e} \cdot \mathbf{r}(\sqrt{\varepsilon}/c) - t), \tag{23}$$

where $\mathbf{e}$ is a unit vector in the direction of propagation. In an absorbing medium, the dielectric constant $\varepsilon(\omega)$ and hence the refractive index represent complex quantities

$$n^* = \sqrt{\varepsilon(\omega)} = n + ik$$

and

$$\varepsilon(\omega) = \varepsilon_1 + i\varepsilon_2 = (n^2 - k^2) + 2ink, \tag{24}$$

where n and k are refractive index and extinction coefficient, respectively, and ε_1 and ε_2 real and imaginary parts of the dielectric constant. Equation (23) accordingly becomes

$$\mathbf{E} = \mathbf{E}_0 \exp\left(-\frac{\omega k}{c} \mathbf{e} \cdot \mathbf{r}\right) \exp\left[i\omega\left(\frac{n}{c} \mathbf{e} \cdot \mathbf{r} - t\right)\right] \tag{25}$$

and describes the simultaneous effects of refraction and attenuation. In terms of the absorption coefficient α, the attenuation is given by the Beer–Lambert law

$$|E^2| = |E_0^2| \exp(-\alpha r). \tag{26}$$

Comparison of Eqs. (25) and (26) give the relation between α and k

$$\alpha = 2\omega k/c = 4\pi k/\lambda.$$

In terms of the real and the imaginary parts of $\varepsilon(\omega)$ [Eq. (22)], one has

$$\varepsilon_1 = n^2 - k^2 = \varepsilon_\infty + \frac{(\varepsilon_0 - \varepsilon_\infty)[1 - (\omega/\omega_0)^2]}{[1 - (\omega/\omega_0)^2]^2 + (\gamma/\omega_0)^2(\omega/\omega_0)^2} \tag{27}$$

and

$$\varepsilon_2 = 2nk = \frac{(\varepsilon_0 - \varepsilon_\infty)(\gamma/\omega_0)(\omega/\omega_0)}{[1 - (\omega/\omega_0)^2]^2 + (\gamma/\omega_0)^2(\omega/\omega_0)^2}. \tag{28}$$

The reflectance R of an absorbing medium is now given by

$$R = \frac{(n-1)^2 + k^2}{(n+1)^2 + k^2} \tag{29}$$

instead of Eq. (20). The reststrahlen spectrum of a damped oscillator for several values of the damping factor γ/ω_0 is shown in Fig. 2.

The optical constants n and k, or ε_1 and ε_2, are obtained by fitting the reststrahlen band by Eqs. (27)–(29). Usually, a best set of oscillator parameters ω_0, γ/ω_0, ε_0, and ε_∞ are chosen so as to minimize the difference between calculated and observed reflectivity as a function of frequency. The fit to the experimental data is obtained by minimizing the quantity

$$S = \sum_{i=1}^{M} [R_{i,\,exp} - R_{i,\,calc}]^2, \tag{30}$$

where S is the square of the difference between the calculated and experimental reflectivity and the summation extends over the number of experimental points M. The computations may be performed utilizing a computer program based on the method of steepest descent outlined by Verleur [5].

Figure 4 gives the calculated optical constants for AlSb from its reststrahlen spectrum shown in Fig. 3 [4]. Although the maximum of k occurs very near ω_0, it is ε_2 and the conductivity $(nk\omega)$ that undergo maxima precisely at $\omega = \omega_0$. Obviously, one identifies ω_0 with the long-wavelength ω_{TO} and the corresponding ω_{LO} may be obtained from the LST relation [Eq. (19)]. In the case of diatomic cubic crystals, ω_{LO} may also be identified with the minimum in ε_2 versus ω curves.

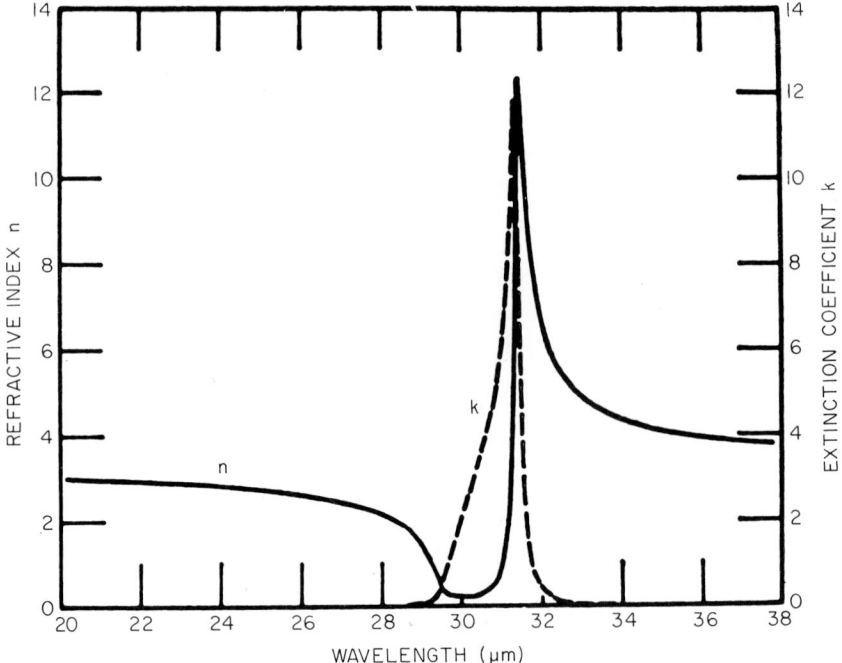

Fig. 4. Optical constants of AlSb in the one-phonon region. (From Turner and Reese [4].)

Crystals with more than two particles per unit cell may possess more than one reflection maximum, and analysis of such data requires a multioscillator fit of the observed data. Additional resonances may also occur as a result of a strong two-phonon or a strong impurity process. For the multiresonance case, Eq. (22) is replaced by the phenomenological relation

$$\varepsilon(\omega) = \varepsilon_\infty + \sum_{j=1}^{n} \frac{4\pi\rho_j}{1 - (\omega/\omega_j)^2 - i(\gamma_j/\omega_j)(\omega/\omega_j)}, \tag{31}$$

where $\varepsilon_0 = \varepsilon_\infty + \sum_{j=1}^{n} 4\pi\rho_j$. The sum is intended to include all resonances necessary to represent the reststrahlen spectrum and the dielectric response function. An example of such a reflection spectrum is shown in Fig. 5 for single-crystal GeO_2 with light polarized perpendicular to the unique axis of the crystal [6]. This spectrum was fitted with three oscillators.

For crystals with more than two atoms per unit cell a modified LST relation holds [7] involving the transverse and longitudinal phonons of the same symmetry types

$$\prod_i (\omega_{LO_j}/\omega_{TO_j}) = (\varepsilon_0/\varepsilon_\infty)^{1/2}. \tag{32}$$

In the presence of damped phonons, Eqs. (19) and (32) undergo slight modifications as pointed out by Barker [8] and Chang et al. [9]. It is also realized

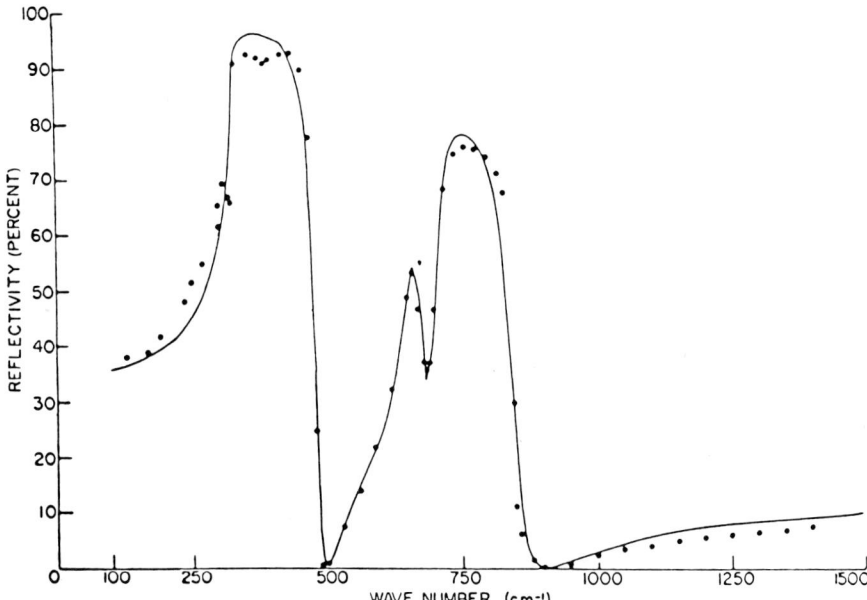

Fig. 5. Reststrahlen spectrum of tetragonal GeO_2 with $E \perp c$. Solid line is classical oscillator fit to experimental data (open circle). (From Kahan *et al.* [6].)

that in the case of multimode crystals, LO frequencies are not readily available from the TO frequencies through Eq. (32). Chang *et al.* [9] have given analytical relations for ω_{LO}'s in terms of the ω_{TO}'s and other oscillator parameters including damping. They also suggest that in the case of crystals with multimode damped oscillators, the ω_{TO}'s may be identified with the maxima of the modulus of dielectric response function $|\varepsilon(\omega)| = (\varepsilon_1^2 + \varepsilon_2^2)^{1/2}$, while ω_{LO}'s with the minima in $|\varepsilon(\omega)|$ to a very good approximation. Barker [10] suggests identification of ω_{TO}'s with maxima in ε_2 and ω_{LO}'s with maxima in $Im(-1/\varepsilon)$. Either method gives good agreement with the analytical solution or other measurements of ω_{LO}'s if the damping terms are not too large.

The long-wavelength transverse optical-mode frequency of a polar crystal may also be determined directly by transmission measurements. However, for an ionic crystal, the strong absorption in the neighborhood of the dispersion frequencies calls for the use of thin samples. For very thin (< 1-μm) samples, it may be shown that the minima in transmission occur exactly at the infrared-active ($k = 0$), TO-mode frequencies. If transmission measurements on thin films are done with p-polarized light incident obliquely to the surface, transmission minima occur also at the corresponding LO mode frequencies as shown by Berreman [11]. The three different infrared measurement configurations described and the information they reveal are illustrated in Fig. 6 for a crystal with a single damped oscillator.

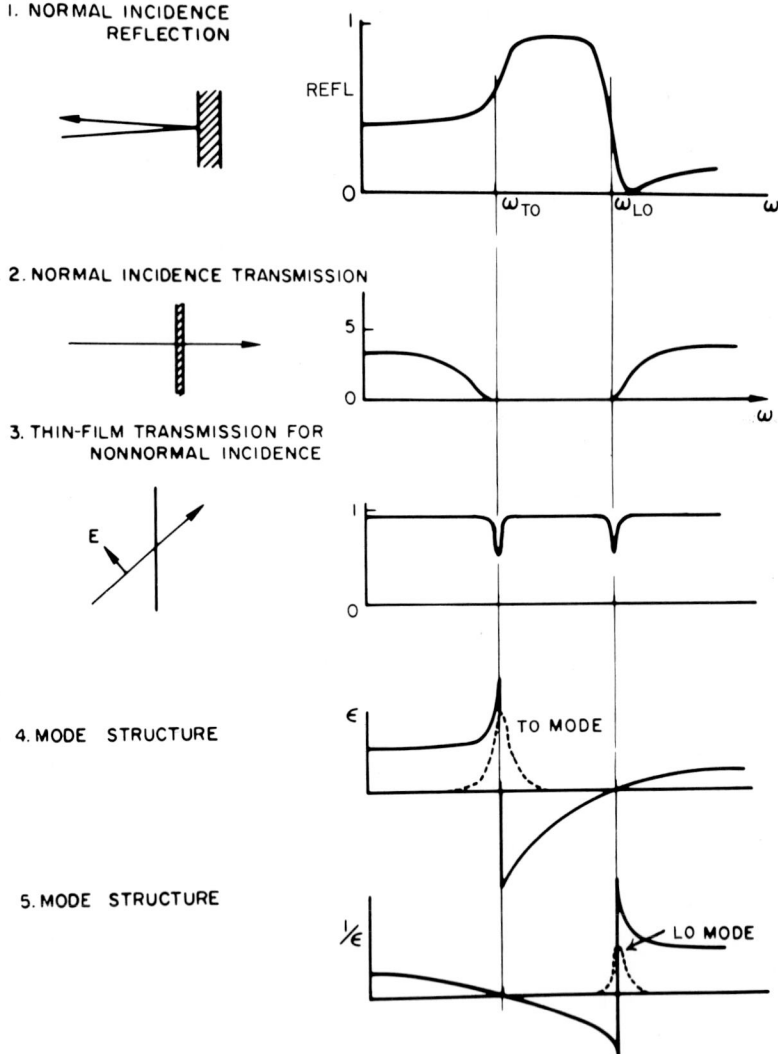

Fig. 6. Infrared spectra of three types of sample configuration. The sample is taken to have the idealized one-oscillator-mode structure given at the bottom of the figure with one TO frequency, one LO frequency and a constant (frequency-independent) damping factor. In part 4, the solid line represents real (ε) and the broken line imaginary (ε); in part 5, real $(1/\varepsilon)$ and imaginary $(-1/\varepsilon)$. (From Barker [10].)

As as example of the usefulness of Eq. (31) in obtaining optical constants, we present the case of MgO studied by Jasperse *et al.* [12]. The reflection spectra at several temperatures ranging from 8 to 1950 K and their fit by Eq. (31) requiring two oscillators are presented in Fig. 7, with the

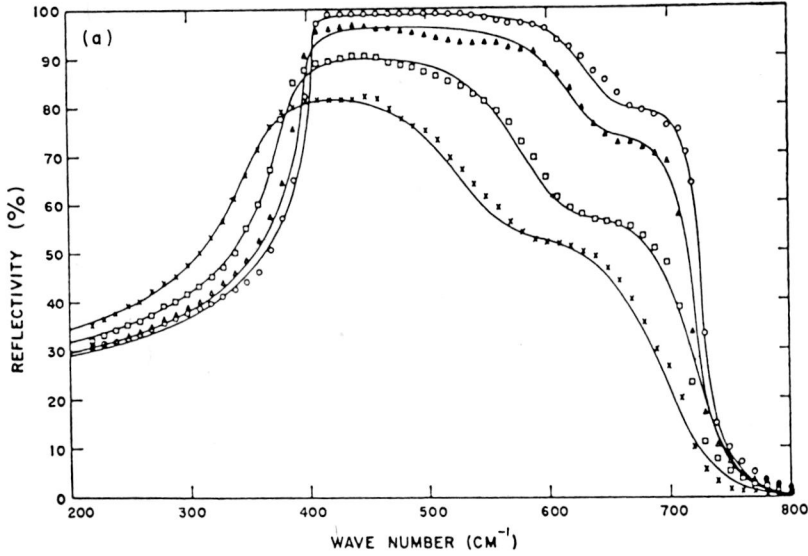

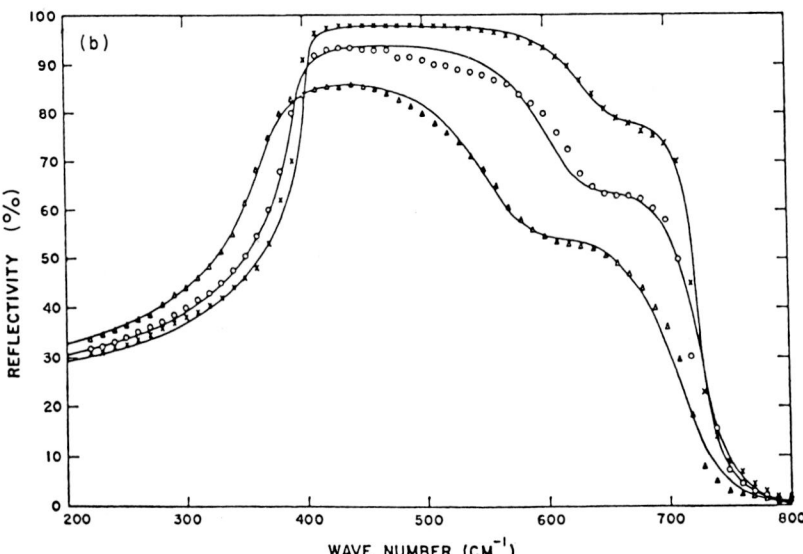

Fig. 7. Infrared reflectivity of MgO as a function of wave number and temperature. The various symbols are experimental points, and the solid lines are the theoretical curves calculated from the dispersion parameters of Table I. (a) ○, 8 K; ▲, 295 K; □, 950 K; ×, 1950 K; (b) ×, 85 K; ○, 545 K; ▲, 1410 K. (From Jasperse *et al.* [12].)

TABLE I

Infrared-Dispersion Parameters and Related Quantities for MgO, Temperature Independent[a,b]

Temperature (K)	v_1 cm^{-1}	v_2 cm^{-1}	γ_1/v_1	γ_2/v_2	$4\pi\rho_1$	$4\pi\rho_2$	ε_0	v_{LO} (cm^{-1})	LST
8	408	653	0.0045	0.140	6.30	0.025	9.34	724	719
85	406	650	0.0100	0.145	6.40	0.030	9.44	725	719
295	401	640	0.0190	0.160	6.60	0.045	9.64	725	718
545	394	630	0.0325	0.170	6.75	0.075	9.84	724	712
950	382	610	0.0570	0.195	7.10	0.120	10.23	720	705
1410	368	590	0.0850	0.225	7.45	0.175	10.64	711	692
1950	355	566	0.1200	0.260	7.95	0.225	11.19	703	685

[a] From Jasperse et al. [12].

[b] v_{LO} is determined from the zero in the dielectric response function ε' and from the LST relation, and $\varepsilon_\infty = 3.01$.

corresponding fitting parameters given in Table I. The index of refraction and extinction coefficient as functions of frequency and temperature can be easily obtained from these parameters as shown in Fig. 8 and 9.

It should be emphasized that the classical dispersion theory, even with the inclusion of damping factors and several poles, is only an approximation of the true frequency dependence of the dielectric response function. It is useful because it has a simple form, can represent the frequency dependence of reflectivity reasonably well, and enables one to obtain the various optical constants as functions of frequency in a fairly straightforward manner.

In the pseudoharmonic approximation a damping constant was introduced in Eq. (21) in an arbitrary manner. In an equally arbitrary manner a number of discrete poles were introduced to obtain Eq. (31). In reality, the phonon spectrum of a crystal is quasi-continuous extending from $\omega = 0$ to $\omega = \omega_{LO}$. Potentially any and all combinations of these phonons obeying certain conservation rules may contribute to the dielectric and optical response. Thus, γ is no longer frequency or temperature independent and the summation of Eq. (31) should be replaced by integrations involving joint densities of phonon states. A complete analytical expression for $\varepsilon(\omega)$ that includes all possible combinations of phonons and all possible decay channels is well nigh impossible. However, several approximations to include the factors previously described have been attempted [13–17] with varying success. It is interesting to note that in the vicinity of a resonance frequency (ω_j) all of these treatments reduce to the expression one obtains for the classical case. Thus, it is safe to conclude that Eq. (31) is useful in determining optical constants in the vicinities of resonances only where α has high values.

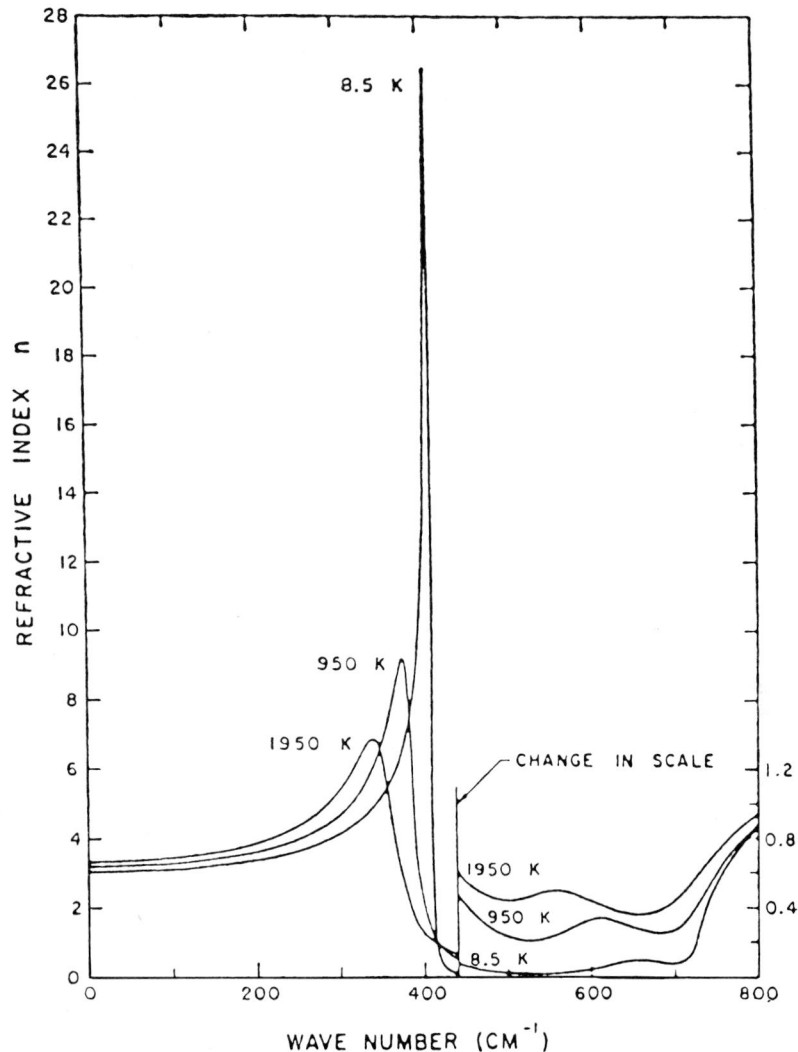

Fig. 8. Calculated refractive indices of MgO as functions of wave number and temperature (note change in vertical scale). (From Jasperse *et al.* [12].)

KRAMERS–KRONIG DISPERSION RELATIONS   III

The optical constants and hence the long-wavelength optical phonon frequencies may also be obtained from the observed reflection spectrum by performing a Kramers–Kronig (KK) dispersion analysis [18]. The KK relations

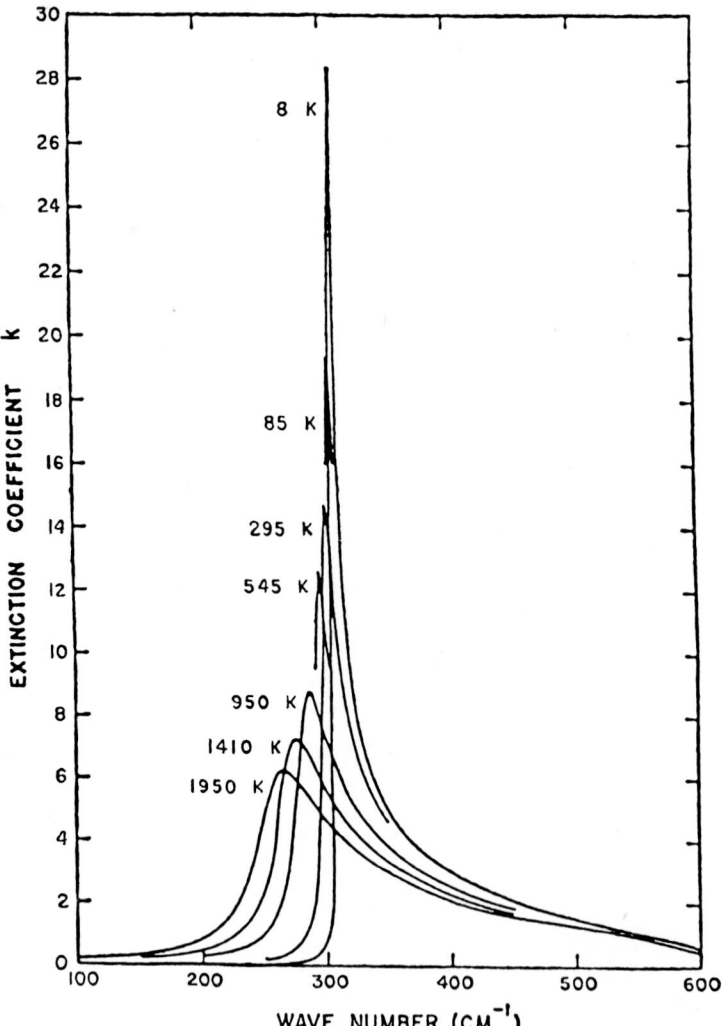

Fig. 9. Computed extinction coefficients of MgO as functions of wave number and temperature. (From Jasperse *et al.* [12].)

express an exact relationship between the real and imaginary parts of the dielectric constant and do not assume a model for calculating ε. This can be derived as a direct consequence of the principles of causality and the fact that the dielectric constant must be an analytically continuous function of the complex frequency [19]. The optical constants are given by

$$n = \frac{1 - r^2}{1 + r^2 - 2r \cos \theta}, \qquad k = \frac{-2r \sin \theta}{1 + r^2 - 2r \cos \theta}, \tag{33}$$

where the quantity $R = r^2$ is the experimentally observed reflectivity (single-surface reflection). The phase difference θ between the incident and the reflected waves is obtained from the relation

$$\theta_c = \frac{1}{\pi} \int_0^\infty \frac{d \ln r}{d\omega} \ln \left| \frac{\omega + \omega_c}{\omega - \omega_c} \right| d\omega$$

$$= 2 \frac{\omega_c}{\pi} \int_0^\infty \frac{\ln r(\omega) - \ln r(\omega_c)}{\omega^2 - \omega_c^2} d\omega. \tag{34}$$

Here, θ_c is the value of the phase difference at frequency ω_c. The integral is evaluated numerically. The limits of integration are, in practice, the two extremities of the spectrum where the reflection approaches constant values. Compared to the oscillator-fit technique described in Section II, the KK relations impose much less restrictive conditions on ε. In practice, however, the damped-oscillator representation of ε actually fits the experimental reflectivity data to a reasonable degree of accuracy, although one may often have to use a large number of oscillators [20]. The basic problem with the KK method is the limits of integration in Eq. (34) that necessitates extrapolation of data, often with large inaccuracies in the optical constants at frequencies away from the region of high reflectivity. Some of these difficulties have been discussed in detail in the literature [21–24] and will not be further discussed here. Procedures that combine the oscillator fit with KK analysis have also been tried with some success [25].

Thus, in principle, the absorption coefficient as a function of frequency can be determined from only one type of measurement, e.g., reflection or transmission, over a large spectral region. This could be accomplished by the use of KK analysis. However, since an error at any wavelength may influence the results over the entire spectral region of consideration, this method (in conjunction with pole fit or not) is only useful in the strong absorption $(\alpha \geq 10^3 \text{ cm}^{-1})$ regime also.

In fact, the KK integrals exist [18] for pairs of complementary optical constants, e.g., ε_1 and ε_2, n and k, and so on. In principle, if one has experimental values of one of these optical constants as a function of frequency, all others could be obtained.

DETERMINATION OF ABSORPTION COEFFICIENT IV
IN THE INTERMEDIATE REGION

Usually the absorption coefficient in the intermediate region ($10^0 \text{ cm}^{-1} < \alpha < 10^3 \text{ cm}^{-1}$) as a function of frequency is determined by two independent measurements, e.g., reflection and transmission at each frequency.

Considering the multiple reflection and interference, transmittance of a monochromatic beam incident normally on a slab of thickness d and with parallel faces is given by [26]

$$T = \frac{I}{I_0} = \frac{(1 - R)^2 + 4R \sin^2 \phi}{\exp(\alpha d) + R^2 \exp(-\alpha d) - 2R \cos 2(\psi + \phi)}, \tag{35}$$

where R is given by Eq. (29),

$$\tan \phi = 2k/(n^2 + k^2 - 1), \tag{36}$$

and

$$\psi = 2\pi nd. \tag{37}$$

Equation (35) is simplified in the case of low resolution [27], when interference fringes cannot be resolved, to

$$T = \frac{(1 - R)^2 + 4R \sin^2 \phi}{\exp(\alpha d) + R^2 \exp(-\alpha d)}. \tag{38}$$

At infrared wavelengths, the second term of the numerator can be neglected compared to the first one; thus

$$T = \frac{(1 - R)^2}{\exp(\alpha d) - R^2 \exp(-\alpha d)}, \tag{39}$$

or solving for αd,

$$\alpha d = \log\left\{\frac{(1 - R)^2}{2T} + \left[R^2 + \frac{(1 - R)^2}{4T^2}\right]^{1/2}\right\}. \tag{40}$$

Tables are available [28, 29] for the determination of α by using Eq. (39) or (40) from measured values of R and T. Provided that samples of different thicknesses varying from a few tens of micrometers to a few millimeters are available, measured α for ionic and partially ionic solids could vary from a few reciprocal centimeters to a few hundred reciprocal centimeters.

V   ABSORPTION COEFFICIENT IN THE TRANSPARENT REGIME

The transparent regime could be broadly defined as the spectral region for which $\alpha < 10^0$ cm^{-1}, although Barker et al. [30] have defined the onset of transparency as the frequency (ω_{onset}) at which α becomes 10^{-2} cm^{-1} and thus for $\omega > \omega_{onset}$, $\alpha < 10^{-2}$ cm^{-1}. They have also shown that this transparent regime starts around $\omega_{onset} \simeq 3\,\omega_{LO}$ as shown in Table II. The

TABLE II

**Frequency of Onset of Transparency Compared
with the Three-Phonon Cutoff Frequency[a]**

Material	$\omega_{max} = \omega_{LO}$ (cm^{-1})	$3\omega_{LO}$	ω_{onset}[b]
LiF	665	1995	2000
NaF	422	1226	1350
NaCl	265	795	700
KCl	205	615	580
KBr	163	489	430
CsBr	114	342	330
CsI	90	270	240
MgO	728	2184	1750
AgCl	189	567	500
CuCl	216	648	620
CuBr	169	507	500
ZnS	347	1041	1000
AlSb	340	1020	1110
InSb	197	591	550
GaP	403	1209	1200
GaAs	286	858	810
CaF$_2$	479	1437	1470
SrF$_2$	389	1167	1250
BaF$_2$	338	1014	1070
Si	508	1524	1520
Ge	300	900	880
C (diamond)	1332	3996	3950

[a] From Barker et al. [30].
[b] For $\alpha = 0.01$ cm^{-1} (see text).

measurement of absorption coefficient in this regime by straightforward transmission measurement utilizing thick slabs is not always possible. The chief uncertainties in the measured values come from surface effects and scattering, because these contributions are often comparable to the absorption coefficient itself. The surface effects could be minimized by difference measurements involving slabs of two different thicknesses as is usually done with attenuation measurements in optical fibers.

A number of different techniques have evolved in the past several years for the measurement of very low absorption coefficients (10^{-2} cm$^{-1} > \alpha > 10^{-5}$ cm^{-1}). They include laser calorimetry, band-edge-shift calorimetry, surface-acoustic-wave calorimetry, differential spectrophotometry, and so on. The various experimental methods for the measurement of low absorption coefficients have been thoroughly reviewed by Skolnik [32] and Hordvik [33]; we refer the reader to these articles for further information. The various methods and their range of suitability are summarized in Table III (reproduced from Skolnik [32]).

TABLE III

"Readily" Measurable Optical-Loss Levels[a]

Method	Absorption loss (cm^{-1})	Reference(s)
Thermal methods		
Thermocouple laser calorimetry	1×10^{-3}–1×10^{-4}	Many [1–7]
Laser Doppler calorimetry	5×10^{-5}	Skolnik [24]
Band-edge-shift calorimetry	1×10^{-4}–5×10^{-5}	Nurmikko [27]
Photoacoustic calorimetry	1×10^{-6}	Kerr [33]
		Rosencwaig [31, 32]
Thermoacoustic calorimetry	5×10^{-6}	Hordvik [34]
Surface acoustic wave calorimetry	1×10^{-5}–5×10^{-6}	Parks [35]
Thermal lens (IR-solids)	1×10^{-2}–5×10^{-3}	Hordvik [41]
Direct loss methods		
Laser differential loss	1×10^{-4}–1×10^{-5}	Rehn [46]
Optical bridge Q-meter	1×10^{-5}	Birnbaum [47]
Differential spectrophotometry	5×10^{-3}–2×10^{-3}	Deutsch [48, 49]
Emittance spectroscopy	1×10^{-5}	Stierwalt [51, 52]
		Skolnik [54]

[a] From Skolnik [32]. For references in the last column, see the original reference [32].

VI   MULTIPHONON ABSORPTION

A   General Remarks

In the harmonic approximation, electric-dipole selection rules permitting only the long-wavelength ($k = 0$) optical phonons display resonances in the infrared spectrum. Combinations and overtones do not arise. Furthermore, the phonon resonances should appear as sharp lines. The appearance of multiphonon spectra, the line broadening, and the phenomenon of thermal expansion of solids, however, indicate the presence of cubic and higher-order terms in the vibrational potential energy of a solid. Inclusion of anharmonicity in the general theory of the vibrational spectra of crystals not only allows combinations and overtones but, in fact, puts very little restriction on the selection rules applicable to such transitions.

For polar crystals, the mechanism of the interaction of electromagnetic radiation with phonons is due to the anharmonic (mechanical) part of the potential energy associated with the lattice vibrations. The interaction takes place through the dipole moment associated with the lattice mode. In homopolar crystals the long-wavelength optical phonons are infrared inactive because the associated electric-dipole moment is zero, and thus the above mechanism does not apply. For such crystals Lax and Burstein [34] have shown that two or more phonons can interact directly with the radiation through terms in the electric moment of second or higher order in the atomic displacements. For a semiconductor such as GaAs, which is partially co-valent and partially ionic, both mechanisms contribute. Multiphonon spectra of either origin will be allowed, space-group selection rules permitting, if the energy and wave vectors are conserved:

$$\hbar\omega = \sum_i (\pm \hbar\omega_i) \tag{41}$$

and

$$\mathbf{k} + n\mathbf{K} = \sum_i (\pm \mathbf{q}_i), \tag{42}$$

where $\hbar\omega$ and $\mathbf{k}$ are the energy and wave vector of the absorbed photon; $\hbar\omega_i$ and $\mathbf{q}_i$ are the energy and the wave vector of the ith phonon; $\mathbf{K}$ is a reciprocal lattice vector; and n is an integer. The positive and negative signs indicate emission and absorption of phonons, respectively. Since the wave vector of the photon in the infrared region is small compared to the phonon wave vectors, the second condition becomes

$$n\mathbf{K} = \sum_i (\pm \mathbf{q}_i). \tag{43}$$

Strictly speaking, the multiphonon process should give rise to a continuous spectrum. Features in multiphonon spectra, however, occur because of singu-larities in the phonon frequency distribution. Any structure in the spectrum thus reflects structure in the frequency dependence of the combined phonon density of states. The regions of high phonon concentration in a phonon dispersion occur at critical points (CP). The singularities corresponding to the CPs occur where the dispersion curves for the individual branches are flat. A CP in a phonon branch occurs where $\nabla_q\omega(q)$ (gradient of ω with respect to q) either is zero or changes sign discontinuously. In the case of the cubic diamond and zincblende structures, the relevant CPs, e.g., are

Γ,   the center of the Brillouin zone $(0, 0, 0)$,
L,   the center of a face with coordinates $(\frac{1}{2}, \frac{1}{2}, \frac{1}{2})$,
X,   the center of a face with coordinates $(1, 0, 0)$,
W,   the density peak at coordinates $(1, \frac{1}{2}, 0)$, and
K,   the edge of the zone with coordinates $(\frac{3}{4}, \frac{3}{4}, 0)$

Each of the CP phonons is related to one of the irreducible representa-tions of the cystal space group. With the use of group theory, one may, in principle, enumerate the selection rules governing the infrared spectra of a crystal allowing for the possibility of creation and/or annihilation of one or more of these CP phonons. One derives the reduction coefficients of the so-called Clebsch–Gordon coefficients for the Kronecker products of space-group irreducible representations. This corresponds to the problem of finding whether a combination of two, three, or many phonons from different or same critical points can interact with a photon through an electric-dipole, an induced electric-dipole, an electric-quadrupole, or a magnetic-dipole change. Birman [35] has worked out the space-group selection rules for the diamond and zincblende structures. Mitra [36] has applied these selection rules to the understanding of the infrared spectrum of GaAs . Johnson and Loudon [37] have applied them to the interpretation of the spectra of diamond, Si, and Ge. Burstein *et al.* [38] have obtained the selection rules for the NaCl structure, and Ganesan and Burstein [39] have done so for the fluorite structure.

The temperature dependence of the infrared absorption at any frequency depends on the number of phonons available to participate in a proposed multiphonon process at a given temperature. The occupation number of a phonon of frequency ω_i is given by

$$n_i = [\exp(h\omega_i/kT) - 1]^{-1}. \tag{44}$$

The probability of emission or absorption of one of these phonons is proportional to the square of the matrix elements of the phonon creation a^+ or annihilation a^- operator, respectively. These matrix elements are given by

$$\langle \psi(n_i)|a^+|\psi(n_i + 1)\rangle = (n_i + 1)^{1/2} \tag{45}$$

for emission and

$$\langle \psi(n_i)|a^-|\psi(n_i - 1)\rangle = n_i^{1/2}. \tag{46}$$

for absorption. The probability of absorption of a phonon is thus proportional to n_i, and that of emission of a phonon is proportional to $(1 + n_i)$. Let us suppose that a feature in the infrared absorption spectrum is assigned to a two-phonon process such that the absorption of a photon is accompanied by the emission of two phonons $(i = 1, 2)$ in accordance with the conservation principles Eqs. (41) and (42). The probability of the process is proportional to $(1 + n_1)(1 + n_2)$. The net absorption is obtained by correcting for spontaneous emission of the photon by the reverse process in which two phonons are absorbed. Then the temperature dependence of the net absorption is proportional to $(1 + n_1)(1 + n_2) - n_1 n_2 = 1 + n_1 + n_2$. The temperature-dependence factors for some two- and three-phonon processes are given in Table IV. The two conservation laws and the temperature dependence of the absorption intensity subject the phenomenological analysis of infrared spectral data to rather stringent self-consistency requirements.

TABLE IV

**Temperature-Dependence Factors for Two- and
Three-Phonon Absorption Processes**

Combination	Description	Infrared absorption intensity
2ν	First overtone	$1 + 2n$
$\nu_1 + \nu_2$	Sum	$1 + n_1 + n_2$
$\nu_1 - \nu_2$	Difference	$n_2 - n_1$
3ν	Second overtone	$1 + 3n + 3n^2$
$\nu_1 + \nu_2 + \nu_3$	Sum	$1 + n_1 n_2 + n_2 n_3 + n_3 n_1 + n_1 + n_2 + n_3$
$\nu_1 + \nu_2 - \nu_3$	Sum and difference	$(1 + n_1)(1 + n_2)n_3 - n_1 n_2(1 + n_3)$

It is implicit from the foregoing discussion that the absorption coefficient for a multiphonon event will depend on three factors, viz., the strength of the multiphonon interaction (the dipole-moment matrix element contributing to the oscillator strength), the joint density of states at the frequency of consideration, and the phonon-occupation-number function at the temperature of interest. There are many different ways of expressing these dependences; we shall outline here a well-understood treatment due to Szigeti [40]. When the anharmonic contributions are included, the vibrational potential energy of a solid has the form

$$V = \sum_j^q f_j^q (Q_j^q)^2 + \sum_{i,j,k}^{q_1 q_2 q_3} b_{ijk}^{q_1, q_2, q_3} Q_i^{q_1} Q_j^{q_2} Q_k^{q_3}$$
$$+ \sum_{ijkl}^{q_1 q_2 q_3 q_4} c_{ijkl}^{q_1 q_2 q_3 q_4} Q_1^{q_1} Q_j^{q_2} Q_k^{q_3} Q_l^{q_4} + \cdots. \tag{47}$$

Here, f, b, c, and so on, are harmonic, cubic, quartic, and so on, force constants in suitable units [see Eq. (1), for example], and Q_i is a normal coordinate. The superscripts refer to wave vectors, and the subscripts refer to branches. Similarly, the dipole moment M [see Eq. (2)] can be written as

$$\mathbf{M} = \alpha_0 \mathbf{Q}_0 + \sum_{ij}^{q_1 q_2} \beta_{ij}^{q_1 q_2} Q_i^{q_1} Q_j^{q_2} + \cdots. \tag{48}$$

The subscript 0 refers to the resonance absorption or ω_0 denotes the frequency of resonance ω_{TO} and $\alpha_0 Q_0$ the associated first-order dipole moment [b_2 of Eq. (2)].

For a region of spectrum sufficiently away from the resonance frequency, Szigeti [40] has derived the combined contributions from anharmonicity and higher-order electric moment to the imaginary part of the dielectric constant $\varepsilon_2(\omega)$ as

$$\varepsilon_2(\omega) = \frac{h\pi^2}{V\omega \, \delta\omega} \sum_{ij} \frac{\omega}{\omega_i \omega_j} D_{ij}^2 \left\{ \frac{1 + n_i + n_j}{|n_i - n_j|} \right\} + \cdots, \tag{49}$$

where

$$D_{ij}^2 = \left[\beta_{ij}^{q_1 q_2} - \frac{\alpha_0 b_{0ij}^{q_0 q_i q_j}}{\omega_0^2 - \omega^2} \right]^2.$$

The summation goes over pairs of modes for which $\omega_i \pm \omega_j$ is between ω and $\omega + \delta\omega$ so that $\omega = \omega_{ij}$ in D_{ij}. The population factors $(1 + n_i + n_j)$ and $(n_i - n_j)$ refer to sum and difference processes as given in Table IV. An expression for $\varepsilon_1(\omega)$, the real part of the dielectric function, could also be written in an analogous manner. However, at $\omega \gg \omega_0$, ε_1 is not much different from ε_∞. The absorption coefficient $\alpha(\omega)$ is easily obtained through manipulation of Eqs. (24) and (28). The terms described in Eq. (49) are almost impossible to obtain from a realistic solid, except under certain very special approximations, as we shall see in Section B.

B   Absorption in the Region $\omega > \omega_{LO}$

1   Two- and Three-Phonon Processes

In ionic crystals like NaCl the absorption in the frequency region $\omega_{LO} < \omega < 3\omega_{LO}$ is still dominated by the tail of the strong one-phonon absorption often accompanied by some broad features due to two-phonon processes [41] as illustrated in Fig. 10. Such a one-phonon absorption process is absent in a homopolar crystal such as silicon and germanium. In the latter case, for pure crystals (without free carriers), the entire absorption process originates from multiphonon interactions. Figure 11 shows the experimental data [42] for Si at three temperatures along with a CP analysis of the multiphonon singularities due to Bilz and Genzel [43].

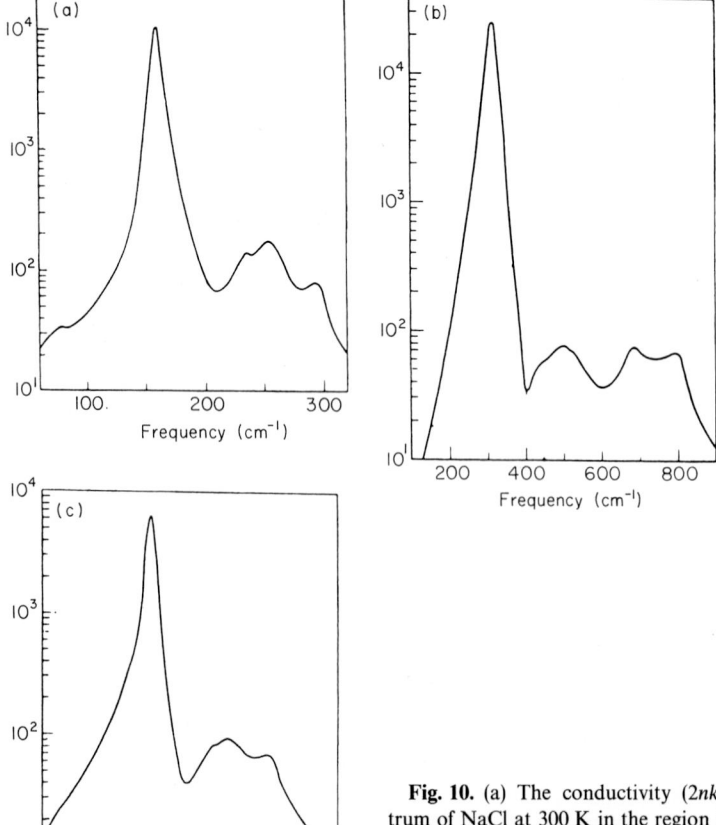

Fig. 10. (a) The conductivity ($2nkv$) spectrum of NaCl at 300 K in the region immediately above ω_{TO}. (b) The conductivity of LiF at 100 K. (c) The conductivity of KCl at 300 K. (From Smart *et al.* [41].)

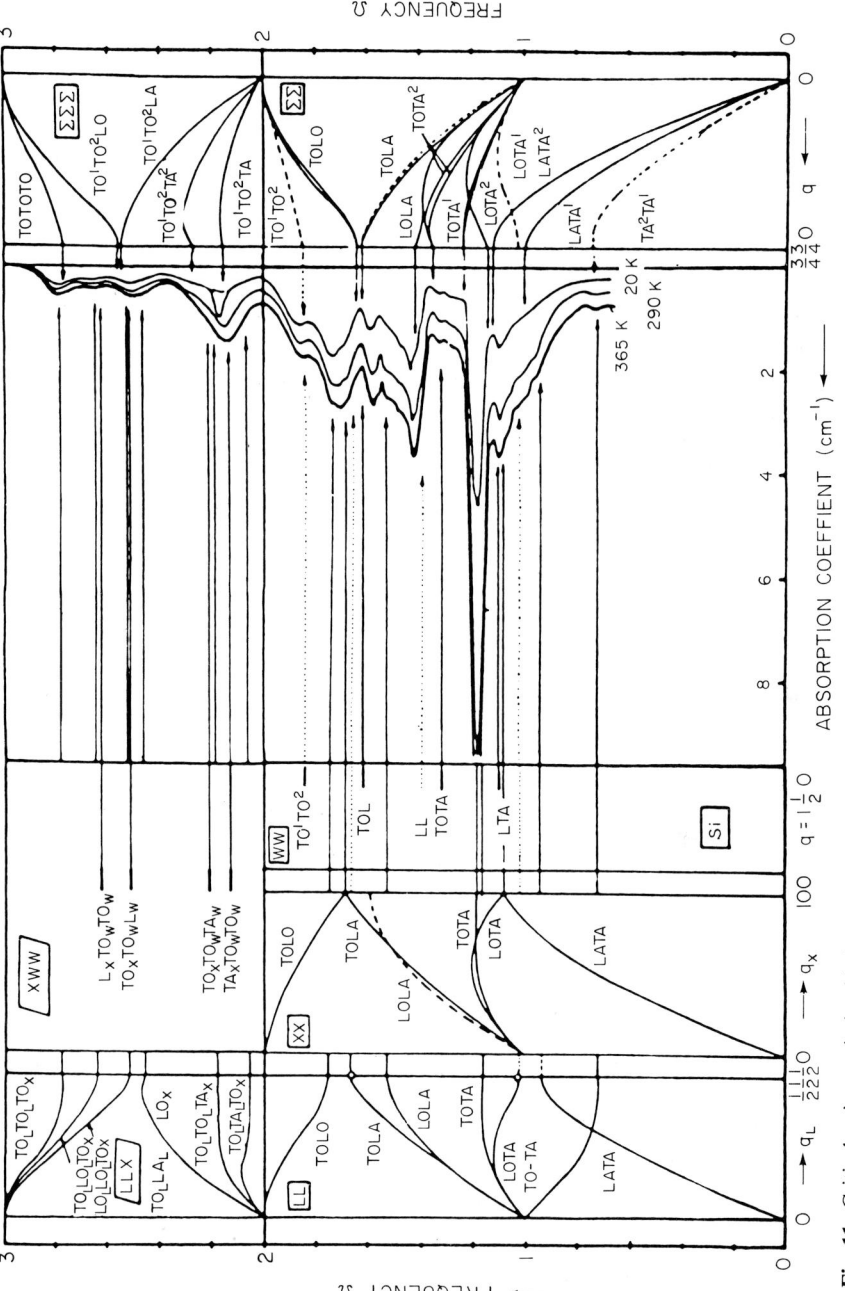

Fig. 11. Critical point analysis of infrared spectra of Si (experimental data from Johnson [42] and analysis from Bilz and Genzel [43]).

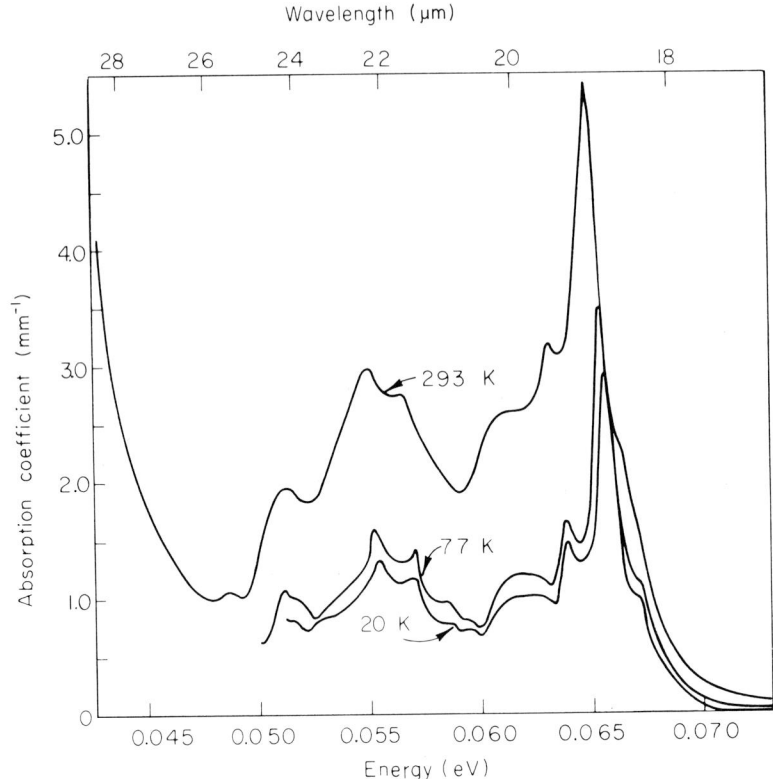

Fig. 12. Infrared absorption spectrum of high-resistivity *n*-type GaAs in the two-phonon region. (From Cochran *et al.* [44].)

Partially ionic crystals such as the II–VI and III–V semiconductors combine both of the preceding features, viz., tail from the one-phonon absorption process superimposed with rich features from two- and three-phonon processes. Examples of this are GaAs (Cochran *et al.* [44]), AlSb (Turner and Reese [4]), and ZnSe (Mitra [45]). We shall discuss the case of GaAs and AlSb (shown in Figs. 12 and 13) in somewhat greater detail. The reststrahlen spectrum of AlSb was described in Section II. The analysis of this spectrum yields $\omega_{TO} = 319$ cm^{-1} and $\omega_{LO} = 340$ cm^{-1}. The spectrum in the region $\omega > 340$ cm^{-1} was assigned by Turner and Reese [4] in terms of combinations of four Brillouin zone-boundary characteristic phonons. The positions of these zone-boundary characteristic phonons are estimated from the seven two-phonon band assignments as indicated in Fig. 13. The triplet 2LO, LO + TO, and 2TO is better resolved at 77 K. The peaks shift toward higher frequencies by 5 cm^{-1} at 77 K. The assignment is confirmed by a comparison of the net power absorbed in a two-phonon summation at two temperatures. A temperature factor f is defined as the ratio of the magnitudes of absorption

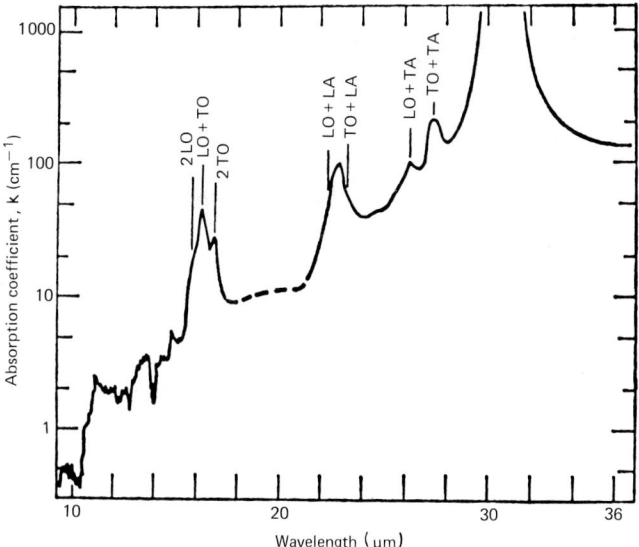

Fig. 13. High-frequency tail of the infrared absorption in AlSb. (From Turner and Reese [4].)

coefficient of a feature at two temperatures. Thus

$$f_{\text{calc}} = \frac{(1 + n_1 + n_2)_{300\,\text{K}}}{(1 + n_1 + n_2)_{77\,\text{K}}} \tag{50}$$

in the present case for a two-phonon sum process. The observed temperature factor f_{obs} was obtained after correction for the background due to free-carrier absorption, impurity absorption, and the tail of the fundamental absorptions. The assignment for the two-phonon features along with their temperature dependence is given in Table V. Thirty-eight additional features at the $\omega > 2\omega_{\text{LO}}$ cutoff (680 cm^{-1}) were assigned to three- and four-phonon processes.

A more thorough analysis [36] is possible for GaAs (Fig. 12) for which neutron-scattering data on critical-point phonons are available. The assignment of the CP phonons at X and L is straightforward from the dispersion relations given by Waugh and Dolling [46]. The CP at W presents some problems. The W is located at $(1, \frac{1}{2}, 0)$ and is not in a (110) plane, for which the experimental values are available. For reasons given by Phillips [47] it is a good approximation to assume $\omega(\text{W}) = \omega(\Sigma)$, where Σ denotes a CP along the (110) direction near $k = (0.75, 0.75, 0)$. There is, however, a lack of clear separation at W of branches into transverse and longitudinal. The CP phonon assignment of GaAs is given in Table VI [36]. There is a total of 16 different phonon frequencies that may take part in infrared-active combinations.

TABLE V

Two-Phonon Combination Bands in AlSb [a,b]

λ (μm)	ν_{obs} (cm^{-1})	f_{obs}	Assignment	ν_{calc} (cm^{-1})	f_{calc}
27.50	363	2.83	TO + TA	362	2.84
26.30	380	2.86	LO + TA	381	2.82
23.29	429	2.27	TO + LA	429	2.23
22.33	448	1.91	LO + LA	448	2.20
16.81	595	1.70	2TO	594	1.62
16.31	613	1.57	LO + TO	613	1.59
15.80	633	1.56	2LO	632	1.55

[a] From Turner and Reese [4].
[b] Data calculated from the four characteristic zone-boundary phonon frequencies: LO = 316 cm^{-1}, TO = 297 cm^{-1}, LA = 132 cm^{-1}, and TA = 65 cm^{-1}.

TABLE VI

Critical Points, Coordinates, and Phonon Modes for GaAs
(Space Group T_d^2) [a]

Critical point	Coordinate	Phonon mode	Frequency[b] (cm^{-1})
Γ	$(0, 0, 0)(1/a)$	TO(Γ)	267 (273)[c]
		LO(Γ)	285 (297)[c]
X	$(2\pi, 0, 0)(1/a)$	TO(X)	252
		LO(X)	241
		LA(X)	227
		TA(X)	79
L	$(\pi, \pi, \pi)(1/a)$	TO(L)	261
		LO(L)	238
		LA(L)	209
		TA(L)	62
W(Σ)[d]	$(2\pi, \pi, 0)(1/a)$	O$_1$(W)	263
		O$_2$(W)	215
		O$_3$(W)	250
		A$_1$(W)	116
		A$_2$(W)	188
		A$_3$(W)	79

[a] Based on Mitra [36].
[b] Data from Waugh and Dolling [46] if not stated otherwise.
[c] The values obtained by Hass and Henvis [48] from infrared reflection measurements. A value of 290 cm^{-1} has been obtained by Iwasa et al. [49] for the LO(Γ) mode from infrared transmission measurements on thin films.
[d] See text for explanation.

TABLE VII

Infrared-Active Two-Phonon Processes in GaAs[a]

Assignment	Calculated (cm^{-1})	Observed[b]
2TO(Γ)[c]	534 (546)[c]	
2LO(Γ)	570 (594)	578; 593
TO(Γ) + LO(Γ)	552 (570)	578
2TO(X)	504	509
2TA(X)	158	
2TO(L)	522	533
2TA(L)	124	
2LO(L)	476	468
2LA(L)	418	411
TO(X) + LO(X)	493	494
TO(X) + LA(X)	479	
TA(X) + LO(X)	320	321
TA(X) + LA(X)	306	307
TO(X) + TA(X)	331	333
LO(X) + LA(X)	468	468
TO(L) + LO(L)	499	494
TO(L) + LA(L)	470	468
TA(L) + LO(L)	300	307
TA(L) + LA(L)	271	
TO(L) + TA(L)	323	321
LO(L) + LA(L)	447	442
O_1(W) + O_2(W)	478	
$O_1 + O_3$[d]	513	509
$O_1 + A_1$	379	387
$O_1 + A_2$	451	456
$O_1 + A_3$	342	
$O_2 + O_3$	465	468
$O_2 + A_1$	331	333
$O_2 + A_2$	403	411
$O_2 + A_3$	294	
$O_3 + A_1$	366	(360)
$O_3 + A_2$	438	442
$O_3 + A_3$	329	333
$A_1 + A_2$	304	307
$A_1 + A_3$	195	
$A_2 + A_3$	267	

[a] Based on Mitra [36].

[b] Data from Cocharan et al. [44]. The values in the parentheses are not listed by Cochran et al. but are estimated from their diagrams.

[c] Based on Hass and Henvis [48] assignment.

[d] In the rest of this column, O_i and A_i (where $i = 1, 2, 3$) mean O_i(W) and A_i(W).

All infrared-active two-phonon combinations are listed in Table VII. It is evident that the total number of possible combinations far exceeds the absorption maxima recorded by Cochran et al. [44]. However, all observed features below 600 cm^{-1} can be assigned to one or the other of the multitude of binary combinations. The infrared-active three-phonon processes are even more numerous. In Table VIII, some possible assignments of the high-frequency bands are shown.

Although the observed infrared spectra of zincblende and diamond types of crystals can be completely understood in terms of the CP phonons known from the neutron scattering, the reverse, i.e., an unambiguous assignment of the CP phonons from the observed infrared absorption spectrum may not always be possible. In the absence of detailed phonon dispersion curves, however, it is still possible to obtain approximate Brillouin zone-boundary values of longitudinal and transverse optical and acoustic phonons in the so-called spherical Brillouin-zone-approximation, often consistent with the requirements of the space-group selection rules, as was demonstrated in the case of AlSb.

TABLE VIII

Some Infrared-Active Three-Phonon Processes in GaAs[a]

Observed[b] (cm^{-1})	Assignment	Calculated (cm^{-1})
578	TA(X) + TO(Γ) + LA(X)	573 (579)[c]
	TA(X) + TO(X) + LO(X)	572
	TO(L) + LO(L) + TA(X)	578
593	TA(X) + LO(Γ) + LA(X)	591 (603)
	TA(X) + TO(Γ) + LO(X)	587 (593)
	2TO(L) + TA(X)	601
694	TO(L) + LA(L) + LA(X)	697
	LO(L) + LA(L) + TO(X)	699
714	TO(L) + LA(L) + LO(X)	711
	TO(X) + LA(X) + LO(X)	720
(752)	TO(L) + LO(L) + TO(X)	751
	LO(X) + LA(X) + LO(X)	753 (765)
	TO(X) + TO(Γ) + LA(X)	746 (752)
	3TO(X)	756
770	2TO(L) + TO(X)	774
	TO(X) + LO(Γ) + LA(X)	764 (776)

[a] Based on Mitra [36].

[b] Data from Cocharan et al. [44]. The values in the parentheses are not listed by Cochran et al. but are estimated from their diagrams.

[c] Based on Hass and Henvis [48] assignment.

As described in the preceding subsection, the two- and three-phonon processes usually produce a structure in the absorption, particularly in the high-frequency side of the one-phonon absorption band (reststrahlen region) of ionic solids [41], and the absorption coefficient is still so substantial that the crystal can hardly be considered transparent. An operational definition of the range of transparency may be the frequency region in which approximately four- to eight-phonon processes are operative. In this subsection we consider the problem of absorption due to higher-order multiphonon processes. In ionic solids such as alkali halides (Fig. 14 and Sparks [50]), the absorption spectrum in this region is already nearly structureless, although a considerable number of features persist in a covalent solid such as Si [Fig. 15], and some structure is evident in a partially ionic solid such as ZnSe [51].

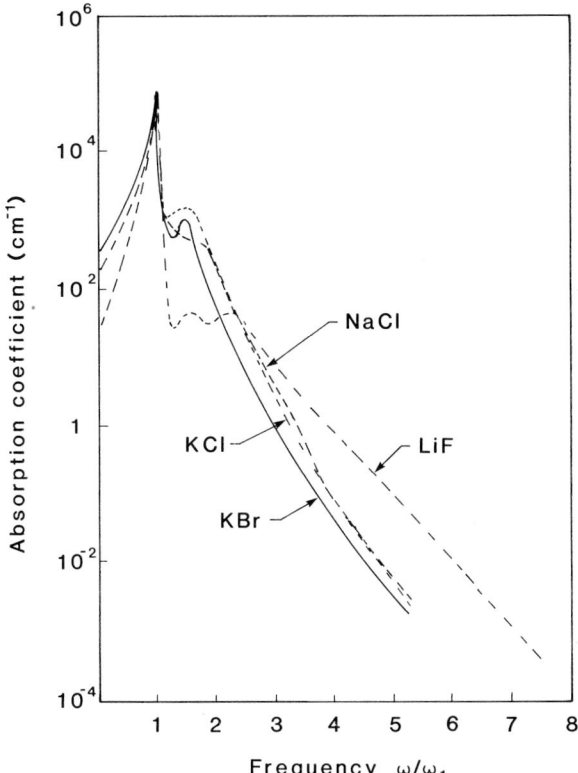

Fig. 14. Experimentally measured absorption versus photon energy at room temperature for several alkali halides. (From Sparks [50].)

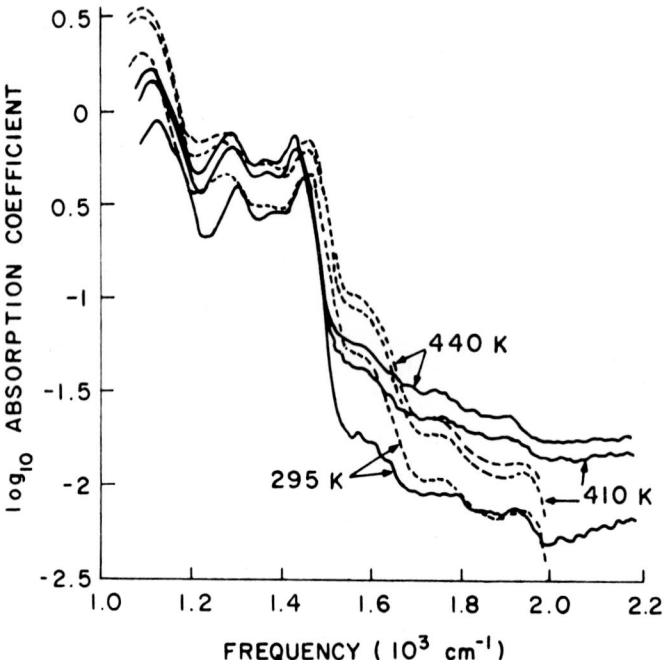

Fig. 15. Multiphonon absorption spectrum of Si in the highly transparent regime. (——, experiment; ––––, theory.) (From Bendow *et al.* [51].)

Recent compilation [52] of experimental data on the absorption coefficient of a large number of dielectric solids in the transparent regime has revealed a striking, almost universal, dependence on frequency. In a wide variety of solids, including alkali halides, alkaline earth fluorides, oxides, and semi-conductors, the absorption coefficient is observed [53] to vary exponentially with frequency as

$$\alpha = A \exp(-\gamma\omega), \tag{51}$$

where A and γ are characteristic material parameters. The temperature dependence of α has also been measured for a number of solids [52], and α increases with increasing temperature.

A number of theoretical attempts [54–61] have been made to show that the absorption coefficient indeed varies as Eq. (51), starting from the basic equation for multiphonon interaction [see Eq. (49)]. The most successful of these treatments are the so-called oscillator models in which the solid is approximated by a series of diatomic molecules. The interatomic interaction is approximated by empirical potential functions such as Born–Mayer or Morse potentials, and the molecular dipole moment is assumed to be either linear or nonlinear. Here we shall briefly describe a theoretical model due to Namjoshi and Mitra [55]. In terms of ε_1 and ε_2, α is given by

$$\alpha = \varepsilon_2 \omega / c \varepsilon_1^{1/2} \qquad (52)$$

where

$$\varepsilon = \varepsilon_1 + i\varepsilon_2$$

$$= \varepsilon_\infty + \frac{A_0}{[1 - (\omega/\omega_0)^2] + i(\gamma_0 \omega/\omega_0^2)} + \sum_j \frac{A_j}{[1 - (\omega/\omega_j)^2] + i(\gamma_j \omega/\omega_j^2)} \qquad (53)$$

and γ_j's are the damping factors, ω_j's the oscillator frequencies, and A_j's the oscillator-strength parameters. The subscript zero stands for the fundamental process and the subscript j stands for a multiphonon process. In the limit when γ_j's go to zero [55], the imaginary part of the dielectric constant reduces to

$$\varepsilon_2 = \frac{A_0 \gamma_0 \omega}{[1 - (\omega/\omega_0)^2]^2 + (\gamma_0 \omega/\omega_0^2)^2} - \sum_j A_j(\omega)\rho_j(\omega)\frac{\omega\pi}{2}. \qquad (54)$$

Here $\rho_j(\omega)$ corresponds to the j-phonon density of states. Contributions to α from the multiphonon processes of the order higher than two are shown [55] to be equal to

$$\alpha = \sum_{j \geq 2} A_j \rho_j(\omega) \pi \omega^2 / 2 c \varepsilon_1^{1/2}. \qquad (55)$$

For the calculation of higher-order-phonon densities of states, Namjoshi and Mitra [55] have shown that the wave vector conservation rule is not important and that as the order increases, the combined density-of-states function resembles more and more a normal distribution function. Therefore, higher-order density-of-states functions do not show any structure. From three phonons onward, the density-of-states function is calculated from the relation

$$\rho_n(\omega) = \int_{\omega_1 = 0}^{\omega} \int_{\omega_2 = 0}^{\omega - \omega_1} \cdots \int_{\omega_{n-1} = 0}^{\omega - \omega_1 - \omega_2 \cdots \omega_{n-2}} \rho(\omega_1)\rho(\omega_2) \cdots \rho(\omega_{n-1})$$
$$\cdot \rho(\omega - \omega_1 - \omega_2 \cdots \omega_{n-1}) \, d\omega_1 \, d\omega_2 \cdots d\omega_{n-1}, \qquad (56)$$

where $\rho(\omega)$ is the one-phonon density of state and $\rho_n(\omega)$ is the n-phonon density of state. Figure 16 shows the result of these iterations for the one- to six-phonon processes for KBr. It may be noted that higher-order-phonon densities of states gradually approach a normal distribution.

The oscillator strengths were assumed [55] to depend on the corresponding derivatives of the crystal potential energy. Namjoshi and Mitra [55] had originally used a Born–Mayer-type potential for this purpose, although other authors [30, 62] have used a Morse-type potential. The calculated and experimental absorption coefficient as a function of frequency is shown in Fig. 17 for three alkaline earth fluorides [63].

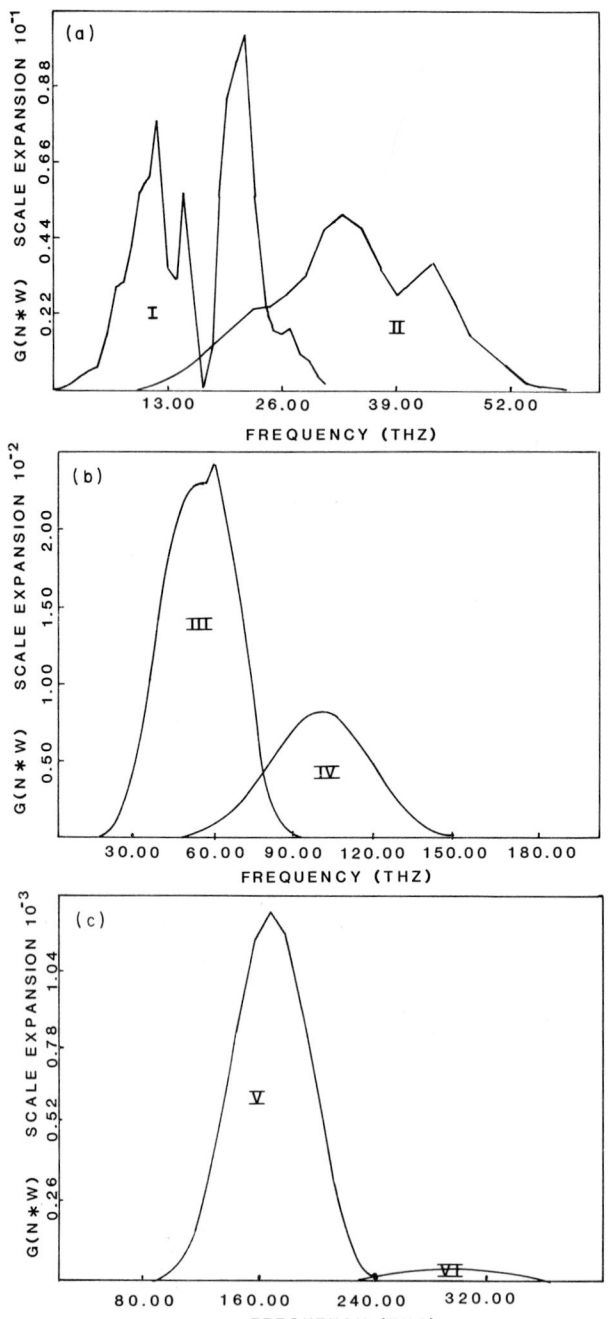

Fig. 16. One- to six-phonon densities of states for KBr: (a) one (I) and two (II), (b) three (III) and four (IV), and (c) five (V) and six (VI). Note that the vertical scale has been expanded to show the normal distribution behavior of higher-order processes. One-phonon density of states has been normalized to one. (From Barker *et al.* [30].)

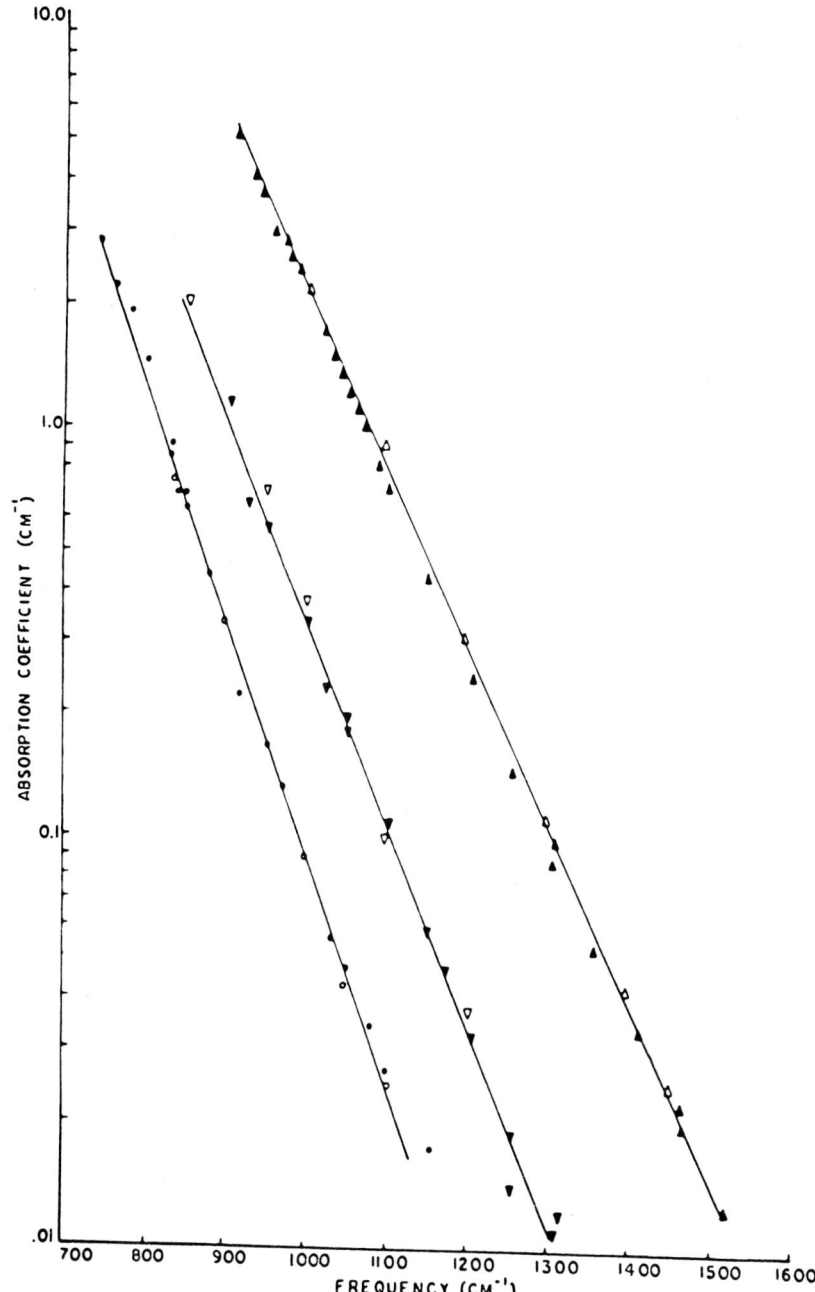

Fig 17. The room-temperature (293-K) absorption coefficient as a function of frequency for the three alkaline earth fluorides. Symbols represent experimental data and lines calculation using the Namjoshi–Mitra model. ▲, △: CaF_2; ▼, ▽: SrF_2; ●, ○: BeF_2. (From Lipson *et al.* [63].)

Namjoshi and Mitra [55] have also done first-principle calculation of the temperature dependence of α in the transparent regime by separately considering the temperature dependence of the oscillator strength A_j and the joint density of state function $\rho_j(\omega)$. They have shown that A_j varies as T^{j-1} for $j > 2$, indeed a strong temperature dependence. The temperature dependence of $\rho_j(\omega)$, in principle, is easy to estimate. In general crystal-mode frequencies decrease with temperature with a different temperature dependence for each mode. Hence, a one-phonon density-of-states function will shift to lower frequencies with increasing temperature, in addition to undergoing any deformation in its shape. Accordingly, the higher-order-phonon density-of-states functions also shift toward lower frequencies. Therefore, the order of the processes making significant contribution to α increases. Since the higher-order processes are weaker than the corresponding lower-order processes, these two effects [i.e., the temperature dependence of A_j and the temperature dependence of $\rho_j(\omega)$] have opposite contributions to the temperature dependence of α. One-phonon density-of-states function at different temperatures can be obtained by solving the lattice dynamics for different temperatures, provided that the temperature dependence of input parameters is known. Namjoshi and Mitra [55] made the simplifying approximation that all the normal-mode frequencies have the same temperature dependence and that it is equal to the corresponding temperature dependence of the bulk modulus of the solid. Under this assumption, one-phonon density of states will shift to lower frequencies as the temperature increases and thus can be obtained from its room-temperature values through multiplication of a suitable scaling factor. Neglecting any deformation in shape, higher phonon density-of-states functions can be obtained from their room-temperature values in an analogous manner. Table IX gives results for the absorption coefficient of NaCl at 10.6 μm for several temperatures that show surprisingly good agreement with the experimental results [64].

The preceding approach to the understanding to the temperature dependence is absolute, albeit approximate, and requires tedious calculations. The same considerations could be incorporated into the framework of a model due to Mitra [30, 63] that aims at calculating relative values and has proved

TABLE IX

Temperature Dependence of α for NaCl at 10.6 μm [a]

Temperature (K)	α_{calc} $(10^{-3} \text{ cm}^{-1})$	α_{exp} $(10^{-3} \text{ cm}^{-1})$
300	1.5	2.7
600	8.2	6.5
900	17.0	20
1000	18.2	22

[a] From Namjoshi and Mitra [55].

extremely useful. Including temperature dependence, the observed behavior of $\alpha(\omega)$ in the transparent regime as given in Eq. (51) can be rewritten [65] as

$$\alpha(\omega, T) = \alpha_0 \frac{\{n[\omega_0(T)] + 1\}^{\omega/\omega_0(T)}}{n(\omega) + 1} \exp[-\beta\omega/\omega_0(T)] \qquad (57)$$

in terms of an effective phonon frequency ω_0. Here n is the phonon occupation number defined in Eq. (44) and α_0 and β depend weakly on ω_0 and/or T. It is clear that α is enhanced at high T according to the T^{j-1} dependence mentioned earlier for a j ($= \omega/\omega_0$) phonon process, provided that ω_0 is independent of temperature. In general, $\omega_0(T)$ decreases with increasing T, which tends to suppress the T dependence of α as remarked before.

To apply Eq. (57) we must specify the effective oscillator frequency $\omega_0(T)$. Because the contributions of acoustic phonons are suppressed due to energy conservation, the average frequency ω defined as the first moment of the density of states should not be an appropriate choice. On the other hand, an average optical phonon frequency, such as Brout frequency [66] ω_B, should be more appropriate [30, 63].

It turns out that ω_B is close to the Debye frequency ω_D for most poly-atomic crystals [67] and that both are related to the bulk modulus [68]. Regarding the T dependence of ω_0, one notes that for ionic solids the principal T dependence arises from thermal-expansion effects [69]. A small contribution arises from anharmonicity of the crystal potential and may be neglected [69] in many cases because of cancellation between the contributions of cubic and quartic terms.

Absorption in the Region $\omega < \omega_{TO}$ C

So far our discussion of optical absorption processes in a solid has been confined to allowed one-phonon, and multiphonon sum processes in the high-frequency side of ω_{LO}. As is evident from Fig. 1, there is a considerable amount of optical absorption present in the low-frequency side of the reststrahlen region. This spectral region for a typical solid ranges from the far-infrared to submillimeter- and millimeter-wave regions. Experimental work on far-infrared absorption in solids dates back to the beginning of the century. For example, as early as 1912, Rubens and Hertz [70] measured the absorption coefficient of several ionic solids at 33 cm^{-1} as a function of temperature and found it to vary linearly. More recently, careful measurements of the frequency and temperature dependence of absorption coefficient have been conducted and the materials range from ionic solids [71], to partially ionic solids such as GaAs [72] and CdTe [73] to a host of refractory, amorphous, and polymeric materials. The primary source of absorption in this region is expected to be two-phonon difference processes, although in general one also expects contribution from three-phonon difference processes and sum processes involving low-energy acoustical phonons, and one-phonon absorption processes in the case of structurally disordered solids and solids containing imperfections and impurities.

That the far-infrared spectrum in the region immediately below the reststrahlen region could consist of two-phonon difference and/or sum processes is illustrated by the example of CdTe. The real and imaginary parts of the dielectric response of CdTe were determined by Birch and Murray [73] as a function of frequency ($40-240$ cm^{-1}) at room temperature from measurements of the amplitude and phase reflection spectra by Fourier transform spectroscopy. This experimental result was used by Parker *et al.* [74] to obtain the frequency dependence of the imaginary part of the self-energy. Twenty or more features were revealed in the spectrum and were assigned to two- and three-phonon decay processes involving CP phonons. Table X shows a typical example of such an analysis.

At frequencies farther removed from the reststrahlen region ($\omega \ll \omega_{TO}$) one will expect a featureless absorption spectrum, and its temperature dependence will be useful in delineating the absorption processes. Advances in far-infrared techniques and the development of high-power microwave

TABLE X

Assignment of Peaks of the Imaginary Part of Self-Energy of CdTe versus Frequency in Terms of Two- and Three-Phonon Combinations[a]

Feature	Assignment	Expected frequency	Observed frequency
1	TO(L) − LA(L)	39	40
3	2TA(L)	62	62.5
4	2TA(X)	69	69
5	LA(L) − TA(L)	73	73
8	LA(X) − TA(X)	91	91
9	LO(X) − TA(X)	102	102
10	TO(L) − TA(L)	112	112
11	TO(X) − TA(X)	115	116
12	LA(L) + TA(L)	135	134
16	LA(X) + TA(X)	160	160
17	LO(X) + TA(X)	171	171
18	TO(L) + TA(L)	174	173
19	TO(X) + TA(X)	184	184
21	2LA(L)	208	208
2	TO(L) + TA(L) − LA(X)	48.5	49.5
13	TA(L) + LO(L) − TA(X)	140.5	140.5
15	TO(X) + TO(Γ) − LO(X)	153.5	152.5
22	TO(L) + LA(L) − TA(X)	212.5	212.5
23	2TA(X) + TO(X)	218.5	219

[a] From Billard [75].

sources and millimeter-wave lasers have helped produce a plethora of such data. For example, the temperature dependence (up to 450 K) of the far-infrared absorption in GaAs measured by Stolen [72] indicated that both two- and three-phonon difference processes contribute to the absorption. He further speculated that the two-phonon part of the absorption is dominated by processes that involve optical and longitudinal acoustic phonons rather than longitudinal and transverse acoustic phonons. The far-infrared absorption in Al_2O_3 and MgO has been measured by Billard et al. [75] in the region covering 1 to 300 cm^{-1} and the temperature range of 200 to 2000 K. Their result is shown in Fig. 18 for Al_2O_3. At high temperatures, the absorption coefficient essentially vary as T^{n-1}, where n is the order of the difference process. The data of Fig. 18 show a temperature dependence of the type

$$\alpha = AT + BT^2, \tag{58}$$

thus indicating the presence of both two- and three-phonon difference mechanisms.

Considerable experimental data also exist on amorphous solids, glasses, and polymers [76–79]. The absorption coefficient of partially crystalline solids in the far-infrared region has more complicated frequency and temperature dependence. The long-wavelength (4–8-mm) region is still dominated by temperature-dependent multiphonon difference processes, whereas the shorter-wavelength (< 1-mm) region is primarily dominated by disorder-induced, one-phonon absorption (see Section VII) involving the entire phonon density of states and is essentially temperature independent. In general, the magnitude of the absorption in amorphous solids in this region is an order of magnitude higher than that in their crystalline counterparts. Some amorphous materials exhibit an additional anomalous behavior. For example, silica-based glasses display [79] a distinct minimum in the temperature dependence of the very-far-infrared (1–10-cm^{-1}) absorption coefficient near 10 K. This anomalous behavior is interpreted [79] by a two-level tunneling model.

The frequency and temperature dependence of the absorption coefficient in the millimeter and submillimeter regions can, in general, be understood in terms of the multiphonon theory outlined in Subsection VI B. In particular, this absorption will consist of (a) multiphonon difference processes, (b) less significantly, multiphonon sum processes, (c) disorder-induced, one-phonon processes, and (d) a host of anomalous behavior at very long wavelengths and very low temperatures. For example, Hardy and Karo [80] explained the long-wavelength absorption in a number of alkali halides in terms of difference processes caused by third-order anharmonic terms. They found that a nearest-neighbor approximation to the anharmonic potential and deformation-dipole-model-generated eigenvectors and eigenfrequencies can account for most of the absorption. The discrepancies between theory and experiment were understood in terms of three-phonon processes. At

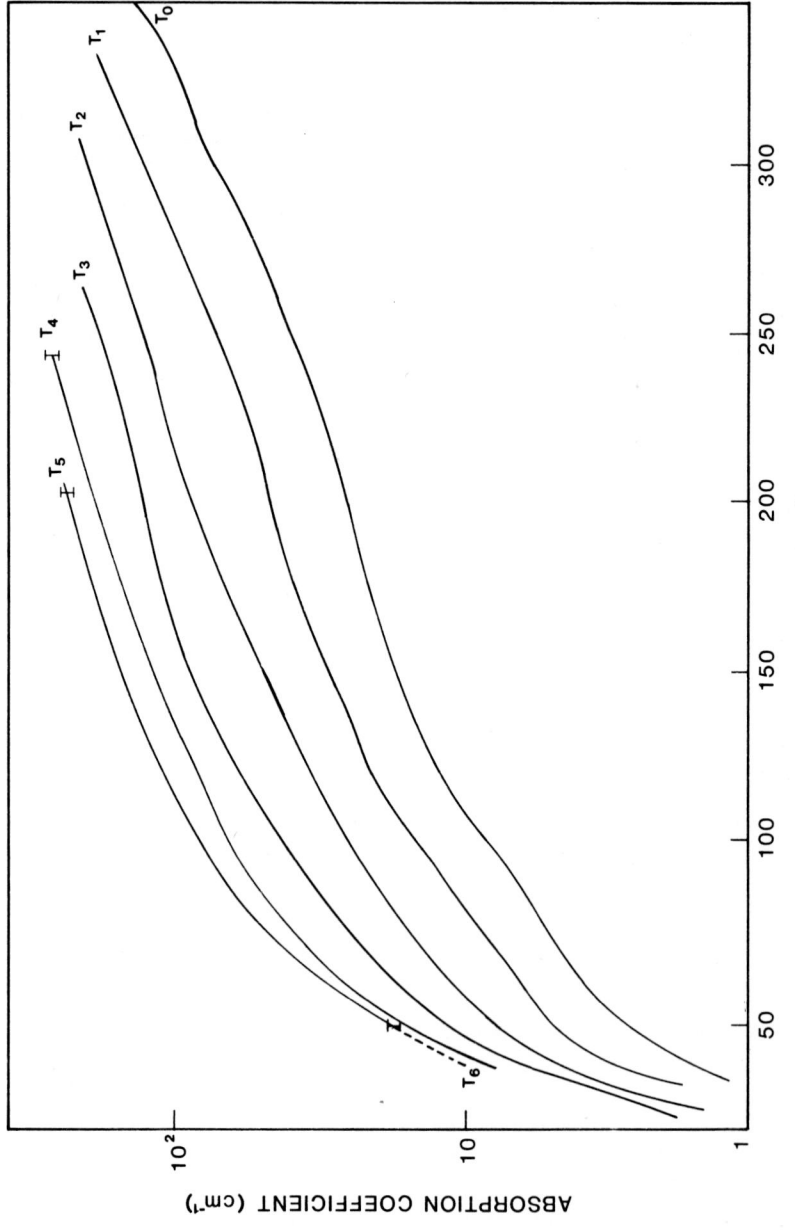

Fig. 18a. Far-infrared absorption spectrum of crystalline Al_2O_3 with $E \perp c$ at several temperatures. ($T_0 = 295$ K; $T_1 = 400$ K; $T_2 = 555$ K; $T_3 = 795$ K; $T_4 = 1050$ K; $T_5 = 1090$ K; and $T_6 = 1310$ K.) (From Billard [75].)

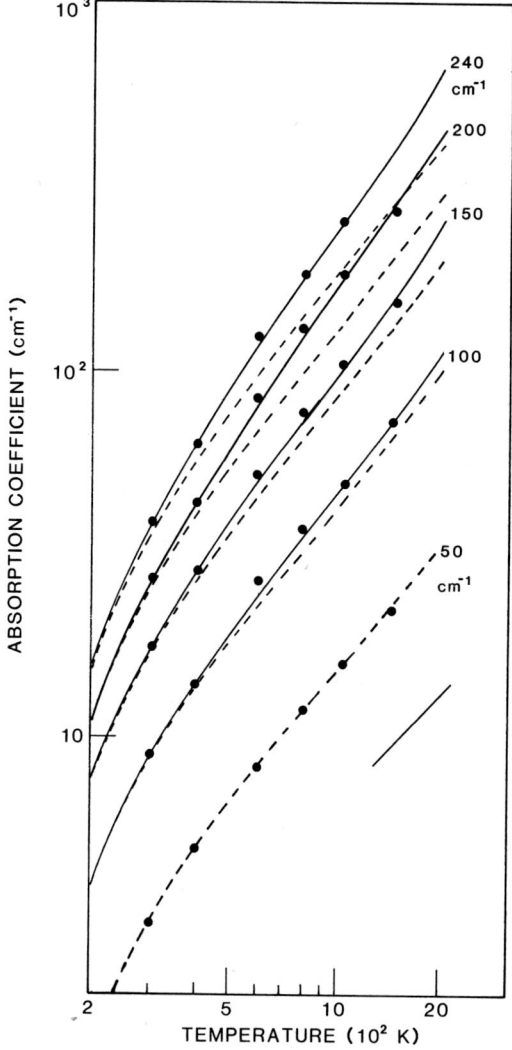

Fig. 18b. Temperature dependence of absorption coefficient of Al_2O_3 with $E \perp c$. (From Billard [75].)

longer wavelengths (~ 3 mm) further discrepancies occurred and were explained by invoking the effects of lifetime broadening of the final-state phonons. The latter mechanism is further emphasized by Sparks *et al.* [81]. They show that at long wavelengths lifetime-broadened, two-phonon processes dominate in contrast to energy-conserving, two- and three-phonon processes. Defining a critical frequency ω_c as originally proposed by Stolen and Dransfeld [71] as the difference between an optical phonon frequency and an acoustical phonon frequency at a critical point in the Brillouin zone boundary. Sparks *et al.* [81] claim that their lifetime-broadened results agree with previous energy-conserving results and experimental results in the region $\omega > \omega_c + \gamma$, where γ is the combined-phonon inverse lifetime at room temperature. The purely energy-conserving results were shown to disagree with experimental results for both the magnitude and the temperature and frequency dependence of absorption coefficients in the region $\omega < \omega_c$.

VII INFRARED ABSORPTION BY DEFECTS AND DISORDERS

A General Remarks

Impurities and defects in an otherwise perfect crystal are widely known to give rise to localized, gap, and/or resonance modes of oscillation. Furthermore, because of their presence, the translational symmetry of the lattice breaks down and the $q \simeq 0$ selection rule is relaxed, making it potentially possible to obtain optical excitation of the entire phonon density of states. Of particular interest is the case of silicon or germanium. Because of their homopolar nature, the perfect crystal does not absorb infrared radiation by a one-phonon process. However, the presence of defects modifies the situation and leads to the excitation of one-phonon infrared absorption as was observed by Fan and Ramdas [82] in neutron-irradiated Si. Simple theoretical considerations indicate that the absorption in the allowed frequency range, in addition to being dependent on the defect concentration, will, in general, follow the density of states modified by a frequency-dependent, dipole-moment matrix element that depends on the detailed nature of the defect. The defects may be chemical in nature (e.g., an impurity atom) or mechanical in nature (e.g., vacancies at particular lattice sites), or they may be extended defects (e.g., dislocation). If the impurity content of a solid is increased to such an extent that interactions between impurity atoms become significant, the system is then termed a disordered solid rather than an impure crystal. Disordered lattices are of two main types: the disordered alloy (mixed crystals) and the glasslike substance in which the disorder is spatial rather than configurational.

Absorption of infrared radiation by a crystal containing defects as compared to a perfect crystal will have the following characteristics: (a) appearance of new absorption peaks, electric-dipole selection rule permitting, due to localized modes of vibration occurring in regions outside of the normal frequency spectrum of the perfect crystal; (b) appearance of features from the entire phonon frequency spectrum due to relaxation of selection rules; and (c) weakening and broadening of allowed one-phonon characteristics. Thus, one often encounters lower absorption near the one-phonon peaks and increased absorption in the wings.

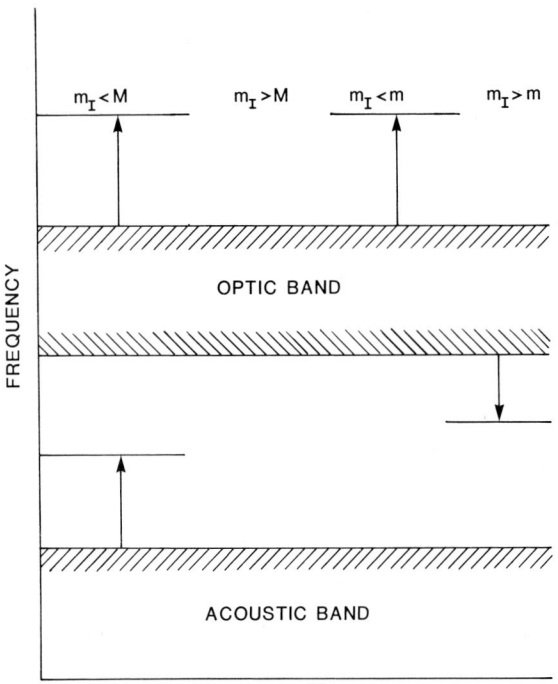

Fig. 19. Impurity modes in a linear diatomic chain ($m < M$). Four cases shown are a light impurity atom replaces the heavy atom of the host lattice ($m_1 < M$); a heavy impurity atom replaces the heavy atom ($m_1 > M$); a light impurity atom replaces the light atom ($m_1 < m$); a heavy impurity atom replaces the light atom ($m_1 > m$).

Crystals with Substitutional Impurities   B

The introduction of impurity atoms modifies the potential-energy function in a lattice. Hence, the phonon spectrum of the crystal is modified as is any physical property of the crystal in which the lattice vibrations play a central role. In the case of a substitutional impurity whose mass is different from that of the atom it replaces, the modification of the phonon spectrum is usually manifested in the appearance of localized frequencies. These frequencies may appear within the allowable frequency range of the perfect crystal, in which case they are known as resonant modes, or outside the frequency spectrum, in which case they are known as gap or local modes, as shown in Fig. 19.

The first theoretical discovery of impurity-induced localized vibrational modes in crystals was made by Lifshitz [83] in 1943. This work was followed by the work of Montroll and Potts [84] and Mazur *et al.* [85]. However, the early work was restricted on one- and two-dimensional lattices and, therefore, was only of qualitative value. The direct experimental observation, in 1960, of the existence of localized modes due to U centers (H⁻ ions) in alkali

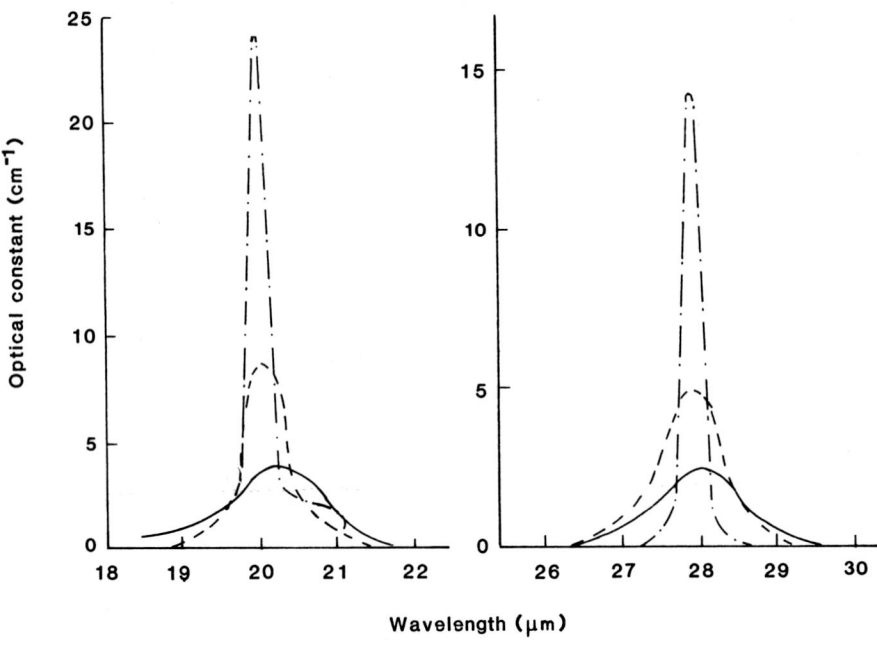

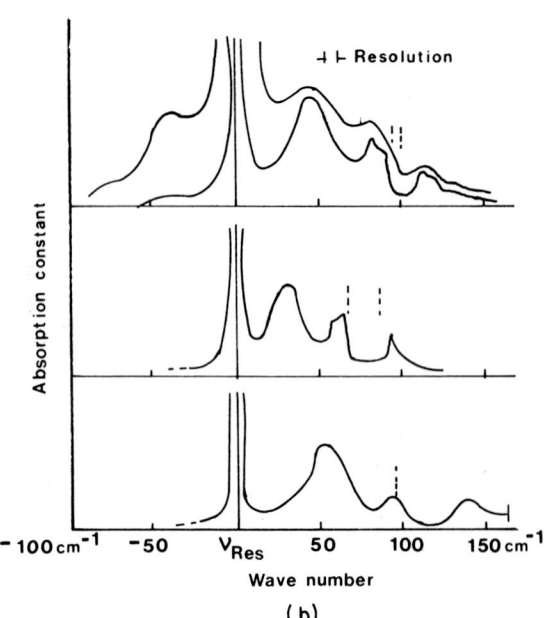

Fig. 20. (a) U-center local-mode absorption due to H^- and D^- impurities in KCl and its temperature dependence. For $KCl \cdot KH$, $N = 8.7 \times 10^{17}$ cm^{-3} and for $KCl \cdot KD$, $N = 5.1 \times 10^{17}$ cm^{-3}. The lines represent various temperatures:———, 300 K; ————, 200 K; —·—, 100 K. (b) One-phonon side bands of U-center absorption in KBr, KI, and NaBr. For $KBr:H^-$, $T = 90$ K and 20 K, $\nu_{Res} = 446$ cm^{-1}, for $KI:H^-$, $T = 9$ K, $\nu_{Res} = 380$ cm^{-1}; for $NaBr:H^-$, $T = 9$ K, $\nu_{Res} = 498$ cm^{-1}. (From Klein [96].)

halides by Schaefer [86] paved the way for a serious study of three-dimensional lattices. The U-center absorption spectra in KCl due to H^- and D^- ions are shown in Fig. 20.

The dynamical behavior of the impurities has been described theoretically by either a Green's function technique and by a molecular model. The Green's function technique introduced by Lifshitz [87] has been applied extensively to describe the dynamical behavior of an impurity atom in a lattice (for example, see Dawber and Elliot [88], Lifshitz [89], Izyumov [90], and Gaur et al. [91]). In this model, it is usually assumed that the defect-induced excitations are localized. This enables one to express the defect problem in terms of the elementary excitations of the unperturbed host crystal. Thus, for a simple mass defect perturbing the phonon spectrum of the host lattice, knowledge of the complete phonon dispersion relation and the eigenvectors of the host-lattice phonons are essential. Under such conditions, this method has an advantage over the conventional perturbation techniques in that it provides, in principle, an exact solution to the defect problem. It is expected that in addition to the change in mass, a softening (or hardening, depending on the impurity) of short-range forces around the impurity ion will also occur. The Green's function technique does not allow one to incorporate such effects easily. A molecular model, introduced by Jaswal [92] to study the U-center local mode in alkali halides, on the other hand, is capable of determining the changes in short-range forces around the impurity. The molecular model was adapted for the impurities in zincblende-type semiconductors by Singh and Mitra [93]. A large body of data on point-defect-induced local, gap, and, band-mode absorption and theoretical interpretation exist in the literature (see Genzel [94], Maradudin [95], Klein [96], Newman [97], and Vetelino and Mitra [98]) and will not be elaborated here.

Before leaving the subject of defect-induced absorption, it is worthwhile to comment on at least two interesting aspects of it. Absorption of infrared radiation resulting in the simultaneous creation (or annihilation) of a local-mode phonon and a lattice phonon giving rise to sidebands around a local-mode resonance are not an uncommon occurrence. An example of this is shown in Fig. 20 for the case of U centers. Resonant absorption of infrared radiation by molecular impurities and defect clusters have also been observed extensively (see, for example, Newman [97]), and although the latter are qualitatively understood, an exact theory is not straightforward. As an example, the absorption spectrum from boron–lithium pairs in silicon is shown in Fig. 21.

If the impurity content of a solid is increased to such an extent that inter-action between impurity atoms begins to play an effective role, the system should then be termed a mixed crystal rather than an impure crystal. The study of mixed crystals dates back to as early as 1928 (see Krueger et al. [99]). Dyson [100] was the first to obtain a formal solution for the frequency spectrum of the disordered linear chain. Several phenomenological models

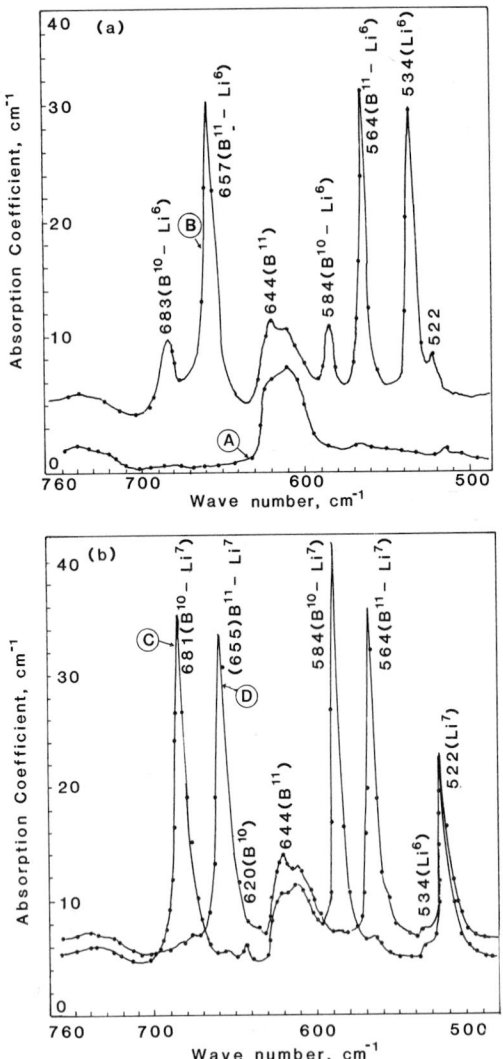

Fig. 21. The absorption spectrum from B–Li pairs in Si. (a) High-purity Si (A) and ^{6}Li doping (B); (b) ^{10}B and Li (C) and ^{11}B and Li (D). (From Newman [97].)

[101–107] have been proposed to explain the vibrations of mixed-crystal systems and their interaction with infrared radiation. As far as the behavior of the long-wavelength optical phonons is concerned, there seem to exist two types of mixed crystals. In one class of mixed systems, termed here the one-mode-behavior type, each of the $q = 0$ optical-mode frequencies varies continuously and approximately linearly with concentration from the frequency characteristic of one end member to that of the other end member. Furthermore, the strength of the resonant absorption remains approximately constant. The examples of this behavior are most of the mixed alkali halides, $GaAs_xSb_{1-x}$, $Zn_xCd_{1-x}S$, $ZnSe_xTe_{1-x}$, and so on. In the other class of mixed-crystal systems, termed here the two-mode behavior type, two phonon frequencies for each of the allowed optical modes of the pure crystal are observed to occur at frequencies close to those of the pure end members. In addition, the strength of absorption by each phonon mode of the mixed crystal is approximately proportional to the mole fraction of the component it represents. Examples of the two-mode type of behavior are in $InAs_xP_{1-x}$, $GaAs_xP_{1-x}$, CdS_xSe_{1-x}, ZnS_xSe_{1-x}, and so on. As examples, the reststrahlen spectra of $Zn_xCd_{1-x}S$ and ZnS_xSe_{1-x} are shown in Fig. 22.

It has been observed that the model using the concept of a unit cell does provide some basic understanding of mixed-crystal behavior. Verleur and Barker [103] considered a cluster model to account for the two-mode behavior. This model assumed that like negative ions cluster around positive ions or vice versa, depending on whether the ad-atom is an anion or a cation. In the random element isodisplacement (REI) model, Chen et al. [104] assumed that in a mixed crystal $AB_{1-x}C_x$ the B and C atoms are distributed on the anion sublattice and the anions of like species vibrate in phase with identical amplitudes against the cations, which also vibrate as a rigid unit. Chang and Mitra [105] modified the REI model to include the polarization field. Subsequently, Chang and Mitra [108] proposed the pseudo-unit-cell model and predicted the zone-boundary phonons of systems exhibiting the two-mode behavior at the zone center. Kutty [109] applied the Green's function technique to derive the phonon dispersion relations in mixed crystals as a function of wave vector. Sen and Hartman [110] explained the switching from one- to two-mode-type behavior observed in some III–V mixed semiconductors by using the coherent-potential-approximation technique in one dimension. Varshney et al. [111] have given a complete lattice dynamics of a mixed linear diatomic chain by using the pseudo-unit-cell model of Chang and Mitra [108]. Massa et al. [112] have extended this model to three dimensions and successfully explained the optical absorption and lattice dynamics of mixed alkali halides.

Amorphous and Glassy Systems   C

In contrast to the crystalline state, in which the positions of atoms are fixed into a definite structure except for small thermal vibrations, the amorphous state of the same material displays varying degrees of departure from this fixed structure. The amorphous state almost always shows no long-range order. Short-range order, up to several neighbors, may often be retained, although averaged considerably around their crystalline values. In the discussion of

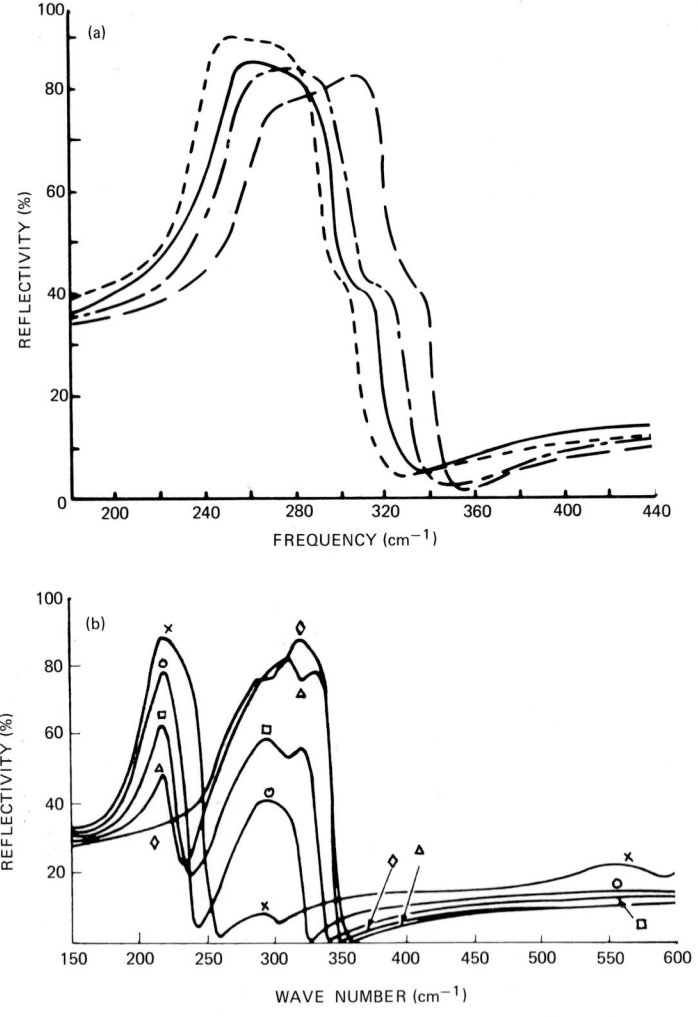

Fig. 22. Reststrahlen spectra of (a) $Zn_xCd_{1-x}S$ for $E \perp c$ (one-mode type) at 300 K [106]. (----, $Zn_{0.19}Cd_{0.81}S$; ——, $Zn_{0.36}Cd_{0.64}S$; –––, $Zn_{0.55}Cd_{0.45}S$; and ——, $Zn_{0.88}Cd_{0.12}S$.) (b) ZnS_xSe_{1-x} (two-mode type). ($\diamond$, $x = 1.00$; $\triangle$, $x = 0.82$; $\square$, $x = 0.60$; $\bigcirc$, $x = 0.33$; and $\times$, $x = 0.015$.) (From Brafman *et al.* [107].)

absorption of light by the vibrational modes of a glassy or an amorphous material, it is useful to distinguish between modes that are allowed or forbidden in a one-phonon absorption process. For a crystal of the diamond structure, e.g., Si or Ge, there is no one-phonon infrared absorption as could be seen from the lower curve of Fig. 21a. On the other hand, the absorption spectrum of a neutron-irradiated sample shows [113] (Fig. 23) a rich absorption spectrum almost mimicking the phonon density of states of the crystalline

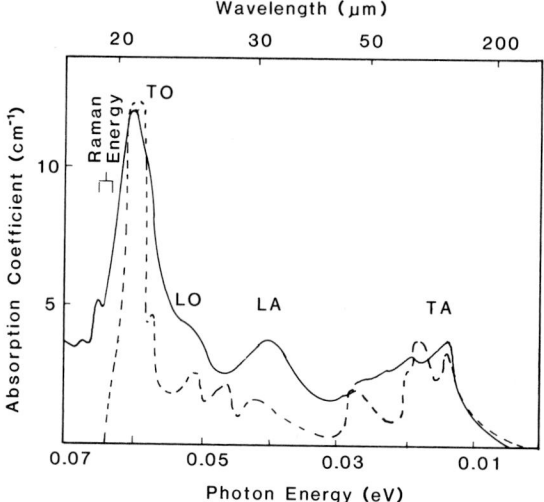

Fig. 23. One-phonon absorption in Si irradiated with 9.3×10^{18} fast neutrons/cm^2. The dashed line represents a calculated density of states curve. (From Angress *et al.* [113].)

counterpart but somewhat broadened. On the other hand, a crystal like SiC displays a strong one-phonon infrared absorption. Spectrum of its amorphous form [114] is still dominated by the one-phonon process, albeit considerably broadened, as shown in Fig. 24. It is worth noting that the areas under the absorption spectra for the crystalline and amorphous materials are almost equal and that while the peak absorption of the crystalline material is more than an order of magnitude higher than its amorphous counterpart, the opposite is true in the wings.

These two contrasting examples (viz., Si and SiC) emphasize the necessity for distinguishing between crystalline-forbidden and crystalline-allowed analog processes. While in the former case the entire density of states plays an important role and a nonvanishing dipole-moment matrix element appears in the entire spectral range [88], a statistical approach to the prediction of infrared absorption in amorphous solids is used by Mitra *et al.* [115] in the latter crystalline-allowed analog processes. In the statistical model of Mitra, a small degree of microscopic density disorder is introduced in which the properties of the disordered solid are related to those of its crystalline counterpart. The infrared response is treated by determining the changes in the phonon spectrum as functions of local density. As an example of application of this model, the crystal-allowed case of SiC is shown in Fig. 24, in which the calculated amorphous one-phonon absorption spectrum consists of nothing but a broadened version of the allowed crystalline case broadened by the radial distribution function of the amorphous phase. This kind of analysis provides a quantitative gauge of the departure from crystallinity in the optical

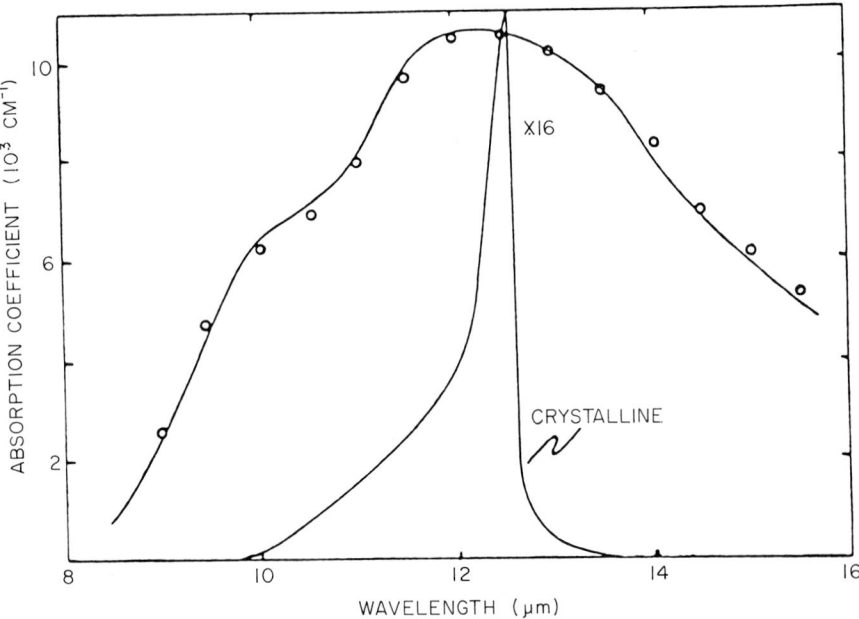

Fig. 24. Infrared absorption in amorphous SiC. Circles represent experimental data [114]. Solid line is a calculation on the basis of local-density-disorder model due to Mitra [115]. The spectrum of the crystalline phase is also indicated (note the multiplicative factor in the vertical scale). (From Vetelino and Mitra [98].)

spectrum, i.e., the relative contribution of processes that do not conserve crystal momentum.

In the preceding examples, both Si and SiC are tetrahedrally bonded semiconductors, and in structure their amorphous counterparts are not that drastically different from their crystalline phases. We should conclude this section with a more dramatic example of disorder, viz., silica glass and quartz. In both their crystalline and vitreous forms, SiO_2 is composed of tetrahedra with SiO_4 units, although in the vitreous phase significant variations in bond angles and distances have been reported and also an almost complete absence of order beyond a few near neighbors. Yet the basic SiO_4 tetrahedron offers a plethora of infrared-active absorption bonds in the range of 300 to 1400 cm^{-1}. In Fig. 25 we compare the infrared reflection spectra of crystalline quartz with that of vitreous silica and neutron-irradiated quartz and silica glass [116]. The following points are worth noting: (a) the dielectric response in quartz and glass [116] shows resonances approximately in the same regions of the spectrum, indicating the integrity of SiO_4 units; (b) the glass spectra, as expected, are smeared out; (c) with increasing neutron dosage, the spectrum of crystalline quartz approaches that of silica glass; and (d) irradiation has very little or no effect on the spectrum of vitreous silica, indicating an equilibrium disordered state.

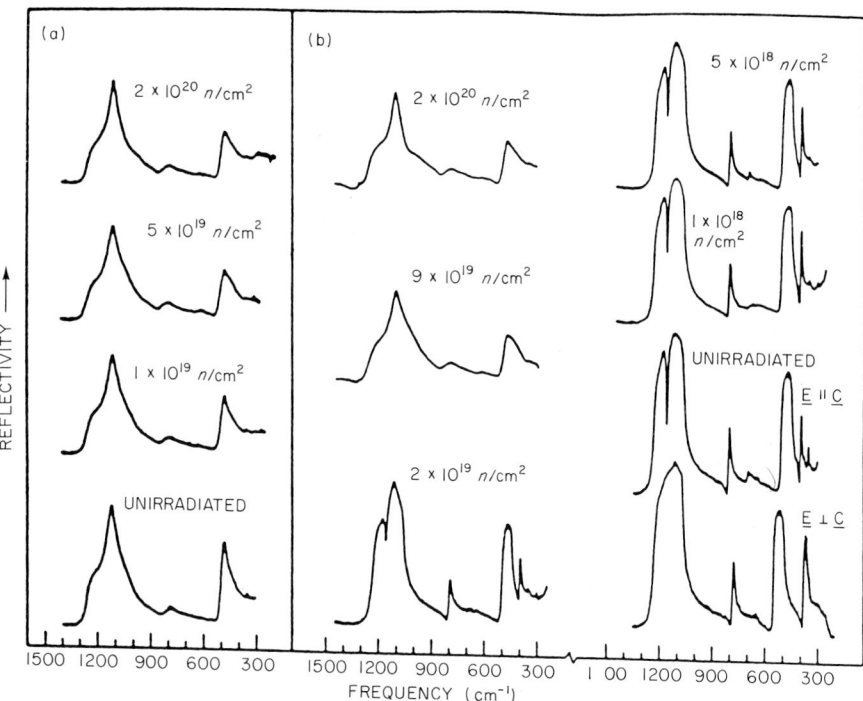

Fig. 25. Infrared reflection spectra of neutron-irradiated (a) vitreous silica and (b) quartz. (From Bates *et al.* [116].)

INFRARED DISPERSION BY PLASMONS   VIII

In the preceding discussion of absorption of infrared radiation by phonons in a solid, it was tacitly assumed that the solid was an insulator. This assumption does not hold well for a narrow-gap semiconductor at ordinary temperatures or for a doped semiconductor with a partially filled conduction or valence band. The collective excitation of this free-carrier electron gas (plasma) in such a crystal will modify the infrared absorption by phonons, as discussed earlier. The dispersion mechanisms through which electromagnetic radiation interacts with a solid should include the contribution of free-charge carriers in solids for which their numbers are significant, in addition to contributions from bound electrons and phonons. The dielectric response function of such a solid can now be written as

$$\varepsilon = 1 + 4\pi(\chi_{BE} + \chi_{L} + \chi_{FC}), \qquad (59)$$

where χ_{BE}, χ_{L}, and χ_{FC}, respectively, represent the bound electron, lattice, and free-carrier contributions to the electrical susceptibility. For the spectral region of interest, we are not concerned with the bound-electron dispersion; thus this term, as usual, will be represented by a dispersion-free, high-frequency dielectric-constant term

$$\varepsilon_\infty = 1 + 4\pi\chi_{BE}. \tag{60}$$

An approximate expression for the dielectric response function for a free-electron gas in a solid can be obtained from the classical Drude model in which a free electron of effective mass m^* and charge e is displaced by an amount $\mathbf{x}$ as a result of interaction with the electric field $\mathbf{E}$, with the equation of motion

$$m^*\ddot{\mathbf{x}} + m^*\gamma\dot{\mathbf{x}} = e\mathbf{E}_0 e^{-i\omega t}. \tag{61}$$

The damping term proportional to the velocity obviously represents the electron–phonon scattering in a phenomenological manner. Solving for $\mathbf{x}$, one obtains

$$\mathbf{x} = -e\mathbf{E}/m^*\omega(\omega + i\gamma). \tag{62}$$

The polarization, defined as the electric-dipole moment per unit volume, is given by

$$\mathbf{P} = N e\mathbf{x}, \tag{63}$$

where N is the carrier concentration. Recalling that

$$\mathbf{P} = [(\varepsilon - 1)/4\pi]\mathbf{E}, \tag{64}$$

one readily obtains

$$\varepsilon_{FC} = \varepsilon_\infty - \{4\pi N e^2/[m^*\omega(\omega + i\gamma)]\} \tag{65}$$

for the dielectric response function due to a single-component plasma. In terms of the plasma frequency defined as

$$\omega_p = (4\pi N e^2/\varepsilon_\infty m^*)^{1/2} \tag{66}$$

ε_{FC} becomes

$$\varepsilon_{FC} = \varepsilon_\infty\{1 - [\omega_p^2/\omega(\omega + i\gamma)]\}. \tag{67}$$

For a number of semiconductors there exists a region of the infrared spectrum in which the free-carrier contribution dominates, and both the bound-electron and lattice contributions to the dielectric response function are negligible. For a semiconductor, such a situation prevails in a region of the infrared spectrum in which $\omega_g \gg \omega_p \gg \omega_{TO}$, where $\hbar\omega_g$ is the electronic band gap. The analysis of optical response is also simpler in a region in which the reflection coefficient is independent of carrier scattering. For such a situation to prevail, one usually finds $\gamma \ll \omega$ and $\varepsilon_1 \gg \varepsilon_2$. In a few semiconductors for which all these conditions are met, the reflectivity spectrum in the appropriate $(\sim\omega_p)$ infrared region becomes a function of carrier concentration. Here

PbTe satisfies the preceding specified conditions, and its infrared reflection spectrum [117] as a function of hole carrier concentration is shown in Fig. 26. Since this case satisfies the preceding outlined conditions, the reflectivity minimum is a measure of the effective mass

$$m^* \cong [1/(\varepsilon_\infty - 1)](Ne^2/\pi c^2)\lambda_{min}^2, \tag{68}$$

and the real and imaginary part of ε are given by

$$\varepsilon_1 = \varepsilon_\infty - \frac{e^2}{\pi c^2}\left(\frac{N}{m^*}\right)\lambda^2 \quad \text{and} \quad \varepsilon_2 = \frac{e^2}{2\pi c^3}\left(\frac{N}{m^*}\right)\gamma\lambda^3. \tag{69}$$

The λ^2 dependence of ε_1 is confirmed experimentally [117].

For most polar semiconductors, however, lattice dispersion contributes significantly. For such cases,

$$\varepsilon(\omega) = \varepsilon_\infty\left[1 - \frac{\omega_p^2}{\omega(\omega + i\gamma)}\right] + \frac{(\varepsilon_0 - \varepsilon_\infty)\omega_0^2}{\omega_0^2 - \omega^2}, \tag{70}$$

where the last term represents a single undamped oscillator representing the long-wavelength optical phonons. It may be recalled from Subsection B that

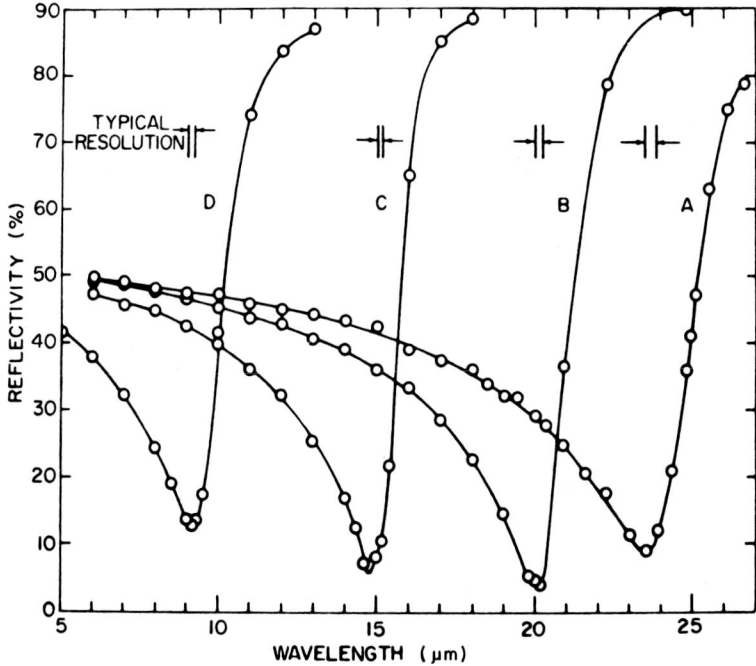

Fig. 26. Wavelength dependence of the normal reflectivity at $(81 \pm 2$ K) for p-type PbTe samples having carrier concentration ranging from 3.5×10^{18} cm^{-3} (A) to 4.8×10^{19} cm^{-3} (D). (From Dixon and Riedl [117].)

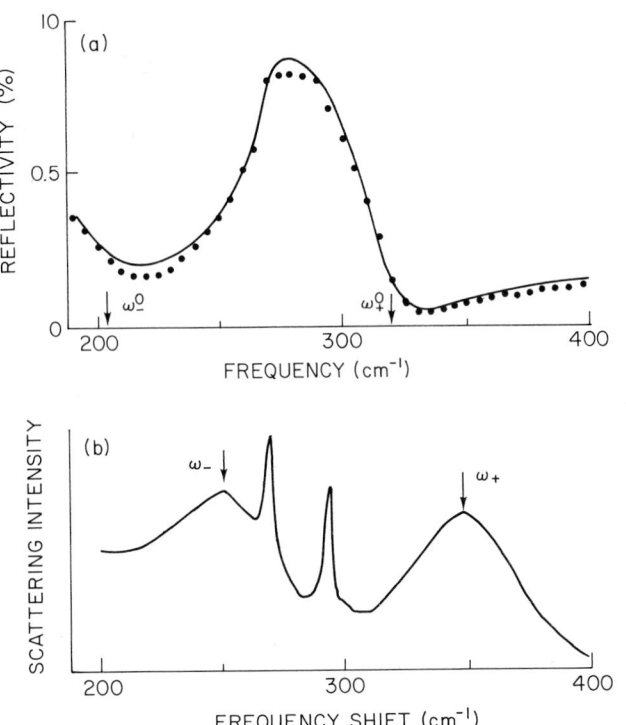

Fig. 27. (a) Infrared reflection spectrum of n-GaAs ($n = 5 \times 10^{17}$ cm^{-3}). The solid curve is the calculated spectrum and the dots are experimental data. The calculated ω_+^0 and ω_-^0 are marked by arrows. (b) Raman scattering spectrum of the same sample. Notice the difference in the position of the arrows, indicating a wave vector dependence of ω_+ and ω_-. (From Zemski et al. [121].)

$\omega_{TO} = \omega_0$ was identified with the maximum in ε, whereas ω_{LO} with its zero. Because of the first term, the second condition is now altered: i.e., ω_{TO} can still be identified with the maximum in ε but not ω_{LO} with its minimum. Thus for a polar semiconductor with an appreciable carrier concentration, the resonance corresponding to the long-wavelength LO frequency will no longer correspond to the real LO phonon frequency. In the limit of negligible damping for both electrons and phonons, the solution of Eq. (70) is given by

$$\omega_\pm^2 = \tfrac{1}{2}(\omega_{LO}^2 + \omega_p^2) \pm \tfrac{1}{2}[(\omega_{LO}^2 + \omega_p^2) - 4\omega_p^2\omega_{TO}^2]^{1/2} \qquad (71)$$

where ω_+ and ω_- represent new resonances above and below ω_{LO}.

The plasmon–phonon interaction obviously will be modified by the inclusion of both plasmon and phonon damping terms. The influence of damping and the wave vector dependence on the phonon–plasmon interaction in semiconductors have been theoretically treated by Gurevich et al. [118],

Varga [119], and Singwi and Tosi [120]. Such effects have indeed been observed experimentally. The wave vector dependence of the plasmon-phonon interaction has been investigated by Zemski et al. [121], who studied n-InP and n-GaAs by both infrared reflection spectroscopy and Raman light scattering. Whereas the former measurement gives coupled LO plasmon–phonon modes with $q \simeq 10^3$ cm^{-1}, the latter measurement gives modes with $q \simeq 10^6$ cm^{-1}. This is illustrated in Fig. 27 for GaAs. Note the difference in the positions of ω_+ and ω_- in the two spectra.

REFERENCES

1. K. Huang, *Proc. R. Soc. London Ser. A* **208**, 352 (1951).
2. M. Born and K. Huang, "Dynamical Theory of Crystal Lattices," p. 82, Oxford Univ. Press, New York, 1954.
3. R. H. Lyddane, R. G. Sachs, and E. Teller, *Phys. Rev.* **59**, 673 (1941).
4. W. J. Turner and W. E. Reese, *Phys. Rev.* **127**, 126 (1962).
5. H. W. Verleur, *J. Opt. Soc. Am.* **58**, 1356 (1968).
6. A. Kahan, J. W. Goodman, R. S. Singh, and S. S. Mitra, *J. Appl. Phys.* **42**, 4444 (1971).
7. W. Cochran, *Z. Krist.* **112**, 465 (1959).
8. A. S. Barker, *Phys. Rev.* **136A**, 1290 (1964).
9. I. F. Chang, S. S. Mitra, J. N. Plendl, and L. C. Mansur, *Phys. Status Solidi* **28**, 663 (1968).
10. A. S. Barker, *in* "Far Infrared Properties of Solids" (S. S. Mitra and S. Nudelman eds.), p. 247, Plenum, New York, 1970.
11. D. W. Berreman, *Phys. Rev.* **130**, 2193 (1963).
12. J. R. Jasperse, A. Kahan, J. N. Plendl, and S. S. Mitra, *Phys. Rev.* **146**, 526 (1966).
13. M. Born and K. Huang, "Dynamical Theory of Crystal Lattices," p. 341, Oxford Univ. Press, New York, 1954.
14. A. A. Maradudin and R. F. Wallis, *Phys. Rev.* **125**, 4 (1962).
15. V. V. Mitskevich, *Fiz. Tverd. Tela* **4**, 3035 (1962); V. V. Mitskevich, [English transl.: *Soviet Phys.—Solid State* **4**, 2224 (1963)].
16. J. Neuberger and R. D. Hatcher, *J. Chem. Phys.* **34**, 5 (1961).
17. R. A. Cowley, *Adv. Phys.* **12**, 421 (1963).
18. R. de L. Kronig *J. Opt. Soc. Am.* **12**, 547 (1926); H. A. Kramers, *Atti Congr. Intern. Fis. Como.* **2**, 545 (1927); M. Cardona, *in* "Optical Properties of Solids" (S. Nudelman and S. S. Mitra. eds.), p.137, Plenum, New York, 1969.
19. M. Sharnoff, *Am. J. Phys.* **32**, 1 (1964).
20. W. G. Spitzer and D. A. Kleinman, *Phys. Rev.* **121**, 1324 (1961).
21. P. N. Schatz, S. Maeda, and K. Kozima, *J. Chem. Phys.* **38**, 2658 (1963).
22. H. J. Bowlden and J. K. Wilmshurst, *J. Opt. Soc. Am.* **53**, 1073 (1963).
23. M. Gottlieb, *J. Opt. Soc. Am.* **50**, 343 (1960).
24. R. B. Sanderson, *J. Phys. Chem. Solids* **26**, 803 (1965).
25. G. Andermann, A. Caron, and D. A. Dows, *J. Opt. Soc. Am.* **55**, 1210 (1965).
26. L. Kellner, *Z. Phys.* **56**, 212 (1929).
27. H. Y. Fan and M. Becker, *in* "Proceedings of Conference on Semiconducting Materials: Reading, U.K.," p. 132, Academic Press, New York, 1951.
28. A. Kahan and H. G. Lipson, Air Force Cambridge Research Laboratory Report AFCRL-63-325, Bedford, Massachusetts [limited circulation].
29. W. Nazarewicz, P. Rolland, E. da Silva, and M. Balkanski, *Appl. Opt.* **1**, 369 (1962).
30. A. J. Barker, G. R. Wilkinson, N. E. Massa, and S. S. Mitra, *in* "Optical Properties of Highly Transparent Solids" (S. S. Mitra and B. Bendow, eds.), p. 45, Plenum, New York, 1975.

31. A. J. Barker, G. R. Wilkinson, N. E. Massa, and S. S. Mitra, *in* "Optical Properties of Highly Transparent Solids" (S. S. Mitra and B. Bendow, eds.), p. 405–525, Plenum, New York, 1975.

32. L. H. Skolnik, *in* "Optical Properties of Highly Transparent Solids" (S. S. Mitra and B. Bendow, eds.), pp. 405–433, Plenum, New York, 1975.

33. A. Hordvik, *Appl. Opt.* **16**, 2827 (1977).

34. M. Lax and E. Burstein, *Phys. Rev.* **97**, 39 (1955).

35. J. L. Birman, *Phys. Rev.* **127**, 1093 (1962); J. L. Birman, *Phys. Rev.* **131**, 1489 (1963).

36. S. S. Mitra, *Phys. Lett.* **11**, 119 (1964).

37. F. A. Johnson and R. Loudon, *Proc. R. Soc. London Ser.* **A 231**, 274 (1964).

38. E. Burstein, F. A. Johnson, and R. Loudon, *Phys. Rev.* **139A**, 1239 (1965).

39. S. Ganesan and E. Burstein, *J. Phys. Paris* **26**, 645 (1965).

40. B. Szigeti, *in* "Lattice Dynamics" (R. F. Wallis ed.), p. 405, Pergamon, Oxford, 1965.

41. C. Smart, G. R. Wilkinson, A. M. Karo, and J. R. Hardy, *in* "Lattice Dynamics" (R. F. Wallis, ed.), p. 387, Pergamon, Oxford, 1965.

42. F. A. Johnson, *Proc. Phys. Soc. London* **73**, 265 (1959).

43. H. Bilz and I Genzel, *Z. Phys.* **169**, 53 (1962).

44. W. Cochran, S. J. Fray, F. A. Johnson, J. E. Quarrington, and N. Williams, *J. Appl. Phys.* **32**, 2102 (1961).

45. S. S. Mitra, *J. Phys. Soc. Jpn.* **21** (suppl.), 61 (1966).

46. J. L. T. Waugh and G. Dolling, *Phys. Rev.* **132**, 2410 (1963).

47. J. C. Phillips, *Phys. Rev.* **113**, 147 (1959).

48. M. Hass and B. W. Henvis, *J. Phys. Chem. Solids* **23**, 1099 (1962).

49. S. Iwasa, I. Balslev, and E. Burstein, *Bull. Am. Phys. Soc.* **9**, 237 (1964).

50. M. Sparks, *Phys. Rev. B* **9**, 827 (1974).

51. B. Bendow, H. G. Lipson, and S. P. Yukon, *Phys. Rev. B* **16**, 2684 (1977).

52. B. Bendow, *Solid State Phys.* **33**, 249 (1978).

53. G. Ruprecht, *Phys. Rev. Lett.* **12**, 580 (1964).

54. T. C. McGill, R. W. Hellwarth, M. Mangir, and H. V. Winston, *J. Phys. Chem. Solids.* **34**, 2105 (1973).

55. K. V. Namjoshi and S. S. Mitra, *Phys. Rev. B* **9**, 815 (1974); K. V. Namjoshi and S. S. Mitra, *Solid State Commun.* **15**, 317 (1974).

56. D. L. Mills and A. A. Maradudin, *Phys. Rev. B* **8**, 1617 (1973).

57. B. Bendow, S. C. Ying, and S. P. Yukon, *Phys. Rev. B* **8**, 1679 (1973).

58. L. J. Sham and M. Sparks, *Phys. Rev. B* **9**, 827 (1974).

59. H. B. Rosenstock, *Phys. Rev. B* **9**, 1963 (1974).

60. D. L. Mills and A. A. Maradudin, *Phys. Rev. B* **10**, 1713 (1974).

61. B. Bendow, S. P. Yukon, and S. C. Ying, *Phys. Rev. B* **10**, 2286 (1974).

62. L. L. Boyer, J. A., Harrington, M. Hass, and H. B. Rosenstock, *in* "Optical Properties of Highly Transparent Solids" (S. S. Mitra and B. Bendow, eds.), p. 59, Plenum, New York, 1975.

63. H. G. Lipson, B. Bendow, N. E. Massa, and S. S. Mitra, *Phys. Rev. B* **13**, 2614 (1976).

64. J. A. Harrington and M. Hass, *Phys. Rev. Lett.* **11**, 710 (1973); K. V. Namjoshi and S. S. Mitra, *Solid State Commun.* **15**, 317 (1974).

65. B. Bendow, *Appl. Phys. Lett.* **23**, 133 (1973).

66. R. Brout, *Phys. Rev.* **113**, 43 (1959); S. S. Mitra and R. Marshall, *J. Chem. Phys.* **41**, 3158 (1964).

67. S. S. Mitra, unpublished.

68. S. S. Mitra and S. K. Joshi, *Physica* **26**, 825 (1960); S. S. Mitra and S. K. Joshi, *Physica* **27**, 345 (1961).

69. S. S. Mitra, *in* "Optical Properties of Solids" (S. Nudelman and S. S. Mitra, eds.), p. 333, Plenum, New York, 1969.

70. H. Rubens and G. Hertz, *Preuss. Akad. Wiss. Berlin* **14**, 256 (1912).
71. H. Dotsch and H. Happ. *Z. Phys*, **177**, 360 (1964); J. E. Eldridge and P. R. Staal, *Phys. Rev. B* **16**, 4608 (1977); R. H. Stolen and K. Dransfeld, *Phys. Rev.* **139A**, 1295 (1965).
72. R. H. Stolen, *Phys. Rev. B* **11**, 767 (1975).
73. J. R. Birch and D. K. Murray, *Infrared Phys.* **18**, 283 (1978).
74. T. J. Parker, W. G. Chambers, and J. F. Angress, *Infrared Phys.* **14**, 207 (1974).
75. D. Billard, private communication, 1980; D. Billard, seminar given at the University of Rhode Island, Kingston, Rhode Island, December 19, 1979.
76. R. H. Stolen, *Phys. Chem. Glasses*, **11**, 83 (1970).
77. E. M. Armhein and H. Heil, *J. Phys. Chem. Solids*, **32**, 1925 (1971).
78. U. Strom and P. C. Taylor, *Phys. Rev. B* **16**, 5512 (1977).
79. M. A. Bosch, *Phys. Rev. Lett.* **40**, 879 (1978).
80. J. R. Hardy and A. M. Karo, *Phys. Rev. B* **26**, 3327 (1982).
81. M. Sparks, D. F. King, and D. L. Mills, *Phys. Rev.* **26**, 6987 (1982).
82. H. Y. Fan and A. K. Ramdas, *J. Appl. Phys.* **30**. 1127 (1959).
83. I. M. Lifshitz, *J. Phys. (USSR)* **7**, 215 (1943); I. M. Lifshitz, *J. Phys (USSR)* **7**, 249 (1943); I. M. Lifshitz, *J. Phys. (USSR)* **8**, 89 (1944).
84. E. W. Montroll and R. B. Potts, *Phys. Rev.* **100**, 525 (1955).
85. P. Mazur, E. W. Montroll, and R. E. Potts, *J. Wash, Acad. Sci.* **46**, 2 (1956).
86. G. Schaefer, *J. Phys. Chem. Solids* **12**, 233 (1960).
87. I. M. Lifshitz, *Nuovo Cimento Suppl.* **3**, 716 (1956).
88. P. G. Dawber and R. J. Elliot, *Proc. R. Soc. London* **272**, 222 (1963); P. G. Dawber and R. J. Elliot, *Proc. Phys. Soc. London* **81**, 453 (1963).
89. I. M. Lifshitz, *Adv. Phys.* **13**, 485 (1964).
90. Yu. A. Izyumov. *Adv. Phys.* **14**, 569 (1965).
91. S. P. Gaur, J. F. Vetelino, and S. S. Mitra, *J. Phys. Chem. Solids* **32**, 2737 (1971).
92. S. S. Jaswal. *Phys. Rev.* **137A**, 302 (1965).
93. R. S. Singh and S. S. Mitra, *Phys. Rev. B* **5**, 733 (1972).
94. L. Genzel, *in* "Optical Properties of Solids" (S. Nudelman and S. S. Mitra, eds.), pp. 453–487, Plenum, New York, 1969.
95. A. A. Maradudin, *Solid State Phys.* **18**, 273 (1966); A. A. Maradudin, *Solid State Phys.* **19**, 2 (1966).
96. M. V. Klein, *in* "Physics of Color Centers" (W. B. Fowler, ed.), pp. 430–531, Academic Press, New York, 1968.
97. R. C. Newman, "Infrared Studies of Crystal Defects," Taylor and Francis, London, 1974.
98. J. F. Vetelino and S. S. Mitra, *in* "Physics of Structurally Disordered Solids" (S. S. Mitra, ed.), pp. 541–622, New York, 1974.
99. F. Krueger, O. Reinkober, and E. Koch-Holm, *Ann. Phys.* **85**, 110 (1928).
100. F. J. Dyson, *Phys. Rev.* **92**, 133 (1953).
101. R. H. Braunstein, R. Moore, and F. Herman, *Phys. Rev.* **109**, 695 (1958).
102. J. S. Langer, *J. Math. Phys.* **2**, 584 (1961).
103. H. W. Verleur and A. S. Barker, *Phys. Rev.* **149**, 715 (1966).
104. Y. S. Chen, W. Shockley, and G. L. Pearson, *Phys. Rev.* **151**, 648 (1966).
105. I. F. Chang and S. S. Mitra, *Phys. Rev.* **172**, 924 (1968).
106. G. Lucovsky, E. Lind, and E. A. Davis, *in* "II–VI Semiconducting Compounds" (D. G. Thomas, ed.), p. 1150, Benjamin, New York, 1967.
107. O. Brafman, I. F. Chang, G. Lengyel, S. S. Mitra, and E. Carnall, *Phys. Rev. Lett.* **19**, 1120 (1967).
108. I. F. Chang and S. S. Mitra, *Adv. Phys.* **20**, 359 (1971).
109. A. P. G. Kutty, *Solid State Commun.* **14**, 213 (1974).
110. P. N. Sen and W. M. Hartman, *Phys. Rev. B* **9**, 367 (1974).
111. S. C. Varshney, J. F. Vetelino, S. S. Mitra, and I. F. Chang, *Phys. Rev. B* **12**, 5912 (1975).

112. N. E. Massa, S. S. Mitra, and J. F. Vetelino, *Phys. Rev. B* **26**, 4579, 4606 (1982).
113. J. F. Angress, S. D. Smith, and K. F. Renk, *in* "Lattice Dynamics" (R. F. Wallis, ed.), p. 467, Pergamon, Oxford, 1965.
114. E. A. Fagan, *Proc. Int. Conf. Amorphous Liq. Semicond. 5th, Garmisch-Partenkirchen,* 6701 (1973).
115. S. S. Mitra, Y. F. Tsay, D. K. Paul, and B. Bendow, *Phys. Stat. Solidi,* **72**, 475 (1975).
116. J. D. Bates, R. W. Hendricks, and L. B. Shaffer, *J. Chem. Phys.* **61**, 4163 (1974).
117. J. E. Dixon and H. R. Riedl, *Phys. Rev.* **138**, A873 (1965).
118. V. I. Gurevich. A. I. Larkin, and Yu. A. Firsov, *Fiz. Tverd. Tela,* **4**, 185 (1962).
119. B. H. Varga, *Phys. Rev.* **137A**, 1896 (1965).
120. K. B. Singwi and M. P. Tosi, *Phys. Rev.* **147**, 658 (1966).
121. V. I. Zemski, E. L. Ivchenko, D. N. Mirlin, and I. I. Reshina, *Solid State Commun.* **16**, 221 (1975).

Part **II**
Critiques

1
Metals

Comments on the Optical
Constants of Metals
and an Introduction
to the Data
for Several Metals

DAVID W. LYNCH

Department of Physics and Ames Laboratory, U.S. Department of Energy
Iowa State University
Ames, Iowa

W. R. HUNTER*
Naval Research Laboratory
Washington, D.C.

INTRODUCTION I

Tables and plots of the optical constants $N = n - ik$ for the metals Ag, Au, Cu, Ir, Mo, Ni, Os, Pt, Rh, W, and Al, are presented in the following sections as a function of wavelength from the soft x-ray region to the near infrared. Preceding each data set is a short critique outlining the procedures used to obtain the constants and the conditions under which they were measured. Most measurements were made at room temperature (297–300 K).

Each critique has its own independent set of references. We tried to choose metals that are of practical value to the designer of optical filters as well as of interest for their electronic properties. Of these metals, aluminum appears to have the most widespread spectral application, being used in optical components from the infrared to the soft x-ray region. It also has the distinction of having been measured and studied more thoroughly than the other metals; therefore more space has been allotted for its critique than for the other

* Present address: Sachs/Freeman Associates, Inc., Bowie, Maryland.

275

metals. Aluminum is discussed further by Smith *et al.* in the section on metallic aluminum in this handbook.

The following paragraphs serve to remind the reader of the methods (discussed in more detail elsewhere in this volume) used to obtain the optical constants. There are also some general comments on the data. Comments germane to a particular metal are included in the individual critique for that metal. We finish with some warnings about attempting to use infrared optical data for free-electron-like metals (here, Cu, Au, Ag, and, to some extent, Al) to calculate accurately some expected optical characteristic, e.g., a reflectance value for a given sample. There is an inherent difficulty in defining optical constants in such cases because of the anomalous skin effect.

The optical constants for the metals listed have been measured by a number of investigators, and there is a large scatter in the reported values for any one metal. Instrumental sources beyond the investigator's control, e.g., second-order radiation or scattered light from the monochromator, are responsible for some of the scatter. Other contributing factors arise from errors in the quantities measured and from genuine differences in the samples caused by the methods of preparation. Methods of sample preparation are given, when known, in the individual critiques.

To plot n and k over such a large wavelength range, data from a number of investigators have been used. Agreement at the junctions of the data sets is rare. Rather than interpolate values to form a smooth junction between disjointed data sets, which sometimes requires an exercise of judgment bordering on the omniscient, we usually give only the data. Readers may create their own smooth junction if one is required.

There are four general methods for obtaining experimental data from which the optical constants can be deduced. The first three are based on the Fresnel relations for the reflection and transmission of radiation at a flat, smooth interface between two media. The first method that is widely used is known as the Kramers–Kronig method and is discussed by Smith in Chapter 3. The second method, sometimes referred to as the reflectance method, requires measuring reflectances at two or more angles of incidence. A description of this method is given by Hunter in Chapter 4. Ellipsometry is the third method for obtaining n and k and is described by Aspnes in Chapter 5.

The fourth method is the use of the electron-loss spectrum [1, 2]. The passage of fast (10-keV) electrons through a thin-film sample yields a spectrum of energy loss proportional to $\text{Im}(-1/\tilde{\varepsilon})$. This spectrum can be measured over a wide energy range and analyzed by using the Kramers–Kronig relations to give $\text{Re}(-1/\tilde{\varepsilon})$, hence $\tilde{\varepsilon}$ from which $\tilde{N} = \sqrt{\tilde{\varepsilon}}$ can be obtained. Such an analysis is subject to the usual restrictions inherent in the Kramers–Kronig method. Furthermore, the absolute magnitude of $\tilde{\varepsilon}$ is dependent on either a determination of the film thickness or an adjustment to fit some independently determined optical datum. The spectra obtained this way can be in good accord with those determined by genuine optical methods, but usually data at energies less than 1 eV (wavelengths longer than about 1 μm) cannot be obtained in this way.

Optical constants in the soft x-ray spectral region (10–100 Å) have attracted much interest in recent years because they are required to design grazing incidence optical systems for both x-ray astronomy and synchrotron radiation instrumentation. Because of this interest, the plots and tables have been extended to 2 keV (6.2 Å) and for three metals (Au, Ag, and Ni) to approximately 9.9 keV (1.25 Å). Many of these values are not measured but are calculated from atomic scattering factors, based on absorption measurements, that have been presented by Henke et al. [3]. This method contains no adjustable parameters, depending only on the density of atoms used to construct the solid. The values for Au and Ni were taken from tables presented by Zombeck et al. [4] that contain some measured values originally obtained by Ershov et al. [5] and to which Van Speybroeck [6] has fitted calculated values obtained by using the Kallman–Mark dispersion formalism described by Compton and Allison [7]. The data for Ag were obtained by Hagemann et al. [8] and are described in the section on Ag. Where overlap or near overlap occurs with values measured at longer wavelengths, the agreement with the calculated soft x-ray data is often very good.

The accuracy with which optical constants can be obtained from reflectance or ellipsometric measurements is not great. In the tables we have usually retained three significant figures, but this is probably too generous. When compiling the indices of refraction for the soft x-ray region, we have retained the first three digits after the repeating nines, rounding the third digit. This is essentially quoting δ, where $n = 1 - \delta$, to three significant figures.

ANOMALOUS SKIN EFFECT   II

A problem with free-electron metals in the infrared is that the optical constants may not be well defined. All of the optical constants of such metals arise from the response of the electrons to the electromagnetic field of the wave. As an example, consider the conductivity, which, for a cubic crystal or random polycrystal, relates the current density $\mathbf{J}$ to the electric field $\mathbf{E}$ as $\mathbf{J} = \sigma\mathbf{E}$. This really means that $\mathbf{J}$ at each point in space and moment in time is directly proportional to the electric field at that point and moment, i.e., $\mathbf{J}(\mathbf{r}, t) = \sigma\mathbf{E}(\mathbf{r}, t)$. We know that because of the inertia of the electrons, this response must decrease if $\mathbf{E}$ varies rapidly in time. We then Fourier analyze the fields with respect to time and treat one component at a time: $\mathbf{J}_\omega(\mathbf{r}) = \sigma(\omega)\mathbf{E}_\omega(\mathbf{r})$, and $\sigma(\omega)$ becomes smaller as ω increases. Now $\mathbf{J}$ arises from the driven motion of the electrons and is proportional to the velocity of the electrons at the point $\mathbf{r}$. The constitutive relations used so far imply that the velocity of an electron at $\mathbf{r}$ is proportional to the electric field at $\mathbf{r}$. This is often a good approximation, but it is not always valid. When it is not valid, we cannot relate fields and currents algebraically unless we are willing to Fourier analyze the fields with respect to space as well. This leads to the replacement of $\sigma(\omega)$ by $\sigma(\omega, \mathbf{k})$, so that a spectrum of σ versus ω would be needed for each component $\mathbf{k}$.

Classically, the electric field decreases exponentially with depth in a metal, with a decay length equal to the skin depth $\delta_0 = c/(2\pi\sigma_0\omega)^{1/2}$, where σ_0 is the dc conductivity. (Note that the skin depth is derived by assuming that $\mathbf{J} = \sigma_0\mathbf{E}$.) For a free-electron gas at high frequency, the conductivity becomes $\sigma_0/(1 + \omega^2\tau^2)$, and so we use $\delta = \delta_0(1 + \omega^2\tau^2)^{1/2}$ as the length over which electric fields in a metal are approximately constant. If the electron mean free path $l = v_F\tau$, where v_F is the average speed of an electron on the Fermi surface, becomes of the order of δ or larger, the electron then samples a nonconstant electric field between its last scattering event and its arrival at $\mathbf{r}$. The velocity at $\mathbf{r}$ no longer is proportional to the field at $\mathbf{r}$ but is proportional to an integral of the varying electric field over the electron's path before it reached $\mathbf{r}$. We then cannot write $J_\omega(\mathbf{r}) = \sigma(\omega)E_\omega(\mathbf{r})$. We are left with an integral constitutive relation or, if we Fourier analyze with respect to $\mathbf{r}$, the quantity $\sigma(\omega, \mathbf{k})$ already mentioned. In using the Fresnel relations to obtain the reflected electric field at some frequency, we must integrate over all spatial frequencies or wave numbers $\mathbf{k}$. Even though only one component $\mathbf{k}$ may be present in the wave in the vacuum, many components will appear in the fields induced in the metal when the response is nonlocal. The term $\sigma(\omega, \mathbf{k})$ can be obtained from a solution of the Boltzmann equation, and under some ranges of the parameters, closed-form solutions exist. Often, instead of relating bulk quantities, the surface impedance $\tilde{Z} = (4\pi/c)(E/H)_0$ is calculated, where the field magnitudes are evaluated on the surface. From this, reflectances can be found.

The nonlocal nature of the response of the electrons is crucial whenever the mean free path becomes long compared with the classical skin depth, e.g., for pure metal single crystals at low temperatures. In such cases the classical picture can lead to calculated results in error by orders of magnitude. At room temperature, if the film grain size is large enough so that the scattering is nearly all by phonons, i.e., if the film and single-crystal samples have the same electrical and optical properties, the nonlocality is marginal and causes small but nonnegligible corrections. For Cu at 300 K and for a frequency corresponding to 10-μm wavelength, we have $\omega\tau = 5.1$, $\delta_0 = 1.2 \times 10^{-6}$ cm, $\delta = 5.9 \times 10^{-6}$ cm, and $l = 4.2 \times 10^{-6}$ cm. If we assume that $\omega\tau \gg 1$, the classical expression for the reflectance at normal incidence is $R = 1 - A$ with absorptance $A = 2/(\omega_p\tau)$. For Cu at 300 K, $A = 0.0045$. The anomalous-skin-effect theory gives two possible corrections to this, one for the fraction p of electrons that scatter specularly from the metal surface and one for the fraction $(1 - p)$ that scatter diffusely. These corrections are simply additive to A; they are not calculated via optical constants. If we assume only diffuse scattering, the correction to A is the addition of a term $\frac{3}{4}(v_F/c)$, which thus reduces R. For copper this term is 0.0039, about as large as the classical departure of R from unity. Thus, the anomalous skin effect makes a small correction to R but a correction of the order of 100% to A. This large departure from the classical value will have a large effect on any values of the optical constants determined from the reflectance. The corrections are much larger at low temperatures, for upon cooling τ gets very large and the classical result for A may become very small, 10^{-5} or less. The term $\frac{3}{4}(v_F/c)$ remains and dominates R. (In fact, at low temperatures there are other mechanisms that also contribute to A, which we neglect here.)

In addition to the mathematical complexity of the anomalous skin effect [9–12], one also must contend with the problem of specular versus diffuse scattering of the conduction electrons from the sample surface, i.e., deciding a value for p. Perhaps the best experimental study of this problem to date is that on Ag at 300 K by Bennett *et al.* [13], to which the reader is referred. Gold, too, requires consideration of the anomalous skin effect in the infrared region [14]. Other general references to the anomalous skin effect are also listed [15–20].

REFERENCES

1. H. Raether, *Springer Tracts Mod. Phys.* **38**, 84 (1965).
2. J. Daniels, C. V. Festenburg, H. Raether, and K. Zeppenfeld, *Springer Tracts Mod. Phys.* **54**, 77 (1970).
3. B. L. Henke, P. Lee, T. J. Tanaka, R. L. Shimabukuro, and B. K. Fujikawa, "Low Energy X-Ray Diagnostics—1981" (D. T. Attwood and B. L. Henke, eds.), p. 340, *AIP Conf. Proc.* No. 75, American Institute of Physics, New York, 1981.
4. M. V. Zombeck, G. K. Austin, and D. T. Torgerson, Smithsonian Astrophysical Observatory Report SAO-AXAF-80-003, Appendix B, Tables B1 (Au) and B2 (Ni), Cambridge, Massachusetts, 1980.
5. O. A. Ershov, I. A. Brytov, and A. P. Lukirskii, *Opt. Spectros.* **22**, 66 (1967).
6. L. Van Speybroeck, Smithsonian Astrophysical Observatory, Cambridge, Massachusetts, private communication (1982).
7. A. H. Compton and S. K. Allison, "X-Rays in Theory and Experiment," p. 263, Van Nostrand-Reinhold, Princeton, New Jersey, 1935.
8. H. J. Hagemann, W. Gudat, and C. Kunz, DESY Report SR-74/7, Hamburg, 1974; H. J. Hagemann, W. Gudat and C. Kunz, *J. Opt. Soc. Am.* **65**, 742 (1975).
9. G. E. H. Reuter and E. H. Sondheimer, *Proc. R. Soc. Ser. A* **195**, 336 (1948).
10. R. B. Dingle, *Physica* **19**, 311 (1953).
11. K. L. Kliewer and R. Fuchs, *Phys. Rev.* **172**, 607 (1968).
12. K. L. Kliewer and R. Fuchs, *Phys. Rev. B* **2**, 2923 (1970).
13. H. E. Bennett, J. M. Bennett, E. J. Ashley, and R. J. Motyka, *Phys. Rev.* **165**, 755 (1968).
14. M. L. Thèye, *Phys. Lett.* **25A**, 764 (1967); M. L. Thèye, *Phys. Rev. B* **2**, 3061 (1970).
15. A. B. Pippard, "*Dynamics of Conduction Electrons,*" p. 58, Gordon and Breach, New York, 1965.
16. A. V. Sokolov, "*Optional Properties of Metals,*" Chapter 8, Blackie, Glasgow and London, 1967.
17. F. Stern, *Solid State Physics* **15**, 299 (1963).
18. H. Jones, *in* "Handbuch der Physik," Vol. 19 (S. Flugge, ed.), p. 308, Springer-Verlag, Berlin, 1956.
19. F. Wooten, "Optical Properties of Solids," Chapter 4, Academic Press, New York, 1972.
20. H. B. G. Casimir and J. Ubbink, *Philips Tech. Rev.* **28**, 300 (1967).

III COPPER (Cu)

Although there have been many measurements of the optical properties of Cu [1–34], there are limited data in the infrared and vacuum ultraviolet spectral regions.

Table I lists the values of n and k chosen and the pertinent references. These data are plotted in Fig. 1 as smooth curves. The data used from 1.38 Å to 1.237 μm are those of Hagemann et al. [22], and those of Dold and Mecke [14] are used in the infrared.

Hagemann et al. [22] prepared their samples by evaporating thin films of copper onto substrates of collodion that were supported on copper screen of the type used for electron microscopy. The evaporation was done from resistance-heated boats at a pressure of about 5×10^{-7} torr at rates of 10–50 Å/sec. The plastic substrates were dissolved away, leaving copper films on copper screens. These processes required exposing the copper films to air before measurements. Transmittance measurements were made from 954 to 83 Å (13–150 eV) to obtain an absorption spectrum that was extended by using the data of others [11, 15, 17, 20] to provide an absorption spectrum large enough for a Kramers–Kronig analysis.

Dold and Mecke [14] evaporated copper onto polished glass substrates from molybdenum boats. The films were measured in air by using an ellipsometric technique described by Beattie and Conn [35]. If these data are used to calculate the normal-incidence reflectance R, a spectrum with a broad peak at 8 μm results. The reflectance should increase monotonically with increasing wavelength. Neglect of the anomalous skin effect causes a calculated R to be too high at longer wavelengths, not lower. The cause of the peak probably is erroneously low k values at the longer wavelengths.

REFERENCES

1. G. Joos and A. Klopfer, *Z. Physik* **138**, 251 (1954).
2. L. G. Schulz, *J. Opt. Soc. Am.* **44**, 357, 540 (1954).
3. L. G. Schulz and D. R. Tangherlini, *J. Opt. Soc. Am.* **44**, 362 (1954).
4. J. R. Beattie and G. K. T. Conn, *Philos. Mag.* **46**, 989 (1955).
5. J. N. Hodgson, *Proc. Phys. Soc. London Ser. B* **68**, 593 (1955).
6. M. A. Biondi, *Phys. Rev.* **102**, 964 (1956).
7. L. G. Schulz, *Adv. Phys.* **6**, 102 (1957).
8. M. A. Biondi and J. R. Rayne, *Phys. Rev.* **115**, 1522 (1959).
9. S. Roberts, *Phys. Rev.* **118**, 1509 (1960).
10. M. Otter, *Z. Phyzik* **161**, 163 (1961).
11. H. Ehrenreich and H. R. Philipp, *Phys. Rev.* **128**, 1622 (1962).
12. W. T. Spencer and M. P. Givens, *J. Opt. Soc. Am.* **54**, 1337 (1964).
13. D. Beaglehole, *Proc. Phys. Soc. London* **85**, 1007 (1965).
14. B. Dold and R. Mecke, *Optik* **22**, 435 (1965).
15. L. R. Canfield and G. Hass, *J. Opt. Soc. Am.* **55**, 61 (1965).
16. M. A. Biondi and A. I. Guobadia, *Phys. Rev.* **166**, 667 (1968).

17. R. Haensel, C. Kunz, T. Sasaki, and B. Sonntag, *Appl. Opt.* **7**, 301 (1968).
18. G. P. Pells and M. Shiga, *J. Phys. C* **2**, 1835 (1969).
19. D. H. Seib and W. E. Spicer, *Phys. Rev. B* **2**, 1676 (1970).
20. P. B. Johnson and R. W. Christy, *Phys. Rev. B* **6**, 4370 (1972).
21. I. I. Sasovskaya and M. M. Noskov, *Fiz. Tverd. Tela* **14**, 999 (1972); I. I. Sasovskaya and M. M. Noskov, *Sov. Phys. Solid State* **14**, 857 (1972).
22. H. J. Hagemann, W. Gudat, and C. Kunz, DESY Report SR-74/7, Hamburg, 1974; H. J. Hagemann, W. Gudat, and C. Kunz, *J. Opt. Soc. Am.* **65**, 742 (1975).
23. P. B. Johnson and R. W. Christy, *Phys. Rev. B* **11**, 1315 (1975).
24. J. Rivory and M. L. Thèye, *J. Physique* **36**, L129 (1975).
25. H. J. Hagemann, W. Gudat, and C. Kunz, *Phys. Stat. Solidi B* **74**, 507 (1976).
26. J. A. MacKay and J. R. Rayne, *Phys. Rev. B* **13**, 673 (1976).
27. P. Winsemius, F. F. van Kampen, H. P. Lengkeek, and C. G. van Went, *J. Phys. F* **6**, 1583 (1976).
28. D. Beaglehole and B. Thiéblemont, *Il Nuovo Cimento B* **39**, 477 (1977).
29. L. A. Feldkamp, L. C. Davis, and M. B. Stearns, *Phys. Rev. B* **15**, 5535 (1977).
30. K. G. Ramanathan, S. H. Yen, and E. A. Estalote, *Appl. Opt.* **16**, 2810 (1977).
31. T. Smith, *J. Opt. Soc. Am.* **67**, 48 (1977).
32. R. Smalley and A. J. Sievers, *J. Opt. Soc. Am.* **68**, 1516 (1978).
33. D. Beaglehole, M. DeCrescenzi, M. L. Thèye, and G. Vuye, *Phys. Rev. B* **19**, 6303 (1979).
34. B. Window and G. Harding, *J. Opt. Soc. Am.* **71**, 354 (1981).
35. J. R. Beattie and G. K. T. Conn, *Philos. Mag.* **46**, 222 (1955).

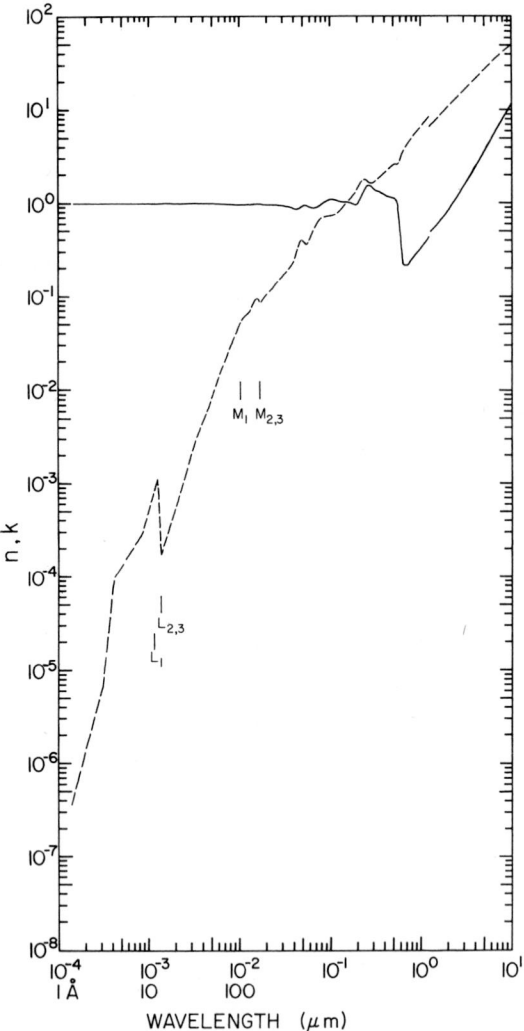

Fig. 1. Log–log plot of n (——) and k (----) versus wavelength in micrometers for copper.

TABLE I

Values of n and k Obtained from Various References for Copper[a]

eV	cm^{-1}	Å	n	k
9000		1.378	1.0010 [22]	3.65 x 10^{-7} [22]
8900		1.393	1.0010	3.56
8000		1.550	1.0010	5.38 x 10^{-7}
4000		3.099	1.0000	6.62 x 10^{-6}
3000		4.133	1.0000	9.49 x 10^{-5}
1500		8.265	1.0000	2.73 x 10^{-4}
1200		10.33	0.999	6.24
1100		11.27	0.999	8.23
1000		12.40	0.999	8.51
900		13.78	1.000	1.74
800		15.50	0.999	2.63
700		17.71	0.998	3.95
600		20.66	0.997	6.51 x 10^{-4}
500		24.80	0.997	1.11 x 10^{-3}
400		30.99	0.993	2.45
300		41.33	0.988	5.13 x 10^{-3}
200		61.99	0.978	1.59 x 10^{-2}
150		82.65	0.969	3.10
120		103.3	0.965	5.40
110		112.7	0.968	5.99
105		118.1	0.968	6.25
100		124.0	0.967	6.53
95		130.5	0.965	6.93
92		134.8	0.963	7.44
90		137.8	0.963	7.88
88		140.9	0.963	8.34
86		144.2	0.966	8.77
84		147.6	0.969	9.11
83		149.4	0.971	9.23
82		151.2	0.973	9.34
81		153.1	0.974	9.43
80		155.0	0.977	9.46
79		156.9	0.979	9.41
78		158.9	0.980	9.36
77		161.0	0.982	9.21
76		163.1	0.984	9.10
75		165.3	0.983	8.72
74		167.5	0.979	8.50
73		169.8	0.977	8.57
72		172.2	0.975	8.68
71		174.6	0.974	8.81
70		177.1	0.972	8.93
68		182.3	0.968	9.27
66		187.8	0.967	9.63 x 10^{-2}
64		193.7	0.965	0.102
62		200.0	0.965	0.107
60		206.6	0.966	0.109
58		213.8	0.964	0.110
56		221.4	0.960	0.114
54		229.6	0.957	0.120
52		238.4	0.954	0.126
50		248.0	0.953	0.134
48		258.3	0.952	0.140
46		269.5	0.951	0.146

(*continued*)

[a] The references from which the values were extracted are given in brackets.

TABLE I (*Continued*)

Copper

eV	cm^{-1}	Å	n	k
44		281.8	0.948	0.151
42		295.2	0.942	0.157
40		309.9	0.934	0.167
38		326.3	0.927	0.180
36		344.4	0.922	0.193
34		364.6	0.910	0.205
32		387.4	0.891	0.223
30		413.3	0.856	0.262
29		427.5	0.849	0.302
28		442.8	0.862	0.346
27		459.2	0.884	0.382
26		476.8	0.921	0.403
25		495.9	0.960	0.396
24		516.6	0.961	0.373
23		539.0	0.943	0.366
22		563.5	0.917	0.381
21		590.4	0.897	0.411
20		619.9	0.882	0.455
19		652.5	0.883	0.507
18		688.8	0.890	0.561
17		729.3	0.913	0.620
16		774.9	0.954	0.673
15.5		799.9	0.981	0.695
15		826.5	1.01	0.707
14.5		855.0	1.03	0.717
14		885.6	1.06	0.724
13		953.7	1.08	0.724

eV	cm^{-1}	μm	n	k
12		0.1033	1.09	0.731
11		0.1127	1.07	0.754
10		0.1240	1.04	0.818
9.5		0.1305	1.03	0.867
9.0		0.1378	1.03	0.921
8.5		0.1459	1.03	0.979
8.0		0.1550	1.03	1.03
7.5		0.1653	1.01	1.09
7.0		0.1771	0.972	1.20
6.5		0.1907	0.958	1.37
6.0	48,390	0.2066	1.04	1.59
5.8	46,780	0.2138	1.10	1.67
5.6	45,170	0.2114	1.18	1.74
5.4	43,550	0.2296	1.28	1.78
5.2	41,940	0.2384	1.38	1.80
5.0	40,330	0.2480	1.47	1.78
4.8	38,710	0.2583	1.53	1.71
4.6	37,100	0.2695	1.52	1.67
4.4	35,490	0.2818	1.49	1.64
4.2	33,880	0.2952	1.42	1.64
4.0	32,260	0.3099	1.34	1.72
3.8	30,650	0.3263	1.34	1.81
3.6	29,040	0.3444	1.31	1.87
3.4	27,420	0.3646	1.27	1.95

TABLE I (*Continued*)

Copper

eV	cm^{-1}	μm	n	k
3.2	25,810	0.3874	1.18	2.21
3.0	24,200	0.4133	1.18	2.21
2.8	22,580	0.4428	1.17	2.36
2.6	20,970	0.4768	1.15	2.50
2.4	19,360	0.5166	1.12	2.60
2.3	18,550	0.5390	1.04	2.59
2.2	17,740	0.5635	0.826	2.60
2.1	16,940	0.5904	0.468	2.81
2.0	16,130	0.6199	0.272	3.24
1.9	15,320	0.6525	0.214	3.67
1.85	14,920	0.6702	0.215	3.86
1.80	14,520	0.688	0.213	4.05
1.75	14,110	0.7084	0.214	4.24
1.70	13,710	0.7293	0.223	4.43
1.50	12,100	0.8265	0.260	5.26
1.00	8,065	1.240	0.433	8.46
0.98	7,904	1.265	0.496 [14]	6.78 [14]
0.96	7,743	1.291	0.505	6.92
0.94	7,582	1.319	0.515	7.06
0.92	7,420	1.348	0.525	7.21
0.90	7,259	1.378	0.536	7.36
0.88	7,098	1.409	0.547	7.53
0.86	6,937	1.442	0.559	7.70
0.84	6,775	1.476	0.572	7.88
0.82	6,614	1.512	0.586	8.06
0.80	6,453	1.550	0.606	8.26
0.78	6,291	1.589	0.627	8.47
0.76	6,130	1.631	0.649	8.69
0.74	5,969	1.675	0.672	8.92
0.72	5,807	1.722	0.697	9.16
0.70	5,646	1.771	0.723	9.41
0.68	5,485	1.823	0.752	9.68
0.66	5,323	1.878	0.782	9.97
0.64	5,162	1.937	0.815	10.3
0.62	5,001	2.000	0.850	10.6
0.60	4,839	2.066	0.890	11.0
0.58	4,678	2.138	0.933	11.3
0.56	4,517	2.214	0.980	11.8
0.54	4,355	2.296	1.03	12.2
0.52	4,194	2.384	1.09	12.7
0.50	4,033	2.480	1.15	13.2
0.48	3,872	2.583	1.22	13.7
0.46	3,710	2.695	1.29	14.4
0.44	3,549	2.818	1.37	15.0
0.42	3,388	2.952	1.47	15.7
0.40	3,226	3.100	1.59	16.5
0.38	3,065	3.263	1.73	17.4
0.36	2,904	3.444	1.90	18.4
0.34	2,742	3.646	2.09	19.6
0.32	2,581	3.874	2.32	20.8
0.30	2,420	4.133	2.59	22.2
0.28	2,258	4.428	2.92	23.7
0.26	2,097	4.768	3.32	25.4
0.24	1,936	5.166	3.81	27.5

(*continued*)

TABLE I (*Continued*)

Copper

eV	cm^{-1}	μm	n	k
0.22	1,774	5.635	4.44	30.0
0.20	1,613	6.199	5.23	33.0
0.18	1,452	6.888	6.23	36.3
0.16	1,291	7.749	7.66	40.3
0.15	1,210	8.265	8.57	42.6
0.14	1,129	8.856	9.64	45.1
0.13	1,049	9.537	10.8	47.5

IV GOLD (Au)

There have been more measurements of the optical properties of Au [1–53] than of any other metal, but most of them have been made in the visible region. As with other metals that show free-electron gas behavior in the infrared, the infrared data are very limited and agreement in the n spectra is not good. Moreover, the scatter in the vacuum ultraviolet spectra is far larger than one would expect from a relatively inert metal.

Table II lists the values of n and k chosen and the pertinent references. These data are plotted in Fig. 2 as smooth curves. The data from 1.25 to 124 Å are those reported by Zombeck et al. [53], from 140 to 470 Å those reported by Hagemann et al. [39], from 470 to 2000 Å those of Canfield et al. [10], from 1000 Å to 2 μm those reported by Thèye [27], and in the infrared those of Dold and Mecke [14].

All investigators except Hagemann et al. [39] stated that the gold used in their investigations was better than 99.99% pure. Hagemann et al. [39] prepared their samples by evaporating thin films of gold onto substrates of collodion that were supported on copper screen of the type used for electron microscopy. The evaporation was done from resistance-heated boats at a pressure of about 5×10^{-7} torr at rates of 10–50 Å/sec. The plastic substrates were dissolved away, leaving gold films on copper screens. These processes required exposing the gold films to air before measurements were performed. Transmittance measurements were made from 13 to 150 eV to obtain an absorption spectrum that was extended by using the data of others [19, 26, 31, 33] to provide an absorption spectrum large enough for a Kramers–Kronig analysis.

Canfield et al. [10] used the reflectance method to obtain n and k from gold films vacuum evaporated onto polished glass substrates at room temperature (30–40°C) in a conventional vacuum system. The pressures during evaporation were approximately 10^{-6} torr and deposition rates were on the order of hundreds of angstroms per second. Some reflectance measurements were made without exposing the films to air, and others were made after exposure. Exposure had no effect on reflectance.

Thèye [27] prepared her films by evaporation in ultrahigh vacuum (10^{-10}–10^{-11} torr) onto superpolished fused silica substrates. The substrate temperature ranged from room temperature to 100°C. Annealing at 100 to 150°C was done. She measured reflectance and transmittance at normal incidence in air for known film thicknesses (100–250 Å) from which she obtained n and k.

Dold and Mecke [14] evaporated gold onto polished glass substrates from tantalum boats. The films were measured in air by using an ellipsometric technique described by Beattie and Conn [54]. If these data are used to calculate the normal-incidence reflectance R, a spectrum with a broad peak at 5–6 μm results. The reflectance should increase monotonically with increasing wavelength. Neglect of the anomalous skin effect causes a calculated R to be too high at longer wavelengths, not lower. The cause of the peak probably is erroneously low k values at the longer wavelengths.

Aspnes et al. [52] made an extensive study of the optical properties of gold films. In the region of interband absorption they that found the most important cause of disagreement among different sets of data results not from experimental errors, but from genuine sample differences caused by different sample preparation methods. Of these differences, voids are the most important. The best films are those that have the highest value of ε_2 in this spectral region. The data we have chosen, those of Thèye [27], meet this criterion, which is discussed further by Aspnes in Chapter 5 of this handbook. In the intraband region, just below the interband threshold, Aspnes et al. [52] show that the most important differences in data sets arise from crystallite size in the films, with large-grained films giving the more bulklike results. In this case the criterion for quality is the smallest value of ε_2. Further in the infrared, measurement errors increase, and this criterion becomes less valid. In comparing the limited amount of data for Cu, Ag, and Au for wavelengths longer than 2 μm, there is no one investigator or group of investigators whose data yield consistently lower values of ε_2 for all three metals. We therefore did not use this criterion.

REFERENCES

1. G. Joos and A. Klopfer, Z. Physik **138**, 251 (1954).
2. L. G. Schulz, J. Opt. Soc. Am. **44**, 357, 540 (1954).
3. L. G. Schulz and F. R. Tangherlini, J. Opt. Soc. Am. **44**, 362 (1954).
4. J. R. Beattie and G. K. T. Conn, Philos. Mag. **46**, 989 (1955).
5. J. N. Hodgson, Proc. Phys. Soc. London Ser. B **68**, 593 (1955).
6. L. G. Schulz, Adv. Phys. **6**, 102 (1957).

7. M. Otter, *Z. Physik* **161**, 163 (1961).
8. P. R. Wessel, *Phys. Rev.* **132**, 2062 (1963).
9. H. Mayer and H. Bohme, *J. Physique* **25**, 81 (1964).
10. L. R. Canfield, G. Hass, and W. R. Hunter, *J. Physique* **25**, 124 (1964).
11. A. P. Lukirskii, E. P. Savinov, O. A. Ershov, and Y. F. Shepelev, *Opt. Spektrosk.* **16**, 310 (1964); A. P. Lukirskii, E. P. Savinov, O. A. Ershov, and Y. F. Shepelev, *Opt. Spectrosc.* **16**, 168 (1964).
12. D. Beaglehole, *Proc. Phys. Soc. London* **85**, 1007 (1965).
13. B. R. Cooper, H. Ehrenreich, and H. R. Philipp, *Phys. Rev.* **138**, A494 (1965).
14. B. Dold and R. Mecke, *Optik* **22**, 435 (1965).
15. H. E. Bennett and J. M. Bennett, *in* "Optical Properties and Electronic Structure of Metals and Alloys" (F. Abélès, ed.), p. 175, North-Holland Publ., Amsterdam, 1966.
16. H. Fukutani and O. Sueoka, *in* "Optical Properties and Electronic Structure of Metals and Alloys" (F. Abélès, ed.), p. 565, North-Holland Publ., Amsterdam, 1966.
17. D. Beaglehole, *in* "Optical Properties and Electronic Structure of Metals and Alloys" (F. Abélès, ed.), p. 154, North-Holland Publ., Amsterdam, 1966.
18. D. Jaegle and G. Missoni, *C. R. Acad. Sci. Paris* **262B**, 71 (1966).
19. E. Meyer, H. Frede, and H. Knof, *J. Appl. Phys.* **38**, 3682 (1967).
20. W. J. Scouler, *Phys. Rev. Lett.* **18**, 445 (1967).
21. W. Koster and R. Stahl, *Z. Metallkd.* **58**, 768 (1967).
22. R. Haensel, C. Kunz, T. Sasaki, and B. Sonntag, *Appl. Opt.* **7**, 301 (1968).
23. J. N. Hodgson, *J. Phys. Chem. Solids* **29**, 2175 (1968).
24. K. Platzöder and W. Steinmann, *J. Opt. Soc. Am.* **58**, 588 (1968).
25. G. P. Pells and M. Shiga, *J. Phys. C* **2**, 1835 (1969).
26. R. Haensel, K. Radler, B. Sonntag, and C. Kunz, *Solid State Commun.* **7**, 1495 (1969).
27. M. L. Thèye, *Phys. Rev. B* **2**, 3060 (1970).
28. P. O. Nilsson, *Phys. Kondens. Mater.* **11**, 1 (1970).
29. G. Jungk, *Phys. Stat. Solidi A* **3**, 965 (1970).
30. W. R. Hunter, *Proc. 3rd Int. Conf. Vac. Ultraviolet Radiation Physics, Tokyo*, p. 2aC2-1 (1971).
31. G. B. Irani, T. Huen, and F. Wooten, *J. Opt. Soc. Am.* **61**, 128 (1971).
32. G. B. Irani, T. Huen, and F. Wooten, *Phys. Rev. B* **6**, 2904 (1972).
33. P. B. Johnson and R. W. Christy, *Phys. Rev. B* **11**, 1315 (1975).
34. R. C. Linton, NASA Tech. Note D-7061 (1972).
35. M. Schlüter, *Z. Physik* **250**, 87 (1972).
36. D. E. Aspnes, *Opt. Commun.* **8**, 222 (1973).
37. O. Hunderi, *Phys. Rev. B* **7**, 3419 (1973).
38. L. N. Aksyutov, *Inzh-Fiz. Zh.* **27**, 197 (1974); L. N. Aksyutov, *J. Eng. Phys.* **27**, 913 (1974).
39. H. J. Hagemann, W. Gudat, and C. Kunz, DESY Report SR-74/7, Hamburg, 1974; H. J. Hagemann, W. Gudat, and C. Kunz, *J. Opt. Soc. Am.* **65**, 742 (1975).
40. C. Wehenkel and B. Gauthé, *Opt. Commun.* **11**, 62 (1974).
41. N. Fuschillo, B. Lalevic, W. Slusark, and A. Delahoy. *J. Vac. Sci. Technol.* **12**, 84 (1975).
42. C. G. Granquist and O. Hunderi, *Solid State Commun.* **19**, 939 (1976).
43. J. A. MacKay and J. A. Rayne, *Phys. Rev. B* **13**, 673 (1976).
44. P. Winsemius, F. F. van Kampen, H. P. Lengkeek, and C. G. van Went, *J. Phys. F* **6**, 1583 (1976).
45. G. Jungk and R. Grundler, *Phys. Stat. Solidi B* **76**, 541 (1976).
46. D. Beaglehole and B. Thiéblemont, *Il Nuovo Cimento B* **39**, 477 (1977).
47. T. Hollstein, V. Kreibig, and F. Leis, *Phys. Stat. Solidi B* **83**, K49 (1977).
48. T. A. McMath, R. A. D. Hewko, O. Singh, A. E. Curzon, and J. C. Irwin, *J. Opt. Soc. Am.* **67**, 630 (1977).
49. R. Kirsch, *Phys. Stat. Solidi A* **46**, 459 (1978).
50. D. Beaglehole, M. DeCrescenzi, M. L. Thèye, and G. Vuye, *Phys. Rev. B* **19**, 6303 (1979).

51. M. L. Scott and G. T. Johnston, *Appl. Opt.* **18**, 2905 (1979).
52. D. E. Aspnes, E. Kinsbron, and D. D. Bacon, *Phys. Rev. B* **21**, 3290 (1980).
53. M. V. Zombeck, G. K. Austin, and D. T. Torgerson, Smithsonian Astrophysical Observatory Report SAO-AXAF-80-003, Appendix B, Table B1, Cambridge, Massachusetts, 1980.
54. J. R. Beattie and G. K. T. Conn, *Philos. Mag.* **46**, 222 (1955).

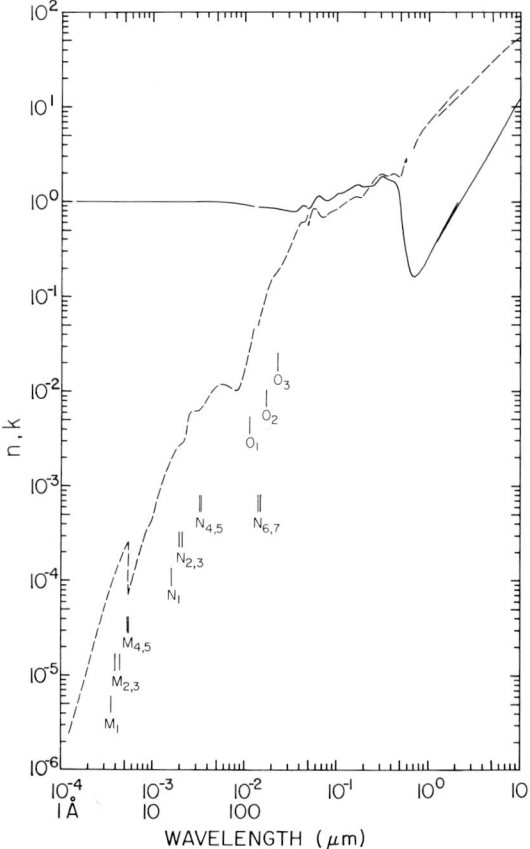

Fig. 2. Log–log plot of n (——) and k (----) versus wavelength in micrometers for gold.

TABLE II

Values of _n_ and _k_ Obtained from Various References for Gold[a]

eV	cm^{-1}	Å	n	k
9919		1.25	0.9999719 [28]	2.39 x 10^{-6} [53]
9184		1.35	0.9999665	3.13
8551		1.45	0.9999608	4.02
8266		1.50	0.9999578	4.52
7085		1.75	0.9999415	7.78 x 10^{-6}
6199		2.00	0.9999227	1.25 x 10^{-5}
4959		2.50	0.999879	2.75
4133		3.00	0.999831	5.25
3542		3.50	0.999794	9.07 x 10^{-5}
3100		4.00	0.999769	1.20 x 10^{-4}
2480		5.00	0.999706	2.36 x 10^{-4}
2066		6.00	0.999637	9.88 x 10^{-5}
1771		7.00	0.999395	1.68 x 10^{-4}
1550		8.00	0.999156	2.67
1487		8.34	0.999085	3.10
1254		9.89	0.99885	4.40
1188		10.44	0.99871	5.33
1012		12.25	0.99820	9.40 x 10^{-4}
929.4		13.34	0.99788	1.23 x 10^{-3}
851.5		14.56	0.99759	1.57
776.4		15.97	0.99734	1.86
704.9		17.59	0.99708	2.34
637.5		19.45	0.99665	2.68
572.9		21.64	0.99652	2.85
524.9		23.62	0.99624	3.76
500.3		24.78	0.99600	5.60
452.2		27.42	0.99592	6.10
395.4		31.36	0.99525	6.15
300.0		41.33	0.99100	8.95 x 10^{-3}
260.0		47.69	0.988	1.06 x 10^{-2}
220.0		56.36	0.986	1.18
200		61.99	0.983	1.15
180		68.88	0.979	1.07
150		82.65	0.963	1.01
120		103.3	0.929	1.78
110		112.7	0.911	2.83
105		118.1	0.900	3.53
100		124.0	0.889	4.54
88		140.9	0.889 [39]	5.54 x 10^{-2} [39]
87.5		141.7	0.888	5.58
87		142.5	0.887	5.76
86		144.2	0.886	6.13
85		145.9	0.887	6.48
84.5		146.7	0.888	6.54
84		147.6	0.888	6.33
83.5		148.5	0.885	6.37
83		149.4	0.884	6.53
82		151.2	0.881	6.82
80		155.0	0.878	7.52
78		158.9	0.875	8.18
76		163.1	0.873	8.71
74		167.5	0.869	9.27
72		172.2	0.862	9.97 x 10^{-2}
70		177.1	0.857	0.112

[a] The references from which the values were extracted are given in brackets.

TABLE II (*Continued*)

Gold

eV	cm^{-1}	Å	n	k
68		182.3	0.855	0.124
66		187.8	0.856	0.136
64		193.7	0.857	0.147
62		200.0	0.859	0.156
60		206.6	0.863	0.162
58		213.8	0.861	0.166
56		221.4	0.856	0.171
54		229.6	0.848	0.180
52		238.4	0.839	0.190
50		248.0	0.828	0.204
49		253.0	0.821	0.213
48		258.3	0.814	0.225
47		263.8	0.809	0.238
46		269.5	0.805	0.254
45		275.5	0.805	0.270
44		281.8	0.807	0.282
43		288.3	0.807	0.292
42		295.2	0.804	0.303
41		302.4	0.799	0.317
40		309.9	0.795	0.334
39		317.9	0.793	0.353
38		326.3	0.791	0.371
37		335.1	0.788	0.391
36		344.4	0.786	0.414
35		354.2	0.783	0.439
34		364.6	0.779	0.473
33		375.7	0.781	0.515
32		387.4	0.805	0.570
31		399.9	0.851	0.596
30.5		406.5	0.870	0.601
30		413.3	0.887	0.603
29.5		420.3	0.900	0.601
29		427.5	0.905	0.599
28		442.8	0.901	0.607
27		459.2	0.895	0.635

eV	cm^{-1}	µm	n	k
26.38		0.0470	0.855 [10]	0.548 [10]
26		0.04768	0.896 [39]	0.679 [39]
25.83		0.0480	0.846 [10]	0.565 [10]
25.30		0.0490	0.846	0.600
24.80		0.0500	0.850	0.645
24.31		0.0510	0.860	0.695
23.84		0.0520	0.872	0.740
23.39		0.0530	0.890	0.795
22.96		0.0540	0.915	0.825
22.54		0.0550	0.950	0.840
22.14		0.0560	0.985	0.848
21.75		0.0570	1.022	0.850
21.38		0.0580	1.055	0.842
21.01		0.0590	1.085	0.830
20.66		0.0600	1.113	0.813
20.33		0.0610	1.134	0.795

(*continued*)

TABLE II (*Continued*)

Gold

eV	cm^{-1}	μm	n	k
20.00		0.0620	1.146	0.770
19.68		0.0630	1.153	0.750
19.37		0.0640	1.157	0.730
19.07		0.0650	1.155	0.710
18.79		0.0660	1.140	0.700
18.51		0.0670	1.125	0.694
18.23		0.0680	1.107	0.687
17.97		0.0690	1.088	0.680
17.71		0.0700	1.075	0.678
17.46		0.0710	1.060	0.680
17.22		0.0720	1.050	0.685
16.98		0.0730	1.042	0.690
16.75		0.0740	1,038	0.697
16.53		0.0750	1.033	0.704
16.31		0.0760	1.030	0.713
16.10		0.0770	1.029	0.720
15.90		0.0780	1.028	0.730
15.69		0.0790	1.028	0.739
15.50		0.0800	1.029	0.745
15.31		0.0810	1.030	0.752
15.12		0.0820	1.033	0.759
14.94		0.0830	1.037	0.765
14.76		0.0840	1.041	0.770
14.59		0.0850	1.048	0.775
14.42		0.0860	1.053	0.780
14.25		0.0870	1.061	0.784
14.09		0.0880	1.070	0.789
13.93		0.0890	1.080	0.793
13.78		0.0900	1.090	0.798
13.62		0.0910	1.100	0.801
13.48		0.0920	1.110	0.806
13.33		0.0930	1.121	0.809
13.19		0.0940	1.133	0.812
13.05		0.0950	1.146	0.815
12.92		0.0960	1.159	0.819
12.78		0.0970	1.170	0.823
12.65		0.0980	1.180	0.826
12.52		0.0990	1.190	0.831
12.40		0.100	1.200	0.836
12.28		0.101	1.207	0.842
12.16		0.102	1.210	0.848
12.04		0.103	1.213	0.853
11.92		0.104	1.215	0.860
11.81		0.105	1.217	0.865
11.70		0.106	1.217	0.870
11.59		0.107	1.218	0.878
11.48		0.108	1.218	0.885
11.37		0.109	1.220	0.893
11.27		0.110	1.222	0.900
11.17		0.111	1.223	0.907
11.07		0.112	1.225	0.914
10.97		0.113	1.228	0.922
10.88		0.114	1.232	0.927
10.78		0.115	1.237	0.932

TABLE II (*Continued*)

Gold

eV	cm^{-1}	μm	n	k
10.69		0.116	1.242	0.935
10.60		0.117	1.247	0.942
10.51		0.118	1.250	0.950
10.42		0.119	1.255	0.955
10.33		0.120	1.260	0.962
10.25		0.121	1.265	0.967
10.16		0.122	1.270	0.975
10.08		0.123	1.275	0.982
10.00		0.124	1.280	0.987
9.919		0.125	1.285	0.994
9.840		0.126	1.290	1.000
9.763		9,127	1,295	1.005
9.686		0.128	1.300	1.012
9.611		0.129	1.304	1.017
9.537		0.130	1.308	1.020
9.465		0.131	1.313	1.027
9.393		0.132	1.318	1.032
9.322		0.133	1.323	1.038
9.253		0.134	1.328	1.045
9.184		0.135	1.333	1.050
9.117		0.136	1.338	1.053
9.050		0.137	1.345	1.058
8.984		0.138	1.350	1.063
8.920		0.139	1.355	1.067
8.856		0.140	1.360	1.072
8.551		0.145	1.386	1.089
8.266		0.150	1.419	1.102
7.999		0.155	1.450	1.108
7.749		0.160	1.483	1.106
7.514		0.165	1.512	1.093
7.293		0.170	1.519	1.070
7.085		0.175	1.500	1.070
6.888		0.180	1.470	1.085
6.702		0.185	1.442	1.107
6.526		0.190	1.427	1.135
6.358		0.195	1.424	1.170
6.199		0.200	1.427	1.215
6.0	48,390	0.2066	1.422 [27]	1.306 [27]
5.9	47,590	0.2101	1.430	1.334
5.8	46,780	0.2138	1.432	1.364
5.7	45,970	0.2175	1.438	1.388
5.6	45,170	0.2214	1.442	1.418
5.5	44,360	0.2254	1.452	1.442
5.4	43,550	0.2296	1.454	1.478
5.3	42,750	0.2339	1.462	1.510
5.2	41,940	0.2384	1.470	1.550
5.1	41,130	0.2431	1.478	1.590
5.0	40,330	0.2480	1.484	1.636
4.9	39,520	0.2530	1.490	1.698
4.8	38,710	0.2583	1.504	1.748
4.7	37,910	0.2638	1.546	1.784
4.6	37,100	0.2695	1.598	1.822
4.5	36,290	0.2755	1.648	1.852
4.4	34,590	0.2818	1.690	1.882

(*continued*)

TABLE II (*Continued*)

Gold

eV	cm^{-1}	μm	n	k
4.3	34,680	0.2883	1.742	1.900
4.2	33,880	0.2952	1.776	1.918
4.1	33,070	0.3024	1.812	1.920
4.0	32,260	0.3100	1.830	1.916
3.9	31,460	0.3179	1.840	1.904
3.8	30,650	0.3263	1.824	1.878
3.7	29,840	0.3351	1.798	1.860
3.6	29,040	0.3444	1.766	1.846
3.5	28,230	0.3542	1.740	1.848
3.4	27,420	0.3647	1.716	1.862
3.3	26,620	0.3757	1.696	1.906
3.2	25,810	0.3875	1.674	1.936
3.1	25,000	0.4000	1.658	1.956
3.0	24,200	0.4133	1.636	1.958
2.9	23,390	0.4275	1.616	1.940
2.8	22,580	0.4428	1.562	1.904
2.7	21,780	0.4592	1.426	1.846
2.6	20,970	0.4769	1.242	1.796
2.5	20,160	0.4959	0.916	1.840
2.4	19,360	0.5166	0.608	2.120
2.3	18,550	0.5391	0.402	2.540
2.2	17,740	0.5636	0.306	2.88
2.1	16,940	0.5904	0.236	
2.0	16,130	0.6199	0.194	
1.9	15,320	0.6526	0.166	3.15
1.8	14,520	0.6888	0.160	3.80
1.7	13,710	0.7293	0.164	4.35
1.6	12,900	0.7749	0.174	4.86
1.5	12,100	0.8266	0.188	5.39
1.4	11,290	0.8856	0.210	5.88
1.3	10,490	0.9537	0.236	6.47
1.2	9,679	1.033	0.272	7.07
1.1	8,872	1.127	0.312	7.93
1.0	8,065	1.240	0.372	8.77
0.98	7,904	1.265	0.389 [14]	8.09 [14]
0.96	7,743	1.291	0.403	8.25
0.94	7,582	1.319	0.419	8.42
0.92	7,420	1.384	0.436	8.59
0.90	7,259	1.378	0.454	8.77
			0.458 [27]	9.72 [27]
0.88	7,098	1.409	0.473 [14]	8.96 [14]
0.86	6,936	1.442	0.493	9.15
0.84	6,775	1.476	0.515	9.36
0.82	6,614	1.512	0.537	9.58
0.80	6,452	1.550	0.559	9.81
			0.550 [27]	11.5 [27]
0.78	6,291	1.590	0.583 [14]	10.1 [14]
0.76	6,130	1.631	0.609	10.3
0.74	5,968	1.675	0.636	10.6
0.72	5,807	1.722	0.665	10.9
0.70	5,646	1.771	0.696	11.2
			0.726	12.9
0.68	5,485	1.823	0.730	11.5
0.66	5,323	1.879	0.767	11.9

TABLE II *(Continued)*

Gold

eV	cm^{-1}	μm	n	k
0.64	5,162	1.937	0.807	12.2
0.62	5,001	2.000	0.850	12.6
0.60	4,839	2.066	0.896	13.0
			0.976 [27]	15.1 [27]
0.58	4,678	2.138	0.947 [14]	13.4 [14]
0.56	4,517	2.214	1.002	13.9
0.54	4,355	2.296	1.063	14.4
0.52	4,194	2.384	1.130	14.9
0.50	4,033	2.480	1.205	15.5
0.48	3,871	2.583	1.287	16.1
0.46	3,710	2.695	1.379	16.8
0.44	3,549	2.818	1.482	17.5
0.42	3,388	2.952	1.598	18.3
0.40	3,226	3.100	1.728	19.2
0.39	3,146	3.179	1.800	19.7
0.38	3,065	3.263	1.876	20.2
0.37	2,984	3.351	1.958	20.7
0.36	2,904	3.444	2.046	21.3
0.35	2,823	3.542	2.141	21.9
0.34	2,742	3.647	2.242	22.5
0.33	2,662	3.757	2.352	23.1
0.32	2,581	3.875	2.471	23.9
0.31	2,500	4.000	2.600	24.6
0.30	2,420	4.133	2.749	25.4
0.29	2,339	4.275	2.912	26.3
0.28	2,258	4.428	3.091	27.2
0.27	2,178	4.592	3.289	28.2
0.26	2,097	4.769	3.507	29.3
0.25	2,016	4.959	3.748	30.5
0.24	1,936	5.166	4.007	31.7
0.23	1,855	5.391	4.292	32.9
0.22	1,774	5.636	4.611	34.3
0.21	1,694	5.904	4.971	35.9
0.20	1,613	6.199	5.423	37.5
0.195	1,573	6.358	5.684	38.3
0.19	1,532	6.526	5.966	39.1
0.185	1,492	6.702	6.270	40.1
0.18	1,452	6.888	6.598	41.0
0.175	1,411	7.085	6.937	42.0
0.17	1,371	7.293	7.282	43.0
0.165	1,331	7.514	7.655	44.1
0.16	1,290	7.749	8.060	45.2
0.155	1,250	7.999	8.500	46.4
0.15	1,210	8.266	9.016	47.6
0.145	1,169	8.551	9.582	48.8
0.14	1,129	8.856	10.21	50.2
0.135	1,089	9.184	10.84	51.6
0.13	1,049	9.537	11.51	53.1
0.125	1,008	9.919	12.24	54.7

V IRIDIUM (Ir)

There are not many optical studies of Ir[1–7]. Table III lists the n and k values of Weaver *et al.* [7] from 310 Å to approximately 12.4 μm and of Henke [8] from 6.2–124 Å. These data are presented as smooth curves in Fig. 3. It is tempting to interpolate between the data set with smooth curves; however, there is an hiatus of 186 Å, where both O and N absorption edges are located so such an interpolation would be highly speculative. Further measurements are necessary to fill the gap.

Weaver *et al.* [7] used large polycrystalline samples cut from a crystal rod of iridium of unstated purity. The samples were mechanically polished, boiled in aqua regia, and heated *in vacuo* (10^{-7} torr) at temperatures of 1200–1500°C for several hours. The samples were then transferred in air to ion-pumped experimental chambers. From 6.2 μm to 2818 Å, a calorimetric technique [9] was used and at the shorter wavelengths (4312–310 Å) the reflectance at 10° angle of incidence was measured. These data were combined with other data [4], Drude infrared extrapolations, and a power-law decay in R extending to 10^5 eV to obtain n and k by using the Kramers–Kronig technique.

REFERENCES

1. B. T. Barnes, *J. Opt. Soc. Am.* **56**, 1546 (1966).
2. G. Hass, G. F. Jacobus, and W. R. Hunter, *J. Opt. Soc. Am.* **57**, 758 (1967).
3. J. A. R. Samson, J. P. Padur, and A. Sharma, *J. Opt. Soc. Am.* **57**, 966 (1967).
4. R. Haensel, K. Radler, B. Sonntag, and C. Kunz, *Solid State Commun.* **7**, 1495 (1972).
5. M. M. Kirillova, L. V. Nomerovannaya, and M. M. Noskov, *Fiz. Met. Metalloved.* **34**, 291 (1972); M. M. Kirillova, L. V. Nomerovannaya, and M. M. Noskov, *Phys. Met. Metall.* **34** (2), 61 (1972).
6. J. H. Weaver, *Phys. Rev. B* **11**, 1416 (1975).
7. J. H. Weaver, C. G. Olson, and D. W. Lynch, *Phys. Rev. B* **15**, 4115 (1977).
8. B. L. Henke, P. Lee, T. J. Tanaka, R. L. Shimabukuro, and B. K. Fujikawa, "Low Energy X-Ray Diagnostics—1981" (D. T. Attwood and B. L. Henke, eds.), p. 340, *AIP Conf. Proc.* No. 75, American Institute of Physics, New York, 1981.
9. L. W. Bos and D. W. Lynch, *Phys. Rev. B* **2**, 4567 (1970).

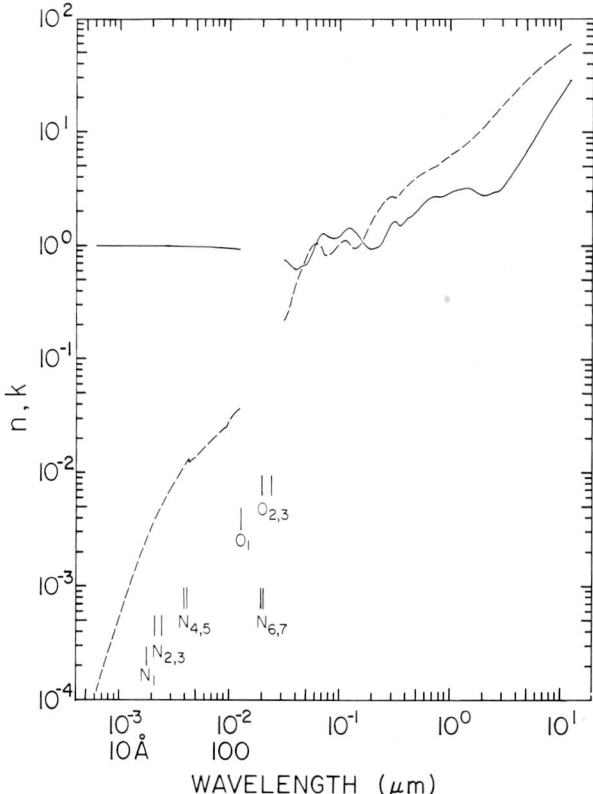

Fig. 3. Log–log plot of n (——) and k (----) versus wavelength in micrometers for iridium.

TABLE III

Values of n and k Obtained from Various References for Iridium[a]

eV	cm^{-1}	Å	n	k
2000		6.199	0.999710 [8]	1.22 x 10^{-4} [8]
1952		6.351	0.999585	1.31
1905		6.508	0.999451	1.42
1860		6.665	0.999388	1.53
1815		6.831	0.999335	1.65
1772		6.997	0.999284	1.78
1730		7.166	0.999235	1.92
1688		7.345	0.999185	2.07
1648		7.523	0.999134	2.23
1609		7.705	0.999082	2.40
1570		7.897	0.999028	2.59
1533		8.087	0.99897	2.79
1496		8.287	0.99892	3.00
1460		8.492	0.99886	3.23
1426		8.694	0.99880	3.48
1392		8.906	0.99873	3.74
1358		9.129	0.99867	4.03
1326		9.350	0.99860	4.34
1294		9.581	0.99853	4.67
1263		9.816	0.99846	5.03
1233		10.05	0.99838	5.41
1204		10.30	0.99830	5.80
1175		10.55	0.99822	6.24
1147		10.81	0.99814	6.71
1119		11.08	0.99805	7.22
1093		11.34	0.99796	7.75
1067		11.62	0.99787	8.33
1041		11.91	0.99777	8.97
1016		12.20	0.99768	9.64 x 10^{-4}
992		12.50	0.99758	1.03 x 10^{-3}
968		12.81	0.99747	1.11
945		13.12	0.99737	1.19
923		13.43	0.99726	1.27
901		13.76	0.99715	1.36
879		14.10	0.99703	1.45
858		14.45	0.99692	1.55
838		14.79	0.99680	1.66
818		15.16	0.99668	1.77
798		15.54	0.99655	1.89
779		15.92	0.99643	2.02
760		16.31	0.99630	2.16
742		16.71	0.99618	2.30
725		17.10	0.99607	2.45
707		17.54	0.99598	2.62
690		17.97	0.99587	2.70
674		18.39	0.99571	2.79
658		18.84	0.99553	2.96
642		19.31	0.99536	3.19
627		19.77	0.99522	3.34
612		20.26	0.99507	3.53
597		20.77	0.99491	3.74
583		21.27	0.99477	3.95

[a] The references from which the values were extracted are given in brackets.

TABLE III (*Continued*)

Iridium

eV	cm^{-1}	Å	n	k
569		21.79	0.99463	4.17
555		22.34	0.99449	4.40
542		22.87	0.99436	4.62
529		23.44	0.99423	4.85
516		24.03	0.99411	5.07
504		24.60	0.99397	5.26
492		25.20	0.99380	5.46
480		25.83	0.99362	5.68
469		26.43	0.99343	5.89
458		27.07	0.99322	6.11
447		27.74	0.99297	6.38
436		28.44	0.99273	6.71
426		29.10	0.99252	7.03
415		29.87	0.99229	7.41
406		30.54	0.99212	7.74
396		31.31	0.99193	8.14
386		32.12	0.99176	8.51
377		32.89	0.99159	8.86
368		33.69	0.99140	9.24
359		34.53	0.99121	9.64 x 10^{-3}
351		35.32	0.99104	1.00 x 10^{-2}
342		36.25	0.99086	1.05
334		37.12	0.99070	1.09
326		38.03	0.99056	1.14
318		38.99	0.99045	1.19
311		39.86	0.99040	1.23
303		40.92	0.99045	1.29
296		41.88	0.99072	1.34
289		42.90	0.99109	1.26
282		43.96	0.99044	1.29
275		45.08	0.99006	1.32
269		46.09	0.9898	1.33
262		47.32	0.9894	1.35
256		48.43	0.9889	1.38
250		49.59	0.9885	1.41
244		50.81	0.9880	1.45
238		52.09	0.9876	1.49
232		53.44	0.9870	1.52
227		54.62	0.9866	1.56
221		56.10	0.9860	1.60
216		57.40	0.9855	1.64
211		58.76	0.9850	1.68
206		60.18	0.9843	1.72
201		61.68	0.9836	1.76
196		63.25	0.9828	1.82
191		64.91	0.9823	1.90
187		66.30	0.9819	1.93
182		68.12	0.9812	1.95
178		69.65	0.9804	1.98
174		71.25	0.9795	2.02
170		72.93	0.9786	2.05
166		74.69	0.9776	2.10

(*continued*)

TABLE III (*Continued*)

Iridium

eV	cm^{-1}	Å	n	k
162		76.53	0.9764	2.15
158		78.47	0.9752	2.22
154		80.51	0.9739	2.30
150		82.65	0.9729	2.39
147		84.34	0.9720	2.42
143		86.70	0.9708	2.46
140		88.56	0.9696	2.49
136		91.16	0.9676	2.52
133		93.22	0.9658	2.54
130		95.37	0.9634	2.57
127		97.62	0.9609	2.72
124		99.98	0.9588	2.88
121		102.5	0.9570	3.05
118		105.1	0.9555	3.18
115		107.8	0.9536	3.27
112		110.7	0.9513	3.37
110		112.7	0.9498	3.45
107		115.9	0.9473	3.53
104		119.2	0.9443	3.62
102		121.5	0.9419	3.68
100		124.0	0.9394	3.74

eV	cm^{-1}	µm	n	k
40		0.03100	0.76 [7]	0.22 [7]
38		0.03263	0.73	0.24
36		0.03444	0.69	0.27
34		0.03647	0.64	0.35
32		0.03875	0.62	0.44
30		0.04133	0.64	0.53
29.5		0.04203	0.64	0.55
29		0.04275	0.65	0.57
28.5		0.04350	0.66	0.59
28		0.04428	0.66	0.61
27.5		0.04509	0.66	0.63
27		0.04592	0.66	0.66
26.5		0.04678	0.67	0.69
26		0.04769	0.67	0.72
25.5		0.04862	0.68	0.76
25		0.04959	0.69	0.79
24.5		0.05060	0.70	0.83
24		0.05166	0.73	0.87
23.5		0.05276	0.76	0.92
23		0.05391	0.79	0.96
22.5		0.05510	0.84	0.99
22		0.05636	0.89	1.00
21.5		0.05766	0.94	1.02
21		0.05904	0.99	1.04
20.5		0.06048	1.04	1.05
20		0.06199	1.10	1.06
19.6		0.06326	1.15	1.05
19.2		0.06458	1.20	1.03

TABLE III (*Continued*)

Iridium

eV	cm^{-1}	μm	n	k
18.8		0.06595	1.24	1.00
18.4		0.06738	1.27	0.97
18		0.06888	1.30	0.93
17.6		0.07045	1.30	0.87
17.2		0.07208	1.28	0.83
16.8		0.07380	1.25	0.82
16.4		0.07560	1.23	0.82
16		0.07749	1.21	0.83
15.6		0.07948	1.20	0.83
15.2		0.08157	1.19	0.84
14.8		0.08377	1.18	0.87
14.4		0.08610	1.17	0.88
14		0.08856	1.16	0.91
13.6		0.09117	1.17	0.95
13.2		0.09393	1.18	0.96
12.8		0.09686	1.19	1.01
12.4		0.09999	1.21	1.05
12		0.1033	1.24	1.08
11.8		0.1051	1.25	1.10
11.6		0.1069	1.28	1.12
11.4		0.1088	1.31	1.13
11.2		0.1107	1.34	1.14
11.0		0.1127	1.38	1.13
10.8		0.1148	1.41	1.12
10.6		0.1170	1.43	1.09
10.4		0.1192	1.44	1.07
10.2		0.1216	1.45	1.04
10.0		0.1240	1.45	1.01
9.9		0.1252	1.44	1.00
9.8		0.1265	1.44	0.99
9.7		0.1278	1.43	0.98
9.6		0.1291	1.42	0.97
9.5		0.1305	1.40	0.96
9.4		0.1319	1.39	0.95
9.3		0.1333	1.38	0.95
9.2		0.1348	1.36	0.95
9.1		0.1362	1.35	0.94
9.0		0.1378	1.33	0.94
8.9		0.1393	1.31	0.94
8.8		0.1409	1.29	0.95
8.7		0.1425	1.27	0.95
8.6		0.1442	1.26	0.96
8.5		0.1459	1.24	0.97
8.4		0.1476	1.22	0.98
8.3		0.1494	1.20	0.99
8.2		0.1512	1.18	1.00
8.1		0.1531	1.15	1.01
8.0		0.1550	1.13	1.03
7.9		0.1569	1.11	1.05
7.8		0.1590	1.08	1.06
7.7		0.1610	1.05	1.10
7.6		0.1631	1.03	1.14

(*continued*)

TABLE III (*Continued*)

Iridium

eV	cm^{-1}	μm	n	k
7.5		0.1653	1.03	1.18
7.4		0.1675	1.02	1.20
7.3		0.1698	1.00	1.23
7.2		0.1722	0.99	1.27
7.1		0.1746	0.98	1.30
7.0		0.1771	0.97	1.34
6.9		0.1797	0.96	1.38
6.8		0.1823	0.95	1.42
6.7		0.1851	0.95	1.46
6.6		0.1879	0.94	1.50
6.5		0.1907	0.95	1.54
6.4		0.1937	0.94	1.59
6.3		0.1968	0.94	1.63
6.2		0.2000	0.94	1.68
6.1		0.2033	0.94	1.73
6.0	48,390	0.2066	0.95	1.78
5.9	47,590	0.2101	0.97	1.82
5.8	46,780	0.2138	0.96	1.86
5.7	45,970	0.2175	0.97	1.92
5.6	45,170	0.2214	0.98	1.98
5.5	44,360	0.2254	0.99	2.03
5.4	43,550	0.2296	1.00	2.09
5.3	42,750	0.2339	1.02	2.15
5.2	41,940	0.2384	1.04	2.22
5.1	41,130	0.2431	1.07	2.29
5.0	40,330	0.2480	1.10	2.35
4.9	39,520	0.2530	1.13	2.42
4.8	38,710	0.2583	1.18	2.49
4.7	37,910	0.2638	1.24	2.56
4.6	37,100	0.2695	1.31	2.60
4.5	36,290	0.2755	1.37	2.65
4.4	35,490	0.2818	1.45	2.68
4.3	34,680	0.2883	1.51	2.70
4.2	33,880	0.2952	1.58	2.71
4.1	33,070	0.3024	1.62	2.70
4.0	32,260	0.3100	1.64	2.68
3.9	31,460	0.3179	1.64	2.67
3.8	30,650	0.3263	1.61	2.69
3.7	29,840	0.3351	1.57	2.72
3.6	29,040	0.3444	1.52	2.81
3.5	28,230	0.3542	1.50	2.93
3.4	27,420	0.3647	1.53	3.05
3.3	26,620	0.3757	1.57	3.15
3.2	25,810	0.3875	1.62	3.26
3.1	25,000	0.4000	1.68	3.35
3.0	24,200	0.4133	1.73	3.43
2.9	23,390	0.4275	1.77	3.51
2.8	22,580	0.4428	1.81	3.61
2.7	21,780	0.4592	1.85	3.73
2.6	20,970	0.4769	1.91	3.86
2.5	20,160	0.4959	1.98	4.00
2.4	19,360	0.5166	2.07	4.14
2.3	18,550	0.5391	2.18	4.26
2.2	17,740	0.5636	2.29	4.38

TABLE III (*Continued*)

Iridium

eV	cm^{-1}	μm	n	k
2.1	16,940	0.5904	2.40	4.48
2.0	16,130	0.6199	2.50	4.57
1.9	15,320	0.6526	2.57	4.68
1.8	14,520	0.6888	2.64	4.81
1.7	13,710	0.7293	2.69	4.92
1.6	12,900	0.7749	2.68	5.08
1.5	12,100	0.8266	2.65	5.39
1.4	11,290	0.8856	2.72	5.74
1.3	10,490	0.9537	2.85	6.07
1.2	9,679	1.033	2.96	6.41
1.1	8,872	1.127	3.04	6.84
1.0	8,065	1.240	3.15	7.31
0.9	7,259	1.378	3.19	7.88
0.8	6,452	1.550	3.14	8.61
0.7	5,646	1.771	2.93	9.78
0.6	4,839	2.066	2.79	11.6
0.5	4,033	2.480	2.98	14.1
0.45	3,629	2.755	3.05	15.8
0.4	3,226	3.100	3.42	18.1
0.35	2,823	3.542	4.11	20.8
0.3	2,420	4.133	5.16	24.3
0.25	2,016	4.959	6.85	28.8
0.2	1,613	6.199	9.69	35.3
0.15	1,210	8.266	15.3	45.2
0.1	806.5	12.40	28.5	60.6

MOLYBDENUM (Mo)   VI

The optical properties of Mo [1–30] have been measured many times and the results generally agree rather well. Table IV lists the *n* and *k* values of Weaver *et al.* [19] and Henke *et al.* [31]. These data are plotted in Fig. 4 as smooth curves.

Weaver *et al.* [19] used samples spark-cut from Mo buttons of high purity that had been melted with an electron-beam gun. The samples were macro-etched, mechanically polished, electropolished, and annealed. After anneal-ing, they were recleaned, electropolished, washed in acetone and then in ethyl alcohol, and transferred in air to the measuring apparatus. In the infrared region measurements were made by using a calorimetric technique [32], and at shorter wavelengths reflectance measurements at near-normal incidence were made. These data were then analyzed by using the Kramers–Kronig method to obtain n and k.

There may be some influence of surface oxides on the vacuum ultraviolet data, but truly clean surfaces have not been measured, nor have oxidation studies been carried out. For wavelengths longer than about 3 μm, molybdenum resembles a free-electron metal, but, in principle, there are interband transitions at longer wavelengths that would affect the fit to a free-electron model for wavelengths shorter than about 10 μm.

REFERENCES

1. L. J. LeBlanc, J. S. Farrell, and D. W. Juenker, *J. Opt. Soc. Am.* **54**, 956 (1964).
2. J. P. Waldron and D. W. Juenker, *J. Opt. Soc. Am.* **54**, 204 (1964).
3. M. M. Kirillova and B. A. Charikov, *Fiz. Met. Metalloved.* **19**, 495 (1965); M. M. Kirillova and B. A. Charikov, *Phys. Met. Metallogr.* **19**(4), 13 (1965).
4. H. R. Apholte and K. Ulmer, *Phys. Lett.* **22**, 552 (1966).
5. B. T. Barnes, *J. Opt. Soc. Am.* **56**, 1546 (1966).
6. A. P. Lenham and D. M. Treherne, *J. Opt. Soc. Am.* **56**, 1137 (1966).
7. A. P. Lenham and D. M. Treherne, *in* "Optical Properties and Electronic Structure of Metals and Alloys" (F. Abélès, ed.), p. 196, North-Holland, Publ., Amsterdam, 1966.
8. M. M. Kirillova, G. A. Bolotin, and M. V. Mayevskiy, *Fiz. Met. Metalloved.* **24**, 95 (1967); M. M. Kirillova, G. A. Bolotin, and M. V. Mayevskiy, *Phys. Met. Metallogr.* **24**(1), 91 (1967).
9. A. P. Lenham, *J. Opt. Soc. Am.* **57**, 473 (1967).
10. D. W. Juenker, L. J. LeBlanc, and C. R. Martin, *J. Opt. Soc. Am.* **58**, 164 (1968).
11. M. L. Kapitsa, Y. P. Udoev, and E. I. Shirokikh, *Fiz. Tverd. Tela.* **11**, 814 (1969); M. L. Kapitsa, Y. P. Udoev, and E. I. Shirokikh, *Sov. Solid State Phys.* **11**, 665 (1969).
12. A. Cezairliyan, M. S. Morse, H. A. Berman, and C. W. Beckett, *J. Res. National Bureau of Standards* **74A**, 65 (1970).
13. K. A. Kress and G. J. Lapeyre, *J. Opt. Soc. Am.* **60**, 1681 (1970).
14. M. M. Kirillova, L. V. Nomerovannaya, and M. M. Noskov, *Zh. Eksp. Teor. Fiz.* **60**, 2252 (1971); M. M. Kirillova, L. V. Nomerovannaya, and M. M. Noskov, *Sov. Phys. JETP* **33**, 1210 (1971).
15. Y. P. Udoyev, N. S. Koz'yakova and M. L. Kapitsa, *Fit. Met. Metalloved.* **31**, 439 (1971); Y. P. Udoyev, N. S. Koz'yakova and M. L. Kapitsa, *Phys. Met. Metallogr.* **31**(2), 229 (1971).
16. P. Gravier, *Thin Solid Films* **11**, 135 (1972).
17. V. Vujnovic and V. Grzeta, *Fizika (Yugoslavia)* **4**, 173 (1972).
18. D. N. Baria, T. S. Kin, and R. G. Bautista, *High Temp.-High Press.* **5**, 545 (1973).
19. J. H. Weaver, D. W. Lynch, and C. G. Olson, *Phys. Rev. B* **10**, 501 (1974).
20. B. W. Veal and A. P. Paulikas, *Phys. Rev. B* **10**, 1280 (1974).
21. Y. Ballu, J. Lecante, and H. Rousseau, *Phys. Rev. B* **14**, 3201 (1976).
22. G. Chassaing, P. Granier, and M. Sigrist, *Thin Solid Films* **35**, L25 (1976).
23. J. H. Weaver and C. G. Olson, *Phys. Rev. B* **14**, 3251 (1976).
24. E. S. Black, D. W. Lynch, and C. G. Olson, *Phys. Rev. B* **16**, 2337 (1977).
25. H. S. Gurev and C. Selvage, *Proc. Soc. Photo-Opt. Instrum. Eng.* **85**, 32 (1977).
26. A. S. Siddiqui and D. M. Treherne, *Infrared Phys.* **17**, 33 (1977).
27. P. Gravier, G. Chassaing, and M. Sigrist, *Thin Solid Films* **57**, 93 (1979).
28. G. E. Carver and B. O. Seraphin, *Appl. Phys. Lett.* **34**, 279 (1979).
29. R. Manzke, *Phys. Stat. Solidi B* **97**, 157 (1980).
30. J. E. Nestell, Jr. and R. W. Christy, *Phys. Rev. B,* **21**, 3173 (1980).

31. B. L. Henke, P. Lee, T. J. Tanaka, R. L. Shimabukuro, and B. K. Fujikawa, "Low Energy
    X-Ray Diagnostics—1981" (D. T. Attwood and B. L. Henke, eds.), p. 340, *AIP Conf. Proc.*
    No. 75, American Institute of Physics, New York 1981.
32. L. W. Bos and D. W. Lynch, *Phys. Rev. B* **2**, 4567 (1970).

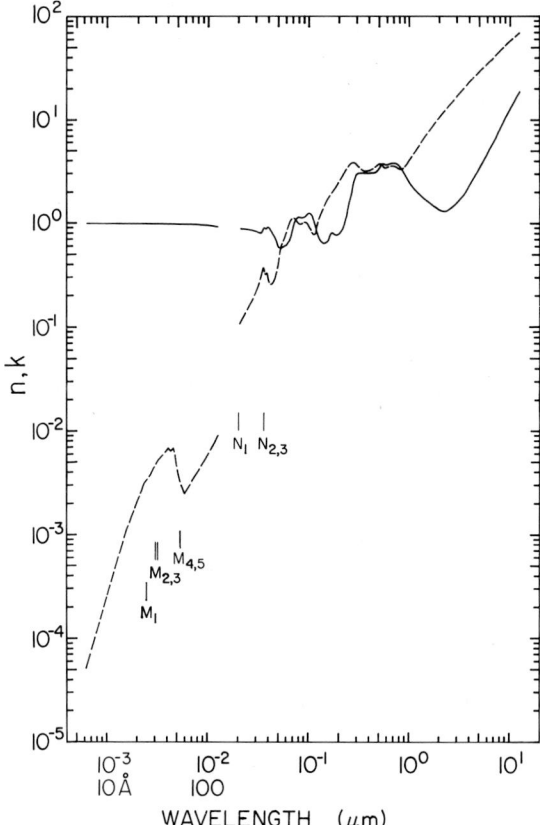

Fig. 4. Log–log plot of n (——) and k (––––) versus wavelength in micrometers for
molybdenum.

TABLE IV

Values of _n_ and _k_ Obtained from Various References for Molybdenum[a]

eV	cm^{-1}	Å	n	k
2000		6.199	0.999622 [31]	4.84x10^{-5} [31]
1952		6.351	0.999600	5.26
1905		6.508	0.999577	5.73
1860		6.665	0.999555	6.22
1851		6.831	0.999530	6.77
1772		6.997	0.999505	7.36
1730		7.166	0.999479	7.99
1688		7.345	0.999451	8.69
1648		7.523	0.999423	9.44x10^{-5}
1609		7.705	0.999393	1.02x10^{-4}
1570		7.897	0.999361	1.11
1533		8.087	0.999329	1.21
1496		8.287	0.999294	1.31
1460		8.492	0.999258	1.43
1426		8.694	0.999222	1.54
1392		8.906	0.999183	1.68
1358		9.129	0.999142	1.82
1326		9.350	0.999100	1.97
1294		9.581	0.999055	2.14
1263		9.816	0.999008	2.32
1233		10.05	0.99896	2.52
1204		10.30	0.99891	2.72
1175		10.55	0.99886	2.95
1147		10.81	0.99880	3.19
1119		11.08	0.99875	3.47
1093		11.34	0.99869	3.75
1067		11.62	0.99863	4.06
1041		11.91	0.99856	4.40
1016		12.20	0.99850	4.77
992		12.50	0.99843	5.15
968		12.81	0.99836	5.57
945		13.12	0.99829	6.02
923		13.43	0.99821	6.49
901		13.76	0.99813	7.01
879		14.10	0.99805	7.59
858		14.45	0.99797	8.19
838		14.79	0.99789	8.83
818		15.16	0.99781	9.53x10^{-4}
798		15.54	0.99772	1.03x10^{-3}
779		15.92	0.99764	1.11
760		16.31	0.99755	1.19
742		16.71	0.99747	1.27
725		17.10	0.99738	1.35
707		17.54	0.99728	1.44
690		17.97	0.99718	1.54
674		18.39	0.99708	1.64
658		18.84	0.99699	1.75
642		19.31	0.99688	1.87
627		19.77	0.99678	1.99
612		20.26	0.99668	2.13
597		20.77	0.99658	2.28
583		21.27	0.99649	2.43

[a] The references from which the values were extracted are given in brackets.

TABLE IV (*Continued*)

Molybdenum

eV	cm^{-1}	Å	n	k
569		21.79	0.99639	2.59
555		22.34	0.99631	2.77
542		22.87	0.99625	2.95
529		23.44	0.99621	3.15
516		24.03	0.99622	3.28
504		24.60	0.99618	3.38
492		25.20	0.99612	3.48
480		25.83	0.99604	3.59
469		26.43	0.99595	3.69
458		27.07	0.99583	3.80
447		27.74	0.99568	3.95
436		28.44	0.99554	4.16
426		29.10	0.99544	4.37
415		29.87	0.99535	4.61
406		30.54	0.99528	4.83
396		31.31	0.99526	5.09
386		32.12	0.99528	5.28
377		32.89	0.99529	5.41
368		33.69	0.99526	5.55
359		34.53	0.99522	5.70
351		35.32	0.99518	5.84
342		36.25	0.99514	6.00
334		37.12	0.99512	6.15
326		38.03	0.99510	6.31
318		38.99	0.99511	6.48
311		39.86	0.99517	6.63
303		40.92	0.99541	6.82
296		41.88	0.99558	6.63
289		42.90	0.99569	6.39
282		43.96	0.99568	6.15
275		45.08	0.99565	5.85
269		46.09	0.99552	5.43
262		47.32	0.99520	4.98
256		48.43	0.99480	4.61
250		49.59	0.99433	4.26
244		50.81	0.99379	3.93
238		52.09	0.99315	3.62
232		53.44	0.99243	3.33
227		54.62	0.99176	3.10
221		56.10	0.99085	2.83
216		57.40	0.9900	2.60
211		58.76	0.9890	2.40
206		60.18	0.9879	2.58
201		61.68	0.9869	2.69
196		63.25	0.9859	2.80
191		64.91	0.9848	2.91
187		66.30	0.9839	3.00
182		68.12	0.9828	3.12
178		69.65	0.9818	3.22
174		71.25	0.9807	3.33
170		72.93	0.9795	3.45
166		74.69	0.9783	3.60

(*continued*)

TABLE IV (*Continued*)

Molybdenum

eV	cm^{-1}	Å	n	k
162		76.53	0.9770	3.75
158		78.47	0.9757	3.92
154		80.51	0.9742	4.10
150		82.65	0.9726	4.28
147		84.34	0.9713	4.42
143		86.70	0.9695	4.62
140		88.56	0.9680	4.78
136		91.16	0.9659	5.01
133		93.22	0.9641	5.19
130		95.37	0.9623	5.38
127		97.62	0.9603	5.59
124		99.98	0.9581	5.80
121		102.5	0.9557	6.03
118		105.1	0.9532	6.28
115		107.8	0.9504	6.54
112		110.7	0.9473	6.90
110		112.7	0.9451	7.19
107		115.9	0.9415	7.64
104		119.2	0.9377	8.14
102		121.5	0.9349	8.50
100		124.0	0.9318	8.89×10^{-3}

eV	cm^{-1}	μm	n	k
60.0		0.02066	0.897 [19]	0.106 [10]
50.0		0.02480	0.868	0.151
45.0		0.02755	0.850	0.187
40.0		0.03100	0.828	0.235
39.5		0.03139	0.823	0.241
39.0		0.03179	0.818	0.250
38.5		0.03220	0.813	0.261
38.0		0.03263	0.810	0.273
37.75		0.03284	0.809	0.281
37.5		0.03306	0.808	0.289
37.25		0.03328	0.809	0.297
37.0		0.03351	0.810	0.303
36.75		0.03374	0.811	0.312
36.5		0.03397	0.812	0.321
36.25		0.03420	0.815	0.331
36.0		0.03444	0.819	0.341
35.8		0.03463	0.825	0.350
35.6		0.03483	0.831	0.358
35.4		0.03502	0.841	0.367
35.2		0.03522	0.853	0.373
35.0		0.03542	0.867	0.375
34.8		0.03563	0.880	0.373
34.6		0.03583	0.893	0.368
34.4		0.03604	0.901	0.360
34.2		0.03625	0.907	0.352
34.0		0.03647	0.911	0.344
33.8		0.03668	0.912	0.334
33.6		0.03690	0.910	0.326
33.5		0.03701	0.907	0.322

TABLE IV (Continued)

Molybdenum

eV	cm^{-1}	μm	n	k
33.4		0.03712	0.903	0.321
33.2		0.03734	0.897	0.323
33.0		0.03757	0.896	0.327
32.8		0.03780	0.898	0.330
32.6		0.03803	0.901	0.331
32.4		0.03827	0.907	0.331
32.2		0.03850	0.912	0.327
32.0		0.03875	0.915	0.323
31.75		0.03905	0.918	0.315
31.5		0.03936	0.919	0.307
31.25		0.03968	0.918	0.298
31.0		0.04000	0.915	0.288
30.75		0.04032	0.909	0.279
30.5		0.04065	0.902	0.271
30.25		0.04099	0.892	0.263
30.0		0.04133	0.879	0.259
29.75		0.04168	0.868	0.256
29.5		0.04203	0.855	0.257
29.25		0.04239	0.844	0.256
29.0		0.04275	0.832	0.259
28.75		0.04313	0.822	0.261
28.5		0.04350	0.811	0.263
28.25		0.04389	0.799	0.265
28.0		0.04428	0.788	0.269
27.75		0.04468	0.776	0.272
27.5		0.04509	0.764	0.276
27.25		0.04550	0.750	0.281
27.0		0.04592	0.734	0.289
26.75		0.04635	0.721	0.298
26.5		0.04679	0.706	0.308
26.25		0.04723	0.694	0.319
26.0		0.04769	0.680	0.329
25.8		0.04806	0.666	0.341
25.6		0.04843	0.656	0.353
25.4		0.04881	0.645	0.367
25.2		0.04920	0.636	0.380
25.0		0.04959	0.624	0.393
24.8		0.04999	0.612	0.410
24.6		0.05040	0.603	0.429
24.4		0.05081	0.596	0.447
24.2		0.05123	0.588	0.468
24.0		0.05166	0.583	0.491
23.8		0.05209	0.581	0.511
23.6		0.05254	0.579	0.532
23.4		0.05299	0.579	0.553
23.2		0.05344	0.579	0.575
23.0		0.05391	0.582	0.595
22.8		0.05438	0.586	0.614
22.6		0.05486	0.589	0.633
22.4		0.05535	0.594	0.650
22.2		0.05585	0.596	0.668
22.0		0.05636	0.602	0.685

(continued)

TABLE IV (*Continued*)

Molybdenum

eV	cm^{-1}	μm	n	k
21.8		0.05687	0.607	0.700
21.6		0.05740	0.608	0.715
21.4		0.05794	0.609	0.733
21.2		0.05848	0.610	0.752
21.0		0.05904	0.612	0.773
20.8		0.05961	0.616	0.794
20.6		0.06019	0.621	0.815
20.4		0.06078	0.626	0.838
20.2		0.06138	0.632	0.862
20.0		0.06199	0.639	0.887
19.8		0.06262	0.648	0.913
19.6		0.06326	0.660	0.940
19.4		0.07391	0.674	0.968
19.2		0.06458	0.691	0.996
19.0		0.06526	0.712	1.02
18.8		0.06595	0.737	1.05
18.6		0.06666	0.766	1.08
18.4		0.06738	0.799	1.10
18.2		0.06812	0.833	1.11
18.0		0.06888	0.866	1.12
17.8		0.06965	0.903	1.13
17.6		0.07045	0.940	1.14
17.4		0.07126	0.978	1.14
17.2		0.07208	1.01	1.13
17.0		0.07293	1.04	1.12
16.8		0.07280	1.07	1.11
16.6		0.07469	1.10	1.10
16.4		0.07560	1.12	1.08
16.2		0.07653	1.13	1.06
16.0		0.07749	1.14	1.04
15.8		0.07847	1.15	1.02
15.6		0.07948	1.15	1.01
15.4		0.08051	1.15	0.994
15.2		0.08157	1.14	0.989
15.0		0.08266	1.14	0.987
14.8		0.08377	1.14	0.987
14.6		0.08492	1.13	0.990
14.4		0.08610	1.13	0.997
14.2		0.08731	1.14	1.01
14.0		0.08856	1.15	1.01
13.8		0.08984	1.15	1.02
13.6		0.09117	1.17	1.02
13.4		0.09253	1.18	1.02
13.2		0.09393	1.20	1.03
13.0		0.09537	1.21	1.02
12.8		0.09686	1.23	1.01
12.6		0.09840	1.24	0.993
12.4		0.09999	1.25	0.975
12.2		0.1016	1.26	0.953
12.0		0.1033	1.26	0.921
11.8		0.1051	1.25	0.886
11.6		0.1069	1.23	0.845

TABLE IV (*Continued*)

Molybdenum

eV	cm^{-1}	μm	n	k
11.4		0.1088	1.18	0.802
11.2		0.1107	1.12	0.782
11.0		0.1127	1.05	0.774
10.8		0.1148	0.977	0.791
10.6		0.1170	0.909	0.835
10.4		0.1192	0.859	0.881
10.2		0.1216	0.812	0.930
10.0		0.1240	0.770	0.987
9.8		0.1265	0.735	1.05
9.6		0.1291	0.707	1.12
9.4		0.1319	0.685	1.19
9.2		0.1348	0.670	1.25
9.0		0.1378	0.652	1.33
8.8		0.1409	0.647	1.41
8.6		0.1442	0.649	1.49
8.4		0.1476	0.659	1.57
8.2		0.1512	0.672	1.65
8.0		0.1550	0.685	1.73
7.9		0.1569	0.697	1.77
7.8		0.1590	0.711	1.81
7.7		0.1610	0.726	1.85
7.6		0.1631	0.745	1.90
7.5		0.1653	0.781	1.93
7.4		0.1675	0.808	1.95
7.3		0.1698	0.816	1.96
7.2		0.1722	0.813	1.98
7.1		0.1746	0.808	2.00
7.0		0.1771	0.798	2.04
6.9		0.1797	0.791	2.08
6.8		0.1823	0.784	2.12
6.7		0.1851	0.777	2.18
6.6		0.1879	0.779	2.24
6.5		0.1907	0.783	2.30
6.4		0.1937	0.789	2.36
6.3		0.1968	0.798	2.43
6.2		0.2000	0.813	2.50
6.1	49,200	0.2033	0.836	2.57
6.0	48,390	0.2066	0.850	2.64
5.9	47,590	0.2101	0.873	2.72
5.8	46,780	0.2138	0.895	2.80
5.7	45,970	0.2175	0.924	2.89
5.6	45,170	0.2214	0.959	2.99
5.5	44,360	0.2254	1.00	3.09
5.4	43,550	0.2296	1.06	3.20
5.3	42,750	0.2339	1.13	3.31
5.2	41,940	0.2384	1.22	3.42
5.1	41,130	0.2431	1.33	3.53
5.0	40,330	0.2480	1.46	3.62
4.9	39,520	0.2530	1.61	3.70
4.8	38,710	0.2583	1.75	3.76
4.7	37,910	0.2638	1.90	3.81
4.6	37,100	0.2695	2.06	3.84

(*continued*)

TABLE IV (*Continued*)

Molybdenum

eV	cm^{-1}	μm	n	k
4.5	36,290	0.2755	2.21	3.87
4.4	35,490	0.2818	2.39	3.88
4.3	34,680	0.2883	2.59	3.86
4.2	33,880	0.2952	2.77	3.77
4.1	33,070	0.3024	2.91	3.67
4.0	32,260	0.3100	3.01	3.51
3.9	31,460	0.3179	3.04	3.40
3.8	30,650	0.3263	3.04	3.31
3.7	29,840	0.3351	3.04	3.27
3.6	29,040	0.3444	3.05	3.24
3.5	28,230	0.3542	3.06	3.21
3.4	27,420	0.3647	3.06	3.19
3.3	26,620	0.3757	3.06	3.18
3.2	25,810	0.3875	3.05	3.18
3.1	25,000	0.4000	3.03	3.22
3.0	24,200	0.4133	3.04	3.27
2.9	23,390	0.4275	3.05	3.33
2.8	22,580	0.4428	3.08	3.42
2.7	21,780	0.4592	3.13	3.51
2.6	20,970	0.4769	3.22	3.61
2.5	20,160	0.4959	3.36	3.73
2.4	19,360	0.5166	3.59	3.78
2.3	18,550	0.5391	3.79	3.61
2.2	17,740	0.5636	3.76	3.41
2.1	16,940	0.5904	3.68	3.45
2.0	16,130	0.6199	3.68	3.52
1.9	15,320	0.6526	3.74	3.58
1.8	14,520	0.6888	3.81	3.58
1.7	13,710	0.7293	3.84	3.51
1.6	12,900	0.7749	3.77	3.41
1.5	12,100	0.8266	3.53	3.30
1.4	11,290	0.8856	3.15	3.40
1.3	10,490	0.9537	2.77	3.74
1.2	9,679	1.033	2.44	4.22
1.1	8,872	1.127	2.16	4.85
1.0	8,065	1.240	1.94	5.58
0.9	7,259	1.378	1.74	6.48
0.86	6,936	1.442	1.70	6.90
0.82	6,614	1.512	1.64	7.35
0.78	6,291	1.590	1.60	7.83
0.74	5,968	1.675	1.52	8.38
0.70	5,646	1.771	1.48	8.99
0.66	5,323	1.879	1.43	9.67
0.62	5,001	2.000	1.38	10.4
0.58	4,678	2.138	1.34	11.3
0.54	4,355	2.296	1.35	12.4
0.50	4,033	2.480	1.37	13.5
0.46	3,710	2.695	1.46	14.9
0.42	3,388	2.952	1.57	16.5
0.38	3,065	3.263	1.70	18.4
0.34	2,742	3.647	1.99	20.8
0.30	2,420	4.133	2.44	23.8
0.29	2,339	4.275	2.58	24.7
0.28	2,258	4.428	2.74	25.6

TABLE IV (*Continued*)

Molybdenum

eV	cm⁻¹	μm	n	k
0.27	2,178	4.592	2.92	26.6
0.26	2,097	4.769	3.14	27.6
0.25	2,016	4.959	3.36	28.8
0.24	1,936	5.166	3.61	30.0
0.23	1,855	5.391	3.92	31.3
0.22	1,774	5.636	4.26	32.7
0.21	1,694	5.904	4.65	34.3
0.20	1,613	6.109	5.10	36.0
0.19	1,532	6.526	5.61	37.8
0.18	1,452	6.888	6.21	39.9
0.17	1,371	7.293	6.90	42.2
0.16	1,291	7.749	7.74	44.7
0.15	1,210	8.266	8.78	47.5
0.14	1,129	8.856	9.96	50.7
0.13	1,049	9.537	11.4	54.4
0.12	967.9	10.33	13.4	58.4
0.11	887.2	11.27	15.6	63.1
0.10	806.5	12.40	18.5	68.5

NICKEL (Ni)   **VII**

Overall agreement among the published n and k spectra of Ni [1–42] is good. Table V lists n and k values chosen and the pertinent references. These data are plotted in Fig. 5 as smooth curves.

The data extending from 1.3 to 124 Å are those of Zombeck *et al.* [40] determined as described in the introduction. Data calculated from a model of Henke *et al.* [43] overlap in the 6.1- to 124- Å region. The data from 88 to 248 Å are those of Bartlett *et al.* [42] (extinction coefficient) and Feldkamp *et al.* [39] (index of refraction). The data from 248 to 495 Å (Weaver *et al.* [41]), from 495 to 3540 Å (Vehse and Arakawa [13]), and from 4132 Å to 15.5 μm (Lynch *et al.* [17]) were combined for a Kramers–Kronig (KK) analysis, and the results are listed in the table and shown in Fig. 5 for 248 Å to 12.4 μm. The data of Henke [41] agree well with those of Feldkamp *et al.* [39] and Bartlett and Olson [42] in the overlap region.

Bartlett and Olson [42] used evaporated Ni films to obtain an absorption spectrum of Ni. They evaporated Ni onto thin-film substrates of carbon (1600 Å thick) or aluminum (900 Å thick). The pressure during deposition was less than 10^{-8} torr. The Ni was of high purity (resistivity ratio of 100 to 200).

Feldkamp *et al.* [39] obtained the electron energy loss spectrum of un-backed Ni films and used a KK analysis to obtain n and k. They evaporated Ni of unspecified purity from an alumina crucible at a deposition rate of 100 Å/min at a pressure of 10^{-7} torr. The substrate was a cleaned and polished {001} surface of NaCl and was maintained at 150°C during deposition. The film was floated off the substrate in water and collected on a 75-mesh electron-microscope grid. Film thicknesses were 500–1000 Å; the films were poly-crystalline and continuous (hole free).

Vehse and Arakawa [13] used evaporated Ni films for their reflectance measurements. The Ni had a purity of 99.9995%. The evaporation and measurements took place in the same vacuum system with no exposure to air. Nickel was evaporated from resistance-heated filaments at a pressure of about 10^{-8} torr onto microscope slides. The films were 1800–2200 Å thick and were probably deposited onto the slides at room temperature. Near-normal-incidence reflectance data were obtained from 495 to 3540 Å, and some reflectance versus angle of incidence measurements were also made at selected wavelengths within this range. A KK analysis was then done by using the n, k values obtained at the selected wavelengths as guiding values.

Lynch *et al.* [17] used a single crystal at 4 K ({111} face). The crystal was spark-cut, polished, and etched. A calorimetric technique [44] was used.

A number of sharp structures have been reported in the visible and in the near infrared that are suspected of being spurious and are not reported here. The remaining infrared structures have been determined to be present by all investigators. Nickel is not a free-electron-like metal, except at wavelengths longer than about 20 μm. This accounts for the good agreement among investigators of Ni in the infrared. A number of VUV measurements are all in general agreement. The structure in the spectra at about 187 Å arises from the excitation of 3p electrons, and there is considerable scatter in the mag-nitudes of adsorption coefficient values reported in this region.

REFERENCES

1. S. Roberts, *Phys. Rev.* **114**, 104 (1959).
2. H. Ehrenreich, H. R. Philipp, and D. J. Olechna, *Phys. Rev.* **131**, 2469 (1963).
3. B. Dold and R. Mecke, *Optik* **22**, 435 (1965).
4. A. P. Lenham and D. M. Treherne, *J. Opt. Soc. Am.* **56**, 1137 (1966).
5. A. P. Lenham, *J. Opt. Soc. Am.* **57**, 473 (1967).
6. A. P. Lenham and D. M. Treherne, *in* "Optical Properties and Electronic Structure of Metals and Alloys" (F. Abélès, ed.), p. 196, North-Holland Publ., Amsterdam, 1966.
7. M. M. Noskov and I. I. Sasovskaya, *Opt. Spektrosk.* **22**, 358 (1966); M. M. Noskov and I. I. Sasovskaya, *Opt. Spectrosc.* **22**, 657 (1966).
8. R. H. W. Graves and A. P. Lenham, *J. Opt. Soc. Am.* **58**, 884 (1968).
9. M. A. Biondi and A. I. Guobadia, *Phys. Rev.* **166**, 667 (1968).
10. J. Feinleib, W. J. Scouler, and J. Hanus, *J. Appl. Phys.* **40**, 1400 (1969).
11. B. Sonntag, R. Haensel, and C. Kunz, *Solid State Commun.* **7**, 597 (1969).
12. M. Shiga and G. P. Pells, *J. Phys. C* **2**, 1847 (1969).
13. R. C. Vehse and E. T. Arakawa, *Phys. Rev.* **180**, 695 (1969).

14. D. H. Seib and W. E. Spicer, *Phys. Rev. B* **2**, 1694 (1970).
15. M. Ph. Stoll, *Solid State Commun.* **8**, 1207 (1970).
16. F. C. Brown, C. Gahwiller, and A. B. Kunz, *Solid State Commun.* **9**, 487 (1971).
17. D. W. Lynch, R. Rosei, and J. H. Weaver, *Solid State Commun.* **9**, 2195 (1971).
18. I. I. Sasovskaya and M. M. Noskov, *Fiz. Met. Metalloved.* **32**, 723 (1971); I. I. Sasovskaya and M. M. Noskov, *Phys. Met. Metallogr.* **32**(4), 48 (1971).
19. M. P. Stoll, *J. Appl. Phys.* **42**, 1717 (1971).
20. G. L. Zuppardo and K. G. Ramanatham, *J. Opt. Soc. Am.* **61**, 1607 (1971).
21. J. C. Jones, D. C. Palmer, and C. L. Tien, *J. Opt. Soc. Am.* **62**, 353 (1972).
22. M. M. Kirillova, *Zh. Eksp. Teor. Fiz.* **61**, 336 (1971); M. M. Kirillova, *Sov. Phys. JEPT* **34**, 178 (1972).
23. N. Ya. Gorban, V. S. Stashchuk, A. V. Shirin, and A. A. Shishlovskii, *Opt. Spektrosk.* **35**, 508 (1973); N. Ya. Gorban, V. S. Stashchuk, A. V. Shirin, and A. A. Shishlovskii, *Opt. Spectrosc.* **35**, 295 (1973).
24. B. W. Veal and A. P. Paulikas, *Int. J. Magn.* **4**, 57 (1973).
25. P. B. Johnson and R. W. Christy, *Phys. Rev. B* **9**, 5056 (1974).
26. C. Wehenkel and B. Gauthe, *Phys. Lett.* **47A**, 253 (1974).
27. C. Wehenkel and B. Gauthe, *Phys. Stat. Solidi B* **64**, 515 (1974).
28. B. K. Reddy and T. C. Goel, *Indian J. Appl. Phys.* **13**, 138 (1975).
29. I. I. Sasovskaya and M. M. Noskov, *Opt. Spektrosk.* **39**, 111 (1975); I. I. Sasovskaya and M. M. Noskov, *Opt. Spectrosc.* **39**, 64 (1975).
30. A. A. Studna, *Solid State Commun.* **16**, 1063 (1975).
31. S. Moritani, K. Kondo, and J. Nakai, *Jpn. J. Appl. Phys.* **15**, 1549 (1976).
32. T. J. Moravec, J. C. Rife, and R. N. Dexter, *Phys. Rev. B* **13**, 3297 (1976).
33. T. Smith, *J. Opt. Soc. Am.* **67**, 48 (1977).
34. A. S. Siddiqui and D. M. Treherne, *Infrared Phys.* **17**, 33 (1977).
35. M. Tokumoto, H. D. Drew, and A. Bagchi, *Phys. Rev. B* **16**, 3497 (1977).
36. V. S. Gushchin, K. M. Shvarev, B. A. Baum, and P. V. Gel'd, *Fiz. Tverd. Tela* **20**, 1637 (1978); V. S. Gushchin, K. M. Shvarev, B. A. Baum, and P. V. Gel'd, *Sov. Phys. Solid State* **20**, 948 (1978).
37. M. P. Stoll and C. Jung, *J. Physique* **39**, 389 (1978).
38. M. P. Stoll and C. Jung, *J. Phys. F* **9**, 2491 (1979).
39. L. A. Feldkamp, M. B. Stearns, and S. S. Shinozaki, *Phys. Rev. B* **20**, 1310 (1979).
40. M. V. Zombeck, G. K. Austin, and D. T. Torgerson, Smithsonian Astrophysica Observatory Report SAO-AXAF-80-003, Appendix B, Table B1, Cambridge, Massachusetts, 1980.
41. J. H. Weaver, C. Krafka, D. W. Lynch, and E. E. Kohn, "Physik Daten, Physics Data, Optical Properties of Metals," Part 1, Fach-information-zentrum, Karlsruhe, 1981.
42. R. J. Bartlett and C. G. Olson, *Phys. Stat. Solidi B* **111**, K33 (1982).
43. B. L. Henke, P. Lee, T. J. Tanaka, R. L. Shimabukuro, and B. K. Fujikawa, "Low Energy X-Ray Diagnostics—1981" (D. T. Attwood and B. L. Henke, eds.), p. 340, *AIP Conf. Proc.* no. 75, American Institute of Physics, New York, 1981.
44. L. W. Bos and D. W. Lynch, *Phys. Rev. B* **2**, 4567 (1970).

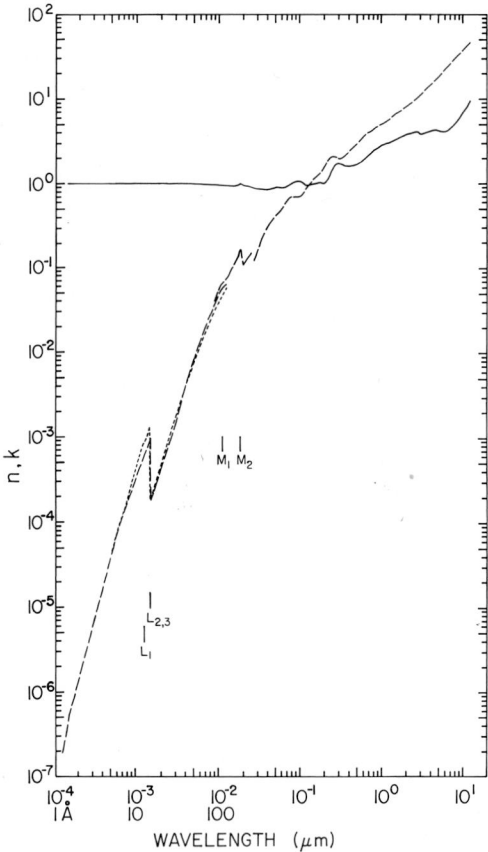

Fig. 5. Log–log plot of n (——) and k (––––) versus wavelength in micrometers for nickel.

TABLE V

Values of n and k Obtained from Various References for Nickel[a]

eV	cm^{-1}	Å	n		k	
9919		1.25	0.9999827	[40]	1.92 x 10^{-6}	[40]
9184		1.35	0.9999808		2.54	
8551		1.45	0.9999804		3.29 x 10^{-6}	
8266		1.50	0.9999816		4.98 x 10^{-7}	
7085		1.75	0.9999687		8.74 x 10^{-7}	
6199		2.00	0.9999579		1.43 x 10^{-6}	
4959		2.50	0.9999325		3.23	
4133		3.00	0.9999012		6.35 x 10^{-6}	
3542		3.50	0.999864		1.12 x 10^{-5}	
3100		4.00	0.999820		1.84	
2480		5.00	0.999715		4.19	
2066		6.00	0.999585		8.23	
2000		6.199	0.999559	[43]	8.81	[43]
1952		6.351	0.999538		9.61 x 10^{-5}	
1905		6.508	0.999516		1.05 x 10^{-4}	
1860		6.665	0.999493		1.14	
1815		6.831	0.999468		1.25	
1772		6.997	0.999444		1.36	
1771		7.00	0.999436	[40]	1.46	[40]
1730		7.166	0.999418	[43]	1.48	[43]
1688		7.345	0.999391		1.62	
1648		7.523	0.999363		1.76	
1609		7.705	0.999335		1.91	
1570		7.897	0.999304		2.09	
1550		8.00	0.999415	[40]	1.78	[40]
1533		8.087	0.999274	[43]	2.27	[43]
1496		8.287	0.999242		2.48	
1487		8.340	0.999409	[40]	1.90	[40]
1460		8.492	0.999210	[43]	2.70	[43]
1426		8.694	0.999178		2.92	
1392		8.906	0.999145		3.18	
1358		9.129	0.999109		3.46	
1326		9.350	0.999075		3.76	
1294		9.581	0.999040		4.09	
1263		9.816	0.999005		4.45	
1254		9.890	0.999200	[40]	3.20	[40]
1233		10.05	0.99897	[43]	4.83	[43]
1204		10.30	0.99894		5.24	
1188		10.44	0.999130	[40]	4.40	[40]
1175		10.55	0.99891	[43]	5.69	[43]
1147		10.81	0.99888		6.18	
1119		11.08	0.99885		6.72	
1093		11.34	0.99883		7.28	
1067		11.62	0.99881		7.91	
1041		11.91	0.99881		8.60	
1016		12.20	0.99883		9.33	
1012		12.25	0.999115	[40]	5.30	[40]
992		12.50	0.99886		9.09	[43]
968		12.81	0.99885		9.69 x 10^{-4}	
945		13.12	0.99886		1.05 x 10^{-3}	
929.4		13.34	0.99898	[40]	7.20 x 10^{-4}	[40]
923		13.43	0.99892	[43]	1.14 x 10^{-3}	[43]

(*continued*)

[a] The references from which the values were extracted are given in brackets.

TABLE V (*Continued*)

Nickel

eV	cm^{-1}	Å	n		k	
901		13.76	0.999081		1.23	
879		14.10	0.999565		1.34	
858		14.45	0.999820		6.30 x 10^{-4}	
851.5		14.56	0.99886	[40]	9.60	[40]
838		14.79	0.99299	[43]	1.95	[43]
818		15.16	0.99890		2.09	
798		15.54	0.99870		2.25	
779		15.92	0.99853		2.42	
776.4		15.97	0.99877	[40]	2.10	[40]
760		16.31	0.99837	[43]	2.63	[43]
742		16.71	0.99822		2.84	
725		17.10	0.99809		3.06	
707		17.54	0.99793		3.32	
704.9		17.59	0.99825	[40]	2.80	[40]
690		17.97	0.99779	[43]	3.59	[43]
674		18.39	0.99764		3.87	
658		18.84	0.99749		4.19	
642		19.31	0.99733		4.53	
637.5		19.45	0.99783	[40]	3.70	[40]
627		19.77	0.99717	[43]	4.89	[43]
612		20.26	0.99701		5.28	
597		20.77	0.99683		5.71	
583		21.27	0.99665		6.16	
572.9		21.64	0.99695	[40]	6.10	[40]
569		21.79	0.99646	[43]	6.65	[43]
555		22.34	0.99626		7.19	
542		22.87	0.99607		7.75	
529		23.44	0.99585		8.36	
524.9		23.62	0.99667	[40]	7.30	[40]
516		24.03	0.99563	[43]	9.03	[43]
504		24.60	0.99540		9.72	
500.3		24.78	0.99606	[40]	9.10 x 10^{-4}	[40]
492		25.20	0.99517	[43]	1.05 x 10^{-3}	[43]
480		25.83	0.99491		1.13	
469		26.43	0.99466		1.21	
458		27.07	0.99440		1.31	
452.2		27.42	0.99527	[40]	1.00	[40]
447		27.74	0.99411	[43]	1.41	[43]
436		28.44	0.99381		1.52	
426		29.10	0.99352		1.63	
415		29.87	0.99317		1.77	
406		30.54	0.99287		1.89	
396		31.31	0.99251		2.03	
395.4		31.36	0.99400	[40]	1.50	[40]
386		32.12	0.99213	[43]	2.20	[43]
377		32.89	0.99177		2.36	
368		33.69	0.99138		2.53	
359		34.53	0.99096		2.72	
351		35.32	0.99056		2.91	
342		36.25	0.99009		3.14	
334		37.12	0.9896		3.37	
326		38.03	0.9892		3.61	

TABLE V (*Continued*)

Nickel

eV	cm^{-1}	Å	n		k	
308		38.99	0.9887		3.88	
311		39.86	0.9882		4.14	
303		40.92	0.9876		4.47	
300.0		41.33	0.9880	[40]	5.13	[40]
296		41.88	0.9871	[43]	4.77	[43]
289		42.90	0.9865		5.11	
282		43.96	0.9859		5.47	
275		45.08	0.9853		5.88	
269		46.09	0.9847		6.25	
262		47.32	0.9840		6.72	
260.0		47.69	0.9851	[40]	7.65	[40]
256		48.43	0.9834	[43]	7.16	[43]
250		49.59	0.9827		7.64	
244		50.81	0.9820		8.17	
238		52.09	0.9812		8.74	
232		53.44	0.9804		9.35	
227		54.62	0.9797		9.91 x 10^{-3}	
221		56.10	0.9788		1.06 x 10^{-2}	
220		56.35	0.9809	[40]	1.22	[40]
216		57.40	0.9781	[43]	1.13	[43]
211		58.76	0.9772		1.20	
206		60.18	0.9764		1.28	
201		61.68	0.9755		1.36	
200		61.99	0.9780	[40]	1.59	[40]
196		63.25	0.9746	[43]	1.45	[43]
191		64.91	0.9736		1.55	
187		66.30	0.9728		1.63	
182		68.12	0.9717		1.74	
180		68.88	0.9751	[40]	2.03	[40]
178		69.65	0.9708	[43]	1.84	[43]
174		71.25	0.9699		1.94	
170		72.93	0.9689		2.05	
166		74.69	0.9679		2.17	
162		76.53	0.9668		2.30	
158		78.47	0.9658		2.44	
154		80.51	0.9647		2.58	
150		82.65	0.9635		2.73	
			0.9690	[40]	3.10	[40]
147		84.34	0.9625	[43]	2.86	[43]
143		86.70	0.9612		3.04	
140		88.56	0.9603		3.18	
			0.951	[39]	4.16	[42]
136		91.16	0.9589	[43]	3.39	[43]
133		93.22	0.9578		3.57	
130		95.37	0.9570		3.75	
			0.948	[39]	5.13	[42]
127		97.62	0.9560	[43]	3.91	[43]
124		99.98	0.9549		4.09	
121		102.5	0.9538		4.27	
120		103.3	0.9650	[40]	5.40	[40]
			0.948	[39]	6.21	[42]
118		105.1	0.9525	[43]	4.47	[43]

(continued)

David W. Lynch and W. R. Hunter

TABLE V (*Continued*)

Nickel

eV	cm^{-1}	Å	n		k	
115		107.8	0.9511		4.69	
112		110.7	0.9497		4.92	
110		112.7	0.9487		5.08	
			0.9680	[40]	5.99	[40]
			0.950	[39]	7.03	[42]
107		115.9	0.9471	[43]	5.34	[43]
105		118.1	0.9680	[40]	6.25	[40]
104		119.2	0.9454	[43]	5.63	[43]
102		121.5	0.9443		5.83	
100		124.0	0.9430		6.04	
			0.9670	[40]	6.53	[40]
			0.943	[39]	7.75	[42]
98		126.5	0.939		8.01	
96		129.1	0.938		8.36	
94		131.9	0.936		8.73	
92		134.8	0.935		9.11	
90		137.8	0.936		9.44	
88		140.9	0.936		9.81 x 10^{-2}	
86		144.2	0.934		0.102	
84		147.6	0.932		0.107	
82		151.2	0.933		0.114	
80		155.0	0.936		0.121	
78		159.0	0.942		0.128	
76		163.1	0.944		0.135	
74		167.5	0.947		0.137	
72		172.2	0.952		0.144	
70		177.1	0.952		0.156	
69		179.7	0.960		0.161	
68		182.3	0.975		0.168	
67		185.1	0.986		0.166	
66		187.9	1.009		0.163	
65		190.7	1.002		0.124	
64		193.7	0.980		0.106	
63		196.8	0.971		0.108	
62		200.0	0.964		0.112	
61		203.3	0.960		0.115	
60		206.6	0.955		0.118	
58		213.8	0.948		0.123	
56		221.4	0.942		0.129	
54		229.6	0.937		0.135	
52		238.4	0.931		0.143	
50		248.0	0.926	[41]	0.151	[41]
45		275.5	0.884		0.129	
40		310.0	0.865		0.179	
35		354.2	0.858		0.241	
30		413.3	0.860		0.318	
29		427.5	0.863		0.336	
28		442.8	0.868		0.354	
27		459.2	0.873		0.372	
26		476.9	0.877		0.394	
25		495.9	0.888	[13]	0.418	[13]
24.5		506.1	0.897		0.427	

TABLE V (*Continued*)

Nickel

eV	cm^{-1}	Å	n	k
24		516.6	0.905	0.431
23.5		527.6	0.911	0.438
23		539.1	0.915	0.436
22.5		551.0	0.911	0.443
22		563.6	0.909	0.451
21.5		576.7	0.906	0.458
21		590.4	0.898	0.471
20.5		604.8	0.893	0.493
20		619.9	0.894	0.513
19.5		635.8	0.896	0.537
19		652.6	0.904	0.560
18.5		670.2	0.912	0.582
18		688.8	0.924	0.606
17.5		708.5	0.941	0.628
17		729.3	0.959	0.643
16.5		751.4	0.975	0.660
16		774.9	0.993	0.674
15.75		787.2	1.00	0.680
15.5		799.9	1.01	0.687
15.25		813.0	1.02	0.694
15		826.6	1.03	0.700
14.75		840.5	1.04	0.704
14.5		855.1	1.05	0.703
14.25		870.0	1.05	0.705
14		885.6	1.07	0.707
13.75		901.7	1.07	0.704
13.5		918.4	1.07	0.705
13.25		935.7	1.08	0.706
13		953.7	1.08	0.706
12.75		972.4	1.08	0.709
12.5		991.9	1.08	0.709

eV	cm^{-1}	μm	n	k
12.25		0.1012	1.07	0.712
12		0.1033	1.07	0.715
11.75		0.1055	1.07	0.714
11.5		0.1078	1.05	0.715
11.25		0.1102	1.04	0.717
11		0.1127	1.01	0.727
10.8		0.1148	0.993	0.747
10.6		0.1170	0.974	0.759
10.4		0.1192	0.951	0.800
10.2		0.1216	0.948	0.833
10		0.1240	0.946	0.867
9.8		0.1265	0.948	0.895
9.6		0.1291	0.949	0.927
9.4		0.1319	0.954	0.957
9.2		0.1348	0.959	0.987
9		0.1378	0.966	1.01
8.8		0.1409	0.972	1.05
8.6		0.1442	0.981	1.08

(*continued*)

TABLE V (*Continued*)

Nickel

eV	cm^{-1}	μm	n	k
8.4		0.1476	0.992	1.11
8.2		0.1512	1.00	1.13
8.0		0.1550	1.01	1.15
7.8		0.1590	1.02	1.18
7.6		0.1631	1.02	1.21
7.4		0.1675	1.03	1.24
7.2		0.1722	1.03	1.27
7.0		0.1771	1.03	1.30
6.8		0.1823	1.02	1.35
6.6		0.1879	1.01	1.40
6.4		0.1937	1.01	1.46
6.2		0.2000	1.00	1.54
5.9	47,590	0.2101	1.02	1.67
5.8	46,780	0.2138	1.04	1.73
5.7	45,970	0.2175	1.06	1.78
5.6	45,170	0.2214	1.09	1.83
5.5	44,360	0.2254	1.12	1.88
5.4	43,550	0.2296	1.16	1.94
5.3	42,750	0.2339	1.21	1.99
5.2	41,940	0.2384	1.27	2.04
5.1	41,130	0.2431	1.33	2.07
5.0	40,330	0.2480	1.40	2.10
4.9	39,520	0.2530	1.47	2.11
4.8	38,710	0.2583	1.53	2.11
4.7	37,910	0.2638	1.59	2.10
4.6	37,100	0.2695	1.63	2.09
4.5	36,290	0.2755	1.67	2.07
4.4	35,490	0.2818	1.71	2.06
4.3	34,680	0.2883	1.73	2.03
4.2	33,880	0.2952	1.74	2.01
4.1	33,070	0.3024	1.74	1.99
4.0	32,260	0.3100	1.73	1.98
3.9	31,460	0.3179	1.72	1.98
3.8	30,650	0.3263	1.69	1.99
3.7	29,840	0.3351	1.66	2.02
3.6	29,040	0.3444	1.64	2.07
3.5	28,230	0.3542	1.63	2.11
3.4	27,420	0.3647	1.62	2.17
3.3	26,620	0.3757	1.61	2.23
3.2	25,810	0.3875	1.61	2.30
3.1	25,000	0.4000	1.61	2.36
3.0	24,200	0.4133	1.61 [17]	2.44 [17]
2.9	23,390	0.4275	1.62	2.52
2.8	22,580	0.4428	1.62	2.61
2.7	21,780	0.4592	1.64	2.71
2.6	20,970	0.4769	1.66	2.81
2.5	20,160	0.4959	1.67	2.93
2.4	19,360	0.5166	1.71	3.06
2.3	18,550	0.5391	1.75	3.19
2.2	17,740	0.5636	1.80	3.33
2.1	16,940	0.5904	1.85	3.48
2.0	16,130	0.6199	1.93	3.65

TABLE V (*Continued*)

Nickel

eV	cm^{-1}	μm	n	k
1.95	15,730	0.6358	1.98	3.74
1.90	15,320	0.6526	2.02	3.82
1.85	14,920	0.6702	2.08	3.91
1.80	14,520	0.6888	2.14	4.00
1.75	14,110	0.7085	2.21	4.09
1.70	13,710	0.7293	2.28	4.18
1.65	13,310	0.7514	2.36	4.25
1.60	12,900	0.7749	2.43	4.31
1.55	12,500	0.7999	2.48	4.38
1.50	12,100	0.8266	2.53	4.47
1.45	11,690	0.8551	2.59	4.55
1.40	11,290	0.8856	2.65	4.63
1.35	10,890	0.9184	2.69	4.73
1.30	10,490	0.9537	2.74	4.85
1.25	10,080	0.9919	2.80	4.97
1.20	9,679	1.033	2.85	5.10
1.15	9,275	1.078	2.91	5.24
1.10	8,872	1.127	2.97	5.38
1.05	8,469	1.181	3.01	5.55
1.00	8,065	1.240	3.06	5.74
0.95	7,662	1.305	3.11	5.98
0.90	7,259	1.378	3.18	6.23
0.85	6,856	1.459	3.27	6.51
0.80	6,452	1.550	3.38	6.82
0.75	6,049	1.653	3.49	7.13
0.70	5,646	1.771	3.59	7.48
0.65	5,243	1.907	3.69	7.92
0.60	4,839	2.066	3.84	8.35
0.55	4,436	2.254	3.90	8.92
0.50	4,033	2.480	4.03	9.64
0.45	3,629	2.755	4.20	10.2
0.40	3,226	3.100	3.84	11.4
0.38	3,065	3.263	3.92	12.1
0.36	2,904	3.444	4.00	12.7
0.34	2,742	3.647	4.07	13.4
0.32	2,581	3.875	4.12	14.2
0.30	2,420	4.133	4.19	15.0
0.28	2,258	4.428	4.30	16.0
0.26	2,097	4.769	4.29	17.1
0.24	1,936	5.166	4.16	18.4
0.22	1,774	5.636	4.11	20.2
0.20	1,613	6.199	4.12	22.5
0.19	1,533	6.526	4.30	23.8
0.18	1,452	6.888	4.45	25.2
0.17	1,371	7.293	4.68	26.8
0.16	1,290	7.749	5.00	28.6
0.15	1,210	8.266	5.45	30.6
0.14	1,129	8.856	5.83	32.8
0.13	1,049	9.537	6.44	35.3
0.12	967.9	10.33	7.11	38.3
0.11	887.2	11.27	8.12	41.8
0.10	806.5	12.40	9.54	45.8

VIII   OSMIUM (Os)

There have been only two measurements of the optical constants of Os that have been reported, and they do not agree very well in magnitude although their spectral dependences are approximately the same. Because one set of measurements used evaporated films and the other bulk samples, the differences in the results may be caused by differences in sample density.

Cox et al. [1] measured the reflectance of osmium films prepared in a conventional vacuum system. An electron beam was used to evaporate the osmium of 99.8% purity, and the films were condensed on both hot (300°C) and room-temperature (40°C), superpolished, fused-silica substrates. The deposition rates were approximately 50 Å/sec. The coated substrates were transferred in air to a reflectometer for measurements. There was very little difference in the reflectance values of films deposited on hot and cold substrates. The optical constants were obtained by using the reflectance method.

Lynch et al. [2] used a polycrystalline bulk sample of unknown crystallite orientation that was optically polished, boiled in aqua regia, and then heated in vacuo to 1200–1500°C for several hours. After cooling, the samples were transferred in air to a reflectometer. The optical constants were obtained from reflectance measurements at near-normal incidence by using the Kramers–Kronig method.

The data of Lynch et al. [2] are presented on Table VI and in Fig. 6, along with the soft x-ray data of Henke et al. [37]. Some of the data of Cox et al. [1] are also given in the table, so that the results obtained from films can be compared with those obtained from the bulk samples.

Osmium is a noncubic metal; therefore its optical properties are anisotropic. Presumably, the data of Lynch et al. [2] are characteristic of a random crystallite orientation. An evaporated film may not be similarly random in that the film often consists of crystallites with their basal planes in the plane of the film. Thus, differences between n and k for bulk and film samples are to be expected but predominantly for wavelengths longer than 0.1 to 0.2 μm.

REFERENCES

1. J. T. Cox, G. Hass, J. B. Ramsey, and W. R. Hunter, J. Opt. Soc. Am. **63**, 435 (1973).
2. D. W. Lynch, C. G. Olson, and J. H. Weaver, in "Physics Data" (J. H. Weaver, C. Krafka, D. W. Lynch, and E. E. Koch, eds.), Part 1, p. 253, Fach-informations-zentrum, Karlsruhe, 1981.
3. B. L. Henke, P. Lee, T. J. Tanaka, R. L. Shimabukuro, and B. K. Fujikawa, "Low Energy X-Ray Diagnostics—1981" (D. T. Attwood and B. L. Henke, eds.), p. 340, AIP Conf. Proc. No. 75, American Institute of Physics, New York, 1981.

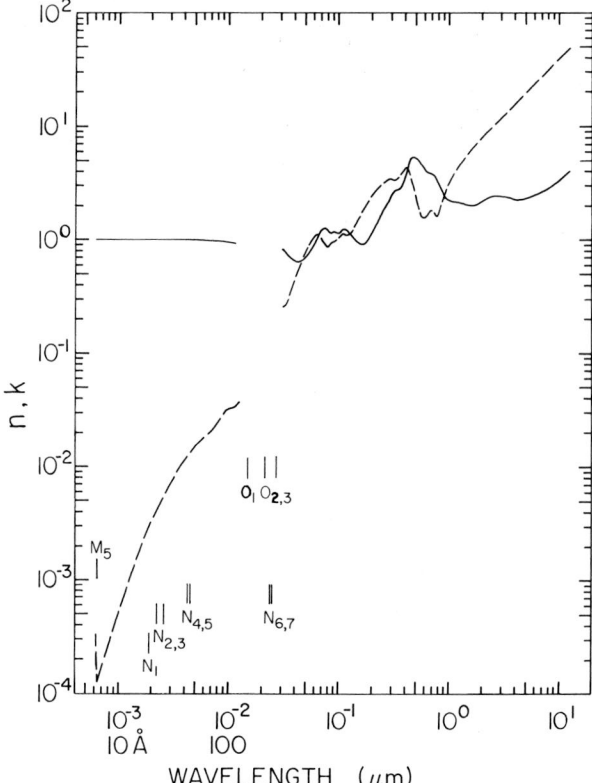

Fig. 6. Log–log plot of n (——) and k (----) versus wavelength in micrometers for osmium.

TABLE VI

Values of *n* and *k* Obtained from Various References for Osmium[a]

eV	cm^{-1}	Å	n	k
2000		6.199	0.999626 [3]	3.36 x 10^{-4} [3]
1952		6.351	0.999529	1.25
1905		6.508	0.999467	1.35
1860		6.665	0.999405	1.45
1815		6.831	0.999350	1.57
1772		6.997	0.999297	1.69
1730		7.166	0.999245	1.82
1688		7.345	0.999192	1.96
1648		7.523	0.999140	2.11
1609		7.705	0.999087	2.28
1570		7.897	0.999031	2.45
1533		8.087	0.99898	2.64
1496		8.287	0.99892	2.85
1460		8.492	0.99886	3.06
1426		8.694	0.99880	3.29
1392		8.906	0.99873	3.54
1358		9.129	0.99866	3.82
1326		9.350	0.99860	4.11
1294		9.581	0.99852	4.42
1263		9.816	0.99845	4.76
1233		10.05	0.99837	5.12
1204		10.30	0.99830	5.49
1175		10.55	0.99821	5.91
1147		10.81	0.99813	6.35
1119		11.08	0.99804	6.84
1093		11.34	0.99795	7.34
1067		11.62	0.99785	7.88
1041		11.91	0.99775	8.49
1016		12.20	0.99765	9.12
992		12.50	0.99755	9.77 x 10^{-4}
968		12.81	0.99744	1.05 x 10^{-3}
945		13.12	0.99733	1.12
923		13.43	0.99722	1.20
901		13.76	0.99710	1.29
879		14.10	0.99698	1.38
858		14.45	0.99686	1.47
838		14.79	0.99674	1.57
818		15.16	0.99661	1.68
798		15.54	0.99647	1.80
779		15.92	0.99634	1.92
760		16.31	0.99620	2.05
741		16.71	0.99607	2.18
725		17.10	0.99594	2.32
707		17.54	0.99579	2.48
690		17.97	0.99567	2.64
674		18.39	0.99556	2.80
658		18.84	0.99543	2.91
642		19.31	0.99527	3.03
627		19.77	0.99509	3.19
612		20.26	0.99491	3.38
597		20.77	0.99473	3.58
583		21.27	0.99456	3.79

[a] The references from which the values were extracted are given in brackets.

TABLE VI (*Continued*)

Osmium

eV	cm $^{-1}$	Å	n	k
569		21.79	0.99440	4.01
555		22.34	0.99422	4.25
542		22.87	0.99408	4.48
529		23.44	0.99395	4.72
516		24.03	0.99383	4.93
504		24.60	0.99369	5.11
492		25.20	0.99351	5.29
480		25.83	0.99331	5.49
469		26.43	0.99310	5.68
458		27.07	0.99286	5.88
447		27.74	0.99257	6.14
436		28.44	0.99230	6.46
426		29.10	0.99205	6.78
415		29.87	0.99179	7.16
406		30.54	0.99159	7.49
396		31.31	0.99139	7.89
386		32.12	0.99122	8.23
377		32.89	0.99101	8.53
368		33.69	0.99078	8.85
359		34.53	0.99054	9.20
351		35.32	0.99030	9.52
342		36.25	0.99002	9.91 x 10^{-3}
334		37.12	0.9897	1.03 x 10^{-2}
326		38.03	0.9895	1.07
318		38.99	0.9892	1.11
311		39.86	0.9889	1.15
303		40.92	0.9886	1.19
296		41.88	0.9883	1.24
289		42.90	0.9880	1.28
282		43.96	0.9878	1.33
275		45.08	0.9876	1.38
269		46.09	0.9873	1.41
262		47.32	0.9870	1.45
256		48.43	0.9867	1.49
250		49.59	0.9863	1.53
244		50.81	0.9860	1.57
238		52.09	0.9855	1.61
232		53.44	0.9851	1.65
227		54.62	0.9846	1.69
221		56.10	0.9841	1.74
216		57.40	0.9836	1.78
211		58.76	0.9832	1.82
206		60.18	0.9827	1.85
201		61.68	0.9820	1.89
196		63.25	0.9812	1.92
190		64.91	0.9804	1.97
187		66.30	0.9796	2.00
182		68.12	0.9785	2.05
178		69.65	0.9775	2.09
174		71.25	0.9764	2.13
170		72.93	0.9752	2.20
166		74.69	0.9741	2.28

(*continued*)

TABLE VI (*Continued*)

Osmium

eV	cm⁻¹	Å	n	k
162		76.53	0.9729	2.36
158		78.47	0.9717	2.45
154		80.51	0.9704	2.54
150		82.65	0.9690	2.63
147		84.34	0.9680	2.71
143		86.70	0.9665	2.82
140		88.56	0.9654	2.91
136		91.16	0.9639	3.03
133		93.22	0.9626	3.13
130		95.37	0.9619	3.22
127		97.62	0.9608	3.24
124		99.98	0.9593	3.26
121		102.5	0.9574	3.29
118		105.1	0.9552	3.31
115		107.8	0.9526	3.34
112		110.7	0.9495	3.36
110		112.7	0.9471	3.38
107		115.9	0.9431	3.48
104		119.2	0.9389	3.59
102		121.5	0.9356	3.66
100		124.0	0.9323	3.74

eV	cm⁻¹	μm	n	k	n	k
41.33		0.030	0.833 [1]	0.127 [1]		
40.00		0.031	0.800	0.138		
39		0.03179			0.81 [2]	0.26 [2]
38.75		0.032	0.768	0.155		
38		0.03263			0.79	0.26
37.57		0.033	0.740	0.173		
37		0.03351			0.77	0.27
36.47		0.034	0.710	0.193		
36		0.03444			0.74	0.29
35.42		0.035	0.685	0.222		
35		0.0354			0.72	0.31
34.44		0.036	0.660	0.247		
34		0.03646			0.70	0.34
33.51		0.037	0.638	0.279		
33		0.03757			0.68	0.37
32.63		0.038	0.620	0.315		
32		0.03874			0.66	0.41
31.79		0.039	0.613	0.350		
31		0.04000	0.590	0.390	0.65	0.45
30.24		0.041	0.580	0.432		
30		0.04133			0.65	0.49
29.6		0.04189			0.65	0.51
29.52		0.042	0.572	0.473		
29.2		0.04246			0.65	0.53
28.83		0.043	0.566	0.513		
28.4		0.04365			0.64	0.57
28.18		0.044	0.562	0.552		
28		0.04428			0.64	0.59

TABLE VI (*Continued*)

Osmium

eV	cm^{-1}	Å	n	k	n	k
27.55		0.045	0.557	0.590		
27.2		0.04558			0.65	0.62
26.95		0.046	0.557	0.627		
26.8		0.04626			0.63	0.66
26.4		0.04696			0.65	0.69
26.38		0.047	0.560	0.662		
26		0.04768			0.66	0.72
25.83		0.048	0.563	0.695		
25.6		0.04843			0.67	0.75
25.30		0.049	0.568	0.739		
24.8		0.04999			0.70	0.80
24.80		0.050	0.578	0.760		
24.4		0.05081			0.72	0.82
24.31		0.051	0.588	0.790		
24		0.05166			0.73	0.84
23.84		0.052	0.597	0.817		
23.6		0.05253			0.75	0.86
23.39		0.053	0.603	0.842		
23.2		0.05344			0.75	0.88
22.96		0.054	0.610	0.868		
22.8		0.05438			0.77	0.90
22.54		0.055	0.617	0.890		
22.4		0.05535			0.78	0.93
22.14		0.056	0.627	0.910		
22		0.05635			0.80	0.96
21.75		0.057	0.630	0.932		
21.6		0.05740			0.83	0.99
21.38		0.058	0.642	0.952		
21.2		0.05848			0.86	1.02
21.01		0.059	0.653	0.973	0.87	1.04
20.8		0.05961			0.89	1.05
20.66		0.060	0.667	0.992		
20.6		0.06018			0.91	1.07
20.4		0.06077			0.93	1.09
20.33		0.061	0.687	1.01		
20.2		0.06138			0.96	1.10
20.00		0.062	0.708	1.03		
19.8		0.06262			1.02	1.11
19.68		0.063	0.730	1.05		
19.6		0.06326			1.05	1.11
19.4		0.06391			1.07	1.11
19.37		0.064	0.760	1.06		
19.2		0.06457			1.10	1.10
19.07		0.065	0.788	1.08		
19.0		0.06525			1.12	1.10
18.8		0.06595			1.14	1.10
18.79		0.066	0.824	1.08		
18.6		0.06666			1.17	1.09
18.51		0.067	0.857	1.09		
18.4		0.06738			1.19	1.08
18.23		0.068	0.887	1.08		
18.2		0.06812			1.21	1.06

(*continued*)

TABLE VI (*Continued*)

Osmium

eV	cm^{-1}	μm	n	k	n	k
18.0		0.0688			1.23	1.04
17.97		0.069	0.920	1.08		
17.8		0.06965			1.24	1.03
17.71		0.070	0.857	1.09		
17.6		0.07044			1.26	1.01
17.46		0.071	0.975	1.05		
17.4		0.07125			1.27	0.99
17.22		0.072	0.995	1.03		
17.2		0.07208			1.27	0.97
17.0		0.07293			1.28	0.96
16.98		0.073	1.01	1.01		
16.8		0.07380			1.28	0.94
16.75		0.074	1.03	0.993		
16.6		0.07469			1.28	0.92
16.53		0.075	1.03	0.963		
16.4		0.07560			1.28	0.90
16.31		0.076	1.03	0.942		
16.2		0.7653			1.26	0.88
16.10		0.077	1.03	0.922		
16.0		0.07749			1.25	0.87
15.90		0.078	1.03	0.908		
15.8		0.07847			1.22	0.86
15.69		0.079	1.03	0.902		
15.6		0.07947			1.20	0.86
15.50		0.080	1.02	0.902		
15.4		0.08051			1.18	0.87
15.31		0.081	1.01	0.908		
15.2		0.08157			1.17	0.89
15.12		0.082	1.01	0.918		
15.0		0.08265			1.16	0.90
14.94		0.083	1.01	0.927		
14.8		0.08377			1.16	0.91
14.76		0.084	1.00	0.935		
14.6		0.08492			1.15	0.93
14.59		0.085	0.999	0.945		
14.42		0.086	0.997	0.953		
14.4		0.08610			1.16	0.94
14.25		0.087	1.00	0.963		
14.2		0.08731			1.17	0.96
14.09		0.088	1.00	0.972		
14.0		0.08856			1.17	0.96
13.93		0.089	1.01	0.983		
13.8		0.08984			1.18	0.96
13.78		0.090	1.02	0.993		
13.62		0.091	1.04	1.01		
13.6		0.09116			1.17	0.97
13.48		0.092	1.06	1.01		
13.4		0.09252			1.17	0.97
13.33		0.093	1.07	1.03		
13.2		0.09392			1.16	0.98
13.19		0.094	1.09	1.04		
13.05		0.095	1.10	1.05		

TABLE VI (*Continued*)

Osmium

eV	cm^{-1}	μm	n	k	n	k
13.0		0.09537			1.16	0.99
12.92		0.096	1.12	1.06		
12.8		0.09686			1.15	1.01
12.78		0.097	1.13	1.07		
12.65		0.098	1.14	1.08		
12.6		0.09840			1.17	1.12
12.52		0.099	1.15	1.09		
12.4		0.09998			1.14	1.03
12.40		0.100	1.16	1.10		
12.2		0.1016			1.14	1.06
12.0		0.1033			1.15	1.08
11.81		0.105	1.21	1.14		
11.8		0.1051			1.16	1.10
11.6		0.1069			1.17	1.12
11.4		0.1088			1.19	1.15
11.27		0.110	1.24	1.17		
11.2		0.1107			1.23	1.14
11.0		0.1127			1.24	1.13
10.8		0.1148			1.25	1.11
10.78		0.115	1.25	1.18		
10.6		0.1170			1.24	1.10
10.4		0.1192			1.22	1.08
10.33		0.120	1.26	1.18		
10.3		0.1204			1.20	1.08
10.2		0.1216			1.19	1.08
10.0		0.1240			1.16	1.10
9.919		0.125	1.25	1.19		
9.8		0.1265			1.13	1.11
9.6		0.1292			1.10	1.14
9.537		0.130	1.23	1.23		
9.4		0.1319			1.08	1.16
9.2		0.1348			1.04	1.19
9.184		0.135	1.20	1.27		
9.0		0.1378			1.01	1.24
8.856		0.140	1.17	1.33		
8.8		0.1409			0.98	1.29
8.6		0.1442			0.96	1.34
8.551		0.145	1.12	1.41		
8.4		0.1476			0.94	1.40
8.266		0.150	1.08	1.49		
8.2		0.1512			0.91	1.48
8.0		0.1550			0.91	1.55
7.999		0.155	1.05	1.59		
7.8		0.1590			0.90	1.63
7.749		0.160	1.03	1.69		
7.6		0.1631			0.90	1.72
7.514		0.165	1.02	1.78		
7.4		0.1675			0.91	1.81
7.293		0.170	1.02	1.88		
7.2		0.1722			0.92	1.91
7.085		0.175	1.03	1.97		
7.0		0.1771			0.95	2.00

(*continued*)

TABLE VI (*Continued*)

Osmium

eV	cm^{-1}	μm	n	k	n	k
6.888		0.180	1.05	1.05		
6.8		0.1823			0.97	2.11
6.702		0.185	1.08	2.13		
6.6		0.1879			1.01	2.21
6.526		0.190	1.12	2.20		
6.4		0.1937			1.06	2.33
6.358		0.195	1.16	2.27		
6.2		0.2000			1.13	2.44
6.199		0.200	1.18	2.33		
6.0	48,390	0.2066			1.20	2.54
5.9	47,590	0.2101			1.24	2.60
5.8	46,780	0.2138			1.27	2.65
5.7	45,970	0.2175			1.32	2.71
5.6	45,170	0.2214			1.36	2.77
5.5	44,360	0.2254			1.41	2.83
5.4	43,550	0.2296			1.46	2.88
5.3	42,750	0.2339			1.52	2.94
5.2	41,940	0.2384			1.58	3.00
5.1	41,140	0.2431			1.65	3.07
5.0	40,330	0.2480			1.74	3.12
4.9	39,520	0.2530			1.82	3.15
4.8	38,720	0.2583			1.88	3.19
4.7	37,910	0.2638			1.94	3.24
4.6	37,100	0.2695			2.01	3.31
4.5	36,300	0.2755			2.11	3.38
4.4	35,490	0.2818			2.24	3.44
4.3	34,680	0.2883			2.39	3.47
4.2	33,880	0.2952			2.53	3.44
4.1	33,070	0.3024			2.64	3.40
4.0	32,260	0.3100			2.71	3.34
3.9	31,460	0.3179			2.73	3.31
3.8	30,650	0.3263			2.73	3.32
3.7	29,840	0.3351			2.73	3.37
3.6	29,040	0.3444			2.75	3.45
3.5	28,230	0.3542			2.79	3.59
3.4	27,420	0.3647			2.93	3.79
3.3	26,620	0.3757			3.15	3.88
3.2	25,810	0.3874			3.29	3.96
3.1	25,000	0.3999			3.51	4.21
3.0	24,200	0.4133			4.05	4.40
2.9	23,390	0.4275			4.65	4.18
2.8	22,580	0.4428			5.05	3.78
2.7	21,780	0.4592			5.30	3.38
2.6	20,970	0.4769			5.36	2.82
2.5	20,160	0.4959			5.28	2.38
2.4	19,360	0.5166			5.10	2.01
2.3	18,550	0.5390			4.84	1.76
2.2	17,740	0.5636			4.58	1.62
2.1	16,940	0.5904			4.26	1.54
2.0	16,130	0.6199			3.98	1.60
1.95	15,730	0.6358			3.88	1.67
1.90	15,320	0.6525			3.81	1.75

TABLE VI (*Continued*)

Osmium

eV	cm^{-1}	μm	n	k	n	k
1.85	14,920	0.6702			3.79	1.81
1.80	14,520	0.6888			3.78	1.83
1.75	14,110	0.7085			3.76	1.80
1.70	13,710	0.7293			3.70	1.75
1.65	13,310	0.7514			3.57	1.66
1.60	12,910	0.7749			3.36	1.62
1.57	12,660	0.7897			3.21	1.63
1.53	12,340	0.8103			2.99	1.70
1.50	12,100	0.8265			2.84	1.80
1.45	11,700	0.8550			2.65	2.01
1.40	11,290	0.8856			2.49	2.23
1.35	10,890	0.9184			2.35	2.48
1.30	10,490	0.9537			2.25	2.77
1.25	10,080	0.9918			2.19	3.04
1.20	9,679	1.033			2.16	3.35
1.15	9,276	1.078			2.17	3.59
1.10	8,872	1.127			2.15	3.84
1.05	8,469	1.181			2.12	4.11
1.00	8,065	1.240			2.09	4.41
0.95	7,662	1.305			2.05	4.74
0.90	7,259	1.378			2.03	5.10
0.85	6,856	1.459			2.01	5.51
0.80	6,452	1.550			2.00	5.95
0.75	6,049	1.653			2.00	6.46
0.70	5,646	1.771			2.02	7.04
0.65	5,243	1.907			2.11	7.68
0.60	4,839	2.066			2.21	8.37
0.55	4,436	2.254			2.33	9.12
0.50	4,033	2.480			2.41	9.97
0.45	3,630	2.755			2.43	11.0
0.40	3,226	3.100			2.45	12.3
0.35	2,823	3.542			2.33	14.1
0.30	2,420	4.133			2.23	16.6
0.25	2,016	4.959			2.35	20.0
0.20	1,613	6.199			2.44	25.1
0.15	1,210	8.265			2.90	33.6
0.10	806.5	12.40			4.08	50.2

PLATINUM (Pt)　IX

Measurements of the optical properties of Pt [1–24] range over the vacuum ultraviolet to the infrared. Table VII lists the *n* and *k* values chosen and the pertinent references. These data are plotted in Fig. 7 as smooth curves. From 150–2200 Å the data are those of Hunter *et al.* [23] and from 2000 Å to about

12.4 μm the data are those of Weaver [21], and Henke et al.'s [25] calculated data cover the wavelength region from 6.2–124 Å.

Weaver [21] used reflectance [9, 16, 18] and transmittance [11] data from a number of sources to obtain n and k by the Kramers–Kronig technique.

Hunter et al. [23] made reflectance-versus-angle-of-incidence measurements on evaporated films of platinum. They used a conventional evaporation system with an electron-beam gun to evaporate the platinum, substrates heated to 300°C, and deposition rates of about 5–10 Å/sec. During deposition the pressure was about 10^{-6} torr. The purity of the Pt was 99.9%.

Except for the n spectra, Weaver's data [2] and those of Hunter et al. [23] show good agreement at 2200 Å. Although Henke et al.'s [25] data and those of Hunter et al. [23] do not meet or overlap, interpolations connecting the n and k spectra appear to be quite reasonable. It should be borne in mind that platinum films condensed on room-temperature substrates will have different values of n and k than those shown here and different reflectance values.

As with other transition metals, platinum is not expected to exhibit the optical properties of a free-electron metal in the infrared region.

REFERENCES

1. G. Hass and R. Tousey, *J. Opt. Soc. Am.* **49**, 593 (1959).
2. R. P. Madden and L. R. Canfield, *J. Opt. Soc. Am.* **51**, 838 (1961).
3. G. F. Jacobus, R. P. Madden, and L. R. Canfield, *J. Opt. Soc. Am.* **53**, 1084 (1963).
4. B. T. Barnes, *J. Opt. Soc. Am.* **56**, 1546 (1966).
5. A. P. Lenham and D. M. Treherne, *J. Opt. Soc. Am.* **56**, 1137 (1966).
6. A. P. Lenham and D. M. Treherne, *in* "Optical Properties and Electronic Structure of Metal Alloys" (F. Abélès, ed.), p. 196, North-Holland Publ., Amsterdam, 1966.
7. A. P. Lenham, *J. Opt. Soc. Am.* **57**, 473 (1967).
8. P. Jaegle, F. Combet Farnoux, P. Dhez, M. Cremonese, and G. Oroni, *Phys. Lett.* **26A**, 364 (1968).
9. A. Y.-C. Yu, W. E. Spicer, and G. Hass, *Phys. Rev.* **171**, 834 (1968).
10. G. Hass, J. B. Ramsey, and W. R. Hunter, *Appl. Opt.* **8**, 2255 (1969).
11. R. Haensel, K. Radler, B. Sonntag, and C. Kunz, *Sol. State Commun.* **7**, 1495 (1969).
12. W. R. Hunter, *Proc. Int. Conf. Vac. Ultraviolet Rad. Physics, 3rd. Tokyo* p. 2aC2-1 (1971).
13. J. C. Jones, D. C. Palmer, and C. L. Tien, *J. Opt. Soc. Am.* **62**, 353 (1972).
14. M. M. Kirillova, L. V. Nomerovannaya, and M. M. Noskov, *Phys. Met. Metallogr.* **34**, 291 (1972); M. M. Kirillova, L. V. Nomerovannaya, and M. M. Noskov, *Phys. Met. Metallogr.* **34**(2), 61 (1972).
15. R. C. Linton, NASA Technical Note D-7061 (1972).
16. A. Seignac and S. Robin, *Solid State Commun.* **11**, 217 (1972).
17. L. N. Aksyutov, *Inzh.-Fiz. Zh.* **27**, 197 (1974); L. N. Aksyutov, *J. Eng. Phys.* **27**, 913 (1974).
18. G. Hass and W. R. Hunter, *in* "Space Optics" (B. J. Thompson and R. R. Shannon, eds.), p. 525, National Academy of Sciences, Washington, D.C., 1974.
19. A. H. Madjid, R. L. Stover, and J. M. Martinez, *Phys. Kond. Matter.* **17**, 125 (1974).
20. C. Wehenkel and B. Gauthé, *Opt. Commun.* **11**, 62 (1974).
21. J. H. Weaver, *Phys. Rev. B* **11**, 1416 (1975).
22. A. S. Siddiqui and D. M. Treherne, *Infrared Phys.* **17**, 33 (1977).

23. W. R. Hunter, D. W. Angel, and G. Hass, *J. Opt. Soc. Am.* **69**, 1695 (1979).
24. R. E. Dietz, E. G. McRae, and J. H. Weaver, *Phys. Rev. B* **21**, 2229 (1980).
25. B. L. Henke, P. Lee, T. J. Tanaka, R. L. Shimabukuro, and B. K. Fujikawa, "Low Energy X-Ray Diagnostics—1981" (D. T. Attwood and B. L. Henke, eds.), p. 340, *AIP Conf. Proc.* No. 75, American Institute of Physics, New York, 1981.

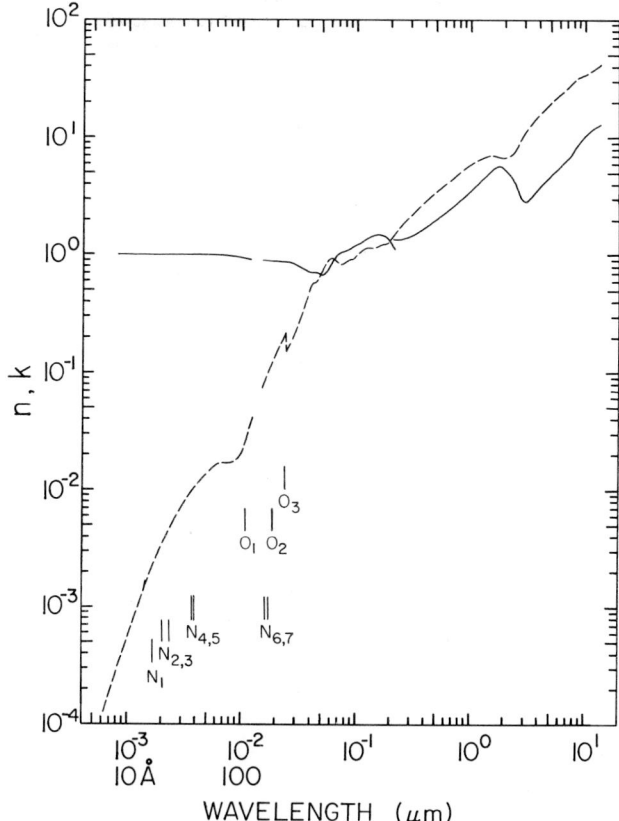

Fig. 7. Log–log plot of *n* (———) and *k* (----) versus wavelength in micrometers for platinum.

TABLE VII

Values of n and k Obtained from Various References for Platinum[a]

eV	cm^{-1}	Å	n	k
2000		6.199	0.999528 [25]	1.23x10^{-4} [25]
1952		6.351	0.999476	1.33
1905		6.508	0.999421	1.43
1860		6.665	0.999377	1.54
1815		6.831	0.999332	1.67
1772		6.997	0.999288	1.80
1730		7.166	0.999243	1.94
1688		7.344	0.999197	2.09
1648		7.523	0.999150	2.25
1609		7.705	0.999102	2.43
1570		7.897	0.999051	2.62
1533		8.087	0.999000	2.82
1496		8.287	0.99895	3.04
1460		8.492	0.99889	3.27
1426		8.694	0.99883	3.51
1392		8.906	0.99877	3.78
1358		9.129	0.99871	4.06
1326		9.350	0.99865	4.38
1294		9.581	0.99858	4.72
1263		9.816	0.99851	5.08
1233		10.05	0.99844	5.46
1204		10.30	0.99837	5.86
1175		10.55	0.99829	6.30
1147		10.81	0.99821	6.77
1119		11.08	0.99813	7.29
1093		11.34	0.99805	7.82
1067		11.62	0.99796	8.40
1041		11.91	0.99787	9.05
1016		12.20	0.99778	9.72x10^{-4}
992		12.50	0.99769	1.04x10^{-3}
968		12.81	0.99759	1.12
945		13.12	0.99749	1.20
923		13.43	0.99739	1.28
901		13.76	0.99729	1.37
879		14.10	0.99718	1.47
858		14.45	0.99707	1.57
838		14.79	0.99697	1.68
818		15.16	0.99686	1.79
798		15.54	0.99675	1.92
779		15.92	0.99666	2.05
760		16.31	0.99656	2.16
742		16.71	0.99646	2.27
725		17.10	0.99634	2.38
707		17.54	0.99620	2.51
690		17.97	0.99606	2.66
674		18.39	0.99592	2.82
658		18.84	0.99579	2.99
642		19.31	0.99566	3.19
627		19.77	0.99555	3.37
612		20.26	0.99544	3.55
597		20.77	0.99532	3.76
583		21.27	0.99521	3.96

[a] The references from which the values were extracted are given in brackets.

TABLE VII (*Continued*)

Platinum

eV	cm^{-1}	Å	n	k
569		21.79	0.99513	4.17
555		22.34	0.99508	4.37
542		22.87	0.99498	4.46
529		23.44	0.99482	4.54
516		24.03	0.99458	4.74
504		24.60	0.99437	4.98
492		25.20	0.99419	5.23
480		25.84	0.99400	5.51
469		26.43	0.99384	5.78
458		27.07	0.99368	6.07
447		27.74	0.99353	6.38
436		28.44	0.99338	6.70
426		29.10	0.99324	7.00
415		29.87	0.99310	7.37
406		30.54	0.99300	7.70
396		31.31	0.99293	8.08
386		32.12	0.99290	8.36
377		32.89	0.99284	8.59
368		33.69	0.99273	8.84
359		34.53	0.99260	9.09
351		35.32	0.99246	9.33
342		36.25	0.99228	9.62
334		37.12	0.99212	9.88×10^{-3}
326		38.03	0.99193	1.02×10^{-2}
318		38.99	0.99174	1.05
311		39.86	0.99155	1.07
303		40.92	0.99130	1.11
296		41.88	0.99108	1.14
289		42.90	0.99085	1.17
282		43.96	0.99060	1.20
275		45.08	0.99034	1.24
269		46.09	0.99010	1.27
262		47.32	0.9898	1.31
256		48.43	0.9896	1.35
250		49.59	0.9893	1.38
244		50.81	0.9891	1.41
238		52.09	0.9888	1.45
232		53.44	0.9885	1.48
227		54.62	0.9881	1.51
221		56.10	0.9877	1.54
216		57.40	0.9874	1.58
211		58.76	0.9870	1.61
206		60.18	0.9866	1.64
201		61.68	0.9863	1.68
196		63.25	0.9859	1.69
191		64.91	0.9855	1.69
187		66.30	0.9850	1.69
182		68.12	0.9841	1.70
178		69.65	0.9834	1.70
174		71.25	0.9826	1.70
170		72.93	0.9816	1.70
166		74.69	0.9805	1.70

(*continued*)

TABLE VII (*Continued*)

Platinum

eV	cm^{-1}	Å	n	k
162		76.53	0.9793	1.70
158		78.47	0.9778	1.71
154		80.51	0.9762	1.72
150		82.65	0.9745	1.74
147		84.34	0.9730	1.75
143		86.70	0.9708	1.77
140		88.56	0.9690	1.79
136		91.16	0.9661	1.81
133		93.22	0.9637	1.86
130		95.37	0.9612	1.92
127		97.62	0.9584	1.99
124		99.98	0.9553	2.06
121		102.5	0.9518	2.18
118		105.1	0.9481	2.34
115		107.8	0.9442	2.50
112		110.7	0.9399	2.71
110		112.7	0.9369	2.87
107		115.9	0.9322	3.14
104		119.2	0.9269	3.50
102		121.5	0.9232	3.77
100		124.0	0.9190	4.08

eV	cm^{-1}	μm	n	k
82.66		0.015	0.885 [23]	7.5 [23]
77.49		0.016	0.892	8.6
72.93		0.017	0.893	9.9×10^{-2}
68.88		0.018	0.892	0.114
65.26		0.019	0.887	0.129
61.99		0.020	0.882	0.145
59.04		0.021	0.877	0.162
56.36		0.022	0.869	0.180
53.91		0.023	0.860	0.198
51.66		0.024	0.881	0.217
45.59		0.025	0.872	0.157
47.69		0.026	0.862	0.166
45.92		0.027	0.850	0.180
44.28		0.028	0.838	0.194
42.75		0.029	0.826	0.210
41.33		0.030	0.810	0.227
40.00		0.031	0.794	0.249
38.75		0.032	0.779	0.272
37.57		0.033	0.762	0.297
36.47		0.034	0.742	0.322
35.42		0.035	0.722	0.350
34.44		0.036	0.696	0.379
33.51		0.037	0.674	0.412
32.63		0.038	0.663	0.447
31.79		0.039	0.675	0.502
31.00		0.040	0.687	0.551
30.24		0.041	0.699	0.581
29.52		0.042	0.709	0.573
28.83		0.043	0.718	0.654

TABLE VII (*Continued*)

Platinum

eV	cm^{-1}	μm	n	k
28.18		0.044	0.722	0.579
27.55		0.045	0.722	0.605
26.95		0.046	0.719	0.622
26.38		0.047	0.717	0.637
25.83		0.048	0.714	0.656
25.30		0.049	0.712	0.679
24.80		0.050	0.709	0.718
24.31		0.051	0.709	0.766
23.84		0.052	0.711	0.797
23.39		0.053	0.713	0.826
22.96		0.054	0.720	0.845
22.54		0.055	0.729	0.867
22.14		0.056	0.744	0.887
21.75		0.057	0.763	0.905
21.38		0.058	0.792	0.924
21.01		0.059	0.812	0.935
20.66		0.060	0.835	0.938
20.33		0.061	0.862	0.937
20.00		0.062	0.887	0.930
19.68		0.063	0.914	0.924
19.37		0.064	0.937	0.915
19.07		0.065	0.957	0.900
18.79		0.066	0.975	0.884
18.51		0.067	0.993	0.869
18.23		0.068	1.01	0.856
17.97		0.069	1.02	0.837
17.71		0.070	1.03	0.826
17.22		0.072	1.04	0.809
16.75		0.074	1.05	0.817
16.31		0.076	1.06	0.835
15.90		0.078	1.07	0.850
15.50		0.080	1.08	0.866
15.12		0.082	1.09	0.878
14.76		0.084	1.10	0.891
14.42		0.086	1.12	0.899
14.09		0.088	1.14	0.903
13.78		0.090	1.16	0.912
13.48		0.092	1.18	0.919
13.19		0.094	1.20	0.930
12.92		0.096	1.19	0.942
12.65		0.098	1.17	0.957
12.40		0.100	1.19	0.981
12.16		0.102	1.22	1.01
11.92		0.104	1.22	1.02
11.70		0.106	1.23	1.04
11.48		0.108	1.24	1.05
11.27		0.110	1.24	1.07
11.07		0.112	1.25	1.09
10.88		0.114	1.27	1.11
10.69		0.116	1.28	1.13
10.51		0.118	1.30	1.15
10.33		0.120	1.32	1.17

(*continued*)

TABLE VII (*Continued*)

Platinum

eV	cm^{-1}	μm	n	k
10.16		0.122	1.35	1.18
9.999		0.124	1.37	1.18
9.840		0.126	1.40	1.17
9.686		0.128	1.42	1.17
9.537		0.130	1.43	1.16
9.393		0.132	1.43	1.15
9.253		0.134	1.43	1.15
9.117		0.136	1.43	1.14
8.984		0.138	1.43	1.14
8.856		0.140	1.45	1.15
8.731		0.142	1.46	1.15
8.610		0.144	1.46	1.15
8.492		0.146	1.47	1.15
8.377		0.148	1.47	1.15
8.266		0.150	1.47	1.15
7.999		0.155	1.46	1.16
7.749		0.160	1.46	1.22
7.514		0.165	1.48	1.21
7.293		0.170	1.49	1.22
7.085		0.175	1.46	1.23
6.888		0.180	1.42	1.23
6.702		0.185	1.38	1.24
6.526		0.190	1.34	1.26
6.358		0.195	1.30	1.29
6.199	50,000	0.200	1.24	1.34
			1.39 [21]	1.35 [21]
6.1	49,200	0.2033	1.38	1.37
6.048	48,780	0.205	1.19 [23]	1.40 [23]
6.0	48,390	0.2066	1.38 [21]	1.40 [21]
5.9	47,590	0.2101	1.37	1.43
5.8	46,780	0.2138	1.36	1.47
5.6	45,170	0.2214	1.36	1.54
5.4	43,550	0.2296	1.36	1.61
5.2	41,940	0.2384	1.36	1.67
5.0	40,330	0.2480	1.36	1.76
4.8	38,720	0.2583	1.38	1.85
4.6	37,100	0.2695	1.39	1.95
4.4	35,490	0.2818	1.43	2.04
4.2	33,880	0.2952	1.45	2.14
4.0	32,260	0.3100	1.49	2.25
3.9	31,460	0.3179	1.51	2.32
3.8	30,650	0.3263	1.53	2.37
3.7	29,840	0.3351	1.56	2.42
3.6	29,040	0.3444	1.58	2.48
3.5	28,230	0.3542	1.60	2.55
3.4	27,420	0.3647	1.62	2.62
3.3	26,620	0.3757	1.65	2.69
3.2	25,810	0.3874	1.68	2.76
3.1	25,000	0.3999	1.72	2.84
3.0	24,200	0.4133	1.75	2.92
2.9	23,390	0.4275	1.79	3.01
2.8	22,580	0.4428	1.83	3.10

TABLE VII (*Continued*)

Platinum

eV	cm^{-1}	μm	*n*	*k*
2.7	21,780	0.4592	1.87	3.20
2.6	20,970	0.4769	1.91	3.30
2.5	20,160	0.4959	1.96	3.42
2.4	19,360	0.5166	2.03	3.54
2.3	18,550	0.5390	2.10	3.67
2.2	17,740	0.5636	2.17	3.77
2.1	16,940	0.5904	2.23	3.92
2.0	16,130	0.6199	2.30	4.07
1.9	15,320	0.6525	2.38	4.26
1.8	14,520	0.6888	2.51	4.43
1.7	13,710	0.7293	2.63	4.63
1.6	12,910	0.7749	2.76	4.84
1.5	12,100	0.8265	2.92	5.07
1.4	11,290	0.8856	3.10	5.32
1.3	10,490	0.9537	3.29	5.61
1.2	9,679	1.033	3.55	5.92
1.15	9,276	1.078	3.70	6.08
1.1	8,872	1.127	3.86	6.24
1.05	8,469	1.181	4.03	6.44
1.0	8,065	1.240	4.25	6.62
0.95	7,662	1.305	4.50	6.77
0.90	7,259	1.378	4.77	6.91
0.88	7,098	1.409	4.91	6.95
0.85	6,856	1.459	5.05	6.98
0.82	6,614	1.512	5.17	7.01
0.80	6,452	1.550	5.31	7.04
0.77	6,210	1.610	5.44	7.04
0.75	6,049	1.653	5.57	7.02
0.73	5,888	1.698	5.67	6.95
0.70	5,646	1.771	5.71	6.83
0.68	5,485	1.823	5.66	6.73
0.65	5,243	1.907	5.52	6.66
0.63	5,081	1.968	5.34	6.70
0.60	4,839	2.066	5.13	6.75
0.57	4,597	2.175	4.86	6.89
0.55	4,436	2.254	4.58	7.14
0.52	4,194	2.384	4.30	7.40
0.50	4,073	2.455	3.91	7.71
0.47	3,791	2.638	3.36	8.40
0.45	3,629	2.755	3.03	9.31
0.43	3,468	2.883	2.92	10.3
0.40	3,226	3.100	2.81	11.4
0.38	3,065	3.263	3.03	12.5
0.35	2,823	3.542	3.28	13.7
0.32	2,581	3.875	3.57	14.9
0.30	2,420	4.133	3.92	16.2
0.28	2,258	4.428	4.24	17.7
0.25	2,016	4.959	4.70	19.4
0.22	1,774	5.636	5.24	21.5
0.20	1,613	6.199	5.90	24.0
0.17	1,371	7.293	6.78	27.2
0.15	1,210	8.266	8.18	31.2
0.13	1,049	9.537	9.91	36.7
0.10	806.5	12.40	13.2	44.7

X RHODIUM (Rh)

Studies of the optical properties of Rh are listed in the references [1–15]. Table VIII lists the n and k values chosen and the pertinent references. These data are plotted in Fig. 8 as smooth curves. The data from 150 to 2000 Å are those of Cox *et al.* [9], from 2000 Å to about 12.4 μm the data are those of Weaver *et al.* [15], and Henke *et al.'s* [16] calculated data extend from 6.2 to 124 Å.

Cox *et al.* [9] obtained n and k values from evaporated films of rhodium by using the reflectance-versus-angle-of-incidence method. The films were evaporated with an electron gun at pressures of 10^{-5} torr in a conventional vacuum system. Deposition rates ranged from 10 to 100 Å/sec and substrate temperatures were 300°C. The purity of the Rh was 99.9%. The films were transferred in air to the reflectometer.

Weaver *et al.* [15] used large polycrystalline samples cut from an electron-beam-melted button of rhodium of unstated purity. The samples were mechanically polished, boiled in aqua regia, and heated *in vacuo* (10^{-7} torr) at temperatures of 1200 to 1500°C for several hours. The samples were then transferred in air to ion-pumped experimental chambers. From 6.2 μm to 2818 Å a calorimetric technique [17] was used, and at the shorter wavelengths the reflectance at 10° angle of incidence was measured. These data were combined with other data [7], Drude infrared extrapolations, and a power-law decay in the reflectance extending to 10^5 eV to obtain n and k by using the Kramers–Kronig method.

REFERENCES

1. G. Hass and R. Tousey, *J. Opt. Soc. Am.* **49**, 593 (1959)
2. R. P. Madden and L. R. Canfield, *J. Opt. Soc. Am.* **51**, 838 (1961).
3. I. Y. Leksina and N. V. Penkina, *Fiz. Met. Metalloved.* **22**, 264 (1966); I. Y. Leksina and N. V. Penkina, *Phys. Met. Metallogr.* **22**(2), 104 (1966).
4. G. A. Bolotin and T. P. Chukina, *Opt. Spektrosk.* **23**, 620 (1967); G. A. Bolotin and T. P. Chukina, *Opt. Spectrosc.* **23**, 333 (1967).
5. F. I. Vilesov, A. A. Azgrubskii, and M. M. Kirillova, *Opt. Spektrosk.* **23**, 153 (1967); F. I. Vileson, A. A. Azgrubskii, and M. M. Kirillova, *Opt. Spectrosc.* **23**, 79 (1967).
6. M. M. Kirillova, L. V. Nomerovannaya, G. A. Bolotin, V. M. Mayevskiy, M. M. Noskov, and M. L. Bolotina, *Fiz. Met. Metalloved.* **25**, 459 (1968); M. M. Kirillova, L. V. Nomerovannaya, G. A. Bolotin, V. M. Mayevkiy, M. M. Noskov, and M. L. Bolotina, *Phys. Met. Metallogr.* **25**(3), 81 (1968).
7. R. Haensel, K. Radler, B. Sonntag, and C. Kunz, *Solid State Commun.* **7**, 1495 (1969).
8. A. Seignac and S. Robin, *C. R. Acad. Sci Ser. B* **271**, 919 (1970).
9. J. T. Cox, G. Hass, and W. R. Hunter, *J. Opt. Soc. Am.* **61**, 360 (1971).
10. W. R. Hunter, *Proc. Int. Conf. Vac. Ultraviolet Rad. Physics, 3rd, Tokyo*, p. 2aC2-1 (1971).
11. A. Seignac and S. Robin, *Solid State Commun.* **11**, 217 (1972).
12. D. T. Pierce and W. E. Spicer, *Phys. Stat. Solidi B* **60**, 689 (1973).
13. J. H. Weaver, *Phys. Rev. B* **11**, 1416 (1975).
14. J. H. Weaver and C. G. Olson, *Phys. Rev. B* **14**, 3251 (1976).

15. J. H. Weaver, C. G. Olson, and D. W. Lynch, *Phys. Rev. B* **15**, 4115 (1977).
16. B. L. Henke, P. Lee, T. J. Tanaka, R. L. Shimabukuro, and B. K. Fujikawa, "Low Energy X-Ray Diagnostics—1981" (D. T. Attwood and B. L. Henke, eds.), p. 340, *AIP Conf. Proc.* No. 75, American Institute of Physics, New York, 1981.
17. L. W. Bos and D. W. Lynch, *Phys. Rev. B* **2**, 4567 (1970).

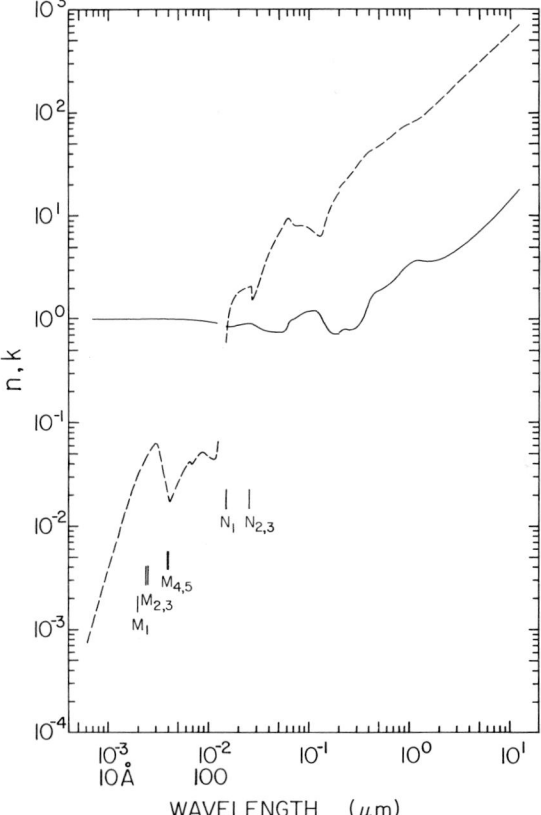

Fig. 8. Log–log plot of n (——) and k (----) versus wavelength in micrometers for rhodium.

TABLE VIII

Values of n and k Obtained from Various References for Rhodium[a]

eV	cm^{-1}	Å	n	k
2000		6.199	0.999512 [16]	7.50x10^{-5} [16]
1952		6.351	0.999486	8.16
1905		6.508	0.999459	8.87
1860		6.665	0.999432	9.63x10^{-5}
1815		6.831	0.999402	1.05x10^{-4}
1772		6.997	0.999372	1.14
1730		7.166	0.999341	1.24
1688		7.345	0.999307	1.34
1648		7.523	0.999272	1.46
1609		7.705	0.999236	1.58
1570		7.897	0.999198	1.71
1533		8.087	0.999158	1.86
1496		8.287	0.999117	2.02
1460		8.492	0.999073	2.19
1426		8.694	0.999029	2.36
1392		8.906	0.99898	2.56
1358		9.129	0.99893	2.78
1326		9.350	0.99888	3.00
1294		9.581	0.99883	3.25
1263		9.816	0.99877	3.52
1233		10.05	0.99871	3.81
1204		10.30	0.99865	4.13
1175		10.55	0.99859	4.47
1147		10.81	0.99852	4.84
1119		11.08	0.99845	5.25
1093		11.34	0.99839	5.67
1067		11.62	0.99831	6.14
1041		11.91	0.99824	6.66
1016		12.20	0.99816	7.21
992		12.50	0.99808	7.79
968		12.81	0.99800	8.42
945		13.12	0.99792	9.10
923		13.43	0.99783	9.81x10^{-4}
901		13.76	0.99774	1.06x10^{-3}
879		14.10	0.99765	1.14
858		14.45	0.99756	1.23
838		14.79	0.99747	1.33
818		15.16	0.99738	1.43
798		15.54	0.99728	1.54
779		15.92	0.99718	1.66
760		16.31	0.99709	1.79
714		16.71	0.99700	1.93
725		17.10	0.99691	2.07
707		17.54	0.99682	2.23
690		17.97	0.99674	2.39
674		18.39	0.99666	2.55
658		18.84	0.99658	2.74
642		19.31	0.99654	2.96
627		19.77	0.99652	3.13
612		20.26	0.99650	3.29
597		20.77	0.99645	3.46
583		21.27	0.99641	3.63
569		21.79	0.99636	3.82

[a] The references from which the values were extracted are given in brackets.

TABLE VIII (*Continued*)

Rhodium

eV	cm^{-1}	Å	n	k
555		22.34	0.99632	4.05
542		22.87	0.99631	4.27
529		23.44	0.99634	4.51
516		24.03	0.99639	4.72
504		24.60	0.99643	4.90
492		25.20	0.99648	5.09
480		25.83	0.99654	5.30
469		26.43	0.99661	5.50
458		27.07	0.99672	5.72
447		27.74	0.99692	6.01
436		28.44	0.99721	6.07
426		29.10	0.99744	6.12
415		29.87	0.99773	6.18
406		30.54	0.99808	6.23
396		31.31	0.99863	5.97
386		32.12	0.99887	5.30
377		32.89	0.99882	4.75
368		33.69	0.99867	4.25
359		34.53	0.99840	3.79
351		35.32	0.99808	3.41
342		36.25	0.99765	3.02
334		37.12	0.99720	2.71
326		38.03	0.99669	2.42
318		38.99	0.99611	2.16
311		39.86	0.99554	1.94
303		40.92	0.99473	1.72
296		41.88	0.99400	1.77
289		42.90	0.99329	1.87
282		43.96	0.99261	1.98
275		45.08	0.99195	2.09
269		46.09	0.99137	2.20
262		47.32	0.99066	2.33
256		48.43	0.99002	2.45
250		49.59	0.9894	2.59
244		50.81	0.9887	2.73
238		52.09	0.9879	2.89
232		53.44	0.9872	3.06
227		54.62	0.9865	3.21
221		56.10	0.9857	3.41
216		57.40	0.9849	3.59
211		58.76	0.9842	3.75
206		60.18	0.9834	3.90
201		61.68	0.9825	4.06
196		63.25	0.9817	4.23
190		64.91	0.9808	4.17
187		66.30	0.9799	4.09
182		68.12	0.9785	3.98
178		69.65	0.9773	4.08
174		71.25	0.9761	4.20
170		72.93	0.9748	4.32
166		74.69	0.9734	4.44
162		76.53	0.9719	4.57

(*continued*)

TABLE VIII (*Continued*)

Rhodium

eV	cm^{-1}	Å	n	k
158		78.47	0.9703	4.71
154		80.51	0.9686	4.86
150		82.65	0.9667	5.01
147		84.34	0.9653	5.14
143		86.70	0.9632	5.20
140		88.56	0.9615	5.12
136		91.16	0.9589	5.02
133		93.22	0.9567	4.93
130		95.37	0.9543	4.85
127		97.62	0.9517	4.77
124		99.98	0.9487	4.69
121		102.5	0.9455	4.61
118		105.1	0.9419	4.52
115		107.8	0.9379	4.44
112		110.7	0.9333	4.36
110		112.7	0.9300	4.39
107		115.9	0.9243	4.43
104		119.2	0.9174	4.73
102		121.5	0.9124	5.59
100		124.0	0.9072	6.62×10^{-3}

eV	cm^{-1}	μm	n	k
82.66		0.015	0.852 [9]	6.0×10^{-2} [9]
77.49		0.016	0.852	0.112
72.93		0.017	0.853	0.140
68.88		0.018	0.854	0.158
65.26		0.019	0.857	0.172
61.99		0.020	0.865	0.183
59.04		0.021	0.875	0.190
56.36		0.022	0.888	0.195
53.91		0.023	0.900	0.199
51.66		0.024	0.910	0.203
49.59		0.025	0.915	0.205
47.69		0.026	0.915	0.208
45.92		0.027	0.907	0.153
44.28		0.028	0.890	0.162
42.75		0.029	0.863	0.177
41.33		0.030	0.838	0.194
40.00		0.031	0.808	0.212
38.75		0.032	0.779	0.237
37.57		0.033	0.753	0.260
36.47		0.034	0.739	0.285
35.42		0.035	0.731	0.312
34.44		0.036	0.727	0.337
33.51		0.037	0.725	0.360
32.63		0.038	0.727	0.390
31.79		0.039	0.729	0.416
31.00		0.040	0.732	0.445
30.24		0.041	0.735	0.468
29.52		0.042	0.737	0.488
28.83		0.043	0.739	0.511

TABLE VIII (*Continued*)

Rhodium

eV	cm^{-1}	μm	n	k
28.18		0.044	0.739	0.536
27.55		0.045	0.739	0.560
26.95		0.046	0.738	0.580
26.38		0.047	0.733	0.603
25.83		0.048	0.728	0.627
25.30		0.049	0.727	0.645
24.80		0.050	0.732	0.673
24.31		0.051	0.742	0.696
23.84		0.052	0.757	0.720
23.39		0.053	0.780	0.743
22.96		0.054	0.795	0.771
22.54		0.055	0.807	0.798
22.14		0.056	0.815	0.832
21.75		0.057	0.821	0.821
21.38		0.058	0.823	0.893
21.01		0.059	0.832	0.925
20.66		0.060	0.875	0.942
20.33		0.061	0.926	0.949
20.00		0.062	0.987	0.950
19.68		0.063	1.02	0.948
19.37		0.064	1.03	0.938
19.07		0.065	1.04	0.913
18.79		0.066	1.05	0.877
18.51		0.067	1.06	0.851
18.23		0.068	1.06	0.834
17.97		0.069	1.07	0.820
17.71		0.070	1.08	0.812
17.46		0.071	1.09	0.805
17.22		0.072	1.10	0.803
16.98		0.073	1.10	0.802
16.75		0.074	1.11	0.802
16.53		0.075	1.12	0.802
16.31		0.076	1.13	0.802
16.10		0.077	1.14	0.802
15.90		0.078	1.15	0.802
15.69		0.079	1.15	0.802
15.50		0.080	1.17	0.802
15.31		0.081	1.17	0.802
15.12		0.082	1.18	0.802
14.94		0.083	1.19	0.802
14.76		0.084	1.20	0.801
14.59		0.085	1.21	0.800
14.42		0.086	1.21	0.799
14.25		0.087	1.22	0.798
14.09		0.088	1.23	0.796
13.93		0.089	1.24	0.793
13.78		0.090	1.24	0.791
13.62		0.091	1.25	0.788
13.48		0.092	1.26	0.787
13.33		0.093	1.26	0.784
13.19		0.094	1.26	0.782
13.05		0.095	1.27	0.778

(*continued*)

TABLE VIII (*Continued*)

Rhodium

eV	cm^{-1}	µm	n	k
12.92		0.096	1.27	0.773
12.78		0.097	1.27	0.769
12.65		0.098	1.27	0.764
12.52		0.099	1.27	0.758
12.40		0.100	1.27	0.755
11.81		0.105	1.26	0.730
11.27		0.110	1.24	0.707
10.78		0.115	1.22	0.682
10.33		0.120	1.18	0.655
9.919		0.125	1.11	0.638
9.537		0.130	1.03	0.638
9.184		0.135	0.953	0.675
8.856		0.140	0.880	0.750
8.551		0.145	0.820	0.843
8.266		0.150	0.766	0.940
7.999		0.155	0.720	1.03
7.749		0.160	0.691	1.12
7.514		0.165	0.681	1.22
7.293		0.170	0.677	1.30
7.085		0.175	0.675	1.38
6.888		0.180	0.672	1.45
6.702		0.185	0.672	1.51
6.526		0.190	0.671	1.57
6.358		0.195	0.672	1.63
6.199		0.200	0.675	1.69
6.2	50,010	0.2000	0.78 [15]	1.85 [15]
6.0	48,390	0.2066	0.76	1.93
5.8	46,780	0.2138	0.79	2.00
5.6	45,170	0.2214	0.80	2.06
5.4	43,550	0.2296	0.80	2.14
5.2	41,490	0.2384	0.79	2.23
5.0	40,330	0.2480	0.79	2.34
4.9	39,520	0.2530	0.79	2.39
4.8	38,710	0.2583	0.79	2.46
4.7	37,910	0.2638	0.79	2.52
4.6	37,100	0.2695	0.78	2.60
4.5	36,290	0.2755	0.79	2.68
4.4	35,490	0.2818	0.80	2.76
4.3	34,680	0.2883	0.83	2.85
4.2	33,880	0.2952	0.84	2.94
4.1	33,070	0.3024	0.84	3.03
4.0	32,260	0.3100	0.86	3.12
3.9	31,460	0.3179	0.88	3.23
3.8	30,650	0.3263	0.91	3.34
3.7	29,840	0.3351	0.95	3.45
3.6	29,040	0.3444	0.99	3.58
3.5	28,230	0.3542	1.04	3.71
3.4	27,420	0.3647	1.11	3.84
3.3	26,620	0.3757	1.20	3.97
3.2	25,810	0.3875	1.30	4.09
3.1	25,000	0.4000	1.41	4.20
3.0	24,200	0.4133	1.53	4.29

TABLE VIII (Continued)

Rhodium

eV	cm^{-1}	μm	n	k
2.9	23,390	0.4275	1.63	4.36
2.7	21,780	0.4592	1.80	4.49
2.6	20,970	0.4769	1.85	4.55
2.5	20,160	0.4959	1.88	4.65
2.4	19,360	0.5166	1.90	4.78
2.3	18,550	0.5391	1.94	4.94
2.2	17,740	0.5636	2.00	5.11
2.1	16,940	0.5904	2.05	5.30
2.0	16,130	0.6199	2.12	5.51
1.9	15,320	0.6526	2.20	5.76
1.8	14,520	0.6888	2.30	6.02
1.7	13,710	0.7293	2.42	6.33
1.6	12,900	0.7749	2.60	6.64
1.5	12,100	0.8266	2.78	6.97
1.4	11,290	0.8856	3.01	7.31
1.3	10,490	0.9537	3.26	7.63
1.25	10,080	0.9919	3.38	7.80
1.2	9,679	1.033	3.51	7.94
1.15	9,275	1.078	3.60	8.09
1.1	8,872	1.127	3.67	8.26
1.05	8,469	1.181	3.72	8.44
1.0	8,065	1.240	3.71	8.67
0.95	7,662	1.305	3.69	8.94
0.90	7,259	1.378	3.62	9.36
0.85	6,856	1.459	3.64	9.81
0.80	6,452	1.550	3.63	10.3
0.75	6,049	1.653	3.63	11.0
0.70	5,646	1.771	3.67	11.7
0.65	5,243	1.907	3.77	12.6
0.60	4,839	2.066	3.87	13.5
0.55	4,436	2.254	3.97	14.7
0.50	4,033	2.480	4.20	16.1
0.45	3,629	2.755	4.45	17.7
0.40	3,226	3.100	4.74	19.8
0.35	2,823	3.542	5.22	22.4
0.30	2,420	4.133	5.85	25.9
0.25	2,016	4.959	7.01	30.7
0.20	1,613	6.199	8.66	37.5
0.15	1,210	8.266	11.5	48.6
0.10	806.5	12.40	18.5	69.4

XI SILVER (Ag)

From among the many studies of the optical properties of Ag [1–36], four have been chosen as representative. They are (1) the data of Hagemann *et al.* [28] from approximately 1.5 to 460 Å, (2) the data of Leveque *et al.* [35] from 460 to 3600 Å, (3) the data of Winsemius *et al.* [33] from 3600 Å to 2.07 μm, and (4) the data of Dold and Mecke [14] from 1.265 to 10 μm. Table IX lists the values of *n* and *k* and the pertinent references. These data are plotted in Fig. 9 as smooth curves.

Hagemann *et al.* [28] prepared their samples by evaporating thin films of silver onto substrates of collodion that were supported on copper screens of the type used for electron microscopy. The evaporation was done from resistance heated boats at a pressure of about 5×10^{-7} torr at rates of 10–50 Å/sec. The plastic substrates were dissolved away, leaving Ag films on copper screens. These processes required exposing the Ag films to air before measurements. Transmission measurements were made from 13 to 150 eV to obtain an absorption spectrum that was extended by using the data of others [10, 15, 22, 24, 26] to provide an absorption spectrum large enough for a Kramers–Kronig analysis.

Leveque *et al.* [35] prepared their samples by the evaporation of silver in a vacuum of about 10^{-7} torr. Reflectance measurements were made in the same chamber, and so there was no exposure to air. A variety of substrates was used, but for the data reported, the substrates were Pyrex plates at room temperature. Reflectance measurements were made in the 3.5–30-eV region by using synchrotron radiation. These data were augmented prior to Kramers–Kronig analysis by using absorptance data, $A = 1 - R$, in the 0.1–2.8-eV region (Weaver [36]), a smooth interpolation to the reflectance at 3.5 eV, a reflectance derived from the absorption coefficient data of Haensel *et al.* [22] between 30 and 150 eV and the data from Hagemann *et al.* [28] for higher energies.

Winsemius *et al.* [33] used polycrystalline silver of 99.999⁺% purity. Samples were spark-cut from larger pieces and electrolytically and chemically polished [37]. The samples were then transferred in air to a reflectometer, where they were vacuum annealed at 700 K for four or more hours. Values of *n* and *k* were obtained by using a polarimetric method described by Beattie [38].

Dold and Mecke [14] evaporated 99.99⁺% pure silver onto polished glass plates from a Mo boat. The substrate temperature was not stated but was probably room temperature. Polarimetric measurements were made in air to obtain *n* and *k*.

Henke *et al.'s* [39] data showed good agreement with that of Hagemann *et al.* [28], especially from 80 to 124 Å. At 460 Å the *n* spectra of Hagemann *et al.* [28] and Leveque *et al.* [35] are in very close agreement, but they diverge to longer wavelengths. The *k* spectra of these two sets of investigators show rather poor agreement at 460 Å but converge and become equal at about 1000 Å. The agreement for both n and k is good at 3600 Å, where the data of Leveque *et al.* [35] and Winsemius *et al.* [33] meet. The data of the latter group show only fair agreement with the data of Dold and Mecke [14] where their data overlap. If the data of Dold and Mecke [14] are used to calculate the normal-incidence reflectance R, a spectrum with a broad peak at 5 μm results. The reflectance should increase monotonically with increasing wavelength. Neglect of the anomalous skin effect causes a calculated R to be too high at longer wavelengths, not lower. The cause of the peak probably is erroneously low k values at longer wavelengths. The measured reflectance of silver is higher than that given by the data presented here, and it increases smoothly to longer wavelengths.

It should be noted that in the longer-wavelength region, the spectra are expected to be somewhat sample dependent, for the electron relaxation time depends on impurity content and crystallite size to an extent sufficient to appear in the spectra. See the discussion of the data for gold.

REFERENCES

1. G. Joos and A. Klopfer, *Z. Physik* **138**, 251 (1954).
2. L. G. Schulz, *J. Opt. Soc. Am.* **44**, 357, 540 (1954).
3. L. G. Schulz and F. R. Tangherlini, *J. Opt. Soc. Am.* **44**, 362 (1954).
4. J. R. Beattie and G. K. T. Conn, *Philos. Mag.* **46**, 989 (1955).
5. J. N. Hodgson, *Proc. Phys. Soc. London Ser. B* **68**, 593 (1955).
6. M. A. Biondi, *Phys. Rev.* **102**, 964 (1956).
7. J. R. Beattie, *Physica* **23**, 898 (1957).
8. L. G. Schulz, *Adv. Phys.* **6**, 102 (1957).
9. M. Otter, *Z. Physik* **161**, 163 (1961).
10. H. Ehrenreich and H. R. Philipp, *Phys. Rev.* **128**, 1622 (1962).
11. P. R. Wessel, *Phys. Rev.* **132**, 2062 (1963).
12. R. H. Huebner, E. T. Arakawa, R. A. MacRae, and R. N. Hamm, *J. Opt. Soc. Am.* **54**, 1434 (1964).
13. A. P. Lukirskii, E. P. Savinov, O. A. Ershov, and Y. F. Shepelev, *Opt. Spektrosk.* **16**, 310 (1964); A. P. Lukirskii, E. P. Savinov, O. A. Ershov, and Y. F. Shepelev, *Opt. Spectrosc.* **16**, 168 (1964).
14. B. Dold and R. Mecke, *Optik* **22**, 435 (1965).
15. L. R. Canfield and G. Hass, *J. Opt. Soc. Am.* **55**, 61 (1965).
16. R. G. Yarovaya and I. N. Shklyarevskii, *Opt. Spektrosk.* **18**, 832 (1964); R. G. Yarovaya and I. N. Shklyarevskii, *Opt. Spectrosc.* **18**, 465 (1965).
17. H. E. Bennett and J. M. Bennett, in "Optical Properties and Electronic Structure of Metals and Alloys" (F. Abélès, ed.), p. 175, North-Holland Publ., Amsterdam, 1966.
18. S. Robin, in "Optical Properties and Electronic Structure of Metals and Alloys" (F. Abélès, ed.), p. 202, North-Holland Publ., Amsterdam, 1966.
19. J. Daniels, *Z. Physik* **203**, 235 (1967).
20. E. Meyer, H. Frede, and H. Knof, *J. Appl. Phys.* **38**, 3682 (1967).
21. H. E. Bennett, J. M. Bennett, E. J. Ashley, and R. J. Motyka, *Phys. Rev.* **165**, 755 (1968).
22. R. Haensel, C. Kunz, T. Sasaki, and B. Sonntag, *Appl. Opt.* **7**, 301 (1968).
23. J. Daniels, *Z. Physik* **277**, 234 (1969).
24. G. B. Irani, T. Huen, and F. Wooten, *J. Opt. Soc. Am.* **61**, 128 (1971).
25. G. B. Irani, T. Huen, and F. Wooten, *Phys. Rev. B* **3**, 2385 (1971).
26. P. B. Johnson and R. W. Christy, *Phys. Rev. B* **6**, 4370 (1972).

27. M. Schlüter, *Z. Physik* **250**, 87 (1972).
28. H. J. Hagemann, W. Gudat, and C. Kunz, DESY Report SR-74/7, Hamburg (1974); H. J. Hagemann, W. Gudat, and C. Kunz, *J. Opt. Soc. Am.* **65**, 742 (1975).
29. C. Wehenkel and B. Gauthé, *Opt. Commun.* **11**, 62 (1974).
30. C. J. Flaten and E. A. Stern, *Phys. Rev. B* **11**, 638 (1975).
31. J. Rivory and M. L. Thèye, *J. Physique* **36**, L129 (1975).
32. J. A. Mackay and J. A. Rayne, *Phys. Rev. B* **13**, 673 (1976).
33. P. Winsemius, F. F. van Kampen, H. P. Lengkeek, and C. G. van Went, *J. Phys. F* **6**, 1583 (1976).
34. K. G. Ramanathan, S. H. Yen, and E. A. Estalote, *Appl. Opt.* **16**, 2810 (1977).
35. G. Leveque, C. G. Olson, and D. W. Lynch, *Phys. Rev. B* **24**, 4654 (1983).
36. J. H. Weaver, private communication (1978).
37. P. Winsemius, H. P. Lengkeek, and F. F. van Kampen, *Physica* **79 B**, 529 (1975).
38. J. R. Beattie, *Philos. Mag.* **46**, 235 (1955).
39. B. L. Henke, P. Lee, T. J. Tanaka, R. L. Shimabukuro and B. K. Fujikawa, "Low Energy X-Ray Diagnostics—1981" (D. T. Attwood and B. L. Henke, eds.), p. 340, *AIP Conf. Proc.* No. 75, American Institute of Physics, New York, 1981.

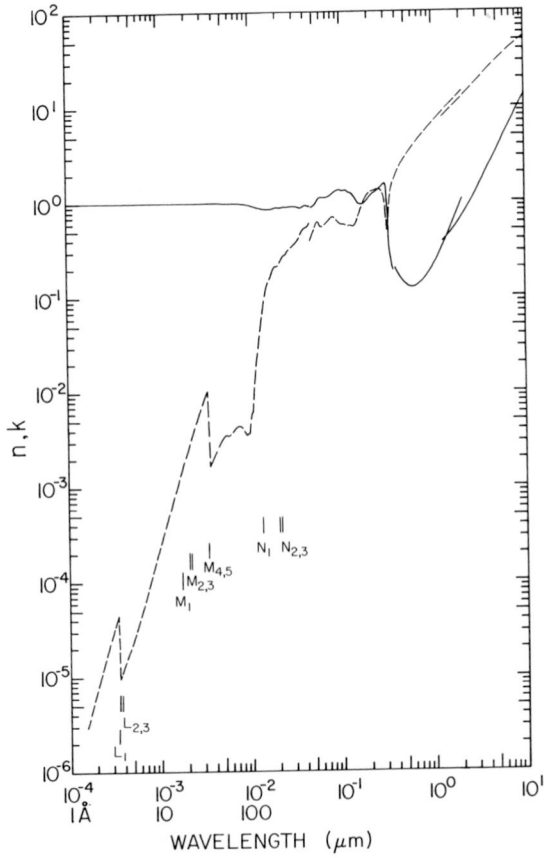

Fig. 9. Log–log plot of n (——) and k (----) versus wavelength in micrometers for silver.

TABLE IX

Values of n and k Obtained from Various References for Silver[a]

eV	cm^{-1}	Å	n	k
10000		1.240	0.9990 [28]	1.29×10^{-6} [28]
8000		1.550	1.000	2.92
6000		2.066	1.002	8.36×10^{-6}
5000		2.480	1.001	1.52×10^{-5}
4500		2.755	1.001	2.21
4000		3.099	1.003	3.57
3600		3.444	1.003	3.24
3500		3.542	1.003	1.72
3000		4.133	1.001	1.43
2500		4.959	1.001	2.40
2000		6.199	1.001	4.72×10^{-5}
1500		8.265	0.999	1.21×10^{-4}
1000		12.40	0.998	5.23×10^{-4}
800		15.50	0.997	1.08×10^{-3}
740		16.75	0.998	1.42
600		20.66	0.997	2.69
500		24.80	0.997	4.57
400		30.99	0.998	8.82
370		33.51	1.007	8.23
350		35.42	1.001	1.68
300		41.33	0.996	2.32
280		44.28	0.994	2.70
260		47.68	0.992	3.11
240		51.66	0.989	3.50
220		56.35	0.987	3.45
200		61.99	0.982	3.77
180		68.88	0.978	4.31
170		72.93	0.973	4.42
160		77.49	0.968	4.38
150		82.65	0.963	4.07
140		88.56	0.955	3.54
130		95.37	0.943	3.66
120		103.3	0.926	6.22×10^{-3}
110		112.7	0.902	1.72×10^{-2}
105		118.1	0.890	2.61
100		124.0	0.876	3.82
95		130.5	0.861	5.65
90		137.8	0.849	8.24×10^{-2}
85		145.9	0.846	0.111
80		155.0	0.848	0.139
78		158.9	0.851	0.149
76		163.1	0.853	0.158
74		167.5	0.853	0.166
72		172.2	0.851	0.181
70		177.1	0.859	0.197
68		182.3	0.871	0.208
66		187.8	0.883	0.211
64		193.7	0.885	0.210
62		200.0	0.881	0.211
60		206.6	0.873	0.221
58		213.8	0.868	0.237
56		221.4	0.871	0.259
54		229.6	0.885	0.274
53		233.9	0.890	0.275

(*continued*)

[a] The references from which the values were extracted are given in brackets.

TABLE IX (*Continued*)

Silver

eV	cm^{-1}	Å	n	k
52		238.4	0.889	0.276
51		243.1	0.886	0.281
50		248.0	0.884	0.290
49		253.0	0.885	0.300
48		258.3	0.888	0.309
47		263.8	0.893	0.317
46		269.5	0.896	0.323
44		281.8	0.899	0.334
42		295.2	0.897	0.349
40		309.9	0.896	0.368
39		317.9	0.895	0.378
38		326.3	0.892	0.388
37		335.1	0.885	0.400
36		344.4	0.876	0.418
35		354.2	0.865	0.454
34		364.6	0.879	0.489
33		375.7	0.899	0.514
32		387.4	0.921	0.528
31		399.9	0.932	0.534
30		413.2	0.931	0.541
29		427.5	0.919	0.557
28.5		435.0	0.911	0.572
28		442.8	0.902	0.590
27.5		450.8	0.851	0.616
27		459.2	0.886	0.650
			0.882 [35]	0.399 [35]
26.5		467.9	0.886	0.421
26		476.9	0.887	0.444
25.5		486.2	0.891	0.468
24.5		506.1	0.906	0.522
23.5		527.6	0.943	0.584
23		539.1	0.980	0.614
22.5		551.0	1.029	0.624
22		563.6	1.075	0.615
21.5		576.7	1.104	0.584
21		590.4	1.109	0.561
20.5		604.8	1.105	0.549
20		619.9	1.098	0.548
19		652.6	1.096	0.565
18		688.8	1.101	0.593
17		729.3	1.121	0.635
16		774.9	1.167	0.676
15.5		799.9	1.202	0.691
15		826.6	1.243	0.693
14.5		855.1	1.281	0.679
14		885.6	1.308	0.656
13.5		918.4	1.322	0.628
13		953.7	1.322	0.603
12.5		991.9	1.315	0.588

eV	cm^{-1}	μm	n	k
12.1		0.1025	1.308	0.581
11.8		0.1051	1.304	0.578

TABLE IX (*Continued*)

Silver

eV	cm^{-1}	μm	n	k
11.5		0.1078	1.300	0.573
11.2		0.1107	1.293	0.566
10.9		0.1137	1.280	0.560
10.6		0.1170	1.265	0.560
10.3		0.1204	1.252	0.564
10.0		0.1240	1.241	0.568
9.7		0.1278	1.229	0.566
9.2		0.1348	1.182	0.550
9.0		0.1378	1.149	0.552
8.8		0.1409	1.112	0.563
8.6		0.1442	1.073	0.581
8.4		0.1476	1.032	0.610
8.2		0.1512	0.993	0.653
8.0		0.1550	0.962	0.706
7.8		0.1590	0.940	0.770
7.6		0.1631	0.935	0.832
7.4		0.1675	0.936	0.892
7.2		0.1722	0.942	0.951
7.0		0.1771	0.953	1.01
6.8		0.1823	0.969	1.07
6.6		0.1879	0.995	1.13
6.4		0.1937	1.028	1.18
6.3		0.1968	1.048	1.21
6.2		0.2000	1.072	1.24
6.1	49,200	0.2033	1.098	1.26
6.0	48,390	0.2066	1.125	1.27
5.8	46,780	0.2138	1.173	1.29
5.6	45,170	0.2214	1.208	1.30
5.4	43,550	0.2296	1.238	1.31
5.2	41,940	0.2384	1.265	1.33
5.0	40,330	0.2480	1.298	1.35
4.9	39,520	0.2530	1.320	1.35
4.8	38,710	0.2583	1.343	1.35
4.7	37,910	0.2638	1.372	1.35
4.6	37,100	0.2695	1.404	1.33
4.5	36,290	0.2755	1.441	1.31
4.4	35,490	0.2818	1.476	1.26
4.3	34,680	0.2883	1.502	1.19
4.2	33,880	0.2952	1.519	1.08
4.15	33,470	0.2988	1.522	0.992
4.1	33,070	0.3024	1.496	0.882
4.05	32,670	0.3061	1.432	0.766
4.0	32,260	0.3100	1.323	0.647
3.98	32,100	0.3115	1.246	0.586
3.95	31,860	0.3139	1.149	0.540
3.93	31,700	0.3155	1.044	0.514
3.9	31,460	0.3179	0.932	0.504
3.88	31,290	0.3195	0.815	0.526
3.85	31,050	0.3220	0.708	0.565
3.83	30,890	0.3237	0.616	0.609
3.8	30,650	0.3263	0.526	0.663
3.75	30,250	0.3306	0.371	0.813
3.73	30,090	0.3324	0.321	0.902
3.7	29,840	0.3351	0.294	0.986

(*continued*)

TABLE IX (*Continued*)

Silver

eV	cm^{-1}	μm	n		k	
3.65	29,440	0.3397	0.259		1.12	
3.6	29,040	0.3444	0.238		1.24	
3.5	28,230	0.3542	0.209		1.44	
3.4	27,420	0.3647	0.186		1.61	
3.3	26,620	0.3757	0.200	[33]	1.67	[33]
3.2	25,810	0.3875	0.192		1.81	
3.1	25,000	0.4000	0.173		1.95	
3.0	24,200	0.4133	0.173		2.11	
2.9	23,390	0.4275	0.160		2.26	
2.8	22,580	0.4428	0.157		2.40	
2.7	21,780	0.4592	0.144		2.56	
2.6	20,970	0.4769	0.132		2.72	
2.5	20,160	0.4959	0.130		2.88	
2.4	19,360	0.5166	0.130		3.07	
2.3	18,550	0.5391	0.129		3.25	
2.2	17,740	0.5636	0.120		3.45	
2.1	16,940	0.5904	0.121		3.66	
2.0	16,130	0.6199	0.131		3.88	
1.9	15,320	0.6526	0.140		4.15	
1.8	14,520	0.6888	0.140		4.44	
1.7	13,710	0.7293	0.148		4.74	
1.6	12,900	0.7749	0.143		5.09	
1.5	12,100	0.8266	0.145		5.50	
1.4	11,290	0.8856	0.163		5.95	
1.3	10,490	0.9537	0.198		6.43	
1.2	9,679	1.033	0.226		6.99	
1.1	8,872	1.127	0.251		7.67	
1.0	8,065	1.240	0.329		8.49	
0.98	7,904	1.265	0.375	[14]	7.78	[14]
0.96	7,743	1.291	0.383		7.92	
0.95	7,662	1.305	0.358		8.95	
0.94	7,582	1.319	0.392		8.06	
0.92	7,420	1.348	0.401		8.21	
0.90	7,259	1.378	0.411		8.37	
			0.396	[33]	9.48	[33]
0.88	7,098	1.409	0.421	[14]	8.37	[14]
0.86	6,936	1.442	0.431		8.70	
0.85	6,856	1.459	0.446	[33]	10.1	[33]
0.84	6,775	1.476	0.442	[14]	8.88	[14]
0.82	6,614	1.512	0.455		9.08	
0.80	6,452	1.550	0.469		9.32	
			0.514	[33]	10.8	[33]
0.78	6,291	1.590	0.485		9.57	
0.76	6,130	1.631	0.501		9.84	
0.75	6,049	1.653	0.624	[33]	11.5	[33]
0.74	5,968	1.675	0.519	[14]	10.1	[14]
0.72	5,807	1.722	0.537		10.4	
0.70	5,646	1.771	0.557		10.7	
			0.844	[33]	12.2	[33]
0.68	5,485	1.823	0.578	[14]	11.1	[14]
0.66	5,323	1.879	0.600		11.4	
0.65	5,243	1.907	0.873	[33]	13.3	[33]
0.64	5,162	1.937	0.624	[14]	11.8	[14]
0.62	5,001	2.000	0.650		12.2	

TABLE IX (*Continued*)

Silver

eV	cm^{-1}	μm	n		k	
0.60	4,839	2.066	0.668		12.6	
			1.064	[33]	14.4	[33]
0.58	4,678	2.138	0.729	[14]	13.0	[14]
0.56	4,517	2.214	0.774		13.5	
0.54	4,355	2.296	0.823		14.0	
0.52	4,194	2.384	0.878		14.5	
0.50	4,033	2.480	0.939		15.1	
0.48	3,871	2.583	1.007		15.7	
0.46	3,710	2.695	1.083		16.4	
0.44	3,549	2.818	1.168		17.1	
0.42	3,388	2.952	1.265		17.9	
0.40	3,226	3.100	1.387		18.8	
0.38	3,065	3.263	1.536		19.8	
0.36	2,904	3.444	1.710		20.9	
0.34	2,742	3.647	1.915		22.1	
0.32	2,581	3.875	2.160		23.5	
0.30	2,420	4.133	2.446		25.1	
0.28	2,258	4.428	2.786		26.9	
0.26	2,097	4.769	3.202		29.0	
0.24	1,936	5.166	3.732		31.3	
0.22	1,774	5.636	4.425		34.0	
0.20	1,631	6.199	5.355		37.0	
0.19	1,532	6.526	5.960		38.6	
0.18	1,452	6.888	6.670		40.4	
0.17	1,371	7.293	7.461		42.5	
0.16	1,290	7.749	8.376		44.8	
0.15	1,210	8.266	9.441		47.1	
0.14	1,129	8.856	10.69		49.4	
0.13	1,049	9.537	12.21		52.2	
0.125	1,008	9.919	13.11		53.7	

TUNGSTEN (W) **XII**

Most measurements of the optical properties of W are in good agreement [1–35]. Table X lists the n and k values of Weaver et al. [26] and Henke et al. [36]. These data are plotted in Fig. 10 as smooth curves.

Weaver et al. [26] used W of unstated purity that had been melted into buttons by an electron beam. Samples were spark-cut from the buttons and after macroetching showed large (3-mm), unoriented, crystal grains. The samples were polished first by using an alumina abrasive and again, immediately prior to loading into the reflectometer, by electropolishing. During loading there was about a 3-min exposure to air. From 2818 Å to 8.3 μm a calorimetric technique [37] was used, and for wavelengths less than 2818 Å, the reflectance at room temperature was measured at $10°$ angle of incidence. These data were used with the absorption spectrum obtained by Haensel *et al.* [12] to do a Kramers–Kronig analysis.

There may be some influence on the reflectance in the VUV because of oxidation, but oxidation effects have never been studied. Above about 5 μm tungsten resembles a free-electron metal, but interband transitions persist to longer wavelengths. Infrared data do not cover a wide enough range to give a definitive limit for the shortest wavelength at which Drude parameters are expected to be free of the influence of interband transitions, but probably it is roughly 10 μm.

REFERENCES

1. S. Roberts, *Phys. Rev.* **114**, 104 (1959).
2. L. J. LeBlanc, J. S. Farrell, and D. W. Juenker, *J. Opt. Soc. Am.* **54**, 956 (1964).
3. C. Tingwaldt, U. Schley, J. Verch, and S. Takata, *Optik* **22**, 48 (1965).
4. H. R. Apholte and K. Ulmer, *Phys, Lett.* **22**, 552 (1966).
5. B. T. Barnes, *J. Opt. Soc. Am.* **56**, 1546 (1966).
6. I. Ye. Leksina and N. V. Penkina, *Fiz. Met. Metalloved.* **22**, 264 (1966); I. Ye. Leksina and N. V. Penkina, *Phys. Met. Metallogr.* **22**(2), 104 (1966).
7. A. P. Lenham and D. M. Treherne, *J. Opt. Soc. Am.* **56**, 1137 (1966).
8. A. P. Lenham and D. M. Treherne, *in* "Optical Properties and Electronic Structure of Metals and Alloys" (F. Abélès, ed.), p. 196, North-Holland, Publ., Amsterdam, 1966.
9. A. P. Lenham, *J. Opt. Soc. Am.* **57**, 473 (1967).
10. D. W. Juenker, L. J. LeBlanc, and C. R. Martin, *J. Opt. Soc. Am.* **58**, 164 (1968).
11. J. J. Carroll and A. J. Melmed, *Surf. Sci.* **16**, 251 (1969).
12. R. Haensel, K. Radler, B. Sonntag, and C. Kunz, *Solid State Commun.* **7**, 1495 (1969).
13. B. A. Konyaev, *Teplo. Vys. Temp.* **7**, 1029 (1969).
14. L. N. Latyev, V. Y. Cheknovskoi, and E. N. Shestakov, *High Temp.-High Press.* **2**, 175 (1970).
15. V. Vujnovic, *J. Opt. Soc. Am.* **60**, 177 (1970).
16. J. J. Carroll and A. J. Melmed, *J. Opt. Soc. Am.* **61**, 470 (1971).
17. W. R. Hunter, *Proc. Int. Conf. Vac. Ultraviolet Rad. Physics, 3rd, Tokyo*, p. 2aC2-1 (1971).
18. Y. P. Udoyev, N. S. Kozyakova, and M. L. Kapitsa, *Fiz. Met Metalloved.* **31**, 439 (1971); Y. P. Udoyev, N. S. Kozyakova, and M. L. Kapitsa, *Phys. Met. Metallogr.* **31**(2), 229 (1971).
19. J. T. Cox, G. Hass, J. B. Ramsey and W. R. Hunter, *J. Opt. Soc. Am.* **62**, 781 (1972).
20. L. N. Latyev, V. Ya. Checkovskoi and E. N. Shestakov, *High Temp.-High Press.* **4**, 679 (1972).
21. L. V. Nomerovannaya, M. M. Kirillova, and M. M. Noskov, *Fiz. Tverd. Tela.* **14**, 633 (1972); L. V. Nomerovannaya, M. M. Kirillova, and M. M. Noskov, *Sov. Phys. Solid State* **14**, 541 (1972).
22. J. J. Rogers, *High Temp.-High Press.* **4**, 271 (1972).
23. T. Smith, *J. Opt. Soc. Am.* **62**, 774 (1972).
24. L. N. Aksyutov. *Inzh.-Fiz. Zh.* **27**, 197 (1974); L. N. Aksyutov, *J. Eng. Phys.* **27**, 913 (1974).
25. G. A. Zhorov, *Teplo. Vys. Temp.* **10**, 1332 (1974).
26. J. H. Weaver, C. G. Olson, and D. W. Lynch, *Phys. Rev. B* **12**, 1293 (1975).
27. J. O. Hylton and R. L. Reid, *AIAA J.* **14**, 1303 (1976).
28. L. K. Thomas and S. Thurm, *J. Phys. F* **6**, 279 (1976).
29. J. H. Weaver and C. G. Olson, *Phys. Rev B* **14**, 3251 (1976).
30. B. V. Habedank, L. K. Thomas, and S. Thurm, *J. Phys. F* **7**, 2217 (1977).
31. H. S. Gurev and C. Selvage, *Proc. Soc. Photo-Opt. Instrum. Eng.* **85**, 32 (1977).

32. P. Gravier, G. Chassaing, and M. Sigrist, *Thin Solid Films* **57**, 93 (1979).
33. J. E. Nestell, Jr., and R. W. Christy, *Phys. Rev. B* **21**, 3173 (1980).
34. J. E. Nestell, Jr., R. W. Christy, M. H. Cohen, and G. C. Ruben, *J. Appl. Phys.* **51**, 655 (1980).
35. L. A. Wojeik, A. J. Sievers, G. W. Graham, and T. N. Rhodin, *J. Opt. Soc. Am.* **70**, 443 (1980).
36. B. L. Henke, P. Lee, T. J. Tanaka, R. L. Shimabukuro, and B. K. Fujikawa, *At. Data Nucl. Data Tables* **27**, 1 (1982).
37. L. W. Bos and D. W. Lynch, *Phys. Rev. B* **2**, 4567 (1970).

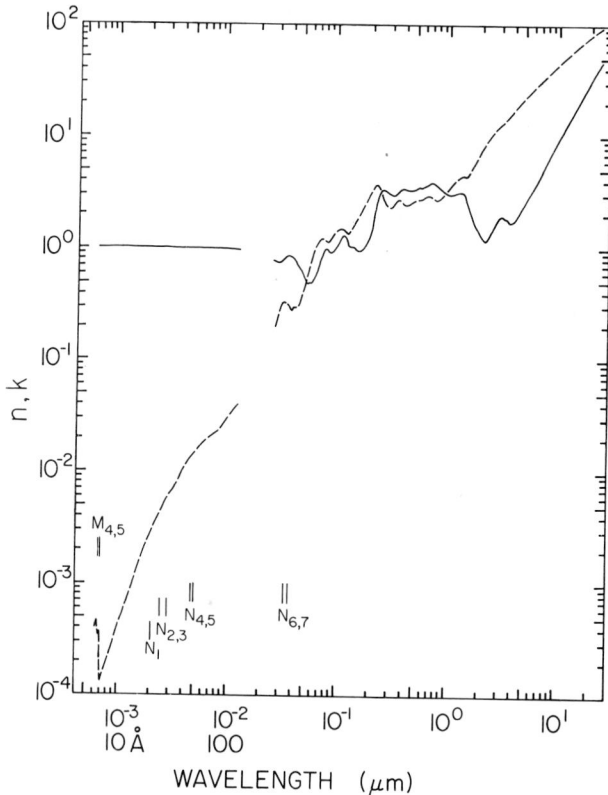

Fig. 10. Log–log plot of *n* (——) and *k* (----) versus wavelength in micrometers for tungsten.

TABLE X

Values of *n* and *k* Obtained from Various References for Tungsten[a]

eV	cm⁻¹	Å	n	k
2000		6.199	0.999560 [37]	4.00 x 10⁻⁴ [36]
1952		6.351	0.999637	4.34
1905		6.508	0.999721	4.71
1860		6.665	0.999810	3.44
1815		6.831	0.999908	3.69
1772		6.997	0.999656	1.31
1730		7.166	0.999536	1.42
1688		7.345	0.999458	1.53
1648		7 523	0.999390	1.65
1609		7.705	0.999330	1.78
1570		7.897	0.999273	1.92
1533		8.088	0.999218	2.07
1496		8.287	0.999161	2.23
1460		8.492	0.999106	2.41
1426		8.694	0.999051	2.59
1392		8.906	0.99899	2.79
1358		9.129	0.99893	3.01
1326		9.350	0.99887	3.24
1294		9.581	0.99881	3.50
1263		9.816	0.99874	3.77
1233		10.05	0.99868	4.05
1204		10.30	0.99861	4.35
1175		10.55	0.99854	4.68
1147		10.81	0.99846	5.03
1119		11.08	0.99838	5.41
1093		11.34	0.99831	5.81
1067		11.62	0.99822	6.24
1041		11.91	0.99814	6.72
1016		12.20	0.99805	7.22
992		12.50	0.99796	7.74
968		12.81	0.99786	8.31
945		13.12	0.99777	8.90
923		13.43	0.99767	9.53 x 10⁻⁴
901		13.76	0.99757	1.02 x 10⁻³
879		14.10	0.99746	1.09
858		14.45	0.99735	1.17
838		14.79	0.99724	1.25
818		15.16	0.99712	1.34
798		15.54	0.99700	1.44
779		15.92	0.99688	1.54
760		16.31	0.99678	1.65
742		16.71	0.99663	1.76
725		17.10	0.99651	1.87
707		17.54	0.99637	2.00
690		17.97	0.99625	2.13
674		18.39	0.99611	2.26
658		18.84	0.99598	2.41
642		19.31	0.99586	2.57
627		19.77	0.99574	2.71
612		20.26	0.99561	2.84
597		20.77	0.99547	2.98
583		21.27	0.99531	3.13
569		21.79	0.99513	3.29
555		22.34	0.99494	3.49

[a] The references from which the values were extracted are given in brackets.

TABLE X (*Continued*)

Tungsten

eV	cm^{-1}	Å	n	k
542		22.87	0.99478	3.70
529		23.44	0.99462	3.91
516		24.03	0.99446	4.14
504		24.60	0.99430	4.35
492		25.20	0.99414	4.58
480		25.83	0.99397	4.83
469		26.43	0.99382	5.07
458		27.07	0.99369	5.34
447		27.74	0.99357	5.57
436		28.44	0.99343	5.78
426		29.10	0.99326	5.97
415		29.87	0.99305	6.20
406		30.54	0.99284	6.40
396		31.31	0.99258	6.63
386		32.12	0.99229	6.95
377		32.89	0.99203	7.26
368		33.69	0.99177	7.59
359		34.53	0.99150	7.95
351		35.32	0.99126	8.29
342		36.25	0.99098	8.70
334		37.12	0.99072	9.10
326		38.03	0.99046	9.52
318		38.99	0.99020	9.97 x 10^{-3}
311		39.86	0.9900	1.04 x 10^{-2}
303		40.92	0.9897	1.09
296		41.88	0.9895	1.14
289		42.90	0.9893	1.19
282		43.96	0.9891	1.25
275		45.08	0.9890	1.30
269		46.09	0.9888	1.34
262		47.32	0.9887	1.38
256		48.43	0.9884	1.42
250		49.59	0.9882	1.46
244		50.81	0.9879	1.50
238		52.09	0.9876	1.55
232		53.44	0.9874	1.60
227		54.62	0.9871	1.64
221		56.10	0.9867	1.70
216		57.40	0.9864	1.75
211		58.76	0.9861	1.80
206		60.18	0.9858	1.84
201		61.68	0.9855	1.89
196		63.25	0.9852	1.94
191		64.91	0.9848	1.99
187		66.30	0.9845	2.02
182		68.12	0.9840	2.05
178		69.65	0.9835	2.09
174		71.25	0.9829	2.12
170		72.93	0.9823	2.15
166		74.69	0.9816	2.19
162		76.53	0.9808	2.23
158		78.47	0.9800	2.26
154		80.51	0.9791	2.30
150		82.65	0.9782	2.34

(*continued*)

TABLE X (*Continued*)

Tungsten

eV	cm^{-1}	Å	n	k
147		84.34	0.9774	2.39
143		86.70	0.9760	2.48
140		88.56	0.9745	2.56
136		91.16	0.9717	2.67
133		93.22	0.9690	2.75
130		95.37	0.9661	2.85
127		97.62	0.9647	2.94
124		99.98	0.9637	3.03
121		102.5	0.9627	3.13
118		105.1	0.9612	3.22
115		107.8	0.9594	3.33
112		110.7	0.9573	3.44
110		112.7	0.9558	3.51
107		115.9	0.9532	3.63
104		119.2	0.9502	3.76
102		121.5	0.9479	3.85
100		124.0	0.9452	3.95

eV	cm^{-1}	μm	n	k
47.0		0.02638	0.774 [26]	0.195 [26]
46.0		0.02695	0.762	0.205
45.0		0.02755	0.750	0.225
44.0		0.02818	0.746	0.245
43.0		0.02883	0.744	0.274
42.0		0.02952	0.756	0.304
41.0		0.03024	0.778	0.319
40.0		0.03100	0.797	0.328
39.0		0.03179	0.814	0.330
38.0		0.03263	0.826	0.330
37.0		0.03351	0.837	0.328
36.0		0.03444	0.846	0.320
35.0		0.03542	0.848	0.306
34.5		0.03594	0.844	0.298
34.0		0.03647	0.838	0.291
33.5		0.03701	0.829	0.285
33.0		0.03757	0.816	0.278
32.5		0.03815	0.797	0.286
32.0		0.03875	0.791	0.295
31.5		0.03936	0.788	0.294
31.0		0.04000	0.778	0.293
30.5		0.04065	0.765	0.292
30.0		0.04133	0.750	0.295
29.75		0.04168	0.742	0.296
29.5		0.04203	0.733	0.297
29.25		0.04239	0.724	0.298
29.0		0.04275	0.712	0.301
28.75		0.04313	0.700	0.306
28.5		0.04350	0.690	0.310
28.25		0.04389	0.679	0.318
28.0		0.04428	0.671	0.322
27.75		0.04468	0.656	0.328
27.5		0.04509	0.643	0.336
27.25		0.04550	0.631	0.346

TABLE X (*Continued*)

Tungsten

eV	cm^{-1}	μm	n	k
27.0		0.04592	0.619	0.357
26.8		0.04626	0.610	0.365
26.6		0.04661	0.600	0.375
26.4		0.04696	0.591	0.383
26.2		0.04732	0.580	0.394
26.0		0.04769	0.571	0.404
25.8		0.04806	0.560	0.415
25.6		0.04843	0.548	0.429
25.4		0.04881	0.537	0.444
25.2		0.04920	0.528	0.461
25.0		0.04959	0.521	0.476
24.8		0.04999	0.511	0.494
24.6		0.05040	0.503	0.513
24.4		0.05081	0.497	0.534
24.2		0.05123	0.491	0.554
24.0		0.05166	0.487	0.575
23.8		0.05209	0.483	0.598
23.6		0.05254	0.484	0.621
23.4		0.05299	0.486	0.643
23.2		0.05344	0.490	0.663
23.0		0.05391	0.494	0.679
22.8		0.05438	0.494	0.695
22.6		0.05486	0.493	0.712
22.4		0.05535	0.492	0.730
22.2		0.05585	0.491	0.751
22.0		0.05636	0.491	0.773
21.8		0.05687	0.492	0.796
21.6		0.05740	0.496	0.819
21.4		0.05794	0.499	0.842
21.2		0.05848	0.503	0.866
21.0		0.05904	0.508	0.892
20.8		0.05961	0.517	0.918
20.6		0.06019	0.525	0.943
20.4		0.06078	0.537	0.969
20.2		0.06138	0.549	0.994
20.0		0.06199	0.563	1.02
19.8		0.06262	0.579	1.05
19.6		0.06326	0.599	1.07
19.4		0.06391	0.621	1.09
19.2		0.06458	0.642	1.11
19.0		0.06526	0.665	1.13
18.8		0.06595	0.688	1.15
18.6		0.06666	0.712	1.16
18.4		0.06738	0.738	1.18
18.2		0.06812	0.766	1.19
18.0		0.06888	0.796	1.20
17.8		0.06965	0.822	1.20
17.6		0.07045	0.847	1.21
17.4		0.07126	0.874	1.21
17.2		0.07208	0.900	1.20
17.0		0.07293	0.922	1.20
16.8		0.07380	0.944	1.19
16.6		0.07469	0.959	1.18
16.4		0.07560	0.973	1.17

(*continued*)

TABLE X (*Continued*)

Tungsten

eV	cm⁻¹	µm	n	k
16.2		0.07653	0.980	1.15
16.0		0.07749	0.981	1.14
15.8		0.07847	0.976	1.13
15.6		0.07948	0.966	1.12
15.4		0.08051	0.951	1.12
15.2		0.08157	0.933	1.13
15.0		0.08266	0.917	1.14
14.8		0.08377	0.904	1.17
14.6		0.08492	0.900	1.20
14.4		0.08610	0.901	1.23
14.2		0.08731	0.906	1.25
14.0		0.08856	0.913	1.28
13.8		0.08984	0.923	1.32
13.6		0.09117	0.941	1.35
13.4		0.09253	0.962	1.37
13.2		0.09393	0.984	1.40
13.0		0.09537	1.01	1.42
12.8		0.09686	1.04	1.44
12.6		0.09840	1.07	1.46
12.4		0.09999	1.10	1.47
12.2		0.1016	1.13	1.48
12.0		0.1033	1.16	1.48
11.9		0.1042	1.18	1.48
11.8		0.1051	1.20	1.48
11.7		0.1060	1.21	1.48
11.6		0.1069	1.22	1.48
11.5		0.1078	1.24	1.47
11.4		0.1088	1.25	1.46
11.3		0.1097	1.26	1.45
11.2		0.1107	1.27	1.44
11.1		0.1117	1.28	1.43
11.0		0.1127	1.28	1.42
10.9		0.1137	1.29	1.41
10.8		0.1148	1.29	1.39
10.7		0.1159	1.28	1.38
10.6		0.1170	1.28	1.36
10.5		0.1181	1.26	1.34
10.4		0.1192	1.24	1.33
10.3		0.1204	1.22	1.33
10.2		0.1216	1.19	1.33
10.1		0.1228	1.16	1.33
10.0		0.1240	1.13	1.34
9.9		0.1252	1.11	1.36
9.8		0.1265	1.09	1.38
9.7		0.1278	1.07	1.41
9.6		0.1291	1.05	1.44
9.5		0.1305	1.04	1.47
9.4		0.1319	1.03	1.50
9.3		0.1333	1.03	1.52
9.2		0.1348	1.02	1.55
9.1		0.1362	1.02	1.57
9.0		0.1378	1.01	1.60
8.9		0.1393	1.02	1.63
8.8		0.1409	1.01	1.65

TABLE X (*Continued*)

Tungsten

eV	cm^{-1}	μm	n	k
8.7		0.1425	0.999	1.67
8.6		0.1442	0.986	1.70
8.5		0.1459	0.973	1.73
8.4		0.1476	0.956	1.76
8.3		0.1494	0.943	1.81
8.2		0.1512	0.935	1.86
8.1		0.1531	0.935	1.91
8.0		0.1550	0.937	1.95
7.9		0.1569	0.932	2.00
7.8		0.1590	0.930	2.06
7.7		0.1610	0.937	2.12
7.6		0.1631	0.950	2.18
7.5		0.1653	0.964	2.24
7.4		0.1675	0.979	2.29
7.3		0.1698	0.993	2.36
7.2		0.1722	1.01	2.43
7.1		0.1746	1.04	2.49
7.0		0.1771	1.06	2.56
6.9		0.1797	1.09	2.63
6.8		0.1823	1.12	2.70
6.7		0.1851	1.16	2.78
6.6		0.1879	1.21	2.87
6.5		0.1907	1.26	2.95
6.4		0.1937	1.31	3.04
6.3		0.1968	1.38	3.14
6.2		0.2000	1.47	3.24
6.1	49,200	0.2033	1.58	3.33
6.0	48,390	0.2066	1.70	3.42
5.9	47,590	0.2101	1.83	3.52
5.8	46,780	0.2138	2.00	3.61
5.7	45,970	0.2175	2.20	3.67
5.6	45,170	0.2214	2.43	3.70
5.5	44,360	0.2254	2.68	3.68
5.4	43,550	0.2296	2.92	3.58
5.3	42,750	0.2339	3.11	3.43
5.2	41,940	0.2384	3.27	3.27
5.1	41,130	0.2431	3.38	3.05
5.0	40,330	0.2480	3.40	2.85
4.9	39,520	0.2530	3.37	2.68
4.8	38,710	0.2583	3.33	2.57
4.7	37,910	0.2638	3.29	2.48
4.6	37,100	0.2695	3.24	2.41
4.5	36,290	0.2755	3.19	2.35
4.4	35,490	0.2818	3.13	2.32
4.3	34,680	0.2883	3.07	2.31
4.2	33,880	0.2952	3.01	2.33
4.1	33,070	0.3024	2.97	2.37
4.0	32,260	0.3100	2.95	2.43
3.95	31,860	0.3139	2.95	2.46
3.9	31,460	0.3179	2.96	2.50
3.85	31,050	0.3220	2.98	2.53
3.8	30,650	0.3263	2.99	2.56
3.75	30,250	0.3306	3.02	2.60
3.7	29,840	0.3351	3.05	2.62

(*continued*)

TABLE X (*Continued*)

Tungsten

eV	cm^{-1}	μm	n	k
3.65	29,440	0.3397	3.09	2.65
3.6	29,040	0.3444	3.13	2.67
3.55	28,630	0.3493	3.18	2.69
3.5	28,230	0.3542	3.24	2.70
3.45	27,830	0.3594	3.32	2.70
3.4	27,420	0.3647	3.39	2.66
3.35	27,020	0.3701	3.43	2.60
3.3	26,620	0.3757	3.45	2.55
3.25	26,210	0.3815	3.45	2.49
3.2	25,810	0.3875	3.43	2.45
3.15	25,410	0.3936	3.41	2.43
3.1	25,000	0.4000	3.39	2.41
3.05	24,600	0.4065	3.37	2.42
3.0	24,200	0.4133	3.35	2.42
2.95	23,790	0.4203	3.33	2.43
2.9	23,390	0.4275	3.32	2.45
2.85	22,990	0.4350	3.31	2.47
2.8	22,580	0.4428	3.30	2.49
2.75	22,180	0.4509	3.31	2.53
2.7	21,780	0.4592	3.31	2.55
2.65	21,370	0.4679	3.32	2.59
2.6	20,970	0.4769	3.34	2.62
2.55	20,570	0.4862	3.35	2.64
2.5	20,160	0.4959	3.38	2.68
2.45	19,760	0.5061	3.42	2.71
2.4	19,360	0.5166	3.45	2.72
2.35	18,950	0.5276	3.48	2.72
2.3	18,550	0.5391	3.50	2.72
2.25	18,150	0.5510	3.50	2.73
2.2	17,740	0.5636	3.49	2.75
2.15	17,340	0.5767	3.51	2.81
2.1	16,940	0.5904	3.54	2.84
2.05	16,530	0.6048	3.57	2.86
2.0	16,130	0.6199	3.60	2.89
1.95	15,730	0.6358	3.65	2.92
1.9	15,320	0.6526	3.70	2.94
1.85	14,920	0.6702	3.76	2.95
1.8	14,520	0.6888	3.82	2.91
1.75	14,110	0.7085	3.85	2.86
1.7	13,710	0.7293	3.84	2.78
1.65	13,310	0.7514	3.78	2.72
1.6	12,900	0.7749	3.67	2.68
1.55	12,500	0.7999	3.56	2.73
1.5	12,100	0.8266	3.48	2.79
1.45	11,690	0.8551	3.38	2.85
1.4	11,290	0.8856	3.29	2.96
1.35	10,900	0.9184	3.20	3.07
1.3	10,490	0.9537	3.12	3.24
1.25	10,080	0.9919	3.05	3.39
1.2	9,679	1.033	3.00	3.64
1.15	9,275	1.078	3.02	3.83
1.1	8,872	1.127	3.05	4.04
1.05	8,469	1.181	3.11	4.19
1.0	8,065	1.240	3.14	4.32

TABLE X (*Continued*)

Tungsten

eV	cm^{-1}	μm	n	k
0.98	7,904	1.265	3.15	4.36
0.96	7,743	1.291	3.15	4.41
0.94	7,582	1.319	3.15	4.43
0.92	7,420	1.348	3.14	4.45
0.90	7,259	1.378	3.11	4.44
0.88	7,098	1.409	3.05	4.42
0.87	7,017	1.425	3.00	4.39
0.86	6,936	1.442	2.92	4.37
0.85	6,856	1.459	2.80	4.33
0.84	6,775	1.476	2.62	4.37
0.83	6,694	1.494	2.47	4.47
0.82	6,614	1.512	2.36	4.61
0.80	6,452	1.550	2.22	4.85
0.78	6,291	1.590	2.12	5.05
0.76	6,131	1.630	1.97	5.27
0.74	5,969	1.675	1.83	5.52
0.72	5,807	1.722	1.70	5.81
0.70	5,646	1.771	1.59	6.13
0.68	5,485	1.823	1.52	6.45
0.66	5,323	1.879	1.45	6.78
0.64	5,162	1.937	1.36	7.14
0.62	5,001	2.000	1.28	7.52
0.60	4,839	2.066	1.21	7.96
0.58	4,678	2.138	1.18	8.44
0.56	4,517	2.214	1.20	8.94
0.54	4,355	2.296	1.23	9.45
0.52	4,194	2.384	1.30	9.98
0.50	4,033	2.480	1.40	10.5
0.48	3,817	2.583	1.54	11.1
0.46	3,710	2.695	1.69	11.6
0.44	3,549	2.818	1.82	12.1
0.42	3,388	2.952	1.92	12.6
0.40	3,226	3.100	1.94	13.2
0.38	3,065	3.263	1.86	13.9
0.36	2,904	3.444	1.82	14.7
0.34	2,742	3.647	1.71	15.7
0.32	2,581	3.875	1.72	16.9
0.30	2,420	4.133	1.83	18.3
0.25	2,016	4.959	2.56	22.4
0.20	1,613	6.199	3.87	28.3
0.18	1,452	6.888	4.72	31.5
0.16	1,290	7.749	5.92	35.3
0.14	1,129	8.856	7.58	40.2
0.12	967.9	10.33	10.1	46.4
0.10	806.5	12.40	14.1	54.7
0.09	725.9	13.78	17.0	60.0
0.08	645.2	15.50	20.9	66.3
0.07	564.6	17.71	26.5	73.8
0.06	483.9	20.66	34.5	82.9
0.05	403.3	24.80	46.5	93.7

The Optical Properties
of Metallic Aluminum*

D. Y. SMITH

Argonne National Laboratory
Argonne, Illinois
and
Max-Planck-Institut für Festkörperforschung
Stuttgart, Federal Republic of Germany

E. SHILES† and MITIO INOKUTI

Argonne National Laboratory
Argonne, Illinois

GENERAL FEATURES   I

The optical properties of metallic aluminum are among the most widely measured and analyzed of any material [1–7]. They are dominated by three practically nonoverlapping groups of electronic transitions corresponding to absorptions by conduction band, L-shell, and K-shell electrons. The absorption spectrum for the crystalline solid is shown in Fig. 11.

The 3 e/at. (electrons/atom) donated to the conduction band by the atomic $3s^2 3p$ valence levels give rise to a typical metallic absorption from zero to ~ 15 eV, the bulk plasmon frequency ω_p. In crystalline samples this portion of the spectrum is dominated by intraband transitions in the far infrared and by two strong interband absorptions at ~ 0.5 eV ($\lambda \approx 2.5$ μm) and ~ 1.5 eV ($\lambda \approx 0.8$ μm). In the noncrystalline or partially crystalline solid state and in the liquid state, the interband spectrum is strongly modified or absent; this effect will be discussed further in Subsection II.C.

Intraband Spectrum   A

The intraband contribution to the optical properties can be described phenomenologically to within experimental error with a Drude-model dielectric

---

* Work supported by the U.S. Department of Energy.
† Present address: Gulf Research and Development Company, Houston, Texas.

369

HANDBOOK OF OPTICAL CONSTANTS OF SOLIDS

function [8, 9] for the electron gas:

$$\varepsilon_{\text{Drude}} = \varepsilon_0 - [\Omega_{\text{p}}^2/\omega(\omega + i/\tau)], \tag{1}$$

where the contribution to the dielectric function from core interband transitions has been included in ε_0. Here Ω_{p} is the (phenomenological) plasma frequency for intraband transitions and τ the intraband relaxation time. The quantity ε_0 is effectively constant throughout the region of conduction–electron absorption (that is, to well above the plasma frequency) with a value [7] of 1.03_5.

In addition to using the quantity Ω_{p}, the strength of the intraband transitions is commonly expressed in terms of the effective number density of electrons participating in intraband transitions

$$n_{\text{eff,intraband}} = m_e\Omega_{\text{p}}^2/4\pi e^2. \tag{2}$$

Here m_e and e are the free-electron mass and charge, respectively. A second alternative is the optical mass defined by [8, 9]

$$m_{\text{opt}} = m_e(n_c/n_{\text{eff,intraband}}), \tag{3}$$

where n_c is the density of conduction electrons.

Similarly, the intraband relaxation time is often given in terms of the damping coefficient

$$\gamma = \tau^{-1}. \tag{4}$$

A quantity often cited for comparison with electrical experiments is the dc conductivity in the Drude model

$$\sigma(0) = \Omega_{\text{p}}^2/4\pi\gamma. \tag{5}$$

A variety of fits of Eq. (1) have been made to infrared data for polycrystalline samples of metallic aluminum [1, 3, 5, 10–15], but early work was incomplete because the interband transitions at ~ 0.5 and ~ 1.5 eV were not always reckoned with. Some of the more recent determinations of the Drude parameters and the experimental bulk dc conductivity are given in Table XI. The "0.5-eV" interband transition was not accounted for by Ehrenreich et al. [1] or by Powell [3], so that some interband absorption strength, as well as damping, is included in their Drude terms. This leads to a conductivity significantly smaller than the measured bulk value. Bennett and Bennett [12] assumed the measured value of $\sigma(0)$ but also neglected the 0.5 eV transition leading to an overestimate of Ω_{p}. Dresselhaus et al. [13], Mathewson and Myers [5], and Benbow and Lynch [14] accounted for both interband transitions, the last two studies with the Ashcroft–Sturm [18] model of the interband spectrum. The values reported by Smith and Segall [15] were derived by fitting the dielectric function of Shiles et al. [7] over the range of 0.04 to 3.0 eV, assuming that it is a linear superposition of a Drude term and the two principal interband transitions. No detailed assumptions as to the form of the interband spectrum were made. The resulting parameters thus yield the least model-dependent description of the room-temperature interband absorption presently available.

Interband Spectrum B

The two strong interband transitions of the conduction-electron spectrum in the crystalline material are a consequence of the "parallel-band" effect [1, 18–20] that occurs in almost free-electron polyvalent metals. In these materials the occupied and unoccupied conduction bands are effectively parallel over substantial regions of k space in the vicinity of high-symmetry planes parallel to the zone faces. Allowed transitions between these parallel bands lead to prominent interband absorptions at energies approximately twice the Fourier component of the effective crystal potential V_K for the reciprocal lattice vector K corresponding to the face in question. In aluminum, absorption between almost parallel bands occurs in the neighborhood of surfaces parallel to the hexagonal (111) and the square (200) zone faces. The corresponding energy gaps predicted by a two-band model [18] using parameters derived by Ashcroft [21] from de Haas–van Alphen measurements below 4 K are $2V_{111} = 0.487$ eV and $2V_{200} = 1.53$ eV. The parallel-band absorption shows sharp rises near these energies followed by long tails toward higher energies that joins smoothly onto the band-to-band absorption arising from the remainder of the Brillouin zone [18, 20]. In addition, theory predicts [22, 23] an accidental degeneracy of the energy bands leading to a weak but finite interband absorption down to zero energy. However, this has not been verified experimentally because the interband component is overwhelmed by the intraband absorption below a few tenths of an electron volt.

The 1.5-eV (0.8-μm) interband absorption is readily apparent as a pronounced drop in the reflectance for crystalline samples (Fig. 12) and as a small peak in the extinction coefficient (Fig. 11). The former appears to have been first reported by Strong [24] and the latter by Schulz [25]. The absorption responsible for these features was identified as a transition between pairs of parallel bands by Ehrenreich et al. [1]. Shklyarevskii and Yarovaya [26] independently identified the absorption as an interband transition but did not recognize the crucial importance of energy-band parallelism near the Fermi energy.

The second major interband absorption at ~0.5 eV is almost completely hidden by the Drude absorption. There is a suggestion of it in the early infrared measurements of Beattie [27] and in the ultrahigh vacuum (uhv) reflectance data of Bennett et al. that show a small drop in reflectance near 25 μm. (see Bennett et al. [28] Fig. 3). This transition was first definitively observed in low-temperature absorbtivity experiments by Bos and Lynch [29]. Figure 13 shows subsequent results of Benbow and Lynch [14] for single-crystal aluminum at 4.2 K together with Smith and Segall's [15] separation of room-temperature data for uhv evaporated films. The results are presented in terms of the real part of the interband conductivity

$$\sigma_{\text{interband}}(\omega) = (-i\omega/4\pi)[\varepsilon(\omega) - \varepsilon_{\text{Drude}}(\omega)]. \tag{6}$$

Below approximately 0.3 eV the uncertainty in the experimental data and in the Drude parameters is so large that the interband contribution cannot be determined reliably from the available measurements.

Note the shift of the interband absorption toward higher energies as the temperature is lowered. Mathewson and Myers [5] studied this temperature dependence in detail for the "1.5-eV" transition (see also Fig. 17).

At low temperatures (~ 20 K) a weak fine structure is superimposed on the tail of the 1.5-eV absorption in the range 2–2.5 eV [30]. Some of these structures appear to arise [31] from transitions near the corners of the Brillouin zone in the vicinity of the intersections of the two groups of planes responsible for the two strong parallel-band absorptions. Along these lines of intersection three bands are almost parallel with band gaps determined by both V_{200} and V_{111}. However, the transitions arising from this "trio" of parallel bands along the intersections are not as prominent as those associated with the "duos" of parallel bands in the planes because of the smaller momentum space involved in the neighborhood of the lines of intersection. The result is two weak absorptions, one near the strong 0.5-eV interband transition and the second just above the strong 1.5-eV transition. These have been predicted in detail by band-structure calculations [23, 31]. (See also Hunderi and Nilsson [30].)

Further interband fine structure has been resolved in recent piezo-modulation experiments by Jiles and Staines [31a]. This includes strain-induced changes around 2–2.5 eV and 4.6 eV, in addition to structure near the 1.5 eV parallel-band transition.

The possibility that excitons may be associated with the interband absorption edges has been investigated theoretically [32]. While excitonic states are predicted, their binding energies are found to be only 5 (singlet) and 20 (triplet) meV as a result of the highly effective screening of the coulomb interaction by the conduction electrons. Such structure is unobservable in conventional optical experiments but may contribute to the broadening of the interband peaks.

II   OPTICAL MEASUREMENTS AND SAMPLE CONDITIONS

A   Polished Polycrystalline Samples

Serious studies of the optical properties of aluminum were begun in 1874 by Quincke [33]. Early studies employed polished bulk or rolled samples [33–40]. Even though bulk aluminum takes a "high polish," it always retains a "hazy white surface" [36] and has a reflectance well below that of an uncontaminated evaporated film. As will become apparent, this is presumably the result of both scattering and absorption by the microscopically rough, oxide-coated surface. A composite of reflectance data for polished surfaces is given in Fig. 12. The curve generally follows the data of Coblentz [36] (IR), Luckiesh [38] (visible + UV), Quincke [33] (visible), and Wulff [40] (visible + UV). The polished bulk-metal data are strongly affected by surface contamination and light scattering, particularly at shorter wavelengths. In the visible, Nutting [35] reports reflectance values $\sim 10\%$ higher than those shown, while values measured by Coblentz and Stair [39] are 10% lower. An even more dramatic range is given by Taylor [41], who quotes reflectances from 33 (mill finish) to 84% (electropolished) at 296.7 nm.

Thin-Film Polycrystalline Samples   B

The key factor in obtaining accurate values of the optical properties was the discovery that aluminum could be sputtered [42] or evaporated [43] to form highly reflecting metal films. Early sputtered films were described as "brilliant" [44, 45]. However, evaporation by the tungsten-coil method [46, 47] has become the preferred method of sample preparation because of its high speed and potential for low contamination. The major impetus for these developments was the production of interferometer and telescope mirrors [24]. (The superior reflectance of evaporated aluminum films over silver at short wavelengths extended stellar spectroscopy some 250 Å into the violet.)

Evaporated films show reflectances much higher than polished surfaces particularly in the ultraviolet. Generally, the higher the vacuum and the faster the evaporation, the higher is the reflectance [48–51]. Two curves for evaporated surfaces are shown in Fig. 12. The upper is for unoxidized, smooth films produced in ultrahigh vacuum (10^{-9}–10^{-10} torr), and the other is for films evaporated in conventional high vacuum ($\sim 10^{-5}$ torr). The first is believed to correspond to a smooth, uncontaminated aluminum surface. The second is for more-or-less contaminated surfaces having oxide coating with no special attempts made to eliminate surface roughness. The uhv curve generally follows the data of Bennett *et al.* [28], Hass and Waylonis [51] (though not uhv films, a high evaporation rate was employed that has been shown to yield the same results as an uhv preparation [52]), and Endriz and Spicer [53]. Beyond 11.8 eV the curve follows the analysis of Shiles *et al.* [7], which relies heavily on electron-energy-loss measurements [54] near ω_p and on transmission and reflectance data of Ditchburn and Freeman [55] and x-ray absorption measurements[†] beyond ω_p.

The high-vacuum curve generally follows the data of Beattie [11], Strong [24], Hass [48, 56], Sabine [57], Banning [58], and Walker *et al.* [59]. The

---

[†] For a list of sources used for x-ray cross sections, see Shiles *et al.* [7].

curve shown is in close agreement from 250 to 2500 nm with the U.S. National Bureau of Standards' reference standard first-surface aluminum mirror #2003a [60], which has an aged oxide coating. The curve also generally has the same reflectance from 200 to 2000 nm as commercially available evaporated mirrors.[†] Films evaporated under higher vacuum or at high deposition rates generally have reflectances between the high-vacuum and uhv curves shown. (For references to additional measurements see Shiles et al. [7] and the sources cited in Schopper [61].)

The differences between the various evaporated film data arises from surface oxide layers [48, 62–66], residual gas incorporated in the film [49, 50], surface roughness [53, 67, 68], and film morphology [52, 69–71].

1 Surface Oxide Layers

A layer of Al_2O_3 forms rapidly on an aluminum surface even in high vacuum [72, 73]. The layer is both adherent and relatively impervious so that once formed it protects the underlying metal from further oxidation [74]. The thickness of the surface layer and its composition depend strongly on sample structure and history, particularly on chemical treatment [24, 52, 69, 71, 74, 75]. Polarimetric studies of evaporated films simply exposed to the atmosphere indicate that the surface oxide layer is from 20 to 55 Å thick [50, 62–66] but that thicker layers readily form in moist environments [74]. The surface layers are probably hydrated to varying degrees depending on formation conditions. Preparation conditions [48, 52, 76], purity [76], structure [69, 71], partial oxidation in vacuum before exposure to the atmosphere [65], etc.; all influence the reported thickness. In addition, exposure of aluminum surfaces in air to ultraviolet light has been found to decrease their reflectance at short wavelengths; this is attributable to an increase in the oxidation rate of the surface [48, 69, 76, 77].

The oxide of aluminum is highly transparent from approximately 6 μm (0.2 eV) [78] to 180 nm (6.8 eV) [6, 79], so that a surface oxide film does not decrease the reflectance significantly over this range. Moreover, in the infrared a light wave does not "see" the film [48]. (The incident and reflected electric waves must be out of phase by approximately π to create the required node in the electric field at the metal surface. Thus, provided that the oxide thickness is much less than the wavelength of light, the electric fields of the incident and reflected waves almost cancel in the oxide layer and coupling to the layer is negligible.) However, at the shorter wavelengths of the vacuum ultraviolet, the oxide layer has a strong effect. For example, at 121.6 nm

[†] The authors are indebted to Professor Wayne Major of the University of Richmond for providing measurements for a commercially supplied, aluminum-film (~1000-Å) reflector on optical glass prepared by D. and E. Technology, Santa Clara, California. Evaporation was performed at ~10^{-6} torr.

(~ 10 eV) an oxide layer 17 Å thick decreases the reflectance by a factor of two [63, 76].

Considerable research has been devoted to developing thin-film dielectric coatings to preserve or enhance the optical qualities of aluminum front-surface mirrors. For many applications, the natural oxide film is either too thin to offer the required protection from chemical change and mechanical damage—particularly when frequent cleaning is needed—or too absorbing for use in the ultraviolet. Coatings of the silicon oxides and Al_2O_3 that offer a high degree of mechanical and chemical protection have been developed [48, 80, 81] for use in the IR, visible, and near UV. At high angles of incidence, silicon oxide coatings show a reduced reflectance in the infrared. For such applications, protective coatings of Y_2O_3 and HfO_2 have been found to be highly successful [82]. At shorter wavelengths, overcoatings of MgF_2 and LiF preserve the high reflectance of aluminum down to ~ 1000 Å by preventing surface oxidation [48, 83]. Multiple dielectric layer coatings to enhance reflectance over a limited wavelength range have also been developed; enhancement of a 90% reflector to 99% reflectance have been reported [84].

Bulk Inclusion of Residual Gas 2

Studies of films evaporated at different rates and pressures indicate that the optical constants are very sensitive to residual gas incorporated into the bulk of the films [49, 50], specifically to the film composition within the penetration depth [50]. This is generally attributed [48–50, 85] to oxide formation arising from the gettering of molecular oxygen or to the reduction of water vapor to form Al_2O_3 and hydrogen. The relevant parameter appears to be the ratio of the residual gas pressure to the deposition rate, which is a measure of the arrival rate of residual gas molecules to that of aluminum atoms [49, 50]. Films deposited from high-purity starting material at terminal pressure-to-rate ratios of the order 10^{-8}–10^{-9} torr min Å^{-1} or less [50] have similar reflectances and appear to be uncontaminated for the purpose of optical property measurements, provided that subsequent oxide formation is avoided.

The presence of surface effects has also been demonstrated in x-ray absorption measurements in the neighborhood of the L edge [86, 87]. This has not been investigated in detail, but it has been associated with an apparent systematic overestimate of the L-shell absorption [7].

Surface Roughness and Morphology 3

An extremely important factor in determining effective optical properties, particularly in the ultraviolet, is the roughness of the sample surfaces [53, 67, 68]. Surface roughness causes both scattering [88] and coupling to surface plasmons [89] (the latter lie at an energy of $\hbar\omega_p/\sqrt{2} \approx 10.6\,\text{eV}$ for an aluminum-vacuum interface). Reflectance curves for two low-contamination films [53, 90, 91] with slight surface roughness are shown in Fig. 12 for comparison with the smooth, uhv surface reflectance. The main reflectance drop lies near the surface-plasmon frequency, but the drop extends well into the visible, largely as a result of scattering. Even a slight surface roughness produces dramatic effects. For example, a film with an rms surface roughness of only 27 Å is reported [53] to exhibit a reflectance drop of 20% at 3 eV (400 nm). These effects account for the wide variation of reflectance found in the UV and often previously attributed to contamination.

Film morphology, particularly grain size and orientation, has a substantial effect on optical properties. This is particularly apparent in aluminum films deposited by evaporation at oblique angles of incidence [92–94]. The optical properties of these films are anisotropic and the films develop diffusely reflecting surfaces with increasing thickness as a result of an increasingly coarse microstructure [80]. During the initial stages of deposition, films prepared in good vacuum systems appear to be compact and to have relatively small grain size, regardless of deposition angle; but beyond a critical film thickness diffuse scattering becomes substantial. For example, aluminum films 300 to 400 Å thick show surface roughness (as measured by scattering in the visible) for deposition angles of 60° or more. Holland [92] showed that the critical film thickness for diffuse reflection decreases sharply as the angle of incidence increases. These properties result from the preferential growth of crystallites in the form of columns parallel to the vapor beam [95] that are "self-shadowing." The amount and nature of the residual gas in the evaporating system have a major effect on the microstructure and, hence, the optical properties both because of gettering [96] and scattering of the metal vapor that alters the angle of incidence on the substrate [94].

C Noncrystalline and Liquid Samples

Vapor-quenched solid and liquid aluminum exhibit typical metallic properties but with a wide range of modifications in the interband spectrum. In the liquid state, aluminum exhibits no interband structure [97]. Rather, the conduction-electron absorption appears to follow a free-electron Drude model to within experimental error. This is in agreement with the expectation that in the liquid state there is no long-range order and, hence, no well-defined band structure nor Brillouin-zone boundaries [98].

Similar properties might be anticipated for samples with a low degree of crystallinity. However, the optical spectrum of vapor-quenched aluminum films deposited at low temperature exhibits a complex behavior [5, 99]. Films deposited at 25, 140, and 198 K show a progressive reduction in the intensity of the 1.5-eV interband absorption with decreasing substrate temperature. Essentially no 1.5-eV absorption remains in films deposited at 25 K [99]. The reduced 1.5-eV absorption does not show a shift in position or broadening, only a decrease in magnitude. Further, annealing of the films at room temperature restores the 1.5-eV absorption. These observations are taken [5] to indicate that deposition from the vapor on a cooled substrate yields a two-phase system that is partly crystalline and partly amorphous but that crystallizes on annealing.

This interpretation cannot be complete since it is not consistent with analysis [99] of the 25-K films. These films show no 1.5-eV interband component, and consequently they would be considered to be noncrystalline in this interpretation. However, their optical properties are not described by the Drude model as those of the liquid can be. Rather, the conduction–electron absorption appears to consist of a Drude term plus a residual absorption near 0.5 eV that is attributed to interband transitions characteristic of a $2V_{111}$ energy gap. A possible explanation [99] is that the film consists of small-diameter, close-packed planar clusters with characteristic (111) translational symmetry but lacking the (200) translational symmetry of bulk aluminum.

TABULATED DATA   III

Room-Temperature Optical Constants   A

The most reliable optical data presently available are for uncontaminated polycrystalline aluminum films prepared in uhv. Measurements generally are made in reflectance at room temperature (18–25°C) and are presently available over the range 0.04 eV (32 μm) to 11.8 eV (105 nm) [28, 53]. In addition, a number of ellipsometric studies [5, 65, 71, 100, 101] of thin films in the near IR and visible have recently become available, particularly through the studies of Mathewson and Myers [5, 100] and of Liljenvall et al. [101]. (The former also includes high- and low-temperature measurements.) Studies [14, 29] of electropolished polycrystalline and single-crystal samples have also been reported, but absolute optical properties were not measured. The results are, however, consistent with absolute measurements on thin films.

Ellipsometric measurements yield the optical constants directly, but reflectance measurements at normal incidence must be analyzed by using dispersion theory. This has recently been done over a wide spectral range for the room-temperature uhv reflectance measurements by Shiles et al. [7], who used a self-consistent Kramers–Kronig analysis. This involved combining the reflectance measurements below 11.8 eV with (1) electron-energy-loss measurements in the vicinity of the plasma frequency and (2) transmission and angular-variation-of-reflectance measurements in the extreme UV and x-ray regions. The energy-loss measurements were mainly those of Gibbons et al. [54]. The higher-energy optical data were taken primarily from Ditchburn and Freeman [55], Haensel et al. [102], Gähwiller and Brown [103], Fomichev et al. [104–106], Singer [107], Ershov et al. [108, 109], Cooke and Stewardson [110], Bearden [111], Henke et al. [112, 113], Singman [114], Lublin et al. [115], Hubbel et al. [116], and Davisson [117]. These additional data were for room temperature but were not for uhv samples and may suffer from surface effects to some degree, particularly in the transmission measurements on thin films near the $L_{II,III}$ edge [86, 87] and to a lesser degree between the plasmon energy and the $L_{II,III}$ edge [104] (see later in this section).

Since the Kramers–Kronig relations require knowledge of one optical function over the entire spectrum, the analysis proceeded by successive approximations. A trial reflectance function was constructed by employing estimates from transmission, electron-energy-loss, and similar data above 11.8 eV. This was then analyzed and the resulting optical functions compared with the input. The trial function was successively modified—primarily by the substitution of experimental data where available—until the calculated and measured optical functions agreed.

Throughout, the results were checked against a number of optical sum rules to ensure agreement with independent theoretical and experimental constraints (see Chapter 3 of this volume). This disclosed that the reported oscillator strength in the L-absorption region was consistently too high by $\sim 14\%$. The absorption coefficient in this region has generally been measured in transmission by using thin films that exhibit a surface component to their absorption [86, 87]. To compensate for this, an *ad hoc* reduction was made in the data based on the thin-film measurements of Lukirskii et al. [118], Fomichev and Lukirskii [104, 105], Haensel et al. [102], and Gähwiller and Brown [103]. This brought the calculated absorption coefficient into agreement with both the f sum rule and later measurements by Balzarotti et al. [87], who separated surface and bulk absorption. While the data of Balzarotti et al. [87] extend only 10 eV above the L edge, the $k(\omega)$ values are consistently some 15% below those used by Shiles et al. [7] in preparing their original input in this range. This is in agreement with the conclusion that previous measurements of the absorption cross section in the neighborhood of the $L_{II,III}$ edge were systematically too large.

While this correction is very important in the soft x-ray region of the spectrum, it was found to have negligible effect on the optical constants below the plasma frequency as determined by the Kramers–Kronig analysis.

The results of the analysis for the refractive index and the reflectance between the infrared and the $L_{II,III}$ edge are listed in Table XII. Given for comparison are the *in situ* uhv ellipsometric measurements of Mathewson and Myers [100], the uhv reflectance data of Bennett *et al.* [28], and the vacuum ultraviolet transmission and reflectance-angular-variation measurements of Ditchburn and Freeman [55]. The index results are graphed in Fig. 14. It will be seen that the Kramers–Kronig analysis is in good agreement with the ellipsometric measurements of Mathewson and Myers [100] in the visible and the IR. The differences—less than 11% in $n(\omega)$ and 8% or less in $k(\omega)$—give an indication of the uncertainty of the data.

A large number of optical constant measurements [10, 11, 25–27, 51, 55, 56, 65, 66, 73, 75, 91, 119–129] have been reported for films prepared on normal substrates in high vacuum and subsequently exposed to the air. The resulting "effective" optical constants describe good but contaminated aluminum surfaces made with no special attempt to produce ultrasmooth films. A number of these measurements are compared with values derived from uhv films in Fig. 14. Over most of the spectrum the effective optical constants for the oxidized surface lie below those for the uhv films. This is most pronounced in the IR, where effective optical constants 25% or more smaller than the uhv values are common. In the visible the effective constants are generally 10 to 15% lower than the uhv values, while in the UV the effective constants are surprisingly close to the uhv values. The exception is for the effective $n(\omega)$ near the surface plasmon frequency ω_s. Here there is a "bump" in the effective index, presumably arising from dispersion associated with the apparent "absorption" of the surface plasma and light scattering by films with rough surfaces.

Reflectance at Normal Incidence   B

The reflectance for smooth unoxidized films at normal incidence as calculated from the results of Shiles *et al.* [7] by using the Fresnel relation

$$R(\omega) = \frac{[n(\omega) - 1]^2 + k^2(\omega)}{[n(\omega) + 1]^2 + k^2(\omega)}, \tag{7}$$

is given in Table XII and Fig. 15. Also given in Table XII are the direct reflectance measurements of Bennett *et al.* [28] and the reflectance calculated from the ellipsometric measurements of Mathewson and Myers [100]. These data cover the infrared and visible and apply to films prepared in uhv. Values of the reflectance in the vacuum ultraviolet are also given; these were taken from the work of Ditchburn and Freeman [55] on films evaporated in high vacuum and are likely to suffer from surface-roughness and surface-oxide effects. (See Fomichev and Lukirskii [104].)

The last column of Table XII gives a composite set of reflectance data for the more common high-quality evaporated aluminum mirrors that are prepared in conventional high vacuum (10^{-5}–10^{-6} torr) and subsequently exposed to the atmosphere. These same data are plotted as the curve marked high-vacuum films in air in Fig. 12. The properties of such films vary widely depending on formation conditions, atmospheric exposure history, and substrate surface roughness. The composite given is for films formed by rapid evaporation in high vacuum on optically polished (but not supersmooth) substrate and subsequently exposed to clean dry air. Such films generally show reflections $\sim 1\%$ lower than smooth oxide-free films in the infrared and visible. In the ultraviolet, their reflectance drops rapidly from the smooth oxide-free film value as a result of scattering and surface-plasmon excitation. A wide range of reflectance is reported at these wavelengths because surface roughness is generally not controlled.

C   Room-Temperature Dielectric Function

The dielectric function covering the complete energy range involved in the Kramers–Kronig analysis is given in Table XIII and Fig. 16. This set of data includes the 14% reduction in the high-energy side of the $L_{II,III}$ edge needed to obtain agreement with the f sum rule and recent transmission measurements [87] taking surface effects into account. The energy intervals used in the table have been chosen to give closely spaced points in regions of pronounced structure such as interband absorption and x-ray edges. Intermediate points can be found by interpolation noting that the dielectric function generally has a power-law dependence on energy in regions far removed from interband edges. Near the edges the energy spacing is too course to show all the x-ray fine structure. For details of this the reader is referred to Balzarotti et al. [87] and Haensel et al. [130] for L-edge structure and to Kiyono et al. [131] for K-edge spectra.

The $\varepsilon_2(\omega)$ curve in Fig. 16 discloses unexpected discontinuities in slope: a small discontinuity at approximately 7 eV and a larger discontinuity near 15 eV. These are not readily apparent in the other optical functions, specifically in $n(\omega)$ and $k(\omega)$. These features may be extraneous, arising from the joining of data from different sources. This is most probably the case for the discontinuity near 7 eV when additional uncertainty was introduced from reading published graphic data.

However, the situation is not as clear for the larger discontinuity near 15 eV (the plasmon energy). Here the ε_2 data fall off significantly faster below 15 eV than in the interval between 15 eV and the $L_{II,III}$ edge. This is contrary to the simple expectation, based on single-particle excitations, that at photon energies high compared to the principal interband transition energies, the valence electrons should behave almost as free electrons so that their contribution to ε_2 should be approximately Drude-like; i.e., $\varepsilon_2(\omega) \sim \omega^{-3}$. A possible explanation is that the change in slope is an experimental artifact: below 15 eV, the $\varepsilon_2(\omega)$ data are based primarily on samples prepared in uhv, whereas from 15 eV to the $L_{II,III}$ edge, the samples were prepared in conventional vacua and the measurements were made in air. Over the latter range, surface contamination could lead to significant errors in the optical constants reported by Ditchburn and Freeman [55]. Such an error cannot be detected as a violation of the f sum rule to within the accuracy of present data, since the contribution to the f sum rule over the region in question is of the order of a few tenths of an electron per atom.

A second and more intriguing possibility is that there is a deviation of the optical properties from the random-phase-approximation values above the plasmon energy. Just such an effect has been proposed by Hopfield [132] as a result of the dynamic screening of the phonon—or disorder—contribution to the optical absorption. A third possibility [133] is the excitation of bulk plasmons by oscillating charges induced at the metal surface by the exciting light; however, studies [134] of the surface photoelectric effect suggest the latter process is negligible. New measurements on well-characterized uhv samples in this energy range are needed to clarify this situation. Such experiments should be possible with the increasing availability of synchrotron radiation sources. Recent inelastic electron-scattering studies by Livins et al. [134a] confirm a deviation from Drude behavior near the plasmon energy that is consistent with Hopfield's suggestion.

Temperature Dependence   D

The effect of temperature on the optical properties of aluminum has been studied directly with ellipsometry by Liljenvall et al. [101] and by Mathewson and Myers [5], who found $\sigma(\omega)$ as a function of energy and temperature as shown in Fig. 17. As the temperature rises, the 1.5-eV interband transition shows a pronounced broadening, shifts toward lower energies, and becomes weaker. The free-electron contribution also shows a broadening but becomes stronger with rising temperature. This is in line with the fact that in the liquid metal interband transitions are not observed [97], all the conduction electron oscillator strength having been transferred to the Drude-like intraband term. The ellipsometric results have been analyzed [5] within the framework of the Ashcroft–Sturm [18] two-band model to obtain the temperature dependence of the Drude parameters, the Fourier components of the crystal field, and the interband relaxation time (see Table XI and Mathewson and Myers [5]).

As a result of the high values of $n(\omega)$ and $k(\omega)$ in the visible and infrared, even a large variation of the optical constants with temperature produces only a small change in the reflectance R. This effect has been investigated qualitatively in thermomodulation studies by Rosei and Lynch [135]. The results are rather featureless except near the "1.5-eV" interband absorption. They show that dR/dT is negative below approximately 1.3 eV (0.95 μm) and positive at higher energies. This is consistent with the ellipsometric results [5] and may be understood as arising primarily from (1) a broadening and shift toward lower energy of the "1.5-eV" interband transition with rising temperature, combined with (2) a decrease in reflectance in the Drude region (below ~ 1 eV) as a result of a shortening of the intraband lifetime at higher temperatures. Although the temperature modulation involved was not determined, the scale of the effect is given by Pudkov's measurements [136] at 1.96 eV (0.633 μm) that yield a value for dR/dT of approximately 3×10^{-5} $(°C)^{-1}$.

The temperature dependence of the reflectance and absorptance is of particular interest for high-power laser applications (see, for example, Pudkov [136] and Konov and Tokarev [137]). A detailed study of the wavelength and temperature dependence of the absolute reflectance of metallic Al (as well as Ag and Cu) from 4000 Å to 10 μm has been made by Decker and Hodgkin [138]. Measurements were made to a precision of a few parts in 10^5 and are in agreement with the qualitative discussion given above. The reader is referred to Decker and Hodgkin [138] for numerical results.

E Pressure Dependence

The effect of pressure on the reflectance of aluminum has been investigated at room temperature by Tups and Syassen [139] in both the infrared and the visible (0.5–3.0 eV). Polycrystalline samples were studied by using the gasketed diamond-anvil technique [140] with the reflectance measured relative to the diamond windows of the pressure cell. Representative results are shown in Fig. 18. Note that the reflectance of the diamond–aluminum interface shown in Fig. 18 for a pressure of 1 kbar is significantly less than that shown in Fig. 12 for the air–aluminum interface at 1 atm because of the higher refractive index of diamond ($n \approx 2.4$). Tups and Syassen [139] found that as pressure is increased, the reflectance drop arising from the strong interband transitions associated with the square (200) faces of the Brillouin zone shifts toward higher energies. From a value of ~ 1.47 eV at 1 atm, the transition energy rises monotonically to ~ 2.34 eV at 310 kbar. This shift is indicated by the upper set of arrows in the figure. The initial rate of charge of the transition energy is 4.35 ± 0.3 meV/kbar.

Near pressures of 100 kbar, a shoulder developed on the reflectance curve at about 1 eV. With increasing pressure this shoulder moved toward higher energies, and a well-defined reflectance minimum develops at low energies. This feature corresponds to the second strong interband transition associated with the hexagonal (111) faces of the Brillouin zone. At normal pressure and temperature, this transition lies at ~ 0.4 eV and is obscured by the intraband absorption, but it shifts toward higher energies with increasing pressure reaching ~ 1.32 eV at 310 kbar, where it is well resolved from lower-energy intraband processes. The peak positions of this low-energy interband transition are indicated by the lower set of arrows.

In a detailed analysis, Tups and Syassen [139] show that both the interband transitions become stronger at the expense of the oscillator strength of the intraband absorption. This exchange of strength reflects the partial f sum rule for conduction-electron absorption [147], which enters the theory of Ashcroft and Sturm [18]. The weakening of intraband processes is evident in the reflectance spectra as the retreat of the high-reflectance almost-free-electron region that lies below ~ 1 eV under normal conditions, toward lower energies with increasing pressure. A fitting of the data to the theory of Ashcroft and Sturm [18] indicates that the optical mass, which is inversely proportional to the intraband strength, increases by approximately a factor of 1.4 in going from atmospheric pressure to 310 kbar. This corresponds to a transfer of $\sim 28\%$ of the intraband strength of approximately 0.5 e/at. to the interband transitions.

ACKNOWLEDGMENTS

The authors would like to express appreciation to innumerable colleagues for discussions and aid during the course of their studies of aluminum. Particular thanks are due Dr. J. M. Bennett, Professor F. C. Brown, Professor H. P. Myers, Dr. E. M. Pell, Professor S. E. Schnatterly, Professor B. Segall, Dr. K. Syassen, and Dr. B. W. Veal, Jr., for counsel and for supplying unpublished or prepublication copies of their results.

REFERENCES

1. H. Ehrenreich, H. R. Philipp, and B. Segall, *Phys. Rev.* **132**, 1918 (1963).
2. H. R. Philipp and H. Ehrenreich, *J. Appl. Phys.* **35**, 1416 (1964).
3. C. J. Powell, *J. Opt. Soc. Am.* **60**, 78 (1970).
4. T. Sasaki and M. Inokuti, "Conference Digest of the Third International Conference on Vacuum Ultraviolet Radiation Physics" (Y. Nakai, ed.), paper 2aC2–2. Phys. Soc. of Japan, Tokyo, 1971.
5. A. G. Mathewson and H. P. Myers, *J. Phys. F* **2**, 403 (1972).
6. H.-J. Hagemann, W. Gudat, and C. Kunz, *J. Opt. Soc. Am.* **65**, 742 (1975); H.-J. Hagemann, W. Gudat, and C. Kunz, DESY Report SR 74/7, Hamburg, 1974 (unpublished).
7. E. Shiles, T. Sasaki, M. Inokuti, and D. Y. Smith, *Phys. Rev. B* **22**, 1612 (1980).
8. F. Wooten, "Optical Properties of Solids," Section 3.2. Academic Press, New York, 1972.
9. M. H. Cohen, *Philos. Mag.* **3**, 762 (1958).
10. J. N. Hodgson, *Proc. Phys. Soc. London Sec. B* **68**, 593 (1955).
11. J. R. Beattie, *Physica (Utrecht)* **23**, 898 (1957).
12. H. E. Bennett and J. M. Bennett, in "Optical Properties and Electronic Structure of Metals and Alloys" (F. Abelès ed.), p. 175. North-Holland Publ., Amsterdam, 1966.
13. G. Dresselhaus, M. S. Dresselhaus, and D. Beaglehole, in "Electronic Density of States" (L. H. Bennett ed.). National Bureau of Standards Special Publication 323, Washington, D.C., 1971.
14. R. L. Benbow and D. W. Lynch, *Phys. Rev. B* **12**, 5615 (1975).
15. D. Y. Smith and B. Segall, *Bull. Am. Phys. Soc.* **26**, 209 (1981); D. Y. Smith and B. Segall, (to be published).
16. "Handbook of Chemistry and Physics," 58th ed. (R. C. Weast, ed.). CRC Press, Cleveland, 1977.
17. G. T. Meaden, "Electrical Resistance of Metals." Plenum, New York, 1965.
18. N. W. Ashcroft and K. Sturm, *Phys. Rev. B* **3**, 1898 (1971).

19. W. A. Harrison, *Phys. Rev.* **147**, 467 (1966).
20. A. I. Golovashkin, A. I. Kopeliovich, and G. P. Motulevich, *Zh. Eksp. Teor. Fiz.* **53**, 2053 (1967); A. I. Golovashkin, A. I. Kopeliovich, and G. P. Motulevich, *Soviet Phys. JETP* **26**, 1161 (1968).
21. N. W. Ashcroft, *Philos. Mag.* **8**, 2055 (1963).
22. D. Brust, *Phys. Rev. B* **2**, 818 (1970).
23. F. Szmulowicz and B. Segall, *Phys. Rev. B* **24**, 892 (1981).
24. J. Strong, *Astrophys. J.* **83**, 401 (1936).
25. L. G. Schulz, *J. Opt. Soc. Am.* **44**, 357 (1954); L. G. Schulz and F. R. Tangherlini, *J. Opt. Soc. Am.* **44**, 362 (1954).
26. I. N. Shklyarevskii and R. G. Yarovaya, *Opt. Spektrosk.* **16**, 85 (1964); I. N. Shklyarevskii and R. G. Yarovaya, *Opt. Spectrosc. (USSR)* **16**, 45 (1964).
27. J. R. Beattie, *Philos. Mag.* **46**, 235 (1955).
28. H. E. Bennett, M. Silver, and E. J. Ashley, *J. Opt. Soc. Am.* **53**, 1089 (1963).
29. L. W. Bos and D. W. Lynch, *Phys. Rev. Lett.* **25**, 156 (1970).
30. O. Hunderi and P. O. Nilsson, *Solid State Commun.* **19**, 921 (1976); O Hunderi and P. O. Nilsson, *Nuovo Cimento* **39 B**, 459 (1977).
31. D. Y. Smith, D. D. Koelling, and B. Segall, *Bull. Am. Phys. Soc.* **28**, 387 (1983); D. Y. Smith, D. D. Koelling, and B. Segall, to be published.
31a. D. C. Jiles and M. P. Staines, *Solid State Commun.* **47**, 37 (1983).
32. M. S. Miller and N. D. Drew, *J. Phys. F*, **13**, 1885 (1983).
33. G. Quincke, *Ann. Physik, Series 2, Jubelband*, p. 336 (1874); W. Voigt, *Ann. Physik, Series 3*, **23**, 142 (1884); F. F. Martens, "Landolt-Börnstein," 5th ed., Vol. 2, Section 165, p. 906. Springer, Berlin, 1923.
34. P. Drude, *Ann. Physik, Series 3*, **39**, 481 (1890).
35. P. G. Nutting, *Phys. Rev.* **13**, 193 (1901).
36. W. W. Coblentz, *Bull. U.S. Bureau Standards* **2**, 457 (1906).
37. W. v. Uljain, *Phys. Z.* **11**, 784 (1910).
38. M. Luckiesh, *J. Opt. Soc. Am.* **19**, 1 (1929).
39. W. W. Coblentz and R. Stair, *U. S. Bureau Standards J. Res.* **4**, 189 (1930).
40. J. Wulff, *J. Opt. Soc. Am.* **24**, 223 (1934).
41. A. H. Taylor, *J. Opt. Soc. Am.* **24**, 192 (1934).
42. W. R. Grove, *Philos. Trans. R. Soc. London* **142**, 87 (1852); W. Crookes, *Proc. R. Soc. London* **50**, 88 (1891).
43. P. Pringsheim and R. Pohl, *Verh. Dtsh. Phys. Ges.* **14**, 506 (1912).
44. E. O. Hulburt, *Astrophys. J.* **42**, 203 (1915).
45. P. R. Gleason, *Proc. Nat. Acad. Sci. (USA)* **15**, 551 (1929).
46. R. Ritschl, *Z. Instrumentenkunde* **41**, 158 (1929); R. Ritschl, *Z. Instrumentenkunde* **50**, 230 (1930); R. Ritschl, *Z. Phys.* **69**, 578 (1931); H. S. Jones, *Nature* **134**, 522 (1934).
47. J. Strong, *Phys. Rev.* **43**, 498 (1933).
48. G. Hass, *J. Opt. Soc. Am.* **45**, 945 (1955).
49. J. C. Burridge, H. Kuhn, and A. Pery, *Proc. Phys. Soc. London Ser. B* **66**, 963 (1953).
50. J. Halford, F. K. Chin, and J. E. Norman, *J. Opt. Soc. Am.* **63**, 786 (1973).
51. G. Hass and J. E. Waylonis, *J. Opt. Soc. Am.* **51**, 719 (1961).
52. E. T. Hutcheson, G. Hass, and J. K. Coulter, *Opt. Commun.* **3**, 213 (1971).
53. J. G. Endriz and W. E. Spicer, *Phys. Rev. B* **4**, 4144 (1971).
54. P. C. Gibbons, S. E. Schnatterly, J. J. Ritsko, and J. R. Fields, *Phys. Rev. B* **13**, 2451 (1976).
55. R. W. Ditchburn and G. H. Freeman, *Proc. R. Soc. London Sec. A* **294**, 20 (1966).
56. G. Hass, *Optik* **1**, 8 (1946).
57. G. B. Sabine, *Phys. Rev.* **55**, 1064 (1939).
58. M. Banning, *Phys. Rev.* **59**, 914 (1941); M. Banning, *J. Opt. Soc. Am.* **32**, 98 (1942).
59. W. C. Walker, J. A. R. Samson, and O. P. Rustig, *J. Opt. Soc. Am.* **48**, 71 (1958); W. C. Walker, O. P. Rustig, and G. L. Weissler, *J. Opt. Soc. Am.* **49**, 471 (1959).

60. V. R. Weidner and J. J. Hsia, National Bureau of Standards Certificate for Standard Reference Material 2003a, Washington, D.C., 1981; V. R. Weidner and J. J. Hsia, National Bureau of Standards Report NBS-SP-260-75, Washington, D.C., 1982.

61. H. Schopper, in "Landolt-Börnstein Zahlenwerte and Funktionen," 6th ed., (K.-H. Hellwege and A. M. Hellwege, eds.), Vol. 2, Part 8, Section 281, pp. 1-1-1-41. Springer Verlag, Berlin, 1962; R. C. Willams and G. B. Sabine, *Astrophys. J.* **77**, 316 (1933); H. S. Jones, *Nature* **133**, 552 (1934); B. K. Johnson, *Nature* **134**, 216 (1934); M. Auwärter, *Z. Tech. Phys.* **18**, 457 (1937); B. K. Johnson, *Proc. Phys. Soc. London* **53**, 258 (1941); A. Boettcher, *Z. Angew. Phys.* **2**, 340 (1950).

62. G. Hass, *Z. Anorg. Allg. Chem.* **254**, 96 (1947).

63. P. H. Berning, G. Hass, and R. P. Madden, *J. Opt. Soc. Am.* **50**, 586 (1960).

64. I. N. Shklyarevskii and R. G. Yarovaya, *Opt. Spektrosk.* **14**, 252 (1963); I. N. Shklyarevskii and R. G. Yarovaya, *Opt. Spectrosc. (USSR)* **14**, 130 (1963).

65. R. W. Fane and W. E. J. Neal, *J. Opt. Soc. Am.* **60**, 790 (1970).

66. J. Shewchun and E. C. Rowe, *J. Appl. Phys.* **41**, 4128 (1970).

67. B. P. Feuerbacher and W. Steinmann, *Opt. Commun.* **1**, 81 (1969).

68. A. Daude, A. Savary, and S. Robin, *Thin Solid Films* **13**, 255 (1972); A. Daude, A. Savary, and S. Robin, *J. Opt. Soc. Am.* **62**, 1 (1972).

69. W. Walkenhorst, *Z. Tech. Phys.* **22**, 14 (1941).

70. R. S. Sennett and G. D. Scott, *J. Opt. Soc. Am.* **40**, 203 (1950); M. F. Crawford, W. M. Gray, A. L. Schawlow, and F. M. Kelly, *J. Opt. Soc. Am.* **39**, 888 (1949).

71. W. E. J. Neal, R. W. Fane, and N. W. Grimes, *Philos. Mag.* **21**, 167 (1970).

72. R. P. Madden and L. R. Canfield, *J. Opt. Soc. Am.* **51**, 838 (1961).

73. R. P. Madden, L. R. Canfield, and G. Hass, *J. Opt. Soc. Am.* **53**, 620 (1963).

74. R. K. Hart, *Proc. R. Soc. London Sec. A* **236**, 68 (1956).

75. G. Hass, *Ann. Phys. (Leipzig)* **31**, 245 (1938).

76. G. Hass, W. R. Hunter, and R. Tousey, *J. Opt. Soc. Am.* **46**, 1009 (1956); G. Hass, W. R. Hunter, and R. Tousey, *J. Opt. Soc. Am.* **47**, 1070 (1957).

77. N. Cabrera, J. Terrien, and J. Hamon, *Compt. Rend.* **224**, 1558 (1947); N. Cabrera, *Philos. Mag.* **40**, 175 (1949).

78. R. P. Chasmar, J. L. Craston, G. Isaacs, and A. S. Young, *J. Sci. Instrum.* **28**, 206 (1951).

79. G. H. C. Freeman, *Br. J. Appl. Phys.* **16**, 927 (1965); E. T. Arakawa and M. W. Williams, *J. Phys. Chem. Sol.* **29**, 735 (1968).

80. G. Hass and N. W. Scott, *J. Opt. Soc. Am.* **39**, 179 (1949).

81. A. P. Bradford and G. Hass, *J. Opt. Soc. Am.* **53**, 1096 (1963); J. T. Cox, G. Hass, and J. B. Ramsey, *J. de Phys. (Paris)* **25**, 250 (1964); H. Herzig, R. S. Spencer, and J. J. Zaniewski, *Appl. Opt.* **17**, 3031 (1978); J. T. Cox and G. Hass, *Appl. Opt.* **17**, 333 (1978); R. J. Francis, *UV Spectrom. Group Bull.*, No. 6, 35 (December, 1978); and G. Hass, *J. Opt. Soc. Am.* **72**, 27 (1982).

82. J. T. Cox and G. Hass, *Appl. Opt.* **17**, 2125 (1978).

83. G. Hass and R. Tousey, *J. Opt. Soc. Am.* **49**, 593 (1959); P. H. Berning, G. Hass, and R. P. Madden, *J. Opt. Soc. Am.* **50**, 586 (1960); D. W. Angel, W. R. Hunter, R. Tousey, and G. Hass, *J. Opt. Soc. Am.* **51**, 913 (1961); W. R. Hunter, *Opt. Acta* **9**, 255 (1962); L. R. Canfield, G. Hass, and J. E. Waylonis, *Appl. Opt.* **5**, 45 (1966); J. T. Cox, G. Hass, and J. E. Waylonis, *Appl. Opt.* **7**, 1535 (1968); E. T. Hutcheson, G. Hass, and J. T. Cox, *Appl. Opt.* **11**, 2245 (1972); G. Hass and W. R. Hunter, in "Space Optics" (B. J. Thompson and R. R. Shannon, eds.), p. 525, National Academy of Sciences, Washington, D.C., 1974.

84. J. D. Armitage and M. A. Grimm, *IBM Tech. Disclosure Bull.* **23**, 3580 (1981).

85. L. Holland, "Vacuum Deposition of Thin Films." Wiley, New York, 1958.

86. D. H. Tomboulian and E. M. Pell, *Phys. Rev.* **83**, 1196 (1951).

87. A. Balzarotti, A Bianconi, and E. Burattini, *Phys. Rev. B* **9**, 5003 (1974).

88. J. M. Elson and R. H. Ritchie, *Phys. Rev. B* **4**, 4129 (1971).

89. R. H. Ritchie and R. E. Wilems, *Phys. Rev.* **178**, 372 (1969); J. Crowell and R. H. Ritchie, *J. Opt. Soc. Am.* **60**, 794 (1970); J. M. Elson and R. H. Ritchie, *Phys. Lett.* **33A**, 255 (1970).

90. R. C. Vehse, E. T. Arakawa, and J. L. Stanford, *J. Opt. Soc. Am.* **57**, 551 (1967).
91. M. W. Williams, E. T. Arakawa, and L. C. Emerson, *Surf. Sci.* **6**, 127 (1967).
92. L. Holland, *J. Opt. Soc. Am.* **43**, 376 (1953).
93. N. G. Nakhodkin and A. I. Shaldervan, *Thin Solid Films* **10**, 109 (1972).
94. R. T. Kivaisi, *Thin Solid Films* **97**, 153 (1982).
95. J. M. Nieuwenhuizen and H. B. Haanstra, *Philips Tech. Rev.* **27**, 87 (1966).
96. L. Holland, *B. J. Appl. Phys.* **9**, 336 (1958).
97. J. C. Miller, *Philos. Mag.* **20**, 1115 (1969).
98. T. E. Faber, *in* "Optical Properties and Electronic Structure of Metals and Alloys" (F. Abelès ed.), p. 259. North-Holland Publ., Amsterdam, 1966.
99. L. G. Bernland, O. Hunderi, and H. P. Myers, *Phys. Rev. Lett.* **31**, 363 (1973).
100. A.G. Mathewson and H. P. Myers, *Phys. Scr.* **4**, 291 (1971).
101. H. G. Liljenvall, A. G. Mathewson, and H. P. Myers, *Solid State Commun.* **9**, 169 (1971).
102. R. Haensel, B. Sonntag, C. Kunz, and T. Sasaki, *J. Appl. Phys.* **40**, 3046 (1969).
103. C. Gähwiller and F. C. Brown, *Phys. Rev. B* **2**, 1918 (1970).
104. V. A. Fomichev and A. P. Lukirskii, *Opt. Spektrosk.* **22**, 796 (1967); V. A. Fomichev and A. P. Lukirskii, *Opt. Spectrosc. (USSR)* **22**, 432 (1967).
105. V. A. Fomichev and A. P. Lukirskii, *Fiz. Tverd. Tela* **8**, 2104 (1966); V. A. Fomichev and A. P. Lukirskii, *Sov. Phys.-Solid State* **8**, 1674 (1967).
106. V. A. Fomichev, *Fiz. Tverd. Tela* **8**, 2892 (1966); V. A. Fomichev, *Sov. Phys.-Solid State* **8**, 2312 (1967).
107. S. Singer, *J. Appl. Phys.* **38**, 2897 (1967).
108. O. A. Ershov, I. A. Brytov, and A. P. Lukirskii, *Opt. Spektrosk.* **22**, 127 (1967); O. A. Ershov, I. A. Brytov, and A. P. Lukirskii, *Opt. Spectrosc. (USSR)* **22**, 66 (1967).
109. O. A. Ershov and I. A. Brytov, *Opt. Spektrosk.* **22**, 305 (1967); O. A. Ershov and I. A. Brytov, *Opt. Spectrosc. (USSR)* **22**, 165 (1967).
110. B. A. Cooke and E. A. Stewardson, *Br. J. Appl. Phys.* **15**, 1315 (1964).
111. A. J. Bearden, *J. Appl. Phys.* **37**, 1681 (1966).
112. B. L. Henke and R. L. Elgin, *in* "Advances in X-Ray Analysis" (B. L. Henke, J. B. Newkirk, and G. R. Mallett, eds.), Vol. 13, p. 639. Plenum, New York, 1970.
113. B. L. Henke and E. S. Ebisu, *in* "Advances in X-Ray Analysis" (C. L. Grant, C. S. Barrett, J. B. Newkirk, and C. O. Ruud, eds.), Vol. 17, p. 150. Plenum, New York, 1974.
114. L. Singman, *J. Appl. Phys.* **45**, 1885 (1974).
115. P. Lublin, P. Cukor, and R. J. Jaworowski, *in* "Advances in X-Ray Analysis" (B. L. Henke, J. B. Newkirk, and G. R. Mallett, eds.), Vol. 13, p. 632. Plenum, New York, 1970.
116. J. H. Hubbell, W. H. McMaster, N. K. Del Grande, and J. H. Mallett, *in* "International Tables for X-Ray Crystallography" (J. A. Ibers and W. C. Hamilton, eds.), Vol. IV, p. 47. Kynoch, Birmingham, England, 1974.
117. C. M. Davisson, *in* "Alpha-, Beta-, and Gamma-Ray Spectroscopy" (K. Siegbahn, ed.), Vol. I, p. 37. North-Holland Publ., Amsterdam, 1965.
118. A. P. Lukirskii, E. P. Savinov, O. A. Ershov, and Yu. F. Shepelev, *Opt. Spektrosk.* **16**, 310 (1964); A. P. Lukirskii, E. P. Savinov, O. A. Ershov, and Yu. F. Shepelev, *Opt. Spectrosc. (USSR)* **16**, 168 (1964).
119. A. Smakula, *Z. Phys.* **88**, 114 (1934).
120. W. Woltersdorff, *Z. Phys.* **91**, 230 (1934).
121. H. M. O'Bryan, *J. Opt. Soc. Am.* **26**, 122 (1936).
122. K. B. Hunt, Investigations of the Reflectivity and Transmissivity of Selected Materials in the Infrared Region. M.S. thesis, Purdue University, West Lafayette, Indiana, 1945 (unpublished).
123. A. I. Golovashkin, G. P. Motulevich, and A. A. Shubin, *Zh. Eksp. Teor. Fiz.* **38**, 51 (1960); A. I. Golovashkin, G. P. Motulevich, and A. A. Shubin, *Sov. Phys.-JETP* **11**, 38 (1960).
124. T. T. Cole and F. Oppenheimer, *Appl. Opt.* **1**, 709 (1962).
125. W. R. Hunter, *J. Appl. Phys.* **34**, 1565 (1963); W. R. Hunter, *J. Opt. Soc. Am.* **54**, 208 (1964); W. R. Hunter, *J. Phys. (Paris)* **25**, 154 (1964).

126. G. P. Motulevich, A. A. Shubin, and O. F. Shustova, *Zh. Eksp. Teor. Fiz.* **49**, 1431 (1965); G. P. Motulevich, A. A. Shubin, and O. F. Shustova, *Sov. Phys.-JETP* **22**, 984 (1966).

127. A. P. Lenham and D. M. Terherne, *J. Opt. Soc. Am.* **56**, 752 (1966).

128. A. Daude, M. Priol, and S. Robin, *C. R. Acad. Sci. (Paris) B* **263**, 1178 (1966).

129. A. Daude, A. Savary, G. Jezequel, and S. Robin, *C. R. Acad. Sci. (Paris) B* **269**, 901 (1969).

130. R. Haensel, G. Keitel, B. Sonntag, C. Kunz, and P. Schreiber, *Phys. Status Solidi A* **2**, 85 (1970); J. J. Ritsko, S. E. Schnatterly, and P. C. Gibbons, *Phys. Rev. Lett.* **32**, 671 (1974); H. Peterson and C. Kunz, *Phys. Rev. Lett.* **35**, 863 (1975); C. G. Olson and D. W. Lynch, *Solid State Commun.* **31**, 601 (1979); and C. G. Olson and D. W. Lynch, *Solid State Commun.* **36**, 513 (1980).

131. S. Kiyono, S. Chiba, Y. Hayasi, S. Kato, and S. Mochimaru, *Jpn. J. Appl. Phys.* **17** (Suppl. 17-2), 212 (1978); and M. H. Heinonen and J. A. Leiro, *Philos. Mag. B* **46**, 669 (1982).

132. J. J. Hopfield, *Phys. Rev.* **139**, A419 (1965).

133. K. L. Kliewer and R. Fuchs, *Phys. Rev.* **172**, 607 (1968).

134. H. J. Levinson and E. W. Plummer, *Phys. Rev. B* **24**, 628 (1981).

134a. P. Livins, S. E. Schnatterly, T. Aton, and A. Cafolla, *Bull. Am. Phys. Soc.* **29**, 517 (1984).

135. R. Rosei and D. W. Lynch, *Phys. Rev. B* **5**, 3883 (1972).

136. S. D. Pudkov, *Zh. Tekh. Fiz.* **47**, 649 (1977); S. D. Pudkov, *Sov. Phys.-Tech. Phys.* **22**, 389 (1977).

137. V. I. Konov and V. N. Tokarev, *Kvantovaya Elektron.* **10**, 327 (1983); V. I. Konov and V. N. Tokarev, *Sov. J. Quantum Electron.* **13**, 177 (1983).

138. D. L. Decker and V. A. Hodgkin, National Bureau of Standards Special Publication NBS-SP-620, Washington, D.C. 1981, p. 190.

139. H. Tups and K. Syassen, *J. Phys. F* **14**, 2753 (1984).

140. A. Jayaraman, *Rev. Mod. Phys.* **55**, 65 (1983).

141. D. Y. Smith and E. Shiles, *Phys. Rev. B* **17**, 4689 (1978).

142. D. Y. Smith, E. Shiles, and M. Inokuti, "The Optical Properties and Complex Dielectric Function of Metallic Aluminum from 0.04 to 10^4 eV." Argonne National Laboratory Report ANL-83-24, Argonne, Illinois, March 1983.

143. A. R. P. Rau and V. Fano, *Phys. Rev.* **162**, 68 (1967).

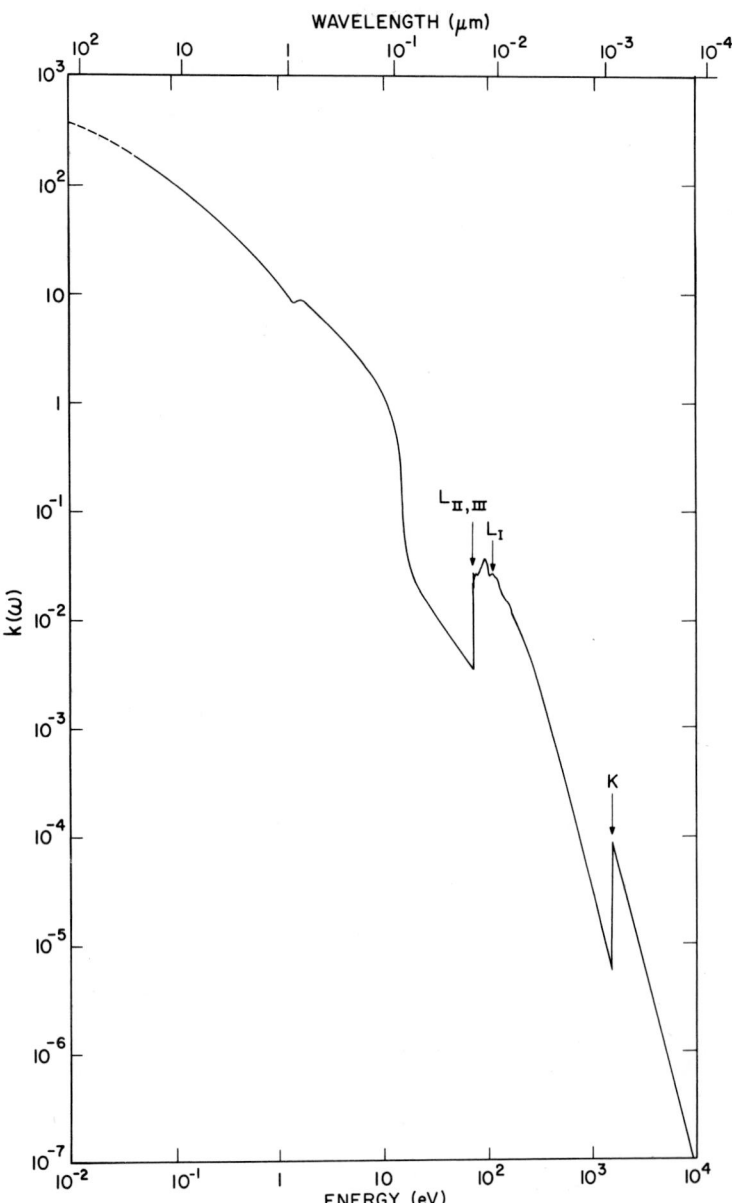

Fig. 11. The extinction coefficient $k(\omega)$; i.e., the imaginary part of the refractive index, for metallic aluminum at room temperature. (After Shiles *et al.* [7].)

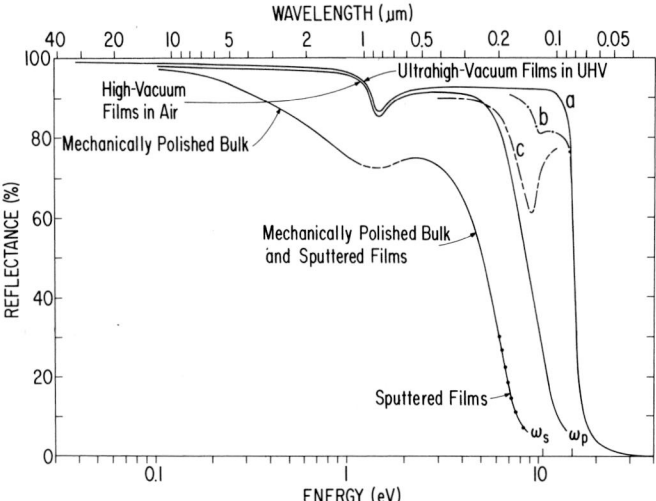

Fig. 12. The reflectance of metallic aluminum at room temperature. The uppermost curve gives the reflectance of opaque evaporated aluminum films prepared in ultrahigh vacuum. These films are presumed to be free of bulk and surface oxide contamination. The effect of light scattering and coupling to surface plasmons (at $\omega_s \approx 10.6$ eV) is illustrated by curves b and c. Curve b was obtained by Vehse *et al.* [90] for a film on a microscope slide cleaned by vapor decreasing and evaporated in a vacuum of $\sim 3 \times 10^{-8}$ torr; the "oxide-free" reflectance was obtained by extrapolation of the reflectance back to the time that the evaporation was completed. Curve c is for a uhv film with rms surface roughness σ of 18 Å, as reported by Endriz and Spicer [53]. The curve labeled high vacuum is an average for films evaporated in vacuum of the order of 10^{-5} torr. At this pressure surface and bulk oxidation occurs. The values reported for such films vary widely, especially in the ultraviolet, presumably because of surface roughness. The third curve is representative of mechanically polished samples, but the reported range of measurements is extremely large. Data for sputtered films are available from 380 nm well into the UV [44, 45]; curiously enough, they track the data for polished samples very closely.

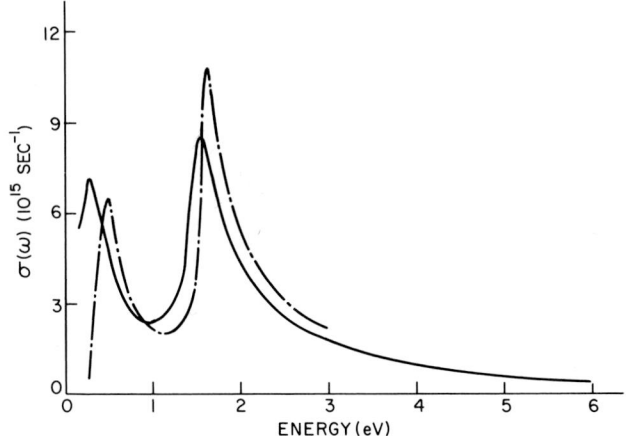

Fig. 13. The real part of the interband conductivity for aluminum obtained by subtracting the intraband contribution in the Drude model from the measured optical conductivity. The solid curve is for room temperature and was calculated by Smith and Segall [15], who employed Shiles *et al.*'s [7] analysis of uhv reflectance data. The dashed curve is taken from Benbow and Lynch [14] and applies to electropolished samples at 4.2 K.

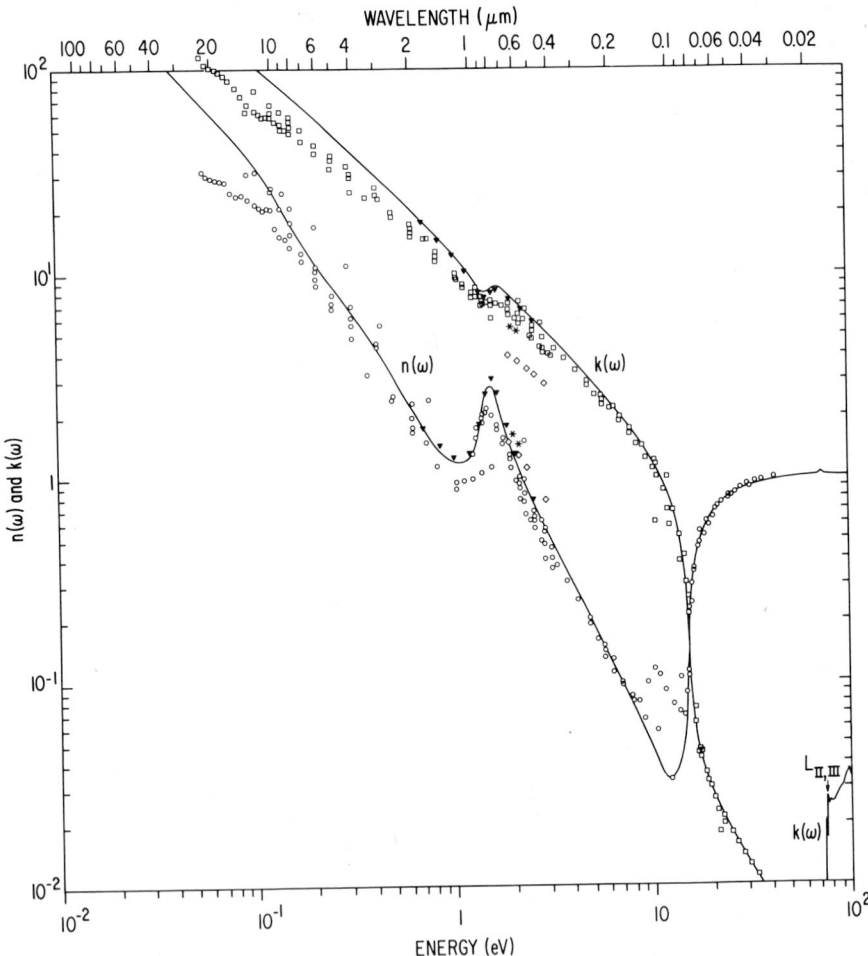

Fig. 14. The complex refractive index $n(\omega) + ik(\omega)$ for aluminum. The curves are taken from Shiles *et al.*'s [7] analysis of uhv reflectance data. A portion of the uhv ellipsometric data of Mathewson and Myers [100] is given for comparison. Quincke's [33] and Drude's [34] results for polished bulk samples are shown for historical interest; considering the materials and techniques available, these early measurements are remarkably good, especially for $n(\omega)$. The remainder of the data points are given to show the range of values of $n(\omega)$ (○) and $k(\omega)$ (□) reported in the literature. Most refer to evaporated films prepared in conventional or high vacuum and measured by using polarimetric, interferometric, and like methods. The sources of these data are given in the references [10, 11, 25, 26, 27, 51, 55 56, 65, 66, 73, 75, 91, and 121–129]. Note that the curves for $n(\omega)$ and $k(\omega)$ curve cross each other at roughly 15 eV, the plasmon energy. This corresponds to the plasmon condition $\varepsilon_1(\omega_p) = n^2(\omega_p) - k^2(\omega_p) \approx 0$. The onset of the L-shell absorption appears in the lower-right-hand corner. The corresponding dispersion is visible as a little "pimple" on the index curve near 72 eV. (Mathewson and Myers [100], ▼; Quincke [33], ◇; Drude [34], *.)

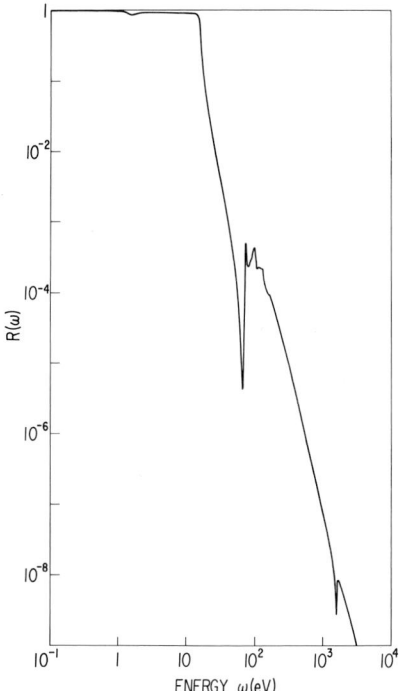

Fig. 15. The reflectance $R(\omega)$ at normal incidence of a smooth oxide-free metallic aluminum surface in vacuum. (After Shiles *et al.* [7].)

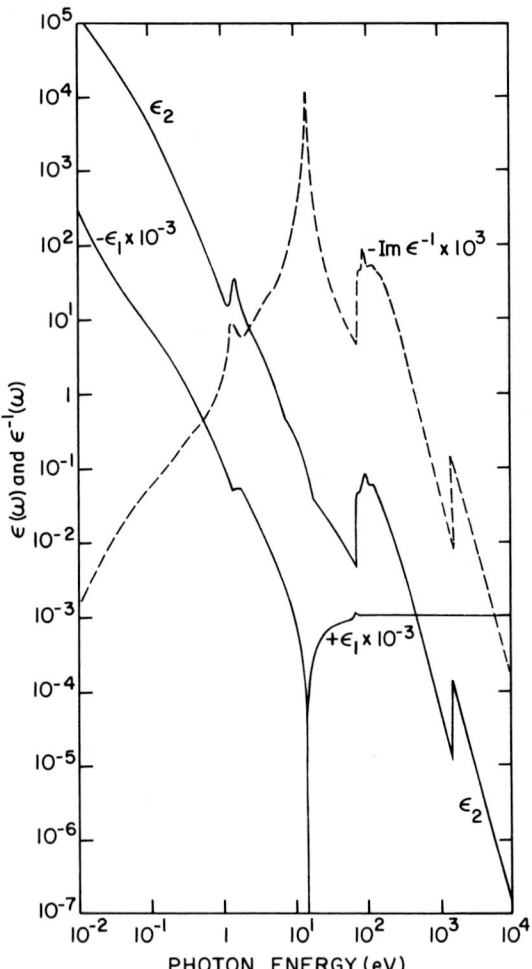

Fig. 16. The dielectric-response function of metallic aluminum. The real part $\varepsilon_1(\omega)$ is negative with large absolute values at low energies, vanishes at $\hbar\omega_p \approx 15.0$ eV (the plasmon energy for conduction electrons), and is positive and approaches unity at higher energies. The imaginary part $\varepsilon_2(\omega)$ is also largest at low energies and decreases with energy except at thresholds for newer modes of excitation (first at the beginning of interband excitation of valence electrons, next at the L-shell threshold, and finally at the K-shell threshold). The quantity $\mathrm{Im}[-1/\varepsilon(\omega)]$, which governs the energy transfer from fast charged particles, shows a prominent maximum near $\hbar\omega_p$ and is small at both lower and higher energies. (After Shiles *et al.* [7].)

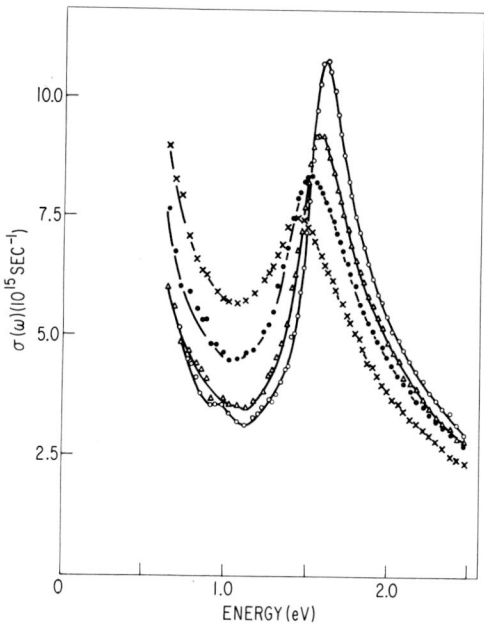

Fig. 17. The real part of the optical conductivity as a function of temperature for clean poly-crystalline aluminum films deposited at room temperature. (198 K, ○; 298 K, △; 404 K, ●; and 552 K, ×.) (After Mathewson and Myers [5].) (Copyright © 1972 by the Institute of Physics).

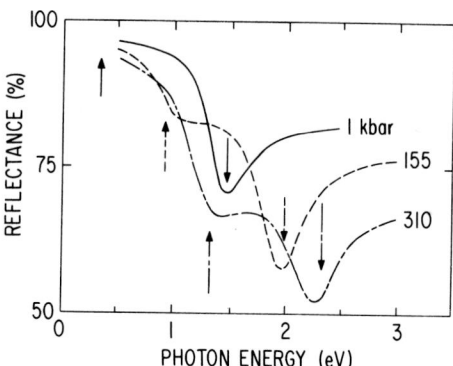

Fig. 18. The reflectance of aluminum at an interface with diamond as a function of pressure and photon energy. Pressures are given in kilobars to the right of each curve. The positions of the principal interband transitions, as determined by a fitting to the theory of Ashcroft and Sturm [18], are indicated by arrows coded to the corresponding reflectance curve. The position of the high-energy transition $2V_{200}$ is indicated by the upper set of arrows. That of the low-energy transition $2V_{111}$ is indicated by the lower set of arrows. (After Tups and Syassen [139].)

TABLE XI

Drude Parameters for the Intraband Absorption of Metallic Aluminum[a]

Source	Temperature (K)	Strength[b] $h\Omega_p$ (eV)	n_{eff} (e/at.)	m_{opt} (m_e)	Damping τ (10^{-14} sec)	$h\gamma = h\tau^{-1}$ (meV)	DC conductivity $\sigma(0)$ (10^{17} sec^{-1}) Optical	Electrical
Ehrenreich et al. [1]	RT	12.7	1.94	1.55	0.512	129	1.52	3.18 [16]–3.28 [17]
Bennett et al. [12]	RT	14.7	2.60	1.15	0.801	82.2	3.18 (input)[c]	3.18 [16]–3.28 [17]
Powell [3]	RT	12.2	1.80	1.67	0.66	100	1.81	3.18 [16]–3.28 [17]
Dresselhaus et al. [13]	RT	12.9 ± 0.7	2.0 ± 0.2	1.5 ∓ 0.15	0.5 ± 0.2	160 ∓ 60	1.60 ± 0.8	3.18 [16]–3.28 [17]
Mathewson and Myers [5]	198	12.8	1.99	1.51[d]	1.18	55.8	3.58	5.49 [5]
Mathewson and Myers [5]	298	13.0	2.03	1.48[d]	1.02	64.5	3.14	3.21 [5]
Mathewson and Myers [5]	404	13.2	2.10	1.43[d]	0.62	105	1.96	2.30 [5]
Mathewson and Myers [5]	552	13.3	2.13	1.41[d]	0.52	128	1.65	1.63 [5]
Benbow and Lynch [14]	4.2	12.7	1.94	1.55 (input)[e]	1.10	60	3.25	
Smith and Segall [15]	RT	12.5 ± 0.3	1.88 ± 0.09	1.60 ∓ 0.08	1.06 ± 0.12	63 ∓ 7	3.00 ± 0.3	3.18 [16]–3.28 [17]

[a] See Eqs. (1)–(4) in the text.

[b] The effective density of electrons n_{eff} is given in electrons per atom (e/at.). In aluminum the actual density of conduction electrons is 3 e/at. The optical conductivity at $\omega = 0$ has been calculated from Eq. (5); the electrical values of $\sigma(0)$ are for dc measurements on bulk samples.

[c] The measured bulk value of $\sigma(0)$ was used to fix the ratio Ω_p^2/γ via Eq. (5).

[d] Evaluated by fitting the interband absorption and use of Eq. (42) of Ashcroft and Sturm [18].

[e] Assumed on the basis of a fit by Ashcroft and Sturm [18].

TABLE XII

The Optical Constants n and k and Reflectivity of Evaporated Metallic Aluminum Films at Room Temperature from 32 μm to the $L_{II,III}$ Edge[a,b]

eV	cm⁻¹	μm	n	n	k	k	R	R	R	$R_{h\nu}$
75		0.01653	1.01 [7]		2.40×10^{-2} [7]		1.74×10^{-4} [7]			
74		0.01676	1.01		2.42×10^{-2}		1.97			
73		0.01698	1.02		1.91×10^{-2}		2.33×10^{-4}			
72		0.01722	1.02		3.46×10^{-3}		6.85×10^{-5}			
71		0.01746	1.01		3.46		2.75			
70		0.01771	1.01		3.52		1.27×10^{-5}			
65		0.01908	0.995		4.17		9.85×10^{-6}			
60		0.02066	0.987		4.41		4.64×10^{-5}			
55		0.02254	0.979		5.08		1.21×10^{-4}			
50		0.02480	0.969		5.87		2.57			
45		0.02755	0.957		6.82		5.01			
40		0.03100	0.940		8.16		9.89×10^{-4}			
38		0.03263	0.931		8.71		1.30×10^{-3}			
36		0.03444	0.921	0.96 [55]	9.32×10^{-3} [7]	9.5×10^{-3} [55]	1.73	4.40×10^{-4} [55]		
34		0.03647	0.909		1.02×10^{-2}		2.33			
33		0.03757	0.902	0.943	1.07	1.10×10^{-2}	2.71	8.93×10^{-4}		
32		0.03875	0.894		1.11		3.17			
31		0.04000	0.885		1.16		3.75			

(continued)

[a] Columns labeled [7] are the composite data of Shiles et al. [7]; the values tabulated are based primarily on the Kramers-Kronig analysis of reflectance measurements of uhv films from 32 to 0.1 μm. The uhv reflectance data of Bennett et al. [28], are given in columns labeled [28]. The in situ uhv ellipsometric data of Mathewson and Myers [100] are given in columns labeled [100], and the vacuum-UV transmission and angular-dependence-of-reflection measurements of Ditchburn and Freeman [55] are given in columns labeled [55]. Composite reflectance values for films evaporated in high vacuum and measured in air, $R_{h\nu}$, are given in columns labeled [142]. These data are taken from Smith et al. [142] and are also shown as the "high vacuum" curve in Fig. 12. The reflectance of these oxidized films is not unique. However, films prepared by rapid evaporation in high vacuum and exposed to a normal laboratory environment are generally reported to have reflectances throughout the IR and visible which are roughly 1% below the values for smooth, unoxidized films prepared in uhv. At higher energies, the high-vacuum films show a sharp fall-off in reflectance because of surface-oxide absorption and because of the scattering and surface-plasmon absorption caused by surface roughness. A wide range of optical properties is generally observed in the ultraviolet unless special precautions are taken to control the film's surface roughness.

[b] A tabulation of optical constants at energies higher than those given here is available in Reference 142.

TABLE XII (Continued)
Metallic Aluminum

eV	cm⁻¹	μm	n	n	k	k	R	R	R_hv
30		0.04133	0.876	0.912	1.25	1.25×10^{-2}	4.44	2.16×10^{-3}	
29		0.04275	0.865		1.35		5.28		
28		0.04428	0.854	0.880	1.45	1.41	6.30	4.13	
27		0.04592	0.841		1.55		7.58		
26		0.04769	0.826	0.838	1.65	1.59	9.20×10^{-3}	7.84×10^{-3}	
25		0.04959	0.809		1.77		1.13×10^{-2}		
24		0.05166	0.789	0.785	1.90	1.82	1.40	1.46×10^{-2}	
23		0.05391	0.766		2.05		1.77		
22		0.05636	0.740	0.718	2.22	2.13	2.26	2.71	
21		0.05904	0.707		2.42		2.96		
20		0.06199	0.668	0.635	2.68	2.67	3.98	5.01	
19.5		0.06358	0.646		2.84		4.67		
19		0.06526	0.620	0.580	3.02	3.07	5.54	7.10×10^{-2}	
18.5		0.06702	0.591		3.24		6.64		
18		0.06888	0.558	0.520	3.48	3.55	8.09×10^{-2}	0.10	
17.5		0.07085	0.520		3.81		0.100		
17		0.07293	0.474	0.445	4.23	4.24	0.128	0.148	
16.5		0.07514	0.419		4.87		0.168		
16		0.07749	0.351	0.345	5.95	6.32×10^{-2}	0.233	0.239	
15.5		0.07999	0.258	0.225	7.77×10^{-2}	0.220	0.350	0.419	0.050 [142]
15		0.08266	0.125		0.153		0.612		0.057
14.5		0.08551	0.0616		0.301		0.798		0.064
14		0.08856	0.0481	0.104	0.416	0.390	0.849	0.697	0.074
13.5		0.09184	0.0409		0.517		0.879		0.086
13		0.09537	0.0376		0.609		0.896		0.102
12.5		0.09919	0.0344		0.700		0.912		0.127
12	96,790	0.1033	0.0328	0.033	0.791	0.580	0.922	0.906	0.165
11.5	92,750	0.1078	0.0331		0.883		0.928		

11	88,720	0.1127	0.0356	0.978	0.930		0.215
10.5	84,690	0.1181	0.0396	1.08	0.929		0.270
10	80,660	0.1240	0.0442	1.18	0.929		0.328
9.5	76,620	0.1305	0.0495	1.29	0.928		0.388
9	72,590	0.1378	0.0557	1.40	0.928		0.448
8.75	70,570	0.1417	0.0592	1.46	0.927		0.478
8.5	68,560	0.1459	0.0630	1.53	0.927		0.510
8.25	66,540	0.1503	0.0671	1.59	0.927		0.543
8	64,520	0.1550	0.0716	1.66	0.927		0.579
7.75	62,510	0.1600	0.0765	1.74	0.927		0.616
7.5	60,490	0.1653	0.0820	1.81	0.927		0.656
7.25	58,480	0.1710	0.0880	1.90	0.926		0.698
7	56,460	0.1771	0.0946	1.98	0.926		0.735
6.75	54,440	0.1837	0.102	2.07	0.926		0.766
6.5	52,430	0.1907	0.110	2.17	0.926		0.793
6.25	50,410	0.1984	0.119	2.28	0.926		0.820
6	48,390	0.2066	0.130	2.39	0.926		0.841
5.75	46,380	0.2156	0.141	2.51	0.926		0.858
5.5	44,360	0.2254	0.155	2.64	0.926		0.873
5.25	42,340	0.2362	0.172	2.79	0.925		0.884
5	40,330	0.2480	0.190	2.94	0.924		0.894
4.75	38,310	0.2610	0.209	3.11	0.925		0.901
4.5	36,300	0.2755	0.233	3.30	0.925		0.905
4.25	34,280	0.2917	0.261	3.51	0.925		0.908
4.133	33,330	0.300	0.276	3.61	0.925	0.9208 [28]	0.909
4	32,260	0.3100	0.294	3.74	0.925		0.910
3.9	31,460	0.3179	0.310	3.84	0.925		0.911
3.8	30,650	0.3263	0.326	3.95	0.925		0.912
3.7	29,840	0.3351	0.344	4.06	0.925		0.913
3.6	29,040	0.3444	0.364	4.17	0.925		0.913
3.542	28,570	0.350	0.375	4.24	0.925	0.9205	0.914
3.5	28,230	0.3542	0.385	4.30	0.925		0.914

(continued)

TABLE XII (*Continued*)

Metallic Aluminum

eV	cm⁻¹	μm	n	n	k	k	R	R	R	R_hv
3.4	27,420	0.3647	0.407		4.43		0.924			0.914
3.3	26,620	0.3757	0.432		4.56		0.924			0.915
3.2	25,810	0.3875	0.460		4.71		0.924			0.915
3.1	25,000	0.400	0.490		4.86		0.924	0.9194		0.915
3.0	24,200	0.4133	0.523		5.02		0.924			0.914
2.9	23,390	0.4275	0.558		5.20		0.924			0.914
2.8	22,580	0.4428	0.598		5.38		0.924			0.913
2.755	22,220	0.450	0.618		5.47		0.924			0.913
2.7	21,780	0.4592	0.644		5.58		0.924	0.9175		0.913
2.6	20,970	0.4769	0.695		5.80		0.924			0.912
2.5	20,160	0.4959	0.755	0.779 [100]	6.03	5.84 [100]	0.923	0.9162	0.916 [100]	0.911
2.480	20,000	0.500	0.769		6.08		0.923			0.911
2.45	19,760	0.5061	0.789	0.818	6.15	5.93	0.923		0.915	0.911
2.4	19,360	0.5166	0.826	0.852	6.28	6.10	0.923		0.916	0.910
2.35	18,950	0.5276	0.867	0.891	6.42	6.23	0.922		0.916	0.909
2.3	18,550	0.5391	0.912	0.944	6.55	6.36	0.922	0.9157	0.915	0.909
2.254	18,180	0.550	0.958		6.69		0.921			0.908
2.25	18,150	0.5510	0.963	1.00	6.70	6.52	0.921		0.914	0.908
2.2	17,740	0.5636	1.02	1.07	6.85	6.64	0.920		0.912	0.907
2.15	17,340	0.5767	1.08	1.15	7.00	6.77	0.919		0.909	0.906
2.1	16,940	0.5904	1.15	1.22	7.15	6.93	0.918		0.908	0.905
2.066	16,670	0.600	1.20		7.26		0.917	0.9117		0.904
2.05	16,530	0.6048	1.22	1.31	7.31	7.09	0.916		0.906	0.904
2.0	16,130	0.6199	1.30	1.40	7.48	7.21	0.915		0.903	0.903
1.95	15,730	0.6358	1.39	1.51	7.65	7.37	0.913	0.9057	0.900	0.901
1.908	15,390	0.650	1.47		7.79		0.912			0.899
1.9	15,320	0.6526	1.49	1.63	7.82	7.49	0.912		0.897	0.899

1.85	14,920	0.6702	1.60	1.78	8.01	7.66	0.910		0.893	0.896
1.8	14,520	0.6888	1.74	1.94	8.21	7.83	0.907		0.889	0.893
1.771	14,290	0.700	1.83		8.31		0.905	0.8977		0.890
1.75	14,120	0.7085	1.91	2.13	8.39	8.00	0.903		0.885	0.888
1.7	13,710	0.7293	2.14	2.34	8.57	8.15	0.897		0.880	0.882
1.653	13,330	0.750	2.40		8.62		0.888	0.8862		0.876
1.65	13,310	0.7514	2.41	2.57	8.62	8.25	0.888		0.873	0.876
1.6	12,905	0.7749	2.63	2.87	8.60	8.23	0.879		0.861	0.868
1.55	12,500	0.7999	2.80	3.04	8.45	8.08	0.869	0.8773	0.851	0.860
1.503	12,120	0.825	2.75		8.31		0.868	0.8676		0.858
1.5	12,100	0.8266	2.74	2.94	8.31	7.76	0.868	0.8657	0.845	0.858
1.459	11,770	0.850	2.61		8.22		0.870	0.8677		0.859
1.45	11,700	0.8551	2.58	2.53	8.21	7.61	0.872		0.856	0.860
1.417	11,430	0.875	2.38		8.18			0.878	0.8744	0.868
1.4	11,290	0.8856	2.24	2.14	8.21	7.75	0.885		0.878	0.874
1.378	11,110	0.900	2.06		8.30		0.895	0.8908		0.884
1.35	10,890	0.9184	1.86	1.82	8.44	8.14	0.906		0.902	0.897
1.340	10,810	0.925	1.77		8.49		0.911	0.9075		0.902
1.305	10,530	0.950	1.49		8.88		0.930	0.9243		0.916
1.3	10,490	0.9537	1.47	1.58	8.95	8.69	0.932		0.923	0.918
1.25	10,080	0.9919	1.37	1.41	9.49	9.19	0.943		0.937	0.932
1.240	10,000	1.000	1.35		9.58		0.944	0.9402		0.934
1.2	9,679	1.033	1.26	1.32	10.0	9.73	0.952		0.947	0.942
1.15	9,275	1.078	1.21	1.25	10.6	10.3	0.958		0.955	0.948
1.1	8,872	1.127	1.20	1.23	11.2	10.9	0.963		0.960	0.953
1.05	8,469	1.181	1.21	1.22	11.8	11.6	0.967		0.965	0.957
1.033	8,333	1.200	1.21		12.0		0.968	0.9637		0.958
1	8,066	1.240	1.21	1.25	12.5	12.2	0.970		0.967	0.960
0.95	7,662	1.305	1.23	1.30	13.2	12.9	0.973		0.970	0.963
0.9	7,259	1.378	1.26	1.35	14.0	13.7	0.975		0.972	0.964
0.85	6,856	1.459	1.33	1.43	14.9	14.5	0.977		0.974	0.966
0.8266	6,667	1.500	1.38		15.4		0.977	0.9742		0.967

(continued)

TABLE XII (*Continued*)

Metallic Aluminum

eV	cm^{-1}	µm	n	n	k	k	R	R	R	R$_{hv}$
0.8	6,452	1.550	1.44	1.50	16.0	15.5	0.978		0.976	0.967
0.75	6,049	1.653	1.59	1.59	17.1	16.5	0.979		0.977	0.968
0.7	5,646	1.771	1.77	1.75	18.3	17.8	0.979		0.978	0.969
0.65	5,243	1.907	1.99		19.8		0.980			0.970
0.6199	5,000	2.0	2.15		20.7		0.980	0.9779		0.970
0.6	4,839	2.065	2.27		21.4		0.981			0.970
0.55	4,436	2.254	2.62		23.3		0.981			0.971
0.5	4,033	2.480	3.07		25.6		0.982			0.972
0.45	3,629	2.755	3.68		28.3		0.982			0.973
0.4133	3,333	3.0	4.24		30.6		0.982	0.9805		0.973
0.4	3,226	3.100	4.45		31.5		0.983			0.974
0.375	3,025	3.306	4.88		33.4		0.983			0.974
0.35	2,823	3.542	5.44		35.6		0.983			0.975
0.325	2,621	3.815	6.00		38.1		0.984			0.975
0.3100	2,500	4.0	6.43		39.8		0.984	0.9826		0.975
0.3	2,420	4.133	6.76		41.0		0.984			0.976
0.275	2,218	4.509	7.61		44.3		0.985			0.976
0.25	2,016	4.959	8.59		48.2		0.986			0.977
0.2480	2,000	5.0	8.67		48.6		0.986	0.9843		0.977
0.225	1,815	5.510	9.85		53.2		0.987			0.977
0.2066	1,667	6.0	11.1		57.6		0.987	0.9856		0.978
0.2	1,613	6.199	11.7		59.4		0.987			0.978
0.19	1,532	6.526	12.7		62.2		0.987			0.978
0.18	1,452	6.888	13.7		65.2		0.988			0.979
0.1771	1,429	7.0	14.0		66.2		0.988	0.9866		0.979
0.17	1,371	7.293	14.9		68.8		0.988			0.979
0.16	1,290	7.749	16.5		72.7		0.988			0.980
0.1550	1,250	8.0	17.5		74.9		0.988	0.9872		0.980

0.15	1,210	8.266	18.6	77.0
0.14	1,129	8.856	20.9	81.5
0.1378	1,111	9.0	21.5	82.6
0.13	1,049	9.537	23.5	86.5
0.1240	1,000	10.0	25.3	89.8
0.12	967.9	10.33	26.6	92.2
0.1127	909.1	11.0	29.2	96.6
0.11	887.2	11.27	30.2	98.4
0.1033	833.3	12.0	33.0	103
0.1	806.5	12.40	34.5	106
0.09537	769.2	13.0	36.6	109
0.095	766.2	13.05	36.8	110
0.09	725.9	13.78	39.7	114
0.08856	714.3	14.0	40.5	116
0.085	685.6	14.59	42.8	119
0.08	645.2	15.50	46.0	124
0.07749	625.0	16.0	47.7	127
0.075	604.9	16.53	49.7	129
0.07	564.6	17.71	53.8	136
0.06888	555.6	18.0	54.7	137
0.065	524.3	19.07	57.6	143
0.06199	500.0	20.0	60.7	147
0.06	483.9	20.66	62.9	151
0.05636	454.6	22.0	66.9	157
0.055	443.6	22.54	68.3	160
0.05166	416.7	24.0	72.2	168
0.05	403.3	24.80	75.0	172
0.04769	384.6	26.0	79.1	178
0.045	362.9	27.55	84.7	186
0.04428	357.1	28.0	86.3	189
0.04133	333.3	30.0	94.2	199
0.04	322.6	31.00	98.6	204
0.03875	312.5	32.0	103	208

0.988	0.9874	0.980
0.988	0.9876	0.980
0.988	0.9879	0.980
0.988	0.9882	0.981
0.988	0.9884	0.981
0.989	0.9886	0.981
0.989	0.9892	0.982
0.989	0.9896	0.982
0.989	0.9902	0.982
0.989	0.9907	0.982
0.989	0.9912	0.982
0.989	0.9918	0.982
0.989	0.9923	0.982
0.989	0.9928	0.983
0.989	0.992	0.983
0.990	0.9933	0.983
0.990		0.983
0.990		0.983
0.990		0.983
0.990		0.983
0.990		0.983
0.990		0.983
0.991		0.983
0.991		0.983
0.991		0.984
0.991		0.984
0.992		0.984
0.992		0.984
0.992		0.984
0.992		0.984
0.992		0.984
0.992		0.984
0.992		0.985
		0.985
		0.985
		0.985
		0.985
		0.985

TABLE XIII

The Dielectric Function for Evaporated Metallic Aluminum Films at Room Temperature from 0.04 to 10,000 eV^{a-d}

eV	Å	ε_1	ε_2	eV	Å	ε_1	ε_2
10,000	1.240	0.99999	1.42×10^{-7}	1,556	7.968	0.99982	6.209
9,000	1.378	0.99999	2.16	1,555	7.973	0.99983	5.909
8,000	1.550	0.99998	3.40	1,550	7.999	0.99978	1.200
7,000	1.771	0.99998	5.80×10^{-7}	1,540	8.051	0.99974	1.230
6,000	2.066	0.99997	1.046×10^{-6}	1,530	8.104	0.99971	1.260
5,000	2.480	0.99996	2.08	1,520	8.157	0.99969	1.290
4,000	3.100	0.99993	4.900	1,510	8.211	0.99968	1.324
3,500	3.542	0.99991	8.006×10^{-6}	1,500	8.266	0.99966	1.356
3,000	4.133	0.99988	1.420×10^{-5}	1,450	8.551	0.99961	1.560
2,500	4.959	0.99983	2.718	1,400	8.856	0.99957	1.780
2,000	6.199	0.99974	6.406	1,300	9.537	0.99947	2.319
1,900	6.526	0.99972	7.691	1,200	10.33	0.99936	3.119
1,800	6.888	0.99970	9.426×10^{-5}	1,100	11.27	0.99922	4.312
1,700	7.293	0.99968	1.159×10^{-4}	1,000	12.40	0.99904	6.222
1,650	7.514	0.99968	1.309	900	13.78	0.99879	9.355×10^{-5}
1,600	7.749	0.99970	1.476	800	15.50	0.99845	1.462×10^{-4}
1,590	7.798	0.99971	1.520	700	17.71	0.99796	2.363
1,580	7.847	0.99973	1.570	600	20.66	0.99717	4.408
1,570	7.897	0.99976	1.622	500	24.80	0.99590	8.695×10^{-4}
1,565	7.922	0.99979	1.686	400	31.00	0.99378	1.939×10^{-3}
1,564	7.927	0.99981	1.709	300	41.33	0.9898	4.684
1,563	7.933	0.99983	1.636	250	49.59	0.9862	8.325×10^{-3}
1,562	7.938	0.99985	1.437	200	61.99	0.9820	1.490×10^{-2}
1,561	7.943	0.99985	1.220	190	65.26	0.9810	1.680
1,560	7.948	0.99985	1.000×10^{-4}	180	68.88	0.9797	1.913
1,559	7.953	0.99985	8.229×10^{-5}	170	72.93	0.9776	2.174
1,558	7.958	0.99984	6.583	160	77.49	0.9785	2.718
1,557	7.963	0.99983	5.860	150	82.68	0.9791	2.925

145	85.51	0.9782	3.055×10^{-2}	88	140.9	1.010	5.823×10^{-2}
140	88.56	0.9772	3.225	86	144.2	1.011	5.680
135	91.84	0.9756	3.511	84	147.6	1.013	5.402
130	95.37	0.9746	4.070	82	151.2	1.014	5.129
125	99.19	0.9780	4.634	80	155.0	1.013	4.928
120	103.3	0.9814	4.769	79	156.9	1.013	4.933
115	107.8	0.9840	4.947	78	159.0	1.014	4.987
110	112.7	0.9871	5.059	77	161.0	1.018	5.082
105	118.1	0.9860	4.850	76	163.1	1.020	4.818
100	124.0	0.9804	5.927	75.5	164.2	1.021	4.841
98	126.5	0.9832	6.559	75	165.3	1.022	4.856
96	129.2	0.9915	7.150	74.5	166.4	1.025	4.928
94	131.9	0.9998	7.066	74	167.5	1.029	4.904
92	134.8	1.007	6.706	73.9	167.8	1.030	4.915
90	137.8	1.010	6.218	73.8	168.0	1.031	4.928

(continued)

ᵃ Based on Shiles et al. [7].

ᵇ These data are based primarily on the analysis of reflectance measurements below the $L_{II,III}$ edge and on transmission measurements above the edge. They are in close agreement with direct ellipsometric measurements on oxide-free films by Mathewson and Myers [100] in the range 0.7 to 2.5 eV (see Table XII). Suggested extrapolations are given in the footnotes c and d.

ᶜ Low-energy extrapolations. The room-temperature dielectric function presented here may be extrapolated to lower energies by using the Drude model [Eq. (1)] with $\hbar\Omega_p = 11.5$ eV and $\hbar\gamma = 50.6$ meV [$\sigma(0) = 3.16 \times 10^{17}$ sec^{-1}]. This extrapolation fits on continuously at 0.04 eV, but with a small discontinuity in slope. However, it should be recognized that, while the dc conductivity is in excellent agreement with the measured value ($3.18 - 3.28 \times 10^{17}$ sec^{-1}) [16, 17], Ω_p, and γ are somewhat outside the range recommended by Smith and Segall [5] (see Table XI). This represents a negligible conflict since, for the high reflectances ($>99\%$) beyond 20 μm, the experimental error ($\pm 0.1\%$) in $\varepsilon(\omega)$ in the reflectance measurements used to obtain the dielectric function translates into an uncertainty of the order of $\pm 25\%$ in $\varepsilon(\omega)$ even in favorable cases. The values of Smith and Segall [15] arise from fits over the range 0.04–3 eV, not just the lowest energy point, which is subject to considerable uncertainty.

ᵈ High-energy extrapolations. For $\varepsilon_1(\omega)$: at frequencies well above the K edge the real part of the dielectric function approaches $\varepsilon_1 \approx 1 - \omega_{p,t}^2/\omega^2$, where $\omega_{p,t}$ is the plasma frequency for the total electron density (13 e/at.). For aluminum at room temperature $\omega_{p,t} \approx 32.86$ eV and the asymptotic region is achieved by $\omega \simeq 3000$ eV. For $\varepsilon_2(\omega)$: beyond the K edge $\varepsilon_2(\omega)$ falls off very nearly as a power of energy $\varepsilon_2(\omega) \sim \omega^{-\delta}$. The exponent varies slowly with energy. At 5000 eV, $\delta \sim 3.8$; at 10^4 eV, $\delta \sim 4.0$; and from 10^4 to 10^5 eV, it increases to $\delta = 4.2$. An extrapolation from 10^4 eV using $\delta = 4.1$ reproduces photoelectric data up to 10^5 eV reasonably well. Theory predicts an asymptotic value of $\delta = 4.5$ [143].

TABLE XIII (Continued)

Metallic Aluminum

eV	Å	ε_1	ε_2	eV	Å	ε_1	ε_2
73.7	168.2	1.032	4.949×10^{-2}	55	225.4	0.9581	9.942×10^{-3}
73.6	168.5	1.033	4.976	50	248.0	0.9389	1.138×10^{-2}
73.5	168.7	1.035	5.048	45	275.5	0.9153	1.304
73.4	168.9	1.038	5.177	40	310.0	0.8827	1.533
73.3	169.1	1.043	5.308	35	354.2	0.8368	1.797
73.2	169.4	1.050	4.955	30	413.3	0.7665	2.197
73.1	169.6	1.052	3.998	25	495.9	0.6536	2.863
73	169.8	1.049	3.910	20	619.9	0.4460	3.582
72.9	170.1	1.051	4.089	19	652.6	0.3835	3.745
72.8	170.3	1.061	4.118	18	688.8	0.3103	3.885
72.7	170.5	1.069	2.549×10^{-2}	17	729.3	0.2232	4.013
72.6	170.8	1.061	8.139×10^{-3}	16	774.9	0.1195	4.174
72.5	171.0	1.050	7.264	15.8	784.7	0.09693	4.161
72.4	171.3	1.044	6.789	15.6	794.8	0.07327	4.081
72.3	171.5	1.040	6.871	15.4	805.1	0.04769	3.941
72.2	171.7	1.037	6.940	15.2	815.7	0.02019	3.855
72.1	172.0	1.035	6.949	15.0	826.6	−0.007825	3.834
72	172.2	1.033	7.042	14.8	837.7	−0.03727	3.613
71	174.6	1.020	6.982	14.6	849.2	−0.07038	3.644
70	177.1	1.012	7.086	14.4	861.0	−0.1033	3.802
68	182.3	1.002	7.339	14.2	873.1	−0.1365	3.928
66	187.9	0.9940	7.776	14.0	885.6	−0.1712	4.011
64	193.7	0.9876	8.204	13.5	918.4	−0.2656	4.229
62	200.0	0.9810	8.457	13.0	953.7	−0.3701	4.588
60	206.6	0.9746	8.709	12.5	991.9	−0.4887	4.814

eV	µm	ε_1	ε_2	eV	µm	ε_1	ε_2
12.0	0.1033	−0.6241	5.190×10^{-2}	1.90	0.6526	−58.96	23.28
11.5	0.1078	−0.7794	5.854	1.85	0.6702	−61.65	25.72

11.0	0.1127	−0.9562	6.971×10^{-2}	1.80	0.6888	−64.29	28.57
10.5	0.1181	−1.157	8.517×10^{-2}	1.75	0.7085	−66.68	32.00
10.0	0.1240	−1.386	0.1042	1.70	0.7293	−68.91	36.74
9.5	0.1305	−1.651	0.1273	1.65	0.7514	−68.52	41.61
9.0	0.1378	−1.962	0.1560	1.60	0.7749	−67.03	45.13
8.5	0.1459	−2.327	0.1922	1.58	0.7847	−66.27	46.67
8.0	0.1550	−2.762	0.2381	1.56	0.7948	−64.31	47.38
7.5	0.1653	−3.285	0.2974	1.54	0.8051	−62.95	47.06
7.0	0.1771	−3.923	0.3753	1.52	0.8157	−62.05	46.39
6.5	0.1907	−4.711	0.4791	1.50	0.8266	−61.50	45.61
6.0	0.2066	−5.700	0.6208	1.48	0.8377	−60.96	44.56
5.5	0.2254	−6.965	0.8192	1.46	0.8492	−60.71	43.08
5.0	0.2480	−8.617	1.120	1.44	0.8610	−60.94	41.51
4.5	0.2755	−10.83	1.538	1.42	0.8731	−61.07	39.28
4.0	0.3100	−13.90	2.203	1.40	0.8856	−62.42	36.74
3.8	0.3263	−15.47	2.575	1.38	0.8984	−64.45	34.50
3.6	0.3444	−17.29	3.035	1.36	0.9117	−66.70	32.54
3.4	0.3647	−19.42	3.606	1.34	0.9253	−69.01	30.03
3.2	0.3875	−21.95	4.327	1.32	0.9393	−72.93	27.48
3.0	0.4133	−24.97	5.258	1.30	0.9537	−77.92	26.28
2.9	0.4275	−26.71	5.800	1.28	0.9686	−82.27	26.11
2.8	0.4428	−28.64	6.440	1.26	0.9840	−86.20	26.10
2.7	0.4592	−30.78	7.188	1.24	0.9999	−89.97	25.92
2.6	0.4769	−33.15	8.066	1.22	1.016	−93.97	25.47
2.5	0.4959	−35.81	9.111	1.20	1.033	−98.61	25.23
2.4	0.5166	−38.79	10.38	1.15	1.078	−110.5	25.66
2.3	0.5391	−42.13	11.96	1.10	1.127	−123.6	26.86
2.2	0.5636	−45.83	13.94	1.05	1.181	−137.7	28.45
2.1	0.5904	−49.86	16.43	1.00	1.240	−153.9	30.21
2.0	0.6199	−54.24	19.50	0.95	1.305	−172.8	32.45
1.95	0.6358	−56.52	21.25	0.90	1.378	−195.0	35.45

(continued)

TABLE XIII (Continued)

Metallic Aluminum

eV	μm	ε_1	ε_2	eV	μm	ε_1	ε_2
0.85	1.459	−221.5	39.82	0.23	5.391	−2,622	994.9
0.80	1.550	−252.5	46.07	0.22	5.636	−2,845	1,104
0.75	1.653	−289.0	54.33	0.21	5.904	−3,096	1,231
0.70	1.771	−332.8	64.88	0.20	6.199	−3,387	1,393
0.65	1.907	−386.4	78.53	0.19	6.526	−3,703	1,578
0.60	2.066	−452.9	97.30	0.18	6.888	−4,065	1,791
0.55	2.254	−536.6	122.2	0.17	7.293	−4,512	2,054
0.50	2.480	−645.0	157.2	0.16	7.749	−5,016	2,407
0.45	2.755	−785.7	208.3	0.15	8.266	−5,578	2,859
0.40	3.100	−971.5	280.4	0.14	8.856	−6,205	3,410
0.39	3.179	−1,018	296.8	0.13	9.537	−6,930	4,065
0.38	3.263	−1,068	315.8	0.12	10.33	−7,787	4,899
0.37	3.351	−1,122	337.7	0.11	11.27	−8,769	5,940
0.36	3.444	−1,178	361.8	0.10	12.40	−9,964	7,279
0.35	3.542	−1,238	387.1	0.095	13.05	−10,720	8,080
0.34	3.647	−1,302	413.6	0.090	13.78	−11,450	9,049
0.33	3.757	−1,374	441.6	0.085	14.59	−12,260	10,160
0.32	3.875	−1,454	473.8	0.080	15.50	−13,210	11,380
0.31	4.000	−1,540	511.5	0.075	16.53	−14,290	12,860
0.30	4.133	−1,632	553.7	0.070	17.71	−15,470	14,580
0.29	4.275	−1,731	598.7	0.065	19.07	−16,990	16,420
0.28	4.428	−1,841	647.0	0.060	20.66	−18,790	18,960
0.27	4.592	−1,964	701.3	0.055	22.54	−20,840	21,810
0.26	4.769	−2,098	760.6	0.050	24.80	−24,030	25,830
0.25	4.959	−2,253	828.3	0.045	27.55	−27,580	31,590
0.24	5.166	−2,424	905.2	0.040	31.00	−31,770	40,170

2
Semiconductors

Cadmium Telluride
(CdTe)

EDWARD D. PALIK

Naval Research Laboratory

Washington, D.C. 20375

The UV–visible spectrum has been measured by wavelength ellipsometry by Marple and Ehrenreich [1] in the 5–1.5-eV region. Samples were high-resistivity, single crystals with cleaved surfaces. The measurements were completed in less than an hour after cleavage in air. The temperature was 300 K. The reflectivity has been measured to higher photon energies, but no Kramers–Kronig analysis was done [2]. The positions of peaks in ε_1 and ε_2 were in agreement with reflectivity peaks. Table I lists n and k as determined from reading the graph in Marple and Ehrenreich [1]. Figure 1 shows n and k.

Myers *et al.* [3] have measured n and k for both single crystals and poly-crystalline film samples in the region 0.45–1.9 μm. The films were made by sublimation of the compound with deposition on heated fused-quartz substrates. Their resistivities were $> 10^8 \, \Omega$ cm. Films of cubic structure as well as films with mixed cubic-hexagonal structure were produced. However, the percentage mixture was not clearly determined. The thin-film constants were determined directly from measurements of R (at 6° by using an absolute reflectance attachment) and T. As expected, the constants varied with substrate preparation temperature and probably with cubic-hexagonal mixing. Roughly, the higher the temperature, the larger was n, while the reverse was true for k. The single-crystal material was mechanically polished with 0.25-μm grit. Only the reflectance was measured directly. In this case average values of k for the films were assumed, and then n was calculated. This was a reasonable assumption, since $n > k$ (~ 3 compared to ~ 0.4 at 0.5 μm). For the deposited films k ranged from 0.35 to 0.45 at 0.5 μm, for example, while its value from Marple and Ehrenreich [1] is 0.5. We have chosen to list the single-crystal data of Myers *et al.* [3] for n for comparison with the data of Marple and Ehrenreich [1]. In the transparent region to 1.9 μm, the n values [3] agree with Marple's values [4] to ~ 0.01. In the opaque region, they are lower than the values of Marple and Ehrenreich [1] by ~ 0.1. We also list an average k determined for films because of the peculiar behavior of k of

409

HANDBOOK OF OPTICAL CONSTANTS OF SOLIDS

ISBN 0-12-544420-6

Marple and Ehrenreich [1] near the fundamental band gap; also the film data tend to extrapolate to agree with data of Marple [5].

The absorption at the fundamental band edge was measured by Marple [5] and Konak et al. [6]. In Marple's [5] research a series of slab samples were mechanically polished with Linde A to thicknesses of 3 mm to 100 μm and then chemically polished to remove 10 μm of additional material. Temperature was 297 K. The results of Marple's [5] findings, which are tabulated, and those of Konak et al. [6] were reasonably close at 1.47 eV within 20%, but below 1.43 eV they deviated to a factor of 4. Konak et al.'s [6] k value remains excessively high, suggesting impurity- and defect-absorption problems.

Marple [4] measured the refractive index by minimum deviation with a prism made of melt-grown material with mechanically polished surfaces. The data in the 1.45–0.5-eV region were fitted with the formula

$$n^2 = A + [B\lambda^2]/[\lambda^2 - C^2], \tag{1}$$

with $A = 5.68$, $B = 1.53$, and $C^2 = 0.366$ μm^2. Notice that at large λ, $n^2 \rightarrow A + B = 7.21 = \varepsilon_\infty$, the high-frequency dielectric constant often used in the Lorentz- oscillator-model equation given later. Random error in n was quoted as ± 0.001, while systematic error was ± 0.002.

Pikhtin and Yas'kov [7] have refitted these data with a more detailed oscillator-model formula of the form

$$n^2 = 1 + \frac{A}{\pi} \ln \frac{E_1^2 - (\hbar\omega)^2}{E_0^2 - (\hbar\omega)^2} + \frac{G_1}{E_1^2 - (\hbar\omega)^2} + \frac{G_2}{E_2^2 - (\hbar\omega)^2} + \frac{G_3}{E_3^2 - (\hbar\omega)^2}, \tag{2}$$

with $E_0 = 1.5$ eV, $E_1 = 3.5$ eV, $E_2 = 5.5$ eV, $E_3 = 17.4 \times 10^{-3}$ eV, $G_1 = 64.375$ eV^2, $G_2 = 17.664$ eV^2, $G_3 = 1.08 \times 10^{-3}$ eV^2, $A = 0.7\sqrt{E_0} = 0.857$. A lattice-vibration term G_3 presumably improves the fit at lower energy. We have listed these values of n for the spectral region 1.45–0.05 eV. Where they overlap Marple's data [4], they are in agreement to ± 0.003.

Danielewicz and Coleman [8] quote n data of Ladd [9] for hot-pressed CdTe in the near infrared at 300 K and have fitted it with an oscillator model together with their single-crystal n data in the 15–45-cm^{-1} region. We list n for some near-infrared wavelengths for comparison with the results of Pikhtin and Yas'kov [7]. Agreement is usually to one digit or better in the third significant figure. However, the agreement among three types of CdTe samples in Ladd [9], Harvey and Wolfe [10], and De Bell et al. [11] at 8 μm, for example, suggests that the n values of Pihktin and Yas'kov [7] are about 0.01 too small at the longer wavelengths.

The far IR has been studied by Danielewicz and Coleman [8], Mitsuishi et al. [12, 13], Manabe et al. [14, 15], Vodop'yanov et al. [16], and Johnson et al. [17]. The reststrahlen reflectivity spectrum has been measured at 300 K at an angle of incidence of 12° [12, 14]. The surface was mechanically polished. The reflectivity was fitted by using both KK analysis and an oscillator-model fit [12, 14]. The resultant values of n and k agree to $\pm 30\%$ in the spectral

region 50–75 μm. The oscillator-model fit is listed in Table I. It is obtained
with the formula

$$(n - ik)^2 = \varepsilon_\infty \left[1 + \frac{(\omega_L^2 - \omega_T^2)}{\omega_T^2 - \omega^2 + i\Gamma\omega} \right], \tag{3}$$

with the parameters $\varepsilon_\infty = 7.1$, $\omega_T = 141$ cm^{-1}, $\omega_L = 169$ cm^{-1}, and $\Gamma = 6.6$ cm^{-1}; T = transverse and L = longitudinal. We have allowed some overlap with the n of Pikhtin and Yas'kov [7] for comparison.

While we have concentrated on the oscillator fit to reflectivity, which is in
reasonable agreement with the KK analysis, Birch and Murray [18] have
carried out asymmetric Fourier-transform spectroscopy at 290 K to obtain
directly the complex dielectric function (or complex refractive index). Their n
and k data show ripples that have been analyzed in detail as multiphonon
structure below and above the resonant frequency [19]. The complex
frequency-dependent dielectric function is given as

$$\varepsilon(v) = \varepsilon_\infty + \frac{(\varepsilon_0 - \varepsilon_\infty)v(oj)^2}{v(oj)^2 - v^2 + 2v(oj)[\Delta(oj, v) - i\Gamma(oj, v)]}, \tag{4}$$

where $v(oj)$ is the harmonic frequency of the TO phonon and Δ and Γ are
the real and imaginary parts, respectively, of the irreducible self-energy of the
TO phonon. After determining ε experimentally, analysis gives Γ, which
suppresses the TO resonance but shows much weak structure that can be
assigned to a large number of multiphonon modes through which the TO
phonon decays. We have determined the values of n and k in the reststrahlen
region by reading a graph but smoothing out the ripples and therefore losing
the details. These are listed in Table I and are probably the best numbers
in this spectral region. While we have plotted the oscillator fit in the region
180–120 cm^{-1} (55.56–83.33 μm), there is no difference on the coarse scale
of the graph. However, at longer wavelengths to 125 μm, the values are larger
than the oscillator fit and begin to approach the data of Manabe et al. [15]
somewhat better than the oscillator fit.

Danielewicz and Coleman [8] and Johnson et al. [17] fitted n data on
both sides of ω_T with similar Lorentz oscillators as Manabe et al. [14] and
obtained similar parameters. In view of the fact that the data of Manabe et
al. [14] fit the experimental reflectivity over a wider range and that we do
not know how well the other two fits of n actually fit the reflectivity, we chose
to tabulate the data of Manabe et al. [14].

Vodop'yanov et al. [16] did a more detailed study of the reststrahlen spectrum that demonstrated changes in R with surface preparation and analyzed R by KK analysis. Unfortunately, they did not present numerical data.
They also applied an oscillator model and discussed the frequency dependence of the lattice damping constant, which is neglected in the usual fitting
procedure.

Multiphonon absorption straddles the resonant frequency ω_T for many tens of reciprocal centimeters. The two-phonon summation band region from 33 to 52 μm has been measured by transmission at 300 K [15], and the k data as read from a graph are tabulated. The data of Johnson et al. [17] and Stafsudd et al. [20] are in reasonable agreement with this result. Note that the oscillator-model fit for k is larger than the experimental values at 49 μm, indicating a poor fit. The 20–26-μm region (three-phonon absorption) has been measured by Deutsch [21], who also quotes the absorption coefficient at 10.6 μm. Data of Sherman [22] are quoted in Danielewicz and Coleman [8] for the 22–30-μm region.

The low-frequency region below ω_T has been measured in some detail [8, 15, 17] with transmission samples at 300 K. Some of the k values determined are listed and are to be compared with the oscillator-model values. The samples of Johnson et al. [17] had resistivities of $8.5 \times 10^5 \ \Omega$ cm. Danielewicz and Coleman [8] made measurements by using the channel-spectrum technique with both Fourier-transform and grating spectrometers. Samples were single crystals with resistivities of $10^8 \ \Omega$ cm polished with alumina. They also quote k values that were presumably obtained from thicker samples. The data from different laboratories differ by factors of 2, and it is not clear what structural differences would cause such differences in the optical-phonon features. At longer wavelengths of $> 100 \ \mu$m, some free-carrier absorption might be present.

We have included n data for two other types of CdTe, hot-pressed [9, 10] and chemical vapor deposition (CVD) [11] in the transparent IR region. These measurements, done by a modified minimum-deviation technique [23], give six significant figures. Accuracies are about one part in the fourth decimal place, so that the change in index as a function of temperature could be closely followed.

For the hot-pressed material of Harvey and Wolfe [10] the temperature was 297.5 K. The mean value of dn/dT at 10 μm was 0.951×10^{-4}/K. For the hot-pressed material of Ladd [9] room temperature was not specified. For CVD material the temperature was 300 K. A Sellmeier equation of the following form was fitted to the data:

$$n^2 - 1 = A_1\lambda^2/(\lambda^2 - \lambda_1^2) + A_2\lambda^2/(\lambda^2 - \lambda_2^2), \qquad (5)$$

with $A_1 = 6.1977889$, $A_2 = 3.2243821$, $\lambda_1^2 = 0.1005326 \ \mu$m^2, and $\lambda_2^2 = 5279.518 \ \mu$m^2. A Herzberger equation gave a similar fit with the residual no more than one part in the fourth decimal place. The temperature coefficient in the range 80–300 K is 1.0×10^{-4}/K.

Table I shows a little systematic variation in n from single crystal to hot pressed to CVD material. Since the temperature varies about three degrees kelvin among these materials the fourth decimal place is touchy.

The far-IR (50–10-cm^{-1}) properties of hot-pressed Irtran VI have been measured by Randall and Rawcliffe [24]. Room temperature was not specified. The n values are slightly larger in the third significant figure than those of Danielewicz and Coleman [8], but the k values are consistently about a factor of 2 larger than for crystalline material.

REFERENCES

1. D. T. F. Marple and H. Ehrenreich, *Phys. Rev. Lett.* **8**, 87 (1962).
2. M. Cardona and D. C. Greenaway, *Phys. Rev.* **131**, 98 (1963).
3. T. H. Myers, S. W. Edwards, and J. F. Schetzina, *J. Appl. Phys.* **52**, 4231 (1981); T. H. Myers, S. W. Edwards, and J. F. Schetzina, private communication (1983).
4. D. T. F. Marple, *J. Appl. Phys.* **35**, 539 (Part 1) (1964).
5. D. T. F. Marple, *Phys. Rev.* **150**, 728 (1966).
6. C. Konak, J. Dillinger, and V. Prosser, *Proc. II–VI Semiconducting Compounds, 1967 Int. Conf.* (D. G. Thomas, ed.), p. 850, Benjamin, New York, 1967.
7. A. N. Pikhtin and A. D. Yas'kov, *Sov. Phys. Semicond.* **12**, 622 (1978); A. N. Pikhtin and A. D. Yas'kov, *Fiz. Tekh. Poluprovodn.* **12**, 1047 (1978).
8. E. J. Danielewicz and P. D. Coleman, *Appl. Opt.* **13**, 1164 (1974).
9. L. S. Ladd, *Infrared Phys.* **6**, 145 (1966).
10. J. E. Harvey and W. L. Wolfe, *J. Opt. Soc. Am.* **65**, 1267 (1975).
11. A. G. DeBell, E. L. Dereniak, J. Harvey, J. Nissley, J. Palmer, A. Selvarajin, and W. L. Wolfe, *Appl. Opt.* **18**, 3114 (1979).
12. A. Mitsuishi, *J. Phys. Soc. Jpn.* **16**, 533 (1961).
13. A. Mitsuishi, H. Yoshinaga, K. Yata, and A. Manabe, *Jpn. J. Appl. Phys.* **4**, 581 (1965).
14. A. Manabe, A. Mitsuishi, and H. Yoshinga, *Jpn. J. Appl. Phys.* **6**, 593 (1967).
15. A. Manabe, A. Mitsuishi, H. Yoshinaga, Y. Ueda, and H. Sei, *Technol. Rep. Osaka Univ. Jpn.* **17**, 263 (1967).
16. L. K. Vodop'yanov, E. A. Vinogradov, V. V. Kolotkov, and Y. A. Mityagin, *Sov. Phys.-Solid State* **16**, 912 (1974); L. K. Vodop'yanov, E. A. Vinogradov, V. V. Kolotkov, and Y. A. Mityagin, *Fiz. Tverd. Tela.* **16**, 1419 (1974).
17. C. J. Johnson, G. H. Sherman, and R. Weil, *Appl. Opt.* **8**, 1667 (1969).
18. J. R. Birch and D. K. Murray, *Infrared Phys.* **18**, 283 (1978).
19. T. J. Parker, J. R. Birch, and C. L. Mok, *Solid State Commun.* **36**, 581 (1980).
20. O. M. Stafsudd, F. A. Haak, and K. Radisavljevic, *J. Opt. Soc. Am.* **57**, 1475 (1967).
21. T. F. Deutsch, *J. Phys. Chem. Solids* **34**, 2091 (1973).
22. G. H. Sherman, Ph.D. dissertation, University of Illinois, Urbana, Illinois (1972).
23. B. C. Platt, H. W. Jeenogle, J. E. Harvey, R. Korinski, and W. Wolfe, *J. Opt. Soc. Am.* **654**, 1264 (1975).
24. C. M. Randall and R. D. Rawcliffe, *Appl. Opt.* **7**, 213 (1968).

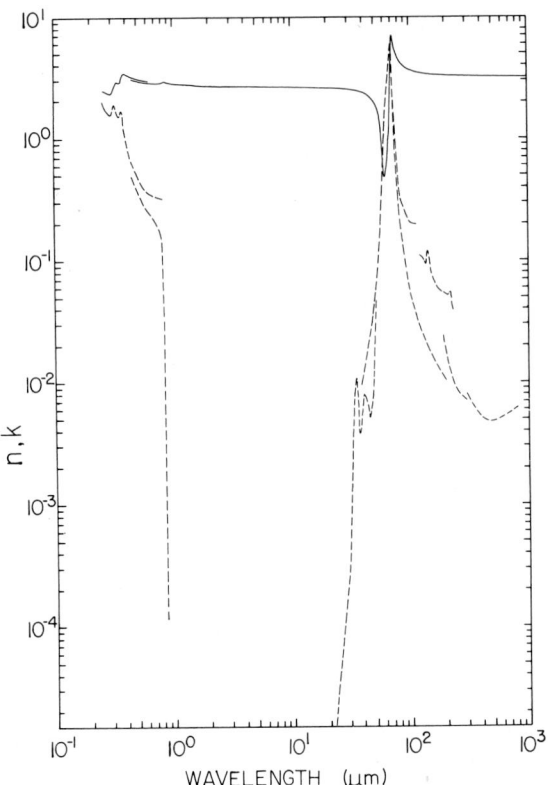

Fig. 1. Log–log plot of n (——) and k (----) versus wavelength in micrometers for cadmium telluride. Data from various references sometimes do not join together in overlap regions.

TABLE I

Values of n and k for Cadmium Telluride Obtained from Various References[a]

eV	cm^{-1}	μm	n	n	k	k
4.9	39,520	0.2530	2.48 [1]			2.04 [1]
4.8	38,710	0.2583	2.49			1.89
4.7	37,910	0.2638	2.48			1.80
4.6	37,100	0.2695	2.45			1.73
4.5	36,290	0.2755	2.43			1.67
4.4	35,490	0.2818	2.41			1.64
4.3	34,680	0.2883	2.38			1.60
4.2	33,880	0.2952	2.36			1.57
4.1	33,070	0.3024	2.33			1.59
4.0	32,260	0.3100	2.39			1.71
3.9	31,460	0.3179	2.57			1.90
3.8	30,650	0.3263	2.80			1.77
3.7	29,840	0.3351	2.92			1.61
3.6	29,040	0.3444	2.93			1.54
3.5	28,230	0.3542	2.89			1.52
3.4	27,420	0.3647	2.91			1.67
3.3	26,620	0.3757	3.30			1.67
3.2	25,810	0.3875	3.44			1.16
3.1	25,000	0.4000	3.43			1.02
3.0	24,200	0.4133	3.37			0.861
2.755	22,220	0.45	3.080 [3]			0.485 [3]
2.75	22,180	0.4509	3.23 [1]			0.636 [1]
2.610	21,050	0.475	3.045 [3]			
2.5	20,160	0.4959	3.14 [1]			0.525

[a] The reference from which the values were extracted is given in brackets next to the first n and first k value. Reading down, the reference is the same until another set of brackets appears for a new reference.

(continued)

TABLE I (*Continued*)

Cadmium Telluride

eV	cm⁻¹	μm	n	n	n	k	k	k
2.480	20,000	0.50	3.010 [3]				0.380 [3]	
2.362	19,050	0.525	2.970				0.300	
2.254	18,180	0.55	2.945					
2.25	18,150	0.5510	3.05 [1]				0.411 [1]	
2.156	17,390	0.575	2.917 [3]					
2.066	16,670	0.60	2.894				0.261 [3]	
2.0	16,130	0.6199	2.99 [1]				0.351 [1]	
1.984	16,000	0.625	2.873 [3]					
1.907	15,380	0.65	2.860				0.238 [3]	
1.837	14,810	0.675	2.860					
1.771	14,290	0.70	2.861				0.210	
1.75	14,110	0.7085	2.95 [1]				0.338 [1]	
1.710	13,790	0.725	2.868 [3]					
1.653	13,330	0.75	2.870				0.185 [3]	
1.631	13,160	0.76					0.179	
1.610	12,990	0.77					0.174	
1.600	12,900	0.775	2.872					
1.590	12,820	0.78					0.165	
1.569	12,740	0.79					0.155	
1.550	12,500	0.80	2.876				0.125	
1.531	12,350	0.81					8.0×10^{-2}	
1.512	12,200	0.82	2.880				4.0×10^{-2}	
1.50	12,100	0.8266	2.98 [1]				0.319 [1]	
1.494	12,050	0.83					2.0×10^{-2} [3]	
1.475	11,900	0.840	2.905 [3]				1.34×10^{-3} [5]	
1.47	11,860	0.8434					6.71×10^{-4}	
1.465	11,820	0.8463					3.37	

1.46	11,780	0.8492	2.948	1.89
1.459	11,760	0.850		
1.455	11,740	0.8521	2.9565 [7]	
1.45	11,690	0.8551		1.08×10^{-4}
1.445	11,650	0.8580		5.10×10^{-5}
1.442	11,630	0.860		2.73
1.44	11,610	0.8610	2.952 [3]	
1.43	11,530	0.8670	2.9479 [7]	1.37
1.42	11,450	0.8731	2.9402	
1.41	11,370	0.8793	2.9331	
1.409	11,360	0.880	2.9266	
1.40	11,290	0.8856	2.920 [3]	
1.39	11,210	0.8920	2.9205 [7]	
1.38	11,130	0.8984	2.9147	
1.378	11,110	0.900	2.9092	
1.36	10,970	0.9116	2.899 [3]	
1.35	10,890	0.9184	2.8989 [7]	
1.34	10,810	0.9252	2.8940	
1.340	10,810	0.925	2.8894	
1.33	10,730	0.9322	2.878 [3]	
1.32	10,650	0.9393	2.8848 [7]	
1.31	10,570	0.9464	2.8804	
1.305	10,530	0.950	2.8761	
1.30	10,490	0.9537	2.860 [3]	
1.29	10,400	0.9611	2.8720 [7]	
1.28	10,320	0.9686	2.8679	
1.272	10,260	0.975	2.8640	
1.27	10,240	0.9762	2.850 [3]	
1.26	10,160	0.9840	2.8601 [7]	
1.25	10,080	0.9919	2.8563	
1.24	10,000	0.9999	2.8527	
			2.8490	

(continued)

TABLE I (*Continued*)

Cadmium Telluride

eV	cm^{-1}	μm	n	n	n	k	k	k
1.240	10,000	1.0	2.840 [3]					
1.23	9,921	1.008	2.8455 [7]					
1.22	9,840	1.016	2.8420					
1.21	9,759	1.025	2.8386					
1.20	9,679	1.033	2.8353					
1.19	9,598	1.042	2.8320					
1.18	9,518	1.051	2.8288					
1.17	9,437	1.060	2.8257					
1.16	9,356	1.069	2.8226					
1.15	9,275	1.078	2.8195					
1.14	9,195	1.088	2.8165					
1.13	9,114	1.097	2.8136					
1.12	9,034	1.107	2.8107					
1.11	8,953	1.117	2.8078					
1.10	8,872	1.127	2.8050					
1.09	8,792	1.137	2.8023					
1.08	8,711	1.148	2.7996					
1.07	8,630	1.159	2.7969					
1.06	8,550	1.170	2.7943					
1.05	8,469	1.181	2.7917					
1.04	8,388	1.192	2.7891					
1.03	8,308	1.204	2.7866					
1.02	8,227	1.216	2.7842					
1.01	8,146	1.228	2.7817					
1.00	8,065	1.240	2.7793					
0.99	7,985	1.252	2.7770					
0.98	7,904	1.265	2.7747					

0.97	7,824	1.278	2.7724
0.96	7,743	1.291	2.7701
0.95	7,662	1.305	2.7679
0.94	7,582	1.319	2.7657
0.93	7,501	1.333	2.7636
0.92	7,420	1.348	2.7615
0.91	7,340	1.362	2.7594
0.90	7,259	1.378	2.7573
0.89	7,170	1.393	2.7553
0.88	7,098	1.409	2.7533
0.87	7,017	1.425	2.7514
0.86	6,936	1.442	2.7494
0.85	6,856	1.459	2.7475
0.84	6,775	1.476	2.7457
0.83	6,694	1.494	2.7438
0.82	6,614	1.512	2.7420
0.81	6,533	1.531	2.7402
0.80	6,452	1.550	2.7384
0.79	6,372	1.569	2.7367
0.78	6,291	1.589	2.7350
0.77	6,210	1.610	2.7333
0.76	6,130	1.631	2.7317
0.75	6,049	1.653	2.7301
0.74	5,968	1.675	2.7285
0.73	5,888	1.698	2.7269
0.72	5,807	1.722	2.7253
0.71	5,726	1.746	2.7238
0.70	5,646	1.771	2.7223
0.69	5,565	1.797	2.7208
0.68	5,485	1.823	2.7194
0.67	5,404	1.851	2.7180

(continued)

TABLE I (*Continued*)

Cadmium Telluride

eV	cm^{-1}	μm	n	n	n	k	k	k
0.66	5,323	1.879	2.7166					
0.65	5,243	1.907	2.7152					
0.64	5,162	1.937	2.7138					
0.63	5,081	1.968	2.7125					
0.62	5,001	2.000	2.7112					
0.6199	5,000	2.0	2.713 [9]	2.71772 [10]				
0.61	4,920	2.033	2.7099 [7]					
0.60	4,839	2.066	2.7086					
0.59	4,759	2.101	2.7074					
0.58	4,678	2.138	2.7062					
0.57	4,597	2.175	2.7050					
0.56	4,517	2.214	2.7038					
0.55	4,436	2.254	2.7027					
0.54	4,355	2.296	2.7015					
0.53	4,275	2.339	2.7004					
0.52	4,194	2.384	2.6993					
0.51	4,113	2.431	2.6983					
0.50	4,033	2.480	2.6972					
0.4959	4,000	2.5	2.702 [9]					
0.49	3,952	2.530	2.6962 [7]					
0.48	3,871	2.583	2.6952					
0.47	3,791	2.638	2.6942					
0.46	3,710	2.695	2.6932					
0.45	3,629	2.755	2.6923					
0.44	3,549	2.818	2.6914					
0.43	3,468	2.883	2.6904					
0.42	3,388	2.952	2.6895					

0.4133	3,333	3.0	2.695 [9]	2.69933	
0.41	3,307	3.024	2.6887 [7]		
0.40	3,226	3.100	2.6878		
0.39	3,146	3.179	2.6870		
0.38	3,065	3.263	2.6861		
0.37	2,984	3.351	2.6853		
0.36	2,904	3.444	2.6845		
0.3542	2,857.1	3.5	2.691 [9]		
0.35	2,823	3.542	2.6837 [7]		
0.34	2,742	3.647	2.6829		
0.33	2,662	3.757	2.6822		
0.32	2,581	3.875	2.6814		
0.31	2,500	4.000	2.6807	2.69258 [10]	
			2.688 [9]		
0.30	2,420	4.133	2.6800 [7]		
0.29	2,339	4.275	2.6792		
0.28	2,258	4.428	2.6785		
0.27	2,178	4.592	2.6778		
0.26	2,097	4.769	2.6770		
0.25	2,016	4.959	2.6763		
0.2480	2,000	5.0	2.684 [9]		
0.24	1,936	5.166	2.6755 [7]		
0.23	1,855	5.391	2.6747		
0.22	1,774	5.636	2.6739		
0.21	1,694	5.904	2.6731		2.68198 [11]
0.2066	1,667	6.0	2.681 [9]	2.68570	
0.20	1,613	6.199	2.6722 [7]		
0.19	1,532	6.526	2.6713		
0.18	1,452	6.888	2.6703		
0.1771	1,428.6	7.0	2.679 [9]		
0.17	1,371	7.293	2.6692 [7]		

(continued)

TABLE I (*Continued*)

Cadmium Telluride

eV	cm⁻¹	μm	n	n	n	k	k
0.16	1,290	7.749	2.6679				
0.1550	1,250	8.0	2.677 [9]	2.68044	2.67730		
0.15	1,210	8.266	2.6665 [7]				
0.14	1,129	8.856	2.6649				
0.13	1,049	9.537	2.6628				
0.1240	1,000	10.0	2.653 [14] 2.672 [9]	2.67513 [10]	2.67242 [11]		
0.12	967.9	10.33	2.6605 [7]				
0.1178	950.0	10.53	2.651 [14]				
0.1169	943	10.6				5.1 × 10⁻⁸ [21]	
0.1116	900	11.11	2.650				
0.11	887.2	11.27	2.6574 [7]				
0.1054	850	11.76	2.648 [14]				
0.1033	833.3	12.0	2.6535 [7]	2.66932	2.66677		
0.10	806.5	12.40	2.646 [14]				
0.09919	800	12.50	2.643				
0.09299	750	13.33					
0.09	725.9	13.78	2.6482 [7]	2.66280	2.66020		
0.08856	714.3	14.0	2.640 [14]				
0.08679	700	14.29	2.636				
0.08059	650	15.38	2.6407 [7]				
0.08	645.2	15.50	2.630 [14]				
0.07749	625.0	16.0	2.6296 [7]	2.65451	2.65253		
0.07439	600	16.67					
0.07	564.6	17.71					
0.06888	555.6	18.0		2.64492	2.64366		

0.06819	550	18.18	2.623 [14]			3.9×10^{-7}
0.0625	504	19.8	2.614			
0.06199	500.0	20.0	2.6118 [7]	2.63428	2.63343	
0.06	483.9	20.66				
0.0573	462	21.6				3.8×10^{-6}
0.05636	454.5	22.0	2.601 [14]	2.62265	2.62177	
0.05579	450	22.22				
0.0521	420	23.8				3.4×10^{-5}
0.05166	416.7	24.0		2.60928		
0.05061	408.2	24.5				
0.05	403.3	24.80	2.5801 [7]			3.51×10^{-5} [22]
0.04959	400	25	2.581 [14]			
0.04835	390	25.64	2.576			
0.04769	384.6	26		2.59466		
0.04711	380	26.32	2.570			8.0×10^{-5} [21]
0.0469	378	26.5				9.88×10^{-5} [22]
0.04592	370.3	27				
0.04587	370	27.03	2.564			
0.04463	360	27.78	2.557			
0.04428	357.1	28		2.57724		
0.04215	340	29.41	2.541			2.97×10^{-3} [14]
0.04133	333.3	30		2.55916		2.86×10^{-4} [22]
0.04092	330	30.30	2.531			3.34 [14]
0.03968	320	31.25	2.521			3.79
0.03844	310	32.26	2.508			4.32
0.03757	303.0	33.00				6.57 [15]
0.03720	300	33.33	2.494			4.97 [14]
0.03701	298.5	33.50				8.13 [15]
0.03647	294.1	34.00				8.93
0.03596	290	34.48	2.478			5.77×10^{-3} [14]

(continued)

TABLE I (*Continued*)

Cadmium Telluride

eV	cm⁻¹	μm	n	n	n	k	k	k
0.03594	289.9	34.50						1.08×10^{-2} [15]
0.03542	285.7	35.00						9.19×10^{-3}
0.03493	281.7	35.5						7.91
0.03472	280	35.71			2.459		6.76 [14]	
0.03444	277.8	36.0						4.58 [15]
0.03397	274.0	36.5						3.92
0.03351	270.3	37.0						3.83
0.03348	270	37.04			2.437		8.01 [14]	
0.03306	266.7	37.5						4.03 [15]
0.03263	263.2	38.0						4.54
0.03224	260	38.46			2.410		9.62 [14]	
0.03220	259.7	38.5						5.06 [15]
0.03179	256.4	39.0						6.21
0.03139	253.2	39.5						7.23
0.03100	250	40			2.378		1.18×10^{-2} [14]	7.96×10^{-3} [15]
0.03061	246.9	40.5						7.57
0.03024	243.9	41.0						7.83
0.02988	241.0	41.5						7.43
0.02976	240	41.67			2.339		1.46×10^{-2} [14]	
0.02952	238.1	42.0						7.35×10^{-3} [15]
0.02917	235.3	42.5						6.93
0.02883	232.6	43.0						6.50
0.02852	230	43.48		2.28 [18]	2.289		1.87×10^{-2} [14]	
0.02850	229.9	43.5						6.23 [15]
0.02818	227.3	44.0						5.95
0.02786	224.7	44.5						5.31

0.02755	222.2	45.0			5.19	
0.02728	220	45.45	2.224	2.23	2.47×10^{-2} [14]	
0.02725	219.8	45.5			5.43×10^{-3} [15]	
0.02695	217.4	46.0			5.86	
0.02666	215.1	46.5			5.92	
0.02638	212.8	47.0			6.54	
0.02610	210.5	47.5			6.80	
0.02604	210	47.62	2.137	2.18	3.40×10^{-2} [14]	
0.02583	208.3	48.0			8.02×10^{-3} [15]	
0.02556	206.2	48.5			1.12×10^{-2}	
0.02530	204.1	49.0			1.44	
0.02505	202.0	49.5			1.60	
0.02480	200	50.00	2.013	2.05	4.97×10^{-2} [14] 2.07 [15]	
0.02455	198.0	50.5			2.65	
0.02431	196.1	51.0			3.69	
0.02407	194.2	51.5			5.33	
0.02384	192.3	52.0			6.21	
0.02356	190	52.63	1.824	1.9	8.00×10^{-2} [14]	
0.02232	180	55.56	1.498	1.5	0.154	0.1 [18]
0.02108	170	58.82	0.8021	0.80	0.522	0.5
0.02046	165	60.61	0.5183	0.40	1.17	0.9
0.01984	160	62.50	0.5085	0.20	1.89	1.8
0.01922	155	64.52	0.6392	0.40	2.71	2.75
0.01860	150	66.67	1.021	0.80	3.81	3.7
0.01835	148	67.57	1.353		4.41	
0.01810	146	68.49	1.939		5.12	
0.01798	145	68.97		1.8		
0.01785	144	69.44	3.053		5.87	5.2
0.01767	142.5	70.18		3.8		
0.01760	142	70.42	5.009		6.02	6.2

(continued)

TABLE I (*Continued*)

Cadmium Telluride

eV	cm⁻¹	μm	n	n	k	k
0.01736	140	71.43	6.778	5.5	4.50	5.5
0.01711	138	72.46	6.917		2.65	
0.01686	136	73.53	6.399		1.59	2.2
0.01674	135	74.07		6.75	1.03	
0.01661	134	74.63	5.882		0.722	
0.01636	132	75.76	5.468	5.1	0.534	
0.01612	130	76.92	5.146	4.7	0.294	0.8
0.01550	125	80.00	4.598	4.4	0.187	0.55
0.01488	120	83.33	4.260		0.129	0.3
0.01426	115	86.96	4.032	3.9	9.47×10^{-2}	
0.01364	110	90.91	3.868		7.23	0.25
0.01302	105	95.24	3.745		5.68	
0.01240	100	100	3.649	3.62	3.75×10^{-2}	0.2 [18]
0.01118	90	111.1	3.510	3.45	0.111 [15]	0.2
0.01033	83.33	120			2.62×10^{-2} [14]	
0.009919	80	125	3.415	3.32	0.103 [15]	0.2
0.009537	76.92	130			9.34×10^{-2}	
0.009184	74.07	135			0.116	
0.008856	71.43	140			1.89×10^{-2} [14]	
0.008679	70	142.9	3.348	3.3	0.108 [15]	0.2
0.008551	68.97	145			8.27×10^{-2}	
0.008266	66.67	150			6.86	
0.007749	62.50	160			1.39 [14]	
0.007439	60	166.7	3.299	3.3	6.05 [15]	
0.007293	58.82	170			5.67	
0.006888	55.56	180			5.56	
0.006526	52.63	190			2.42 [17]	

0.006199	50	200	3.263	3.3	
					1.03 [14]
0.005904	47.62	210			1.83 [17]
0.005767	46.51	215			5.32 [15]
0.0056956	45.938	217.68	3.2733 [8]		5.18
0.005636	45.45	220			5.37
0.0055904	45.089	221.78	3.2720		
0.005510	44.44	225			5.46
0.005391	43.48	230			
0.0052828	42.608	234.70	3.2628		4.24
0.0051798	41.778	239.36	3.2597		3.96 × 10^{-2}
0.0050756	40.937	244.28	3.2574		
0.0049712	40.095	249.41	3.2550		
0.004959	40	250.0	3.236 [14]		7.52 × 10^{-3} [14]
0.004133	33.33	300			8.55 [17]
0.0042395	34.194	292.45	3.2360 [8]		7.17
0.0041293	33.305	300.25	3.2371		8.12 [8]
0.0038128	30.752	325.18	3.2294		
0.003720	30	333.3			
0.0031678	25.550	391.39	3.217 [14]		
0.003100	25.00	400	3.2203 [8]		5.09
0.0028443	22.941	435.90	3.2155		
0.0025176	20.306	492.46	3.2136		
0.002480	20	500.0			
0.0023015	18.563	538.71	3.2096		4.77
0.002066	16.67	600			
0.0019749	15.929	627.78	3.2062		5.16
0.001771	14.29	700			5.57
0.001550	12.50	800			6.18

Gallium Arsenide (GaAs)

EDWARD D. PALIK

Naval Research Laboratory
Washington, D.C.

The optical constants of pure (semi-insulating) GaAs are derived from a number of papers including the far-infrared reststrahlen work of Johnson et al. [1], Kachare et al. [2], and Holm et al. [3]; the mid-IR work of Cochran et al. [4]; the near-IR work of Pikhtin and Yas'kov [5]; the calorimetry work of Christensen et al. [6]; the fundamental band gap work of Casey et al. [7]; the visible–UV ellipsometry work of Theeten et al. [8]; the UV reflection work of Philipp and Ehrenreich [9]; and the synchrotron transmission work of Cardona et al. [10]. A smooth plot of the n and k data, given in Fig. 2, is obtained from the data of Table II.

The measurements of Johnson et al. [1] to determine n and k were 300-K transmission measurements of thin samples of resistivity $3.5 \times 10^8 \ \Omega$ cm. Slab thickness was known to $\pm 1\%$. The transmittance was measured to $\pm 5\%$ with a Beckmann IR-11 spectrometer system. Surface preparation was described as optical polish. If this were a mechanical grit polish, then the surfaces were damaged, as discussed by Holm and Palik [11]. The effect of a damage layer thousands of angstroms thick can probably be neglected except at the reststrahlen peak itself. The n data listed were obtained from a graph of a calculated oscillator fit to the experimental points, the scatter of the points being ± 0.02 from the fit. The k data listed were read directly off a smooth-curve graph (no fit in this case). Similar less-detailed measurements for k have been made by Stolen [12], and his 294-K data (as read from a graph) are also listed for comparison.

In the reststrahlen region reflectivity data have been fitted by an oscillator model of the form

$$(n - ik)^2 = \varepsilon_\infty \left[1 + \frac{\omega_L^2 - \omega_T^2}{\omega_T^2 - \omega^2 + i\Gamma\omega} \right] \tag{1}$$

used by Kachare et al. [2] and Holm et al. [3]. We averaged these parameters to obtain $\omega_L = 292.1$ cm^{-1}, $\omega_T = 268.7$ cm^{-1}, $\Gamma = 2.4$ cm^{-1}, and $\varepsilon_\infty = 11.0$. Surfaces were chemically polished to remove saw-cut and grinding damage. The reference was an aluminum mirror whose reflectivity is given Bennett

429

ISBN 0-12-544420-6

et al. [13]. A comparison can be made in the longer-wavelength overlap region with the data of Johnson *et al.* [1] showing agreement to ~0.03 and better in *n*. The calculated values of *k* are almost a factor of 2 smaller than those measured by Johnson *et al.* [1] probably because on the low-frequency side of ω_T difference bands occur in the 80–200-μm region that would make the experimental values of *k* larger than the single-oscillator model, as is observed. However, *k* measurements of Iwasa *et al.* [14] on thin mechanically polished and epitaxial films are only ~10% smaller than those calculated in the 240–320-cm^{-1} region (after a decimal point correction in Fig. 2. of Iwasa *et al.* [14], so we chose the oscillator-model fit through this region. The oscillator model can be modified to include a frequency-dependent damping constant to account for some of the structure away from ω_T [15]. The static dielectric constant is found to be 13.00.

The oscillator fit for *n* is extended up to 480 cm^{-1}. However, several multiphonon bands occur above ω_L that have been measured directly by Cochran *et al.* [4] in the 347–807-cm^{-1} region. They measured transmission and reflection of thin slabs. Surface preparation was not described. Room temperature varies among the samples and is usually 297 or 300 K, except for these samples for which it was 292 K (a cold English laboratory or ice water in the dewar?). Dopant impurities such as Si, Li, and B give rise to local vibration modes that often fall in this spectral region [16].

Calorimetry measurements at CO_2 laser wavelengths [6] produced the small *k* values listed in the 1000-cm^{-1} region for Czochrolski-grown, undoped, high-resistivity GaAs. No surface preparation was given, although surely a chemical polish was used. These numbers varied from sample to sample by a factor of 2 and with other measurements also [17]. Temperature-dependence studies suggest that this absorption is primarily intrinsic due to multiphonon absorption, although the variation from sample to sample suggests that residual impurities and surface effects may still be present.

Pikhtin and Yas'kov [5] have measured the index of refraction in the near IR by using prism (minimum deviation) and interference-fringe techniques. The authors fitted theirs and Marple's data [18] to an oscillator formula of the form

$$n^2 = 1 + \frac{A}{\pi} \ln \frac{E_1^2 - (\hbar\omega)^2}{E_0^2 - (\hbar\omega)^2} + \frac{G_1}{E_1^2 - (\hbar\omega)^2} + \frac{G_2}{E_2^2 - (\hbar\omega)^2} + \frac{G_3}{E_3^2 - (\hbar\omega)^2}, \quad (2)$$

with the parameters $E_0 = 1.428$ eV, $E_1 = 3.0$ eV, $E_2 = 5.1$ eV, $E_3 = 0.0333$ eV, $G_1 = 39.194$ eV^2, $G_2 = 136.08$ eV^2, $G_3 = 0.00218$ eV^2, and $A = 0.7/\sqrt{E_0} = 0.5858$. No damping was assumed. These results for *n* are listed from 0.1 to 1.4 eV and fit the experimental values to a few units in the third decimal place. This fit of Eq. (2) agrees with the reststrahlen oscillator fit of Eq. (1) to within 0.02 at 0.06 eV, but *n* is smaller primarily due to the fact that in Eq. (1) $\varepsilon_\infty = 11.0$, while in Eq. (2) the first four terms give 10.9 to three figures from 5000 to 0 cm^{-1} when $G_3 = 0$. Usually, ε_∞ is the high-frequency background dielectric constant conveniently defined, where n^2 in Eq. (2) is equal to its low-frequency value when $G_3 = 0$. The second term in Eq. (1) could replace (without ε_∞) the last term in Eq. (2) for better description of n over the entire IR region along with k in the reststrahlen region.

At the fundamental band gap, Casey et al. [7] have measured k for "pure" and doped thin films by transmission and Kramers–Kronig (KK) analysis of reflectivity. A 1-m Jarrell Ash monochromator was used. The Burstein–Moss effect, the change in interband absorption due to free carriers blocking interband transitions, does change k significantly, but this effect is not discussed here. Transmission samples were bulk samples thinned down to appropriate thicknesses. Surface preparation was not described. For near-normal-incidence reflection, epitaxial films of GaAs on $Al_xGa_{1-x}As$ were used. The surfaces in this case would be undamaged. Uncertainty in absorption coefficient $\alpha = 4\pi k/\lambda$ was quoted to $\pm 5\%$ for the transmission data for $\alpha \leq 10^3$ cm^{-1}, primarily due to uncertainty in thickness measurements.

Theeten et al. [8] have performed continuous wavelength ellipsometry measurements on chemically polished samples and their results are presented in the 1.5–6.0-eV region. We have overlapped the Casey et al. [7] and the Theeten et al. [8] data to demonstrate how these data can differ; at 1.5 eV the respective k's are 0.079 and 0.041, while are 1.6 eV they are 0.095 and 0.066. The transmission measurements are preferred if the film thickness is well determined and the surface quality is good. However, in the Casey [7]/Theeten [8] overlap region the KK analysis was done by Casey et al. [7], so we prefer the more direct ellipsometric data.

The ellipsometric data [8] are considered better than the KK analyzed reflection data of Philipp and Ehrenreich [9], since it is a more direct determination with few approximations and assumptions. At 6 eV n is 1.264 compared to 1.395, a difference of 0.1, while k is 2.472 compared to 2.048, a difference of 0.4. Above 6 eV we list the results of Philipp and Ehrenreich [9] as tabulated by Seraphin and Bennett [19]. These were near-normal-incidence reflection measurements taken over the range 1–25 eV in a vacuum spectrometer system. Surfaces were chemically prepared.

Transmission of synchrotron radiation was measured by Cardona et al. [10] in the 15–155-eV region to determine k only. Normal incidence and grazing incidence vacuum spectrometers were used. Films were flash evaporated on carbon foils or on microscope slides coated with KCl. The films could be floated on water and picked up on a copper screen. To obtain polycrystalline rather than amorphous samples, the substrate temperature was $\geq 250°$C. The k data listed were obtained from a graph readable to three figures. At 15 eV the KK-analyzed reflection data [9] give $k = 0.41$ while the transmission data give $k = 0.28$.

Throughout the visible–UV region we are plagued with values of k from various laboratories that differ by as much as a factor of 2, although the usual differences are $< 50\%$. Differences in n are usually less than 10%. Often, absolute reflectivity is quoted to $\pm 3\%$, which could account for much of the differences between KK analysis, between two laboratories, or between KK analysis and ellipsometric measurements.

Except for the spectral regions in which the minimum deviation and interference-fringe techniques are used, we are hard-pressed to quote n to better than two significant figures. Things are even worse for k that is generally not determined better than $\pm 30\%$. Physicists, who did most of the measurements cited, were more interested in positions of peaks to identify vibrational frequencies and interband transitions, so relative values of k were typically good but absolute values varied as has been discussed.

The temperature dependence of n in the near IR between 5 and 20 μm has been found by Cardona [20] to be

$$(1/n)(dn/dT) = (4.5 \pm 0.2) \times 10^{-5} \, (\mathrm{C}^{-1}). \tag{3}$$

Near strong absorption features like band edges and lattice vibrations, the effects are no doubt even stronger. Data are usually quoted at 297 or 300 K, which would mean a variation of a few units in the fourth decimal place. Therefore, we cannot use this decimal place with confidence, although it is sometimes quoted in Table II.

It is clear that all samples probably have a native oxide in the form of Ga_2O_3 and As_2O_3 of $\sim$20-Å thickness. This layer presents little or no problems in the IR region but must be considered in the region above the band gap of this mixed oxide ($\sim$5 eV). Theeten et al. [8] and Philipp and Ehrenreich [9] prepared surfaces by chemical etching and polishing, but the samples may have been exposed to air before being put into the inert gas system or the vacuum system. Native oxides typically grow to their final thickness in a few hours in air but grow rapidly at first exposure. Therefore, we must assume that an oxide was growing on the samples. This may cause uncertainties in the determination of k in the UV. Only work with cleaved or evaporated surfaces in UHV would be free of oxides for a few hours.

REFERENCES

1. C. J. Johnson, G. H. Sherman, and R. Weil, *Appl. Opt.* **8**, 1667 (1969).
2. A. H. Kachare, W. G. Spitzer, F. K. Euler, and A. Kahan, *J Appl. Phys.* **45**, 2938 (1974).
3. R. T. Holm, J. W. Gibson, and E. D. Palik, *J. Appl. Phys.* **48**, 212 (1977).
4. W. Cochran, S. J. Fray, F. A. Johnson, J. E. Quarrington, and N. Williams, *J. Appl. Phys. Suppl.* **32**, 2102 (1961).
5. A. N. Pikhtin and A. D. Yas'kov, *Sov. Phys. Semicond.* **12**, 622 (1978); A. N. Pikhtin and A. D. Yas'kov, *Fiz. Tekh. Poluprovodn.* **12**, 1047 (1978).
6. C. P. Christensen, R. Joiner, S. K. T. Nieh, and W. H. Steier, *J. Appl. Phys.* **45**, 4957 (1974).
7. H. C. Casey, D. D. Sell, and K. W. Wecht, *J. Appl. Phys.* **46**, 250 (1975).
8. J. B. Theeten, D. E. Aspnes, and R. P. H. Chang, *J. Appl. Phys.* **49**, 6097 (1978); J. B. Theeten, D. E Aspnes, and R. P. H. Chang, private communication (1982).
9. H. R. Philipp and H. Ehrenreich, *Phys. Rev.* **129**, 1550 (1963).
10. M. Cardona, W. Gudat, B. Sonntag, and P. Y. Yu, in *Proc. Int. Conf. Phys. Semicond., 10th Cambridge*, 1970, p. 208. U.S. Atomic Energy Commission, Oak Ridge, Tennessee, 1970.

11. R. T. Holm and E. D. Palik, *J. Vac. Sci. Technol.* **13**, 889 (1976).

12. R. H. Stolen, *Phys. Rev. B* **11**, 767 (1975); R. H. Stolen, *Appl. Phys. Lett.* **15**, 74 (1969).

13. H. E. Bennett, M. Silver, and E. J. Ashley, *J. Opt. Soc. Am.* **53**, 1089 (1963).

14. S. Iwasa, I. Balslev, and E. Burstein, *in Int. Conf. Phys. Semicond., Paris*, 1964, p. 1077. Dunod, Paris, and Academic Press, New York, 1964.

15. A. S. Barker, Jr., *Phys. Rev.* **165**, 917 (1968).

16. W. G. Spitzer, *in* "Semiconductors and Semimetals" (R. K. Willardson and A. C. Beer, eds.), Vol. 3, p. 17. Academic Press, New York, 1967.

17. L. H. Skolnik, H. G. Lipson, and B. Bendow, *Appl. Phys. Lett.* **25**, 442 (1974).

18. D. T. F. Marple, *J. Appl. Phys.* **35**, 1241 (1964).

19. B. O. Seraphin and H. E. Bennett, *in* "Semiconductors and Semimetals" (R. K. Willardson and A. C. Beer, eds.), Vol. 3, p. 499. Academic Press, New York, 1967.

20. M. Cardona, *Proc. Int. Conf. Semicond. Phys., Prague*, 1960, p. 388. Czech. Acad. Sci., Prague, and Academic Press, New York, 1961.

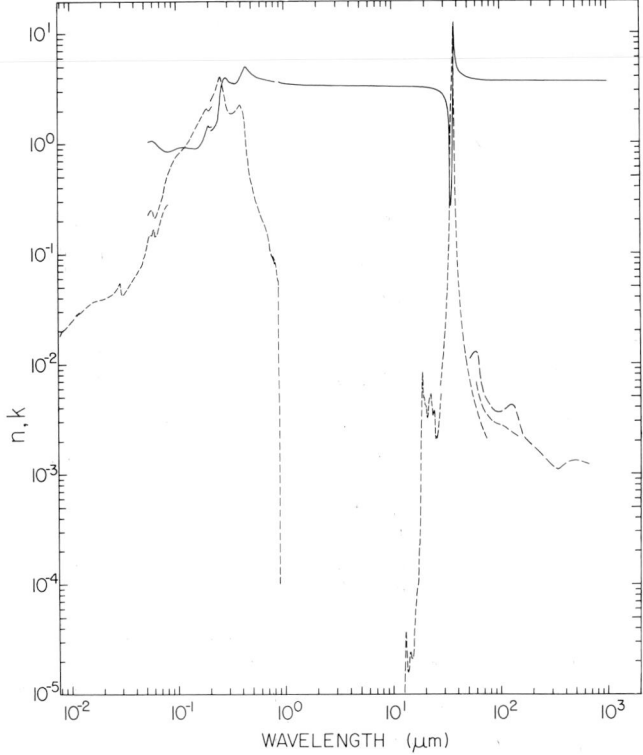

Fig. 2. Log–log plot of n (———) and k (----) versus wavelength in micrometers for gallium arsenide. Data from various references sometimes do not join together in overlap regions.

TABLE II

Values of *n* and *k* for Gallium Arsenide Obtained from Various References[a]

eV	cm^{-1}	μm	n	k
155		0.007999		0.0181 [10]
150		0.008266		0.0193
145		0.008551		0.0203
144		0.008610		0.0204
143		0.008670		0.0206
142		0.008731		0.0204
141		0.008793		0.0205
140		0.008856		0.0206
135		0.009184		0.0213
130		0.009537		0.0224
125		0.009919		0.0234
120		0.01033		0.0245
115		0.01078		0.0256
113		0.01097		0.0263
111		0.01117		0.0274
110		0.01127		0.0278
109		0.01137		0.0281
108		0.01148		0.0283
107		0.01159		0.0276
106		0.01170		0.0288
105		0.01181		0.0293
104		0.01192		0.0284
103		0.01204		0.0294
102		0.01215		0.0289
100		0.01240		0.0294
95		0.01305		0.0308
90		0.01378		0.0323
85		0.01459		0.0338
80		0.01550		0.0353
75		0.01653		0.0368
70		0.01771		0.0376
65		0.01907		0.0387
60		0.02066		0.0389
55		0.02254		0.0407
50		0.02480		0.0430
48		0.02583		0.0448

[a] The data of Philipp and ·Ehrenreich [9] given in Seraphin and Bennett [19] are specified by the wavelength (in micrometers). However, the original data were given in electron volts, and it is obvious that most of the *n*, *k* values were determined for energies in whole numbers or tenths of a whole number. They were rounded off to three significant figures and then given the wavelength listed in Seraphin and Bennett [19]. Sometimes we list these wavelengths and then recalculate energy (in electron volts), thus producing roundoff error in election volts. The reference is given at the beginning of a tabulation of *n* and *k*, for example, and is understood to refer to all *n*'s and *k*'s below it until a new reference appears in brackets. The value of *k* is frequently given as an exponent of 10. This value of exponent is understood to refer to all *k*'s below it in the table until a new exponent appears or the number changes from 10^{-2} to 10^{-1} when a decimal only is used.

TABLE II (*Continued*)
Gallium Arsenide

eV	cm^{-1}	μm	n	k
46		0.02695		0.0476
44		0.02818		0.0518
43		0.02883		0.0546
42		0.02952		0.0470
41		0.03024		0.0426
40		0.03100		0.0426
38		0.03263		0.0452
35		0.03542		0.0513
30		0.04133		0.0648
29		0.04275		0.0683
28		0.04428		0.0740
27		0.04592		0.0785
26		0.04769		0.0872
25		0.04959		0.0986
24		0.05166		0.115
23.5		0.05276		0.126
23		0.05391	1.037 [9, 19]	0.141
22.5		0.05510		0.228 [9, 19] 0.146 [10]
22		0.05636	1.043	0.148
21.5		0.05767		0.243 [9, 19] 0.147 [10]
21		0.05904	1.058	0.162
20.5		0.06048		0.245 [9, 19] 0.164 [10]
20		0.06199	1.025	0.144
19.5		0.06358		0.212 [9, 19] 0.147 [10]
19		0.06526	0.981	0.152
18		0.06888	0.936	0.234 [9, 19] 0.178 [10]
17		0.07293	0.889	0.267 [9, 19] 0.223 [10]
16		0.07749	0.850	0.324 [9, 19] 0.261 [10]
15		0.08266	0.836	0.411 [9, 19] 0.282 [10]
14		0.08856	0.840	0.503 [9, 19] 0.602
13		0.09537	0.864	0.709
12.0		0.1033	0.895	0.791
10.8		0.1148	0.923	0.881
10.0		0.1240	0.913	0.974
9.0		0.1378	0.901	1.136
8.0		0.1550	0.899	1.435
7.0		0.1771	1.063	1.838

(*continued*)

TABLE II (*Continued*)

Gallium Arsenide

eV	cm^{-1}	μm	n	k
6.6		0.1879	1.247	2.047
6.3		0.1968	1.441	1.988
6.2		0.2000	1.424	1.976
6.0	48,390	0.2066	1.264 [8]	2.472 [8]
			1.395 [9, 19]	2.048 [9, 19]
5.94	47,910	0.2087	1.287 [8]	2.523 [8]
5.90	47,590	0.2101	1.288	2.556
5.84	47,100	0.2123	1.304	2.592
5.80	46,780	0.2138	1.311	2.625
5.74	46,300	0.2160	1.321	2.673
5.70	45,970	0.2175	1.325	2.710
5.64	45,490	0.2198	1.339	2.772
5.60	45,170	0.2214	1.349	2.815
5.54	44,680	0.2238	1.367	2.885
5.50	44,360	0.2254	1.383	2.936
5.44	43,880	0.2279	1.410	3.019
5.40	43,550	0.2296	1.430	3.079
5.34	43,070	0.2322	1.468	3.181
5.30	42,750	0.2339	1.499	3.255
5.24	42,260	0.2366	1.552	3.384
5.20	41,940	0.2384	1.599	3.484
5.14	41,460	0.2412	1.699	3.660
5.10	41,130	0.2431	1.802	3.795
5.04	40,650	0.2460	2.044	3.992
5.00	40,330	0.2480	2.273	4.084
4.94	39,840	0.2510	2.654	4.106
4.90	39,520	0.2530	2.890	4.047
4.84	39,040	0.2562	3.187	3.898
4.80	38,710	0.2583	3.342	3.770
4.74	38,230	0.2616	3.511	3.574
4.70	37,910	0.2638	3.598	3.452
4.64	37,420	0.2672	3.708	3.280
4.60	37,100	0.2695	3.769	3.169
4.54	36,620	0.2731	3.850	3.018
4.50	36,290	0.2755	3.913	2.919
4.44	35,810	0.2792	4.004	2.715
4.40	35,490	0.2818	4.015	2.563
4.34	35,000	0.2857	3.981	2.368
4.30	34,680	0.2883	3.939	2.260
4.24	34,200	0.2924	3.864	2.132
4.20	33,880	0.2952	3.810	2.069
4.14	33,390	0.2995	3.736	2.001
4.10	33,070	0.3024	3.692	1.969
4.04	32,580	0.3069	3.634	1.935
4.00	32,260	0.3100	3.601	1.920
3.94	31,780	0.3147	3,559	1.907
3.90	31,460	0.3179	3.538	1.904

TABLE II (*Continued*)

Gallium Arsenide

eV	cm^{-1}	μm	n	k
3.84	30,970	0.3229	3.512	1.905
3.80	30,650	0.3263	3.501	1.909
3.74	30,160	0.3315	3.488	1.920
3.70	29,840	0.3351	3.485	1.931
3.64	29,360	0.3406	3.489	1.950
3.60	29,040	0.3444	3.495	1.965
3.54	28,550	0.3502	3.513	1.992
3.50	28,230	0.3542	3.531	2.013
3.48	28,070	0.3563	3.541	2.024
3.46	27,910	0.3583	3.553	2.036
3.44	27,750	0.3604	3.566	2.049
3.42	27,580	0.3625	3.580	2.062
3.40	27,420	0.3647	3.596	2.076
3.38	27,260	0.3668	3.614	2.091
3.36	27,100	0.3690	3.635	2.107
3.34	26,940	0.3712	3.657	2.123
3.32	26,780	0.3734	3.681	2.142
3.30	26,620	0.3757	3.709	2.162
3.28	26,450	0.3780	3.740	2.183
3.26	26,290	0.3803	3.776	2.207
3.24	26,130	0.3827	3.818	2.232
3.22	25,970	0.3850	3.871	2.260
3.20	25,810	0.3875	3.938	2.288
3.18	25,650	0.3899	4.023	2.307
3.16	25,490	0.3924	4.126	2.304
3.14	25,330	0.3949	4.229	2.270
3.12	25,160	0.3974	4.313	2.212
3.10	25,000	0.4000	4.373	2.146
3.08	24,840	0.4025	4.413	2.082
3.06	24,680	0.4052	4.439	2.029
3.04	24,520	0.4078	4.462	1.988
3.02	24,360	0.4105	4.483	1.961
3.00	24,200	0.4133	4.509	1.948
2.98	24,040	0.4161	4.550	1.952
2.96	23,870	0.4189	4.626	1.967
2.94	23,710	0.4217	4.755	1.960
2.92	23,550	0.4246	4.917	1.885
2.90	23,390	0.4275	5.052	1.721
2.88	23,230	0.4305	5.107	1.529
2.86	23,070	0.4335	5.102	1.353
2.84	22,910	0.4366	5.065	1.206
2.82	22,740	0.4397	5.015	1.088
2.80	22,580	0.4428	4.959	0.991
2.78	22,420	0.4460	4.902	0.912
2.76	22,260	0.4492	4.845	0.846
2.74	22,100	0.4525	4.793	0.789

(*continued*)

TABLE II (*Continued*)

Gallium Arsenide

eV	cm^{-1}	μm	n	k
2.72	21,940	0.4558	4.741	0.739
2.70	21,780	0.4592	4.694	0.696
2.68	21,660	0.4626	4.649	0.659
2.66	21,450	0.4661	4.605	0.626
2.64	21,290	0.4696	4.567	0.595
2.62	21,130	0.4732	4.525	0.569
2.60	20,970	0.4769	4.492	0.539
2.58	20,810	0.4806	4.456	0.517
2.56	20,650	0.4843	4.423	0.497
2.54	20,490	0.4881	4.392	0.476
2.52	20,330	0.4920	4.362	0.458
2.50	20,160	0.4959	4.333	0.441
2.48	20,000	0.4999	4.305	0.426
2.46	19,840	0.5040	4.279	0.411
2.44	19,680	0.5081	4.254	0.398
2.42	19,520	0.5123	4.229	0.385
2.40	19,360	0.5166	4.205	0.371
2.38	19,200	0.5209	4.183	0.359
2.36	19,030	0.5254	4.162	0.347
2.34	18,870	0.5299	4.141	0.337
2.32	18,710	0.5344	4.120	0.327
2.30	18,550	0.5391	4.100	0.320
2.28	18,390	0.5438	4.082	0.308
2.26	18,230	0.5486	4.063	0.301
2.24	18,070	0.5535	4.045	0.294
2.22	17,910	0.5585	4.029	0.285
2.20	17,740	0.5636	4.013	0.276
2.18	17,580	0.5687	3.998	0.266
2.16	17,420	0.5740	3.983	0.257
2.14	17,260	0.5794	3.968	0.251
2.12	17,100	0.5848	3.954	0.245
2.10	16,940	0.5904	3.940	0.240
2.08	16,780	0.5961	3.927	0.232
2.06	16,610	0.6019	3.914	0.228
2.04	16,450	0.6078	3.902	0.223
2.02	16,290	0.6138	3.890	0.213
2.00	16,130	0.6199	3.878	0.211
1.98	15,970	0.6262	3.867	0.203
1.96	15,810	0.6326	3.856	0.196
1.94	15,650	0.6391	3.846	0.187
1.92	15,490	0.6458	3.836	0.183
1.90	15,320	0.6526	3.826	0.179
1.88	15,160	0.6595	3.817	0.173
1.86	15,000	0.6666	3.809	0.173
1.84	14,840	0.6738	3.799	0.168
1.82	14,680	0.6812	3.792	0.158
1.80	14,520	6.8888	3.785	0.151

TABLE II (*Continued*)

Gallium Arsenide

eV	cm^{-1}	μm	n	k
1.78	14,360	0.6965	3.779	0.152
1.76	14,200	0.7045	3.772	0.134
1.74	14,030	0.7126	3.762	0.127
1.72	13,870	0.7208	3.752	0.118
1.70	13,710	0.7293	3.742	0.112
1.68	13,550	0.7380	3.734	0.105
1.66	13,390	0.7469	3.725	0.101
1.64	13,230	0.7560	3.716	0.097
1.62	13,070	0.7653	3.707	0.093
1.60	12,900	0.7749	3.700	0.091
1.58	12,740	0.7847	3.693	0.089
1.56	12,580	0.7948	3.685	0.087
1.54	12,420	0.8051	3.679	0.085
1.52	12,260	0.8157	3.672	0.083
1.50	12,100	0.8266	3.666	0.080
				7.89×10^{-2} [7]
1.49	12,020	0.8321		7.28
1.48	11,940	0.8377		6.86
1.47	11,860	0.8434		6.64
1.46	11,780	0.8492		6.28
1.45	11,690	0.8551		6.12
1.44	11,610	0.8610		5.68
1.435	11,570	0.8640		5.57
1.43	11,530	0.8670		5.72
1.425	11,490	0.8700		5.54
1.42	11,450	0.8731		2.71×10^{-2}
1.41	11,370	0.8793		5.95×10^{-3}
1.40	11,290	0.8856	3.6140 [5]	1.69×10^{-3}
1.39	11,210	0.8920		5.68×10^{-4}
1.38	11,130	0.8984		2.50
1.37	11,050	0.9050		1.33×10^{-4}
1.35	10,890	0.9184	3.5690	
1.30	10,490	0.9537	3.5388	
1.25	10,080	0.9919	3.5138	
1.20	9,679	1.033	3.4920	
1.15	9,275	1.078	3.4724	
1.10	8,872	1.127	3.4546	
1.05	8,469	1.181	3.4383	
1.00	8,065	1.240	3.4232	
0.95	7,662	1.305	3.4094	
0.90	7,259	1.378	3.3965	
0.85	6,856	1.459	3.3847	
0.80	6,452	1.550	3.3737	
0.75	6,049	1.653	3.3636	
0.70	5,646	1.771	3.3543	
0.65	5,243	1.907	3.3457	

(*continued*)

TABLE II (*Continued*)

Gallium Arsenide

eV	cm⁻¹	μm	n	k
0.60	4,839	2.066	3.3378	
0.55	4,436	2.254	3.3306	
0.50	4,033	2.480	3.3240	
0.45	3,629	2.755	3.3180	
0.40	3,226	3.100	3.3125	
0.35	2,823	3.542	3.3075	
0.30	2,420	4.133	3.3027	
0.29	2,339	4.275	3.3017	
0.28	2,258	4.428	3.3008	
0.27	2,178	4.592	3.2998	
0.26	2,097	4.768	3.2988	
0.25	2,016	4.959	3.2978	
0.24	1,936	5.166	3.2968	
0.23	1,855	5.391	3.2954	
0.22	1,774	5.636	3.2946	
0.21	1,694	5.904	3.2934	
0.20	1,613	6.199	3.2921	
0.19	1,532	6.526	3.2907	
0.18	1,452	6.888	3.2891	
0.17	1,371	7.293	3.2874	
0.16	1,290	7.749	3.2854	
0.15	1,210	8.266	3.2831	
0.14	1,129	8.856	3.2803	
0.1347	1,087	9.2		4.4×10^{-7} [6]
0.13	1,048	9.537	3.2769	
0.1239	1,042	9.6		6.1
0.12	967.9	10.33	3.2727	
0.1169	943.4	10.6		7.6×10^{-7}
0.11	887.2	11.27	3.2671	
0.100	806.5	12.40	3.2597	4.93×10^{-6} [4]
0.098	790.4	12.65		1.61×10^{-5}
0.096	774.3	12.91		3.49
0.095	766.2	13.05		3.63
0.094	758.2	13.19		2.72
0.092	742.0	13.48		1.61
0.090	725.9	13.78	3.2493	1.64
0.088	709.8	14.09		2.24
0.086	693.6	14.42		2.40
0.084	677.5	14.76		2.11
0.082	661.4	15.12		2.16
0.080	645.2	15.50	3.2336	2.83
0.078	629.1	15.90		4.55
0.076	613.0	16.31		7.66
0.074	596.8	16.75		9.46×10^{-5}
0.073	588.8	16.98		1.00×10^{-4}
0.072	580.7	17.22		1.50
0.071	572.6	17.46		2.05

TABLE II (*Continued*)

Gallium Arsenide

eV	cm^{-1}	μm	n	k
0.070	564.6	17.71	3.2081	2.32
0.069	556.5	17.97		6.00×10^{-4}
0.068	548.5	18.23		1.17×10^{-3}
0.067	540.4	18.51		2.59
0.066	532.3	18.79		4.18
0.0655	528.3	18.93		6.10
0.065	524.3	19.07	3.1886	8.27
0.0645	520.2	19.22	3.205 [2, 3]	6.27
0.064	516.2	19.37		4.85
0.0635	512.2	19.53		4.83
0.063	508.1	19.68		4.85
0.0625	504.1	19.84		4.23
0.062	500.1	20.00		4.17
0.0615	496.0	20.16		4.18
0.061	492.0	20.33		4.09
0.060	483.9	20.66	3.1609 [5]	3.45
0.0595	480.0	20.84	3.176 [2, 3]	3.18
0.059	475.9	21.01		3.19
0.058	467.8	21.38		3.57
0.057	459.7	21.75	3.157	4.50
0.0565	455.7	21.94		4.78
0.056	451.7	22.14		4.84
0.0555	447.6	22.34		4.97
0.055	443.6	22.54		5.32
0.054	435.5	22.96	3.126	4.56
0.053	427.5	23.39		3.68
0.0525	423.4	23.62		3.38
0.052	419.4	23.84	3.100	3.43
0.051	411.3	24.31		3.71
0.050	403.3	24.80		2.74
0.0495	399.2	25.05	3.058	2.07
0.049	395.2	25.30		2.07
0.0485	391.2	25.56		2.17
0.048	387.1	25.83		2.10
0.047	379.1	26.38	2.997	2.33
0.046	371.0	26.95		2.91
0.045	362.9	27.55		3.94
0.04463	360	27.78	2.913	6.50 [2, 3]
0.044	354.9	28.18		5.15 [4]
0.04339	350	28.57	2.851	8.40 [2, 3]
0.043	346.8	28.83		8.95×10^{-3} [4]
0.04215	340	29.41	2.770	1.13×10^{-2} [2, 3]
0.04092	330	30.30	2.659	1.59
0.03968	320	31.25	2.495	2.43
0.03906	315	31.75	2.380	3.14

(*continued*)

TABLE II (*Continued*)

Gallium Arsenide

eV	cm^{-1}	μm	n	k
0.03844	310	32.26	2.229	4.21
0.03782	305	32.79	2.020	6.02
0.03720	300	33.33	1.707	9.59×10^{-2}
0.03707	299	33.44	1.623	0.107
0.03695	298	33.56	1.529	0.122
0.03682	297	33.67	1.422	0.141
0.03670	296	33.78	1.298	0.166
0.03658	295	33.90	1.151	0.201
0.03645	294	34.01	0.975	0.257
0.03633	293	34.13	0.761	0.357
0.03620	292	34.25	0.536	0.552
0.03608	291	34.36	0.391	0.826
0.03596	290	34.48	0.323	1.09
0.03583	289	34.60	0.291	1.34
0.03571	288	34.72	0.275	1.57
0.03558	287	34.84	0.267	1.79
0.03546	286	34.96	0.266	2.01
0.03534	285	35.09	0.270	2.23
0.03521	284	35.21	0.278	2.46
0.03509	283	35.33	0.290	2.69
0.03496	282	35.46	0.307	2.94
0.03484	281	35.59	0.329	3.20
0.03472	280	35.71	0.358	3.48
0.03459	279	35.84	0.395	3.79
0.03447	278	35.97	0.443	4.14
0.03434	277	36.10	0.506	4.52
0.03422	276	36.23	0.592	4.97
0.03409	275	36.36	0.713	5.49
0.03397	274	36.50	0.890	6.12
0.03385	273	36.63	1.17	6.91
0.03372	272	36.76	1.64	7.94
0.03360	271	36.90	2.56	9.32
0.03348	270	37.04	4.66	11.0
0.03341	269.5	37.10	6.63	11.6
0.03335	269	37.17	9.30	11.3
0.03329	268.5	37.24	11.6	9.37
0.03323	268	37.31	12.4	6.74
0.03317	267.5	37.38	12.0	4.66
0.03310	267	37.45	11.3	3.30
0.03304	266.5	37.52	10.6	2.43
0.03298	266	37.59	9.90	1.87
0.03286	265	37.74	8.87	1.20
0.03273	264	37.88	8.12	0.844
0.03261	263	38.02	7.55	0.629
0.03248	262	38.17	7.11	0.489
0.03236	261	38.31	6.76	0.393
0.03224	260	38.46	6.47	0.323

TABLE II (*Continued*)

Gallium Arsenide

eV	cm^{-1}	μm	n	k
0.03162	255	39.22	5.57	0.153
0.03100	250	40.00	5.08	9.02×10^{-2}
0.02976	240	41.67	4.57	4.26
0.02852	230	43.48	4.30	2.49
0.02728	220	45.45	4.13	1.63
0.02604	210	47.62	4.02	1.15×10^{-2}
0.02480	200	50.00	3.93	8.49×10^{-3}
			3.87 [1]	
0.02356	190	52.63		1.09×10^{-2} [12]
0.02232	180	55.56		1.22
0.02108	170	58.82		1.25×10^{-2}
0.02066	166.7	60	3.77 [2, 3]	3.89×10^{-3} [2, 3]
			3.74 [1]	6.92×10^{-3} [1]
0.01984	160	62.50		1.15×10^{-2} [12]
0.01860	150	66.67		6.4×10^{-3}
0.01771	142.9	70	3.71 [2, 3]	3.79 [1]
			3.68 [1]	
0.01736	140	71.43		5.0 [12]
0.01612	130	76.92		4.4
0.01550	125	80	3.681 [2, 3]	1.84 [2, 3]
			3.65 [1]	3.18 [1]
0.01488	120	83.33		3.9 [12]
0.01378	111.1	90	3.662 [2, 3]	
			3.635 [1]	2.86 [1]
0.01364	110	90.91		3.6 [12]
0.01240	100	100	3.650 [2, 3]	3.6
			3.625 [1]	2.78 [1]
0.01116	90	111.1		3.8 [12]
0.009919	80	125.0		4.2
0.008679	70	142.9		3.5
0.008266	66.67	150	3.623 [2, 3]	2.14 [1]
			3.62 [1]	
0.007439	60	166.7		2.1 [12]
0.006199	50.00	200	3.615 [2, 3]	1.8 [12]
			3.61 [1]	
0.004959	40	250.0		1.4
0.004133	33.33	300	3.611 [2, 3]	
			3.60 [1]	
0.003720	30	333.3		1.1
0.003100	25	400		1.2
0.002480	20	500	3.607 [2, 3]	1.3
0.001860	15	667		1.2
0.001240	10	1000	3.606	

Gallium Phosphide (GaP)

A. BORGHESI and G. GUIZZETTI

Dipartimento di Fisica "A. Volta" and
Gruppo Nazionale di Struttura della Materia del CNR
Università di Pavia
Pavia, Italy

The room-temperature values of n and k tabulated here were obtained from the following works and references therein: Kleinman and Spitzer [1], Barker [2], Parsons and Coleman [3], Abagyan et al. [4, 5], Giehler and Jahne [6], and Pikhtin et al. [7] for the medium and far infrared; Spitzer et al. [8], Bond [9], Dean and Thomas [10], Dean et al. [11], Remenyuk et al. [12], Nelson and Turner [13], and Pikhtin and Yas'kov [7, 14, 15] for the near infrared and visible; and Philipp and Ehrenreich [16], Cardona et al. [17, 18], Stokowski and Sell [19], Gudat et al. [20], and Jungk [21] for the visible–ultraviolet and the vacuum ultraviolet.

A composite smooth-curve plot is given in Fig. 3. Numerical values are given in Table IV with appropriate references.

In the infrared near the transverse optic mode ($v_T \simeq 27.5$ μm), where the absorption coefficient α is too large to allow transmission T measurements using the thinnest samples available, bulk reflectance R measurements at near-normal incidence were performed [1, 6]. Values for n and k were obtained from a Kramers–Kronig (KK) analysis [2] or by fitting R (from 13 to 42 μm) within the experimental error by classical dispersion theory [1]. As a by-product of this procedure, the optical constants were calculated and checked against T measurements in the regions outside the fundamental band. At longer wavelengths, transmission samples of thicknesses from 30 μm to 1 mm were used; k was derived from T and R, including multiple-reflection effects; n was deduced from the interference transmission pattern.

All the samples, except those in Kleinman and Spitzer [1], were single crystals grown by different techniques [floating zone, Czochralski, vapor transport, and liquid encapsulation Czochralski (LEC)], both of n- and p-type with free-carrier density N ranging from $<5 \times 10^{15}$ to 10^{18} cm^{-3}. Some transmission samples [1, 4] were copper compensated to reduce the background absorption, and their surfaces were mechanically polished and etched in concentrated bromine. In the other cases no description of the surface treatment was given.

The R values in the infrared from various labs agree to within the experimental uncertainty. It was determined that differences in the free-carrier concentration had no effect on R. The n values for $\lambda \geq 50$ μm from Parsons and Coleman [3] and Abagyan et al. [4] present a difference of 0.28% at $\lambda = 50$ μm, which lowers to 0.2% at $\lambda = 1000$ μm. In Abagyan et al. [4] the error Δn in n, resulting from the measurement of the sample thickness and of wavelength λ, is quoted not exceeding 0.005, while in Parsons and Coleman [3] n is quoted with four decimal places and its values are those reported in Table IV for 50 μm $\leq \lambda \leq 400$ μm.

Parsons and Coleman [3] fit the infrared data and the accurate n data from Bond [9] and Nelson and Turner [13] in the 0.5–4.0-μm range by using for the complex dielectric function $\tilde{\varepsilon}$ the following form proposed by Barker [2]:

$$\tilde{\varepsilon}(v) = 1 + \sum_{i=1}^{3} \frac{S_{ei}v_{ei}^2}{v_{ei}^2 - v^2} + \frac{S_0\left(1 - \sum_{j=1}^{2} S_j\right)}{v_0^2 - v^2 + iv\gamma_0 - v_0^2 \sum_{j=1}^{2} [S_j v_j^2/(v_j^2 - v^2 + iv\gamma_j)]}, \tag{1}$$

i.e., a classical-oscillator equation, where S_i, v_i^2, and γ_i are the oscillator strength, restoring force, and linewidth, respectively. The second term in Eq. (1) represents the electronic contribution and the third one the lattice contribution with frequency-dependent damping. The fit parameters, given in Table III, are the same used by Barker [2] to fit α, R, and Raman data near the transverse optic mode, except S_0, which had to be increased from 2.01 to 2.056 for best fit. The net result of this fit is that the refractive index of GaP can be calculated from Eq. (1) to an accuracy somewhat better than four figures from microwave to visible frequencies. In particular, in the range 30–200 μm the fit agrees with the experimental n to a few units in the fourth decimal place ($\pm 0.03\%$) and yields the precise value $\varepsilon_0 = 11.147$ for the static relative dielectric constant. In Table IV we have labeled this fit with n_f for the data of Barker [2].

Also, Pikhtin et al. [7] proposed an analytic dispersion relationship for n in the 0.5–1000-μm spectral range; it has only six parameters and is simpler than Eq. (1) but less accurate ($\pm 0.8\%$). However, it is useful to report it, since by means of a relation in Pikhtin et al. [7] it has been possible also to fit spectra at 105 K. The fit parameters are from Pikhtin et al. [7].

For frequencies above or below $v_T \equiv v_0$, the k spectrum has considerable structure and values much higher than those predicted from Eq. (1). Since the same bands are observed in compensated and uncompensated samples, they are attributed to absorption by the lattice (combination bands) [1, 2]. However, the α spectra from different labs are in qualitative but not in quantitative agreement. In the 12–20-μm region we have tabulated the data from Pikhtin and Yas'kov [15] because the n-type sample was mono-crystalline, not intentionally doped, with the lowest carrier concentration (1.15×10^{17} cm^{-3}). Besides, T spectra were obtained with a double-beam spectrophotometer, and R was checked against accurate measurements of n. Near v_T the reported values (labeled k_f) were obtained from Eq. (1).

As for the 4–23-μm range, Table IV contains experimental values of n deduced from the positions of interference peaks in T [5, 7]. The thickness of the samples and the interference order were accurately determined by comparing, in the 0.5–4-μm range, the results obtained with this method and with the more precise results obtained by using the prism method. The data from Abagyan et al. [5] and Pikhtin et al. [7] agree to within the 0.1%, with an error [7] of $\pm 0.2\%$ at 4 μm and of $\pm 0.47\%$ at 23 μm. The samples were n type, undoped, and high resistivity ($N < 5 \times 10^{16}$ cm^{-3}). Measurements performed on low-resistivity samples, doped both n- and p-type up to 1.7×10^{18} cm^{-3} have shown that for $\lambda \geq 3$-μm doping reduces n [5, 7]. For this reason we have disregraded older data for n [22, 23] that are $\sim 2\%$ lower. The absorption coefficient between 1 and 12 μm is free-carrier-like; it typically increases with wavelength and with free-carrier concentration. Besides, all the n-type samples show a broad absorption band in the 1–4-μm region that does not occur in p-type material. In this region there are many measurements of α [8, 12, 15] on crystals doped with different impurities and with carrier densities ranging from 10^{17} to 10^{18} cm^{-3}. We have tabulated those from Pikhtin and Yas'kov [15] on the basis of the same reasons that determined the choice in the 12–20-μm region.

In the 0.53–4-μm range, where GaP is nearly transparent, n measurements were carried out by the method of prism minimum deviation [7, 9, 13] and by interference spacing [11] at 2.24 eV (0.5536 μm) and 2.62 eV (0.4733 μm). The scattering of n data from different authors is within ± 0.0025. In Pikhtin et al. [7] the absolute values of n were determined to within ± 0.0025; in Bond [9] n was measured to the fourth decimal. In Nelson and Turner [13] the uncertainty in the individual value was ± 0.0002; however, the value obtained for different prisms differed by substantially more than this. The differences are believed to be real and to reflect the variation that can be expected from different samples of GaP. To include the value for all of the prisms, an uncertainty of ± 0.0012 would have to be assigned to the n values from 0.545 to 0.70 μm in Table IV. These variations were not correlated with the level of doping, which varied from no intentional doping to 10^{18} cm^{-3} with the type of dopant (Te, Sn, Zn) or the method of crystal growth (high-pressure boat growth, floating zone, or solution growth).

The intrinsic electronic absorption edge of single crystals was presented and analyzed by Spitzer et al. [8] at 300 K, by Dean and Thomas [10] between 1.6 and 300 K, and by Pikhtin and Yas'kov [14] between 4.2 and 500 K. The results agree qualitatively, although there is some quantitative difference. In Table IV we have reported in the 0.54–0.57-μm range the values from Dean and Thomas [10] that are a little lower than those reported by Spitzer *et al.* [8] and Pikhtin and Yas'kov [14] and were obtained with exceptionally perfect single crystals grown by a modified wet-hydrogen transport process. All crystals had very high resistivities and three different thicknesses (0.0212, 0.0724, and 0.362 cm).

The 0.8–0.2-μm region straddling the fundamental band gap has been measured more recently by ellipsometry on surfaces containing native oxides that are mathematically "removed" in the analysis, or samples are chemically polished to remove this oxide and then put into an inert atmosphere while the measurement of the ellipsometric parameters ψ and Δ are performed [24]. We have included these values and at the ends of this spectral region have overlapped other data for comparison.

The optical properties from the fundamental absorption edge (0.53 μm) to about 0.08 μm can be determined by KK analysis of normal-incidence reflectance data. The R data between 0.53 and 0.21 μm obtained with a sensitive double-beam spectrophotometer by Stokowski and Sell [19] provide a considerable improvement in sensitivity over the older measurements reported by Philipp and Ehrenreich [16] and Seraphin and Bennett [23]. The samples used in Stokowski and Sell [19] were grown by the Czochralski method, were sliced parallel to a {111} face, and a weak ($\sim$0.2%) bromine–methanol solution was used as a polishing etch. The absolute-reflectance values were accurate to within $\pm 3\%$ of the reflectivity. The data were about 10% higher than those in Philipp and Ehrenreich [16], in agreement within 1% with those calculated from n at 0.4733 μm as measured by Dean *et al.* [11]. In the 0.21–0.08-μm range no other R data are available except those from Philipp and Ehrenreich [16].

Absorption and reflection measurements in the 0.08–0.03-μm range [18, 20] and only absorption measurements in the 0.08–0.008-μm range [17] were performed with synchrotron radiation and with a typical resolution of 2 Å over the whole spectral range. The crystalline samples for transmission were obtained by flash evaporation and samples of at least three different thicknesses between 500 and 1500 Å were used. The thicknesses were determined with a Tolansky interferometer with an accuracy of about 10% and also during evaporation with a calibrated quartz-crystal monitor. The absolute values determined for the absorption coefficient were affected by an error of less than 20%, which arose mainly from uncertainties in the thicknesses. The R measurements [20] were performed under nearly normal incidence [15] on polished bulk samples, etched in bromine–methanol solution. Since the incoming and reflected intensities were not determined simultaneously, no accurate absolute values of R were obtained ($\pm 50\%$); therefore the measured reflectivities were scaled so as to coincide around the first maximum at 0.059 μm with the ones calculated by the KK analysis transforms from α data. Some discrepancies between scaled R and R from KK analysis may be due to a structural difference between the thin films used to measure T and the bulk samples for R. Also, there may be some influence of the surface treatment and of light scattering. Larger discrepancies between the α experimental data from Cardona *et al.* [17, 18] and the α values obtained from KK processing of early R data [16] can be attributed to R-limited range of R spectra and to the inadequacy of the conventional gas-discharge spectroscopic sources; in fact, hot-cathode argon lamps have typical separation of about 1 eV between adjacent lines, equal to the expected spin-orbit splitting of core levels (0.53 eV for Ga) and to separation between peaks in the density of conduction states.

In Table IV n and k values in the 0.44–0.08-μm range have been obtained by KK transform on R spectra from Philipp and Ehrenreich [16] and Stokowski and Sell [19], extended to shorter wavelengths with the data of Gudat *et al.* [20] and longer wavelengths (visible and infrared) with those of Kleiman and Spitzer [1], Giehler and Jahne [6], and Bond [9]. The results are in fair agreement with the dielectric function ellipsometrically determined between 0.36 and 0.31 μm [21].

REFERENCES

1. D. A. Kleinman and W. G. Spitzer, *Phys. Rev.* **118**, 110 (1960).
2. A. S. Barker, Jr., *Phys. Rev.* **165**, 917 (1968).
3. D. F. Parsons and P. D. Coleman, *Appl. Opt.* **10**, 1683 (1971).
4. S. A. Abagyan, G. A. Ivanov, Y. E. Shanurin, and V. I. Amosov, *Sov. Phys. Semicond.* **5**, 889 (1971).
5. S. A. Abagyan, G. A. Ivanov, A. P. Izergin, and Y. E. Shanurin, *Sov. Phys. Semicond.* **6**, 985 (1972).
6. M. Giehler and E. Jahne, *Phys. Status Solidi B* **73**, 503 (1976).
7. A. N. Pikhtin, V. T. Prokopenko, and A. D. Yas'kov, *Sov. Phys. Semicond.* **10**, 1224 (1976).
8. W. G. Spitzer, M. Gershenzon, C. J. Forsch, and D. F. Gibbs, *J. Phys. Chem. Sol.* **11**, 339 (1959).
9. W. L. Bond, *J. Appl. Phys.* **36**, 1674 (1965).
10. P. J. Dean and D. G. Thomas, *Phys. Rev.* **150**, 690 (1966).
11. P. J. Dean, G. Kaminsky, and R. B. Zetterstrom, *J. Appl. Phys.* **38**, 3551 (1967).
12. A. D. Remenyuk, L. G. Zabelina, Yu. I. Ukhanov, and Y. V. Shmartzev, *Sov. Phys. Semicond.* **2**, 557 (1968).
13. D. F. Nelson and E. H. Turner, *J. Appl. Phys.* **39**, 3337 (1968).
14. A. N. Pikhtin and D. A. Yas'kov. *Sov. Phys. Solid State* **11**, 455 (1969).
15. A. N. Pikhtin and D. A. Yas'kov. *Phys. Status Solidi* **34**, 815 (1969).
16. H. R. Philipp and H. Ehrenreich, *in* "Semiconductors and Semimetals" (R. K. Willardson and A. C. Beer, eds.), Vol. 3, Chapter 4. Academic, New York, 1967.
17. M. Cardona, W. Gudat, B. Sonntag, and P. Y. Yu, *Proc. Int. Conf. Phys. Semicond. Cambridge, 1970,* p. 208. U.S. Atomic Energy Commission, Oak Ridge, Tennessee, 1970.
18. M. Cardona, W. Gudat, E. E. Koch, M. Skibowski, B. Sonntag, and P. Yu, *Phys. Rev. Lett.* **25**, 659 (1970).
19. S. E. Stokowski and D. D. Sell, *Phys. Rev. B* **5**, 1636 (1972).
20. W. Gudat, E. E. Koch, P. Y. Yu, M. Cardona, and C. M. Penchina, *Phys. Status Solidi B* **52**, 505 (1972).
21. G. Jungk, *Phys. Status Solidi B* **67**, 85 (1975).
22. H. Welker, *J. Electron.* **1**, 181 (1955).
23. B. O. Seraphin and H. E. Bennett, *in* "Semiconductors and Semimetals" (R. K. Willardson and A. C. Beer, eds.), Vol. 3, p. 219. Academic, New York, 1967.
24. D. E. Aspnes and A. A. Studna, *Phys. Rev. B* **27**, 985 (1983); D. E. Aspnes and A. A. Studna, private communication (1982).

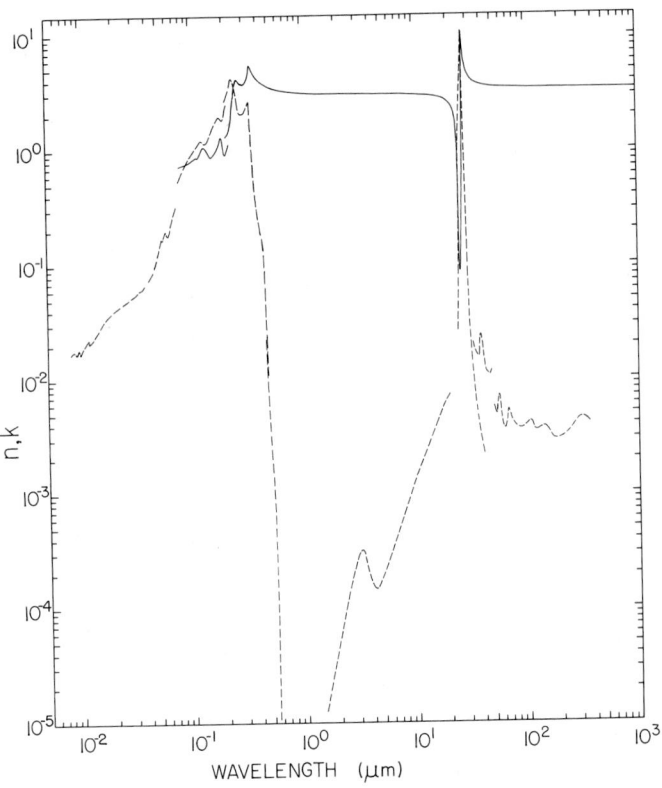

Fig 3. Log–log plot of n (——) and k(----) versus wavelength in micrometers for gallium phosphide.

TABLE III

Dielectric Parameters for Gallium Phosphide at 300 K[a]

Electronic		Ionic	
S_{e1}	2.570	ε_∞	9.091
v_{e1}	29,000 cm^{-1}	S_0	2.056
S_{e2}	4.131	v_0	363.4 cm^{-1}
v_{e2}	42,700 cm^{-1}	γ_0	1.1 cm^{-1}
S_{e3}	1.390	S_1	7.0×10^{-4}
v_{e3}	58,000 cm^{-1}	v_1	349.4 cm^{-1}
		γ_1	21 cm^{-1}
		S_2	3.5×10^{-4}
		v_2	358.4 cm^{-1}
		γ_2	12.6 cm^{-1}

[a] Data from Parsons and Coleman [3].

TABLE IV

Values of n and k for Gallium Phosphide Obtained from Various References[a]

eV	cm⁻¹	μm	n	n_f	k	k_f
154.0		0.00805			1.7×10^{-2} [17]	
144.0		0.00861			1.8	
134.0		0.00925			1.7	
131.2		0.00945			1.85	
126.0		0.00985			1.7	
122.0		0.0102			1.8	
119.0		0.0104			1.9	
115.0		0.0108			2.0	
110.0		0.0113			2.15	
109.2		0.0114			2.2	
107.0		0.0116			2.2	
106.1		0.0117			2.25	
105.0		0.0118			2.25	
104.0		0.0119			2.1	
100.0		0.0124			2.15	
96.0		0.0129			2.25	
90.0		0.0138			2.5	
80.0		0.0155			3.0	
70.0		0.0177			3.6	
66.0		0.0188			3.8	
60.0		0.0207			4.15	
50.0		0.0248			4.7	
45.0		0.0275			5.0	
40.0		0.0310			5.5	
37.0		0.0335			6.1	

(continued)

[a] References are indicated in brackets. Here n_f and k_f are fits to experimental data obtained with Eq. (1) and the parameters of Table III.

TABLE IV (*Continued*)

Gallium Phosphide

eV	cm^{-1}	μm	n	n_f	k	k_f
35.0		0.0354			6.1	
33.0		0.0376			6.7	
31.0		0.0400			7.3 [20]	
29.0		0.0427			8.0	
27.0		0.0459			9.3×10^{-2}	
25.0		0.0496			0.122	
24.2		0.0512			0.141	
23.5		0.0528			0.162	
23.1		0.0537			0.169	
22.6		0.0549			−0.167	
22.2		0.0558			0.174	
21.8		0.0569			0.187	
21.4		0.0579			0.193	
21.2		0.0585			0.197	
21.0		0.0590			0.195	
20.6		0.0602			0.190	
20.5		0.0605			0.188	
20.0		0.0620			0.180	
19.5		0.0636			0.192	
19.0		0.0652			0.213	
18.0		0.0689			0.269	
17.0		0.0729			0.323	
16.0		0.0775	0.730 [16–20]		0.537 [16–20]	
15.0		0.0826	0.748		0.628	
14.0		0.0886	0.759		0.718	
13.0		0.0954	0.780		0.825	
12.0		0.1033	0.822		0.938	

11.0		0.1127	0.873	1.040
10.5		0.1181	0.875	1.104
9.95		0.1246	0.998	1.224
9.5		0.1305	1.076	1.183
9.05		0.1370	1.047	1.151
8.5		0.1459	0.943	1.224
8.05		0.1540	0.872	1.401
7.5		0.1653	0.944	1.652
7.0		0.1771	1.054	1.876
6.75		0.1837	1.200	1.968
6.55		0.1893	1.301	1.896
6.4		0.1937	1.207	1.836
6.0	48,390	0.2066	0.906	2.281
5.94	47,910	0.2087	1.309 [24]	2.690 [24]
5.9	47,590	0.2101	1.327	2.766
5.84	47,100	0.2123	1.327	2.803
5.8	46,780	0.2138	1.342	2.882
			1.348	2.934
5.74	46,300	0.2160	0.988 [16–20]	2.657 [16–20]
			1.368 [24]	3.028 [24]
5.7	45,970	0.2175	1.385	3.096
5.64	45,490	0.2198	1.418	3.211
5.6	45,170	0.2214	1.444	3.297
5.54	44,680	0.2238	1.494	3.441
5.5	44,360	0.2254	1.543	3.556
5.44	43,880	0.2279	1.660	3.752
5.4	43,550	0.2296	1.778	3.889
5.38	43,390	0.2305	1.851	3.955
5.36	43,230	0.2313	1.936	4.019
5.34	43,070	0.2322	2.028	4.077

(continued)

TABLE IV (*Continued*)

Gallium Phosphide

eV	cm⁻¹	μm	n	n_r	k	k_r
5.32	42,910	0.2331	2.134		4.128	
5.30	42,750	0.2339	2.248		4.168	
5.28	42,590	0.2348	2.368		4.192	
5.26	42,420	0.2357	2.490		4.201	
5.24	42,260	0.2366	2.607		4.197	
5.22	42,100	0.2375	2.718		4.185	
5.20	41,940	0.2384	2.825		4.170	
5.18	41,780	0.2394	2.930		4.150	
5.16	41,620	0.2403	3.036		4.123	
5.14	41,460	0.2412	3.144		4.086	
5.12	41,300	0.2422	3.248		4.036	
5.10	41,130	0.2431	3.342		3.975	
5.08	40,970	0.2441	3.424		3.909	
5.06	40,810	0.2450	3.498		3.839	
5.04	40,650	0.2460	3.561		3.768	
5.02	40,490	0.2470	3.615		3.698	
5.00	40,330	0.2480	3.661		3.631	
4.98	40,170	0.2490	3.701		3.567	
4.96	40,000	0.2500	3.739		3.509	
4.94	39,840	0.2510	3.774		3.454	
4.92	39,680	0.2552	3.806		3.404	
4.90	39,520	0.2530	3.844		3.358	
4.88	39,360	0.2541	3.880		3.316	
4.86	39,200	0.2551	3.923		3.269	
4.84	39,040	0.2562	3.970		3.218	
4.82	38,880	0.2572	4.018		3.162	

4.80	38,710	0.2583	4.062	3.096
4.78	38,550	0.2594	4.100	3.024
4.76	38,390	0.2605	4.132	2.949
4.74	38,230	0.2616	4.157	2.871
4.72	38,070	0.2627	4.173	2.791
4.70	37,910	0.2638	4.181	2.712
4.68	37,750	0.2649	4.181	2.634
4.66	37,590	0.2661	4.173	2.560
4.64	37,420	0.2672	4.158	2.491
4.62	37,260	0.2684	4.137	2.427
4.60	37,100	0.2695	4.113	2.371
4.54	36,620	0.2731	4.031	2.241
4.50	36,290	0.2755	3.978	2.180
4.44	35,810	0.2792	3.907	2.117
4.40	35,490	0.2818	3.867	2.090
4.34	35,000	0.2857	3.817	2.066
4.30	34,680	0.2883	3.792	2.058
4.24	34,200	0.2924	3.765	2.057
4.20	33,880	0.2952	3.754	2.063
4.14	33,390	0.2995	3.748	2.082
4.10	33,070	0.3024	3.752	2.100
4.04	32,580	0.3069	3.769	2.137
4.00	32,260	0.3100	3.790	2.171
3.94	31,780	0.3147	3.840	2.240
3.90	31,460	0.3179	3.890	2.303
3.88	31,290	0.3195	3.923	2.343
3.86	31,130	0.3212	3.966	2.389
3.84	30,970	0.3229	4.021	2.443
3.82	30,810	0.3246	4.095	2.502
3.80	30,650	0.3263	4.196	2.562

(continued)

TABLE IV (*Continued*)

Gallium Phosphide

eV	cm^{-1}	μm	n	n_f	k	k_f
3.78	30,490	0.3280	4.328		2.615	
3.76	30,330	0.3297	4.497		2.649	
3.74	30,160	0.3315	4.700		2.646	
3.72	30,000	0.3333	4.927		2.585	
3.70	29,840	0.3351	5.149		2.451	
3.68	29,680	0.3369	5.328		2.250	
3.66	29,520	0.3388	5.437		2.010	
3.64	29,360	0.3406	5.472		1.766	
3.62	29,200	0.3425	5.454		1.550	
3.60	29,040	0.3444	5.406		1.368	
3.58	28,870	0.3463	5.339		1.217	
3.56	28,710	0.3483	5.268		1.089	
3.54	28,550	0.3502	5.194		0.982	
3.52	28,390	0.3522	5.121		0.893	
3.50	28,230	0.3542	5.050		0.819	
3.48	28,070	0.3563	4.983		0.752	
3.46	27,910	0.3583	4.920		0.697	
3.44	27,750	0.3604	4.861		0.649	
3.42	27,580	0.3625	4.805		0.605	
3.40	27,420	0.3647	4.751		0.568	
3.38	27,260	0.3668	4.700		0.534	
3.36	27,100	0.3690	4.651		0.503	
3.34	26,940	0.3712	4.604		0.475	
3.32	26,780	0.3734	4.560		0.449	
3.30	26,620	0.3757	4.518		0.426	
3.28	26,450	0.3780	4.479		0.407	

3.26	26,290	0.3803	4.442	0.388
3.24	26,130	0.3827	4.406	0.369
3.22	25,970	0.3850	4.372	0.353
3.20	25,810	0.3875	4.339	0.337
3.18	25,650	0.3899	4.308	0.323
3.16	25,490	0.3924	4.278	0.310
3.14	25,330	0.3949	4.249	0.298
3.12	25,160	0.3974	4.222	0.285
3.10	25,000	0.4000	4.196	0.275
3.08	24,840	0.4025	4.171	0.264
3.06	24,680	0.4052	4.147	0.253
3.04	24,520	0.4078	4.124	0.244
3.02	24,360	0.4105	4.102	0.233
3.00	24,200	0.4133	4.081	0.224
2.98	24,040	0.4161	4.060	0.218
2.96	23,870	0.4189	4.041	0.208
2.94	23,710	0.4217	4.023	0.202
2.92	23,550	0.4246	4.006	0.194
2.90	23,390	0.4275	3.990	0.183
2.88	23,230	0.4305	3.976	0.176
2.86	23,070	0.4335	3.964	0.165
2.84	22,910	0.4366	3.952	0.152
2.82	22,740	0.4397	3.936	0.135
2.80	22,580	0.4428	3.919	0.118
2.78	22,420	0.4460	3.904	0.103
2.76	22,260	0.4492	3.896	8.5×10^{-2}
2.743	22,120	0.452		3.12 [8, 23]
2.74	22,100	0.4525	3.869	5.7 [24]
2.731	22,030	0.454		2.36 [8, 23]
2.72	21,940	0.4558	3.835	3.5 [24]

(continued)

TABLE IV (Continued)

Gallium Phosphide

eV	cm⁻¹	μm	n	n_f	k	k_f
2.719	21,930	0.456			1.87 [8, 23]	
2.713	21,880	0.457			1.59	
2.701	21,780	0.459			1.43	
2.70	21,780	0.4592	3.805		2.7 [24]	
2.678	21,600	0.463			1.16×10^{-2} [8, 23]	
2.672	21,550	0.464			9.49×10^{-3}	
2.64	21,290	0.4696	3.730		1.5×10^{-2} [24]	
2.638	21,280	0.470			7.68×10^{-3} [8, 23]	
2.621	21,140	0.473	3.73 [11]		6.37	
2.60	20,970	0.4769	3.691 [24]			
2.599	20,960	0.477			5.47	
2.583	20,830	0.480			4.66	
2.572	20,750	0.432			4.29	
2.541	20,490	0.488			3.41	
2.54	20,490	0.4881	3.638 [24]			
2.52	20,320	0.492			3.06	
2.500	20,160	0.496	3.605		2.54	
2.480	20,000	0.500	3.590		2.47	
2.475	19,960	0.501	3.4595 [9]		2.18	
2.460	19,840	0.504			1.85	
2.44	19,680	0.5081	3.561 [24]		1.42	
2.431	19,610	0.510			1.19	
2.412	19,450	0.514				
2.40	19,360	0.5166	3.535			
2.389	19,270	0.519			1.04×10^{-3}	
2.380	19,190	0.521			8.57×10^{-4}	

E (eV)	cm⁻¹	µm	n	n	k
2.34	18,870	0.5299	3.497		5.34
2.339	18,870	0.530			3.34
2.322	18,730	0.534			3.34
2.322	18,730	0.534			3.34
2.30	18,550	0.5391	3.474		2.48 [10]
2.29	18,470	0.5414			1.88 [10]
2.28	18,390	0.5438	3.463		1.3×10^{-4}
2.275	18,350	0.545	3.4522 [13]	3.4491	
2.27	18,310	0.5462			9.3×10^{-5}
2.26	18,230	0.5486	3.452 [24]		6.2
2.254	18,180	0.55	3.4411 [13]	3.4382	4.2
2.24	18,070	0.5535	3.441 [24]		2.7
2.23	17,990	0.5560			1.5×10^{-5}
2.22	17,910	0.5585	3.430		3.1×10^{-6}
2.214	17,860	0.56	3.4203 [13]		3.0
2.21	17,820	0.5610			1.4×10^{-6}
2.20	17,740	0.5636	3.421 [24]		6.5×10^{-7}
2.19	17,660	0.5661			2.8
2.18	17,580	0.5687	3.411		
2.175	17,540	0.57	3.4012 [13]		
2.14	17,260	0.5794	3.393 [24]		
2.138	17,240	0.58	3.3837 [13]		
2.101	16,950	0.59	3.3675		
2.10	16,940	0.5904	3.375 [24]		
2.066	16,670	0.60	3.3524 [13]		
2.04	16,450	0.6078	3.350 [24]		
2.033	16,390	0.61	3.3384 [13]		
2.000	16,130	0.62	3.3254		
1.968	15,870	0.63	3.334 [24], 3.3132 [13]		

(continued)

A. Borghesi and G. Guizzetti

TABLE IV (*Continued*)

Gallium Phosphide

eV	cm⁻¹	μm	n	n_f	k	k_f
1.94	15,650	0.6391	3.311 [24]			
1.937	15,630	0.64	3.3018 [13]			
1.907	15,380	0.65	3.2912	3.2891		
1.90	15,320	0.6526	3.295 [24]			
1.879	15,150	0.66	3.2811 [13]			
1.851	14,930	0.67	3.2716			
1.84	14,840	0.6738	3.275 [24]			
1.823	14,710	0.68	3.2626 [13]			
1.80	14,520	0.6888	3.262 [24]			
1.797	14,490	0.69	3.2541 [13]			
1.771	14,290	0.70	3.2462			
1.74	14,030	0.7126	3.245 [24]			
1.70	13,710	0.7293	3.234			
1.64	13,230	0.7560	3.219			
1.60	12,900	0.7749	3.209			
1.550	12,500	0.80	3.1830 [9]			
1.54	12,420	0.8051	3.191 [24]			
1.50	12,100	0.8266	3.178			
1.378	11,110	0.90	3.1430 [9]			
1.240	10,000	1.0	3.1192	3.1173		
1.127	9,091	1.1	3.0981			
1.033	8,333	1.2	3.0844			
0.8856	7,143	1.4	3.0646	3.0646	1.27×10^{-5} [15]	
0.8551	6,897	1.45				
0.7749	6,250	1.6	3.0509	3.0522	3.27	
0.7085	5,714	1.75				

0.6888	5,556	1.8	3.0439		
0.6199	5,000	2.0	3.0379		
0.5767	4,651	2.15			7.70×10^{-5}
0.5636	4,545	2.2	3.0331		
0.5166	4,167	2.4	3.0296	3.0292	
0.4959	4,000	2.5			1.71×10^{-4}
0.4769	3,846	2.6	3.0271		
0.4428	3,571	2.8	3.0236		
0.4275	3,448	2.9			2.88
0.4133	3,333	3.0	3.0215	3.0217	
0.3875	3,125	3.2			3.18
0.3647	2,941	3.4	3.0181		
0.3444	2,778	3.6	3.0166	3.0166	
0.3263	2,632	3.8	3.0159		1.69
0.3100	2,500	4.0	3.0137	3.0137	1.53
0.2952	2,381	4.2			1.47
0.2583	2,083	4.8			1.91
0.2480	2,000	5.0	3.004 [5, 7]	3.007	
0.2254	1,818	5.5	3.001	2.95	
0.2175	1,754	5.7		3.001	
0.2066	1,667	6.0	2.998		
0.1907	1,538	6.5	2.995	2.993	4.29
0.1771	1,429	7.0	2.992		
0.1653	1,333	7.5	2.98		
0.1590	1,282	7.8			
0.1550	1,250	8.0	2.984	2.985	7.14×10^{-4}
0.1459	1,176	8.5	2.980		
0.1378	1,111	9.0	2.975	2.975	
0.1305	1,053	9.5	2.970		
0.1292	1,042	9.6			1.3×10^{-3}
0.1240	1,000	10	2.964	2.964	2.964

(continued)

TABLE IV (*Continued*)

Gallium Phosphide

eV	cm^{-1}	μm	n	n_f	k	k_f
0.1127	909.1	11	2.951			
0.1033	833.3	12	2.936		2.29	
0.09537	769.2	13	2.918			
0.08856	714.3	14	2.896			
0.08266	666.7	15	2.871	2.868		
0.07749	625.0	16	2.841		4.49	
0.07293	588.2	17	2.803			
0.06888	555.6	18	2.755			
0.06526	526.3	19	2.694 [7]			
0.06199	500.0	20	2.615	2.606	7.16 × 10^{-3}	
0.05904	476.2	21	2.511			
0.05636	454.5	22	2.354			
0.05391	434.8	23		2.076		2.6 × 10^{-2}
0.05166	416.7	24.0		1.585		3.3
0.05123	413.2	24.2		1.421		4.6
0.05081	409.8	24.4		1.208		7.2 × 10^{-2}
0.05040	406.5	24.6		0.908		0.22
0.04999	403.2	24.8		0.360		0.89
0.04959	400.0	25.0		0.108		1.74
0.04881	393.7	25.4		0.086		2.52
0.04805	387.6	25.8		0.103		2.96
0.04769	384.6	26.0		0.123		4.05
0.04696	378.8	26.4		0.208		5.86
0.04626	373.1	26.8		0.515		7.48
0.04592	370.4	27.0		1.064		8.69
0.04575	369.0	27.1		1.740		10.22
0.04558	367.6	27.2		3.231		

0.04541	366.3	27.3		6.580		11.19
0.04525	365.0	27.4		10.314		8.62
0.04509	363.6	27.5		10.483		5.15
0.04492	362.3	27.6		9.501		3.40
0.04460	359.7	27.8		8.060		2.10
0.04428	357.1	28.0		7.348		1.54
0.04366	352.1	28.4		6.431		0.78
0.04275	344.8	29.0		5.578		0.31
0.04217	340.1	29.4		5.182		0.16
0.04133	333.3	30	4.732 [4]	4.765		6.5×10^{-2}
0.04000	322.6	31	4.339			
0.03875	312.5	32	4.115			
0.03815	307.7	32.5			2.07×10^{-2} [2]	
0.03757	303.0	33	3.971			
0.03647	294.1	34	3.869			
0.03542	285.7	35	3.793	3.809	1.67	5.18×10^{-3}
0.03397	274.0	36.5			1.54	
0.03306	266.7	37.5			2.24	
0.03163	255.1	39.2			2.09	
0.03100	250.0	40	3.594	3.606	1.81	2.31
0.02917	235.3	42.5			1.15	
0.02755	222.2	45	3.508	3.518	1.11	1.46
0.02695	217.4	46			1.06	
0.02610	210.5	47.5			1.17×10^{-2}	
0.02480	200.0	50	3.461	3.470	5.77×10^{-3} [3]	1.06×10^{-3}
0.02356	190.0	52.63	3.4440 [3]	3.4435	4.90	
0.02280	183.93	54.37				
0.02244	181.0	55.25			7.03	
0.02173	175.29	57.05	3.4302		5.67	
0.02080	167.78	59.60	3.4238		4.03	

(continued)

TABLE IV (*Continued*)

Gallium Phosphide

eV	cm⁻¹	μm	n	n_f	k	k_f
0.02066	166.7	60		3.4200		6.9×10^{-4}
0.02009	162.00	61.73	3.4134		3.67	
0.01859	149.96	66.68	3.4010	3.4015	5.31	5.2
0.01771	142.9	70		3.3947		
0.01727	139.27	71.80	3.3922	3.3915	4.34	
0.01594	128.54	77.80	3.3830	3.3826	3.84	
0.01459	117.70	84.96	3.3754	3.3747	3.65	
0.01363	109.90	90.99	3.3706	3.3697	3.77	
0.01240	100.0	100		3.3639		3.0
0.01168	94.21	106.1	3.3621	3.3609	4.26	
0.01090	87.92	113.7	3.3584	3.3579	3.53	
0.008935	72.07	138.8	3.3522	3.3513	3.81	
0.008266	66.67	150		3.3494		
0.007360	59.36	168.5	3.3463	3.3472	2.95	
0.006567	52.97	188.8	3.3450	3.3454	3.00	
0.006199	50.00	200		3.3447		1.3×10^{-4}
0.005724	46.17	216.6	3.3438	3.3438	3.19	
0.004186	33.76	296.2	3.3399	3.3414	4.53	
0.004133	33.33	300		3.3413		
0.003387	27.32	366.0	3.3407	3.3405	3.93	7.2×10^{-5}
0.003100	25.00	400	3.3331 [4]	3.3402		
0.002480	20.00	500	3.3326	3.3397		
0.002066	16.67	600	3.3323	3.3394		
0.001771	14.29	700	3.3321	3.3392		
0.001550	12.50	800	3.3320	3.3391		
0.001378	11.11	900	3.33195	3.3390		
0.001240	10.00	1000	3.3319	3.3389		

Germanium (Ge)

ROY F. POTTER

Department of Physics and Astronomy
Western Washington University
Bellingham, Washington

Germanium transistor technology in the 1950s was the basis for extensive research activity. Good-quality single crystals became available for studies of the electrical, elastic, and optical properties. Because infrared systems were also developing, Ge was investigated for its photoelectric characteristics as well as its optical properties. Much of the work during this period concerned impurity or doping effects in terms of free-carrier absorption and forbidden-gap determinations. Early optical measurements made on evaporated Ge films had widely scattered results, depending strongly on the conditions of preparation. However, when quality bulk samples became available, measurement results were obtained that have withstood the passage of time remarkably well. Briggs's tabular data [1] for refractive indices between 1.8 and 2.6 μm are within 0.25% of the present-day accepted values. We include these values since they have stood the test of time. Salzberg and Villa [2] measured single-crystal and polycrystalline material almost a decade later for wavelengths between 2 and 16 μm. Their data served as the basis for IR optical design parameters until present day and are within 0.025% of present-day measurements. Cardona *et al.* [3] published results for Ge for several temperatures between 87 and 197 K. An examination of their figures indicates a similar degree of agreement with today's values.

In 1976, Icenogle *et al.* [4] reported refractive-index values for a series of temperatures and wavelengths between 2.5 and 13 μm. These authors claimed a conservative error of approximately 6×10^{-4} in index n for a percentage error of $\sim 0.0125\%$. A few years later, Edwin *et al.* [5] reported very precise results for wavelengths between 8 and 14 μm. The value of their results lies in the fact that two different experimental setups, each in a separate laboratory, were used on the same crystal sample of Ge. The differences between the two laboratories were between 1×10^{-4} and 3×10^{-4} (10^{-2} and $3 \times 10^{-2}\%$). This represents the apparent level of precision currently available for measuring refractive indices of high-index materials. All the experiments discussed so far are based on high-quality, single-crystal (in most instances) germanium

465

cut as polished prisms having included angles ranging between 5 and 18°. The angle of minimum deviation was measured to obtain the refractive index. Data of Icenogle et al. [4] at 297 K and some data of Edwin et al. [5] at 25°C, originally given in tabular form, are listed in Table V. The tabulated data are plotted in Fig. 4.

The precision of the data of Icenogle et al. [4] was such that Barnes and Piltch [6] used them as a basis for evaluating the coefficients of a temperature-dependent, Sellmeier-type equation

$$n^2 = A + B\lambda^2/(\lambda^2 - C) + D\lambda^2(\lambda^2 - E).$$

Using the temperature-dependent values of n, they found the following values for the coefficients:

$$A = -6.040 \times 10^{-3}T + 11.05128,$$

$$B = 9.295 \times 10^{-3}T + 4.00536,$$

$$C = -5.392 \times 10^{-4}T + 0.5999034,$$

$$D = 4.151 \times 10^{-4}T + 0.09145,$$

$$E = 1.51408T + 3426.5.$$

We have evaluated this expression for "room temperture" of 18°C (291 K), over a large infrared wavelength range, and these values are included in Table V. Some comparison with the results from the several experiments outlined here can be made in the table. The Sellmeier equation represents the accurate values for high-grade optical-quality germanium within 0.025%, or better, over the infrared spectral range 2.5–14 μm and can be extrapolated to 2.0 and 40 μm.

Many more measurements have been reported for this spectral range, but in most cases the specimens were thin films, amorphous material, or heavily doped. Evaporated films do not, in general, have properties representative of bulk material, so such results are applicable only to material evaporated or sputtered by a given recipe. Li [7] tabulates most such measurements that have appeared in the literature in the past half-century. Li [7] also gives a table of recommended values for the infrared at various temperatures.

In the infrared spectral region there are two contributions to the intrinsic absorption properties of germanium. When the infrared radiation field interacts with the crystal lattice, it is described as a photon–phonon interaction. Since selection rules include the conservation of energy and wave vector or momentum, the spectrum will include features related to the photon excitation of multiple phonons. Such "phonon" bands have been observed for Ge. Collins and Fan [8] reported absorption coefficients for phonon bands for 300 K. These have been read from a graph and are listed in Table V. Values of Lax and Burstein [9] are comparable. More recently, Aronson et al. [10] have measured the absorption coefficient for high-resistivity Ge ($\sim 50\ \Omega$ cm) at 300 K. They used samples of different thickness measuring the transmittance. We have read values from a graph and tabulate these. Their results overlap the data of Collins and Fan [8]. The band at 300 cm^{-1} in the data of Collins and Fan [8] occurs at 280 cm^{-1} in the data of Aronsen et al. [10]. Stierwalt and Potter [11] measured the spectral emittance of Ge at 50°C from 24 μm > λ > 10 μm. The reader should note that $k \sim 10^{-5}$ at 10 μm and becomes very much smaller at shorter wavelengths corresponding to three and more phonon interactions.

Randall and Rawcliffe [12] used channel spectra, while Afsar et al. [13] used asymmetric Fourier-transform spectroscopy to measure n and k for high-resistivity Ge (10–50 Ω cm^{-1}). Their results are in good agreement and the data of Randall and Rawcliffe [12] are in Table V, although room temperature was not defined. At λ > 200-μm free-carrier absorption is beginning to appear.

As the photon energy is increased, the values of k due to phonon bands will decrease to insignificant values until the onset of electronic transitions in the near IR. Absorption increases for frequencies less than the first "edge" in Ge, again due to phonon or lattice interactions with electrons. Some of the data of MacFarlane and Roberts [14] obtained on ultrathin bulk samples by using transmittance at 291 K, as read from a graph, have been tabulated to show that k has measured values in the 10^{-7} range.

At wavelengths shorter than 2 μm, Ge becomes increasingly absorbing because of the dominance of electronic interband transitions; hence it is no longer possible to use the angle-of-deviation method for determining n. Transmittance measurements can be used to measure the absorption coefficient α (or k), but data are not valid for λ < 0.6 μm. Data of Dash and Newman [15] at 300 K, as read from a graph, are given for 2 μm > λ > 0.7 μm. The agreement in the wavelength overlap with MacFarlane and Roberts [14] is good, except for the shift of the indirect absorption edge due to the temperature difference.

The optical-measurement methods remaining for shorter wavelengths are related to aspects of the Fresnel reflectivity coefficients, i.e., ellipsometry and reflectance. Because the Fresnel reflectivity is a complex quantity obeying causality laws, it can be expressed as a modulus $R^{1/2}$ and a phase function. The Kramers–Kronig (KK) relation yields the spectral phase function, given R over the entire frequency spectrum [16]. Philipp and Taft [17] determined the reflectance R over a wide spectral region, applying the KK relations to give n and k. Some of their results as extracted from a graph are tabulated for λ < 0.25 μm for 300 K.

Another type of reflectance method is to measure reflectances at nonzero angles of incidence. Potter [18] measured the ratio R_p/R_s at several angles of incidence to get the minimum value R_p/R_s and the associated pseudo-Brewster angle. Some room-temperature values of n and k for Ge are given for the spectral range 1.77 μm > λ > 0.6 μm.

Ellipsometry has been used to measure optical properties from the visible spectral region to 6 eV (0.207 μm). The tabulated data of Aspnes and Studna

[19] are listed in Table V. Room temperature was not stated. Archer [20] also measured the optical properties over a reduced spectral range with less spectral detail. Cardona et al. [21] prepared crystalline and amorphous Ge films on heated and room-temperature KCl substrates, respectively. Samples were floated off and picked up with a copper mesh. Transmission measurements were done with synchrotron radiation. The data in Table V were read from a graph. Room temperature was not stated.

Lukirskii et al. [22] have measured the unpolarized reflectance for Ge as a function of angle of incidence at four x-ray wavelengths (since $\varepsilon < 1$, this gives a type of external attenuated total reflection). Samples were deposited on glass substrates. Room temperature was not specified. Note that the angles as given in Lukirskii et al. [22] are grazing angles, i.e., $\pi/2 - \phi$, where ϕ is angle of incidence. These data are obtained from a table and are listed as four points but plotted in Fig. 4 as a continuous line. Note in Fig. 4 that the k data tend to join the longer-wavelength data of Cardona et al. [21] and that these data, in turn, tend to join the data of Philipp and Taft [17].

Aspens and Studna's data [19] overlap Potter's [18] for 0.83 μm $> \lambda >$ 0.41 μm (1.5–3eV). When the extinction coefficient is low, the agreement in n between the two sets is quite good (better than 0.25% for 0.83 μm $> \lambda >$ 0.74 μm). As k becomes larger, the discrepancy increases to about 1% at 2 eV and 4% at 3 eV. At the longer wavelengths Archer's data [20] are about 4% below those of Aspnes and Studna [19], coming into reasonable agreement (less than 1% difference) for 0.54 μm $> \lambda > 0.36$ μm. Thus, the data of Aspnes and Studna [19] for 0.82 μm $> \lambda > 0.2$ μm appear to be the best choice for refractive-index values. Data of Philipp and Taft [17] appear to be too low.

In the spectral region of interband transitions the extinction coefficient becomes quite large ($k > 4$), corresponding to absorption coefficients of the order of 10^6 cm^{-1}. Reflectance-type measurements must be used. However, some overlap with transmittance measurements exists for comparison. Measurements of Dash and Newman [15] extend to 0.69 μm, which permits comparison with Potter [18] as well as with Aspnes and Studna [19]. The former appears to be low by about a factor of 2, whilte the agreement with the latter is within 10–20%, which, given sample treatment and technique differences, must be considered good. As k becomes larger, Potter's data [18] come closer to agreeing with those of Aspnes and Studna [19]. For 0.62 μm $> \lambda > 0.4$ μm, Potter's data [18] are consistently larger than the data of Aspnes and Studna [19] by 10–20%. Again, it seems that the data of Aspnes and Studna [19] would be the choice for values of k.

For values of n overlap regions occur for the data of Briggs [1], Potter [18], and the Sellmeier fit. Given that the data of Briggs [1] and Potter [18] were taken at 20°C or more, the agreement is good. Below about 2 μm, the Sellmeier results would increasingly deviate from the measured values. That the reflectance method agrees with the angle-of-deviation method better than 0.25% in this spectral region establishes the compatibility of the two methods.

Reflectivity-type experiments can presents serious problems to the unwary for reducing the experimental data to intrinsic material parameters such as n and k. The condition of the reflecting surface must be well established. Any damage layer present due to grinding, polishing, etc., must be removed. Chemical polishes can avoid this problem. Another incipient problem is the likely presence of a native oxide. A stable oxide grows on germanium to thicknesses of 50 to 100 Å in ambient air. For high-index materials such as Ge, such a layer must be removed or accounted for. Aspnes et al. [19] stripped the oxide and took steps to inhibit the oxide growth during the experimental run. While ellipsometry is very sensitive to the presence of a low-index layer, reflectance measurements also become quite sensitive to such a layer as k becomes significant. Both Potter [18] and Archer [20] assumed the presence of an oxide on their specimens and corrected for it. Archer [20] assumed a thickness of 10 to 20 Å, while Potter [18] assumed a layer with thickness of 50 Å and an index of 2.0. (See Aspnes and Stunda [19] for further discussion.)

REFERENCES

1. H. B. Briggs, *Phys. Rev.* **77**, 286 (1950).
2. C. D. Salzberg and J. J. Villa, *J. Opt. Soc. Am.* **47**, 244 (1957).
3. M. Cardona, W. Paul, and H. Brooks, *J. Phys. Chem. Solids* **8**, 204 (1959).
4. H. W. Icenogle, B. C. Platt, and W. L. Wolfe, *Appl. Opt.* **15**, 2348 (1976).
5. R. P. Edwin, M. T. Dudermel, and M. Lamare, *Appl. Opt* **17**, 1066 (1978).
6. N. P. Barnes and M. S. Piltch, *J. Opt. Soc. Am.* **69**, 178 (1979).
7. H. H. Li, *J. Phys. Chem. Ref. Data*, **9**, 561 (1980).
8. R. J. Collins and H. Y. Fan, *Phys. Rev.* **93**, 674 (1954).
9. M. Lax and E. Burstein, *Phys. Rev.* **97**, 39 (1955).
10. J. R. Aronson, H. G. McLinden, and P. J. Gielisse, *Phys, Rev.* **135**, A785 (1964).
11. D. L. Stierwalt and Roy F. Potter, *Rep. Int. Conf. Phys. Semicond., Exter*, 1962, p. 513. Institute of Physics and Physical Society, London, 1962.
12. C. M. Randall and R. D. Rawcliffe, *Appl. Opt.* **6**, 1889 (1967).
13. M. N. Afsar, D. D. Honijk, W. F. Passchier, and J. Goulon, *IEEE Trans. Micro. Theor. Tech.* **MTT-25**, 505 (1977).
14. G. G. MacFarlane and V. Roberts, *Phys. Rev.* **98**, 1865 (1955).
15. W. C. Dash and R. Newman, *Phys. Rev.* **99**, 1151 (1955).
16. M. Cardona, in "Optical Properties of Solids" (S. Nudelman and S. S. Mitra, eds.), p. 137. Plenum, New York, 1969.
17. H. R. Philipp and E. A. Taft, *Phys. Rev.* **113**, 1002 (1959).
18. Roy F. Potter, *Phys. Rev.* **150**, 562 (1966).
19. D. E. Aspnes and A. A. Studna, *Phys. Rev. B* **27**, 985 (1983); D. E. Aspnes, private communication (1983).
20. R. J. Archer, *Phys. Rev*, **110**, 354 (1958).
21. M. Cardona, W. Gudat, B. Sonntag, and P. Y. Yu, *Proc. 10th Int. Conf. Phys. Semicond., Cambridge, 1970*, p. 208. U.S. Atomic Energy Commission, Oak Ridge, Tennessee, 1970.
22. A. P. Lukirskii, E. P. Savinov, O. A. Ershov, and Y. F. Shepelov, *Opt. Spectrosc.* **16**, 168 (1964); A. P. Lukirskii, E. P. Savinov, O. A. Ershov, and Y. F. Shepelov, *Opt. Spektrosk.* **16**, 310 (1964).

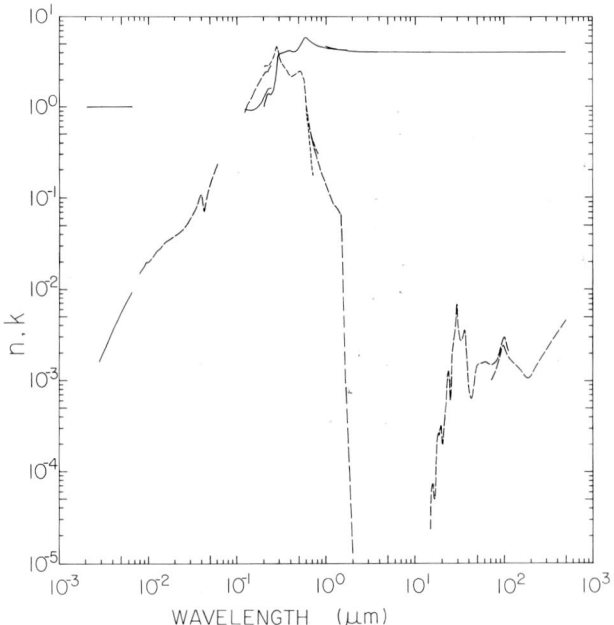

Fig. 4. Log–log plot of n (——) and k (––––) versus wavelength in micrometers for germanium. Data from various references sometimes do not join together in overlap regions.

TABLE V

Values of n and k for Germanium Obtained from Various References[a]

eV	cm^{-1}	μm	n	k	k
525.3		0.00236	0.99741 [22]	9.8×10^{-4} [22]	
394.8		0.00314	0.99590	2.05×10^{-3}	
281.8		0.0044	0.99385	4.30	
185.1		0.0067	0.99050	9.50×10^{-3}	
155		0.007999		1.47×10^{-2} [21]	
150		0.008266		1.55	
145		0.008550		1.63	
140		0.008856		1.71	
135		0.009184		1.77	
130		0.009537		1.98	
127.5		0.009725		1.94	
125		0.009919		1.98	
122.5		0.01012		2.02	
120		0.01033		1.99	
115		0.01078		2.06	
110		0.01127		2.21	
105		0.01181		2.37	
100		0.01240		2.57	
97.5		0.01272		2.65	
95		0.01305		2.71	
90		0.01378		2.79	
85		0.01459		3.03	
80		0.01550		3.26	
75		0.01653		3.39	
70		0.01771		3.52	
65		0.01907		3.67	
60		0.02066		3.94	
55		0.02254		4.23	
50		0.02480		4.52	
45		0.02755		5.04	
40		0.03100		6.04	
38		0.03263		6.51	
36		0.03444		7.40	
34		0.03647		8.56	
32		0.03875		9.99×10^{-2}	
31		0.04000		0.106	
30		0.04133		0.102	
29		0.04275		7.14×10^{-2}	
28		0.04428		7.47×10^{-2}	
26		0.04769		0.110	
24		0.05166		0.144	
22		0.05636		0.179	
20		0.06199		0.237	
10		0.1240	0.93 [17]	0.86 [17]	
9.5		0.1305		1.0	

(continued)

[a] References are indicated in brackets. Data from Stierwalt and Potter [11] obtained at 50°C are included because no room-temperature data were available in the 10–15-μm region.

TABLE V (*Continued*)

Germanium

eV	cm^{-1}	μm	n	k	k
9		0.1378	0.92	1.14	
8.5		0.1459	0.92	1.2	
8		0.1550	0.92	1.4	
7.5		0.1653		1.6	
7		0.1771	1.0	1.8	
6.5		0.1907	1.1	2.05	
6	48,390	0.2066	1.3	2.34	
			1.022 [19]	2.774 [19]	
5.94	47,910	0.2087	1.073	2.801	
5.9	47,590	0.2101	1.108	2.831	
5.84	47,100	0.2123	1.167	2.862	
5.8	46,780	0.2138	1.209	2.873	
5.75	46,380	0.2156	1.48 [17]	2.4 [17]	
5.74	46,300	0.2160	1.273 [19]	2.874 [19]	
5.7	45,970	0.2175	1.293	2.163	
5.64	45,490	0.2198	1.345	2.852	
5.6	45,170	0.2214	1.360	2.846	
5.54	44,680	0.2238	1.372	2.842	
5.5	44,360	0.2254	1.380	2.842	
			1.55 [17]	2.4 [17]	
5.44	43,880	0.2279	1.387 [19]	2.849 [19]	
5.4	43,550	0.2296	1.383	2.854	
5.34	43,070	0.2322	1.377	2.877	
5.3	42,750	0.2339	1.371	2.897	
5.25	42,340	0.2362	1.57 [17]	2.7 [17]	
5.24	42,260	0.2366	1.366 [19]	2.938 [19]	
5.2	41,940	0.2384	1.364	2.973	
5.14	41,460	0.2412	1.366	3.031	
5.1	41,130	0.2431	1.370	3.073	
5.04	40,650	0.2460	1.382	3.146	
5.0	40,330	0.2480	1.394	3.197	
			1.6 [17]	2.86 [17]	
4.94	39,840	0.2510	1.417 [19]	3.282 [19]	
4.9	39,520	0.2530	1.435	3.342	
4.84	39,040	0.2562	1.470	3.439	
4.8	38,710	0.2583	1.498	3.509	
4.74	38,230	0.2616	1.546	3.624	
4.7	37,910	0.2638	1.586	3.709	
4.64	37,420	0.2672	1.659	3.852	
4.6	37,100	0.2695	1.720	3.960	
4.54	36,620	0.2731	1.839	4.147	
4.5	36,290	0.2755	1.953	4.297	
4.44	35,910	0.2792	2.226	4.543	
4.4	35,080	0.2818	2.516	4.669	
4.34	35,000	0.2857	3.038	4.653	
4.3	34,680	0.2883	3.338	4.507	
4.24	34,200	0.2924	3.633	4.207	
4.2	33,880	0.2952	3.745	4.009	
4.14	33,390	0.2995	3.837	3.756	

TABLE V (*Continued*)

Germanium

eV	cm^{-1}	μm	n	k	k
4.1	33,070	0.3024	3.869	3.614	
4.04	32,580	0.3069	3.896	3.436	
4.0	32,260	0.3100	3.905	3.336	
3.94	31,780	0.3147	3.916	3.209	
3.9	31,460	0.3179	3.920	3.137	
3.84	30,970	0.3229	3.930	3.043	
3.8	30,650	0.3263	3.936	2.986	
3.74	30,160	0.3315	3.949	2.909	
3.7	29,840	0.3351	3.958	2.863	
3.64	29,360	0.3406	3.974	2.799	
3.6	29,040	0.3444	3.985	2.759	
3.54	28,550	0.3502	4.005	2.702	
3.5	28,230	0.3542	4.020	2.667	
3.44	27,750	0.3604	4.048	2.615	
3.4	27,420	0.3647	4.070	2.579	
3.34	26,940	0.3712	4.106	2.517	
3.3	26,620	0.3757	4.128	2.469	
3.24	26,130	0.3827	4.150	2.392	
3.2	25,810	0.3875	4.157	2.340	
3.14	25,330	0.3949	4.153	2.262	
3.1	25,000	0.4000	4.141	2.215	
3.04	24,520	0.4078	4.107	2.164	
3.0	24,200	0.4133	4.082	2.145	
2.94	23,710	0.4217	4.052	2.135	
2.9	23,390	0.4275	4.037	2.140	
2.84	22,910	0.4366	4.030	2.162	
2.8	22,580	0.4428	4.035	2.181	
2.74	22,100	0.4525	4.058	2.215	
2.7	21,780	0.4592	4.082	2.240	
2.64	21,290	0.4696	4.133	2.281	
2.6	20,970	0.4769	4.180	2.309	
2.54	20,490	0.4881	4.267	2.353	
2.5	20,160	0.4959	4.340	2.384	
2.44	19,680	0.5081	4.482	2.429	
2.4	19,360	0.5166	4.610	2.455	
2.34	18,870	0.5299	4.883	2.434	
2.3	18,550	0.5391	5.062	2.318	
2.24	18,070	0.5535	5,198	2.125	
2.2	17,740	0.5636	5.283	2.049	
2.14	17,260	0.5794	5.554	1.924	
2.1	16,940	0.5904	5.748	1.634	
2.05	16,530	0.6048	5.9 [18]		1.1 [18]
2.04	16,450	0.6078	5.708 [19]	1.150 [19]	
2.0	16,130	0.6199	5.588	0.933	
			5.64 [18]		0.8
1.95	15,730	0.6358	5.5		0.66
1.9	15,320	0.6526	5.38		0.54
			5.294 [19]	0.638	

(*continued*)

TABLE V (*Continued*)

Germanium

eV	cm^{-1}	μm	n	k	k
1.85	14,920	0.6702	5.21 [18]		0.4
1.84	14,840	0.6738	5.152 [19]	0.537	
1.8	14,520	0.6888	5.067	0.500	
			5.08 [18]		0.26
				0.466 [15]	
1.75	14,110	0.7085	4.99		0.2
1.74	14,030	0.7126	4.961 [19]	0.430 [19]	
1.7	13,710	0.7293	4.897	0.401	
			4.95 [18]		0.18
				0.389 [15]	
1.64	13,230	0.7560	4.816 [19]	0.364 [19]	
1.6	12,900	0.7749	4.763	0.345	
			4.75 [18]	0.308 [15]	
1.54	12,420	0.8051	4.684 [19]	0.316 [19]	
1.5	12,100	0.8266	4.653	0.298	
			4.64 [18]	0.237 [15]	
1.4	11,280	0.8856	4.56	0.190	
1.3	10,490	0.9537	4.495	0.167	
1.240	10,000	1.0	4.61733 [6]		
1.2	9,679	1.033	4.42 [18]	0.123	
1.1	8,872	1.127	4.385	0.103	
1.033	8,333	1.2	4.35673 [6]		
1.0	8,065	1.240	4.325 [18]	8.09 × 10^{-2}	
0.9	7,259	1.378	4.285	7.45	
0.8856	7,143	1.4	4.23840 [6]		
0.88	7,098	1.409		6.73	
0.86	6,936	1.442		6.88	
0.84	6,775	1.476		6.58	
0.82	6,614	1.512		5.66 × 10^{-2}	
0.8	6,452	1.550	4.275 [16]	5.67 × 10^{-3}	
0.78	6,291	1.590		3.42 × 10^{-3}	
0.775	6,251	1.600	4.17262 [6]		8.15 × 10^{-4} [19]
0.76	6,130	1.631	4.17262 [6]	7.79 × 10^{-4}	
0.75	6,049	1.653			5.06
0.74	5,968	1.675		6.67	
0.725	5,847	1.710			3.01
0.72	5,807	1.722		4.66	
0.7	5,646	1.771	4.18 [18]	2.82	1.27 × 10^{-4}
0.6888	5,556	1.8	4.13164 [6]		
			4.143 [1]		
0.68	5,485	1.823		1.31 × 10^{-4}	
0.675	5,444	1.837			4.22 × 10^{-5}
0.6702	5,405	1.85	4.135		
0.66	5,323	1.879		6.73 × 10^{-5}	
0.6525	5,263	1.90	4.129		
0.65	5,243	1.907			9.71 × 10^{-6}
0.64	5,162	1.937		7.71 × 10^{-6}	
0.625	5,041	1.984			1.42 × 10^{-6}
0.6199	5,000	2.0	4.116		

TABLE V (*Continued*)

Germanium

eV	cm^{-1}	μm	n	k	k
			4.10415 [6]		
0.6	4,839	2.066			6.58×10^{-7}
0.5904	4,762	2.1	4.104 [1]		
0.5636	4,545	2.2	4.092		
			4.08469 [6]		
0.5391	4,348	2.3	4.085 [1]		
0.5166	4,167	2.4	4.078		
			4.07038 [6]		
0.4959	4,000	2.5	4.072 [1]		
0.4854	3,915	2.554	4.06230 [4]		
0.4769	3,846	2.6	4.068 [1]		
			4.05951 [6]		
0.4675	3,771	2.652	4.05754 [4]		
0.4538	3,660	2.732	4.05310		
0.4428	3,571	2.8	4.05105 [6]		
0.4341	3,501	2.856	4.04947 [4]		
0.4191	3,381	2.958	4.04595		
0.4133	3,333	3.0	4.04433 [6]		
0.4012	3,236	3.09	4.04292 [4]		
0.3874	3,125	3.2	4.03889 [6]		
0.3647	2,941	3.4	4.03443		
0.3444	2,778	3.6	4.03072		
0.3263	2,632	3.8	4.02760		
0.3100	2,500	4.0	4.02495		
0.3009	2,427	4.12	4.02457 [4]		
0.2952	2,381	4.2	402268 [6]		
0.2818	2,273	4.4	4.02072		
0.2695	2,174	4.6	4.01901		
0.2583	2,083	4.8	4.01751		
0.2480	2,000	5.0	4.01619		
0.2389	1,927	5.19	4.01617 [4]		
0.2384	1,923	5.2	4.01503 [6]		
0.2296	1,852	5.4	4.01399		
0.2214	1,786	5.6	4.01305		
0.2138	1,724	5.8	4.01222		
0.2066	1,667	6.0	4.01146		
0.2000	1,613	6.2	4.01078		
0.1937	1,563	6.4	4.01015		
0.1879	1,515	6.6	4.00958		
0.1823	1,471	6.8	4.00906		
0.1771	1,429	7.0	4.00858		
0.1722	1,389	7.2	4.00814		
0.1675	1,351	7.4	4.00773		
0.1631	1,316	7.6	4.00736		
0.1590	1,282	7.8	4.00701		
0.1550	1,250	8.0	4.00668		
			4.00748 [5]		
0.1512	1,220	8.2	4.00637 [6]		

(*continued*)

TABLE V (*Continued*)

Germanium

eV	cm^{-1}	μm	n	k	k
0.1506	1,215	8.23	4.00743 [4]		
0.1476	1,190	8.4	4.00609 [6]		
0.1442	1,163	8.6	4.00582		
0.1409	1,136	8.8	4.00556		
0.1378	1,111	9.0	4.00533		
			4.00620 [5]		
0.1348	1,087	9.2	4.00510 [6]		
0.1319	1,064	9.4	4.00489		
0.1292	1,042	9.6	4.00469		
0.1265	1,020	9.8	4.00449		
0.1240	1,000	10.0	4.00431		
			4.00525 [5]		1.43×10^{-5} [11]
0.1207	973.7	10.27	4.00571 [4]		
0.1181	952.4	10.5	4.00389 [6]		
0.1137	917.4	10.9			1.30
0.1127	909.1	11.0	4.00351		
			4.00436 [5]		
0.1078	869.6	11.5	4.00316 [6]		
0.1033	833.3	12.0	4.00285		3.82
			4.00398 [5]		
0.1003	809.1	12.36	4.00627 [4]		
0.09919	800.0	12.5	4.00255 [6]		
0.09840	793.7	12.6			3.51
0.09537	769.2	13.0	4.00228		
			4.00352 [5]		
0.09460	763.4	13.1			4.59
0.09184	740.7	13.5	4.00202 [6]		
0.08920	719.4	13.9			3.98
0.08856	714.3	14.0	4.00177		
			4.00315 [5]		
0.08551	689.7	14.5	4.00153 [6]		5.77
0.08431	680	14.71		2.34×10^{-5} [8]	
0.08307	670	14.93		3.56	
0.08266	666.7	15.0	4.00130		9.78×10^{-5}
0.08183	660	15.15		5.30	
0.08157	657.9	15.2			1.2×10^{-4}
0.08059	650	15.38		6.24	
0.07999	645.2	15.5	4.00107		1.71×10^{-4}
0.07935	640	15.63		7.34	
0.07811	630	15.87		7.33	
0.07749	625.0	16.0	4.00086		9.55×10^{-5}
0.07687	620	16.13		6.54	
0.07563	610	16.39		5.61	
0.07514	606.1	16.5			1.58×10^{-4}
0.07439	600	16.67		4.91	
0.07377	595	16.81		6.02	
0.07315	590	16.95		9.44×10^{-5}	
0.07293	588.2	17.0	4.00043		2.57
0.07191	580	17.24		1.64×10^{-4}	

TABLE V (*Continued*)

Germanium

eV	cm^{-1}	μm	n	k	k
0.07085	571.4	17.5			3.90
0.07067	570	17.54		2.09	
0.06943	560	17.86		2.56	
0.06888	555.6	18.0	4.00001		3.01
0.06819	550	18.18		2.68	
0.06757	545	18.35		2.63	
0.06702	540.5	18.5			4.27
0.06695	540	18.52		2.51	
0.06633	535	18.69		2.68	
0.06571	530	18.87		3.00	
0.06526	526.3	19.0	3.99958		4.53
0.06447	520	19.23		3.21	
0.06358	512.8	19.5			4.03
0.06323	510	19.61		3.12	
0.06199	500	20.00	3.99915	2.39	3.98
0.06075	490	20.41		1.94	
0.06048	487.8	20.5			3.92
0.05951	480	20.83		2.49	
0.05904	476.2	21			5.01
0.05827	470	21.28		3.39	
0.05767	465.1	21.5			7.70×10^{-4}
0.05703	460	21.74		4.67	
0.05636	454.5	22.0	3.99825		1.02×10^{-3}
0.05579	450	22.22		5.84	6.0×10^{-4} [10]
0.05455	440	22.73		7.24×10^{-4}	8.0×10^{-4}
0.05331	430	23.26		1.11×10^{-3}	1.1×10^{-3}
0.05269	425	23.53		1.27	
0.05207	420	23.81		1.10×10^{-3}	1.35
0.05166	416.7	24.0	3.99728		
0.05083	410	24.39		7.77×10^{-4}	1.4×10^{-3}
0.05021	405	24.69			9.0×10^{-4}
0.04959	400	25.00		6.17	8.0×10^{-4}
0.04897	395	25.32		8.06×10^{-4}	1.5×10^{-3}
0.04835	390	25.64		1.73×10^{-3}	2.0
0.04769	384.6	26.0	3.99621		
0.04711	380	26.32		2.30	2.1
0.04587	370	27.03		2.80	2.5
0.04463	360	27.78		3.76	3.4
0.04428	357.1	28.0	3.99500		
0.04401	355	28.17			5.5
0.04370	352.5	28.37			6.7
0.04339	350	28.57		5.23	6.3
0.04308	347.5	28.78			6.7
0.04277	345	28.99		6.92	6.5
0.04215	340	29.41		4.68	3.2
0.04133	333.3	30.0	3.99363		
0.04092	330	30.30		3.62	2.95
0.03968	320	31.25		2.98	2.9

(*continued*)

TABLE V (*Continued*)

Germanium

eV	cm^{-1}	μm	n	k	k
0.03844	310	32.26		2.82	2.7
0.03782	305	32.79		3.00	
0.03720	300	33.33		3.18	2.75
0.03658	295	33.90		2.83	
0.03596	290	34.48		2.74	3.05
0.03542	285.7	35	3.98925		
0.03534	285	35.09			3.45
0.03472	280	35.71			3.55
0.03410	275	36.36			3.3
0.03348	270	37.04			2.75
0.03224	260	38.46			1.55
0.03100	250	40.00	3.98272		0.85
0.02976	240	41.67			0.7
0.02852	230	43.48			0.65
0.02728	220	45.45			0.85
0.02604	210	47.62			1.25
0.02480	200	50.00			1.5
0.02356	190	52.63			1.53
0.02232	180	55.56			1.55
0.02108	170	58.82			1.6
0.01984	160	62.50			1.6
0.1860	150	66.67			1.5
0.01736	140	71.43	4.0060 [12]	1.02 [12]	1.5
0.01643	132.5	75.47	4.0060	1.14	
0.01612	130	76.92			1.55
0.01550	125	80.00	4.0060	1.27	
0.01488	120	83.33			1.7
0.01457	117.05	85.11	4.0060	1.62	
0.01364	110	90.91	4.0063	2.10	2.4
0.01302	105	95.24			2.8
0.01271	102.5	97.56	4.0064	2.41	
0.01240	100	100.0			3.0
0.01178	95	105.3	4.0067	2.01	2.5
0.01116	90	111.1			2.1 × 10^{-3}
0.01085	87.5	114.3	4.0068	1.73	
0.009919	80	125.0	4.0067	1.59	
0.008989	72.5	137.9	4.0067	1.43	
0.008059	65	153.8	4.0065	1.34	
0.007129	57.5	173.9	4.0061	1.10	
0.006199	50	200.0	4.0059	1.11	
0.005269	42.5	235.3	4.0056	1.50	
0.004339	35	285.7	4.0052	2.04	
0.003410	27.5	363.6	4.0047	2.89	
0.002480	20	500.0	4.0043	4.77 × 10^{-3}	

Indium Arsenide (InAs)

EDWARD D. PALIK and R. T. HOLM

Naval Research Laboratory
Washington, D.C.

The UV–visible reflectivity of InAs was measured by Philipp and Ehrenreich [1] in the spectral region from 1 to 25 eV. Chemically polished surfaces were used. Kramers–Kronig (KK) analysis yielded n and k, which were tabulated by Seraphin and Bennett [2]. Measurements by Morrison [3] in the 1–6-eV region are significantly different, probably because of poorer surface preparation and/or a wider range of extrapolation for the KK analysis. For 6 to 1.5 eV the ellipsometric results of Aspnes and Studna [4] are used. The absorption of deposited films was measured from 20 to 150 eV but these data were not given [5]; only peaks due to interband transitions were listed. The data of Seraphin and Bennett [2] and Aspnes and Stunda [4] are listed in Table VI. Figure 5 is a smooth-curve plot of all the data.

Dixon and Ellis [6] measured the band-edge absorption at 300 K for thin samples mechanically polished. Samples were polished on a beeswax lap with aluminum oxide powder. They were unsupported, or for samples less than 30 μm thick, they were glued on sapphire with glycol phthalate. Their lowest carrier density, 3.6×10^{16} cm^{-3}, did not show signs of the Burstein–Moss effect. The k values listed in Table VI were obtained from our reading of a graph in Dixon and Ellis [6] to three figures.

Lorimor and Spitzer [7] measured transmission interference fringes at room temperature in the near IR and fitted the data with the formula

$$n^2 = A + \frac{B}{1 - (v/3920)^2} + \frac{C}{1 - (v/219)^2} + \frac{D}{v^2}. \tag{1}$$

The B term is the contribution from the band-edge resonance, the C term is the contribution from the reststrahlen band, and the D term comes from the free-carrier susceptibility since the free-carrier effects were significant for $N = 2 \times 10^{16}$ cm^{-3}. The values were $A = 11.1 \pm 0.1$, $B = 0.71 \pm 0.1$, $C = 2.75 \pm 0.2$, and $D = 5.64 \times 10^4$ cm^{-2}. It is obvious that n is uncertain in the third significant figure by a digit or two.

HANDBOOK OF OPTICAL CONSTANTS OF SOLIDS

ISBN 0-12-544420-6

Pikhtin and Yas'kov [8] have refitted these data with a more detailed formula

$$n^2 = 1 + \frac{A}{\pi} \ln \frac{E_1^2 - (\hbar\omega)^2}{E_0^2 - (\hbar\omega)^2} + \frac{G_1}{E_1^2 - (\hbar\omega)^2} + \frac{G_2}{E_2^2 - (\hbar\omega)^2} + \frac{G_3}{E_3^2 - (\hbar\omega)^2}, \quad (2)$$

with $A = 0.7/\sqrt{E_0}$, $E_0 = 0.356$ eV, $E_1 = 2.2$ eV, $E_2 = 4.9$ eV, $E_3 = 2.714 \times 10^{-2}$ eV, $G_1 = 28.748$ eV^2, $G_2 = 79.354$ eV^2, and $G_3 = 2.01 \times 10^{-3}$ eV^2. The second term is the contribution of the fundamental band gap. The third and fourth terms are contributions from two higher-lying band gaps treated as oscillators, while the fifth term is the contribution of the lattice oscillator in the far infrared. These values are tabulated in the near IR. While four figures are given, the third figure is probably uncertain to a digit or two.

Multiphonon absorption in the 450–250-cm^{-1} region has been measured by Lorimor and Spitzer [7]. Their sample showed free-carrier absorption ($N = 2 \times 10^{16}$ cm^{-3}) that was substracted off, assuming a λ^2 dependence. The tabulated data were read off a graph. Absorption in the same spectral region was measured at 15 K for a sample with $N = 3 \times 10^{16}$ cm^{-3}, but no quantitative numbers were given [9].

The reststrahlen reflectivity data are scant [10], and we have elected to take values of $\omega_T = 218$ cm^{-1}, $\omega_L = 240$ cm^{-1}, and $\Gamma = 4$ cm^{-1} from Raman data [11, 12] with $\varepsilon_\infty = 11.7$ [7, 13] and calculate n and k from the formula

$$(n - ik)^2 = \varepsilon_\infty \left[1 + \frac{\omega_L^2 - \omega_T^2}{\omega_T^2 - \omega^2 + i\Gamma\omega} \right]. \quad (3)$$

These are indicated in Table VI as footnote a. No free carriers are assumed. The n's for Pikhtin and Yas'kov [8] and from the oscillator model [Eq. (3)] are compared, starting at 10 μm, and vary by one digit in the third significant figure. This agreement worsens to three digits by 25 μm. We prefer the oscillator model here because of the more precise parameters used from the Raman data.

Direct experimental measurement of n and k at 300 K has been made by Memon et al. [14], who used the asymmetric Fourier-transform spectroscopy technique. While the sample was described as undoped, it showed strong free-carrier effects below 200 cm^{-1}. They analyzed a dispersion equation by using experimental values of n and k to obtain the frequency-dependent damping parameter. We have read a graph to obtain n and k and list them in the 290–140-cm^{-1} region in Table VI. We have plotted in Fig. 5 the data of Memon et al. [14] in the vicinity of ω_T but have included some oscillator model data to lower frequency for n and k since no free-carrier effects are included. Note that the peak in k is $\sim$9 and occurs at 219 cm^{-1} for values from Eq. (3) and is $\sim$7 and occurs at 221 cm^{-1} for Memon et al. [14]. The data of Memon et al. [14] are preferred.

At low frequency, $n^2 = 14.18$ is smaller than, but in reasonable agreement with, the static dielectric constant 14.55 ± 0.3 [7, 13].

REFERENCES

1. H. R. Philipp and H. Ehrenreich, *Phys. Rev.* **129**, 1550 (1963).
2. B. O. Seraphin and H. E. Bennett, *in* "Semiconductors and Semimetals" (R. K. Willardson and A. C. Beer, eds.), Vol. 3, p. 499. Academic Press, New York, 1967.
3. R. E. Morrison, *Phys. Rev.* **124**, 1314 (1961).
4. D. E. Aspnes and A. A. Studna, *Phys. Rev. B* **27**, 985 (1983); D. E. Aspnes and A. A. Studna, private communication (1982).
5. M. Cardona, W. Gudat, B. Sonntag, and P. Y. Yu, *Proc. Int. Conf. Phys. Semicond., 10th Cambridge, 1970*, p. 208. U.S. Atomic Energy Commission, Oak Ridge, Tennessee, 1970.
6. J. R. Dixon and J. M. Ellis, *Phys. Rev.* **123**, 1560 (1961).
7. O. G. Lorimor and W. G. Spitzer, *J. Appl. Phys.* **36**, 1841 (1965).
8. A. N. Pikhtin and A. D. Yas'kov, *Sov. Phys. Semicond.* **12**, 622 (1978); A. N. Pikhtin and A. D. Yas'kov, *Fiz. Tekh. Poluprovodu.* **12**, 1047 (1978).
9. E. S. Koteles and W. R. Datars, *Can. J. Phys.* **54**, 1676 (1976).
10. M. Hass and B. W. Henvis, *J. Phys. Chem. Solids* **23**, 1099 (1962).
11. R. C. C. Leite and J. F. Scott, *Phys. Rev. Lett.* **22**, 130 (1969).
12. E. L. Ivchenko, D. N. Mirlin, and I. I. Reshina, *Sov. Phys. Solid State* **17**, 1510 (1975); E. L. Ivchenko, D. N. Mirlin, and I. I. Reshina, *Fiz. Tverd. Tela.* **17**, 2282 (1975).
13. M. Hass, *in* "Semiconductors and Semimetals" (R. K. Willardson and A. C. Beer, eds.), Vol. 3, p. 3. Academic Press, New York, 1967.
14. A. Memon, T. J. Parker, and J. R. Birch, *Proc. SPIE*, **289**, 20 (1981).

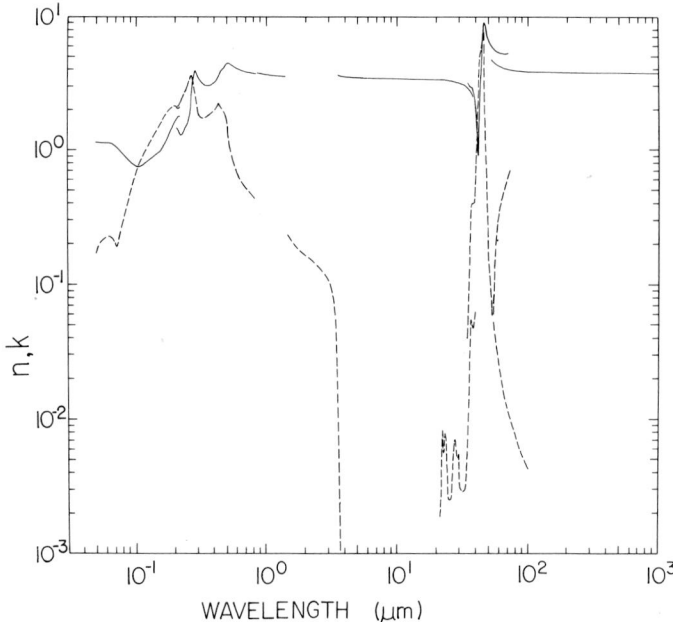

Fig. 5. Log–log plot of n (——) and k (–––) versus wavelength in micrometers for indium arsenide. The transmission data for k at 1.5 μm are probably more reliable than the KK data and can be extrapolated to the first peak in k at 0.45 μm.

TABLE VI

Values of n and k for Indium Arsenide Obtained from Various References[a]

eV	cm^{-1}	μm	n	k
25		0.04959	1.139 [1, 2]	0.168 [1, 2]
24		0.05166	1.135	0.195
23		0.05391	1.135	0.207
22		0.05636	1.133	0.215
21		0.05904	1.131	0.222
20		0.06199	1.125	0.225
19.5		0.06358	1.120	0.224
18.5		0.06702	1.110	0.215
17.5		0.07085	1.047	0.189
16		0.07749	0.948	0.272
15		0.08266	0.894	0.336
14		0.08856	0.829	0.426
13		0.09537	0.766	0.563
12		0.1033	0.745	0.727
11.5		0.1078	0.751	0.830
11		0.1127	0.775	0.905
10		0.1240	0.835	1.071
9.1		0.1362	0.890	1.260
8.1		0.1531	0.967	1.552
7.2		0.1722	1.184	1.889
6.9		0.1797	1.332	1.998
6.6		0.1879	1.483	2.020
6.35		0.1953	1.583	2.120
6.0	48,390	0.2066	1.434 [4]	2.112 [4]
5.4	43,550	0.2296	1.401	2.085
5.94	47,910	0.2087	1.401	2.085
5.9	47,590	0.2101	1.383	2.084
5.84	47,100	0.2123	1.350	2.089
5.8	46,780	0.2138	1.333	2.102
5.74	46,300	0.2160	1.306	2.135
5.7	45,970	0.2175	1.293	2.163
5.64	45,490	0.2198	1.281	2.212
5.6	45,170	0.2214	1.276	2.248
5.54	44,680	0.2238	1.277	2.304
5.5	44,360	0.2254	1.282	2.344

[a] The data of Philipp and Ehrenreich [1] given in Seraphin and Bennett [2] are specified by the wavelength (in micrometers). However, the original data were given in electron volts, and it is obvious that most of the n, k, values were determined for energies in whole numbers or tenths of a whole number. The values were rounded off to three significant figures and then given the wavelength listed in Seraphin and Bennett [2]. Sometimes we list these wavelengths and then recalculate energy (in electron volts), thus producing roundoff error in electron volts. The reference is given at the beginning of the tabulation for n and k and is understood to refer to all n's or all k's below it until a new reference appears. When an exponent of 10 is given, all numbers below it have the same exponent until the power of 10 changes. The power of 10 repeats at the end of a particular set of data or when the next number has 10^{-1}, which is then written as a decimal.

TABLE VI (*Continued*)

Indium Arsenide

eV	cm^{-1}	μm	n	k
			1.765 [1, 2]	2.202 [1, 2]
5.44	43,880	0.2279	1.297 [4]	2.407 [4]
5.4	43,550	0.2296	1.312	2.449
5.34	43,070	0.2322	1.341	2.514
5.3	42,750	0.2339	1.366	2.555
5.24	42,260	0.2366	1.408	2.612
5.2	41,940	0.2384	1.436	2.646
5.14	41,460	0.2412	1.469	2.694
5.1	41,130	0.2431	1.484	2.732
5.04	40,650	0.2460	1.505	2.807
5.0	40,330	0.2480	1.524	2.871
4.98	40,170	0.2490	1.535	2.907
4.96	40,000	0.2500	1.550	2.946
4.94	39,840	0.2510	1.566	2.988
4.92	39,680	0.2520	1.586	3.034
4.9	39,520	0.2530	1.608	3.081
4.88	39,360	0.2541	1.635	3.129
4.86	39,200	0.2551	1.668	3.182
4.84	39,040	0.2562	1.705	3.236
4.82	38,880	0.2572	1.751	3.291
4.8	38,710	0.2583	1.803	3.348
4.78	38,550	0.2594	1.866	3.404
4.76	38,390	0.2605	1.938	3.456
4.74	38,230	0.2616	2.019	3.504
4.72	38,070	0.2627	2.108	3.545
4.7	37,910	0.2638	2.204	3.575
4.68	37,750	0.2649	2.305	3.595
4.66	37,590	0.2661	2.407	3.604
4.64	37,420	0.2672	2.510	3.604
4.62	37,260	0.2684	2.609	3.596
4.6	37,100	0.2695	2.705	3.581
4.58	36,940	0.2707	2.799	3.564
4.56	36,780	0.2719	2.894	3.543
4.54	36,620	0.2731	2.990	3.519
4.52	36,460	0.2743	3.091	3.487
4.5	36,290	0.2755	3.194	3.445
4.48	36,130	0.2768	3.298	3.389
4.46	35,970	0.2780	3.399	3.320
4.44	35,810	0.2792	3.490	3.239
4.42	35,650	0.2805	3.573	3.146
4.4	35,080	0.2818	3.644	3.042
4.38	35,330	0.2831	3.701	2.931
4.36	35,170	0.2844	3.742	2.814
4.34	35,000	0.2857	3.765	2.697
4.32	34,840	0.2870	3.770	2.584
4.3	34,680	0.2883	3.761	2.478

(*continued*)

TABLE VI (*Continued*)

Indium Arsenide

eV	cm^{-1}	μm	n	k
4.28	34,520	0.2897	3.742	2.382
4.26	34,360	0.2910	3.715	2.297
4.24	34,200	0.2924	3.684	2.222
4.22	34,040	0.2938	3.650	2.156
4.2	33,880	0.2952	3.615	2.099
4.18	33,710	0.2966	3.580	2.049
4.16	33,550	0.2980	3.545	2.005
4.14	33,390	0.2995	3.512	1.967
4.12	33,230	0.3009	3.480	1.933
4.1	33,070	0.3024	3.449	1.903
4.08	32,910	0.3039	3.419	1.877
4.06	32,750	0.3054	3.391	1.854
4.04	32,580	0.3069	3.364	1.833
4.02	32,420	0.3084	3.338	1.815
4.0	32,260	0.3100	3.313	1.799
3.94	31,780	0.3147	3.246	1.761
3.9	31,460	0.3179	3.208	1.743
3.84	30,970	0.3229	3.158	1.726
3.8	30,650	0.3263	3.129	1.719
3.74	30,160	0.3315	3.091	1.714
3.7	29,840	0.3351	3.069	1.715
3.64	29,360	0.3406	3.044	1.721
3.6	29,040	0.3444	3.030	1.728
3.54	28,550	0.3502	3.015	1.742
3.5	28,230	0.3542	3.008	1.754
3.44	27,750	0.3604	3.003	1.774
3.4	27,420	0.3647	3.004	1.790
3.34	26,940	0.3712	3.010	1.817
3.3	26,620	0.3757	3.018	1.836
3.24	26,130	0.3827	3.035	1.868
3.2	25,810	0.3875	3.051	1.891
3.14	25,330	0.3949	3.083	1.929
3.1	25,000	0.4000	3.108	1.957
3.04	24,520	0.4078	3.157	2.001
3.0	24,200	0.4133	3.197	2.034
2.98	24,040	0.4161	3.220	2.051
2.96	23,870	0.4189	3.244	2.069
2.94	23,710	0.4217	3.272	2.088
2.92	23,550	0.4246	3.302	2.108
2.9	23,390	0.4275	3.337	2.129
2.88	23,230	0.4305	3.377	2.151
2.86	23,070	0.4335	3.425	2.174
2.84	22,910	0.4366	3.482	2.194
2.82	22,740	0.4397	3.550	2.207
2.8	22,580	0.4428	3.626	2.208
2.78	22,420	0.4460	3.705	2.189
2.76	22,260	0.4492	3.776	2.153
2.74	22,100	0.4525	3.834	2.108

TABLE VI (*Continued*)

Indium Arsenide

eV	cm^{-1}	μm	n	k
2.72	21,940	0.4558	3.878	2.060
2.7	21,780	0.4592	3.911	2.016
2.68	21,620	0.4626	3.935	1.977
2.66	21,450	0.4661	3.956	1.943
2.64	21,290	0.4696	3.976	1.917
2.62	21,130	0.4732	3.996	1.898
2.6	20,970	0.4769	4.021	1.885
2.58	20,810	0.4806	4.052	1.881
2.56	20,650	0.4843	4.098	1.882
2.54	20,490	0.4881	4.168	1.879
2.52	20,330	0.4920	4.265	1.855
2.5	20,160	0.4959	4.364	1.786
2.48	20,000	0.4999	4.437	1.686
2.46	19,840	0.5040	4.476	1.575
2.44	19,680	0.5081	4.489	1.466
2.42	19,520	0.5123	4.484	1.368
2.4	19,360	0.5166	4.466	1.283
2.38	19,200	0.5209	4.443	1.209
2.36	19,030	0.5254	4.417	1.144
2.34	18,870	0.5299	4.389	1.086
2.32	18,710	0.5344	4.360	1.035
2.3	18,550	0.5391	4.331	0.991
2.28	18,390	0.5438	4.303	0.951
2.26	18,230	0.5486	4.276	0.914
2.24	18,070	0.5535	4.249	0.880
2.22	17,910	0.5585	4.224	0.850
2.2	17,740	0.5636	4.199	0.822
2.18	17,580	0.5687	4.176	0.794
2.16	17,420	0.5740	4.153	0.771
2.14	17,260	0.5794	4.131	0.750
2.12	17,100	0.5848	4.109	0.730
2.1	16,940	0.5904	4.088	0.712
2.08	16,780	0.5961	4.068	0.694
2.06	16,610	0.6019	4.050	0.677
2.04	16,450	0.6078	4.031	0.661
2.02	16,290	0.6138	4.012	0.648
2.0	16,130	0.6199	3.995	0.634
1.98	15,970	0.6262	3.979	0.619
1.96	15,810	0.6326	3.962	0.606
1.94	15,650	0.6391	3.947	0.594
1.92	15,490	0.6458	3.932	0.582
1.9	15,320	0.6526	3.917	0.572
1.88	15,160	0.6595	3.902	0.564
1.86	15,000	0.6666	3.889	0.554
1.84	14,840	0.6738	3.875	0.547
1.82	14,680	0.6812	3.862	0.539
1.8	14,520	0.6888	3.851	0.530

(*continued*)

TABLE VI (*Continued*)

Indium Arsenide

eV	cm^{-1}	μm	n	k
1.78	14,360	0.6965	3.839	0.523
1.76	14,200	0.7045	3.828	0.515
1.74	14,030	0.7126	3.817	0.508
1.72	13,870	0.7208	3.807	0.500
1.7	13,710	0.7293	3.798	0.493
1.68	13,550	0.7380	3.788	0.486
1.66	13,390	0.7469	3.780	0.478
1.64	13,230	0.7560	3.770	0.475
1.62	13,070	0.7653	3.761	0.473
1.6	12,900	0.7749	3.755	0.463
1.58	12,740	0.7847	3.746	0.457
1.56	12,580	0.7948	3.735	0.458
1.54	12,420	0.8051	3.729	0.448
1.52	12,260	0.8157	3.720	0.444
1.5	12,100	0.8266	3.714	0.432
1.4	11,290	0.8856	3.696 [1, 2]	
1.2	9,679	1.033	3.613	
1.00	8,065	1.240	3.548	
0.90	7,259	1.378	3.516	
0.85	6,856	1.459		0.232 [6]
0.80	6,452	1.550		0.210
0.75	6,049	1.653		0.191
0.70	5,646	1.771		0.183
0.65	5,243	1.907		0.167
0.60	4,839	2.066		0.161
0.55	4,436	2.254		0.149
0.50	4,033	2.480		0.136
0.45	3,629	2.755		0.120
0.40	3,226	3.100		9.62×10^{-2}
0.37	2,984	3.351		6.4
0.365	2,944	3.397		5.13
0.36	2,904	3.444		3.70
0.355	2,863	3.493		2.17×10^{-2}
0.35	2,823	3.542	3.608 [8]	9.58×10^{-3}
0.345	2,783	3.594		5.14
0.34	2,742	3.647	3.556	2.29
0.335	2,702	3.701		1.12×10^{-3}
0.33	2,662	3.757	3.530	5.08×10^{-4}
0.325	2,621	3.815		2.52
0.32	2,581	3.875	3.512	1.23×10^{-4}
0.31	2,500	4.000	3.499	
0.30	2,420	4.133	3.488	
0.29	2,339	4.275	3.478	
0.28	2,258	4.428	3.470	
0.27	2,178	4.592	3.463	
0.26	2,097	4.769	3.457	
0.25	2,016	4.959	3.451	
0.24	1,936	5.166	3.445	

TABLE VI (*Continued*)

Indium Arsenide

eV	cm⁻¹	μm	n	k
0.23	1,855	5.391	3.441	
0.22	1,774	5.636	3.436	
0.21	1,694	5.904	3.432	
0.20	1,613	6.199	3.427	
0.19	1,532	6.526	3.423	
0.18	1,452	6.888	3.420	
0.17	1,371	7.293	3.416	
0.16	1,290	7.749	3.412	
0.15	1,210	8.266	3.408	
0.14	1,129	8.856	3.404	
0.13	1,049	9.537	3.400	
0.1240	1,000	10.00	3.402[b]	
0.12	967.9	10.33	3.394 [8]	
0.1116	900	11.11	3.398[b]	
0.11	887.2	11.27	3.388 [8]	
0.10	806.5	12.40	3.381	
0.09919	800	12.50	3.391[b]	
0.09	725.9	13.78	3.372 [8]	
0.08679	700	14.29	3.381[b]	
0.08	645.2	15.50	3.358 [8]	
0.07439	600	16.67	3.365[b]	
0.07	564.6	17.71	3.338 [8]	
0.06199	500	20.00	3.334[b]	
0.06	483.9	20.66	3.304 [8]	
0.05641	455	21.98		1.92 × 10⁻³ [7]
0.05579	450	22.22		3.36
0.05517	445	22.47		8.23
0.05455	440	22.73		5.61
0.05393	435	22.99		5.85
0.05331	430	23.26		7.96
0.05269	425	23.53		7.49
0.05207	420	23.81		7.20
0.05145	415	24.10		4.60
0.05083	410	24.39	3.275[b]	3.88
0.05021	405	24.69		2.55
0.05	403.3	24.80	3.237 [8]	
0.04959	400	25.00	3.264[b]	
0.04711	380	26.32		2.72
0.04649	375	26.67		3.40
0.04587	370	27.03		4.73
0.04525	365	27.38		6.54
0.04463	360	27.78		7.07
0.04401	355	28.17		6.73
0.04339	350	28.57	3.182	5.46
0.04277	345	28.99		5.07

[b] Obtained from oscillator model [Eq. (3)].

(*continued*)

TABLE VI (*Continued*)

Indium Arsenide

eV	cm^{-1}	μm	n	k
0.04215	340	29.41		5.15
0.04154	335	29.85		5.46
0.04092	330	30.30		3.38
0.03720	300	33.33	2.988	
0.03658	295	33.90		3.24
0.03596	290	34.48	2.912	4.94 × 10^{-3}
			3.18 [14]	4 × 10^{-2} [14]
0.03534	285	35.09	3.10	6 × 10^{-2}
				7.54 × 10^{-3} [7]
0.03472	280	35.71	2.808^b	1.31 × 10^{-2}
			3.00 [14]	0.17 [14]
0.03410	275	36.36	2.93	0.27
				2.60 × 10^{-2} [7]
0.03348	270	37.04	2.658^b	5.45
			2.92 [14]	0.39 [14]
				3.7 × 10$^{-2\,b}$
0.03286	265	37.74	2.92	0.40 [14]
				5.11 × 10^{-2} [7]
0.03224	260	38.46	2.418^b	4.74
			2.73 [14]	0.40 [14]
				6.3 × 10$^{-2\,b}$
0.03199	258	38.76	2.352^b	7.1
0.03174	256	39.06	2.277	8.1
0.03162	255	39.22	2.37 [14]	0.50 [14]
				5.62 × 10^{-2} [7]
0.03149	254	39.37	2.190^b	9.4^b
0.03124	252	39.68	2.089	0.11
0.03100	250	40.00	1.970^b	6.37 × 10^{-2} [7]
			2.03 [14]	0.83 [14]
				0.13^b
0.03075	248	40.32	1.827^b	0.16
			1.88 [14]	0.97 [14]
0.03050	246	40.65	1.60	1.1
			1.649^b	0.21^b
0.03025	244	40.98	1.424	0.28
			1.24 [14]	1.44 [14]
0.03000	242	41.32	0.98	2.02
			1.128^b	0.41^b
0.02976	240	41.67	0.779	0.71
			0.92 [14]	2.77 [14]
0.02951	238	42.02	1.08	3.50
			0.563^b	1.19^b
0.02926	236	42.37	0.493	1.67
			1.27 [14]	4.20 [14]
0.02901	234	42.74	2.08	4.87
			0.485^b	2.14^b
0.02876	232	43.10	0.514	2.62
			2.79 [14]	5.17 [14]

TABLE VI (*Continued*)

Indium Arsenide

eV	cm^{-1}	μm	n	k
0.02852	230	43.48	3.30	5.36
			0.578[b]	3.15[b]
0.02827	228	43.86	0.690	3.76
			3.85 [14]	5.60 [14]
0.02802	226	44.25	4.35	5.83
			0.882[b]	4.50[b]
0.02777	224	44.64	1.24	5.45
			4.90 [14]	6.12 [14]
0.02765	222	44.84	5.90	6.53
			1.99[b]	6.76[b]
0.02728	220	45.45	3.99	8.44
			7.50 [14]	6.40 [14]
0.02703	218	45.87	8.89	5.05
			8.58[b]	7.87[b]
0.02678	216	46.30	9.16	3.53
			9.24 [14]	3.50 [14]
0.02653	214	46.73	9.13	2.40
			8.32[b]	1.63[b]
0.02628	212	47.17	7.33	0.92
			9.13 [14]	1.53 [14]
0.02604	210	47.62	8.32	1.04
			6.67[b]	0.60[b]
0.02542	205	48.78	5.72	0.27
			7.48 [14]	0.48 [14]
0.02480	200	50.00	6.91	0.30
			5.22[b]	0.16[b]
0.02418	195	51.28	6.42 [14]	8×10^{-2} [14]
0.02356	190	52.63	6.14	6
0.02294	185	54.05	5.97	6×10^{-2}
0.02232	180	55.56	5.64	0.17
			4.41[b]	4.2×10^{-2}[b]
0.02170	175	57.14	5.58 [14]	0.20 [14]
0.02108	170	58.82	5.41	0.32
0.02046	165	60.61	5.31	0.40
0.01984	160	62.50	5.27	0.41
			4.13[b]	1.9×10^{-2}[b]
0.01922	155	64.52	5.30 [14]	0.51 [14]
0.01860	150	66.67	5.27	0.51
0.01798	145	68.97	5.26	0.60
0.01736	140	71.43	5.30	0.71
			3.99[b]	1.1×10^{-2}[b]
0.01488	120	83.33	3.91	6.6×10^{-3}
0.01240	100	100.0	3.85	4.3
0.009919	80	125.0	3.817	
0.007439	60	166.7	3.793	
0.004959	40	250.0	3.778	
0.002480	20	500.0	3.769	
0.001240	10	1000	3.766	

Indium Antimonide (InSb)

R. T. HOLM

Naval Research Laboratory
Washington, D.C.

The collected data for n and k in Table VII are plotted in Fig. 6. The high-energy regime 155–14 eV has been investigated by Cardona et al. [1], who measured k by transmission through evaporated thin films either freely suspended on copper mesh or on transparent substrates of carbon. A synchrotron source provided the wide range of photon energies. The films were polycrystalline but no doubt had a native oxide, since some of the preparation was done outside of vacuum. Samples were presumably at room temperature. These data show several peaks, including three in the 20-eV region that were assigned to interband transitions.

However, Philipp and Ehrenreich [2] found no structure in this region after Kramers–Kronig (KK) analysis, probably because the corresponding reflectivity was <2%; there could be problems in surface preparation even though a chemical polish was used; and a thin native oxide was present on the surface. Native oxides on most semiconductors have band gaps in the 4–10-eV region that means that above these energies even thin films (20–30 Å thick) begin to absorb when α becomes greater than 10^5 cm^{-1}. The factor of 2 disparity in k is illustrated in Table VII in the overlap region between the data of Cardona et al. [1] and Philipp and Ehrenreich [2]. The n and k data of Philipp and Ehrenreich [2], as tabulated by Seraphin and Bennett [3], are used down to 0.8 eV except for the region 6–1.5 eV. Here, ellipsometrically determined values of n and k by Aspnes and Studna [4] have been inserted.

At this point, it joins to the k results of Moss et al. [5], also tabulated in Seraphin and Bennett [3]. The near IR measurements were on slabs of intrinsic InSb of various thickness ranging from 5.5 to 40 μm, which were mechanically polished (surface damage present) and prepared so as to be unbacked but mounted on their edges. Although not specified, samples were at room temperature. Because of free-carrier effects at the fundamental band edge and at lower photon energies, we have elected to tabulate IR results for only intrinsic InSb with a free-carrier density of holes and electrons of

491

HANDBOOK OF OPTICAL CONSTANTS OF SOLIDS

ISBN 0-12-544420-6

$N = P = 2 \times 10^{16}$ cm^{-3}. The Burstein–Moss effect at the band gap, originally discovered in heavily doped InSb [6, 7], is not appreciable in intrinsic material.

The near IR values for n were determined by Moss et al. [5] from interference fringe measurements. Values tabulated in Seraphin and Bennett [3] are listed in Table VII. Uncertainty in n is $\pm 0.5\%$. This work was the first to point out problems of fringe counting when there is significant dispersion in the refractive index.

Additional results for k are given by Kurnick and Powell [8] just below the band gap near 0.2 eV for a temperature of 298 K. Again, transmission measurements were used. Elaborate mechanical polishing to keep the slab faces parallel produced free samples of thicknesses in the range 0.08 to 3 mm. Reflectivity R was determined by measuring T as a function of sample thickness. Corrections for wide slit widths in a spectral range of rapidly changing absorption were made. The data for n- and p-type samples with carrier densities in the range 2×10^{15}–2×10^{17} cm^{-3} were presented. We chose the most suitable intrinsic sample that at low temperature had $N = 2 \times 10^{15}$ cm^{-3} and read the values of absorption coefficient α from the graph.

The far IR properties of nearly intrinsic material ($N = 6 \times 10^{15}$ cm^{-3} at low temperature) have been studied by Yoshinaga and Oetjen [9] and Sanderson [10]. Yoshinaga and Oetjen [9] measured reflectance at an angle of incidence of 45° (no light polarization was mentioned) and normal-incidence transmission in the spectral range 20–220 μm. They determined n and k directly in the transparent region 20–45 μm as tabulated Seraphin and Bennett [3]. The sample temperature was 298 K. The experiment involved a grating spectrometer with filters that did not give adequate spectral purity at long wavelengths beyond the reststrahlen as was subsequently shown by Sanderson [10], who used a lamellar grating interferometric modulator to measure reflectivity. He analyzed the results by a combined Drude and Lorentz oscillator model and also by the KK technique. The results of the two analyses were very close except in the region near the reststrahlen peak. We have used the oscillator model to calculate n and k. These are tabulated in the 10–1000-μm region. The formula used is

$$\varepsilon = (n - ik)^2 = \varepsilon_\infty \left[1 + \frac{\omega_L^2 - \omega_T^2}{\omega_T^2 - \omega^2 + i\Gamma\omega} - \frac{\omega_p^2}{\omega(\omega + i\gamma)} \right],$$

with $\varepsilon_\infty = 15.68$, $\omega_T = 179.1$ cm^{-1}, $\omega_L = 190.4$ cm^{-1}, $\Gamma = 2.86$ cm^{-1}, $\omega_p = 81.0$ cm^{-1}, and $\gamma = 10.7$ cm^{-1} and with ω_p the plasma frequency, γ the electron damping constant characterizing the free carriers, ω_T the TO frequency, ω_L the LO frequency, Γ the vibration damping constant, and ε_∞ the high-frequency lattice background dielectric constant. Above ω_L, these n and k values deviate significantly from those of Yoshinaga and Oetjen [9] in the overlap region as Table VII indicates. We have extended the calculated values of n to 1000 cm^{-1} to demonstrate the ultimate good agreement with the experimental values of Moss et al. [5].

We are hard-pressed to specify uncertainties in n and k. The values of Γ and γ are probably uncertain to $\pm 20\%$ and peaks in n and k near ω_T are very sensitive to these values. Here, n and k are probably reliable to one significant figure at best and a factor of 2 at worst. Of course, n and k as given do yield R over the range 20–200 cm^{-1} to within ± 0.02 reflectivity units everywhere except at the reststrahlen minimum. The calculated value of R is probably better here because the surface is mechanically damaged. A perusal of other data [11, 12] suggests that the reflection minimum should be nearer to 5% than to the experimentally measured value of 12% for intrinsic InSb.

It is obvious from the discussion of wavelength dependence of free-carrier absorption as discussed by Jensen for GaAs, InP, and InAs in Chapter 9 of this handbook that the Drude model with fixed damping constant is not expected to give good values of k above ω_L, although the wavelength dependence of α is λ^2 in agreement with a simple Drude model [13]. On the other hand, in the spectral region 30–45 μm the values of n from Yoshinaga and Oetjen [9] appear significantly smaller than those of Sanderson [10], the difference being larger than the usual discrepancy between measured n and calculated n as obtained with a Lorentz–Drude oscillator fit. The Lorentz oscillator parameters of Sanderson [10] (ε_∞, ω_T, ω_L) appear reasonable compared to another determination [11]; likewise for the free-carrier parameters (ω_p, γ). The data of Yoshinaga and Oetjen [9] do indicate some multiple-phonon absorption in the region 20–45 μm, but such absorption has very little effect on n as determined from an oscillator model.

Moss et al. [5] calculated the refractive index in the 5–21-μm region including various contributions to absorption due to higher-energy band gaps, the fundamental band gap, free carriers, and the reststrahlen band. Their resultant values are about 0.01 higher in the 11–21-μm region than the experimental values, indicating some problems with various oscillator models and the definition of ε_∞. Usually, the dielectric function due to all interband transitions (omitting free-carrier and lattice-vibration contributions) has a value at zero frequency that is the same as its value at about $E_g/3$ to one unit in the second decimal place. The practical value of ε_∞ is then easy to estimate. In a small-band-gap material such as InSb with the lattice vibration absorption being so close to the fundamental band edge, the practical definition of ε_∞ becomes harder. Pihktin and Yas'kov [14] have reanalyzed n in terms of dispersion equation for interband and lattice vibration contributions (no free carriers) and their values of n are significantly larger (~ 0.3) than those of Yoshinaga and Oetjen [9] and Sanderson [10] in the 25-μm region, indicating that the free-carrier contribution is important. It follows that the carrier density must be well determined for meaningful comparison.

The temperature dependence of n has both a lattice part and a free-carrier part. In the near IR transparent region, Cardona [15] determined both of these components, the latter being somewhat complicated due to the temperature dependence of the intrinsic free-carrier density. For the lattice part $(1/n)(dn/dT) = (11.9 \pm 0.2) \times 10^{-5}\,^\circ\mathrm{C}^{-1}$.

REFERENCES

1. M. Cardona, W. Gudat, B. Sonntag, and P. Y. Yu, *Proc. Int. Conf. Phys. Semicond., 10th, Cambridge, 1970*, p. 208. U.S. Atomic Energy Commission, Oak Ridge, Tennessee, 1970.
2. H. R. Philipp and H. Ehrenreich, *Phys. Rev.* **129**, 1550 (1963).
3. B. O. Seraphin and H. E. Bennett, *in* "Semiconductors and Semimetals" (R. K. Willardson and A. C. Beer eds.), Vol. 3, p. 499. Academic Press, New York, 1967.
4. D. E. Aspnes and A. A. Studna, *Phys. Rev. B* **27**, 985 (1983); D. E. Aspnes and A. A. Studna, private communication (1982).
5. T. S. Moss, S. D. Smith, and T. D. F. Hawkins, *Proc. Phys. Soc. London* **70B**, 776 (1957).
6. E. Burstein, *Phys. Rev.* **93**, 632 (1954).
7. T. S. Moss, *Proc. Phys. Soc. London* **67**, 775 (1954).
8. S. W. Kurnick and J. M. Powell, *Phys. Rev.* **116**, 597 (1959).
9. H. Yoshinaga and R. A. Oetjen, *Phys. Rev.* **101**, 526 (1956).
10. R. B. Sanderson, *J. Phys. Chem. Solids* **26**, 803 (1965).
11. R. T. Holm and E. D. Palik, *CRC Crit. Rev. Solid State Sci.* **5**, 397 (1975).
12. R. W. Gammon and E. D. Palik, *J. Opt. Soc. Am.* **64**, 350 (1974).
13. W. G. Spitzer and H. Y. Fan, *Phys. Rev.* **106**, 882 (1957).
14. A. N. Pikhtin and A. D. Yas'kov, *Soviet Phys. Semicond.* **12**, 622 (1978); A. N. Pikhtin and A. D. Yas'kov, *Fiz. Tekh. Poluprovodn.* **12**, 1047 (1978).
15. M. Cardona, *Proc. Int. Conf. Semicond. Phys., Prague, 1960*, p. 349. Czech. Acad. Sci., Prague, and Academic Press, New York, 1961.

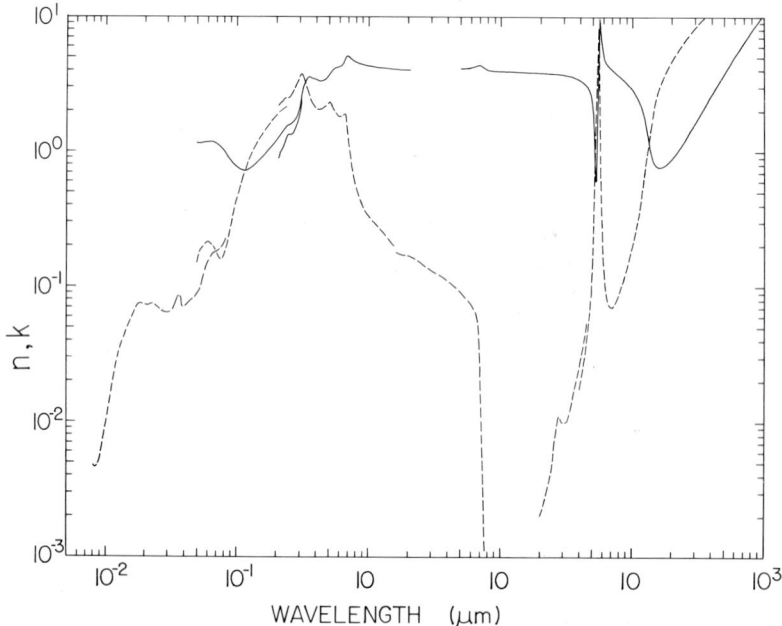

Fig. 6. Log–log plot of n (——) and k (----) versus wavelength in micrometers for indium antimonide. Note that this is intrinsic InSb with strong free-carrier absorption in the far infrared.

TABLE VII

Values of n and k for Indium Antimonide Obtained from Various References[a]

eV	cm^{-1}	μm	n	k
155		0.007999		4.77×10^{-3} [1]
150		0.008266		4.60
145		0.008551		4.83
140		0.008856		5.28
135		0.009184		6.13
130		0.009537		7.59
125		0.009919		9.70×10^{-3}
120		0.01033		1.25×10^{-2}
115		0.01078		1.59
110		0.01127		2.06
105		0.01181		2.63
100		0.01240		3.25
95		0.01305		3.79
90		0.01378		4.45
85		0.01459		5.02
80		0.01550		5.77
75		0.01653		6.60
70		0.01771		7.45
65		0.01907		7.46
60		0.02066		7.30
55		0.02254		7.50
50		0.02480		7.10
45		0.02755		6.51
40		0.03100		6.51
38		0.03263		6.75
37		0.03351		7.06
36		0.03444		7.67
35		0.03542		8.51
34		0.03647		8.56
33		0.03757		8.37
32		0.03875		7.09
31		0.04000		7.16
30		0.04133		7.23
29		0.04275		7.55
28		0.04428		7.89
27		0.04592		8.11
26		0.04769		8.34
25		0.04959	1.15 [2]	8.88×10^{-2}
				0.15 [2]
24		0.05166	1.15	9.37×10^{-2} [1]
				0.18 [2]

(continued)

[a] The value of k is often given in powers of 10 down to 10^{-2}, whereupon we revert to decimals only. The exponent of 10 is understood down the column until the next exponent is about to appear. We have sometimes overlapped data of different references to demonstrate discrepancies. It is not always obvious why the discrepancies exist. The references are given in brackets.

TABLE VII (*Continued*)

Indium Antimonide

eV	cm^{-1}	μm	n	k
23		0.05391	1.15	0.105 [1]
				0.19 [2]
22		0.05636	1.16	0.125 [1]
				0.20 [2]
21		0.05904	1.17	0.145 [1]
				0.21 [2]
20		0.06199	1.17	0.160 [1]
				0.21 [2]
19.5		0.06358		0.163 [1]
19		0.06526	1.18	0.173
				0.20 [2]
18.5		0.06702		0.178 [1]
18		0.06888	1.15	0.178
				0.18 [2]
17.5		0.07085		0.180 [1]
17		0.07293	1.11	0.181
				0.16 [2]
16.5		0.07514		0.191 [1]
16		0.07749		0.205
15.5		0.07999	1.02	0.17 [2]
15		0.08266	0.97	0.230 [1]
				0.19 [2]
14		0.08856	0.88	0.269 [1]
				0.26 [2]
13		0.09537	0.80	0.37
12		0.1033	0.75	0.51
11		0.1127	0.72	0.69
10		0.1240	0.74	0.88
9		0.1378	0.80	1.08
8		0.1550	0.88	1.32
7.6		0.1631	0.94	1.41
6.8		0.1823	1.06	1.61
6.0	48,390	0.2066	1.23	1.91
			0.861 [4]	2.139 [4]
5.94	47,910	0.2087	0.900	2.180
5.90	47,590	0.2101	0.922	2.185
5.84	47,100	0.2123	0.951	2.204
5.80	46,780	0.2138	0.969	2.210
5.74	46,300	0.2160	0.992	2.227
5.70	45,970	0.2175	1.000	2.235
			1.36 [2]	1.97 [2]
5.64	45,490	0.2198	1.016 [4]	2.257 [4]
5.60	45,170	0.2214	1.025	2.275
5.54	44,680	0.2238	1.042	2.307
5.50	44,360	0.2254	1.057	2.333
5.44	43,880	0.2279	1.088	2.369

TABLE VII (*Continued*)

Indium Antimonide

eV	cm^{-1}	μm	n	k
5.40	43,550	0.2296	1.116	2.394
5.34	43,070	0.2322	1.161	2.424
5.30	42,750	0.2339	1.194	2.435
5.24	42,260	0.2366	1.242	2.446
5.20	41,940	0.2384	1.270	2.444
5.14	41,460	0.2412	1.301	2.435
5.10	41,130	0.2431	1.309	2.430
5.04	40,650	0.2460	1.311	2.431
5.00	40,330	0.2480	1.307	2.441
4.94	39,840	0.2510	1.301	2.470
4.90	39,520	0.2530	1.301	2.495
4.84	39,040	0.2562	1.306	2.542
4.80	38,710	0.2583	1.312	2.576
4.74	38,230	0.2616	1.328	2.630
4.70	37,910	0.2638	1.341	2.669
4.64	37,420	0.2672	1.366	2.732
4.60	37,100	0.2695	1.385	2.776
4.54	36,620	0.2731	1.419	2.845
4.50	36,290	0.2755	1.443	2.894
4.44	35,810	0.2792	1.484	2.975
4.40	35,490	0.2818	1.515	3.034
4.34	35,000	0.2857	1.571	3.134
4.30	34,680	0.2883	1.618	3.209
4.24	34,200	0.2924	1.708	3.337
4.20	33,880	0.2952	1.791	3.433
4.14	33,390	0.2995	1.966	3.582
4.10	33,070	0.3024	2.127	3.666
4.04	32,580	0.3069	2.425	3.720
4.00	32,260	0.3100	2.632	3.694
3.94	31,780	0.3147	2.902	3.581
3.90	31,460	0.3179	3.044	3.479
3.84	30,970	0.3229	3.204	3.314
3.80	30,650	0.3263	3.287	3.204
3.74	30,160	0.3315	3.383	3.039
3.70	29,840	0.3351	3.427	2.933
3.64	29,360	0.3406	3.468	2.789
3.60	29,040	0.3444	3.485	2.705
3.54	28,550	0.3502	3.504	2.589
3.50	28,230	0.3542	3.511	2.517
3.44	27,750	0.3604	3.515	2.424
3.40	27,420	0.3647	3.520	2.369
3.34	26,940	0.3712	3.528	2.280
3.30	26,620	0.3757	3.525	2.217
3.24	26,130	0.3827	3.502	2.137
3.20	25,810	0.3875	3.482	2.093

(*continued*)

TABLE VII (*Continued*)

Indium Antimonide

eV	cm^{-1}	μm	n	k
3.14	25,330	0.3949	3.445	2.038
3.10	25,000	0.4000	3.419	2.015
3.04	24,520	0.4078	3.384	1.997
3.00	24,200	0.4133	3.366	1.994
2.94	23,710	0.4217	3.348	1.997
2.90	23,390	0.4275	3.342	2.004
2.84	22,910	0.4366	3.340	2.021
2.80	22,580	0.4428	3.345	2.036
2.74	22,100	0.4525	3.360	2.062
2.70	21,780	0.4592	3.377	2.083
2.64	21,290	0.4696	3.413	2.119
2.60	20,970	0.4769	3.447	2.145
2.54	20,490	0.4881	3.512	2.188
2.50	20,160	0.4959	3.570	2.221
2.48	20,000	0.4999	3.605	2.238
2.46	19,840	0.5040	3.646	2.254
2.44	19,680	0.5081	3.694	2.271
2.42	19,520	0.5123	3.752	2.283
2.40	19,360	0.5166	3.822	2.287
2.38	19,200	0.5209	3.900	2.274
2.36	19,030	0.5254	3.976	2.239
2.34	18,870	0.5299	4.040	2.183
2.32	18,710	0.5344	4.084	2.121
2.30	18,550	0.5391	4.111	2.060
2.28	18,390	0.5438	4.126	2.005
2.26	18,230	0.5486	4.133	1.957
2.24	18,070	0.5535	4.136	1.915
2.22	17,910	0.5585	4.136	1.880
2.20	17,740	0.5636	4.135	1.850
2.18	17,580	0.5687	4.134	1.826
2.16	17,420	0.5740	4.132	1.806
2.14	17,260	0.5794	4.132	1.790
2.12	17,100	0.5848	4.133	1.779
2.10	16,940	0.5904	4.136	1.770
2.08	16,780	0.5961	4.142	1.765
2.06	16,610	0.6019	4.149	1.763
2.04	16,450	0.6078	4.161	1.763
2.02	16,290	0.6138	4.175	1.767
2.00	16,130	0.6199	4.194	1.773
1.98	15,970	0.6262	4.218	1.784
1.96	15,810	0.6326	4.249	1.799
1.94	15,650	0.6391	4.290	1.818
1.92	15,490	0.6458	4.347	1.844
1.90	15,320	0.6526	4.433	1.873
1.88	15,160	0.6595	4.566	1.879
1.86	15,000	0.6666	4.723	1.816

TABLE VII (*Continued*)

Indium Antimonide

eV	cm^{-1}	μm	n	k
1.84	14,840	0.6738	4.841	1.685
1.82	14,680	0.6812	4.898	1.534
1.80	14,520	0.6888	4.909	1.396
			5.13 [2]	1.37 [2]
1.78	14,360	0.6965	4.894 [4]	1.275 [4]
1.76	14,200	0.7045	4.867	1.171
1.74	14,030	0.7126	4.830	1.085
1.72	13,870	0.7208	4.793	1.010
1.70	13,710	0.7293	4.754	0.949
1.68	13,550	0.7380	4.712	0.897
1.66	13,390	0.7469	4.675	0.848
1.64	13,230	0.7560	4.637	0.808
1.62	13,070	0.7653	4.602	0.776
1.60	12,900	0.7749	4.568	0.749
			4.72 [2]	0.60 [2]
1.58	12,740	0.7847	4.534 [4]	0.723 [4]
1.56	12,580	0.7948	4.501	0.701
1.54	12,420	0.8051	4.473	0.675
1.52	12,260	0.8157	4.442	0.659
1.50	12,100	0.8266	4.418	0.643
1.4	11,290	0.8856	4.40 [2]	0.40 [2]
1.2	9,679	1.033	4.24	0.32
1.0	8,065	1.240	4.15	0.26
0.8	6,452	1.550	4.08	0.20
0.7749.	6,250	1.6		0.18 [5]
0.6888	5,556	1.8		0.17
0.6199	5,000	2.0		0.17
0.6	4,839	2.066	4.03	
0.4959	4,000	2.5		0.15
0.4133	3,333	3.0		0.13
0.3542	2,857	3.5		0.12
0.3100	2,500	4.0		0.11
0.2755	2,222	4.5		0.10
0.2480	2,000	5.0	4.14 [5]	9.1 × 10^{-2}
0.2254	1,818	5.5	4.15	
0.2066	1,667	6.0	4.19	7.4
0.2033	1,639	6.1		7.2
0.2000	1,613	6.2		7.0
0.1968	1,587	6.3		6.8
0.1937	1,563	6.4		6.6
0.1907	1,538	6.5	4.30	6.3
0.1879	1,515	6.6		5.9
0.1851	1,493	6.7		5.5
0.1823	1,471	6.8		4.9
0.1797	1,449	6.9		3.7

(*continued*)

TABLE VII (*Continued*)

Indium Antimonide

eV	cm^{-1}	μm	n	k
0.1771	1,429	7.0	4.37	2.8
0.175	1,411	7.085		1.46 × 10^{-2} [8]
0.1722	1,389	7.2		9.2 × 10^{-3} [5]
0.170	1,371	7.293		5.68 [8]
0.1675	1,351	7.4		3.5 [5]
0.1653	1,333	7.5	4.18	2.7 [5]
0.165	1,331	7.5144		2.63 [8]
0.1631	1,316	7.6		1.3 [5]
0.160	1,290	7.749		1.57 [8]
0.1575	1,271	7.87	4.001	
0.155	1,251	7.999		1.18 × 10^{-3}
0.1550	1,250	8.0	3.995	
0.150	1,210	8.266		9.86 × 10^{-4}
0.145	1,169	8.551		9.52
0.140	1,129	8.856		9.15
0.1376	1,110	9.01	3.967	
0.135	1,089	9.184		9.13
0.130	1,049	9.537		9.10 × 10^{-4}
0.1240	1,000	10.00	3.938 [10]	
0.1232	994.0	10.06	3.953 [5]	
0.1126	908.3	11.01	3.937	
0.1116	900	11.11	3.933 [10]	
0.1028	829.2	12.06	3.920 [5]	
0.09919	800	12.50	3.926 [10]	
0.09552	770.4	12.98	3.912 [5]	
0.08920	719.4	13.90	3.902	
0.08679	700	14.29	3.915 [10]	
0.08194	660.9	15.13	3.881 [5]	
0.07852	633.3	15.79	3.873	
0.07439	600	16.67	3.898 [10]	
0.07310	589.6	16.96	3.866 [5]	
0.06946	560.2	17.85	3.850	
0.06577	530.5	18.85	3.843	
0.06205	500.5	19.98	3.826	
0.06199	500	20.00	3.869 [10]	2.0 × 10^{-3} [9]
0.05862	472.8	21.15	3.814 [5]	
0.05585	450.5	22.20	3.805	
0.04959	400.0	25.00	3.78 [9]	5.1
			3.811 [10]	
0.04769	384.6	26.00	3.74 [9]	7.5
0.04592	370.4	27.0	3.66	9.7 × 10^{-3}
0.04463	360	27.78	3.770 [10]	
0.04428	357.1	28.0	3.56 [9]	1.1 × 10^{-2}
0.04275	344.8	29.0	3.50	1.0
0.04215	340	29.41	3.743 [10]	
0.04133	333.3	30.0	3.47 [9]	1.0

TABLE VII (*Continued*)

Indium Antimonide

eV	cm^{-1}	μm	n	k
0.04000	322.6	31.0	3.44	1.0
0.03968	320	31.25	3.708 [10]	
0.03875	312.5	32.0	3.39 [9]	1.0
0.03757	303.0	33.0	3.34	1.1 × 10^{-2}
0.03720	300	33.33	3.662 [10]	7.84 × 10^{-3} [10]
0.03647	294.1	34.0	3.30 [9]	1.2 [9]
0.03596	290	34.48	3.63 [10]	8.96 × 10^{-3} [10]
0.03542	285.7	35.0	3.25 [9]	1.4 × 10^{-2} [9]
0.03472	280	35.71	3.60 [10]	1.03 [10]
0.03348	270	37.04	3.56	1.21
0.03224	260	38.46	3.51	1.44
0.03100	250	40.00	2.98 [9]	2.6 [9]
			3.45 [10]	1.75 [10]
0.02976	240	41.67	3.37	2.20
0.02914	235	42.55	3.32	2.51
0.02852	230	43.48	3.26	2.91
0.02790	225	44.44	3.18	3.44
0.02755	222.2	45.0	2.57 [9]	5.4 [9]
0.02728	220	45.45	3.09 [10]	4.16 [10]
0.02666	215	46.51	2.97	5.23
0.02604	210	47.62	2.82	6.92
0.02542	205	48.78	2.59	9.93 × 10^{-2}
0.02480	200	50.00	2.22	0.165
0.02455	198	50.51	2.00	0.217
0.02430	196	51.02	1.69	0.311
0.02405	194	51.55	1.25	0.525
0.02393	193	51.81	0.989	0.753
0.02380	192	52.08	0.776	1.10
0.02368	191	52.36	0.663	1.49
0.02356	190	52.63	0.618	1.88
0.02343	189	52.91	0.611	2.27
0.02331	188	53.19	0.632	2.67
0.02318	187	53.48	0.680	3.10
0.02306	186	53.76	0.760	3.57
0.02294	185	54.05	0.884	4.10
0.02281	184	54.35	1.08	4.73
0.02269	183	54.64	1.39	5.48
0.02256	182	54.95	1.95	6.43
0.02244	181	55.25	3.05	7.59
0.02232	180	55.56	5.34	8.56
0.02219	179	55.87	8.68	7.34
0.02207	178	56.18	9.61	4.20
0.02194	177	56.50	8.82	2.31
0.02182	176	56.82	7.97	1.42

(*continued*)

TABLE VII (*Continued*)

Indium Antimonide

eV	cm^{-1}	μm	n	k
0.02170	175	57.14	7.32	0.963
0.02157	174	57.47	6.82	0.699
0.02145	173	57.80	6.44	0.534
0.02132	172	58.14	6.14	0.424
0.02120	171	58.48	5.89	0.347
0.02108	170	58.82	5.69	0.291
0.02083	168	59.52	5.37	0.216
0.02058	166	60.24	5.13	0.170
0.02033	164	60.98	4.94	0.140
0.02009	162	61.73	4.79	0.119
0.01984	160	62.50	4.66	0.105
0.01922	155	64.52	4.42	8.38×10^{-2}
0.01860	150	66.67	4.24	7.43
0.01798	145	68.97	4.09	7.10
0.01736	140	71.43	3.96	7.15
0.01674	135	74.07	3.85	7.49
0.01612	130	76.92	3.74	8.08
0.01550	125	80.00	3.62	8.91×10^{-2}
0.01488	120	83.33	3.51	0.100
0.01426	115	86.96	3.38	0.115
0.01364	110	90.91	3.25	0.134
0.01302	105	95.24	3.10	0.159
0.01240	100	100.0	2.92	0.193
0.01178	95	105.3	2.71	0.240
0.01116	90	111.1	2.45	0.310
0.01054	85	117.6	2.12	0.423
0.009919	80	125.0	1.68	0.633
0.009299	75	133.3	1.19	1.09
0.008679	70	142.9	0.881	1.79
0.008059	65	153.8	0.781	2.51
0.007439	60	166.7	0.771	3.21
0.006819	55	181.8	0.811	3.94
0.006199	50	200.0	0.894	4.72
0.005579	45	222.2	1.02	5.59
0.004959	40	250.0	1.21	6.61
0.004339	35	285.7	1.50	7.83
0.003720	30	333.3	1.93	9.35
0.003100	25	400.0	2.63	11.3
0.002480	20	500.0	3.81	14.0
0.001860	15	666.7	6.03	17.9
0.001240	10	1000	10.7	24.0

Indium Phosphide (InP)

O. J. GLEMBOCKI

Naval Research Laboratory
Washington, D.C.

H. PILLER

Department of Physics and Astronomy
Louisiana State University
Baton Rouge, Louisiana

As with other semiconductors, the index of refraction of InP has been reported by various investigators over a wide range of photon energies, 1.2 meV to 20 eV. The number of papers for InP, however, is somewhat limited. Various techniques were employed in these measurements, and there is some overlap in data obtained from different sources. In many cases, values in the overlap regions do not coincide and a choice was made of one set over another based on the technique used and its limitations. We shall discuss the overlap problems when they arise. The values of n and k are listed in Table VIII and a plot is shown in Fig. 7.

Cardona [1] has measured the reflectivity of InP in the 1–20-eV range. The sample surfaces were prepared by mechanical polishing with Linde A-5175 compound. The reflection spectra were then measured and the samples were etched in a 1:1 solution of HNO_3 and HCl to remove any damage and to improve the reflectivity. The reflection spectra were then Kramers–Kronig (KK) analyzed to obtain ε_1 and ε_2 as well as n and k. We read values from a published graph. Above 20 eV Cardona [1] assumed the reflectivity to be proportional to E^{-4} (free-electron gas). In a subsequent work [2] measurements were made to 150 eV, but the data were not given. In Table VIII we list the data of Cardona [1] from 6–20 eV.

The 6-eV cutoff of Cardona's [1] values of n and k results from the availability of better measurements from 1.5 to 6 eV. Recently, Aspnes and Studna [3] have obtained the optical constants of InP, as well as of other semiconductors, through the use of spectroscopic ellipsometry. These measurements allowed them to determine ε_1 and ε_2 or, alternatively, n and k directly without a KK analysis. The values in Table I were obtained from a numerical table. Their samples were of $\langle 100 \rangle$ orientation and undoped. Extensive care

503

was taken to avoid both surface damage and unintentional overlayers such as oxides. Their sample-preparation techniques involved an initial pretreatment with bromine–methanol (0.02-vol % bromine in methanol) chemomechanical polish. Afterward, the final polishing techniques involved a 20-sec polish on a lens paper saturated with the bromine–methanol solution, followed by a-10 sec quench in methanol, and, last, the sample was rinsed in a 1:1 NH_4O_4:OH solution. The final polishing and the measurements were performed in an N_2 atmosphere.

Aspnes and Studna [3] have corrected their values near 1.5 eV so that both ε_1 and ε_2 are KK consistent to within $\pm 0.4\%$. At 5.07 eV they obtain a value of $R = 0.621$, while Cardona [1] measured 0.553. This indicates that the sample surfaces of Cardona [1] were not as good as those of Aspnes and Studna [3].

Below 1.5 eV Pettit and Turner [4] measured n by the minimum-deviation method on an n-type sample of carrier density $N = 5 \times 10^{16}$ cm^{-3}. The surface was mechanically polished to a flatness of 0.5 μm or better. A precision of ± 0.0003 in n is quoted. The data for n between 0.5 and 1.4 eV were found to fit to the expression

$$n^2 = A + B\lambda^2/(\lambda^2 - C^2), \tag{1}$$

where λ is in angstroms and the constants for room temperature are $A = 7.255$, $B = 2.316$, and $C^2 = 0.3922 \times 10^8$. Calculated values obtained from Eq. (1) are tabulated.

The transmission measurements of Newman [5] allow us to obtain k below 1.4 eV. These experiments were performed at 300 K by using an n-type sample with a carrier density of $N = 5 \times 10^{15}$ cm^{-3}. Sample preparation is not discussed. Newman's data [5] for $\lambda > 14$ μm are presented in terms of the transmission spectrum. To obtain k, we used the relationship $T = (1 - R)^2 \exp(-4\pi kd/\lambda)$, which accounts for multiple reflections. The thickness was $d = 0.01$ cm, and R was computed from the oscillator model for n by using $R = [(n - 1)^2 + k^2]/[(n + 1)^2 + k^2]$; k was negligibly small in most cases. Seraphin and Bennett [6] present a tabular form of Newman's data [5] for shorter wavelengths between 0.92 and 0.98 μm.

Overlapping the data of Newman [5] is the work of Pikhtin and Yas'kov [7], in which they fit the n data [4] to a lattice-vibration oscillator model given by

$$n^2 = 1 + \frac{A}{\pi} \ln \frac{E_1^2 - (\hbar\omega)^2}{E_0^2 - (\hbar\omega)^2} + \frac{G_1}{E_1^2 - (\hbar\omega)^2}$$

$$+ \frac{G_2}{E_2^2 - (\hbar\omega)^2} + \frac{G_3}{E_3^2 - (\hbar\omega)^2}, \tag{2}$$

where $E_0 = 1.345$ eV, $E_1 = 3.2$ eV, $E_2 = 5.1$ eV, $E_3 = 37.65 \times 10^{-3}$ eV, $G_1 = 57.889$ eV^2, $G_2 = 65.937$ eV^2, $G_3 = 0.392 \times 10^{-2}$ eV^2, and $A = 0.7\sqrt{E_0}$. A comparison of these values with those of Pettit and Turner [4] reveals some differences, but only in the third decimal place. At longer wavelengths, we expect that the above data will be better than those of Pettit and Turner [4]. This is a consequence of the fact that for $\hbar\omega = 0$ and $G_3 = 0$, $n^2 = 9.52$, corresponding to the high-frequency dielectric constant ε_∞, which will be used in the analysis of Raman scattering data. The value of $n^2 = 9.52$ holds to within three figures, even up to 560 cm^{-1}.

Reynolds et al. [8] measured by transmission the multiphonon absorption in the 12–20-μm range. The temperature used was 293 K. The sample was n-type with a carrier density of $N \approx 10^{15}$ cm^{-3}, and the surfaces were cleaned and etched in hot or cold HCl. The data of Reynolds et al. [8] (read from a graph) and Newman [5] are in good agreement in the ranges of 0.060 to 0.062 eV and 0.083 to 0.085 eV. However, at the vibration peaks near 0.08 eV, they deviate by as much as a factor of 2 and by an order of magnitude in the transparent region near 0.07 eV. We tabulate portions of both data sets; from 12 to 20 μm we plot the values of Reynolds [8], whereas from 20 to 29 μm the Newman [5] data are used.

Koteles and Datars [9] performed absorption measurements at 20 K over the range of 0.025 to 0.10 eV. Their data were rich in structure. However, values of k are not provided, just absorption on a linear scale.

In the far infrared (<0.05 eV), we found little detailed analysis of the re-flectance by a KK inversion or fit to an oscillator model. Newman's [5] data (at 300 K) indicate 100% reflectivity at the reststrahlen peak near 0.04 eV, which is unrealistic. The room-temperature data of Hass and Henvis [10] show too much variation from the simple reststrahlen curve. This may be attributable to free-carrier plasma and mechanical polishing effects. In order to obtain refractive-index data in this region, we used the oscillator model, which yields

$$(n - ik)^2 = \varepsilon_\infty \left[1 + \frac{\omega_L^2 - \omega_T^2}{\omega_T^2 - \omega^2 - i\Gamma\omega} \right], \tag{3}$$

with the parameters $\omega_L = 345.0 \pm 0.3$ cm^{-1}, $\omega_T = 303.7 \pm 0.3$ cm^{-1}, and $\Gamma = 3.5 \pm 0.5$ cm^{-1} from the Raman scattering data of Mooradian and Wright [11], while $\varepsilon_\infty = 9.61$ was obtained from Hass [12]. Other Raman work gave virtually the same values for ω_T and ω_L [13, 14]. In Table VIII we list our values for n and k below 410 cm^{-1} (as footnote b). Near $\omega = \omega_T \pm 30$ cm^{-1}, the n and k values are probably good to $\pm 50\%$. This is based on the systematic variation of the parameters within their probable values. Consequently, the model can only serve as an order-of-magnitude estimate for k far from ω_T. For example, at 0.044 eV the measured k is 2.39×10^{-2}, while the oscillator-model k is 9.9×10^{-2}, a factor of 4 larger. To force the oscillator model to fit better would require increasing Γ significantly, which is not warranted by the Raman linewidth data. We therefore conclude that the oscillator model is a poor approximation away from the resonant frequency.

Jamshidi and Parker [15] have measured n and k by using asymmetric Fourier-transform spectroscopy, which is a more direct technique for obtaining optical constants. The values of n and k that we list have been taken from a graph. We find that their peak reststrahlen reflectivity of 0.897 occurs at 333 cm^{-1}. This is considerably different from the value of $R = 0.924$ at 314 cm^{-1}, which the oscillator model yields and much larger than the experimental value of 0.79 from Hass and Henvis [10]. This experimental value is too low, if anything, and may be due to surface quality and/or stray radiation. Their (Jamshidi and Parker [15]) value of n near 450 cm^{-1} is 2.29, significantly smaller than $n = 2.66$ from Pikhtin and Yas'kov [7]. We feel that the analysis of Pikhtin and Yas'kov [7] cannot be that much in error. In addition, the low-frequency dielectric constant of Jamshidi and Parker [15] is 12.18, much smaller than the value of 12.37 given in the analysis of reststrahlen reflectivity in Hass [12]. It is interesting to note that the values of n and k measured by asymmetric Fourier-transform spectroscopy in the case of CdTe were in good agreement with those obtained by other means, while those for InAs were in poorer agreement (perhaps due to different free-carrier densities).

Jamshidi and Parker [15] suggest that their data for n and k are better in the range 300–380 cm^{-1} than the results from previous reststrahlen analyses. However, outside of this regime, their measured phase is not satisfactorily determined because it is very close to the phase of the reference mirror. This may be the source of the discrepancies in n as noted. Jamshidi and Parker [15] state that the k values are only well determined in the range 300–380 cm^{-1} even though structure outside this region is reproducible. As reported by Newman [5], direct transmission measurements would be better outside the 300–380-cm^{-1} regime. Despite this, we still feel that asymmetric Fourier-transform spectroscopy is a powerful tool because, as in ellipsometry, one directly obtains n and k without a KK analysis.

REFERENCES

1. M. Cardona, *J. Appl. Phys.* **32**, 958 (1961); M. Cardona, *J. Appl. Phys.* **36**, 2181 (1965).
2. M. Cardona, W. Gudat, B. Sonntag, and P. Y. Yu, *Proc. Int. Conf. Phys. Semicond., 10th, Cambridge, 1970*, p. 209. U.S. Atomic Energy Commission, Oak Ridge, Tennessee, 1970.
3. D. E. Aspnes and A. A. Studna, *Phys. Rev. B* **27**, 985 (1983); D. E. Aspnes and A. A. Studna, private communication (1982).
4. G. D. Pettit and W. J. Turner, *J. Appl. Phys.* **36**, 2081 (1965).
5. R. Newman, *Phys. Rev.* **111**, 1518 (1958).
6. B. O. Seraphin and H. E. Bennett, *in* "Semiconductors and Semimetals" (R. K. Willardson and A. C. Beer eds.), Vol. 3, p. 499. Academic Press, New York, 1967.
7. A. N. Pikhtin and A. D. Yas'kov, *Sov. Phys. Semicond.* **12**, 622 (1978); A. N. Pikhtin and A. D. Yas'kov, *Fiz. Tekh. Poluprovodu.* **12**, 1047 (1978).
8. W. N. Reynolds, M. T. Lilburne, and R. M. Dell, *Proc. Phys. Soc. London* **71**, 416 (1958).

9. E. S. Koteles and W. R. Datars, *Solid State Commun.* **19**, 221 (1976).

10. M. Hass and B. W. Henvis, *J. Phys. Chem. Solids* **23**, 1099 (1962).

11. A. Mooradian and G. B. Wright, *Solid State Commun.* **4**, 431 (1966).

12. M. Hass, *in* "Semiconductors and Semimetals" (R. K. Willardson and A. C. Beer, eds.), Vol. 3, p. 3. Academic Press, New York, 1967.

13. V. I. Zemski, E. L. Ivchenko, D. N. Mirlin, and I. I. Reshina, *Solid State Commun.* **16**, 221 (1975).

14. C. Hilsum, S. Fray, and C. Smith, *Solid State Commun.* **7**, 1057 (1969).

15. H. Jamshidi and T. J. Parker, *Int. Meet. Infrared Mm Waves, 7th, Marseilles, 1983*; H. Jamshidi and T. J. Parker, private communication (1983).

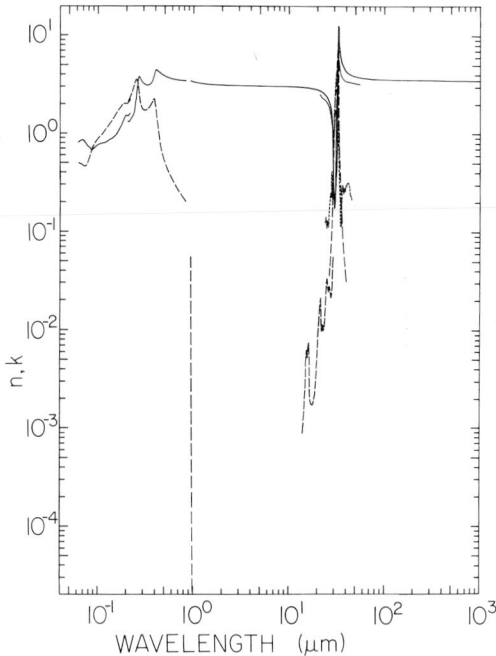

Fig. 7. Log–log plot of n (——) and k (----) versus wavelength in micrometers for indium phosphide.

TABLE VIII

Values of _n_ and _k_ for Indium Phosphide Obtained from Various References[a]

eV	μm	cm^{-1}	n	k	n	k
20	0.06199		0.793	0.494		
19.5	0.06358		0.815	0.499		
19	0.06526		0.834	0.493		
18.5	0.06702		0.843	0.487		
18	0.06888		0.846	0.477		
17.5	0.07085		0.840	0.469		
17	0.07293		0.824	0.454		
16.5	0.07514		0.785	0.457		
16	0.07749		0.742	0.491		
15.5	0.07999		0.719	0.529		
15	0.08266		0.695	0.574		
14.5	0.08851		0.675	0.645		
14	0.08856		0.688	0.706		
13.5	0.09184		0.701	0.765		
13	0.09537		0.726	0.820		
12.5	0.09919		0.754	0.861		
12	0.1033		0.771	0.899		
11.5	0.1078		0.781	0.946		
11	0.1127		0.793	0.996		
10.5	0.1181		0.797	1.056		
10	0.1240		0.806	1.154		
9.9	0.1252		0.820	1.172		
9.8	0.1265		0.832	1.185		
9.7	0.1278		0.840	1.198		
9.6	0.1291		0.847	1.210		
9.5	0.1305		0.852	1.225		
9.4	0.1319		0.859	1.237		
9.3	0.1333		0.861	1.253		
9.2	0.1348		0.865	1.269		
9.1	0.1362		0.868	1.287		
9.0	0.1378		0.872	1.304		
8.9	0.1393		0.874	1.324		
8.8	0.1409		0.875	1.346		
8.7	0.1425		0.877	1.375		
8.6	0.1442		0.885	1.403		
8.5	0.1459		0.894	1.433		
8.4	0.1476		0.909	1.458		
8.3	0.1494		0.919	1.486		
8.2	0.1512		0.934	1.512		
8.1	0.1513		0.947	1.539		
8.0	0.1550		0.960	1.566		
7.9	0.1569		0.973	1.594		
7.8	0.1590		0.984	1.627		
7.7	0.1610		1.000	1.664		
7.6	0.1631		1.022	1.700		
7.5	0.1653		1.046	1.736		
7.4	0.1675		1.072	1.771		

TABLE VIII (*Continued*)

Indium Phosphide

eV	μm	cm⁻¹	n	k	n	k
7.3	0.1698		1.100	1.812		
7.2	0.1722		1.136	1.847		
7.1	0.1746		1.174	1.882		
7.0	0.1771		1.215	1.915		
6.9	0.1797		1.261	1.941		
6.8	0.1823		1.307	1.966		
6.7	0.1851		1.354	1.986		
6.6	0.1879		1.402	2.994		
6.5	0.1907		1.453	2.010		
6.4	0.1937		1.496	2.008		
6.3	0.1968		1.526	1.991		
6.2	0.2000	50,010	1.525	1.982		
6.1	0.2033	49,200	1.508	2.005		
6.0	0.2066	48,390	1.500	2.063		
			1.336 [3]	2.113 [3]		
5.94	0.2087	47,910	1.318	2.151		
5.9	0.2101	47,590	1.301	2.183		
			1.516 [1]	2.130 [1]		
5.84	0.2123	47,100	1.301 [3]	2.239 [3]		
5.8	0.2138	46,780	1.299	2.280		
			1.544 [1]	2.191 [1]		
5.74	0.2160	46,290	1.311 [3]	2.341 [3]		
5.7	0.2175	45,970	1.325	2.383		
			1.573 [1]	2.267 [1]		
5.64	0.2198	45,490	1.352 [3]	2.446 [3]		
5.6	0.2214	45,170	1.375	2.484		
			1.616 [1]	2.349 [1]		
5.54	0.2238	44,680	1.407 [3]	2.532 [3]		
5.5	0.2254	44,360	1.426	2.562		
5.44	0.2279	43,880	1.447	2.611		
5.4	0.2296	43,550	1.455	2.652		
5.34	0.2322	43,070	1.467	2.734		
5.3	0.2339	42,750	1.482	2.802		
5.24	0.2366	42,260	1.519	2.922		
5.2	0.2384	41,940	1.558	3.016		
5.14	0.2412	41,460	1.649	3.176		
5.1	0.2431	41,130	1.745	3.291		
5.04	0.2460	40,650	1.958	3.438		
5.0	0.2480	40,330	2.131	3.495		
4.94	0.2510	39,840	2.391	3.519		
4.9	0.2530	39,520	2.546	3.514		
4.84	0.2562	39,040	2.782	3.523		
4.8	0.2583	38,710	2.984	3.517		
4.74	0.2616	38,230	3.344	3.401		
4.7	0.2638	37,910	3.560	3.223		
4.64	0.2672	37,420	3.757	2.877		

(*continued*)

TABLE VIII (*Continued*)

Indium Phosphide

eV	μm	cm^{-1}	n	k	n	k
4.6	0.2695	37,100	3.800	2.637		
4.54	0.2731	36,620	3.759	2.335		
4.5	0.2755	36,290	3.697	2.186		
4.44	0.2792	35,810	3.593	2.024		
4.4	0.2818	35,490	3.527	1.948		
4.34	0.2857	35,000	3.437	1.865		
4.3	0.2883	34,680	3.384	1.826		
4.24	0.2924	34,200	3.314	1.782		
4.2	0.2952	33,880	3.275	1.762		
4.14	0.2995	33,390	3.224	1.742		
4.1	0.3024	33,070	3.196	1.735		
4.04	0.3069	32,580	3.161	1.730		
4.0	0.3100	32,260	3.141	1.730		
3.94	0.3147	31,780	3.119	1.737		
3.9	0.3179	31,460	3.108	1.744		
3.84	0.3229	30,970	3.097	1.760		
3.8	0.3263	30,650	3.095	1.773		
3.74	0.3315	30,160	3.097	1.797		
3.7	0.3351	29,840	3.103	1.816		
3.64	0.3406	29,360	3.118	1.848		
3.6	0.3444	29,040	3.133	1.872		
3.54	0.3502	28,550	3.165	1.915		
3.5	0.3542	28,230	3.193	1.948		
3.44	0.3604	27,740	3.248	2.009		
3.4	0.3647	27,420	3.299	2.060		
3.34	0.3712	26,940	3.426	2.159		
3.3	0.3757	26,620	3.576	2.209		
3.24	0.3827	26,130	3.819	2.177		
3.2	0.3875	25,810	3.976	2.143		
3.14	0.3948	25,320	4.296	1.966		
3.1	0.4000	25,000	4.415	1.735		
3.04	0.4078	24,520	4.433	1.414		
3.0	0.4133	24,200	4.395	1.247		
2.94	0.4217	23,710	4.314	1.061		
2.9	0.4275	23,390	4.256	0.964		
2.84	0.4366	22,900	4.173	0.847		
2.8	0.4428	22,580	4.121	0.786		
2.74	0.4525	22,100	4.048	0.712		
2.7	0.4592	21,780	4.004	0.667		
2.64	0.4696	21,290	3.940	0.614		
2.6	0.4769	20,970	3.903	0.579		
2.54	0.4881	20,490	3.851	0.536		
2.5	0.4959	20,160	3.818	0.511		
2.44	0.5081	19,680	3.773	0.479		
2.4	0.5166	19,360	3.745	0.457		
2.34	0.5298	18,870	3.706	0.431		
2.3	0.5391	18,550	3.682	0.416		

TABLE VIII (*Continued*)

Indium Phosphide

eV	μm	cm^{-1}	n	k	n	k
2.24	0.5535	18,070	3.649	0.393		
2.2	0.5636	17,740	3.629	0.380		
2.14	0.5794	17,260	3.602	0.358		
2.1	0.5904	16,940	3.585	0.347		
2.04	0.6078	16,450	3.563	0.329		
2.0	0.6199	16,130	3.549	0.317		
1.94	0.6391	15,650	3.530	0.299		
1.9	0.6526	15,320	3.517	0.293		
1.84	0.6738	14,840	3.501	0.278		
1.8	0.6888	14,520	3.492	0.270		
1.74	0.7125	14,030	3.481	0.255		
1.7	0.7293	13,710	3.476	0.242		
1.64	0.7560	13,230	3.469	0.225		
1.6	0.7749	12,900	3.467	0.218.		
1.54	0.8051	12,420	3.459	0.209		
1.5	0.8266	12,100	3.456	0.203		
1.345	0.921	10,850		5.71×10^{-2} [5, 6]		
1.340	0.925	10,810	3.3962 [4]	3.55		
1.335	0.928	10,770		2.04		
1.333	0.930	10,750	3.3905	1.09×10^{-2}		
1.326	0.935	10,700	3.3850	5.90×10^{-3}		
1.319	0.940	10,640	3.3797	3.18		
1.315	0.942	10,610		1.71×10^{-3}		
1.312	0.945	10,580	3.3746	9.67×10^{-4}		
1.305	0.950	10,530	3.3696	5.27		
1.30	0.9534	10,487	3.362 [7]	2.81×10^{-4}		
1.297	0.955	10,470	3.3647 [4]			
1.295	0.957	10,450		1.45×10^{-4}		
1.291	0.960	10,420	3.3600	7.39×10^{-5}		
1.29	0.9608	10,408	3.353 [7]			
1.284	0.965	10,360	3.3555 [4]	3.92		
1.28	0.9683	10,327	3.345 [7]	2.46		
1.277	0.970	10,310	3.3510 [4]			
1.275	0.972	10,290		1.66		
1.271	0.975	10,260	3.3467	1.13		
1.27	0.9759	10,247	3.337 [7]			
1.26	0.9837	10,166	3.330			
1.25	0.9915	10,085	3.324			
1.24	0.9995	10,005	3.318			
1.239	1.000	10,000	3.3269 [4]			
1.23	1.008	9,924	3.312 [7]			
1.22	1.016	9,843	3.307			
1.21	1.024	9,763	3.302			
1.20	1.033	9,682	3.297			
1.15	1.078	9,279	3.274			
1.127	1.10	9,091	3.2687 [4]			

(*continued*)

TABLE VIII (*Continued*)

Indium Phosphide

eV	μm	cm^{-1}	n	k	n	k
1.10	1.127	8,875	3.254 [7]			
1.05	1.180	8,472	3.236			
1.033	1.20	8,333	3.2312 [4]			
1.00	1.239	8,068	3.220 [7]			
0.9534	1.30	7,692	3.2053 [4]			
0.95	1.305	7,665	3.205 [7]			
0.90	1.377	7,261	3.191			
0.8853	1.40	7,143	3.1864 [4]			
0.85	1.458	6,858	3.178 [7]			
0.80	1.549	6,455	3.167			
0.75	1.653	6,051	3.156			
0.70	1.771	5,648	3.146			
0.65	1.907	5,244	3.137			
0.60	2.066	4,841	3.129			
0.55	2.254	4,438	3.121			
0.50	2.479	4,034	3.114			
0.45	2.754	3,631	3.107			
0.40	3.099	3,227	3.101			
0.35	3.541	2,824	3.095			
0.30	4.131	2,420	3.089			
0.25	4.958	2,017	3.083			
0.20	6.197	1,614	3.074			
0.18	6.886	1,452	3.069			
0.16	7.746	1,291	3.062			
0.14	8.853	1,130	3.053			
0.12	10.33	968.2	3.038			
0.1033	12.00	833.3		5.27×10^{-4} [6, 8]		
0.10	12.39	806.8	3.012			
0.09476	13.08	764.5		6.67		
0.09	13.77	726.1	2.990			
0.08853	14.00	714.3		8.86×10^{-4}		
0.08607	14.40	694.4		1.28×10^{-3}		
0.085	14.58	685.8	2.975	1.48×10^{-3} [5]		
0.084	14.75	677.7		1.81		
0.08346	14.85	673.4		3.00 [6, 8]		
0.083	14.93	669.7		6.34 [5]		
0.08263	15.00	666.7		3.71×10^{-3} [6, 8]		
0.082	15.11	661.6		1.15×10^{-2} [5]		
0.08133	15.24	656.2		5.25×10^{-3} [6, 8]		
0.081	15.30	653.5		1.76×10^{-2} [5]		
0.08090	15.32	652.7		6.17×10^{-3} [6, 8]		
0.08022	15.45	647.2		6.26		
0.08	15.49	645.5	2.956	3.90 [5]		
0.07986	15.52	644.3		5.63 [6, 8]		
0.079	15.69	637.4		9.72 [5]		
0.07874	15.74	635.3		5.22 [6, 8]		
0.07820	15.85	630.9		6.13		

TABLE VIII (*Continued*)

Indium Phosphide

eV	μm	cm⁻¹	n	k	n	k
0.078	15.89	629.3		8.83 [5]		
0.07746	16.00	625.0		7.12 [6, 8]		
0.077	16.10	621.2		3.00 [5]		
0.07679	16.14	619.6		7.46 [6, 8]		
0.07646	16.21	616.9		6.67		
0.07613	16.28	614.3		5.16		
0.076	16.31	613.2				
0.07562	16.39	610.1		3.33		
0.075	16.53	605.1	2.932			
0.07489	16.55	604.2		2.31		
0.7291	17.00	588.2		1.77		
0.07	17.71	564.8	2.898			
0.06886	18.00	555.6		1.81		
0.06547	18.93	528.3		2.32		
0.065	19.07	524.4	2.851			
0.064	19.37	516.4				
0.06353	19.51	512.6		3.84		
0.06317	19.62	509.7		4.73		
0.063	19.67	508.3		1.52 [5]		
0.06266	19.78	505.6		6.02 [6, 8]		
0.062	19.99	500.2		8.29 [5]		
0.06197	20.00	500.0		7.94 [6, 8]		
0.06139	20.19	495.3		9.49×10^{-3}		
0.061	20.32	492.2		1.04×10^{-2} [5]		
0.06094	20.34	491.6		1.08×10^{-2} [6, 8]		
0.06070	20.42	489.7		1.15		
0.06025	20.57	486.1		1.3		
0.06	20.66	484.1	2.780	1.46 [5]		
0.05951	20.83	480			2.51 [15]	
0.059	21.01	476.0		1.81		
0.058	21.37	468.0		2.17		
0.05765	21.51	465			2.34	
0.057	21.74	459.9		1.58×10^{-2}		
0.056	22.13	451.8		9.92×10^{-3}		
0.05579	22.22	450			2.24	
0.055	22.53	443.8	2.662	1.15×10^{-2}		
0.054	22.95	435.7		1.01		
0.05393	22.99	435			2.23	
0.053	23.38	427.6		1.15		
0.05207	23.81	420			2.19	0.126 [15]
0.052	23.83	419.6		1.36		
0.05170	23.98	417			2.16	0.135
0.05133	24.15	414			2.14	0.145
0.051	24.30	411.5		2.21		
0.05096	24.33	411			2.14	0.123
0.05083	24.39	410	2.494^b			

(*continued*)

TABLE VIII (*Continued*)

Indium Phosphide

eV	μm	cm^{-1}	n	k	n	k
0.05059	24.51	408			2.12	0.112
0.05021	24.69	405			2.11	0.120
0.050	24.79	403.4	2.429 [7]	3.35		
0.04984	24.88	402			2.09	0.132
0.04959	25.00	400	2.411[b]			
0.04947	25.06	399			2.07	0.129
0.04910	25.25	396			2.04	0.126
0.04897	25.32	395	2.361	1.9[b]		
				2.89 [5]		
0.04873	25.45	393			2.02	0.126
0.04835	25.64	390	2.305	2.1[b]	2.00	0.115
0.048	25.83	387.1		2.49 [5]		
0.04798	25.84	387			1.93	0.129
0.04773	25.97	385	2.239	2.5[b]		
0.04761	26.04	384			1.86	0.182
0.04724	26.25	381			1.84	0.229
0.04711	26.32	380	2.163	2.9		
0.047	26.37	379.2		2.81 [5]		
0.04687	26.46	378			1.84	0.234
0.04649	26.67	375	2.072	3.5[b]	1.76	0.234
0.04612	26.88	372			1.70	0.240
0.046	26.94	371.1		2.51 [5]		
0.04587	27.03	370	1.963	4.3[b]		
0.04575	27.10	369			1.68	0.224
0.04538	27.32	366			1.51	0.234
0.04525	27.40	365	1.827	5.4		
0.045	27.54	363.1		2.15 [5]		
0.04501	27.55	363			1.45	0.316
0.04463	27.78	360	1.653	7.0[b]	1.41	0.398
0.04426	28.01	357			1.41	0.427
0.04401	28.17	355	1.418	9.9		
				2.39×10^{-2} [5]		
0.04389	28.25	354			1.32	0.324
0.04352	28.49	351			1.00	0.240
0.04339	28.57	350	1.069	0.16[b]		
0.04315	28.74	348	0.869	0.22	0.631	0.398
0.043	28.82	346.9		5.62×10^{-2} [5]		
0.04290	28.90	346	0.611	0.34[b]		
0.04277	28.99	345			0.355	0.794
0.04265	29.07	344	0.372	0.61		
0.04240	29.24	342	0.273	0.92	0.282	1.12
0.04215	29.41	340	0.234	1.20		
0.04203	29.50	339			0.224	1.48
0.04191	29.59	338	0.216	1.45		
0.04166	29.76	336	0.209	1.69	0.174	1.86
0.04141	29.94	334	0.209	1.93		
0.04129	30.03	333			0.170	2.29

TABLE VIII (*Continued*)
Indium Phosphide

eV	μm	cm^{-1}	h	k	n	k
0.04116	30.12	332	0.213	2.16		
0.04092	30.30	330	0.221	2.41	0.251	2.75
0.04067	30.49	328	0.234	2.67		
0.04054	30.58	327			0.447	3.31
0.04042	30.67	326	0.251	2.94		
0.04017	30.86	324	0.275	3.24	0.759	3.94
0.03992	31.06	322	0.307	3.57		
0.03980	31.15	321			1.20	4.52
0.03968	31.25	320	0.350	3.94		
0.03943	31.45	318	0.409	4.37	2.09	5.25
0.03918	31.65	316	0.493	4.87		
0.03906	31.75	315			3.24	5.50
0.03893	31.85	314	0.619	5.48		
0.03868	32.05	312	0.822	6.26	4.57	5.56
0.03844	32.26	310	1.81	7.32		
0.03831	32.36	309			5.89	5.25
0.03819	32.47	308	1.94	8.83		
0.03806	32.57	307	2.69	9.85		
0.03794	32.68	306	4.00	11.1	7.41	3.31
0.03782	32.79	305	6.42	12.1		
0.03369	32.89	304	10.2	11.6		
0.03757	33.00	303	12.7	8.24	7.24	1.26
0.03744	33.11	302	12.5	5.00		
0.03732	33.22	301	11.4	3.14		
0.03720	33.33	300	10.4	2.13	6.17	0.447
0.03695	33.56	298	8.93	1.17		
0.03682	33.67	297			5.01	0.162
0.03670	33.78	296	7.95	0.749		
0.03664	33.84	295.5				0.115
0.03645	34.01	294	7.28	0.525	4.68	0.120
0.03620	34.25	292	6.78	0.391		
0.03608	34.36	291			4.37	0.162
0.03596	34.48	290	6.39	0.304		
0.03571	34.72	288			4.12	0.219
0.03534	35.09	285	5.73	0.183	3.98	0.229
0.03496	35.46	282			3.89	0.282
0.03472	35.71	280	5.31	0.124		
0.03459	35.84	279			3.80	0.302
0.03422	36.23	276			3.76	0.282
0.03410	36.36	275	5.01	8.9×10^{-2}		
0.03385	36.63	273			3.72	0.257
0.03366	36.83	271.5				0.245
0.03348	37.04	270	4.79	6.8	3.67	0.263
0.03310	37.45	267			3.55	0.282
0.03273	37.88	264			3.55	0.282
0.03236	38.31	261			3.51	0.257

(*continued*)

TABLE VIII (*Continued*)

Indium Phosphide

eV	μm	cm^{-1}	n	k	n	k
0.03224	38.46	260	4.48	4.3		
0.03199	38.76	258			3.47	0.302
0.03162	39.22	255			3.47	0.295
0.03124	39.68	252			3.47	0.309
0.03100	40.00	250	4.27	3.0		
0.03087	40.16	249			3.47	0.320
0.03050	40.65	246			3.47	0.324
0.03013	41.15	243			3.44	0.320
0.02976	41.67	240	4.13	2.2	3.43	0.316
0.02938	42.19	237			3.43	0.302
0.02901	42.74	234			3.40	0.288
0.02864	43.29	231			3.39	0.234
0.02852	43.48	230	4.02	1.7		
0.02827	43.86	228			3.39	0.234
0.02790	44.48	225			3.35	0.214
0.02728	45.45	220	3.93	1.3		
0.02604	47.62	210	3.87	1.1×10^{-2}	3.31	
0.02480	50.00	200	3.81	8.7×10^{-3}		
0.02418	51.28	195			3.19	
0.02232	55.56	180			3.19	
0.01860	66.67	150	3.65			
0.01240	100.0	100	3.57			
0.009919	125.0	80	3.551			
0.007439	166.7	60	3.538			
0.004959	250.0	40	3.529			
0.002480	500.0	20	3.523			
0.001240	1000.0	10	3.522			

[a] References are indicated in brackets.

[b] Present work: fit by using oscillator model and Raman parameters.

Lead Selenide (PbSe)

G. BAUER and H. KRENN

Institut für Physik
Montanuniversität Leoben
Leoben, Austria

Lead selenide is in many respects similar to the other members of the lead-salt family PbS and PbTe; however, the literature on this compound is not as extensive.

In the far infrared Burstein et $al.$ [1] have determined the transverse optic phonon mode ω_T by transmission experiments by using epitaxial thin PbSe films. Burkhard et $al.$ [2] investigated the reflectivity of bulk n-type material ($N = 10^{17}$ cm^{-3}) in the 20–250-cm^{-1} range by using Fourier-transform spectroscopy. The data were analyzed by a classical oscillator fit (see PbTe), including a Drude term for the free carriers and by Kramers–Kronig (KK) analysis. For room temperature the following parameters were obtained: $\varepsilon_\infty = 23$, $\varepsilon_s = 203 \pm 15$, $\omega_T = 39 \pm 2$ cm^{-1}, and the phonon damping parameter $\Gamma = 24$ cm^{-1}; a plasma frequency $\omega_p = (Ne^2/\varepsilon_0 m_p)^{1/2}$ is 419 cm^{-1} and a free-carrier damping parameter is $\gamma = 20$ cm^{-1}. For frequencies above 40 cm^{-1} the results of the oscillator fit are essentially in agreement with the KK analysis, which was obtained by extrapolating the reflectivity in the low-frequency region according to the Hagen–Rubens law and in the high-frequency region to an asymptotic value given by ε_∞. Another far-infrared investigation on p-PbSe films deposited epitaxially on NaCl substrates ($d \simeq 2$ μm) was performed by Amirtharaj et $al.$ [3] in the frequency range 20–350 cm^{-1}. An oscillator fit to the film/substrate reflectivity spectra yielded slightly different parameters than those found in Burkhard et $al.$ [2]. Thus, both sets of data are shown in Fig. 8, and in Table IX $\varepsilon_\infty = 23.4$, $\varepsilon_s = 212$, $\omega_T = 45$ cm^{-1}, and $\Gamma = 19.6$ cm^{-1} for a sample with a hole concentration of 8.7×10^{17} cm^{-3} with $\omega_p = 880$ cm^{-1} and $\gamma = 170$ cm^{-1}.

In the region of the plasma minimum and at higher frequencies, several investigations have been performed [4–8], and a comparison of the longitudinal-optic-mode frequency from optical data with a value from tunneling experiments [9] is possible. Optical constants below and close to the fundamental absorption edge were obtained by Vyatkin et $al.$ [7], who used

HANDBOOK OF OPTICAL CONSTANTS OF SOLIDS

epitaxial p- and n-type films on BaF_2 substrates in the wavelength region 5–15 μm for the determination of n and 7.5–12 μm for the determination of k. For the absorption constant α an experimental dependence $\alpha \simeq 5 \times 10^{-10} N\lambda^2$ cm^{-1} was found for $\lambda > 8$ μm.

Refractive-index data in the region of the fundamental absorption edge and above (0.1–0.5 eV) were obtained by Zemel et al. [10], who used epitaxially grown films on NaCl and optical techniques described in the critique on PbS. At $T = 300$ K, around 0.3 eV the well-known peak in n (see Table IX) is present. The precision of the n measurements was estimated to be $\pm 3\%$. Further information on the temperature dependence of n in this region is found in Preier [11]. Absorption in this region has been measured by Scanlon [12], and Moss [13] has cited unpublished work by Avery [as cited by Moss [13], p. 187] in the 0.5–5-μm range. Avery measured the reflectance for incident light polarized in as well as perpendicular to the plane of incidence, choosing three angles of incidence per wavelength. Both data sets are given in Table IX.

In the region above the fundamental absorption edge, several authors have studied reflectivity in the 1–6-eV [14–17] and 1–22-eV [15, 17] ranges. Synchrotron radiation was used for a study in the 14–26-eV region [18]. Whereas up to 6 eV the results of a KK analysis of reflectivity data obtained with cleaved bulk samples can be compared with optical constants derived from reflectivity and transmission of thin films ($d \simeq 300$ Å), from 6 to 22 eV only reflectivity data could be analyzed. For the KK analysis and for the extrapolation to energies higher than 25 eV, it was assumed that the reflectivity is given by $R \propto C\omega^{-4}$, where C is calculated so as to have the reflectivity continuous at 25 eV. Actually n and k values were quoted up to 15 eV [17] and are tabulated in Table IX. The structure in the optical constants in the region is due to interband transitions and the effect of the d electrons of lead.

For review papers on the optical and electronic properties of PbSe, we refer the reader to Dalven [19, 20], Ravich et al. [21], and Nimtz [22].

REFERENCES

1. E. Burstein, R. Wheeler, and J. Zemel, *Proc. Int. Conf. Phys. Semicond.* (M. Hulin, ed.), p. 1065. Dunod, Paris, 1964.
2. H. Burkhard, R. Geick, P. Kästner, and K. H. Unkelbach, *Phys. Stat. Solidi B* **63**, 89 (1974).
3. P. M. Amirtharaj, B. L. Bean, and S. Perkowitz, *J. Opt. Soc. Am.* **67**, 939 (1977).
4. A. Aziza, E. Amzallag, and M. Balkanski, *Solid State Commun.* **8**, 873 (1970).
5. A. Mycielski, A. Aziza, M. Balkanski, M. Y. Moulin, and J. Mycielski, *Phys. Stat. Solidi B* **52**, 187 (1972).
6. I. V. Kucherenko, Y. A. Mityagin, L. K. Vodopyanov, and A. P. Shotov, *Sov. Phys. Semicond.* **11**, 282 (1977); I. V. Kucherenko, Y. A. Mityagin, L. K. Vodopyanov, and A. P. Shotov, *Fiz. Tekh. Poluprovodn* **11**, 488 (1977).
7. K. V. Vyatkin and A. P. Shotov, *Sov. Phys. Semicond.* **14**, 785 (1980); K. V. Vyatkin and A. P. Shotov, *Fiz. Tekh. Poluprovodn.* **14**, 1331 (1980).

8. A. Maitre, R. Le Toullec, and M. Balkanski, *Proc. Int. Conf. Phys. Semicond.* (M. Miasek, ed.), p. 826. PWN-Polish Scientific Publishers, Warsaw, 1972.
9. R. N. Hall and J. H. Racette, *J. Appl. Phys.* (Suppl.) **32**, 2078 (1961).
10. J. N. Zemel, J. D. Jensen, and R. B. Schoolar, *Phys. Rev. A* **140**, 330 (1965).
11. H. Preier, *Appl. Phys.* **20**, 189 (1979).
12. W. W. Scanlon, *J. Phys. Chem Solids*, **8**, 423 (1959).
13. T. S. Moss, "Optical Properties of Semiconductors," p. 189. Butterworth, London, 1959.
14. M. L. Belle, *Sov. Phys. Solid State*, **5**, 2401 (1964); M. L. Belle, *Fiz. Tverd. Tela* **5**, 3282 (1963).
15. S. E. Kohn, P. Y. Yu, Y. Petroff, Y. R. Shen, Y. Tsang, and M. L. Cohen, *Phys. Rev. B* **8**, 1477 (1973).
16. F. I. Bogacki, A. K. Sood, C. Y. Yang, S. Rabii, and J. E. Fischer, *Surf. Sci.* **37**, 494 (1973).
17. M. Cardona and D. L. Greenaway, *Phys. Rev. A* **133**, 1685 (1964).
18. M. Cardona, C. M. Penchina, E. E. Koch, and P. Y. Yu, *Phys. Stat. Solidi B* **53**, 327 (1972).
19. R. Dalven, *Infrared Phys.* **9**, 141 (1969).
20. R. Dalven, *Solid State Phys.* **28**, 179 (1973).
21. Y. I. Ravich, B. A. Efimova, and I. A. Smirnov, *in* "Semiconducting Lead Chalcogenides." Plenum, New York, 1979.
22. G. Nimtz, *in* "Landolt-Börnstein, Numerical Data and Functional Relationships in Science and Technology" (K.-H. Hellwege and O. Madelung, eds.), Group III, Vol. 17, Subvolume f, pp. 168, 110. Springer-Verlag, Berlin, 1983.

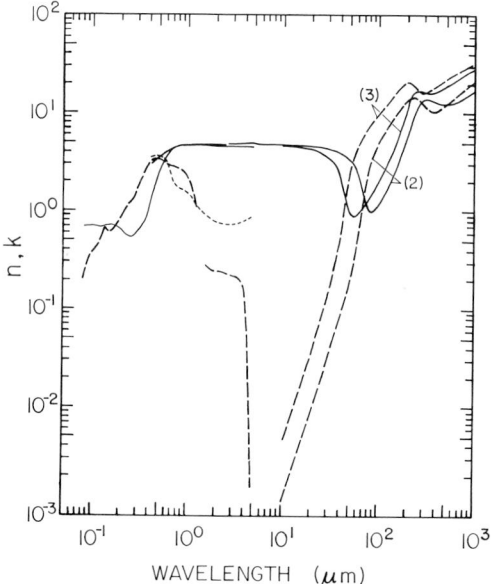

Fig. 8. Log–log plot of n (——) and k (–––) versus wavelength in micrometers for lead selenide. Curves marked (2) and (3) are for samples of two different free-carrier concentrations from Burkbard *et al.* [2] and Amirtharaj *et al.* [3], respectively.

G. Bauer and H. Krenn

TABLE IX

Values of n and k for Lead Selenide Obtained from Various References[a]

eV	cm⁻¹	μm	n	k	n	k
14.5		0.08551	0.72 [17]	0.20 [17]		
14.0		0.08856	0.70	0.25		
13		0.09537	0.70	0.35		
12		0.1033	0.70	0.38		
11.5		0.1078	0.70	0.40		
11		0.1127	0.71	0.44		
10.5		0.1181	0.70	0.46		
10		0.1240	0.68	0.50		
9.5		0.1305	0.68	0.56		
9		0.1378	0.7	0.66		
8.5		0.1459	0.725	0.63		
8		0.1550	0.715	0.62		
7.5		0.1653	0.71	0.66		
7		0.1771	0.70	0.71		
6.5		0.1907	0.68	0.78		
6	48,390	0.2066	0.64	0.86		
5.5	44,360	0.2254	0.58	1.0		
5	40,330	0.2480	0.54	1.2		
4.5	36,290	0.2755	0.56	1.3		
4	32,260	0.3100	0.64	1.9		
3.5	28,230	0.3542	0.74	2.4		
3.0	24,200	0.4133	1.25 [17]	3.20 [17]	3.00 [13]	3.50 [13]
2.0	16,130	0.6199	3.65	2.9	3.90	3.20
1.65	13,310	0.7514			4.51	1.73
1.5	12,100	0.8266	4.64	2.64		
1.24	10,000	1.000			4.67	1.54
1.0	8,065	1.240	4.65	1.1		
0.99	7,985	1.252			4.69	1.14
0.83	6,694	1.494			4.64	0.950
0.80	6,452	1.550		0.284 [12]		
0.75	6,049	1.653		0.269		
0.71	5,726	1.746			4.61	0.830
0.70	5,646	1.771		0.251		
0.65	5,243	1.907		0.240		
0.62	5,001	2.000			4.59	0.770
0.60	4,839	2.066		0.254		
0.55	4,436	2.254		0.238	4.59	0.740
0.50	4,033	2.480		0.235		
			4.70	0.450 [17]		
0.495	3,992	2.505			4.58	0.740
0.48	3,871	2.583	4.90 [10]			
0.46	3,710	2.695	4.88			
0.45	3,629	2.755		0.220 [12]	4.57	0.740
0.44	3,549	2.818	4.885			
0.42	3,388	2.952	4.895			
0.41	3,307	3.024			4.57	0.740

TABLE IX (*Continued*)

Lead Selenide

eV	cm⁻¹	μm	n	k	n	k
0.40	3,226	3.100	4.91			
0.39	3,146	3.179	4.85	0.219		
0.38	3,065	3.263	4.93	0.225	4.57	0.740
0.36	2,904	3.444	4.95			
0.354	2,855	3.502			4.56	0.756
0.35	2,823	3.542		0.199		
0.34	2,742	3.647	4.96	0.194		
0.33	2,662	3.757		0.189	4.54	0.782
0.32	2,581	3.875	4.98	0.173		
0.31	2,500	4.000		0.127	4.52	0.796
0.30	2,420	4.133		0.104	5.05 [12]	
					5.29 [7]	
0.292	2,355	4.246			4.48 [13]	0.836
0.29	2,339	4.275		5.72×10^{-2}		
0.28	2,258	4.428	4.94	3.53×10^{-2} [10]		
0.275	2,218	4.509			5.13 [7]	
0.27	2,178	4.592		9.17×10^{-3} [12]	4.48 [13]	0.836
0.26	2,097	4.769	4.94	2.85	4.47	0.890
0.25	2,016	4.959		1.54	5.02 [7]	
0.248	2,000	5.000			4.45 [13]	
0.24	1,936	5.166	4.90	1.38		
0.23	1,855	5.391	4.83	1.28	4.91 [7]	
0.22	1,774	5.636	4.86			
0.20	1,613	6.199	4.82		4.90	
0.18	1,452	6.888			4.87	
0.17	1,371	7.293			4.86	
0.16	1,290	7.749	4.80		4.85	2.22×10^{-2} [7]
0.15	1,210	8.266	4.78			2.48
0.14	1,129	8.856	4.77			2.69
0.13	1,049	9.537	4.75		4.82	3.4
0.124	1,000	10.00	4.75 [2]	1.06×10^{-3} [2]	4.72 [3]	4.10×10^{-3} [3]
0.1215	980	10.20	4.75	1.13	4.71	4.36
0.1190	960	10.42	4.74	1.20	4.71	4.64
0.1165	940	10.64	4.74	1.28	4.70	4.95
0.1141	920	10.87	4.74	1.37	4.69	5.29
0.1116	900	11.11	4.74	1.46	4.69	5.66
0.1091	880	11.36	4.73	1.57	4.68	6.06
0.1066	860	11.63	4.73	1.68	4.67	6.51
0.1041	840	11.90	4.73	1.80	4.66	6.99
0.1017	820	12.20	4.73	1.94	4.66	7.53
0.09919	800	12.50	4.72	2.09	4.65	8.13
0.09671	780	12.82	4.72	2.26	4.64	8.79
0.09423	760	13.16	4.71	2.44	4.62	9.52×10^{-3}
0.09175	740	13.51	4.71	2.65	4.61	1.03×10^{-2}

(*continued*)

G. Bauer and H. Krenn

TABLE IX (*Continued*)

Lead Selenide

eV	cm^{-1}	μm	n	k	n	k
0.08927	720	13.89	4.70	2.88	4.60	1.13
0.08679	700	14.29	4.70	3.14	4.59	1.23
0.08431	680	14.71	4.69	3.43	4.57	1.34
0.08183	660	15.15	4.69	3.75	4.55	1.48
0.07935	640	15.63	4.68	4.12	4.53	1.63
0.07687	620	16.13	4.67	4.54	4.51	1.80
0.07439	600	16.67	4.66	5.02	4.49	1.99
0.07191	580	17.24	4.65	5.58	4.47	2.22
0.06943	560	17.86	4.64	6.21	4.44	2.48
0.06699	540	18.52	4.63	6.95	4.41	2.78
0.06447	520	19.23	4.62	7.80	4.37	3.14
0.06199	500	20.00	4.60	8.81×10^{-3}	4.33	3.56
0.05951	480	20.83	4.59	1.00×10^{-2}	4.29	4.07
0.05703	460	21.74	4.57	1.14	4.23	4.68
0.05455	440	22.73	4.55	1.31	4.17	5.43
0.05207	420	23.81	4.52	1.52	4.10	6.34
0.04959	400	25.00	4.49	1.77	4.02	7.49
0.04711	380	26.32	4.46	2.08	3.92	8.95×10^{-2}
0.04463	360	27.78	4.42	2.47	3.80	0.11
0.04215	340	29.41	4.37	2.97	3.66	0.13
0.03968	320	31.25	4.31	3.62	3.48	0.17
0.03720	300	33.33	4.24	4.48	3.26	0.22
0.03472	280	35.71	4.15	5.64	2.96	0.30
0.03224	260	38.46	4.04	7.26	2.55	0.43
0.02976	240	41.67	3.89	9.61×10^{-2}	1.97	0.70
0.02728	220	45.45	3.69	0.13	1.27	1.41
0.02480	200	50.00	3.42	0.19	0.96	2.49
0.02232	180	55.56	3.01	0.30	0.91	3.58
0.01984	160	62.50	2.34	0.56	0.99	4.71
0.01736	140	71.43	1.34	1.49	1.15	6.00
0.01488	120	83.33	1.02	3.19	1.45	7.57
0.01240	100	100.0	1.18	5.02	1.97	9.64
0.01215	98	102.0	1.21	5.22	2.04	9.89
0.01190	96	104.2	1.25	5.43	2.12	10.1
0.01165	94	106.4	1.29	5.64	2.21	10.4
0.01141	92	108.7	1.34	5.87	2.30	10.7
0.01116	90	111.1	1.39	6.09	2.39	11.0
0.01091	88	113.6	1.44	6.33	2.50	11.3
0.01066	86	116.3	1.51	6.58	2.61	11.6
0.01041	84	119.0	1.57	6.83	2.73	11.9
0.01017	82	122.0	1.65	7.10	2.87	12.3
0.009919	80	125.0	1.73	7.38	3.02	12.7
0.009671	78	128.2	1.83	7.67	3.18	13.1
0.009423	76	131.6	1.93	7.97	3.36	13.5
0.009175	74	135.1	2.05	8.29	3.56	13.9
0.008927	72	138.9	2.18	8.62	3.79	14.4

TABLE IX (*Continued*)
Lead Selenide

eV	cm⁻¹	μm	n	k	n	k
0.008679	70	142.9	2.33	8.97	4.04	14.9
0.008431	68	147.1	2.50	9.34	4.33	15.4
0.008183	66	151.5	2.69	9.73	4.67	16.0
0.007935	64	156.3	2.91	10.1	5.05	16.6
0.007687	62	161.3	3.16	10.6	5.51	17.2
0.007439	60	166.7	3.46	11.0	6.10	17.9
0.007191	58	172.4	3.80	11.5	6.71	18.6
0.006943	56	178.6	4.20	12.0	7.51	19.3
0.006695	54	185.2	4.68	12.5	8.49	20.0
0.006447	52	192.3	5.25	13.1	9.68	20.6
0.006199	50	200.0	5.93	13.6	11.1	21.1
0.005951	48	208.3	6.7	14.1	12.7	21.2
0.005703	46	217.4	7.7	14.5	14.4	20.9
0.005455	44	227.3	8.8	14.7	15.9	20.1
0.005207	42	238.1	10.0	14.8	16.9	19.0
0.004959	40	250.0	11.2	14.6	17.4	17.8
0.004711	38	263.2	12.3	14.1	17.4	16.9
0.004463	36	277.8	13.2	13.4	17.1	16.4
0.004215	34	294.1	13.8	12.5	16.8	16.3
0.003968	32	312.5	14.1	11.7	16.5	16.5
0.003720	30	333.3	14.0	11.0	16.3	17.0
0.003472	28	357.1	13.8	10.6	16.4	17.7
0.003224	26	384.6	13.4	10.5	16.6	18.1
0.002976	24	416.7	13.0	10.1	17.1	19.7
0.002728	22	454.5	12.7	11.3	17.8	20.9
0.002480	20	500.0	12.6	12.2	18.7	22.3
0.002232	18	555.6	12.8	13.4	19.9	23.1
0.001984	16	625.0	13.2	14.9	21.4	25.4
0.001736	14	714.3	14.1	16.6	23.3	27.3
0.001488	12	833.3	15.5	18.1	25.6	29.6
0.001240	10	1000	17.4	21.1	28.6	32.5

a References are indicated in brackets.

Lead Sulfide (PbS)

G. GUIZZETTI and A. BORGHESI

Dipartimento di Fisica "A. Volta" and
Gruppo Nazionale di Struttura della Materia del CNR
Università di Pavia
Pavia, Italy

The room-temperature values of n and k tabulated here were obtained from the following works and references therein: Riedl and Schoolar [1], Geick [2], Schoolar and Zemel [3], Schoolar and Dixon [4], Zemel et al. [5], Dixon and Riedl [6], and Vakulenko et al. [7] for the medium and far infrared; Schoolar and Dixon [4], Avery [8], Scanlon [9], Wessel [10], and Semenov and Shileika [11] for the near infrared and visible; Wessel [10], Cardona and Greenaway [12], Rossi and Paul [13], Cardona and Haensel [14], Cardona et al. [15], and Heckelmann et al. [16] for the visible–ultraviolet and vacuum–ultraviolet. For the review papers on the optical and electronic properties of PbS we refer to Moss [17], Zemel [18], Dalven [19, 21], and Ravich et al. [20].

A composite smooth-curve plot is given in Fig. 9. Numerical values are given in Table X with appropriate references listed in brackets.

The optical constants were obtained generally by reflectance R and/or transmittance T measurement on single-crystal bulk or film samples. After 1963 almost everyone producing PbS films adopted the technique described by Schoolar and Zemel [3]: single-crystal n-type films were grown in vacuum ($\sim 10^{-6}$ torr) from bulk PbS on heated rocksalt substrates (generally cleaved from synthetic single-crystal of NaCl or KCl). The good crystalline quality of such films, which have very uniform thickness, has been well established [3, 5]. The film's surfaces require on further treatment to make them suitable for either optical T or R measurement. Furthermore, experiments dealing with the electrical, optical, and mechanical properties indicate that the films behave like the best available bulk crystals. No significant thickness dependence of the optical constants has been detected [3–5, 7].

In the far-infrared ($\lambda > 30\ \mu m$) bulk reflectance measurements at near normal incidence (8–10°) [2, 6] and transmission on films [2, 18] were performed. We have disregarded earlier R spectra [22, 23] that were in qualitative agreement with the findings of Geick and Dixon and Riedl [2, 6] but had

HANDBOOK OF OPTICAL CONSTANTS OF SOLIDS

values too low. The n and k values were obtained by fitting R in the 30–2000-μm range by a single classical dispersion oscillator (calculations including a Drude term suggested that the influence of free carriers on the optical constants is negligible for $N \lesssim 10^{16}$ cm^{-3}, at least out to 1000 μm). The R samples were natural or synthetic single crystals with different carrier concentrations N ranging from $\sim 10^{16}$ to $\sim 10^{19}$ cm^{-3}. The optical surfaces were ground mechanically and polished chemically, and it was checked that the damage caused by mechanical polishing does not significantly affect the R spectrum in this spectral region. Experimental and calculated spectra from Geick [2] and Dixon and Riedl [6] were in agreement to within the experimental uncertainties. In Table X n and k values obtained through the fit have been labeled according to Geick [2]. The parameters for the sample with the lowest carrier concentration were [2] $\varepsilon_0 = 150$ for the static relative dielectric constant; $\varepsilon_\infty = 16.8$ for the optical dielectric constant; $v_T = 71$ cm^{-1} and $v_L = 212$ cm^{-1} for the transverse and longitudinal optical phonon frequencies, respectively; and $\gamma = 15$ cm^{-1} for the damping. The fit values for ε_0 and v_T agree within 7% with those derived directly (ε_0 from R value at $\lambda = 2$ mm and v_T from the minimum in T spectra [2, 18]). Besides, the fit values are in good agreement with those deduced from phonon-assisted tunneling by Hall and Racette [24], microwave measurements by Sokoloski and Fang [25], inelastic neutron scattering by Cochran *et al.* [26], and n data in the medium infrared by plotting n^2 versus λ^2 and extrapolating the linear region to zero wavelength [1, 5, 6]. The sight dependence of ε_∞ on carrier density [6] is discussed in the PbTe critique.

As for the 0.95–26-μm range, Table X contains experimental n values deduced from the position of interference peaks in R and/or T for films [3–5]. The thicknesses, ranging from ~ 0.4 to ~ 5 μm, were determined by using the interferometric and weighing techniques described in Riedl and Schoolar [1], with an estimated error less than $\pm 3\%$. The typical carrier concentration was $N = 2 \times 10^{18}$ e/cm^3. At such high carrier density n is a slight function of N and is not expected to join smoothly to the calculated n [2] with no free-carrier contribution. The values of n [2] have been listed to 10 μm and can represent n for pure PbS. Although Geick [2] also lists k values, they are only a qualitative description of the lattice absorption of pure PbS. Phase shifts at the interfaces due to the nonzero values of the extinction coefficients and the optical properties of the substrate were estimated and found to be negligible. The basic precision of the n measurements was estimated to be $\pm 3\%$. The n values for the films are in excellent agreement with those obtained from bulk-reflectivity data [1]; their differences are within the experimental uncertainty. The same agreement exists among the data from [3–5] and the data obtained by other authors [1, 7, 11] with the same technique on films, in narrower spectral ranges. The refractive index in the 2–15-μm region was determined also at 77 and 373 K [5]. We observe that the sharp peak in the n spectrum at ~ 3 μm results from the rapid change in the absorption at the fundamental absorption edge.

The absorption coefficient α between 3 and 15 μm was measured by using a single crystal of n-type galena 0.22 mm thick, with $N = 2 \times 10^{17}$ cm^{-3} [7]. This absorption coefficient, typical of free carriers, was found to vary as $\lambda^{2.8 \pm 0.2}$ giving k to vary as $\lambda^{3.8}$. We would not expect this k to join smoothly on the k calculated for a purer sample in Dixon and Riedl [6]. Multiple-reflection effects were taken into account. Values for α were determined with an error not exceeding 20% and were in agreement, at short wavelengths, with those from Schoolar and Dixon [4] Scanlon [9], while the values from Semiletov et al. [27] were systematically too high (about one order of magnitude over the entire 3–15-μm range).

In the spectral region of the fundamental absorption edge, α has been measured at room temperature several times. The most recent and accurate measurements have been performed by Schoolar and Dixon [4], who, moreover, have discussed the previous results and the large discrepancies. Reflectance (at an angle of incidence less than 15°) and transmittance measurements were made in [4] on n-type films, with $N \simeq 3 \times 10^{18}$ cm^{-3} and thicknesses from 0.37 to 2.36 μm. The experimental error in R and T was estimated to be less than $\pm 4\%$, and the spectral energy resolution was less than 0.01 eV in the vicinity of the edge. Extinction coefficient k was derived from T and R, including multiple-reflection effects and the influence of the film backing. The results were estimated to be uncertain to $\pm 7\%$. In particular, the results approach those of Avery [8] in the higher energy region; in addition, they agree with those of Scanlon [9] in the steep portion of the edge but are greater by amounts up to 30% out to 1.2 μm. This fact and the long absorption plateau reported by Scanlon [9] were probably due to the presence of pinholes in the samples or scattered light in the monochromator, as suggested by Cardona and Greenaway [12]. The room-temperature Burstein–Moss effect is probably negligible at the fundamental band gap [28]. For all shorter wavelengths, the free-carrier effects are probably negligible.

The optical constants from 1.2 μm to about 0.2 μm were determined either by Kramers–Kronig (KK) analysis of near-normal-incidence reflectance data of bulk samples [12, 14] or from systematic and accurate measurement of R and T on thin films [10, 13]. The R values of more than 25 films and bulk agree in both magnitude and structure within about 4% [13]. The same agreement exists between R data from [10] and [13], which are higher than those reported in [12] and [16]. In Wessel [10] the error in R was taken to be $\pm 2\%$ and from 3 to 10% in T. In Rossi and Paul [13] R was repeatable to within $\pm 3\%$ absolute, and the total error was estimated $\pm 5\%$ absolute; the total error in T was estimated to be between 10 and 25%, including errors caused by scattered light from other wavelengths in the thicker samples and errors caused by imperfections in the films and film substrates (NaCl or KCl). No attempt was made to determine the impurity levels, but it was found [12] that the R spectra were not influenced by normally encountered variations in impurity concentrations. The films' thicknesses, ranging from

~ 200 to ~ 800 Å, were determined by assuming the carefully obtained optical constants of Schoolar and Dixon [4] near 1 μm and adding several methods of cross-checking. The thickness was considered to be accurate to approximately $\pm 15\%$. The optical constants obtained from R and T in Wessel [10] and Rossi and Paul [13] agree quite well (within 10%) with those obtained from KK analysis [13], and their error was about the same as the one involved in the R and T measurements, i.e., $\pm 5\%$ or less, if the thickness can be obtained accurately. This is true except in the energy range in which n and k values are such that a small uncertainty in R can lead to a large uncertainty in n [29]; this occurs in PbS near 0.4 μm. In Table X we have reported n and k values derived from thin-film optical data, which are in several respects more reliable than those derived from dispersion relations.

In the region between 0.2 and 0.05 μm only R measurements [12, 15, 16] were performed, either on expitaxial layers or on cleaved single crystals. The optical constants have been obtained from the KK analysis on R spectra from [12] and [15], extended to longer wavelengths with the data from [2] and [6] for the infrared and from [10] and [13] for the visible–ultraviolet. The R values of [12] and [15] were multiplied by a constant factor to make them agree with results from [13] at 0.2 μm and were extrapolated as $R = R_0(\lambda_0/\lambda)^{-P}$ for $\lambda < \lambda_0 = 0.05$ μm. Value P was adjusted to give values of k nearly equal to zero below the energy gap and agreeing with the values of [4, 10, 13] above the gap. The final value of $P = 2.77$ was used.

Transmission measurements on films in the 0.033–0.008-μm range were performed by Cardona and Haensel [14] with synchrotron radiation and with a typical resolution of 2 Å over the whole spectral range. The T technique was chosen over that of reflectance because of the small R or PbS ($< 1\%$) in this region, which increases the error due to scattered light of long wavelengths. The thin carbon substrates of the films made it impossible to extend the measurements to longer wavelengths. The thicknesses were determined with a Tolansky interferometer and also during evaporation with a calibrated quartz-crystal monitor. The absorption coefficient values were affected by a scaling error of about 20% because of uncertainties in the thicknesses.

REFERENCES

1. H. R. Riedl and R. B. Schoolar, *Phys. Rev.* **131**, 2082 (1963).
2. R. Geick, *Phys. Lett.* **10**, 51 (1964).
3. R. B. Schoolar and J. N. Zemel, *J. Appl. Phys.* **35**, 1848 (1964).
4. R. B. Schoolar and J. R. Dixon, *Phys. Rev. A* **137**, 667 (1965).
5. J. N. Zemel, J. D. Jensen, and R. B. Schoolar, *Phys. Rev. A* **140**, 330 (1965).
6. J. R. Dixon and H. R. Riedl, *Phys. Rev. A* **140**, 1283 (1965).
7. O. V. Vakulenko, M. P. Lisitsa, and Ya. F. Kononets, *Sov. Phys. Solid State* **8**, 1356 (1966).
8. D. G. Avery, *Proc. Phys. Soc. London Ser. B* **67**, 2 (1954).
9. W. W. Scanlon, *Solid State Phys.* **9**. 109 (1959).
10. P. R. Wessel, *Phys. Rev.* **153**, 836 (1967).

11. Ya. A. Semenov and A. Yu. Shileika, *Sov. Phys. Solid State* 1419 (1967).

12. M. Cardona and D. L. Greenaway, *Phys. Rev. A* **133**, 1685 (1964).

13. C. E. Rossi and W. Paul, *J. Appl. Phys.* **38**, 1803 (1967).

14. M. Cardona and R. Haensel, *Phys. Rev. B* **1**, 2605 (1970).

15. M. Cardona, C. M. Penchina, E. E. Koch, and P. Y. Yu, *Phys. Stat. Solidi B* **53**, 327 (1972).

16. G. H. Heckelmann, H. H. Landfermann, U. Nielsen, and T. S. Wagner, *Phys. Stat. Solidi B* **69**, K99 (1975).

17. T. S. Moss, *in* "Optical Properties of Semiconductors," p. 183. Butterworth, London, 1959.

18. J. N. Zemel, *Proc. Int. Conf. Phys. Semicond., Paris, 1964*, p. 1061. Dunod, Paris, 1964.

19. R. Dalven, *Infrared Phys.* **9**, 141 (1969).

20. Yu. I. Ravich, B. A. Efimova, and I. A. Smirnov, *in* "Semiconducting Lead Chalcogenides," p. 43. Plenum, New York, 1970.

21. R. Dalven, Solid State Phys. **28**, 179 (1973).

22. J. Strong, *Phys. Rev.* **38**, 1818 (1931).

23. H. Yoshinaga, *Phys. Rev.* **100**, 753 (1955).

24. R. N. Hall and J. H. Racette, *J. Appl. Phys., Suppl.* **32**, 2078 (1961).

25. M. M. Sokoloski and P. H. Fang, *Phys. Lett.* **16**, 222 (1965).

26. W. Cochran, R. Cowley, and G. Dolling, *Proc. R. Soc. London Ser. A* **293**, 433 (1966).

27. S. A. Semiletov, I. P. Voronina, and E. I. Kortukova, *Soviet Phys. Crystallogr.* (*Eng. Trans.*) **10**, 429 (1966).

28. Edward D. Palik, D. L. Mitchell, and J. N. Zemel, *Phys. Rev. A* **135**, 763 (1964).

29. P. M. Grant, *Bull. Am. Phys. Soc., Ser. II* **10**, 546 (1965).

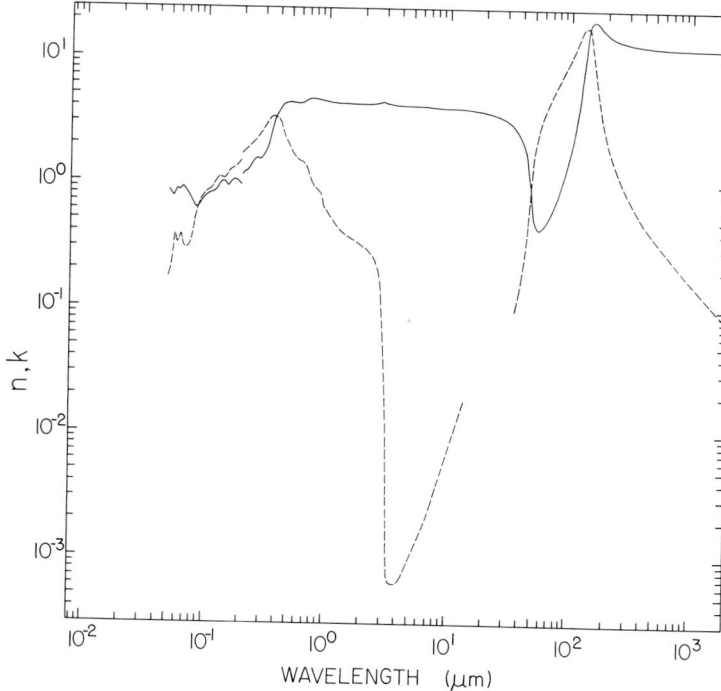

Fig. 9. Log–log plot of n (——) and k (– – – –) versus wavelength in micrometers for lead sulfide. Data from various references do not join together in overlap regions on occasion.

TABLE X

Values of n and k for Lead Sulfide from Various References[a]

eV	cm^{-1}	μm	n	k
150		0.008266		3.86×10^{-3} [14]
145		0.008551		4.09
140		0.008856		4.37
135		0.009184		4.66
130		0.009537		5.10
125		0.009919		5.59
120		0.01033		6.40
115		0.01078		8.96×10^{-3}
110		0.01127		1.05×10^{-2}
105		0.01181		1.26
100		0.01240		1.54
95		0.01305		1.78
90		0.01378		2.06
85		0.01459		2.38
80		0.01550		2.88
75		0.01653		3.72
70		0.01771		4.64
65		0.01907		5.48
60		0.02066		6.17
58		0.02138		6.44
55		0.02254		6.73
50		0.02480		7.05
45		0.02755		5.43
40		0.03100		4.28
38		0.03263		4.23×10^{-2}
25		0.04959	0.845 [12, 15]	0.171 [12, 15]
24		0.05166	0.800	0.196
23.3		0.05321	0.772	0.238
22.7		0.05462	0.760	0.291
21.9		0.05661	0.831	0.372
21.5		0.05767	0.870	0.339
21.3		0.05821	0.858	0.324
20.6		0.06019	0.847	0.339
19.8		0.06262	0.881	0.370
19.2		0.06457	0.911	0.332
18.6		0.06667	0.881	0.300
18.0		0.06888	0.846	0.294
17.5		0.07084	0.814	0.298
17.0		0.07293	0.778	0.307
16.5		0.07514	0.734	0.327
16.0		0.07749	0.687	0.364
15.6		0.07948	0.651	0.412
15.2		0.08157	0.628	0.472
14.8		0.08377	0.619	0.538
14.4		0.08610	0.627	0.606
14.0		0.08856	0.651	0.665
13.5		0.09183	0.687	0.725

TABLE X (*Continued*)

Lead Sulfide

eV	cm^{-1}	μm	n	k
13.0		0.09537	0.725	0.769
12.5		0.09918	0.758	0.807
12.0		0.1033	0.789	0.843
11.5		0.1078	0.811	0.872
11.0		0.1127	0.821	0.912
10.5		0.1181	0.836	0.974
10.0		0.1240	0.879	1.050
9.5		0.1305	0.968	1.096
9.0		0.1378	1.007	1.066
8.6		0.1442	0.997	1.060
8.2		0.1512	0.921	1.140
7.7		0.1610	0.999	1.273
7.2		0.1722	1.037	1.299
6.8		0.1823	1.024	1.361
6.4		0.1937	0.951	1.489
6.2	50,010	0.2000	1.13 [13]	1.70 [13]
5.9	47,590	0.2101	1.22	1.78
5.65	45,570	0.2194	1.22	1.85
5.4	43,550	0.2296	1.33	1.95
5.15	41,540	0.2407	1.43	2.05
4.95	39,920	0.2505	1.52	2.10
4.75	38,310	0.2610	1.53	2.20
4.6	37,100	0.2695	1.55	2.30
4.45	35,890	0.2786	1.53	2.45
4.30	34,680	0.2883	1.57	2.53
4.15	33,470	0.2988	1.67	2.70
4.0	32,260	0.3100	1.73	2.83
3.85	31,050	0.3220	1.90	3.00
3.75	30,250	0.3306	2.07	3.17
3.65	29,440	0.3397	2.33	3.32
3.55	28,630	0.3493	2.58	3.37
3.45	27,830	0.3594	2.98	3.37
3.35	27,020	0.3701	3.17	3.37
3.25	26,210	0.3815	3.32	3.30
3.15	25,410	0.3936	3.62	3.17
3.00	24,200	0.4133	3.88	3.00
2.90	23,390	0.4275	4.12	2.70
2.75	22,180	0.4509	4.25	2.33
2.55	20,570	0.4862	4.35	2.00
2.35	18,950	0.5276	4.33	1.67
2.20	17,740	0.5636	4.30	1.55
2.00	16,130	0.6199	4.29 [10, 13]	1.48 [10, 13]
1.80	14,520	0.6888	4.53 [10]	1.30 [10]
1.60	12,910	0.7749	4.62	0.94
1.40	11,290	0.8856	4.50	0.83
1.30	10,490	0.9537	4.48 [4, 10]	0.627 [4, 10]

(*continued*)

G. Guizzetti and A. Borghesi

TABLE X (*Continued*)

Lead Sulfide

eV	cm^{-1}	μm	n	k
1.24	10,000	1.00	4.43	0.597 [4]
1.18	9,524	1.05	4.385	0.543
1.13	9,091	1.1	4.355	0.508
1.03	8,333	1.2	4.30	0.458
0.953	7,692	1.3	4.275 [4]	0.403
0.885	7,143	1.4	4.26	0.390
0.825	6,667	1.5	4.25	0.370
0.775	6,250	1.6	4.24	0.357
730	5,882	1.7	4.235	0.341
0.690	5,556	1.8	4.235	0.334
0.650	5,263	1.9	4.24	0.318
0.620	5,000	2.0	4.25 [4, 5]	0.310
0.590	4,762	2.1	4.245 [5]	0.304
0.563	4,545	2.2	4.255	0.285
0.539	4,348	2.3	4.26	0.271
0.516	4,167	2.4	4.28	0.255
0.496	4,000	2.5	4.30	0.235
0.477	3,846	2.6	4.335	0.212
0.459	3,704	2.7	4.38	0.182
0.443	3,571	2.8	4.42	0.144
0.432	3,484	2.87	4.435	9.03×10^{-2}
0.413	3,333	3.0	4.345	4.59 [9]
0.400	3,226	3.1	4.30	2.27
0.393	3,175	3.15		1.43
0.387	3,125	3.2		7.9×10^{-3}
0.381	3,077	3.25	4.26	3.46
0.3542	2,857	3.5	4.215	6.48×10^{-4}
0.3306	2,667	3.75	4.185	6.22
0.3100	2,500	4.0	4.16	6.38
0.2917	2,353	4.25	4.14	6.51
0.2755	2,222	4.5	4.125	7.47
0.2480	2,000	5	4.115	9.25×10^{-4}
0.2066	1,667	6	4.10	1.38×10^{-3}
0.1771	1,429	7	4.09	2.06
0.1550	1,250	8	4.07	3.06
0.1378	1,111	9	4.04	4.42
0.1240	1,000	10	4.01	6.32×10^{-3}
			4.02 [2]	
0.1178	950	10.53	4.01	
0.1116	900	11.11	4.00	
0.1054	850	11.76	3.98	
0.1033	833.3	12	3.90 [3]	1.14×10^{-2}
0.09919	800	12.50	3.97 [2]	
0.09299	750	13.33	3.95	
0.08856	714.3	14	3.795 [3]	1.83
0.08679	700	14.29	3.93 [2]	
0.08059	650	15.38	3.90	

TABLE X (*Continued*)

Lead Sulfide

eV	cm^{-1}	μm	n	k
0.07749	625.0	16	3.68 [3]	
0.07439	600	16.67	3.86 [2]	
0.06888	555.6	18	3.545 [3]	
0.06819	550	18.18	3.81 [2]	
0.06199	500.0	20	3.39 [3]	
			3.75 [2]	
0.05636	454.5	22	3.205 [3]	
0.05579	450	22.22	3.66 [2]	
0.05166	416.7	24	2.995 [3]	
0.04959	400	25.00	3.53 [2]	
0.04339	350	28.57	3.33	
0.03720	300	33.33	2.99	
0.03444	277.8	36	2.74	9.77×10^{-2} [2]
0.03263	263.2	38	2.53	0.127
0.03100	250.0	40	2.28	0.167
0.02952	238.1	42	1.98	0.226
0.02818	227.3	44	1.61	0.326
0.02695	217.4	46	1.13	0.540
0.02583	208.3	48	0.688	1.03
0.02480	200.0	50	0.514	1.59
0.02254	181.8	55	0.433	2.67
0.02066	166.7	60	0.452	3.55
0.01771	142.9	70	0.582	5.14
0.01550	125.0	80	0.809	6.74
0.01378	111.1	90	1.175	8.48
0.01240	100.0	100	1.79	10.51
0.01033	83.33	120	5.06	15.98
0.008856	71.43	140	17.41	17.94
0.007749	62.50	160	20.47	7.10
0.006888	55.56	180	17.87	3.47
0.006199	50.00	200	16.27	2.20
0.004959	40.00	250	14.40	1.15
0.004133	33.33	300	13.62	0.785
0.003542	28.57	350	13.21	0.604
0.003100	25.00	400	12.96	0.495
0.002480	20.00	500	12.69	0.367
0.001653	13.33	750	12.44	0.228
0.001240	10.00	1,000	12.35	0.167
0.0008266	6.667	1,500	12.29	0.109
0.0006199	5.000	2,000	12.27	0.0815

[a] Values obtained from the various references are indicated in brackets.

Lead Telluride (PbTe)

G. BAUER and H. KRENN

Institut für Physik
Montanuniversität Leoben
Leoben, Austria

Lead telluride is a member of the IV–VI compound semiconductor family, crystallizing in the cubic rocksalt structure with 10 electrons in the unit cell. The optical direct energy gap is very small, located at the L point of the Brillouin zone, which is the case for all three lead salts: PbS, PbSe, and PbTe [1–3]. The dielectric properties are characterized by unusually high dielectric constants ε_∞ and ε_s, where in PbTe strong temperature dependence of ε_s indicates the inherent tendency of the lattice to undergo a structural phase transition [4].

The room-temperature values of n and k tabulated in Table XI and plotted in Fig. 10 were obtained from the following works and references cited therein: the far-infrared results from 10 cm^{-1} (1000 μm) to 1000 cm^{-1} (10 μm) are from Buss and Kinch [5], Burkhard et al. [6], Perkowitz [7], and Lowney and Senturia [8] (2.5–40 μm). Values of n and k in the mid-infrared region are from Zemel et al. [9] (0.1–0.6 eV), Piccioli et al. [10] (1500–3500 cm^{-1}), Globus et al. [11] (0.15–0.6 eV), McCarthy et al. [21] (1000–2600 cm^{-1}), Walz [31] (0.25–0.35 eV); in the visible and UV region, from Cardona and Greenaway [12] (0.1–17 eV), Korn and Braunstein [13] (0.1–21 eV), and Cardona and Haensel [14] (36–150 eV).

In the low-energy part of the spectrum, the far-infrared results were usually obtained by reflectivity measurements [4–7, 15, 17]. Single-crystal bulk material as well as epitaxially grown films were used. The latter method yield high-quality material with relatively low free-carrier concentration and high carrier mobilities. For more than 10 years the hot wall technique [16] has been used to deposit films of uniform thickness on alkali halide (NaCl, KCl) or BaF_2 substrates. Due to the differences in the lattice constants and the thermal expansion coefficients, stresses appear that influence the electrical and optical properties [29], especially at low temperatures. The PbTe–BaF_2 system is the one in which stresses play the smallest role at room temperature.

535

Epitaxially grown films usually do not require any surface treatment such as grinding, polishing, or etching after growth; bulk materials require electropolishing after surface treatment.

The n and k values in the far infrared were obtained by fitting the reflectivity data [4–8, 15, 17], taken at near normal incidence ($\sim 10°$) with a classical dispersion oscillator for the infrared active phonon and Drude term for the free carriers as given by

$$\varepsilon = (n - ik)^2 = \varepsilon_\infty + \frac{(\varepsilon_s - \varepsilon_\infty)\omega_T^2}{\omega_T^2 - \omega^2 + i\Gamma\omega} - \frac{\omega_p^2}{\omega(\omega + i\gamma)}$$

Here, units of inverse centimeters are used instead of radians per second.

The values given in Table I were obtained by using the following fitting parameters: $\varepsilon_\infty = 32.8$ for the high-frequency dielectric constant [8, 19, 20], $\varepsilon_s = 400 \pm 20$ for the static dielectric constant [4, 6], $\omega_T = 32$ cm^{-1} for the transverse optic phonon frequency (which is also in agreement with inelastic neutron scattering data [18]), and $\Gamma = 10$ cm^{-1} for the phonon damping parameter. The room-temperature electron damping parameter in the Drude term was taken to be 30 cm^{-1}. The plasma frequency given by $\omega_p^2 = Ne^2/\varepsilon_0 m_p$ in mks units is determined by the electron concentration N and effective mass m_p. ($\varepsilon_0 = 8.9 \times 10^{-12}$ C/Nm^2 is the permittivity of free space.) We obtain $\omega_p = 1320$ cm^{-1} from the fit with $N = 10^{18}$ cm^{-3}. The relaxation time τ is the reciprocal of γ if we stay in radians per second and γ (cm^{-1}) $= \gamma$ (rad/sec)/$2\pi c$. Experimental and calculated spectra are usually in agreement within experimental and calculated errors. A detailed discussion on improvements of classical oscillator fits has been given by Perkowitz [7].

At frequencies much greater than the longitudinal optic phonon frequency ω_L and the plasma frequency but less than that of the fundamental absorption edge, n can be expressed by

$$[n(\omega)]^2 = \varepsilon_\infty + \Delta\varepsilon(\omega) - \frac{\omega_p^2}{\omega^2} + \left(1 - \frac{\varepsilon_\infty}{\varepsilon_s}\right)\frac{\omega_L^2 \varepsilon_\infty}{\omega^2}. \tag{1}$$

Here, $\Delta\varepsilon$ is a function that describes additional dispersion due to bound carriers at frequencies near the fundamental absorption edge, and Eq. (1) is valid for $(\omega\tau)^2 > 1$ [8, 19].

Lowney and Senturia [8] describe in detail the determination of the dispersion of n with the aid of thin-film transmission and reflectance data exhibiting interference fringes. A carrier-concentration dependence of ε_∞ is related to Im $\varepsilon(\omega)$ by Dionne and Woolley [19] and Moss [30].

$$\varepsilon_\infty = 1 + \frac{2}{\pi} \int_0^\infty \frac{\text{Im } \varepsilon(\omega)}{\omega} d\omega. \tag{2}$$

It turns out that for carrier densities below 10^{19} cm^{-3}, the values of ε_∞ are also independent of electron concentration N.

In the region below and around the fundamental absorption edge, optical constants have been determined and discussed in a number of works [9–11, 21–28]. Early precise measurements on the refractive index in the 0.1–0.5-eV range were performed by Zemel et al. [9] on epitaxial films by using an interferometric and weighing technique with an experimental error in n of about $\pm 3\%$. There exists a peak in the refractive index close to the fundamental band edge (see Table XI). Piccioli et al. [10] repeated the experiments of Zemel et al. [9] but freed the PbTe films ($p \sim 10^{18}$ cm^{-3}) from the substrate and gave data on absorption as well. The error in refractive index is estimated to be $\pm 1\%$ and in absorption index $\pm 5\%$. The maximum values of n fit Zemel et al.'s [9] results, apart from a small shift in wavelength. McCarthy et al. [21] investigated PbTe films on BaF$_2$ substrates in the 5-12-μm range and showed that a significant reduction in absorption occurs below the band gap as compared with PbTe–NaCl films, presumably due to chlorine surface contamination. Whereas the overall agreement of the refractive-index data with those of Piccioli et al. [10] and Zemel et al. [9] are satisfactory, the absorption coefficient data are smaller than those of Piccioli et al. [10] and of Riedl [22] for samples with carrier concentrations of the order of 10^{17} cm^{-3} by about a factor of 10 to 50.

The most recent data on PbTe–BaF$_2$ films are those by Globus et al. [11]. Reflectivity and transmission were measured by using PbTe films (thickness: 0.6 μm) grown on BaF$_2$ substrates in the wavelength region $\lambda = 2$–8 μm. Reflectivity has been determined with an 8° angle of incidence by using samples of comparatively low carrier concentrations (10^{16} cm^{-3}). With expressions for interference and multiple reflection of the complete film–substrate system, a careful analysis of their data and a comparison with previously published data [9, 10, 24, 27] were performed.

The refractive-index data compare favorably with those of Piccioli et al. [10], Zemel et al. [9], and McCarthy et al. [21]; only in the region near the absorption edge are there some differences (see Table XI). Scanlon's data [24] for k are substantially lower (about 40%) than those of Globus et al. [11]. The results by Globus et al. [11] are not influenced by a Burstein–Moss shift at room temperature. However, there is some fine structure in the absorption spectrum, probably due to phonon-assisted transitions. In addition, the authors suggest that apart from transitions at the L point ($\varepsilon_{gL} \sim 0.32$ eV), transitions originate from other close-lying, valence-band extrema presumably in the direction of Λ. The influence of free carriers on the absorption constant is discussed in the framework of several band models (Kane, two-band, six-band, Cohen models) in [11, 27, 28]. Optical constants of the important Pb$_{1-x}$Sn$_x$Te system can be found [31, 32] in the range of the fundamental absorption edge.

The optical constants above the fundamental absorption gap and in the visible and UV region from 0.5 to 25 eV were determined by Kramers–Kronig (KK) analysis of near-normal-incidence reflectance data of bulk [12, 13, 24, 30] and thin-film samples [12] and from measurements of reflectivity and transmission (below 6 eV) of thin films [12] with thicknesses of the order of 300 Å. Wavelength modulation spectroscopy [13, 35] and thermoreflectance [33, 34] have also been used to detect fine structure in the reflectance spectra, at low temperatures [35] that are related to critical points in the joint density of states. Cardona *et al.* [36] used synchrotron radiation to measure reflectivity in the 14–26-eV range.

Since there are systematic differences between the data of Cardona and Greenaway [12], who used cleaved bulk material and epitaxial films, and those obtained by Korn and Braunstein [13], who worked on polished and electrochemically etched samples, we decided to show both sets of data in Table XI and Fig. 10. Korn and Braunstein [13] cite a maximum absolute reflectance error of about 2% throughout the spectral range 0.1–21 eV. Despite some differences in the extrapolations in the KK analysis between the two papers [12, 13], the discrepancy probably results from the different sample preparation.

In the 36–150-eV region, transmission experiments have been performed with thin PbTe films by Cardona and Haensel [14], who used synchrotron radiation. Whereas below 20 eV the structures are due to valence-band to conduction-band transitions, above 20 eV structures appear to be due to transitions from the outermost d electrons [37].

The measurements were performed on films deposited on carbon substrates, and reflectivity measurements could not be performed because the small reflectivity in the 36–150-eV region ($R < 0.01$) would increase the error due to scattered light. Measurements of the thickness were made by using the Tolansky interferometric method with films deposited on glass substrates next to the carbon substrate used for the transmission experiments in the far-UV range. From uncertainties in the thickness evaluation, the absorption index Cardona and Haensel's [14] data are supposed to be affected by an error of about 20%.

Absorption constant and reflectivity data below and above the fundamental gap were recently compiled by Nimtz and Schlicht [38], and Nimtz [39].

REFERENCES

1. R. Dalven, *Infrared Phys.* **9**, 141 (1969).
2. R. Dalven, *Solid State Phys.* **28**, 179 (1973).
3. Yu. I. Ravich, B. A. Efimova, and I. A. Smirnov, *in* "Semiconducting Lead Chalcogenides." Plenum, New York, 1970.
4. G. Bauer, H. Burkhard, W. Jantsch, F. Unterleitner, A. Lopex-Otero, and G. Schleussner, *in* "Lattice Dynamics" (M. Balkanski ed.), p. 669. Flammarion, Paris, 1978.
5. D. D. Buss and M. A. Kinch, *J. Nonmet.* **1**, 111 (1973).
6. H. Burkhard, G. Bauer, and A. Lopez-Otero, *J. Opt. Soc. Am.* **67**, 943 (1979).
7. S. Perkowitz, *Phys. Rev. B* **12**, 3210 (1975).
8. J. R. Lowney and S. D. Senturia, *J. Appl. Phys.* **47**, 1771 (1976).
9. J. N. Zemel, J. D. Jensen, and R. B. Schoolar, *Phys. Rev.* **140**, A330 (1965).
10. N. Piccioli, J. M. Beson, and M. Balkanski, *J. Phys. Chem. Solids*, **35**, 971 (1974).
11. T. R. Globus, B. L. Gelmont, K. I. Geiman, V. A. Kondrashov, and A. V. Matveenko, *Sov. Phys. JETP* **53**, 1000 (1981); T. R. Globus, B. L. Gelmont, K. I. Geiman, V. A. Kondrashov, and A. V. Matveenko, *Zh. Eksp. Teor. Fiz.* **80**, 1926 (1981).

12. M. Cardona and D. L. Greenaway, *Phys. Rev.* **133**, A1685 (1964).
13. D. M. Korn and R. Braunstein, *Phys. Rev. B* **5**, 4837 (1972).
14. M. Cardona and R. Haensel, *Phys. Rev. B* **1**, 2605 (1970).
15. W. E. Tennant and J. A. Cape, *Phys. Rev. B* **13**, 2540 (1976).
16. A. Lopez-Otero, *Thin Solid Films*, **49**, 1 (1978).
17. E. G. Bylander and M. Hass, *Solid State Commun.* **4**, 51 (1966).
18. W. Cochran, R. A. Cowley, G. Dolling, and M. M. Elcombe, *Proc. R. Soc. London* **293**, 433 (1966).
19. G. Dionne and J. C. Woolley, *Phys. Rev. B* **6**, 3898 (1972).
20. E. Burstein, S. Perkowitz, M. H. Brodsky, *J. Phys. C4* **29**, 78 (1968).
21. S. L. McCarthy, W. H. Weber, and M. Mikkor, *J. Appl. Phys.* **45**, 4907 (1974).
22. H. R. Riedl, *Phys. Rev.* **127**, 162 (1962).
23. J. R. Dixon and H. R. Riedl, *Phys. Rev.* **138**, A873 (1965).
24. W. W. Scanlon, *J. Phys. Chem. Solids* **8**, 423 (1959).
25. R. N. Tauber, A. A. Machonis, and I. B. Cadoff, *J. Appl. Phys.* **37**, 4855 (1966).
26. F. Stern, *Phys. Rev.* **133**, A1653 (1964).
27. D. Genzow, A. G. Mironov, and O. Ziep, *Phys. Status Solidi B* **90**, 535 (1978).
28. W. W. Anderson, *Infrared Phys.* **20**, 363 (1980).
29. E. D. Palik, D. L. Mitchell, and J. N. Zemel, *Phys. Rev.* **135**, A763 (1964).
30. T. S. Moss, "Optical Properties of Solids," p. 181. Butterworth, London, 1959.
31. V. M. Walz, thesis, Naval Postgraduate School, Monterey, California (1972); available from NTIS, Springfield, Virginia, order No. AD 767-677.
32. W. G. Opyd, thesis, Naval Postgrade School, Monterey, California (1973); available from NTIS, Springfield, Virginia, order No. AD 767-677.
33. M. Baleva, *Phys. Status Solidi B* **88**, 335 (1978).
34. R. I. Bogacki, A. K. Sood, C. Y. Yang, S. Rabii, and J. E. Fischer, *Surf. Sci.* **37**, 494 (1973).
35. S. E. Kohn, P. Y. Yu, Y. Petroff, Y. R. Shen, Y. Tsang, and M. L. Cohen, *Phys. Rev. B* **8**, 1477 (1973).
36. M. Cardona, C. M. Penchina, E. E. Koch, and P. Y. Yu, *Phys. Status Solidi B* **53**, 327 (1972).
37. D. E. Aspnes and M. Cardona, *Phys. Rev.* **173**, 714 (1968).
38. G. Nimtz and B. Schlicht, *in* "Springer Tracts in Modern Physics" (G. Höhler, ed.), Vol. 98, p. 1. Springer-Verlag, Berlin, 1983.
39. G. Nimtz, *in* "Landolt-Börnstein, Numerical Data and Functional Relationships in Science and Technology" (K.-H. Hellwege and O. Madelung, eds.), Group III, Vol. 17, Subvolume f, pp. 177 and 473. Springer-Verlag, Berlin, 1983.

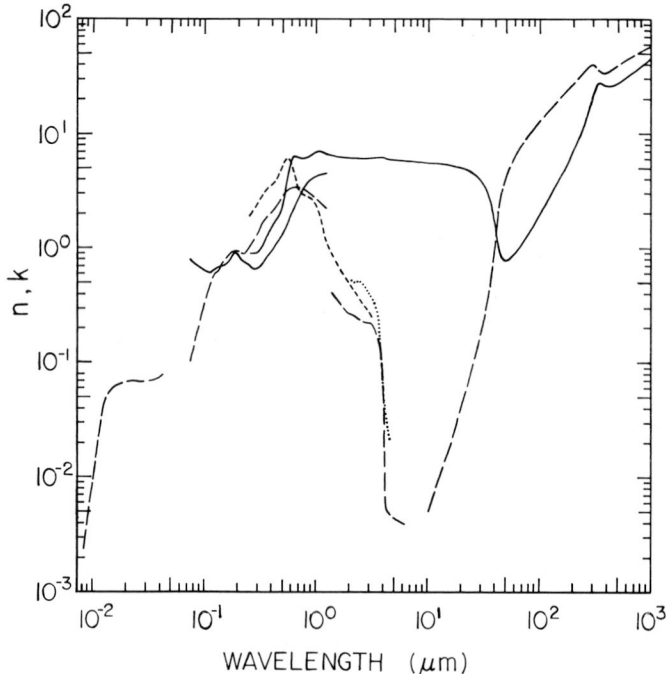

Fig. 10. Log–log plot of n (——) and k (----) versus wavelength in micrometers for lead telluride. Data from various references do not join together in overlap regions on occasion.

TABLE XI

Values of n and k for Lead Telluride Obtained from Various References[a]

eV	cm^{-1}	μm	n	k	n	k
150		0.008266		2.37×10^{-3} [14]		
145		0.008551		3.06		
140		0.008856		4.37		
135		0.009184		5.85		
130		0.009537		7.59		
125		0.009919		9.71×10^{-3}		
120		0.01033		1.31×10^{-2}		
115		0.01078		1.79		
110		0.01127		2.43		
105		0.01181		3.43		
100		0.01240		4.39		
95		0.01305		4.94		
90		0.01378		5.36		
85		0.01459		5.73		
80		0.01550		6.09		
75		0.01653		6.43		
70		0.01771		6.62		
65		0.01907		6.76		
60		0.02066		6.84		
55		0.02254		6.80		
50		0.02480		6.87		
45		0.02755		6.78		
40		0.03100		6.81		
35		0.03542		7.10		
30		0.04133		7.77×10^{-2}		
17		0.07293	0.8 [12]	0.1 [12]		
16.5		0.07514	0.78	0.1		
16		0.07749	0.75	0.1		
15.5		0.07999	0.73	0.15		
15		0.08266	0.72	0.17		
14.5		0.08551	0.70	0.2		
14		0.08856	0.67	0.22		
13.5		0.09184	0.65	0.26		
13		0.09537	0.63	0.3		
12.5		0.09919	0.63	0.35		
12		0.1033	0.62	0.37		
11.5		0.1078	0.60	0.45		
11		0.1127	0.62	0.50		
10.5		0.1181	0.65	0.58		
10		0.1240	0.66	0.60		
9.5		0.1305	0.68	0.62		
9		0.1378	0.68	0.7		
8.5		0.1459	0.7	0.78		
8		0.1550	0.73	0.82		
7.5		0.1653	0.8	0.92		

(*continued*)

TABLE XI (Continued)

Lead Telluride

eV	cm^{-1}	μm	n	k	n	k
7		0.1771	0.9	0.92		
6.5		0.1907	0.9	0.94		
6	48,390	0.2066	0.8	0.92		
5.5	44,360	0.2254	0.75	0.87		
5.0	40,330	0.2480	0.72	1.0	0.90 [13]	1.88 [13]
4.8	38,710	0.2583			0.9	2.02
4.6	37,100	0.2695			0.9	2.16
4.5	36,290	0.2755	0.65	1.22		
4.4	35,490	0.2818			0.9	2.34
4.2	33,880	0.2952			0.91	2.54
4	32,260	0.3100	0.7	1.55	0.96	2.75
3.8	30,650	0.3263			1.05	2.96
3.6	29,040	0.3444			1.2	3.18
3.5	28,230	0.3542	0.86	1.85		
3.4	27,420	0.3647			1.4	3.35
3.2	25,810	0.3875			1.55	3.54
3.0	24,200	0.4133	1.0	2.2	1.65	3.8
2.8	22,580	0.4428			1.8	4.2
2.6	20,970	0.4769			2.14	5.0
2.5	20,160	0.4959	1.35	2.86		
2.4	19,360	0.5166			3.3	5.8
2.2	17,740	0.5636			5.4	5.9
2.0	16,130	0.6199			6.4	4.3
1.8	14,520	0.6888			6.16	3.1
1.6	12,900	0.7749			6.0	2.85
1.5	12,100	0.8266	3.8	3.1		
1.4	11,290	0.8856			6.4	2.74
1.2	9,679	1.033			7	1.8
1.0	8,065	1.240	4.55	2.2	6.5	1.04
0.90	7,259	1.378		0.413 [24]		
0.85	6,856	1.459		0.367		
0.80	6,452	1.550		0.348	6.25	0.71
0.75	6,049	1.653		0.311		
0.70	5,646	1.771		0.289		
0.65	5,243	1.907		0.271		
0.62	5,001	2.000				0.527 [11]
0.61	4,920	2.033				0.529
0.60	4,839	2.066		0.261	6.10 [11]	0.521
					6.10 [13]	0.46 [13]
0.59	4,759	2.101			6.11 [11]	0.516 [11]
0.58	4,678	2.138			6.10	0.508
0.57	4,597	2.175			6.10	0.492
0.56	4,517	2.214			6.10	0.496
0.55	4,436	2.254			6.11	0.505

TABLE XI (*Continued*)

Lead Telluride

eV	cm^{-1}	μm	n	k	n	k
0.54	4,355	2.296			6.10	0.522
0.53	4,275	2.339			6.07	0.523
0.52	4,194	2.384			6.10	0.504
0.51	4,113	2.431			6.10	0.509
0.50	4,033	2.480	5.76	1.4 [12]	6.07	0.507
				0.228 [24]		
0.49	3,952	2.530			6.06	0.505
0.48	3,871	2.583	6.04 [9]		6.05	0.493
0.47	3,791	2.638			6.06	0.465
0.46	3,710	2.695	6.04		6.09	0.447
0.45	3,629	2.755		0.220	6.10	0.438
0.44	3,549	2.818	6.05		6.11	0.427
0.43	3,468	2.883		0.370 [10]	6.14	0.407
0.42	3,388	2.952	6.08	0.374	6.07	0.399
0.41	3,307	3.024	6.06 [10]	0.345	6.07	0.381
0.40	3,226	3.100	6.075	0.331	6.09	0.382
			6.12 [9]	0.220 [24]	6.00 [13]	0.24 [13]
0.39	3,146	3.179			6.10 [11]	0.349 [11]
0.38	3,065	3.263	6.15	0.285 [10]	6.13	0.345
				0.195 [24]		
0.37	2,984	3.351	6.13 [10]	0.284 [10]	6.14	0.320
0.36	2,904	3.444	6.09	0.261	6.15	0.307
			6.16 [9]	0.178 [24]		
0.35	2,823	3.542	6.12 [10]	0.227 [10]	6.17	0.245
				0.188 [24]		0.175 [31]
0.34	2,742	3.647	6.23	0.145	6.17	0.183 [11]
			6.05 [9]			0.151 [31]
0.335	2,702	3.701	6.17 [10]	0.177 [10]		0.134 [11]
0.33	2,662	3.757		0.119 [24]	6.12 [11]	0.149 [31]
0.322	2,597	3.850	6.16	7.81 10^{-2} [10]	6.08	5.50 × 10^{-2} [11]
			5.96 [9]	7.32 [24]		2.75 [31]
0.31	2,500	4.000	6.02 [10]	5.61 [10]	6.02	1.13 [11]
				3.78 [24]		1.13 [31]
				9.55 × 10^{-3} [22]		
0.30	2,420	4.133	5.95	3.55 × 10^{-2} [10]	5.97	7.95 × 10^{-3}
			5.89 [9]	1.23 [24]		
0.29	2,339	4.275		5.2 × 10^{-3}	5.91	7.14
				5.11 [22]		
0.28	2,258	4.428	5.90 [10]	2.49 × 10^{-2} [10]	5.89	1.04 [21]
			5.82 [9]	4.20 × 10^{-3} [24]	5.92 [8]	7.05 [31]
0.27	2,178	4.592	5.88 [10]	2.17 × 10^{-2} [10]	5.86 [11]	6.94
				4.35 × 10^{-3} [24]		
0.26	2,097	4.769	5.85	4.44 [22]	5.85	7.02
			5.82 [9]			

(*continued*)

TABLE XI (*Continued*)

Lead Telluride

eV	cm^{-1}	μm	n	k	n	k
0.25	2,016	4.959	5.84 [10]		5.84	7.18
			5.79 [9]			
0.24	1,936	5.166	5.79		5.83	7.32
0.235	1,895	5.276	5.82 [10]			
0.23	1,855	5.391			5.82	
0.226	1,823	5.486			5.83 [8]	
0.22	1,774	5.636	5.80		5.82 [11]	
			5.76 [9]			6.72 × 10^{-4} [21]
0.21	1,694	5.904	5.78 [10]		5.8	
0.20	1,613	6.199	5.77		5.80	
			5.74 [9]		5.70 [8]	
0.19	1,532	6.526		3.9 × 10^{-3}	5.79 [11]	
0.185	1,492	6.702	5.74			
			5.76 [10]			
0.18	1,452	6.888	5.72 [9]		5.77	1.1 × 10^{-3} [21]
0.175	1,411	7.085			5.69 [8]	
0.17	1,371	7.293			5.77 [11]	
0.16	1,290	7.749	5.70		5.76	8.45 × 10^{-4}
0.155	1,250	7.999	5.72 [9, 10]	5.68 [8]		
0.15	1,210	8.266			5.76 [11]	
0.14	1,129	8.856	5 68 [9]			
0.124	1,000	10.00	5.66 [9, 10]		5.67 [8]	
			5.56 [4–7]	4.97 [15, 17]		
0.1215	980	10.20	5.55	5.29		
0.1190	960	10.42	5.54	5.63		
0.1165	940	10.64	5.54	6.01		
0.1141	920	10.87	5.53	6.42		
0.1116	900	11.11	5.52	6.87		
0.1103	890	11.24	5.66 [8]			
0.1091	880	11.36	5.51 [4–7]	7.37		
0.1066	860	11.63	5.49	7.91		
0.1041	840	11.90	5.48	8.5		
0.1017	820	12.20	5.47	9.16		
0.09199	800	12.50	5.45	9.9 × 10^{-3}		
0.09671	780	12.82	5.44	1.07 × 10^{-2}		
0.09423	760	13.16 ·	5.64 [8]			
			5.51 [9, 10]			
0.09299	750	13.33	5.42 [4–7]	1.16		
0.09175	740	13.51	5.40	1.26		
0.08927	720	13.89	5.38	1.37		
0.08679	700	14.29	5.36	1.50		
0.08431	680	14.71	5.34	1.64		
0.08183	660	15.15	5.31	1.81		
			5.57 [8]			

TABLE XI (*Continued*)

Lead Telluride

eV	cm^{-1}	μm	n	k	n	k
0.07935	640	15.63	5.28 [4–7]	1.99		
0.07687	620	16.13	5.25	2.2		
0.07439	600	16.67	5.22	2.45		
0.07191	580	17.24	5.18	2.73		
0.06943	560	17.86	5.13	3.06		
0.06695	540	18.52	5.09	3.45		
			5.48 [8]			
0.06447	520	19.23	5.03 [4–7]	3.90		
0.06199	500	20.00	4.97	4.44		
0.05951	480	20.83	4.90	5.09		
0.05703	460	21.74	4.82	5.88		
0.05455	440	22.73	4.72	6.85		
0.05207	420	23.81	4.61	8.1		
0.04959	400	25.00	4.50	9.6 × 10^{-2}		
0.04711	380	26.23	4.32	0.12		
0.04463	360	27.78	4.13	0.14		
0.04215	340	29.41	3.89	0.18		
0.03968	320	31.25	3.58	0.23		
0.03720	300	33.33	3.18	0.32		
0.03472	280	35.71	2.62	0.48		
0.03224	260	38.46	1.79	0.87		
0.02976	240	41.67	1.01	1.9		
0.02728	220	45.45	0.80	3.2		
0.02480	200	50.00	0.78	4.4		
0.02232	180	55.56	0.84	5.6		
0.01984	160	62.50	0.95	6.9		
0.1736	140	71.43	1.15	8.5		
0.01488	120	83.33	1.47	10.4		
0.01240	100	100.0	2.00	13.0		
0.01215	98	102.0	2.07	13.3		
0.01190	96	104.2	2.14	13.6		
0.01165	94	106.4	2.22	13.9		
0.01141	92	108.7	2.31	14.3		
0.01116	90	111.1	2.40	14.6		
0.01091	88	113.6	2.50	15.0		
0.01066	86	116.3	2.60	15.4		
0.01041	84	119.0	2.71	15.8		
0.01017	82	122.0	2.83	16.2		
0.009919	80	125.0	2.95	16.6		
0.009671	78	128.2	3.1	17.1		
0.009423	76	131.6	3.2	17.6		
0.009175	74	135.1	3.4	18.1		
0.008927	72	138.9	3.6	18.6		
0.008679	70	142.9	3.7	19.2		

(*continued*)

TABLE XI (*Continued*)

Lead Telluride

eV	cm⁻¹	μm	n	k	n	k
0.008431	68	147.1	3.9	19.8		
0.008183	66	151.5	4.2	20.4		
0.007935	64	156.3	4.4	21.1		
0.007687	62	161.3	4.7	21.8		
0.007439	60	166.7	4.9	22.5		
0.007191	58	172.4	5.3	23.3		
0.006695	54	185.2	6.0	25.1		
0.006447	52	192.3	6.4	26.1		
0.006199	50	200.0	6.9	27.2		
0.005951	48	208.3	7.5	28.4		
0.005703	46	217.4	8.2	29.8		
0.005455	44	227.3	9.1	31.3		
0.005207	42	238.1	10.2	32.9		
0.004959	40	250.0	11.6	34.8		
0.004711	38	263.2	13.6	36.9		
0.004463	36	277.8	16.6	38.8		
0.004215	34	294.1	20.8	39.9		
0.003968	32	312.5	25.3	38.8		
0.003720	30	333.3	27.7	35.7		
0.003472	28	357.1	27.5	33.4		
0.003224	26	384.6	26.5	33.2		
0.002976	24	416.7	26.0	34.5		
0.002728	22	454.5	26.4	36.6		
0.002480	20	500.0	27.6	39.1		
0.002232	18	555.6	29.4	42.0		
0.001984	16	625.0	32.0	45.1		
0.001736	14	714.3	35.4	48.7		
0.001488	12	833.3	39.6	52.8		
0.001240	10	1000	45.1	57.8		

[a] References are indicated in brackets.

Silicon (Si)*

DAVID F. EDWARDS†
University of California
Los Alamos National Laboratory
Los Alamos, New Mexico

Index of refraction values have been reported for the transparent region ($\lambda \geq 1.12$ μm) of silicon by a number of investigators. For these reports the samples have ranged from single cystals to evaporated thin films. The measuring techniques with their inherent differences in accuracy have varied from investigator to investigator as have the measured wavelength ranges. For a given group of investigators using equivalent measurement techniques on samples of about the same quality, the index values disagree by amounts greater than their stated precisions [1–3]. Rather than attempting to resolve this disagreement, we have selected those data by using the criteria given below and present them in Table XII as representative of average values. The data extend from 1.12 to 588 μm and are the results of six investigative groups. Essentially two different measuring techniques were used; angle of minimum deviation for the near infrared and channel spectra for the far infrared. The data-selection criteria are as follows: The samples were single crystal of good or optical quality. The reported refractive-index values were either tabulated or given in a figure which was readable to at least a few places in the third decimal place. We also required agreement in overlap areas to less than five in the third place between different data sets.

Meeting these criteria were the following investigators: Primak [1], who determined the index from 1.12 to 2.16 μm from the angle of minimum deviation measured by an autocollimation technique; Salzberg and Villa [2], who measured the index from 1.3570 to 11.04 μm by a similar autocollimation technique; Edwards and Ochoa [3], who used a channel-spectra technique to measure the index from 2.4373 to 25 μm; Icenogle et al. [4], who used the autocollimator technique to measure the index from 2.554 to 10.270 μm; Loewenstein et al. [5], who determined the far-IR index (28.57–333 μm) from

---

* Work performed under the auspices of the U.S. Department of Energy.
† Present address: Lawrence Livermore National Laboratory, Livermore, California.

547

the channel spectra; and Randall and Rawcliffe [6], who also used the channel-spectra technique to determine the index from 67.8 to 588 μm. Each of these investigators report the index values for room-temperature samples. We have fit these data to both a Herzberger-type and a Sellmeier-type dispersion formula. A total of 184 data points was used.

The Herzberger-type dispersion formula [7] is

$$n = A + BL + CL^2 + D\lambda^2 + E\lambda^4,$$

where $L = 1/(\lambda^2 - 0.028)$ with the wavelength λ in micrometers and 0.028 is the square of the mean asymptote for the short-wavelength abrupt rise in index for 14 materials (silicon included) [7]. For the 184 data points, from 1.12 to 588 μm, the coefficients are $A = 3.41906$, $B = 1.23172 \times 10^{-1}$, $C = 2.65456 \times 10^{-2}$, $D = -2.66511 \times 10^{-8}$, and $E = 5.45852 \times 10^{-14}$. The quality of the fit of reported indices to the dispersion formula is good with differences in the third and fourth decimal places. The pertinent temperature is 26°C.

The Sellmeier-type dispersion formula [8] is

$$n^2 = \varepsilon + A/\lambda^2 + B\lambda_1^2/(\lambda^2 - \lambda_1^2),$$

where $\lambda_1 = 1.1071$ μm for silicon [8]. For the same 184 data points the co-efficients are $\varepsilon = 1.16858 \times 10^1$, $A = 9.39816 \times 10^{-1}$, and $B = 8.10461 \times 10^{-3}$. The quality of the fit is similar to that for the Herzberger-type formula.

In addition to their room-temperature measurements, Icenogle et al. [4] have measured the index for a number of temperatures between about 100 and 296 K for the range from 2.554 to 10.270 μm. For a given wavelength their temperature-dependent indices were fit to a polynomial from which we have evaluated dn/dT. These dn/dT are given in Table XIII for four wavelengths. Loewenstein et al. [5] report the far-infrared index at 1.5 K in addition to 300 K.

Channel spectra used for far-infrared index determination has been described by Randall and Rawcliffe [6] and by Loewenstein et al. [5]. This technique is particularly well suited for the far infrared because of the low order of the interference, typically 20 or 30. For the near infrared, special care must be exercised in using the channel spectra because of the very large fringe order and the difficulty in its determination [9]. A method suggested by Baumeister [10], and Edwards and Ochoa [3] was used to avoid this problem. The Baumeister method has a broad application, and the details are given here.

For the channel-spectra technique, the samples must have faces that are flat to $\lambda/4$ (visible) or better and parallel to a second of arc or better. The flatness and wedge can be the determining factors in the accuracy of the measurement. Fringes of equal chromatic order [11] are produced when radiation undergoes multiple reflections between the parallel surfaces of the sample. The spacing between the fringes depends only on the sample thickness and the real part of the refractive index. Knowing the sample thickness and the wavelength of the fringe, one can, in principle, determine the index. A limitation is knowing the order of the fringe [9]. To overcome this [10], an arbitrary order number is assigned to a fringe and the entire channel spectra is fit to the polynomial

$$m = a_0 + a_1\sigma + a_2\sigma^2 + a_3\sigma^3, \tag{1}$$

where the a_i coefficients are determined by a least-squares fit and m is the order number of the fringes at wave number σ. The slope for this curve at any wave number is

$$dm/d\sigma = a_1 + 2a_2\sigma + 3a_3\sigma^2. \tag{2}$$

In terms of the sample parameters, the order of the fringe is given by

$$m = 2n(\sigma)h\sigma, \tag{3}$$

where $n(\sigma)$ is the refractive index at wave number σ and h the metric thickness. For most materials, such as silicon, with broad transparent regions, the index is a slowly varying function of σ and can accurately be represented as a polynomial

$$n = n_0 + n_1\sigma + n_2\sigma^2, \tag{4}$$

where the n_i are coefficients that can be evaluated from the a_i coefficients and the metric thickness

$$n_0 = a_1/(2h), \qquad n_1 = a_2/(2h), \qquad n_2 = a_3/(2h). \tag{5}$$

The channel spectrum is usually taken in a beam of finite angle, and thus the n_i coefficients must be corrected [6]. In Eq. (5) this beam-angle correction is accomplished by replacing h by $h(1 + \cos\beta_m)/2$, where β_m is the maximum internal angle of a ray with the normal to the surface. The index values evaluated by this Baumeister method are insensitive to the choice of the initial order. The accuracy of the index values is set by the sample thickness, flatness, and parallelism. Good optical practices will ensure sufficient flatness and parallelism. In a metrology laboratory the metric thickness can be determined to a few parts in 10^5, and thus the uncertainty in the index will be determined to first order by the uncertainty in the sample thickness. For index 2 this means an index uncertainty of about $1 - 5 \times 10^{-4}$.

Optical-quality silicon is highly transparent from the band edge at 1.1071 μm to the far infrared with the exception of the 5- to 25-μm lattice absorption band. Outside this lattice band and for $\lambda > 1.1071$ μm, the absorption coefficient has been measured at a few select laser wavelengths. Using a photoacoustic technique, Hordvik and Skolnik [12] have measured the absorption coefficient for silicon at the HF and DF laser wavelengths, 2.7 and 3.8 μm, respectively. Their results are given in Table XII.

Multiphonon absorption limits the transparency in the lattice band of silicon and other wide-gap semiconductors. Because of crystal symmetry, the fundamental vibration in silicon has no dipole moment and is infrared inactive. Lax and Burstein [13] have shown that the phonons can interact directly with the radiation field through terms in the electric moment of second order (or higher) in the atomic displacement. Johnson [14] analyzed the absorption in the 6.7- to 25-μm region for oxygen-free silicon. He verified the absorption to be two phonon. Bendow *et al.* [15] extended these measurements to wider frequency and temperature ranges and found the multiphonon absorption structure to persist to low absorption levels and high temperatures. The work of Johnson [14] and Bendow *et al.* [15] confirms multiphonon interpretation of the 5- to 25-μm band. The absorption coefficient values of Johnson [14] and Bendow *et al.* [15] are given in Table XII for the major absorption peaks and shoulders. These correspond to combinations of four characteristic phonon energies. At wavelengths longer than 28 μm, the values for α of Loewenstein *et al.* [5] are approximately an order of magnitude larger than the extrapolation of the work of Lax and Burstein [13] and Johnson [14]. Therefore, we must omit this work as being anomalous due to surface damage or some other mechanism, as the authors suggest.

Several recent Fourier-transform measurements of n and k (not using Kramers–Kronig analysis) have given smaller values of k [16–19]. The Si used was 10 Ω cm [16–18] or 1000 Ω cm [19] resistivity with carrier type not specified. These showed free-carrier effects below 100 cm^{-1} with n decreasing and k increasing. The plasma frequency for 10 Ω cm n-type material is ~ 4 cm^{-1}, and the carrier density is $\sim 5 \times 10^{14}$ cm^{-3}. The data in the lattice-band region near 600 cm^{-1} in Honijk *et al.* [16] and Passchier *et al.* [17] appear to be in error both in peak value and in shape of bands compared to Lax and Burstein [13] and Johnson [14]. For example, the peak at 606 cm^{-1} is not 17 Np/cm but 9.5 cm^{-1}. Also, there is no band of Si at 650 cm^{-1}. The neper is usually a unit of amplitude attenuation and is $\alpha/2$ in optics language. Therefore, we do not use these data to lower frequency. Since the k values below 100 cm^{-1} varied among the reports of Loewenstein *et al.* [5], Randall and Rawcliffe [6], Honijk *et al.* [16], Passchier *et al.* [17], Birch [18], and Afsar and Hasted [19], we conclude that the free-carrier densities were different. We choose the data of Afsar and Hasted [19] for k below 160 cm^{-1} (measured at 303 K) because the resistivity was the largest, which should minimize the free-carrier effects on k.

Wavelength ellipsometry measurements have been carried out by Aspnes and Theeten [20] to determine n and k in the 1.5–6.0-eV region. Temperature was not stated but was probably 298 K. Samples of resistivity ~ 26 Ω cm with {100}, {111}, and {110} surface orientations were prepared by Syton polishing; and dry thermal oxides were grown, stripped, and regrown so that a study of the transition layer between SiO_2 and Si could be made. The analysis involved mathematical removal of the oxide and its 6-Å transition layer to obtain the final optical constants. The data were combined with ε_2 data of Hulthèn [21] below 2.5 eV to generate the final values we have tabulated.

The Hulthèn work was transmission measurements of epitaxial films of silicon on sapphire substrates, so that surface damage was not a problem as might be the case for Dash and Newman [22], who used mechanically polished thin samples. Temperature was not stated but was probably 298 K. This is fairly important, since the band gap is quite sensitive to temperature [23]. It would also be useful to know whether the entire optical image was focused on the sample (possibly raising the temperature) or if the radiation was predispersed. Strain in the film might also cause a shift in band gap. We obtained n and k from reading graphs [21] and list them so as to get some overlap with data of Aspnes and Theeten [20] near the band edge.

To slightly lower energy (1.0–1.3 eV), the absorption has been measured by McFarlane and Roberts [23] at 290 K for 100-Ω-cm material, mechanically polished. The absorption coefficient is dependent on optical–phonon emission and absorption processes for the indirect transitions, thus being temperature sensitive. Since the data of Hulthèn [21] and McFarlane and Roberts [23] were at 298 and 290 K, respectively, the discrepancy at 1.2 eV is not too surprising (bearing in mind the graph-reading process, also). However, a factor of 5 difference at 1.2 eV is puzzling and leads us to believe that sample surface preparation, the type and magnitude of impurity doping, and possibly strain must produce significant absorption effects near the band edge.

It has been pointed out by Russo [24] through reflection measurements of unpolarized incident radiation that n and k above the fundamental band gap do depend on resistivity (free-carrier density). However, sample resistivity was in the range $10^{-2}\ \Omega$ cm ($n \approx 1 \times 10^{19}\ cm^{-3}$) compared to 95 Ω cm ($\sim 5 \times 10^{13}\ cm^{-3}$) for the effect to be large. For example, at 0.5750 μm k increased by a factor of 3 while n increased by 10%. This is probably a combination of free-carrier absorption coming in from the infrared and the Burstein–Moss effect at the fundamental and other higher-lying band gaps.

Several investigators have measured reflectivity in the visible–ultraviolet region and then determined n and k from a Kramers–Kronig analysis [25–28]. The effect of native oxide on reflectivity measurements in this spectral region is discussed by Philipp [26]. It appears that the samples of Sasaki and Ishiguro [28] were mechanically polished, which would make them suspect at high energies. Philipp [25–27] used chemically prepared surfaces with minimum time spent in air before insertion into the vacuum spectrometer. We have therefore elected to tabulate the Philipp data from 5 to 25 eV overlapping the ellipsometric data for comparison. Interestingly, the data of Philipp [26] and Sasaki and Ishiguro [28] are in reasonable agreement at higher energies, typically $\lesssim 30\%$ for n but $\lesssim 10\%$ for k as well as at lower energy, despite the apparently different surface preparation. The data of Aspnes and Theeten [20] and Philipp [26] are typically within $\lesssim 5\%$ for k and $\lesssim 10\%$ for n in the overlap region 5–6 eV.

For the range 20–95 eV we found data of Hunter [29]. Samples were evaporated Si on glass plates for measurement of n by a critical-angle method; they were freestanding for measurement of k by transmission. While the samples were probably amorphous, we find that the data fit reasonably well between the results of Henke [30] and Philipp [26].

An interpolation has been done between 95 and 99 eV to keep the continuity.

The absorption coefficient near the $L_{2,3}$ absorption edge of Si at ~ 100 eV has been measured [31, 32] with chemically etched and polished samples and with films flash evaporated onto NaCl substrates and floated off; the latter were polycrystalline. A synchroton source was used. Some of the extended fine structure is apparent.

From 99 to 2000 eV we have calculated n and k from a model of Henke et al. [30] as discussed by Lynch and Hunter in the metal section of this handbook. While the calculation cannot account for the fine structure, the magnitude of k is in reasonable agreement with experiment [31, 32].

Very recently, new measurements [33] of ε' and ε'' have been made on n-type Si wafers at 23°C in the millimeter-wave region at 107.3 GHz ($\lambda = 2.796$ mm). The resistivity varied from 5 to 50 Ω cm ($N = 1 \times 10^{15}$ to 1×10^{14} cm^{-3}). A transmission–reflection technique with the sample in free space was used. It was shown that a Drude model fit the data well with

$$\varepsilon' = \varepsilon_\infty \left[1 - \frac{\omega_p^2 \tau^2}{(1 + \omega^2 \tau^2)} \right], \qquad \varepsilon'' = \frac{\varepsilon_\infty \omega_p^2 \tau}{\omega(1 + \omega^2 \tau^2)},$$

where $\omega_p^2 = Ne^2/\varepsilon_0 \varepsilon_\infty m^*$ is the plasma frequency, N is the free-electron density, $\varepsilon_\infty = 11.7$ is the high-frequency dielectric constant, $m^* = 0.26$ is the effective mass, $\tau = 1 \times 10^{-13}$ sec is the constant scattering time, and mks units are used. Since the data fit well over the order-of-magnitude change in N, it is probably safe to extend the calculation of ε' and ε'' at least another order of magnitude in N in both directions to estimate how transparent a sample will be. Also, it is probably safe to calculate the reflectance and transmittance an order of magnitude to either side of 107.3 GHz.

REFERENCES

1. W. Primak, *App. Opt.* **10**, 759 (1971).
2. C. D. Salzberg and J. J. Villa, *J. Opt. Soc. Am.*, **47**, 244 (1957).
3. D. F. Edwards and E. Ochoa, *Appl. Opt.* **19**, 4130 (1980).
4. H. W. Icenogle, B. C. Platt, and W. L. Wolfe, *Appl. Opt.* **15**, 2348 (1976).
5. E. V. Loewenstein, D. R. Smith, and R. L. Morgan, *Appl. Opt.* **12**, 398 (1973).
6. C. M. Randall and R. D. Rawcliffe, *Appl. Opt.* **6**, 1889 (1967).
7. M. Herzberger and C. D. Salzberg, *J. Opt. Soc. Am.* **52**, 420 (1962); M. Herzberger, *Opt. Acta.* **6**, 197 (1959).

8. H. H. Li, *J. Phys. Chem. Ref. Data* **9**, 561 (1980).
9. T. S. Moss, "Optical Properties of Semi-Conductors." Butterworth, London, 1959.
10. P. Baumeister, Optical Coating Laboratory, Inc., private communication (1979).
11. S. Tolansky, *Philos. Mag.* **36**, 225 (1945).
12. A. Hordvik and L. Skolnik, *Appl. Opt.* **16**, 2919 (1977).
13. M. Lax and E. Burstein, *Phys. Rev.* **97**, 39 (1955).
14. F. A. Johnson, *Proc. Phys. Soc. London* **73**, 265 (1959).
15. B. Bendow, H. G. Lipson, and S. P. Yukon, *Appl. Opt.* **16**, 2909 (1977).
16. D. D. Honijk, W. F. Passchier, M. Mandel, and M. N. Afsar, *Infrared Phys.* **17**, 9 (1977).
17. W. F. Passchier, D. D. Honijk, M. Mandel, and M. N. Afsar, *J. Phys. D* **10**, 509 (1977).
18. J. R. Birch, *Infrared Phys.* **18**, 613 (1978).
19. M. N. Afsar and J. B. Hasted, *Infrared Phys.* **18**, 835 (1978).
20. D. E. Aspnes and J. B. Theeten, *J. Electrochem. Soc.* **127**, 1359 (1980); D. E. Aspnes and J. B. Theeten, private communication (1980).
21. R. Hulthèn, *Phys. Scr.* **12**, 342 (1975).
22. W. C. Dash and R. Newman, *Phys. Rev.* **99**, 1151 (1955).
23. G. G. McFarlane and V. Roberts, *Phys. Rev.* **98**, 1865 (1955).
24. O. I. Russo, *J. Electrochem. Soc.* **127**, 953 (1980).
25. H. R. Philipp and E. A. Taft, *Phys. Rev.* **120**, 37 (1960).
26. H. R. Philipp, *J. Appl. Phys.* **43**, 2836 (1972); H. R. Philipp, private communication (1982).
27. H. R. Philipp and H. Ehrenreich, *in* "Semiconductors and Semimetals" (R. K. Willardson and A. C. Beer, eds.), Vol. 2, p. 93. Academic Press, New York, 1967.
28. T. Sasaki and K. Ishiguro, *Phys. Rev.* **127**, 1091 (1962).
29. W. R. Hunter, private communication of unpublished data (1982).
30. B. L. Henke, P. Lee, T. J. Tanaka, R. L. Shimabukuro, and B. K. Fujikawa, "Low Energy X-Ray Diagnostics—1981" (D. T. Attwood and B. L. Henke, eds.), *AIP Conference Proc.*, No. 75, p. 340. American Institute of Physics, New York, 1981.
31. C. Gähwiller and F. C. Brown, *Proc. Int. Conf. Phys. Semicond., 10th, Cambridge, 1970*, p. 213. U.S. Atomic Energy Commission, Oak Ridge, Tennessee, 1970.
32. F. C. Brown, R. Z. Bachrach, and M. Skibowski, *Phys. Rev. B* **15**, 478 (1977).
33. H. T. Kinasewitz and D. Senitzky, *J. Appl. Phys.* **54**, 3394 (1983).

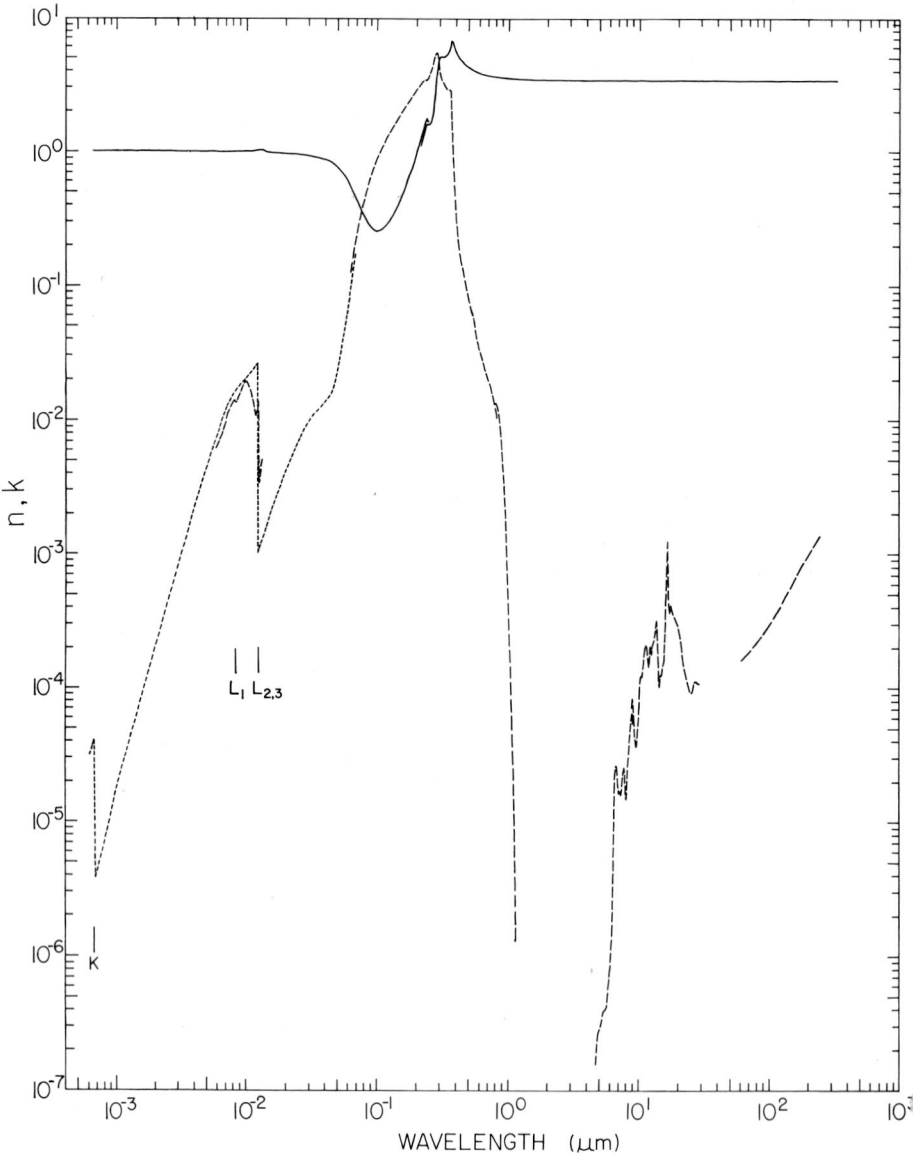

Fig. 11. Log–log plot of n (——) and k (−−−−) versus wavelength in micrometers for silicon. Data from various references sometimes do not join together in overlap regions.

TABLE XII

Values of *n* and *k* for Silicon Obtained from Various References[a]

eV	cm^{-1}	Å	n	k	n	k
2,000		6.199	0.9999048 [30]	3.19×10^{-5} [30]		
1,952		6.351	0.9999131	3.46		
1,905		6.508	0.9999222	3.76		
1,860		6.665	0.9999318	4.08×10^{-5}		
1,815		6.831	0.9999113	3.93×10^{-6}		
1,772		6.997	0.999885	4.32		
1,730		7.166	0.999874	4.74		
1,688		7.345	0.999863	5.20		
1,648		7.523	0.999854	5.70		
1,609		7.705	0.999844	6.24		
1,570		7.897	0.999834	6.85		
1,533		8.087	0.999824	7.49		
1,496		8.287	0.999814	8.22		
1,460		8.492	0.999803	8.99		
1,426		8.694	0.999792	9.81×10^{-6}		
1,392		8.906	0.999781	1.07×10^{-5}		
1,358		9.129	0.999769	1.17		
1,326		9.350	0.999756	1.28		
1,294		9.581	0.999743	1.40		
1,263		9.816	0.999729	1.53		
1,233		10.05	0.999715	1.67		
1,204		10.30	0.999700	1.83		
1,175		10.55	0.999684	2.00		
1,147		10.81	0.999667	2.18		
1,119		11.08	0.999649	2.39		
1,093		11.34	0.999631	2.60		
1,067		11.62	0.999612	2.84		
1,041		11.91	0.999591	3.10		
1,016		12.20	0.999570	3.39		
992		12.50	0.999548	3.70		
968		12.81	0.999524	4.04		
945		13.12	0.999500	4.41		
923		13.43	0.999474	4.80		
901		13.76	0.999447	5.24		
879		14.10	0.999418	5.73		
858		14.45	0.999388	6.25		
838		14.79	0.999358	6.80		
818		15.16	0.999325	7.42		
798		15.54	0.999289	8.11		
779		15.92	0.999253	8.84		
760		16.31	0.999214	9.66×10^{-5}		
742		16.71	0.999174	1.05×10^{-4}		
725		17.10	0.999134	1.14		
707		17.54	0.999088	1.25		
690		17.97	0.999041	1.36		
674		18.39	0.99899	1.48		

(*continued*)

TABLE XII (*Continued*)

Silicon

eV	cm^{-1}	Å	n	k	n	k
658		18.84	0.99894	1.61		
642		19.31	0.99889	1.76		
627		19.77	0.99883	1.91		
612		20.26	0.99878	2.08		
597		20.77	0.99871	2.27		
583		21.27	0.99865	2.47		
569		21.79	0.99858	2.69		
555		22.34	0.99851	2.94		
542		22.87	0.99843	3.19		
529		23.44	0.99836	3.47		
516		24.03	0.99827	3.79		
504		24.60	0.99819	4.11		
492		25.20	0.99810	4.47		
480		25.83	0.99800	4.87		
469		26.43	0.99791	5.27		
458		27.07	0.99781	5.72		
447		27.74	0.99770	6.22		
436		28.44	0.99759	6.77		
426		29.10	0.99748	7.34		
415		29.87	0.99735	8.01		
406		30.54	0.99723	8.63		
396		31.31	0.99710	9.39×10^{-4}		
386		32.12	0.99695	1.02×10^{-3}		
377		32.89	0.99682	1.11		
368		33.69	0.99667	1.20		
359		34.53	0.99651	1.31		
351		35.32	0.99636	1.41		
342		36.25	0.99619	1.53		
334		37.12	0.99602	1.66		
326		38.03	0.99585	1.80		
318		38.99	0.99566	1.95		
311		39.86	0.99549	2.10		
303		40.92	0.99528	2.29		
296		41.88	0.99509	2.47		
289		42.90	0.99489	2.67		
282		43.96	0.99468	2.88		
275		45.08	0.99446	3.13		
269		46.09	0.99427	3.36		
262		47.32	0.99403	3.65		
256		48.43	0.99382	3.93		
250		49.59	0.99361	4.23		
244		50.81	0.99339	4.57		
238		52.09	0.99318	4.94		
232		53.44	0.99296	5.34		
227		54.62	0.99277	5.70		
221		56.10	0.99254	6.19		
216		57.40	0.99236	6.63		

TABLE XII (*Continued*)

Silicon

eV	cm⁻¹	Å	n	k	n	k
211		58.76	0.99219	7.12		
210		59.04				6.30×10^{-3} [31]
206		60.18	0.99206	7.63		
205		60.48				6.59
201		61.68	0.99192	8.18		
200		61.99				6.96
196		63.25	0.99179	8.79		
195		63.58				7.34
191		64.91	0.99169	9.46×10^{-3}		
190		65.26				7.84
187		66.30	0.99167	1.00×10^{-2}		
185		67.02				8.48
182		68.12	0.99164	1.07		
180		68.88				9.21×10^{-3}
178		69.65	0.99164	1.14		
175		70.85				1.01×10^{-2}
174		71.25	0.99167	1.21		
170		72.93	0.99176	1.28		1.14
166		74.69	0.99235	1.45		
165		75.14				1.21
162		76.53	0.99235	1.45		
160		77.49				1.29
158		78.47	0.99287	1.52		
155		79.99				1.38
154		80.51	0.99340	1.57		
152.5		81.30				1.42
150		82.66	0.99384	1.62		1.40
147.5		84.05				1.37
147		84.34	0.99415	1.66		
145		85.51				1.43
143		86.70	0.99460	1.72		
140		88.56	0.99496	1.77		1.54
136		91.16	0.99549	1.83		
135		91.84				1.66
133		93.22	0.99594	1.88		
130		95.37	0.99645	1.94		1.81
127.5		97.24				1.88
127		97.62	0.99705	2.00		
125		99.19				1.91
124		99.98	0.99773	2.06		
121		102.5	0.99856	2.12		
120		103.3				1.85
118		105.1	0.99958	2.19		
115		107.8	1.0008	2.26		1.63
112		110.7	1.0024	2.33		
110		112.7	1.0039	2.39		1.38

(*continued*)

TABLE XII (*Continued*)

Silicon

eV	cm^{-1}	Å	n	k	n	k
107		115.9	1.0070	2.47		
105		118.1				1.08 [32]
104.8		118.3				1.09
104.6		118.5				1.08
104.4		118.8				1.08
104.2		119.0				1.09
104		119.2	1.0132	2.56		1.09
103.8		119.4				1.07
103.6		119.7				1.09
103.4		119.9				1.14
103.2		120.1				1.14
103		120.4				1.12
102.8		120.6				1.16
102.6		120.8				1.17
102.4		121.1				1.13
102.2		121.3				1.13
102		121.6	1.0241	2.62		1.16
101.8		121.8				1.22
101.6		122.0	1.025	2.4×10^{-2}		1.29
101.4		122.3				1.36
101.2		122.5				1.37
101		122.8				1.40
100.8		123.0	1.030	5.0×10^{-3}		1.40
100.6		123.2				1.34
100.4		123.5				1.12
100.2		123.7				1.00×10^{-2}
100		124.0	1.032	1.54		9.7×10^{-3}
99.8		124.2				4.9
99.6		124.5				3.6
99.4		124.7				3.4
99.2		125.0				3.4
99.19		125	1.034 [29]	1.0 [29]		
99.0		125.2				3.4
98.40		126	1.034	1.02		
96.86		128	1.034	1.09		
95.37		130	1.03	1.13		
95		130.5				4.9 [31]
93.93		132	1.022	1.2		
91.17		136	1.008	1.3		
88.56		140	1.000	1.43		
82.66		150	0.993	1.78		
77.49		160	0.991	2.15		
72.93		170	0.988	2.54		
68.88		180	0.985	2.97		
65.26		190	0.982	3.43		
61.99		200	0.978	3.93		
59.04		210	0.976	4.43		

TABLE XII (*Continued*)

Silicon

eV	cm⁻¹	Å	n	k	n	k
56.36		220	0.972	5.0		
53.91		230	0.968	5.53		
51.66		240	0.964	6.1		
49.59		250	0.960	6.7		
47.69		260	0.956	7.3		
45.92		270	0.952	7.85		
44.28		280	0.947	8.4		
42.75		290	0.942	9.0		
41.33		300	0.937	9.5×10^{-3}		
40.00		310	0.930	1.0×10^{-2}		
38.75		320	0.925	1.04		
37.57		330	0.918	1.0		
36.47		340	0.913	1.13		
35.42		350	0.906	1.17		
34.44		360	0.899	1.21		
33.51		370	0.893	1.24		
32.63		380	0.885	1.28		
31.79		390	0.877	1.32		
31.00		400	0.869	1.35		
30.24		410	0.860	1.38		
29.52		420	0.853	1.42		
28.83		430	0.843	1.47		
28.18		440	0.834	1.52		
27.55		450	0.824	1.58		
26.95		460	0.814	1.68		
26.38		470	0.803	1.78		
25.83		480	0.792	1.92		
25.30		490	0.778	2.05		
24.80		500	0.766	2.23		
24.31		510	0.752	2.43		
23.84		520	0.737	2.64		
23.39		530	0.722	2.92		
22.96		540	0.706	3.25		
22.54		550	0.691	3.65		
22.14		560	0.675	4.05		
21.75		570	0.659	4.55		
21.38		580	0.644	5.1		
21.01		590	0.627	5.8		
20.66		600	0.61	6.5		
20.33		610	0.59	7.4		
20.00		620	0.567	8.35×10^{-2}		
			0.569 [26]	0.122 [26]		
19.68		630	0.549 [29]	9.3×10^{-2}		
19.5		635.8	0.542 [26]	0.140 [26]		
10.37		640	0.53 [29]	0.10 [29]		
19.07		650	0.513	0.113		

(*continued*)

TABLE XII (*Continued*)

Silicon

eV	cm⁻¹	μm	n	k	n	k
19.0		0.06526	0.514 [26]	0.163 [26]		
18.5		0.06702	0.485	0.189		
18.0		0.06888	0.455	0.219		
17.75		0.06985	0.440	0.237		
17.5		0.07085	0.426	0.255		
17.25		0.07187	0.411	0.275		
17.0		0.07293	0.397	0.296		
16.80		0.07380	0.386	0.314		
16.7		0.07424	0.379	0.323		
16.6		0.07469	0.374	0.333		
16.5		0.07514	0.369	0.342		
16.25		0.07630	0.357	0.367		
16.0		0.07749	0.345	0.394		
15.75		0.07872	0.333	0.421		
15.5		0.07999	0.323	0.450		
15.25		0.08130	0.313	0.479		
15.0		0.08266	0.304	0.510		
14.75		0.08406	0.296	0.541		
14.5		0.08551	0.288	0.573		
14.25		0.08700	0.281	0.607		
14.0		0.08856	0.275	0.641		
13.75		0.09017	0.269	0.677		
13.5		0.09184	0.265	0.714		
13.25		0.09357	0.261	0.752		
13.0		0.09537	0.258	0.792		
12.75		0.09724	0.256	0.833		
12.5		0.09919	0.255	0.875		
12.25		0.10120	0.256	0.918		
12.0		0.1033	0.257	0.963		
11.75		0.1055	0.259	1.01		
11.5		0.1078	0.263	1.06		
11.25		0.1102	0.267	1.11		
11.0		0.1127	0.272	1.16		
10.75		0.1153	0.278	1.21		
10.5		0.1181	0.286	1.26		
10.25		0.1210	0.295	1.32		
10.0		0.1240	0.306	1.38		
9.75		0.1272	0.318	1.45		
9.5		0.1305	0.332	1.51		
9.25		0.1340	0.348	1.58		
9.0		0.1378	0.367	1.66		
8.75		0.1417	0.389	1.73		
8.5		0.1459	0.414	1.82		
8.25		0.1503	0.444	1.90		
8.0		0.1550	0.478	2.00		
7.75		0.1600	0.517	2.10		
7.5		0.1653	0.563	2.21		
7.25		0.1710	0.618	2.32		

TABLE XII (Continued)

Silicon

eV	cm^{-1}	μm	n	k	n	k
7.0		0.1771	0.682	2.45		
6.75		0.1837	0.756	2.58		
6.5		0.1907	0.847	2.73		
6.25		0.1984	0.968	2.89		
6.0	48,390	0.2066	1.11	3.05	1.010 [20]	2.909 [20]
5.98	48,230	0.2073			1.036	2.928
5.96	48,070	0.2080			1.046	2.944
5.94	47,910	0.2087			1.066	2.937
5.92	47,750	0.2094			1.070	2.963
5.9	47,590	0.2101			1.083	2.982
5.88	47,430	0.2109			1.088	2.987
5.86	47,260	0.2116			1.102	3.005
5.84	47,100	0.2123			1.109	3.015
5.82	46,940	0.2130			1.119	3.025
5.8	46,780	0.2138	1.24	3.18	1.133	3.045
5.78	46,620	0.2145			1.139	3.061
5.76	46,460	0.2153			1.155	3.073
5.74	46,300	0.2160			1.164	3.086
5.72	46,130	0.2168			1.175	3.102
5.7	45,970	0.2175			1.180	3.112
5.68	45,810	0.2183			1.195	3.135
5.66	45,640	0.2191			1.211	3.150
5.64	45,490	0.2198			1.222	3.169
5.62	45,330	0.2206			1.235	3.190
5.6	45,170	0.2214	1.40	3.33	1.247	3.206
5.58	45,010	0.2222			1.265	3.228
5.56	44,840	0.2230			1.280	3.245
5.54	44,680	0.2238			1.299	3.267
5.52	44,520	0.2246			1.319	3.285
5.5	44,360	0.2254	1.51	3.40	1.340	3.302
5.48	44,200	0.2263			1.362	3.319
5.46	44,040	0.2271			1.389	3.334
5.44	43,880	0.2279			1.416	3.350
5.42	43,710	0.2288			1.445	3.359
5.4	43,550	0.2296	1.64	3.44	1.471	3.366
5.38	43,390	0.2305			1.502	3.368
5.36	43,230	0.2313			1.526	3.368
5.34	43,070	0.2322			1.548	3.364
5.32	42,910	0.2331			1.566	3.358
5.3	42,750	0.2339	1.75	3.42	1.579	3.353
5.28	42,590	0.2348			1.585	3.346
5.26	42,420	0.2357			1.590	3.344
5.24	42,260	0.2366			1.591	3.344
5.22	42,100	0.2375			1.592	3.347
5.2	41,940	0.2384	1.78	3.36	1.589	3.354
5.18	41,780	0.2394			1.586	3.363
5.16	41,620	0.2403			1.582	3.376

(continued)

TABLE XII (*Continued*)

Silicon

eV	cm^{-1}	μm	n	k	n	k
					1.579	3.389
5.14	41,460	0.2412			1.573	3.408
5.12	41,300	0.2422			1.571	3.429
5.1	41,130	0.2431	1.72	3.42	1.570	3.451
5.08	40,970	0.2441			1.569	3.477
5.06	40,810	0.2450			1.568	3.504
5.04	40,650	0.2460			1.569	3.533
5.02	40,490	0.2470			1.570	3.565
5.0	40,330	0.2480	1.68	3.58	1.575	3.598
4.98	40,170	0.2490			1.580	3.632
4.96	40,000	0.2500			1.584	3.670
4.94	39,840	0.2510			1.591	3.709
4.92	39,680	0.2520			1.597	3.749
4.9	39,520	0.2530			1.608	3.789
4.88	39,360	0.2541			1.618	3.835
4.86	39,200	0.2551			1.629	3.880
4.84	39,040	0.2562			1.643	3.928
4.82	38,880	0.2572			1.658	3.979
4.8	38,710	0.2583			1.673	4.031
4.78	38,550	0.2594			1.692	4.088
4.76	38,390	0.2605			1.713	4.149
4.74	38,230	0.2616			1.737	4.211
4.72	38,070	0.2627			1.764	4.278
4.7	37,910	0.2638			1.794	4.350
4.68	37,750	0.2649			1.831	4.426
4.66	37,590	0.2661			1.874	4.506
4.64	37,420	0.2672			1.927	4.590
4.62	37,260	0.2684			1.988	4.678
4.6	37,100	0.2695			2.059	4.764
4.58	36,940	0.2707			2.140	4.849
4.56	36,780	0.2719			2.234	4.933
4.54	36,620	0.2731			2.339	5.011
4.52	36,460	0.2743			2.451	5.082
4.5	36,290	0.2755			2.572	5.148
4.48	36,130	0.2768			2.700	5.206
4.46	35,970	0.2780			2.833	5.257
4.44	35,810	0.2792			2.974	5.304
4.42	35,650	0.2805			3.120	5.344
4.4	35,490	0.2818			3.277	5.381
4.38	35,330	0.2831			3.444	5.414
4.36	35,170	0.2844			3.634	5.435
4.34	35,000	0.2857			3.849	5.439
4.32	34,840	0.2870			4.086	5.395
4.3	34,680	0.2883			4.318	5.301
4.28	34,520	0.2897			4.525	5.158
4.26	34,360	0.2910			4.686	4.989
4.24	34,200	0.2924			4.805	4.812
4.22	34,040	0.2938			4.888	4.639
4.2	33,880	0.2952				

TABLE XII (*Continued*)

Silicon

eV	cm^{-1}	μm	n	k	n	k
4.18	33,710	0.2966			4.941	4.480
4.16	33,550	0.2980			4.977	4.335
4.14	33,390	0.2995			4.999	4.204
4.12	33,230	0.3009			5.012	4.086
4.1	33,070	0.3024			5.020	3.979
4.08	32,910	0.3039			5.021	3.885
4.06	32,750	0.3054			5.020	3.798
4.04	32,580	0.3069			5.018	3.720
4.02	32,420	0.3084			5.015	3.650
4.0	32,260	0.3100			5.010	3.587
3.98	32,100	0.3115			5.009	3.529
3.96	31,940	0.3131			5.010	3.477
3.94	31,780	0.3147			5.009	3.429
3.92	31,620	0.3163			5.012	3.386
3.9	31,460	0.3179			5.016	3.346
3.88	31,290	0.3195			5.021	3.310
3.86	31,130	0.3212			5.029	3.275
3.84	30,970	0.3229			5.040	3.242
3.82	30,810	0.3246			5.052	3.211
3.8	30,650	0.3263			5.065	3.182
3.78	30,490	0.3280			5.079	3.154
3.76	30,330	0.3297			5.095	3.128
3.74	30,160	0.3315			5.115	3.103
3.72	30,000	0.3333			5.134	3.079
3.7	29,840	0.3351			5.156	3.058
3.68	29,680	0.3369			5.179	3.039
3.66	29,520	0.3388			5.204	3.021
3.64	29,360	0.3406			5.231	3.007
3.62	29,200	0.3425			5.261	2.995
3.6	29,040	0.3444			5.296	2.987
3.58	28,870	0.3463			5.336	2.983
3.56	28,710	0.3483			5.383	2.984
3.54	28,550	0.3502			5.442	2.989
3.52	28,390	0.3522			5.515	2.999
3.5	28,230	0.3542			5.610	3.014
3.48	28,070	0.3563			5.733	3.026
3.46	27,910	0.3583			5.894	3.023
3.44	27,750	0.3604			6.089	2.982
3.42	27,580	0.3625			6.308	2.881
3.4	27,420	0.3647			6.522	2.705
3.38	27,260	0.3668			6.695	2.456
3.36	27,100	0.3690			6.796	2.169
3.34	26,940	0.3712			6.829	1.870
3.32	26,780	0.3734			6.799	1.577
3.3	26,620	0.3757			6.709	1.321
3.28	26,450	0.3780			6.585	1.110
3.26	26,290	0.3803			6.452	0.945

(*continued*)

TABLE XII (*Continued*)

Silicon

eV	cm^{-1}	μm	n	k	n	k
3.24	26,130	0.3827			6.316	0.815
3.22	25,970	0.3850			6.185	0.714
3.2	25,810	0.3875			6.062	0.630
3.18	25,650	0.3899			5.948	0.561
3.16	25,490	0.3924			5.842	0.505
3.14	25,330	0.3949			5.744	0.456
3.12	25,160	0.3974			5.654	0.416
3.1	25,000	0.4000			5.570	0.387
3.08	24,840	0.4025			5.493	0.355
3.06	24,680	0.4052			5.420	0.329
3.04	24,520	0.4078			5.349	0.313
3.02	24,360	0.4105			5.284	0.291
3.0	24,200	0.4133			5.222	0.269
2.98	24,040	0.4161			5.164	0.255
2.96	23,870	0.4189			5.109	0.244
2.94	23,710	0.4217			5.058	0.228
2.92	23,550	0.4246			5.009	0.211
2.9	23,390	0.4275			4.961	0.203
2.88	23,230	0.4305			4.916	0.194
2.86	23,070	0.4335			4.872	0.185
2.84	22,910	0.4366			4.831	0.185
2.82	22,740	0.4397			4.791	0.170
2.8	22,580	0.4428			4.753	0.163
2.78	22,420	0.4460			4.718	0.149
2.76	22,260	0.4492			4.682	0.149
2.74	22,100	0.4525			4.648	0.133
2.72	21,940	0.4558			4.615	0.131
2.7	21,780	0.4592			4.583	0.130
2.68	21,660	0.4626			4.553	0.131
2.66	21,450	0.4661			4.522	0.134
2.64	21,290	0.4696			4.495	0.120
2.62	21,130	0.4732			4.466	0.120
2.6	20,970	0.4769			4.442	0.090
2.58	20,810	0.4806			4.416	0.094
2.56	20,650	0.4843			4.391	0.083
2.54	20,490	0.4881			4.367	0.079
2.52	20,330	0.4920			4.343	0.077
2.5	20,160	0.4959			4.320	0.073
2.48	20,000	0.4999			4.298	0.073
2.46	19,840	0.5040			4.277	0.066
2.44	19,680	0.5081			4.255	0.072
2.42	19,520	0.5123			4.235	0.060
2.4	19,360	0.5166			4.215	0.060
2.38	19,200	0.5209			4.196	0.056
2.36	19,030	0.5254			4.177	0.053
2.34	18,870	0.5299			4.159	0.043
2.32	18,710	0.5344			4.140	0.045

TABLE XII (*Continued*)

Silicon

eV	cm^{-1}	μm	n	k	n	k
2.3	18,550	0.5391			4.123	0.048
2.28	18,390	0.5438			4.106	0.044
2.26	18,230	0.5486			4.089	0.044
2.24	18,070	0.5535			4.073	0.032
2.22	17,910	0.5585			4.057	0.038
2.2	17,740	0.5636			4.042	0.032
2.18	17,580	0.5687			4.026	0.034
2.16	17,420	0.5740			4.012	0.030
2.14	17,260	0.5794			3.997	0.027
2.12	17,100	0.5848			3.983	0.030
2.1	16,940	0.5904			3.969	0.030
2.08	16,780	0.5961			3.956	0.027
2.06	16,610	0.6019			3.943	0.025
2.04	16,450	0.6078			3.931	0.025
2.02	16,290	0.6138			3.918	0.024
2.0	16,130	0.6199		2.96×10^{-2} [21]	3.906	0.022
1.98	15,970	0.6262			3.893	0.022
1.96	15,810	0.6326			3.882	0.019
1.94	15,650	0.6391			3.870	0.018
1.92	15,490	0.6458			3.858	0.017
1.9	15,320	0.6526		2.4	3.847	0.016
1.88	15,160	0.6595			3.837	0.016
1.86	15,000	0.6666			3.826	0.015
1.84	14,840	0.6738			3.815	0.014
1.82	14,680	0.6812			3.805	0.013
1.8	14,520	0.6888		2.25	3.796	0.013
1.78	14,360	0.6965			3.787	0.013
1.76	14,200	0.7045			3.778	0.012
1.74	14,030	0.7126			3.768	0.011
1.72	13,870	0.7208			3.761	0.011
1.7	13,710	0.7293		1.97	3.752	0.010
1.68	13,550	0.7380			3.745	0.010
1.66	13,390	0.7469			3.736	0.009
1.64	13,230	0.7560			3.728	0.009
1.62	13,070	0.7653			3.721	0.008
1.6	12,900	0.7749		1.42	3.714	0.008
1.58	12,740	0.7847			3.705	0.007
1.56	12,580	0.7948			3.697	0.007
1.54	12,420	0.8051			3.688	0.006
1.52	12,260	0.8157			3.681	0.006
1.5	12,100	0.8266		1.25×10^{-2}	3.673	0.005
1.4	11,290	0.8856		7.75×10^{-3}		
1.3	10,490	0.9537		2.26		
1.28	10,320	0.9686				6.4×10^{-4} [23]
1.26	10,160	0.9840				5.1
1.24	10,000	0.9999				4.0

(*continued*)

TABLE XII (*Continued*)

Silicon

eV	cm^{-1}	μm	n	k	n	k
1.22	9,840	1.016				2.8
1.2	9,679	1.033		1.07		1.8
1.18	9,518	1.051				1.2×10^{-4}
1.16	9,356	1.069				6.7×10^{-5}
1.14	9,195	1.088				4.2
1.12	9,034	1.107				2.5
1.107	8,929	1.12	3.5361 [1, 2, 3, 8]			
1.10	8,872	1.127				1.3×10^{-5}
1.084	8,741	1.144	3.5295			
1.08	8,711	1.148				5.8×10^{-6}
1.06	8,550	1.170				1.5
1.033	8,333	1.20	3.5193			
0.9037	7,289	1.372	3.5007			
0.8856	7,143	1.4	3.4876			
0.8093	6,527	1.532	3.4784			
0.7749	6,250	1.6	3.4710			
0.7310	5,896	1.696	3.4644			
0.6888	5,556	1.8	3.4578			
0.6199	5,000	2.0	3.4490			
0.5087	4,103	2.4373	3.4434			
0.4959	4,000	2.50	3.4424			
0.4568	3,684	2.7144	3.4393	2.5×10^{-9} [12]		
0.4133	3,333	3.00	3.4361			
0.3753	3,027	3.3033	3.4335			
0.3626	2,925	3.4188	3.4327			
0.3542	2,857	3.50	3.4321			
0.3263	2,632	3.80		1.3×10^{-8}		
0.3100	2,500	4.00	3.4294			
0.2480	2,000	5.00	3.4261	1.99×10^{-7} [15]		
0.2418	1,950	5.128		2.82×10^{-7}		
0.2356	1,900	5.263		2.97		
0.2294	1,850	5.405		3.26		
0.2232	1,800	5.556		3.94		
0.2170	1,750	5.714		4.05		
0.2108	1,700	5.882		4.17		
0.2066	1,667	6.00	3.4242			
0.2046	1,650	6.061		5.64		
0.1984	1,600	6.250		8.46×10^{-7}		
0.1922	1,550	6.452		1.74×10^{-6}		
0.1860	1,500	6.667		2.66×10^{-6}		
0.1798	1,450	6.897		2.45×10^{-5}		
0.1771	1,492	7.00	3.4231			
0.1736	1,400	7.143		1.68		
				1.87 [14]		
0.1674	1,350	7.407		1.62 [15]		
0.1612	1,300	7.692		2.49		
				2.82 [14]		

TABLE XII (*Continued*)

Silicon

eV	cm^{-1}	μm	n	k	n	k
0.1550	1,250	8.000	3.4224	1.53 [15]		
0.1488	1,200	8.333		2.41		
0.1426	1,150	8.696		2.52 [14]		
0.1401	1,130	8.850		5.13 [15]		
0.1395	1,125	8.889		7.11 [14]		
0.1389	1,120	8.929		6.44 [15]		
0.1378	1,111	9.00	3.4219	8.38 [14]		
0.1376	1,110	9.009				
0.1364	1,100	9.091		8.46		
				5.49 [15]		
0.1339	1,080	9.259		7.38 [14]		
0.1314	1,060	9.434		4.72		
0.1289	1,040	9.615		3.75		
0.1265	1,020	9.804		3.67		
0.1240	1,000	10.00	3.4215	4.84		
0.1215	980	10.20		6.76 × 10^{-5}		
0.1203	970	10.31		1.09 × 10^{-4}		
0.1190	960	10.42		1.22		
0.1178	950	10.53		1.24		
0.1165	940	10.64		1.22		
0.1141	920	10.87		1.27		
0.1116	900	11.11		1.66		
0.1103	890	11.24		2.02		
0.1091	880	11.36		2.08		
0.1079	870	11.49		2.02		
0.1066	860	11.63		2.01		
0.1054	850	11.76		1.73		
0.1041	840	11.90		1.53		
0.1029	830	12.05		1.44		
0.1017	820	12.20		1.77		
0.1004	810	12.35		2.06		
0.09919	800	12.50		1.97		
0.09795	790	12.66		1.77		
0.09671	780	12.82		2.07		
0.09547	770	12.99		2.14		
0.09423	760	13.16		2.27		
0.09299	750	13.33		2.39		
0.09184	740.7	13.50	3.4209	2.59		
0.09175	740	13.51				
0.09051	730	13.70		3.12		
0.08927	720	13.89		3.03		
0.08803	710	14.08		2.21		
0.08679	700	14.29		1.57		
0.08555	690	14.49		1.02		
0.08550	689.7	14.50	3.4208	1.12		

(*continued*)

TABLE XII (*Continued*)

Silicon

eV	cm^{-1}	μm	n	k	n	k
0.08431	680	14.71		1.25		
0.08307	670	14.93		1.24		
0.08265	666.7	15.00	3.4207			
0.08183	660	15.15		1.41		
0.08059	650	15.38		1.52		
0.07935	640	15.63		1.64		
0.07811	630	15.87		5.05		
0.07687	620	16.13		9.63×10^{-4}		
0.07563	610	16.39		1.11×10^{-3}		
0.07513	608	16.50		1.25×10^{-3}		
0.07439	600	16.67		7.16×10^{-4}		
0.07315	590	16.95		4.18		
0.07191	580	17.24		3.65		
0.07067	570	17.54		4.15		
0.06943	560	17.86		3.84		
0.06695	540	18.52		3.40		
0.06447	520	19.23		3.18		
0.06199	500	20.00	3.4204	2.86		
0.05951	480	20.83		2.22		
0.05703	460	21.74		1.56		
0.05455	440	22.73		1.27		
0.05207	420	23.81		1.06×10^{-4}		
0.04959	400	25.00	3.4201	9.15×10^{-5}		
0.04835	390	25.64		9.39×10^{-5}		
0.04711	380	26.32		1.05×10^{-4}		
0.04587	370	27.03		1.12		
0.04463	360	27.78		1.08		
0.04339	350	28.57				
0.04215	340	29.41		1.08		
0.04092	330	30.30	3.4200 [5, 6]			
0.03720	300	33.33				
0.03100	250	40.00	3.4199			
0.02480	200	50.00	3.4197			
0.01984	160	62.50	3.4195	1.7 [19]		
0.01736	140	71.43	3.4192	1.9		
0.01488	120	83.33	3.4190	2.3		
0.01364	110	90.91	3.4188			
0.01240	100	100.0	3.4185	2.9		
0.01116	90	111.1		3.5		
0.009919	80	125.0	3.4180	4.3		
0.008679	70	142.9		5.5		
0.007430	60	166.7	3.4170	7.2×10^{-4}		
0.006199	50	200.0	3.4165	1.0×10^{-3}		
0.004959	40	250.0	3.4160	1.4		
0.003720	30	333.3	3.4155			

[a] References are indicated in brackets.

TABLE XIII

Temperature Dependence of Index of Refraction[a]

λ (μm)	B_1	B_2	B_3
2.554	-1.1127×10^{-4}	2.0722×10^{-6}	-4.0×10^{-9}
2.732	-1.0255×10^{-5}	9.6748×10^{-7}	-1.2145×10^{-9}
5.190	-4.0089×10^{-5}	1.3368×10^{-6}	-2.3458×10^{-9}
10.270	-6.5198×10^{-5}	1.6358×10^{-6}	-3.2457×10^{-9}

[a] dn/dT for $T = 100$ to 296 K; $dn/dT = B_1 + B_2 T + B_3 T^2$.

Silicon (Amorphous) (a-Si)

H. PILLER
Department of Physics and Astronomy
Louisiana State University
Baton Rouge, Louisiana

The optical properties of amorphous silicon (a-Si) are sensitive to preparation conditions and to doping with hydrogen. Surface conditions and oxide films on the surface affect the optical measurements strongly, especially in the ultraviolet (UV). The optical properties of a-Si are also affected by the amount of disorder in the samples. Disorder is determined by substrate temperature, deposition rate, deposition method, impurities, vacuum conditions, annealing conditions, and environmental conditions. The mismatch of the expansion coefficients of the sample and the substrate introduces stress that can cause grain boundaries and voids in the sample. Annealing of the samples will reduce the strain but will generally increase the surface roughness. The a-Si samples should never be exposed to air before making the optical measurements. A measurement of the surface roughness should be performed, since specular reflectance measurements need to be corrected for imperfect (rough) surfaces. X-ray and low-energy electron-diffraction (LEED) measurements should be made to investigate the amorphous and surface properties of the samples.

Pierce and Spicer [1] measured reflectance R from 0.4 to 11.8 eV. Kramers–Kronig (KK) analysis was used to calculate the optical constants. The samples were evaporated in a vacuum of less than 5×10^{-6} torr by using low evaporation rates of 2 to 5 Å/sec and large evaporator-to-substrate distance (~ 40 cm). Silicon single crystals (1000 Ω cm) were used as substrates. The R measurements were made *in situ* on a 600-Å-thick film. The scattering in the R data indicated a relative measurement accuracy of $\pm 1\%$. The optical constants above 12 eV were derived from an extrapolation of the measured R and transmittance T and are less accurate. A zero-frequency refractive index n of 3.4, identical to that in crystalline Si (c-Si), was obtained. The maximum of the extinction coefficient k occurs at 4 eV compared to a maximum in k at 4.3 eV in c-Si. The data are listed in Table XIV and shown in Fig. 12 for both

571

n and *k*. Weiser *et al.* [2] measured *R* of amorphous glow-discharge Si (g-Si) in the energy range of up to 10 eV. The g-Si films were deposited on quartz substrates in a rf discharge in pure silane at 0.2 mbar. The deposition rate was 2 Å/sec. The H:Si ratio in the samples varied as a function of deposition temperature (room temperature to 400°C) between 50 and 4%. The optical constants quoted in Table XIV and shown in Fig. 12 were determined from the measurements on a sample with a substrate temperature of 400°C. Here, *R* of g-Si shows a strong dependence on deposition temperature but a small dependence on annealing.

Brodsky *et al.* [3] measured the optical constants in a-Si films deposited on single-crystal sapphire substrates. The films were deposited by rf sputtering of c-Si onto a room-temperature substrate in an Ar atmosphere at a pressure of 0.01 torr. The deposition rates varied between 200 and 600 Å/min. Thicknesses of the films varied from 0.3 to 10 μm and were measured to within ±10%. The films were stored in air. The data quoted in the table and shown in the figure were determined on a room-temperature sample. The *k* values agree very well with the data of Pierce *et al.* [1].

Zanzucchi *et al.* [4] measured the absorption in g-Si. Films prepared from silane contain up to an estimated 15–20 at. % of H. The films were deposited by either dc or rf glow discharge in silane at a pressure of 0.5 to 2.0 torr of silane and at deposition rates in the range 0.1–0.5 μm/min. It was found that *n* of g-Si is within 10 to 20% of the c-Si values. Vibration-band absorption due to bonded hydrogen was observed in g-Si associated with two modes, one at 2050 cm^{-1} and the other one at 635 cm^{-1}. These modes are associated with the Si-H stretching mode and the low-energy wagging mode, respectively. The data given in Table XIV and shown in Fig. 12 are for rf g-Si deposited at 215 and 300°C, respectively. The *k* values of rf g-Si, with substrate temperature of 420°C, agree very well with the values of Weiser *et al.* [2] in the band-edge region. Here, the absorption coefficient was determined from the amplitude of interference fringes in 1- to 3-μm-thick films. The measurement accuracy was ±(100–200) cm^{-1}, which at 0.6 μm translates into ±(4.7–9.4) × 10^{-4} for *k*. The experimental data show that the type of discharge is an important factor in determining the optical properties of g-Si. The Si-H bond was shown to be stable to thermal treatments.

Moddel *et al.* [5] determined the low-energy absorption in a-Si:H. The measurements were made on a 10-μm-thick sputtered film. The range of photon energy *hv* was extended to lower energy by making use of photoconductivity (PC) data. The samples were made by rf sputtering of c-Si onto glass substrates in an Ar-H atmosphere. Interference fringes in the PC resulting from front and back surface reflections were not observed because of the low resolution of the spectrometer. Optical constants are shown in Fig. 12 and are listed in Table XIV.

Evangelisti *et al.* [6] determined the absorption coefficient α of a-Si:H from the spectral dependence of the PC over four decades. The samples were grown by rf glow discharge in a capacitive reactor in pure silane or in a silane–argon mixture. Film thicknesses ranged between 0.5 and 1 μm; the hydrogen content varied between 9 and 17%. The exponential dependence of the PC on photon flux and hv was studied, and it was found that the exponent β varies with hv. Value α was then evaluated by means of a recurrent expression. The optical constants presented in Table XIV and shown in Fig. 12 were determined by this procedure. Transmission data were used to normalize α deduced from PC.

Freeman and Paul [7] studied the transmission spectra of a-Si:H as a function of H content and substrate temperature. The samples were prepared by rf sputtering of a c-Si target in an Ar and H plasma. Corning 7059 glass was used as a substrate. The deposition rate was 1 Å/sec and the Ar pressure was 5×10^{-3} torr. The hydrogenated sample (22 at. %) shows a lower n for hv between 0.04 and 0.50 eV; n decreases with increasing H pressure; n of a-Si and a-Si:H show very little hv dependence for $hv < 0.5$ eV. Resonance effects due to Si–H vibrational modes are observed in n of a-Si:H. Here, n of a-Si:H is about 25% less than n of a-Si. The error in n is ± 4%. An increase in the H content in the film shifts the absorption edge to larger hv and also causes a decrease of the density. Refractive indices are shown in Fig. 12 and are listed in Table XIV.

Brodsky and Lurio [8] measured the IR vibrational spectra of a-Si between 35 and 700 cm^{-1}. The absorption spectrum is interpreted in terms of a disorder-induced breakdown of the selection rules for vibronic transitions. Wedged c-Si substrates were used to suppress internal multiple-reflection effects in the substrate. Interference effects within the thin a-Si films are negligibly small because of the low reflection coefficient at the a-Si to c-Si interface (approximately 0.03). Extinction coefficients are shown in Fig. 12 and are listed in Table XIV.

Brown and Rustgi [9] studied the soft x-ray absorption spectra of a-Si. The spectrum for a-Si shows none of the detailed structure observed in c-Si. Extinction coefficients are shown in Fig. 12 and are listed in Table XIV.

Jackson and Amer [10] measured α at the absorption edge in a-Si:H down to $hv = 0.6$ eV by using the photothermal deflection spectroscopy (PDS), which measures directly the optical absorption and is insensitive to scattering. No optical constants are listed here, however.

Additional work is discussed by Chaleravertz and Kaplan [11].

REFERENCES

1. D. T. Pierce and W. E. Spicer, *Phys. Rev. B* **5**, 3017 (1972).
2. G. Weiser, D. Ewald, and M. Milleville, *J. Non-Cryst. Solids* **35/36**, 447 (1980).
3. M. H. Brodsky, R. S. Title, K. Weiser, and G. D. Pettit, *Phys. Rev. B* **1**, 2632 (1970).
4. P. J. Zanzucchi, C. R. Wronski, and D. E. Carlson, *J. Appl. Phys.* **48**, 5227 (1977).
5. G. Moddel, D. A. Anderson, and W. Paul, *Phys. Rev. B* **22**, 1918 (1980).
6. F. Evangelisti, P. Fiorini, G. Fortunato, A. Frova, C. Giovannella, and R. Peruzzi, *J. Non-Cryst. Solids* **55**, 191 (1983).

7. E. C. Freeman and W. Paul, *Phys. Rev. B* **20**, 716 (1979).
8. M. H. Brodsky and A. Lurio, *Phys. Rev. B* **9**, 1646 (1974).
9. F. C. Brown and O. P. Rustgi, *Phys. Rev. Lett.* **28**, 497 (1972).
10. B. Jackson and N. M. Amer, *Phys. Rev. B* **25**, 5559 (1982).
11. B. K. Chakravertz and D. Kaplan, *J. de Phys.* **42**, Colloque C-4 (1981).

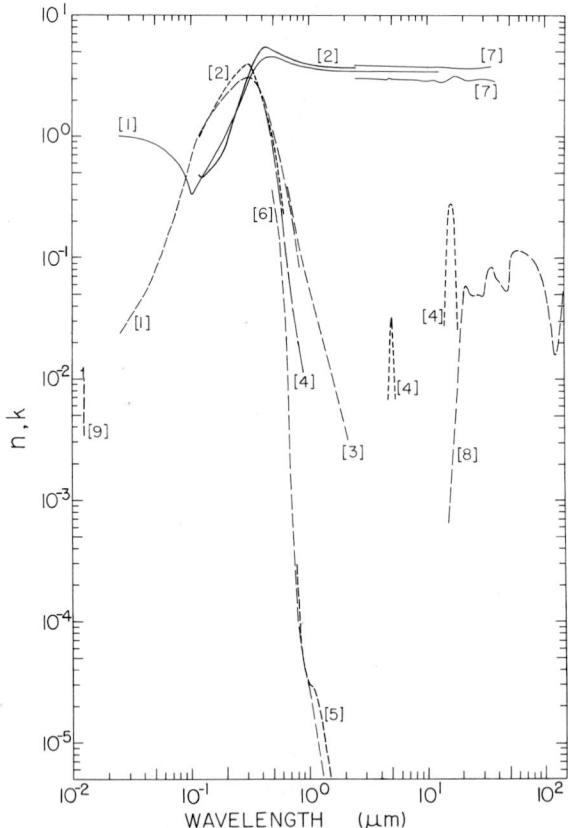

Fig. 12. Log–log plot of n (——) and k (----) versus wavelength in micrometers for amorphous silicon. Various references are indicated in brackets.

TABLE XIV

Values of *n* and *k* for Amorphous Silicon Obtained from Various References[a]

eV	Å	cm⁻¹	n	k
105.0	118.1			1.16×10^{-2} [9]
104.5	118.6			1.13
103.9	119.3			1.12
103.1	120.2			1.11
102.6	120.8			1.11
102.0	121.5			1.13
101.7	122.0			1.14
101.3	122.4			1.16
101.0	122.8			1.19
100.8	123.0			1.22
100.6	123.2			1.23
100.5	123.4			1.19
100.45	123.43			1.12
100.4	123.5			1.05×10^{-2}
100.3	123.6			9.99×10^{-3}
100.2	123.7			9.99
100.1	123.9			9.94
99.93	124.1			9.03
99.83	124.2			7.40
99.72	124.3			5.78
99.68	124.4			4.57
99.45	124.7			3.68
99.13	125.1			3.42
98.86	125.4			3.37
50	248.0		0.995 [1]	2.35×10^{-2} [1]
49	253.0		0.992	2.42
48	258.3		0.990	2.50
47	263.8		0.988	2.58
46	269.5		0.986	2.67
45	275.5		0.984	2.75

(continued)

TABLE XIV (*Continued*)

Amorphous Silicon

eV	Å	cm⁻¹	n	k
44	281.8		0.981	2.85
43	288.3		0.979	2.94
42	295.2		0.976	3.04
41	302.4		0.973	3.14
40	310.0		0.970	3.26
39	317.9		0.967	3.37
38	326.3		0.963	3.50
37	335.1		0.959	3.64
36	344.4		0.955	3.79
35	354.2		0.950	3.95
34	364.7		0.945	4.13
33	375.7		0.939	4.33
32	387.5		0.933	4.55
31	400.0		0.926	4.80
30	413.3		0.918	5.08
29	427.5		0.909	5.41
28	442.8		0.899	5.79
27	459.2		0.888	6.23
26	476.9		0.876	6.75
25	495.9		0.862	7.37
24	516.6		0.846	8.12
23	539.1		0.828	9.03×10^{-2}
22	563.6		0.808	0.101
21	590.4		0.785	0.115
20	619.9		0.758	0.132
19	652.6		0.727	0.154
18	688.8		0.691	0.181
17	729.3		0.651	0.216
16	774.9		0.603	0.261
15	826.6		0.549	0.321
14	885.6		0.485	0.403
13	953.7		0.410	0.519

eV	μm	cm⁻¹	n	k	n(H)	k(H)	k(H)
12	0.1033		0.327	0.726			
11.5	0.1078		0.363	0.847			
11	0.1127		0.392	0.946			
10.5	0.1181		0.423	1.04	0.488 [2]	1.00 [2]	
10	0.1240		0.459	1.14	0.466	1.09	
9.5	0.1305		0.497	1.24	0.467	1.20	
9	0.1378		0.543	1.35	0.499	1.33	
8.5	0.1459		0.597	1.47	0.515	1.49	
8	0.1550		0.660	1.60	0.554	1.66	
7.5	0.1653		0.735	1.74	0.614	1.83	
7	0.1771		0.832	1.89	0.670	2.08	
6.5	0.1907		0.951	2.07	0.774	2.35	
6	0.2066	48,390	1.11	2.28	0.961	2.65	
5.5	0.2254	44,360	1.35	2.51	1.23	2.99	
5	0.2480	40,330	1.69	2.76	1.66	3.38	
4.8	0.2583	38,710	1.86	2.85			
4.6	0.2695	37,100	2.07	2.93			
4.5	0.2755	36,290	2.30	2.99	2.31	3.71	
4.4	0.2818	35,490	2.56	3.04			
4.2	0.2952	33,880	2.87	3.06	3.36	3.92	
4	0.3100	32,260	3.21	3.00			
3.8	0.3263	30,650	3.55	2.88			
3.6	0.3444	29,040	3.73	2.79	4.59	3.38	
3.5	0.3543	28,230	3.90	2.66			
3.4	0.3647	27,420	4.17	2.38			
3.2	0.3875	25,810					

(continued)

TABLE XIV (*Continued*)

Amorphous Silicon

eV	μm	cm^{-1}	n	k	n(H)	k(H)	k(H)
3.0	0.4133	24,200	4.38	2.02	5.43	2.19	
2.897	0.4279	23,370				1.80 [4]	
2.8	0.4428	22,580	4.47	1.64			
2.705	0.4584	21,820				1.40	
2.6	0.4769	20,970	4.49	1.28		0.850	
2.514	0.4932	20,280					0.367 [6]
2.500	0.4960	20,160	4.47	1.12	5.25	0.992 [2]	0.361
2.425	0.5112	19,560					0.288
2.4	0.5166	19,360	4.46	0.969			
2.382	0.5205	19,210				0.620 [4]	
2.318	0.5350	18,690					0.216
2.216	0.5596	17,870					0.175
2.205	0.5623	17,780				0.401	
2.20	0.5636	17,740	4.36	0.690			
2.100	0.5904	16,940				0.253	
2.090	0.5933	16,860					8.92 × 10^{-2}
2.009	0.6171	16,210				0.154	
2.0	0.6199	16,130	4.23	0.461	4.71	0.217 [2]	
1.970	0.6295	15,890					3.48 × 10^{-2}
1.907	0.6503	15,380				9.14 × 10^{-2} [4]	
1.9	0.6526	15,320	4.17	0.363			
1.857	0.6675	14,980		0.384 [3]			8.57 × 10^{-3}
1.856	0.6679	14,970				6.46	
1.804	0.6872	14,550		0.271 [1]			
1.8	0.6888	14,520	4.09				2.13
1.770	0.7006	14,270		0.310 [3]			
1.767	0.7016	14,250					

1.704	0.7275	13,750		0.251			
1.7	0.7293	13,710	4.01	0.199 [1]			6.57×10^{-4}
1.693	0.7323	13,660					
1.643	0.7546	13,250		0.221 [3]			2.32×10^{-4}
1.612	0.7694	13,000				2.37×10^{-2}	
1.6	0.7749	12,900	3.93	0.136 [1]			
1.581	0.7841	12,750		0.174 [3]			
1.533	0.8089	12,360				2.90×10^{-4} [5]	9.60×10^{-5}
1.526	0.8123	12,310					
1.514	0.8190	12,210				2.23	
1.513	0.8194	12,200		0.137			
1.5	0.8266	12,100	3.86	8.12×10^{-2} [1]	4.13		
1.498	0.8276	12,080				1.66×10^{-4}	5.89
1.457	0.8509	11,750				7.99×10^{-5}	
1.440	0.8613	11,610					
1.437	0.8631	11,590		0.110 [3]			
1.409	0.8799	11,360				5.45×10^{-5}	
1.4	0.8856	11,290	3.77	4.01×10^{-2} [1]			
1.397	0.8876	11,270				1.15×10^{-2} [4]	
1.363	0.9094	11,000		8.31 [3]			
1.357	0.9140	10,940				4.09×10^{-5} [5]	3.70
1.333	0.9301	10,750					
1.304	0.9505	10,520				3.41×10^{-5}	
1.3	0.9538	10,480	3.68				
1.291	0.9608	10,410					
1.290	0.9613	10,400		6.80		1.10×10^{-2} [4]	
1.262	0.9824	10,180				3.17×10^{-5} [5]	
1.243	0.9978	10,020					
1.234	1.005	9,950		5.50			2.82
1.211	1.024	9,764				2.94	
1.2	1.033	9,678	3.61				

(continued)

TABLE XIV (Continued)

Amorphous Silicon

eV	μm	cm^{-1}	n	k	n(H)	k(H)	k(H)
1.158	1.070	9,342				2.87	
1.137	1.090	9,172					1.69×10^{-5}
1.130	1.098	9,110		4.20			
1.114	1.113	8,988				2.57×10^{-5}	
1.1	1.127	8,872	3.57				
1.086	1.141	8,762				1.06×10^{-2} [4]	
1.062	1.167	8,567				2.06×10^{-5} [5]	
1.032	1.201	8,324					9.01×10^{-6}
1.023	1.211	8,254		$\cdot$ 2.86			
1.014	1.223	8,176				1.88	
1.0	1.240	8,065	3.54		3.83		
0.9649	1.285	7,782					6.08
0.9634	1.287	7,770				1.30×10^{-5}	
0.9286	1.335	7,489		1.99			
0.9160	1.354	7,388				9.88×10^{-6}	
0.90	1.378	7,259	3.50				
0.8870	1.398	7,154					3.70
0.8637	1.435	6,966				7.19	
0.8561	1.448	6,905		1.40×10^{-2}			
0.8637	1.435						
0.8159	1.520	6,581				4.87	
0.80	1.550	6,452	3.48				
0.7522	1.648	6,067		9.33×10^{-3}			
0.7035	1.762	5,674		6.88			
0.70	1.771	5,646	3.45				
0.6509	1.905	5,250		5.03			
0.6120	2.026	4,936		3.64			

0.60	2.066	4,839	3.44		
0.5647	2.196	4,555	3.88 [7]	2.85	
0.5014	2.473	4,044			3.74
0.50	2.480	4,033	3.87		
0.4920	2.520	3,968			2.99 [7]
0.4918	2.521	3,966			3.00
0.4738	2.617	3,822	3.86		
0.4718	2.628	3,805			2.99
0.4538	2.732	3,660	3.86		
0.4406	2.814	3,553			2.99
0.4344	2.854	3,504	3.85		
0.4202	2.951	3,389			3.00
0.4146	2.990	3,344	3.85		
0.3976	3.119	3,207			3.00
0.3942	3.145	3,180	3.84		
0.3777	3.282	3,047			2.99
0.3751	3.305	3,026	3.85		
0.3628	3.417	2,926			2.99
0.3537	3.506	2,853	3.84		
0.3458	3.586	2,789			2.99
0.3337	3.716	2,691	3.84		
0.3334	3.718	2,689	3.82		
0.3227	3.842	2,603			
0.3142	3.946	2,534	3.84		2.98
0.3093	4.008	2,495			
0.2948	4.206	2,377	3.84		2.98
0.2856	4.341	2,304	3.81		
0.2799	4.430	2,257			
0.2773	4.471	2,237			2.96
0.2702	4.589	2,179	3.81		
0.2664	4.654	2,149			2.93

(continued)

TABLE XIV (Continued)

Amorphous Silicon

eV	μm	cm⁻¹	n	k	n(H)	k(H)	k(H)
0.2654	4.672	2,140				6.63×10^{-3} [4]	
0.2629	4.716	2,120				1.02×10^{-2}	
0.2624	4.725	2,117			2.97		
0.2604	4.761	2,100				1.41	
0.2593	4.782	2,091				1.80	
0.2578	4.809	2,080				2.06	
0.2567	4.831	2,070				2.22	
0.2555	4.853	2,061				2.48	
0.2541	4.880	2,049				2.69	
0.2537	4.887	2,046	3.80				
0.2529	4.904	2,039				2.92	
0.2518	4.925	2,031				3.09	
0.2506	4.948	2,021			3.01		
0.2505	4.949	2,020				3.13	
0.2492	4.975	2,010				3.34	
0.2479	5.001	1,999				3.06	
0.2466	5.028	1,989				2.72	
0.2455	5.051	1,980				2.34	
0.2443	5.075	1,970				2.35	
0.2436	5.090	1,965			2.99		
0.2430	5.102	1,960				1.80	
0.2416	5.132	1,949				1.67	
0.2404	5.157	1,939				1.46×10^{-2}	
0.2379	5.211	1,919				9.66×10^{-3}	
0.2366	5.240	1,909				6.91	
0.2326	5.330	1,876	3.80				
0.2320	5.344	1,871			2.99		

0.2269	5.463	1,830	3.79	2.98	
0.2221	5.582	1,791			
0.2188	5.668	1,764	3.80	2.99	
0.2133	5.812	1,721			
0.2055	6.034	1,657	3.79	2.98	
0.2049	6.051	1,653		2.98	
0.1948	6.365	1,571	3.79		
0.1911	6.489	1,541		2.96	
0.1850	6.701	1,492	3.78		
0.1752	7.079	1,413		2.95	
0.1749	7.089	1,411		2.95	
0.1628	7.618	1,313			
0.1602	7.740	1,292	3.77	2.94	
0.1523	8.142	1,228			
0.1478	8.389	1,192	3.75	2.91	
0.1458	8.503	1,176			
0.1417	8.748	1,143	3.72		
0.1284	9.655	1,036	3.71	2.85	
0.1280	9.683	1,033	3.74		
0.1184	10.47	955.0		2.85	
0.1161	10.68	936.3	3.75		
0.1106	11.21	892.0		2.90	
0.1037	11.95	836.6	3.75		
0.1009	12.29	813.8	3.73		
0.09577	12.95	772.4		2.82	
0.09487	13.07	765.2			
0.08934	13.88	720.6			2.73×10^{-2}
0.08801	14.09	709.9			4.31
0.08795	14.10	709.3	3.73	2.80	
0.08792	14.10	709.1			
0.08666	14.31	698.9			6.85

(continued)

TABLE XIV (Continued)
Amorphous Silicon

eV	μm	cm⁻¹	n	k	n(H)	k(H)	k(H)
0.08567	14.47	690.9				9.82×10^{-2}	
0.08438	14.69	680.6				0.153	
0.08323	14.90	671.3				0.201	
0.08180	15.16	659.8				0.246	
0.08086	15.33	652.1				0.266	
0.08066	15.37	650.6		6.72×10^{-4} [8]			
0.07957	15.58	641.7				0.277	
0.07895	15.70	636.8	3.73				
0.07846	15.80	632.8				0.283	
0.07691	16.12	620.3				0.277	
0.07574	16.37	610.9				0.254	
0.07439	16.67	600.0				0.204	
0.07411	16.73	597.7	3.67	2.83×10^{-3}			
0.07318	16.94	590.2				0.193	
0.07203	17.21	580.9				0.151	
0.07171	17.29	578.4		4.43			
0.07066	17.55	569.9				7.08×10^{-2}	
0.06995	17.72	564.2		6.98×10^{-3}			
0.06935	17.88	559.3				3.19	
0.06908	17.95	557.1	3.62				
0.06811	18.21	549.3				2.52	
0.06808	18.21	549.1		1.06×10^{-2}			
0.06764	18.33	545.6			3.12		
0.06717	18.46	541.7		1.49			
0.06625	18.72	534.3		1.90			
0.06553	18.92	528.5		2.27			
0.06475	19.15	522.2		2.68			

0.06421	19.31	517.8		3.09	
0.06345	19.54	511.7		3.46	
0.06316	19.63	509.4		3.92	
0.06263	19.80	505.1		4.21	
0.06157	20.14	496.5		4.85	
0.06079	20.40	490.3		5.37	
0.05996	20.68	483.6			2.94
0.05953	20.83	480.2	3.62		
0.05917	20.95	477.2		5.94	
0.05745	21.58	463.4		5.76	
0.05657	21.92	456.3		5.47	
0.05531	22.42	446.1		5.20	
0.05385	23.03	434.3		5.02	
0.05318	23.32	428.9			2.87
0.05226	23.73	421.5	3.62		
0.05009	24.75	404.0		4.91	
0.04595	26.98	370.6		5.03	
0.04496	27.57	362.7	3.68		
0.04385	28.27	353.7		4.84	
0.04242	29.23	342.1		4.74	
0.04232	29.30	341.3		4.82	
0.04209	29.46	339.5			2.96
0.04083	30.36	329.3		5.56	
0.04028	30.78	324.9		6.16	
0.04005	30.96	323.0		6.77	
0.03762	32.96	303.4		7.86	
0.03642	34.05	293.7		8.24	
0.03491	35.52	281.5	3.74		
0.03404	36.42	274.6		7.99	
0.03311	37.44	267.1		7.45	
0.03261	38.03	263.0		6.83	

(continued)

TABLE XIV (*Continued*)

Amorphous Silicon

eV	μm	cm⁻¹	n	k	n(H)	k(H)	k(H)
0.03178	39.02	256.3		6.53	2.78		
0.03132	39.59	252.6		6.38			
0.03084	40.21	248.7		6.13			
0.02981	41.60	240.4		5.66			
0.02833	43.77	228.5		5.23			
0.02568	48.27	207.2		7.22			
0.02494	49.71	201.2		7.30			
0.02481	49.98	200.1		8.51			
0.02470	50.20	199.2		9.45×10^{-2}			
0.02454	50.53	197.9		0.110			
0.02333	53.14	188.2		0.114			
0.02078	59.66	167.6		0.112			
0.01884	65.82	151.9		0.104			
0.01725	71.86	139.2		9.63×10^{-2}			
0.01565	79.25	126.2		8.62			
0.01476	84.01	119.0		7.27			
0.01382	89.69	111.5		6.21			
0.01313	94.46	105.9		4.43			
0.01208	102.6	97.44		2.86			
0.01127	110.0	90.90		1.57			
0.01023	121.2	82.54		2.97			
0.009225	134.4	74.41		5.63			
0.008818	140.6	71.12		8.16			
0.008396	147.7	67.72					

a Values of n and k, obtained from various references indicated in brackets. Note the distinction between a-Si and a-Si:H data.

Silicon Carbide (SiC)

W. J. CHOYKE

Westinghouse Research and Development Center
and University of Pittsburgh
Pittsburgh, Pennsylvania

EDWARD. D. PALIK

Naval Research Laboratory
Washington, D.C.

The most widely studied polytype of this material is denoted as SiC II or 6H SiC. We list only room-temperature results for it in Table XV. The samples often used are in the form of hexagonal platelets as indicated in the inset of Fig. 13, with the c axis perpendicular to the surface. Thus, a normal-incidence transmission or reflection experiment measures the ordinary (o) ray properties. A prism cut in the manner shown is often used to determine both n_o and n_e (e for extraordinary ray). A slice parallel to the c axis is used to measure transmission and reflection of both polarizations.

The UV–visible data are a combination of the work of Rehn et al. [1] from 25 to 13 eV and Leveque and Lynch [2] from 30 to 3 eV. In Rehn et al. [1] samples were single crystals heated to 700°C for 30 min before etching in concentrated HF acid for 10 min. The sample was mounted in an absolute reflectometer and reflectance was measured at five angles of incidence (15°, 30°, 45°, 60°, 75°). The p-polarized reflectance was generally measured.

The absolute accuracy in R was quoted as ± 0.03. Analysis of the data finally yielded n and k, although it was not stated whether this was o or e or a mixture. These results are in good agreement with results of Leveque and Lynch [2], who utilized reflectivity of synchrotron radiation and Kramers–Kronig analysis. We quote the data of Leveque and Lynch [2] to three figures, although even two significant figures is probably generous.

At lower energy (4.75–3.2 eV) the band-edge absorption has been measured by Choyke and Patrick [3], who used hexagonal platelets. Thus, they measured k_o. Samples of thickness 3.6–200 μm were unsupported, while thinner ones were prepared by grinding and polishing samples glued to Suprasil II substrates. This work was undertaken to reveal direct and indirect

HANDBOOK OF OPTICAL CONSTANTS OF SOLIDS

transitions in the absorption edge that are not obvious from the data plotted in Fig. 13.

At lower energy the band-edge region has been studied by Groth and Kauer [4], who determined the temperature dependence of the ordinary absorption coefficient. We use their $20°C = 293$ K for k_o as read from a graph in the region 3.12–2.5 eV. At still lower energy in the IR, they measured free-carrier absorption as a function of temperature. This absorption is evident at the low-energy side of the band edge.

Note that the Leveque and Lynch [2] values for k are too large near 3 μm by factors of 2, this points out that transmission measurements of k (where feasible) generally give better values than reflectivity analyses.

In the transparent region just below the band edge, measurements of n have typically been done with minimum-deviation prisms or plates exhibiting interference fringes. A typical prism (see insert of Fig. 13) is cut from a hexagonal plate so that the c axis is parallel to the prism apex edge. Then, both o ($E \perp c$) and e ($E \parallel c$) polarization can be used. Thibault [5] measured n_o and n_e at several wavelengths of H, Na, and Hg. Shaffer [6] fitted his data with the formulas

$$n_o = A' + (B'/\lambda_2), \qquad n_e = A + (B/\lambda_2),$$

with $A' = 2.5531$, $B' = 0.0334 \times 10^6$, $A = 2.5852$, and $B = 0.0368 \times 10^6$ and with wavelength λ in nanometers. The birefringence can, of course, be obtained from these results. Accuracy in n was limited by the estimated error of <1 min of angle to be <0.001.

Choyke and Patrick [7] measured n_o by utilizing interference fringes formed by their plates with the c axis perpendicular to the surface. They fitted their data with the formula

$$n_o^2 - 1 = 5.52/[1 - (h\nu/7.53)^2],$$

with $h\nu$ given in electron volts for the range 4–0.5 eV. These data are listed in Table XV for 20°C.

Spitzer et al. [8] have measured the absorption coefficient α_o from 2 to 10 μm by using plates. The samples were green in color with a resistivity greater than ~ 1 Ω cm, although the individual resistivities were not specified.

Their results for several samples showed free-carrier absorption, and we chose the data for the weakest absorbing sample. While they measured both o and e absorption, there were no significant differences, so we give k_o.

Spitzer et al. [8, 9] also measured the reststrahlen reflection in the 2–22-μm region for o and e polarization. They fitted the reflection with an oscillator model of the form

$$(n - ik)^2 = \varepsilon = \varepsilon_\infty \left[1 + \frac{\omega_L^2 - \omega_T^2}{\omega_T^2 - \omega^2 + i\Gamma\omega} \right],$$

with $\omega_L = 969$ cm^{-1}, $\omega_T = 793$ cm^{-1}, $\Gamma = 4.76$ cm^{-1}, and $\varepsilon_\infty = 6.7$ for the ordinary ray. We list the calculated n_o and k_o. While they also fitted the reflectance for the extraordinary ray, the numbers are very similar, and we are not sure whether the differences are very meaningful. Their experiment was not precise enough to give different values for $\varepsilon_{\infty o}$ and $\varepsilon_{\infty e}$ either. With all the shortcomings of the oscillator model, we still list n_o, k_o but caution that the numbers are probably uncertain in the first significant figure at least in the vicinity of ω_T.

The static dielectric constant has been measured at 20 K to be 10.2 ± 0.2 in the 1–100-kHz region, although conducting samples were a problem [10]. From an analysis of published data using the Lyddane–Sachs–Teller relation, $\varepsilon_o = \varepsilon_\infty(\omega_L/\omega_T)^2$, Patrick and Choyke [11] get $\varepsilon_\infty(\perp, o) = 6.52$, ε_∞ ($\|$, e) = 6.70 and $\varepsilon_s(\perp, o) = 9.66$, $\varepsilon_s(\|, e) = 10.03$ at room temperature.

REFERENCES

1. V. Rehn, J. L. Stanford, and V. O. Jones, *Proc. Int. Conf. Phys. Semicond., 13th*, Rome, *1976* (F. G. Fumi, ed.), p. 985. Typografia Marves, Rome, 1976.
2. G. Leveque and D. A. Lynch, private communication (1982).
3. W. J. Choyke and L. Patrick, *Phys. Rev.* **172**, 769 (1968).
4. R. Groth and E. Kauer, *Phys. Status Solidi* **1**, 445 (1961).
5. N. W. Thibault, *Am. Mineral.* **29**, 327 (1944).
6. P. T. B. Shaffer, *Appl. Opt.* **10**, 1034 (1971).
7. W. J. Choyke and L. Patrick *J. Opt. Soc. Am.* **58**, 377 (1968).
8. W. G. Spitzer, D. A. Kleinman, C. J. Frosch, and D. J. Walsh, *in* "Silicon Carbide, A High Temperature Semiconductor" (J. R. O'Connor and J. Smiltens, eds.), p. 347. Pergamon, New York, 1960.
9. W. G. Spitzer, D. Kleinman, and D. Walsh, *Phys. Rev.* **113**, 127 (1959).
10. D. Hofman, J. A. Lely, and J. Volger, *Physica* **23**, 236 (1957).
11. L. Patrick and W. J. Choyke, *Phys. Rev. B* **2**, 2255 (1970).

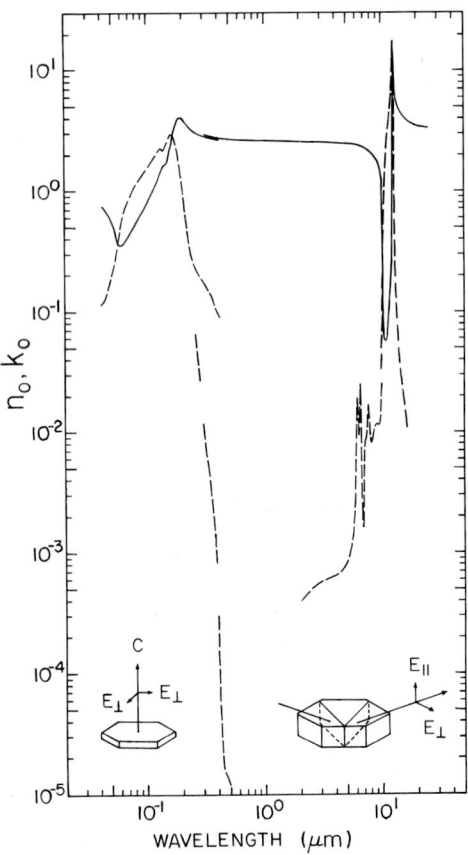

Fig. 13. Log–Log plot of n_o (——) and k_o (----) versus wavelength in micrometers for silicon carbide. Typical sample orientations are shown in the insets.

TABLE XV

Values of n_o and k_o for Silicon Carbide Obtained from Various References[a]

eV	cm^{-1}	μm	n_o	k_o	n_e
30.0		0.04133	0.739 [1, 2]	0.114 [1, 2]	
29.1		0.04261	0.718	0.115	
28.2		0.04397	0.689	0.123	
27.6		0.04492	0.670	0.132	
27.0		0.04592	0.650	0.143	
26.4		0.04696	0.630	0.156	
26.1		0.04750	0.621	0.163	
25.5		0.04862	0.601	0.177	
25.1		0.04940	0.587	0.186	
24.5		0.05061	0.562	0.200	
24.1		0.05145	0.543	0.209	

TABLE XV (*Continued*)

Silicon Carbide

eV	cm^{-1}	μm	n_o	k_o	n_e
23.5		0.05276	0.505	0.225	
23.1		0.05367	0.474	0.243	
22.5		0.05510	0.424	0.285	
22.1		0.05610	0.392	0.326	
21.5		0.05767	0.360	0.402	
21.1		0.05876	0.348	0.481	
20.5		0.06048	0.347	0.532	
20.1		0.06168	0.349	0.581	
19.5		0.06358	0.359	0.653	
19.1		0.06491	0.369	0.699	
18.5		0.06702	0.387	0.766	
18.1		0.06850	0.401	0.810	
17.5		0.07085	0.424	0.873	
17.1		0.07251	0.438	0.915	
16.5		0.07514	0.461	0.983	
15.9		0.07798	0.490	1.05	
15.5		0.07999	0.512	1.10	
15.1		0.08211	0.536	1.15	
14.5		0.08551	0.574	1.23	
14.1		0.08793	0.602	1.28	
13.5		0.09184	0.649	1.36	
13.1		0.09465	0.681	1.41	
12.5		0.09919	0.729 [2]	1.51 [2]	
12.1		0.1025	0.769	1.58	
11.5		0.1078	0.844	1.71	
11.1		0.1117	0.909	1.80	
10.95		0.1132	0.937	1.84	
10.80		0.1148	0.969	1.88	
10.65		0.1164	1.00	1.92	
10.45		0.1186	1.05	1.97	
10.25		0.1210	1.11	2.03	
10.05		0.1234	1.18	2.08	
9.9		0.1252	1.23	2.13	
9.7		0.1278	1.33	2.19	
9.5		0.1305	1.46	2.21	
9.3		0.1333	1.56	2.18	
9.0		0.1378	1.60	2.15	
8.8		0.1409	1.61	2.19	
8.6		0.1442	1.63	2.26	
8.4		0.1476	1.66	2.39	
8.2		0.1512	1.77	2.57	
8.0		0.1550	1.99	2.73	
7.8		0.1590	2.27	2.82	
7.6		0.1631	2.59	2.87	
7.5		0.1653	2.78	2.85	
7.35		0.1687	3.04	2.74	

(*continued*)

TABLE XV (*Continued*)
Silicon Carbide

eV	cm^{-1}	μm	n_o	k_o	n_e
7.15		0.1734	3.33	2.55	
7.0		0.1771	3.51	2.38	
6.8		0.1823	3.69	2.17	
6.6		0.1879	3.97	1.88	
6.4		0.1937	4.05	1.42	
6.2		0.2000	3.96	1.06	
6.0	48,390	0.2066	3.83	0.803	
5.8	46,780	0.2138	3.68	0.609	
5.6	45,170	0.2214	3.54	0.464	
5.4	43,550	0.2296	3.39	0.367	
5.2	41,940	0.2384	3.27	0.304	
5.0	40,330	0.2480	3.16	0.259	
4.75	38,310	0.2610		6.58×10^{-2} [3]	
4.7	37,910	0.2638		5.85	
4.65	37,500	0.2666		5.16	
4.6	37,100	0.2695		4.57	
4.55	36,700	0.2725		4.13	
4.5	36,290	0.2755		3.73×10^{-2}	
			2.96	0.203 [2]	
4.45	35,890	0.2786		3.35×10^{-2} [3]	
4.4	35,490	0.2818		3.02	
4.35	35,080	0.2850		2.72	
4.1	33.070	0.3024		1.18×10^{-2}	
4.0	32,260	0.3100		9.48×10^{-3}	
			2.85	0.174 [2]	
			2.948 [7]		
3.9	31,460	0.3179	2.923	7.65×10^{-3} [3]	
3.8	30,650	0.3263	2.899	5.98	
3.7	29,840	0.3351	2.877	4.93	
3.6	29,040	0.3444	2.856	3.75	
3.5	28,230	0.3542	2.836	2.89	
3.4	27,420	0.3647	2.817	2.04×10^{-3}	
			2.76 [2]	0.119 [2]	
3.3	26,600	0.3757	2.799 [7]	1.38×10^{-3} [3]	
3.2	25,810	0.3875	2.781	8.39×10^{-4} [4]	
3.12	25,160	0.3974		3.00×10^{-4} [4]	
3.10	25,000	0.4000	2.765	1.91	
3.08	24,840	0.4025		1.19×10^{-4}	
3.06	24,680	0.4052		8.06×10^{-5}	
3.04	24,520	0.4078		6.49	
3.02	24,360	0.4105		5.39	
3.0	24,200	0.4133	2.750	4.44×10^{-5}	
			2.69 [2]	0.090 [2]	
2.98	24,040	0.4161		3.74×10^{-5} [4]	
2.96	23,870	0.4189		3.33	
2.94	23,710	0.4217		2.82	
2.92	23,550	0.4246		2.53	

TABLE XV *(Continued)*

Silicon Carbide

eV	cm^{-1}	μm	n_o	k_o	n_e
2.90	23,390	0.4275		2.31	
2.85	22,990	0.4350		1.87	
2.845	22,950	0.4358	2.7305 [5]		2.7824 [5]
2.8	22,580	0.4428		1.55	
2.7	21,780	0.4592		1.46	
2.655	21,410	0.467	2.7074 [6]		2.7553 [6]
2.6	20,970	0.4769		1.37	
2.5	20,160	0.4959	2.684 [7]	1.18×10^{-5}	
2.490	20,080	0.498	2.6870 [6]		2.7331 [6]
2.407	19,420	0.515	2.6789		2.7236
2.270	18,310	0.5461	2.6631 [5]		2.7064 [5]
2.183	17,610	0.568	2.6557 [6]		2.6979 [6]
2.145	17,300	0.5781	2.6511 [5]		2.6933 [5]
2.105	16,980	0.589	2.6488 [6]		2.6911 [6]
2.103	16,960	0.5895	2.6475 [5]		2.6889 [5]
2.013	16,230	0.616	2.6411 [6]		2.6820 [6]
2.0	16,130	0.6199	2.634 [7]		
1.889	15,240	0.6563	2.6296 [5]		2.6696 [5]
1.794	14,470	0.691	2.6243 [6]		2.6639 [6]
1.5	12,100	0.8266	2.598 [7]		
1.0	8,065	1.248	2.573		
0.6199	5,000	2	2.572 [8]	3.98×10^{-4} [8]	
0.5579	4,500	2.222	2.568		
0.5	4,033	2.480	2.558 [7]		
0.4959	4,000	2.5	2.562 [8]	5.17	
0.4339	3,500	2.857	2.554		
0.4133	3,333	3.0		5.73	
0.4	3,226	3.100	2.556 [7]		
0.3720	3,000	3.333	2.540 [8]		
0.3542	2,857	3.5		6.13	
0.3100	2,500	4.0	2.516	6.37	
0.3	2,420	4.133	2.555 [7]		
0.2755	2,222	4.5		7.16	
0.2480	2,000	5.0	2.467 [8]	8.75	
0.2384	1,923	5.2		9.93×10^{-4}	
0.2356	1,900	5.263	2.450		
0.2296	1,852	5.4		1.16×10^{-3}	
0.2232	1,800	5.556	2.430		
0.2214	1,786	5.6		1.56	
0.2138	1,724	5.8		2.68	
0.2108	1,700	5.882	2.404		
0.2066	1,667	6.0		6.21×10^{-3}	
0.2033	1,639	6.1		1.46×10^{-2}	
0.2000	1,613	6.2		1.92	
0.1984	1,600	6.250	2.372		
0.1968	1,587	6.3		9.53×10^{-3}	

(continued)

TABLE XV (*Continued*)

Silicon Carbide

eV	cm^{-1}	μm	n_o	k_o	n_e
0.1937	1,563	6.4		1.02×10^{-2}	
0.1907	1,538	6.5		2.59	
0.1879	1,515	6.6		2.10×10^{-2}	
0.1860	1,500	6.667	2.328		
0.1851	1,493	6.7		3.20×10^{-3}	
0.1823	1,471	6.8		1.62	
0.1797	1,449	6.9		3.07	
0.1771	1,429	7.0		5.57	
0.1746	1,408	7.1		7.91	
0.1736	1,400	7.143	2.267		
0.1722	1,389	7.2		8.59	
0.1698	1,370	7.3		8.71×10^{-3}	
0.1675	1,351	7.4		1.06×10^{-2}	
0.1653	1,333	7.5		1.43	
0.1631	1,316	7.6		1.69	
0.1612	1,300	7.692	2.178		
0.1610	1,299	7.7		1.59	
0.1590	1,282	7.8		1.24×10^{-2}	
0.1569	1,266	7.9		9.43×10^{-3}	
0.1550	1,250	8.0		8.28	
0.1512	1,220	8.2		7.96	
0.1488	1,200	8.333	2.034		
0.1476	1,190	8.4		8.96×10^{-3}	
0.1442	1,163	8.6		1.03×10^{-2}	
0.1409	1,136	8.8		1.11	
0.1378	1,111	9.0		1.15	
0.1364	1,100	9.091	1.768		
0.1348	1,087	9.2		1.16	
0.1339	1,080	9.259	1.684		
0.1319	1,064	9.4		1.14	
0.1314	1,060	9.434	1.581		
0.1292	1,042	9.6		1.15	
0.1289	1,040	9.615	1.453		
0.1265	1,020	9.8	1.286	1.40	
0.1240	1,000	10.0		2.55	
			1.051	3.4 [8]	
0.1227	990	10.10	0.888	4.5	
0.1215	980	10.20	0.663	6.7×10^{-2}	
0.1203	970	10.31	0.274	0.18	
0.1190	960	10.42	0.0872	0.63	
0.1178	950	10.53	0.0663	0.95	
0.1165	940	10.64	0.0593	1.21	
0.1153	930	10.75	0.0569	1.45	
0.1141	920	10.87	0.0569	1.69	
0.1128	910	10.99	0.0587	1.93	
0.1116	900	11.11	0.0621	2.18	
0.1103	890	11.24	0.0672	2.45	

TABLE XV *(Continued)*

Silicon Carbide

eV	cm^{-1}	μm	n_o	k_o	n_e
0.1091	880	11.36	0.0746	2.75	
0.1079	870	11.49	0.0850	3.08	
0.1066	860	11.63	0.0999	3.47	
0.1054	850	11.76	0.122	3.93	
0.1041	840	11.90	0.156	4.51	
0.1029	830	12.05	0.215	5.27	
0.1023	825	12.12	0.262	5.77	
0.1017	820	12.20	0.332	6.38	
0.1010	815	12.27	0.443	7.18	
0.1004	810	12.35	0.639	8.27	
0.09981	805	12.42	1.05	9.93	
0.09919	800	12.50	2.22	12.8	
0.09857	795	12.58	8.74	18.4	
0.09795	790	12.66	17.7	6.03	
0.09733	785	12.74	12.7	1.76	
0.09671	780	12.82	10.3	0.868	
0.09609	775	12.90	8.91	0.531	
0.09547	770	12.99	8.00	0.364	
0.09485	765	13.07	7.35	0.268	
0.09423	760	13.16	6.86	0.208	
0.09299	750	13.33	6.16	0.136	
0.09175	740	13.51	5.68	9.7×10^{-2}	
0.09051	730	13.70	5.32	7.4	
0.08927	720	13.89	5.05	5.8	
0.08803	710	14.08	4.83	4.7	
0.08679	700	14.29	4.65	3.9	
0.08059	650	15.38	4.09	1.8	
0.07439	600	16.67	3.80	1.1	
0.06819	550	18.18	3.61		
0.06199	500	20.00	3.49		
0.05579	450	22.22	3.40		
0.04959	400	25.00	3.34		

[a] References are indicated in brackets.

Zinc Sulfide (ZnS)

EDWARD D. PALIK and A. ADDAMIANO

Naval Research Laboratory

Washington, D.C.

Natural ZnS   The complete phase diagram of the system Zn–S is not known [1–3]. Natural crystal forms include zinc blende (also called sphalerite), and wurtzite. Zinc blende (cubic, space group F$\overline{4}$3m) is the prototype of the B-3 structure [4] (see inset of Fig. 14). Wurtzite (hexagonal, space group C6mc) is the prototype of the B-4 structure [5] (see inset of Fig. 15). Other crystal modifications of natural zinc sulfide are known [6–8]. Substantial amounts of impurities are usually found in zinc sulfide minerals, the principal one being iron. Other common impurities include Cu, Ag, Pb, K, Na, Ga, Hg, Cd, In, Tl, As, Ni, Cr, Ca, Co, Mn, Sn, Ge, and rare earths[9]. These impurities affect the body color of the minerals [10]. The purest crystals are rarely colorless, usually having a light yellow body color [9, 10]. White zinc blende from Franklin, New Jersey, and from Nordmark, Sweden, was named cleiophane [9]. Natural ZnS crystals usually have hydrothermal or metamorphic origin [11].

Synthetic ZnS   A variety of techniques have been developed to obtain large, well-formed, high-purity crystals. These include different "evaporation" or "sublimation" techniques, the high-pressure growth from molten ZnS, hydrothermal techniques, growth from solution, and so on. In addition, thin-layer ZnS coatings on a variety of substrates and sintered hot-pressed polycrystalline ZnS windows (IRTRAN 2 in the USA and KO-2 in the USSR) are being routinely produced for a variety of applications. Owing to differences in composition, their preparative methods, and/or history, all these materials differ somewhat in their physical properties. The differences derive from changes in the impurity background and from changes in the crystal structure. It must be stressed, in particular, that x-ray diffraction work on crystals grown by sublimation has shown that ZnS, like SiC, exists in a very large number of related crystal modifications, usually referred to as polytypes. Alexander *et al.* [12], for instance, observed about 150 new polytypes in their crystals. Approximately a total of some 200 polytypes have been characterized by x-ray diffraction [13–19]. Oftentimes, many polytypes

HANDBOOK OF OPTICAL CONSTANTS OF SOLIDS

ISBN 0-12-544420-6

coexist side by side in minute needles or small plates. Since the preparative method, as well as the history (e.g., cooling rate) of a crystal may influence the fine structure of the material, it is not too surprising to read that the same growing procedure often resulted in different crystals [20]. The optical properties of all the ZnS polytypes have not been measured. The available data are limited to crystals with the zinc blende and the wurtzite structure to thin layers and to sintered ZnS (IRTRAN 2).

Cubic ZnS Data   We shall discuss the data for cubic ZnS, starting in the x-ray region and working to longer wavelength. Because of the success of the model of Henke et al. [21] for calculating optical constants in the x-ray region (6–124 Å), we have included n and k results for ZnS (see the discussion by Lynch and Hunter on metals in this handbook). No attention is paid to the crystal structure, and the data should hold for any form of ZnS, except for scaling the density. The agreement with experimental data [22] in the small overlap region (85–124 Å) encourages us. These results, listed in Table XVI and plotted in Fig. 14, show the characteristic absorption edges due to the Zn and S atoms.

Cardona and Haensel [22] reported the absorption spectra of evaporated thin films in the 150–60-eV region. The experiments were performed by using the Deutsches Elektron Synchrotron (DESY) facility at Hamburg, Germany. The samples were prepared by vacuum deposition at pressures of about 10^{-6} torr *in situ* and *ex situ*. The substrates were carbon films supported by a copper grid and were kept at room temperature during evaporation. While the polycrystallinity of the samples was not discussed, there should be no essential differences between zinc blende and wurtzite spectra. We list k in Table XVI. A double peak in k at 135 Å is suggested to be due to transitions of electrons from 3p levels of Zn into conduction band states.

Hunter [23] has interpolated between 124 and 160 Å to provide the data listed there. Hunter et al. [24] and Cox et al. [25] measured the near-normal-incidence reflectance between 160 and 2200 Å of films vacuum-deposited onto glass. The ZnS was evaporated by using radiant heat. Kramers–Kronig (KK) analysis yielded the n and k values listed in Table XVI. The polycrystallinity of this material was not discussed in detail. However, the interband transitions in the 0.15-μm region are not nearly as sharp as those obtained by Cardona and Harbecke [26] with bulk crystals, suggesting some structure-related differences. In this experiment cubic crystals were freshly cleaved prior to measurement. Below 6 eV the measurements were done with a 50-cm Bausch and Lomb grating spectrometer. At higher photon energies a 1-m Jarrell–Ash vacuum UV spectrometer was used in windowless operation. The reflectivity was analyzed by the KK method. We list n and k from 20 to 2 eV. Reflectivity was measured between 25 and 4 eV and KK analyzed by Balkanski and Petroff [27], also.

While the KK analysis gives values of k at the absorption edge, we cannot trust these over direct transmission measurements such as given by Czyzak *et al.* [20]. As mentioned previously, the color of different ZnS samples varies owing to impurity absorption near the band edge and to structure differences. We have chosen one of the synthetic crystals of low absorption measured by Czyzak *et al.* [20] and tabulated its *k*. These values are much smaller than those obtained from the KK analysis.

The optical constants of thin evaporated films were also obtained by Khawaja and Tomlin [28] from measurements of reflectance and transmittance at normal incidence as suggested by Denton *et al.* [29]. The films of thickness 50 to 300 nm were prepared on quartz substrates at room temperature and 180°C in the form of wedges of 3° angle, by evaporation from an aluminum crucible heated by an external tungsten wire coil in a vacuum of 10^{-5} torr. The films were predominantly cubic crystallites with a small proportion of the hexagonal phase. It was stressed by Khawaja and Tomlin [28] that surface roughness, clearly visible by electron microscopy of Pd-shadowed carbon replicas of the surface, must be taken into account in an interpretation of the data. Results were similar to Czyzak *et al.* [20, 30].

Refractive-index measurements of sphalerite crystals were made over a century ago. These early determinations have been reviewed by Mellor [31] and Gmelin [32]. Routinely, the crystals were described as having a light yellow body color. The refractive index was found to increase with iron content [32]. Mell [33] has measured the refractive indices of 0 and 20°C between 0.4 and 0.8 μm, fitting the data with a dispersion formula. He also fitted the temperature dependence with an empirical formula.

Bond [34] reported refractive index measurements of a zinc blende from Spain that were "not transparent beyond 2.4 μm." He used the minimum-deviation method. His data were fitted by Pikhtin and Yas'kov [35] with a detailed dispersion formula with agreement to one unit in the third decimal place. We list these calculated values in Table XVI along with extrapolation to longer wavelengths.

DeVore [36] reported room-temperature values of the refractive index of a sphalerite prism made from a "clear, water-like natural crystal." The deviation of the mercury light was measured by a photographic technique from 3650 to 15,296 Å. The data were fitted with the dispersion formula

$$n^2 = 5.164 + (1.208 \times 10^7)/(\lambda^2 - 0.732 \times 10^7). \tag{1}$$

Czyzak *et al.* [30] measured the room-temperature refractive index of sublimation-grown "cubic" ZnS by using the prism method. The data from 4000 to 14,000 Å were fitted with the formula

$$n^2 = 5.131 + (1.275 \times 10^7)/(\lambda^2 - 0.732 \times 10^7) \tag{2}$$

and are in good agreement with DeVore's data [36] for natural zinc blende.

The refractive index of sputtered and evaporated films was reported by Burgiel *et al.* [37] in the 4–1-eV region. The data are in close agreement with results of Hall [38] on vacuum-evaporated ZnS deposited on quartz.

In the infrared region, the three- and two-phonon absorption was measured by Klein and Donadio [39] at an unspecified room temperature. Their samples were grown by chemical vapor deposition but were not characterized except as cubic. It is not obvious whether their thin sample ($t = 0.018$ cm) for the two-phonon region (700–400 cm^{-1}) was on a substrate, so we do not calculate the absorption coefficient from their transmission data. For the three-phonon region (1000–750 cm^{-1}), the sample was 1.59 cm thick. Assuming that it was not on a substrate and using the reflectivity as calculated from the n's in Table XVI, we determined k and listed it in Table XVI.

Deutsch [40] measured the room-temperature transmission of crystals from Czyzak et al. [20] and a sample grown elsewhere [41] that had fewer stacking faults. The spectra in the two-phonon region were similar except that the sample of Nitsche [41] had a weak additional band at 19 μm. We list Deutsch's k values in Table XVI. Kwasniewski et al. [42] have measured absorption in the 800–200-cm^{-1} region to locate phonon peaks, but they give no values for absorption coefficient.

The reflectance was measured in the reststrahlen region (15–34 μm) and fitted with an oscillator model by Deutsch [40]. However, we have selected the more recent data of Manabe et al. [43]. They measured the reflectance at a 12° angle of incidence from 20 to 100 μm with a grating spectrometer at 300 K. The cubic sample was mechanically polished. Reflectivity was determined by using an aluminum mirror standard and was quoted to $\pm 1\%$. The reflectivity was fitted with an oscillator model to determine n and k with the formulas

$$R = [(n - 1)^2 + k^2]/[(n + 1)^2 + k^2] \tag{3}$$

and

$$(n - ik)^2 = \varepsilon_\infty \left[1 + \frac{\omega_L^2 - \omega_T^2}{\omega_T^2 - \omega^2 + i\Gamma\omega} \right], \tag{4}$$

with $\varepsilon_\infty = 5.7$, $\omega_T = 282$ cm^{-1}, $\omega_L = 352$ cm^{-1}, and $\Gamma = 6.77$ cm^{-1}. The resulting n and k are listed in Table XVI.

At still-longer wavelengths (133–666 μm) Hattori et al. [44] studied the refractive index and absorption of ZnS crystals grown by the Bridgman method under high temperature and high pressure. These were described as of "mixed cubic and hexagonal type." The deviation of the surface flatness was smaller than 0.4%. The measurements were made at 300, 80, and 2 K with an evacuated Michelson interferometer coupled to a germanium bolometer through a light-pipe system. The refractive index was obtained from analysis of transmission interference maxima. We list values of n and k in Table XVI for 300 K.

Hexagonal ZnS Data   The data for the wurtzite crystal are much sparser. Cardona and Harbecke [26] measured the reflectivity at 298 K of cleaved samples in the spectral region 11–2 eV and performed the KK analysis. They used polarized light (produced by reflection from LiF) to determine both

R_o ($E \perp c$) and R_e ($E \parallel c$). Significant differences were noted. In Table XVII we list their data for n_o, k_o and n_e, k_e as read from a graph and plot them in Fig. 15. Similar results were obtained by Freeouf [45] and Balkanski and Petroff [27]. Baars [46] also measured the reflectivity of several polytypes. Ebina et al. [47] have pointed out that considerable differences in the reflectivity spectra of structurally pure wurtzite and zinc blende crystals exist. They used melt-grown wurtzite crystals stabilized by the addition of 10^{-3}-mol Al per mole ZnS. Crystals grown without Al additions were found to have cubic structure with broadened x-ray diffraction spots on the n layers of the $\langle 111 \rangle$ diffraction patterns where $n \neq 3m$ (m an integer). For Al additions of 10^{-4}-mol Al per mole ZnS, the crystals were found to be mixtures of 6H and 3C (hexagonal and cubic polytypes). The conclusion is made that melt-grown ZnS crystals are superior to other synthetic crystals because of higher crystal perfection, reproducibility, and short production time.

Bieniewski and Czyzak [48] reported refractive indices of synthetic crystals between 0.36 and 1.4 μm for 298 K. Only very small traces ($\leq 0.0001\%$) of B, Mg, Si, Cu, and Fe were detected in these crystals. The crystals had to be heat treated to ensure the hexagonal structure. We list the data in Table XVII. Piper et al. [49] give similar data.

Pikhtin and Yas'kov [50] have fitted a detailed oscillator model to data of Voronkova et al. [51] and discussed the birefringence. Unfortunately, their table of parameters seems incomplete since they refer to work of Sobolev et al. [52] for parameters for three exciton peaks that we could not find. Therefore, we could not calculate the exciton contribution δn, although they did provide a plot of δn versus λ from 0.4 to 1.5 μm from which we might extrapolate $\delta n < 0.03$ for $\lambda > 1.5$ μm. Geidur and Yas'kov [53] had previously discussed the birefringence and given a curve for δn; this suggests that it is significantly smaller. Kulakovskii et al. [54] have also studied the birefringence. We list a few values of n_o for the near infrared as read off a graph in Pikhtin and Yas'kov [50].

Marshall and Mitra [55] give percentage transmission spectra of multiple-phonon absorption (presumably T_o) in the 775–350-cm^{-1} region but no optical constants. Samples were platelet crystals grown from the vapor phase.

Manabe et al. [43] measured R_o ($E \perp c$) in the 20–100-μm spectral region at 300 K and performed an oscillator fit to the data. Their parameters were used in Eq. (3) to obtain n_o and k_o, which are listed in Table XVII. The parameters are $\varepsilon_\infty = 5.7$, $\omega_T = 274$ cm^{-1}, $\omega_L = 356$ cm^{-1}, and $\Gamma = 4.66$ cm^{-1}. We extrapolated the data to 10 μm to compare with the data of Pikhtin and Yas'kov [50]. A significant difference is probably due to the ε_∞ value of Manabe et al. [43] being too large.

Hot-Pressed ZnS Data IRTRAN 2 is the trade name of hot-pressed polycrystalline ZnS manufactured by the Eastman Kodak Company. The material is composed of 95% cubic and 5% hexagonal phases [56]. The room-temperature refractive index, listed in Table XVI, was measured in the region

1–10 μm and the values were extrapolated to 13 μm [57]. The method of measurement was presumably minimum deviation.

Transmittance and reflectance of a 1-mm-thick sample were measured by McCarthy [58]. Reflection at 30° incidence was compared to an aluminum mirror. No optical constants were given. Transmission data for various thicknesses were also reported by Eastman Kodak in the 0.4–15-μm range [57, 59]. Packard [60] compared the measured ("actual") transmittance of a 1-mm-thick sample with the "typical" (i.e., advertised) transmittance of the same sample and stressed the fact that there are considerable differences. For example, for one sample at 10 μm the actual transmittance was 62%, versus 66% reported as typical transmittance.

Considerable variations in the spectral transmission of KO-2 ceramics prepared "by the recrystallization vacuum-molding method from commercially manufactured zinc sulfide powder" were reported by Nosov and Dronova [61]. The KO-2 samples are known to contain some 0.1–1.5% of SO_4^{2-} ions by weight. Inclusions of sulfur and zinc oxide probably are responsible for the measured changes in transmission and for absorption bands appearing in the 9–11-μm region.

Stierwalt *et al.* [62] have reported the spectral absorptance between 3 and 15 μm of 2-mm-thick IRTRAN 2 at 60, 120, and 180°C. Stierwalt [63] measured the spectral emittance at 4.2, 77, and 373 K by using a modified Beckman IR-3 spectrophotometer.

Deposited Thin Films   The optical constants of various types of deposited and grown ZnS films on a substrate have been discussed in some detail, since ZnS is a useful antireflection coating. We mentioned a few examples earlier. The subject is too complicated to organize in a few paragraphs, so we refer the reader to Pulker [64], who discusses the dependence of optical constants on structure, microstructure, packing density, chemical composition, and homogeneity, among other things, with some reference to ZnS [65]. Evaluations of sputtered and molecular-beam-deposited films [66] show wide variations in k and small systematic variations in n. Ever-present problems are the control of the initial constituents, the specification of the deposition process and the nonoptical characterization of the films as to structure and content.

In Jones *et al.* [66] the bulk index of refraction is given in the visible with no description of the measurement (presumably a prism technique). These numbers are consistently ~0.004 less than the numbers given in Table XVI for cubic ZnS.

REFERENCES

1. M. Hansen, *in* "Constitution of Binary Alloys," p. 1171. McGraw-Hill, New York, 1958.
2. R. P. Elliott, *in* "Constitution of Binary Alloys," first suppl., p. 800. McGraw-Hill, New York, 1965.

3. F. A. Shunk, *in* "Constitution of Binary Alloys," second suppl., p. 644. McGraw-Hill, New York, 1969.
4. P. P. Ewald and C. Hermann, "Strukturbericht 1913–28," p. 76. Akademische Verlagsgesellschaft M. B. H., Leipzig, 1931.
5. P. P. Ewald and C. Hermann, "Strukturbericht 1913–28," p. 78. Akademische Verlagsgesellschaft M. B. H., Leipzig, 1931.
6. H. T. Evans and E. T. McKnight, *Am. Mineral* **44**, 210 (1959).
7. C. Frondel and C. Palache, *Am. Mineral* **35**, 29 (1950); C. Frondel and C. Palache, *Science* **107**, 602 (1948).
8. D. C. Buck and L. W. Strock, *Am. Mineral* **39**, 318 (1944); D. C. Buck and L. W. Strock, *Am. Mineral* **40**, 192 (1955).
9. J. W. Mellor, "Inorganic and Theoretical Chemistry," Vol. 4, p. 586. New Impression, New York, 1957.
10. L. Gmelin, "Handbuch der anorganischen Chemie," 8th ed. System-Nummer 32, pp. 123 and 521. Springer-Verlag, Berlin, 1956.
11. L. Gmelin, "Handbuch der anorganischen Chemie," 8th ed. System-Nummer 32, p. 25. Springer-Verlag, Berlin, 1956.
12. E. Alexander, Z., H. Kafman, S. Mardix, and I. T. Steinberger, *Philos. Mag.* **21**, 1237 (1970).
13. Y. Sonnenblick and I. Kiflawi, *Israel J. Chem.* **10**, 7 (1972).
14. I. T. Steinberger, E. Alexander, Y. Brada, Z. H. Kalman, I. Kiflawi, and S. Mardix, *J. Cryst. Growth* **13/14**, 285 (1972).
15. I. Kiflawi, Z. H. Kalman, S. Mardix, and I. T. Steinberger, *Acta Cryst.* **B 28**, 2110 (1972).
16. S. Kume, E. Kodera, T. Aikami, and J. Kakinoki, *J. Phys. Soc. Jpn.* **32**, 288 (1972).
17. I. Kiflawi, S. Mardix, and I. T. Steinberger, *Acta Cryst.* **B 27**, 378 (1971).
18. I. Kiflawi and S. Mardix, *Acta Cryst.* **B 26**, 1192 (1970); I. Kiflawi and S. Mardix, *Acta Cryst,* **B 25**, 1195 (1969).
19. R. W. G. Wyckoff, "Crystal Structures," 2nd ed., Vol. 1, p. 113. Interscience, New York, 1963.
20. S. L. Czyzak, D. C. Reynolds, R. C. Allen, and C. C. Reynolds, *J. Opt. Soc. Am.* **44**, 864 (1954).
21. B. L. Henke, P. L. Lee, T. J. Tanaka, R. L. Shimabukuro, and B. F. Fujikawa, "Low Energy X-Ray Diagnostics—1981" (D. T. Attwood and B. L. Henke, eds.). *AIP Conf. Proc.*, No. 75. American Institute of Physics, New York, 1981.
22. M. Cardona and R. Haensel, *Phys. Rev. B* **1**, 2605 (1970).
23. W. R. Hunter, private communication, (1983).
24. W. R. Hunter, D. W. Angel, and G. Hass, *J. Opt. Soc. Am.* **68**, 1319 (1978).
25. J. T. Cox, J. E. Waylonis, and W. R. Hunter, *J. Opt. Soc. Am.* **59**, 807 (1959).
26. M. Cardona and G. Harbeke, *Phys. Rev.* **137**, A1467 (1965).
27. M. Balkanski and Y. Petroff, *Proc. Conf. Phys. Semicond., 7th, Paris, 1964*, p. 245. Dunod, Paris, 1964.
28. E. Khawaja and S. G. Tomlin, *J. Phys. D.* **8**, 581 (1975).
29. R. E. Denton, R. D. Campbell, and S. G. Tomlin, *J. Phys. D* **5**, 852 (1972).
30. S. J. Czyzak, W. M. Baker, R. C. Crane, and J. B. Home, *J. Opt. Soc. Am.* **47**, 240 (1957).
31. J. W. Mellor, "Inorganic and Theoretical Chemical," Vol. 4, p. 597. New Impression, New York, 1957.
32. L. Gmelin, "Handbuch der anorganischen Chemie," 8th ed. System-Nummer 32, p. 200. Springer-Verlag, Berlin, 1956.
33. M. Mell, *Z. Phys.* **16**, 244 (1923).
34. W. L. Bond, *J. Appl. Phys.* **36**, 1674 (1965).
35. A. N. Pikhtin and A. D. Yas'kov, *Sov. Phys. Semicond.* **12**, 622 (1978).
36. J. R. DeVore, *J. Opt. Soc. Am.* **41**, 416 (1951).
37. J. C. Burgiel, Y. S. Chen, F. Vratny, and G. Smolinsky, *J. Electrochem. Soc.* **115**, 729 (1968).
38. J. F. Hall, Jr., *J. Opt. Soc. Am.* **46**, 1013 (1956).
39. C. A. Klein and R. N. Donadio, *J. Appl. Phys.* **51**, 797 (1980).

40. T. Deutsch, *Proc. Int. Conf. Phys. Semicond., 6th Exeter 1962*, p. 505. The Institute of Physics and the Physical Society, London, 1962.
41. R. Nitsche, *J. Phys. Chem. Solids* **17**, 163 (1960).
42. E. A. Kwasniewski, E. S. Koteles, and W. R. Datars, *Canada J. Phys.* **54**, 1053 (1976).
43. A. Manabe, A. Mitsuishi and H. Yoshinaga, *Jpn. J. Appl. Phys.* **6**, 593 (1967).
44. T. Hattori, Y. Homma, and A. Mitsuishi, *Opt. Commun.* **7**, 229 (1973).
45. J. L. Freeouf, *Phys. Rev. B* **7**, 3810 (1973).
46. J. W. Baars, *Proc. Int. Conf. II–VI Semicond. Comp., Providence, 1968* (D. G. Thomas, ed.), p. 631. Benjamin, New York, 1968.
47. A. Ebina, E. Fukunaga, and T. Takahashi, *Phys. Rev. B* **12**, 687 (1975).
48. T. M. Bieniewski and S. J. Czyzak, *J. Opt. Soc. Am.* **53**, 496 (1963).
49. W. W. Piper, D. T. F. Marple, and P. D. Johnson, *Phys. Rev.* **110**, 323 (1958).
50. A. N. Pikhtin and A. D. Yas'kov, *Sov. Phys. Semicond.* **15**, 8 (1981).
51. E. M. Voronkova, B. N. Grechushnikov, G. I. Distler, and I. P. Petrov, "*Optical Materials for Infrared Technology.*" Nauka, Moscow, 1965 [in Russian].
52. V. V. Sobolev, V. I. Donetskikh, and E. F. Zagaïnov, *Sov. Phys. Semicond.* **12**, 646 (1978).
53. S. A. Geidur and A. D. Yas'kov, *Opt. Spectrosc.* **48**, 618 (1980); S. A. Geidur and A. D. Yas'kov, *Opt. Spektrok.* **48**, 1130 (1980).
54. V. D. Kulakovskii, V. I. Grinev, and M. P. Kulakov, *Sov. Phys. Solid State* **19**, 345 (1977).
55. R. Marshall and S. S. Mitra, *Phys. Rev.* **134**, A1019 (1964).
56. A. J. Moses, "Refractive Index of Optical Materials in the Infrared Region," D. S. 166 (January 1970), p. 5, Air Force Materials Laboratory, Culver City, California; A. J. Moses, A. D. 704555 (1970).
57. Eastman Kodak, publication No. U-72 Rochester, New York (1981).
58. D. E. McCarthy, *Appl. Opt.* **2**, 591 (1963).
59. W. L. Wolfe, *in* "Handbook of Optics," pp. 7–33. McGraw-Hill, New York, 1978.
60. R. D. Packard, *Appl. Opt.* **8**, 1901 (1969).
61. V. B. Nosov and G. N. Dronova, *Sov. J. Opt. Technol.* **47**, 470 (1980).
62. D. L. Stierwalt, J. D. Bernstein, and D. D. Kirth, *Appl. Opt.* **2**, 1169 (1963).
63. D. L. Stierwalt, *Appl. Opt.* **5**, 1911 (1966).
64. H. K. Pulker, *Appl. Opt.* **18**, 1969 (1979).
65. H. K. Pulker, *Thin Solid Films* **34**, 343 (1976).
66. P. L. Jones, D. R. Cotton, and D. Moore, *Thin Solid Films* **88**, 163 (1982).

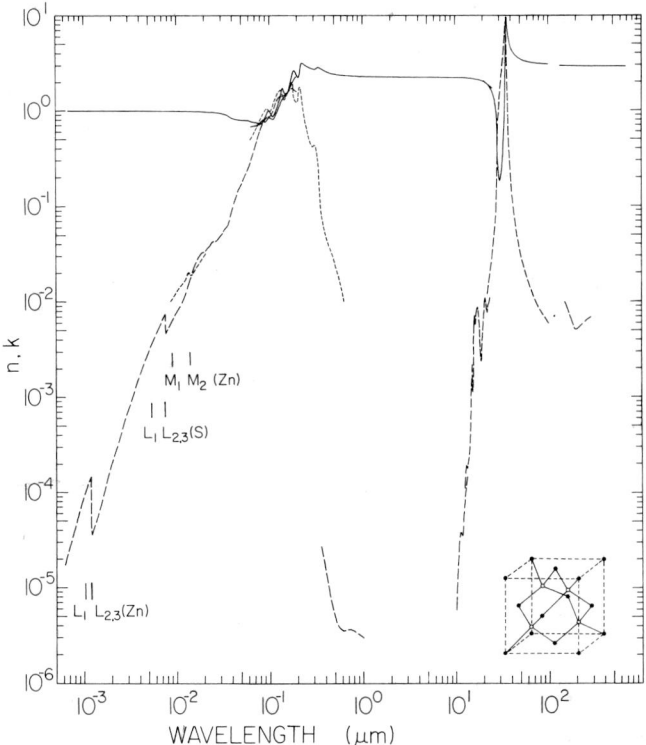

Fig. 14. Log–log plot of n (——) and k (––––) versus wavelength in micrometers for cubic ZnS. (Inset: ●, Zn; ○, S.)

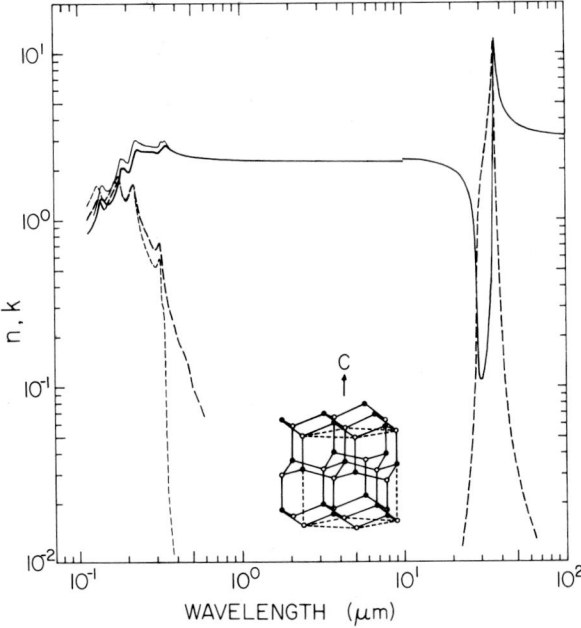

Fig. 15. Log–log plot of n_o (——), k_o (––––), n_e (——), and k_e (––––) versus wavelength in micrometers for hexagonal ZnS. (Inset: ●, Zn; ○, S.)

TABLE XVI

Values of n and k for Cubic Zinc Sulfide Obtained from Various References[a]

eV	Å	n	k	n	k
2000	6.199	0.999904 [21]	1.76×10^{-5} [21]		
1952	6.351	0.999900	1.92		
1905	6.508	0.999894	2.09		
1860	6.665	0.999889	2.27		
1815	6.831	0.999884	2.48		
1772	6.997	0.999878	2.70		
1730	7.166	0.999873	2.94		
1688	7.345	0.999867	3.20		
1648	7.523	0.999861	3.48		
1609	7.705	0.999855	3.78		
1570	7.897	0.999848	4.12		
1533	8.087	0.999841	4.47		
1496	8.287	0.999835	4.87		
1460	8.492	0.999828	5.30		
1426	8.694	0.999821	5.74		
1392	8.906	0.999814	6.24		
1358	9.129	0.999807	6.80		
1326	9.350	0.999800	7.38		
1294	9.581	0.999793	8.03		
1263	9.816	0.999787	8.74		
1233	10.05	0.999781	9.39×10^{-5}		
1204	10.30	0.999777	1.00×10^{-4}		
1175	10.55	0.999774	1.08		
1147	10.81	0.999770	1.15		
1119	11.08	0.999769	1.24		
1093	11.34	0.999769	1.32		
1067	11.62	0.999773	1.41		
1041	11.91	0.999791	1.52		
1038	11.95	0.99979	1.1×10^{-4}		
1033	12.00	0.999805	9.3×10^{-5}		
1029	12.05	0.999813	7.0		
1025	12.10	0.999822	4.4		
1020	12.15	0.999833	3.0		
1016	12.20	0.999838	3.61		
992	12.50	0.999774	3.91		
968	12.81	0.999734	4.24		
945	13.12	0.999694	4.60		
923	13.43	0.999671	4.98		
901	13.76	0.999647	5.42		
879	14.10	0.999622	5.91		
858	14.45	0.999597	6.43		
838	14.79	0.999573	6.99		
818	15.16	0.999547	7.60		
798	15.54	0.999520	8.29		
779	15.92	0.999493	9.01		
760	16.31	0.999464	9.80×10^{-5}		
742	16.71	0.999435	1.06×10^{-4}		

TABLE XVI (*Continued*)

Cubic Zinc Sulfide

eV	Å	n	k	n	k
725	17.10	0.999405	1.15		
707	17.54	0.999372	1.25		
690	17.97	0.999338	1.35		
674	18.39	0.999304	1.46		
658	18.84	0.999269	1.58		
642	19.31	0.999230	1.72		
627	19.77	0.999191	1.87		
612	20.26	0.999150	2.02		
597	20.77	0.999106	2.19		
583	21.27	0.999062	2.37		
569	21.79	0.999015	2.57		
555	22.34	0.99896	2.79		
542	22.87	0.99892	3.02		
529	23.44	0.99886	3.26		
516	24.03	0.99881	3.54		
504	24.60	0.99875	3.82		
492	25.20	0.99869	4.14		
480	25.83	0.99863	4.49		
469	26.43	0.99857	4.84		
458	27.07	0.99850	5.22		
447	27.74	0.99843	5.63		
436	28.44	0.99836	6.09		
426	29.10	0.99829	6.54		
415	29.87	0.99821	7.10		
406	30.54	0.99814	7.60		
396	31.31	0.99806	8.20		
386	32.12	0.99797	8.86		
377	32.89	0.99789	9.50×10^{-4}		
368	33.69	0.99781	1.02×10^{-3}		
359	34.53	0.99772	1.10		
351	35.32	0.99763	1.17		
342	36.25	0.99753	1.26		
334	37.12	0.99744	1.35		
326	38.03	0.99734	1.45		
318	38.99	0.99724	1.55		
311	39.86	0.99715	1.66		
303	40.92	0.99704	1.79		
296	41.88	0.99694	1.91		
289	42.90	0.99685	2.04		
282	43.96	0.99676	2.19		
275	45.08	0.99667	2.34		
269	46.09	0.99660	2.47		
262	47.32	0.99650	2.62		
256	48.43	0.99641	2.76		
250	49.59	0.99632	2.92		
244	50.81	0.99622	3.09		

(*continued*)

TABLE XVI (*Continued*)

Cubic Zinc Sulfide

eV	Å	n	k	n	k
238	52.09	0.99612	3.27		
232	53.44	0.99602	3.46		
227	54.62	0.99594	3.64		
221	56.10	0.99583	3.86		
216	57.40	0.99574	4.07		
211	58.76	0.99566	4.29		
206	60.18	0.99558	4.55		
201	61.68	0.99553	4.82		
196	63.25	0.99549	5.12		
191	64.91	0.99549	5.44		
187	66.30	0.99553	5.71		
182	68.12	0.99561	6.07		
178	69.65	0.99575	6.37		
174	71.25	0.99604	6.69		
170	72.93	0.99678	7.04		
166	74.69	0.99838	7.42		
165.3	75.00	0.9981	7.3		
164.2	75.50	0.99767	6.1		
163.1	76.00	0.99733	5.0		
162	76.53	0.99703	4.62		
158	78.47	0.99611	4.86		
154	80.51	0.99550	5.10		
150	82.65	0.99504	5.36		
147	84.34	0.99476	5.57		
145	85.51				1.00×10^{-2} [22]
143	86.70	0.99439	5.87		
140	88.56	0.99411	6.11		1.04
136	91.16	0.99375	6.45		
135	91.84				1.10
133	93.22	0.99347	6.73		
130	95.37	0.99321	7.02		1.16
127	97.62	0.99293	7.35		
125	99.18				1.23
124	99.98	0.99265	7.69		
121	102.5	0.99237	8.07		
120	103.3				1.32
118	105.1	0.99209	8.47		
115	107.8	0.99180	8.90		1.40
112	110.7	0.99151	9.36		
110	112.7	0.99132	9.69×10^{-3}		1.50
107	115.9	0.99105	1.02×10^{-2}		
105	118.1				1.59
104	119.2	0.99081	1.08		
102	121.5	0.99068	1.12		
100	124.0	0.99061	1.17		1.66
98	126.5				1.72
97	127.8				1.77

TABLE XVI (*Continued*)
Cubic Zinc Sulfide

eV	Å	n	k	n	k
96	129.2				1.85
95.37	130	0.989 [23]	1.37 [23]		
95	130.5				1.95
94	131.9				1.98
93	133.3				1.97
92	134.8				2.03
91.84	135	0.988	1.52		
91	136.2				2.04
90	137.8				1.96
88.56	140	0.987	1.68		
87.5	141.6				1.88
85	145.9				1.93
82.66	150	0.982	1.79		
80	155.0				2.08
77.49	160	0.974 [24]	2.55 [24]		
75	165.3				2.32
72.93	170	0.974	2.81		
70	177.1				2.55
68.88	180	0.973	3.06		
65.26	190	0.969	3.32		
65	190.7				2.87
61.99	200	0.964	3.32		
60	206.6				3.24
59.04	210	0.964	3.57		
56.36	220	0.962	3.83		
53.91	230	0.962	3.83		
51.66	240	0.962	4.34		
49.59	250	0.957	4.34		
47.69	260	0.957	4.34		
45.92	270	0.954	4.34		
44.28	280	0.949	4.59		
42.75	290	0.944	4.85		
41.33	300	0.941	5.10		
38.75	320	0.929	5.61		
36.47	340	0.911	6.12		
34.44	360	0.890	6.89		
32.63	380	0.867	7.91		
31.00	400	0.847	9.95×10^{-2}		
29.52	420	0.819	0.110		
28.18	440	0.809	0.125		
26.95	460	0.801	0.145		
25.83	480	0.798	0.158		
24.80	500	0.796	0.171		
23.84	520	0.793	0.189		
22.96	540	0.793	0.204		
22.14	560	0.788	0.222		

(*continued*)

TABLE XVI (*Continued*)
Cubic Zinc Sulfide

eV	Å	n	k	n	k
21.38	580	0.788	0.242		
20.66	600	0.783	0.260		
20.00	620	0.776	0.286		
				0.68 [26]	0.49 [26]
19.37	640	0.765	0.314		
19.0	652.6			0.68	0.53
18.79	660	0.758	0.347		
18.0	688.8			0.68	0.59
17.71	700	0.747	0.431		
17.22	720	0.747	0.464		
17.0	729.3			0.69	0.67
16.75	740	0.742	0.500		
16.31	760	0.737	0.543		
16.0	774.9			0.70	0.76
15.90	780	0.737	0.584		
15.50	800	0.742	0.620		
15.12	820	0.758	0.679		
15.0	826.6			0.72	0.86
14.76	840	0.791	0.735		
14.42	860	0.798	0.773		
14.09	880	0.783	0.801		
14.0	885.6			0.81	0.99
13.78	900	0.758	0.824		
13.5	918.4			0.89	1.04
13.48	920	0.765	0.842		
13.19	940	0.793	0.857		
13.0	953.7			0.97	1.04
12.92	960	0.827	0.862		
12.65	980	0.852	0.867		
12.5	991.9			1.00	0.99

eV	cm^{-1}	μm	n	k	n	k
12.40		0.1000	0.862	0.876		
12.0		0.1033			0.95	0.98
11.81		0.105	0.834	0.903		
11.5		0.1078			0.87	1.10
11.27		0.110	0.819	1.02		
11.0		0.1127			0.89	1.26
10.78		0.115	0.855	1.14		
10.5		0.1181			0.99	1.41
10.33		0.120	0.929	1.27		
10.0		0.1240			1.13	1.54
9.919		0.125	1.02	1.36		
9.537		0.130	1.13	1.42		
9.5		0.1305			1.40	1.66
9.184		0.135	1.23	1.44		

TABLE XVI (*Continued*)

Cubic Zinc Sulfide

eV	cm^{-1}	μm	n	k	n	k
9.0		0.1378			1.71	1.50
8.856		0.140	1.33	1.44		
8.551		0.145	1.38	1.45		
8.5		0.1459			1.53	1.28
8.266		0.150	1.41	1.47		
7.999		0.155	1.48	1.53	1.49	1.55
7.749		0.160	1.55	1.58	1.58	1.69
7.514		0.165	1.65	1.64		
7.5		0.1653			1.68	1.75
7.293		0.170	1.76	1.70		
7.25		0.1710			2.05	1.91
7.085		0.175	1.88	1.72		
7.0		0.1771			2.40	2.02
6.888		0.180	2.01	1.72		
6.75		0.1837			2.64	1.40
6.702		0.185	2.11	1.68		
6.526		0.190	2.19	1.64		
6.5		0.1907			2.50	1.25
6.358		0.195	2.26	1.62		
6.25		0.1984			2.35	1.21
6.199		0.200	2.32	1.62		
6.0	48,390	0.2066			2.24	1.65
5.75	46,380	0.2156			2.80	1.72
5.5	44,360	0.2254			3.18	1.30
5.25	42,340	0.2362			3.06	0.85
5.0	40,330	0.2480			2.97	0.68
4.75	38,310	0.2610			2.90	0.54
4.5	36,290	0.2755			2.81	0.46
4.25	34,280	0.2917			2.74	0.42
4.0	32,260	0.3100			2.70	0.44
3.75	30,250	0.3306			2.84	0.30
3.542	28,570	0.35		2.78×10^{-5} [20]		
3.5	28,230	0.3542			2.79	0.10
3.25	26,210	0.3815			2.64	6×10^{-2}
3.100	25,000	0.40		1.46×10^{-5}		
3.0	24,200	0.4133			2.54	4
2.755	22,220	0.45		7.52×10^{-6}		
2.75	22,180	0.4509	2.4687 [35]			
2.70	21,780	0.4592	2.4593			
2.65	21,370	0.4679	2.4502			
2.60	20,970	0.4769	2.4414			
2.55	20,570	0.4862	2.4330			
2.50	20,160	0.4959	2.4248		2.42	3
2.480	20,000	0.50		4.77×10^{-6}		
2.45	19,760	0.5061	2.4170			
2.40	19,360	0.5166	2.4094			

(*continued*)

TABLE XVI (*Continued*)

Cubic Zinc Sulfide

eV	cm^{-1}	μm	n	k	n	k
2.35	18,950	0.5276	2.4021			
2.30	18,550	0.5391	2.3950			
2.254	18,180	0.55		3.85		
2.25	18,150	0.5510	2.3882			
2.20	17,740	0.5636	2.3816			
2.15	17,340	0.5767	2.3753			
2.10	16,940	0.5904	2.3692			
2.066	16,670	0.60		3.58		
2.05	16,530	0.6048	2.3633			
2.00	16,130	0.6199	2.3576		2.37	1
1.95	15,730	0.6358	2.3520			
1.907	15,380	0.65		3.62×10^{-6}		
1.90	15,320	0.6526	2.3467			
1.85	14,920	0.6702	2.3416			
1.80	14,520	0.6888	2.3367			
1.771	14,290	0.70		3.68		
1.75	14,110	0.7085	2.3319			
1.70	13,710	0.7293	2.3274			
1.653	13,330	0.75		3.64		
1.65	13,310	0.7514	2.3229			

eV	cm^{-1}	μm	n	k	n_{IRTRAN}	k_{IRTRAN}
1.60	12,900	0.7749	2.3187			
1.55	12,500	0.7999	2.3146	3.50		
1.50	12,100	0.8266	2.3107			
1.459	11,760	0.85		3.38		
1.45	11,690	0.8551	2.3069			
1.40	11,290	0.8856	2.3033			
1.378	11,110	0.90		3.22		
1.35	10,890	0.9184	2.2998			
1.305	10,530	0.95		3.17		
1.30	10,490	0.9537	2.2965			
1.25	10,080	0.9919	2.2933			
1.240	10,000	1.00		3.02	2.2907 [57]	
1.20	9,679	1.033	2.2903			
1.15	9,275	1.078	2.2874			
1.10	8,872	1.127	2.2846			
1.05	8,469	1.181	2.2820			
1.00	8,065	1.240	2.2795			
0.9919	8,000	1.25			2.2777	
0.95	7,662	1.305	2.2771			
0.90	7,259	1.378	2.2748			
0.85	6,856	1.459	2.2727			
0.8266	6,667	1.50			2.2706	
0.80	6,452	1.550	2.2706			
0.75	6,049	1.653	2.2687			

TABLE XVI (*Continued*)

Cubic Zinc Sulfide

eV	cm^{-1}	μm	n	k	n_{IRTRAN}	k_{IRTRAN}
0.7085	5,714	1.75			2.2662	
0.70	5,646	1.771	2.2669			
0.65	5,243	1.907	2.2651			
0.6199	5,000	2.0			2.2631	6.2×10^{-6} [57]
0.60	4,839	2.066	2.2634			
0.5510	4,444	2.25			2.2608	
0.55	4,436	2.254	2.2618			
0.50	4,033	2.480	2.2603			
0.4959	4,000	2.5			2.2589	
0.4509	3,636	2.75			2.2573	
0.45	3,629	2.755	2.2587			
0.4133	3,333	3.0			2.2558	5.3
0.40	3,226	3.100	2.2570			
0.3815	3,077	3.25			2.2544	
0.3542	2,857	3.5			2.2531	
0.35	2,823	3.542	2.2552			
0.3306	2,667	3.75			2.2518	
0.3100	2,500	4.0			2.2504	3.8
0.30	2,420	4.133	2.2529			
0.2917	2,353	4.25			2.2491	
0.2755	2,222	4.5			2.2477	
0.2610	2,105	4.75			2.2462	
0.25	2,016	4.959	2.2497			
0.2480	2,000	5.0			2.2447	2.8
0.2362	1,905	5.25			2.2432	
0.2254	1,818	5.5			2.2416	
0.2156	1,739	5.75			2.2399	
0.2066	1,667	6.0			2.2381	2.9
0.20	1,613	6.199	2.2443			
0.1984	1,600	6.25			2.2363	
0.1907	1,538	6.5			2.2344	
0.1837	1,481	6.75			2.2324	
0.18	1,452	6.888	2.2409			
0.1771	1,429	7.0			2.2304	3.9
0.1710	1,379	7.25			2.2282	
0.1653	1,333	7.5			2.2260	
0.16	1,290	7.749	2.2362			
0.1550	1,250	8.0			2.2237	
0.1503	1,212	8.25			2.2213	4.5
0.1459	1,176	8.5			2.2188	
0.1417	1,143	8.75			2.2162	
0.14	1,129	8.856	2.2290		2.2135	
0.1378	1,111	9.0			2.2107	5.7
0.1340	1,081	9.25			2.2078	
0.1305	1,053	9.5			2.2048	
0.1272	1,026	9.75			2.2018	

(*continued*)

TABLE XVI (*Continued*)

Cubic Zinc Sulfide

eV	cm^{-1}	μm	n	k	n_{IRTRAN}	k_{IRTRAN}
0.1240	1,000	10.00		6.09×10^{-6} [39]	2.1986	8.8×10^{-6}
0.1209	975	10.26		1.14×10^{-5}		
0.12	967.9	10.33	2.2176			
0.1178	950	10.53		1.82		
0.1170	943.4	10.6				2.4×10^{-5}
0.1147	925	10.81		3.26		
0.1127	909.1	11			2.1846	3.9
0.1116	900	11.11		3.78		
0.1085	875	11.43		3.67		
0.1054	850	11.76		3.49		
0.1033	833.3	12			2.1688	4.5
0.1017	825	12.12		5.19		
0.10	806.5	12.40	2.1969			
0.09919	800	12.50		8.19×10^{-5}		
				1.09×10^{-4} [40]		
0.09724	784.3	12.75		1.88		
0.09537	769.2	13.0		1.86	2.1508	
0.09357	754.7	13.25		1.79		
0.09299	750	13.33		1.53 [39]		
0.09184	740.7	13.5		2.04 [40]		
0.09017	727.3	13.75		2.30		
0.09	725.9	13.78	2.1793			
0.08856	714.3	14.0		2.56		
0.08701	701.8	14.25		3.63		
0.08551	689.7	14.5		6.35		
0.08492	684.9	14.6		8.13×10^{-4}		
0.08377	675.7	14.80		1.53×10^{-3}		
0.08266	666.7	15.0		2.27		
0.08157	657.9	15.2		1.11		
0.08051	649.4	15.4		2.21		
0.07999	645.2	15.5	2.1518	3.82		
0.07948	641.0	15.6		5.21		
0.07847	632.9	15.8		7.04		
0.07749	625.0	16.0		5.79		
0.07653	617.3	16.2		5.87		
0.07560	609.8	16.4		7.05		
0.07514	606.1	16.5		7.48		
0.07469	602.4	16.6		7.86		
0.07380	595.2	16.8		8.56		
0.07293	588.2	17.0		8.79		
0.07188	579.7	17.25		8.51		
0.07085	571.4	17.5		7.24		
0.07	564.6	17.71	2.1040			
0.06985	563.4	17.75		5.93		
0.06888	555.6	18.0		4.66		
0.06794	547.9	18.25		3.27		
0.06702	540.5	18.5		2.44		
0.06613	533.3	18.75		2.40		

TABLE XVI *(Continued)*

Cubic Zinc Sulfide

eV	cm^{-1}	μm	n	k	n$_{IRTRAN}$	k$_{IRTRAN}$
0.06526	526.3	19.0		3.02		
0.06441	519.5	19.25		4.01		
0.06358	512.8	19.5		5.04		
0.06278	506.3	19.75		6.00		
0.06199	500.0	20.0		7.56		
			2.05 [43]	7.2 [43]		
0.06123	493.8	20.25		9.35 [40]		
0.06075	490	20.41	2.03	8.0 [43]		
0.06048	487.8	20.5		1.07 × 10^{-2} [40]		
0.06	483.9	20.66	2.0041 [35]			
0.05975	481.9	20.75		1.09		
0.05951	480	20.83	2.01 [43]	9.0 × 10^{-3} [43]		
0.05904	476.2	21.0		1.01 × 10^{-2} [40]		
0.05835	470.6	21.25		9.13 × 10^{-3}		
0.05827	470	21.28	1.98	1.0 × 10^{-2} [43]		
0.05767	465.1	21.5		8.16 × 10^{-3} [40]		
0.05703	460	21.74	1.95	1.2 × 10^{-2} [43]		
0.05700	459.8	21.75		7.65 × 10^{-3} [40]		
0.05636	454.5	22.0		8.05 × 10^{-3}		
0.05579	450	22.22	1.91	1.3 × 10^{-2} [43]		
0.05572	449.4	22.25		9.15 × 10^{-3} [40]		
0.05510	444.4	22.5		9.31		
0.05455	440	22.73	1.87	1.5 × 10^{-2} [43]		
0.05450	439.6	22.75		9.56 × 10^{-3} [40]		
0.05391	434.8	23.0		9.98 × 10^{-3}		
0.05333	430.1	23.25		1.11 × 10^{-2}		
0.05331	430	23.26	1.83	1.82 [43]		
0.05207	420	23.81	1.76	2.2		
0.05083	410	24.39	1.69	2.6		
0.05	403.3	24.80	1.6866 [35]			
0.04959	400	25.00	1.60 [43]	3.3		
0.04835	390	25.64	1.49	4.2		
0.04711	380	26.32	1.35	5.7		
0.04587	370	27.03	1.14	8.4 × 10^{-2}		
0.04463	360	27.78	0.83	0.15		
0.04339	350	28.57	0.31	0.52		
0.04215	340	29.41	0.19	1.1		
0.04092	330	30.30	0.19	1.7		
0.03968	320	31.25	0.23	2.3		
0.03844	310	32.26	0.31	3.1		
0.03720	300	33.33	0.53	4.2		
0.03695	298	33.56	0.62	4.6		
0.03670	296	33.78	0.74	4.9		
0.03645	294	34.01	0.90	5.4		
0.03620	292	34.25	1.15	5.9		
0.03596	290	34.48	1.52	6.6		
0.03571	288	34.72	2.15	7.4		

(continued)

TABLE XVI (*Continued*)

Cubic Zinc Sulfide

eV	cm^{-1}	μm	n	k	n_{IRTRAN}	k_{IRTRAN}
0.03546	286	34.97	3.29	8.3		
0.03521	284	35.21	5.42	9.0		
0.03496	282	35.46	8.32	8.0		
0.03472	280	35.71	9.54	5.2		
0.03447	278	35.97	9.02	3.1		
0.03422	276	36.23	8.20	2.0		
0.03397	274	36.50	7.48	1.3		
0.03372	272	36.76	6.91	0.99		
0.03348	270	37.04	6.46	0.76		
0.03323	268	37.31	6.10	0.60		
0.03298	266	37.59	5.80	0.49		
0.03273	264	37.88	5.55	0.41		
0.03248	262	38.17	5.34	0.35		
0.03224	260	38.46	5.15	0.30		
0.03100	250	40.00	4.52	0.16		
0.02976	240	41.67	4.15	0.10		
0.02852	230	43.48	3.90	7.1×10^{-2}		
0.02728	220	45.45	3.72	5.2		
0.02604	210	47.62	3.58	4.0		
0.02480	200	50.00	3.48	3.1		
0.02356	190	52.63	3.39	2.5		
0.02232	180	55.56	3.33	2.1		
0.02108	170	58.82	3.27	1.7		
0.01984	160	62.50	3.22	1.5		
0.01860	150	66.67	3.18	1.2		
0.01736	140	71.43	3.15	1.0×10^{-2}		
0.01612	130	76.92	3.12	9.1×10^{-3}		
0.01488	120	83.33	3.10	7.8		
0.01364	110	90.91	3.07	6.7		
0.01240	100	100.0	3.06	5.8		
0.009299	75	133.3	2.950 [44]			
0.008679	70	142.9	2.940	1.0×10^{-2} [44]		
0.008369	67.5	148.1		1.0×10^{-2}		
0.008059	65	153.0	2.932	8.6×10^{-3}		
0.007749	62.5	160.0		8.0		
0.007439	60	166.7	2.924	7.4		
0.007129	57.5	173.9		6.1		
0.006819	55	181.8	2.918	5.4		
0.006199	50	200.0	2.913	5.1		
0.005579	45	222.2	2.908	5.6		
0.004959	40	250.0	2.903	6.2		
0.004339	35	285.7	2.899	7.0		
0.003720	30	333.3	2.896			
0.003100	25	400.0	2.894			
0.002480	20	500.0	2.892			
0.001860	15	666.7	2.890			

^{a} References are indicated in brackets. Some values for IRTRAN-2 are also listed. When a value of n or k appears by itself on a line, the pertinent wavelength is on the preceding line.

TABLE XVII

Values of n_o, k_o and n_e, k_e for Hexagonal Zinc Sulfide Obtained from Various References[a]

eV	cm^{-1}	μm	n_o	k_o	n_e	k_e
11		0.1127	0.84 [26]	1.02 [26]		1.23 [26]
10.75		0.1153		1.09		1.30
10.5		0.1181	0.89	1.13		1.36
10.25		0.1210		1.16		1.42
10.0		0.1240	0.99	1.21	1.08 [26]	1.50
9.75		0.1272		1.29		1.59
9.5		0.1305	1.15	1.34	1.30	1.61
9.25		0.1340	1.24	1.33	1.48	1.54
9.0		0.1378	1.37	1.26	1.62	1.44
8.75		0.1417	1.36	1.18	1.62	1.33
8.5		0.1459	1.29	1.18	1.53	1.33
8.25		0.1503	1.27	1.26	1.50	1.39
8.0		0.1550	1.27	1.34	1.52	1.46
7.75		0.1600	1.33	1.45	1.55	1.57
7.5		0.1653	1.42	1.54	1.69	1.72
7.25		0.1710	1.51	1.66	1.83	1.79
7.0		0.1771	1.73	1.75	2.10	1.81
6.75		0.1837	2.08	1.50	2.34	1.49
6.5		0.1907	2.07	1.34	2.30	1.35
6.25		0.1984	2.00	1.33	2.23	1.36
6.0	48,390	0.2066	1.97	1.50	2.28	1.54
5.75	46,380	0.2156	2.30	1.65	2.85	1.66
5.5	44,360	0.2254	2.65	1.37	3.01	1.20
5.25	42,340	0.2362	2.63	1.05	2.94	0.91
5.0	40,330	0.2480	2.60	0.91	2.88	0.76
4.75	38,310	0.2610	2.60	0.84	2.81	0.67
4.5	36,290	0.2755	2.60	0.74	2.80	0.59
4.25	34,280	0.2917	2.58	0.67	2.77	0.53
4.0	32,260	0.3100	2.57	0.72	2.76	0.58
3.75	30,250	0.3306	2.75	0.45	2.97	0.27
3.5	28,230	0.3542	2.69	0.30	2.80	2.5×10^{-2}
3.444	27,780	0.360	2.705 [48]		2.709 [48]	
3.306	26,670	0.375	2.637		2.640	
3.25	26,210	0.3815	2.62 [26]	0.24	2.64 [26]	1.0×10^{-2}
3.100	25,000	0.400	2.560 [48]		2.564 [48]	
3.024	24,390	0.410	2.539		2.544	
3.0	24,200	0.4133	2.54 [26]	0.18	2.54 [26]	
2.952	23,810	0.420	2.522 [48]		2.525 [48]	
2.917	23,530	0.425	2.511		2.514	
2.883	23,260	0.430	2.502		2.505	
2.818	22,730	0.440	2.486		2.488	
2.755	22,220	0.450	2.473		2.477	
2.75	22,180	0.4509	2.49 [26]	0.15	2.49 [26]	
2.695	21,740	0.460	2.459 [48]		2.463 [48]	
2.638	21,280	0.470	2.448		2.453	
2.610	21,050	0.475	2.445		2.449	
2.583	20,830	0.480	2.438		2.443	

(*continued*)

TABLE XVII (*Continued*)

Hexagonal Zinc Sulfide

eV	cm^{-1}	μm	n_o	k_o	n_e	k_e
2.530	20,410	0.490	2.428		2.433	
2.5	20,160	0.4959	2.42 [26]	0.10	2.42 [26]	
2.480	20,000	0.500	2.421 [48]		2.425 [48]	
2.362	19,050	0.525	2.402		2.407	
2.254	18,180	0.550	2.386		2.392	
2.25	18,150	0.5510		8×10^{-2}		
2.156	17,390	0.575	2.375		2.378	
2.066	16,670	0.600	2.363		2.368	
2.0	16,130	0.6199	2.38 [26]	6	2.36 [26]	
1.984	16,000	0.625	2.354 [48]		2.358 [48]	
1.907	15,380	0.650	2.346		2.350	
1.837	14,810	0.675	2.339		2.343	
1.771	14,290	0.700	2.332		2.337	
1.550	12,500	0.800	2.324		2.328	
1.378	11,110	0.900	2.310		2.315	
1.240	10,000	1.00	2.301		2.303	
1.033	8,333	1.20	2.290		2.294	
0.8856	7,143	1.40	2.285		2.288	
0.6199	5,000	2.0	2.27 [50]			
0.4133	3,333	3.0	2.26			
0.3100	2,500	4.0	2.255			
0.2480	2,000	5.0	2.25			
0.2066	1,667	6.0	2.245			
0.1771	1,492	7.0	2.24			
0.1550	1,250	8.0	2.235			
0.1378	1,111	9.0	2.23			
0.1240	1,000	10.0	2.225			
			2.32 [43]			
0.1033	833.3	12.0	2.29			
0.08856	714.3	14.0	2.24			
0.07749	625.0	16.0	2.18			
0.06888	555.6	18.0	2.11			
0.06199	500.0	20.0	2.00	5.6×10^{-3} [43]		
0.06075	490	20.41	1.98	6.2		
0.05951	480	20.83	1.95	7.0		
0.05827	470	21.28	1.92	7.9		
0.05703	460	21.74	1.88	9.0×10^{-3}		
0.05579	450	22.22	1.84	1.0×10^{-2}		
0.05455	440	22.73	1.79	1.2		
0.05331	430	23.26	1.74	1.4		
0.05207	420	23.81	1.67	1.7		
0.05083	410	24.39	1.59	2.0		
0.04959	400	25.00	1.49	2.5		
0.04835	390	25.64	1.37	3.3		
0.04711	380	26.32	1.21	4.5		

TABLE XVII (*Continued*)

Hexagonal Zinc Sulfide

eV	cm^{-1}	μm	n_o	k_o	n_e	k_e
0.04587	370	27.03	0.97	6.8×10^{-2}		
0.04463	360	27.78	0.57	0.15		
0.04339	350	28.57	0.15	0.72		
0.04215	340	29.41	0.11	1.3		
0.04092	330	30.30	0.11	1.7		
0.03968	320	31.25	0.13	2.2		
0.03844	310	32.26	0.17	2.9		
0.03720	300	33.33	0.25	3.7		
0.03695	298	33.56	0.27	4.0		
0.03670	296	33.78	0.30	4.2		
0.03645	294	34.01	0.34	4.5		
0.03620	292	34.25	0.40	4.8		
0.03596	290	34.48	0.46	5.1		
0.03571	288	34.72	0.56	5.6		
0.03546	286	34.97	0.69	6.1		
0.03521	284	35.21	0.88	6.7		
0.03496	282	35.46	1.19	7.5		
0.03472	280	35.71	1.75	8.6		
0.03447	278	35.97	2.88	10.1		
0.03422	276	36.23	5.61	11.8		
0.03397	274	36.50	10.9	10.6		
0.03372	272	36.76	12.3	5.4		
0.03348	270	37.04	10.7	2.7		
0.03323	268	37.31	9.34	1.6		
0.03298	266	37.59	8.36	1.1		
0.03273	264	37.88	7.64	0.78		
0.03248	262	38.17	7.09	0.59		
0.03224	260	38.46	6.56	0.47		
0.03100	250	40.00	5.38	0.20		
0.02976	240	41.67	4.74	0.11		
0.02852	230	43.48	4.35	7.4×10^{-2}		
0.02728	220	45.45	4.09	5.2		
0.02604	210	47.62	3.90	3.8		
0.02480	200	50.00	3.75	3.0		
0.02356	190	52.63	3.64	2.4		
0.02232	180	55.56	3.55	1.9		
0.02108	170	58.82	3.47	1.6		
0.01984	160	62.50	3.41	1.3		
0.01860	150	66.67	3.36	1.1×10^{-2}		
0.01736	140	71.43	3.32	9.4×10^{-3}		
0.01612	130	76.92	3.28	8.0		
0.01488	120	83.33	3.25	6.9		
0.01364	110	90.91	3.22	5.9		
0.01240	100	100.0	3.20	5.1		

[a] References are given in brackets.

3
Insulators

Arsenic Selenide (As₂Se₃)

D. J. TREACY

Physics Department
U.S. Naval Academy
Annapolis, Maryland

The values of n_a, k_a and n_c, k_c listed in Table I and plotted in Fig. 1 are for crystalline material, while the values of n, k listed in Table II and plotted in Fig. 2 are for vitreous (amorphous) material. Values are for room temperature unless otherwise noted.

Crystalline As₂Se₃ In the crystalline phase the results are incomplete, with the major part of the experimental work having been done by reflectance in the one-phonon region, 200 μm (50 cm^{-1}) to 20 μm (500 cm^{-1}) by Zallen et al. [1], and in the electronic absorption region, 1.24 μm (1 eV) to 0.089 μm (14 eV) by Zallen et al. [2]. There are also transmission measurements in limited portions of both of these ranges.

In the infrared region of the spectrum reflectance measurements were made. Since the samples were opaque, this yielded one-surface reflectivity that was fitted with a Lorentz oscillator model. The fitting procedure was supplemented by including transmission measurements on weaker absorption bands to fix the position and strength of these bands. The root-mean-square uncertainty of the reflectance is quoted. Typically, reflectance measurements can be uncertain by about $\pm 3\%$, but this leads to about a 10% uncertainty in the index of refraction. There are limitations on the data obtained in this fashion. When n and k are substantially different, the value of the smaller is very uncertain.

Arsenic selenide crystallizes in a monoclinic structure [3] and possesses three principal indices of refraction. It exhibits micaceous cleavage perpendicular to the (010) axis, and so the plane of the cleaved sample contains the a and c axes. Therefore, it is difficult to obtain data for E $\parallel$ b. The data listed were obtained from the published [1] oscillator parameters for E $\parallel$ a (n_a, k_a) and E $\parallel$ c (n_c, k_c). The experiment was performed as an investigation of the structure of As₂Se₃ rather than as a definitive work on the optical constants of this material.

For the electronic absorption portion of the spectrum, while a Kramers–Kronig (KK) analysis of reflectivity is discussed for the 1–14-eV region [2],

623

HANDBOOK OF OPTICAL CONSTANTS OF SOLIDS

no results for ε_1, ε_2 or n, k were given. Values of k listed were obtained in a limited range from 0.69 μm (1.8 eV) to 0.54 μm (2.3 eV), for which transmission measurements were made by using thin samples in a foreprism double-monochromator instrument.

Vitreous As_2Se_3   The vitreous form of As_2Se_3 has been widely investigated. The material is isotropic and thus possesses only one set of n and k. The low-frequency index of refraction is usually stated in the literature as the square root of the real part of the dielectric constant with values ranging from 3.29 ($\varepsilon_1 = 10.84$) at $\lambda \to \infty$ (dc) [4] to 3.03 $\pm$ 0.02 ($\varepsilon_1 = 9.2 \pm 0.15$) [5] at $\lambda = 400$ μm to 2.97 ($\varepsilon_1 = 8.87$) [6] at $\lambda = 10^8$ μm. The value of 3.03 $\pm$ 0.02 is a good value since it was obtained from a fringe-counting experiment, but it must still be viewed with some caution since k is down by slightly less than an order of magnitude from its value in the reststrahlen region at this wavelength. Since k is comparable to n at this wavelength, the dc and the 400-μm measurements are in substantial agreement.

The k data for the region $\lambda = 8.6 \times 10^4$ to 270 μm are taken from work of Strom and Taylor [7] and were listed in the original paper as $n\alpha$ versus $\bar{v}$. This implies that none of the values listed can be more accurate than the value of n used to obtain k. The value of n used in this region was taken from Austin and Garbett [5].

There have been numerous studies of the reflectivity of amorphous As_2Se_3 in the reststrahlen region [5, 8–14]. The values of $\Delta\varepsilon$ for this entire region range from 1.9 [5] to 2.25 [11]. If either the high or the low estimate of $\Delta\varepsilon$ due to the reststrahlen region is used in conjunction with the high-frequency index of refraction, this projects to a low-frequency index of refraction that ranges from 3.1 to 3.2, in reasonable agreement with [4] and [5].

In the reststrahlen region the main features of the reflectivity spectrum can be fit with a combination of damped oscillators [11]. There is a simplicity to this model in that it uses a minimum number of parameters to fit the observed spectrum. There are some drawbacks to this approach in that it does not fit the observed reflectivity spectrum perfectly on the high-energy side of the reststrahlen region and that it overestimates the value of k as one moves away from the resonant frequency. The values of n and k for the region $\lambda = 200$ μm to $\lambda = 25$ μm are taken from the osillator parameters of Lucovsky [11].

In the relatively transparent region there are two transmission studies [15, 16]. In this region the value of n is slowly varying and ranges from 2.77 at 8 μm [17] to 2.81 around 16.7 μm [18]. Both of these measurements are fringe measurements and agree to almost 1%. The differences between measurements are due to uncertainty in the thickness of the sample and composition (impurity content) of the glass. Both measurements were on nominally pure material, and a 3% variation could be expected, depending on the composition of any particular glass.

In the multiphonon region, sample in–sample out measurements of k on the amorphous material [16] from 25 to 16.7 μm and double-beam measurements [15] from 25 to 11 μm have been made. These measurements are in agreement and both were made on pure materials. Substantial differences can arise, depending on impurity content. Many of these impurity absorptions are referenced in Moynihan et al. [15].

There is one study that uses laser calorimetry on As$_2$Se$_3$ covering the range from 11 to 4.57 μm [15].The values of k listed in this region were taken for samples that approximate intrinsic absorption.

At the electronic absorption edge there is a prism measurement of the index of refraction from 1.5 to 0.82 μm [19]. The data were fit with a one-oscillator Drude-type model (Sellmeier equation) and for reasonably pure material over this spectral region represents better than a 1% fit. Some caution should be exercised as to the purity of the material before applying these indices uncritically.

Absorption data are available from 0.84 to 0.60 μm [2, 20]. In the low k region this material is very sensitive to impurities. The values listed represent pure material.

Reflectivity data are available [2] to 0.1 μm, but the details of the KK analysis done on both crystalline and amorphous materials have not been given.

Vitreous has been used here to describe a material that when cooled down in the molten state solidifies into the amorphous state. There are, of course, other ways to make amorphous materials.

REFERENCES

1. R. Zallen, M. L. Slade, and A. T. Ward, *Phys. Rev. B* **3**, 4257 (1971).
2. R. Zallen, R. E. Drews, R. L. Emerald, and M. L. Slade, *Phys. Rev Lett.* **26**, 1564 (1971).
3. A. A. Vaipolin, *Sov. Phys. Cryst.* **10**, 509 (1966).
4. J. J. Fontenella, private communication, (1975).
5. I. G. Austin and E. S. Garbett, *Philos. Mag.* **23**, 17 (1971).
6. E. B. Ivkin and B. T. Kolomiets, *J. Non-Cryst. Solids* **3**, 41 (1970).
7. U. Strom and P. C. Taylor, *Phys. Rev. B* **16**, 5512 (1977).
8. M. Onomichi, T. Arai, and K. Kudo, *J. Non-Cryst. Solids* **6**, 362 (1971).
9. L. B. Zlatkin and Y. F. Markov. *Phys. Stat. Solidi A* **4**, 391 (1971).
10. G. Lucovsky, *Mater. Res. Bull.* **4**, 505 (1969).
11. G. Lucovsky, *Phys. Rev. B* **6**, 1480 (1972).
12. P. C. Taylor, S. G. Bishop, and D. L. Mitchell, *Solid State Commun.* **8**, 1783 (1970).
13. D. L. Mitchell, S. G. Bishop, and P. C. Taylor, *J. Non-Cryst. Solids* **8–10**, 231 (1972).
14. P. C. Taylor, S. G. Bishop, and D. L. Mitchell, *Proc. Int. Conf. Phonons, Rennes, 1971*, p. 199. Flammarion Science, Paris, 1971.
15. C. T. Moynihan, P. B. Macedo, M. S. Maklad, R. K. Mohr, and R. E. Howard, *J. Non-Cryst. Solids* **17**, 369 (1975).
16. D. J. Treacy and P. C. Taylor, in "Optical Properties of Highly Transparent Solids" (S. S. Mitra and B. Bendow, eds.), p. 261. Plenum, New York, 1975.
17. J. T. Edmond and H. W. Redfern, *Proc. Phys. Soc. London* **81**, 382 (1963).
18. D. J. Treacy, unpublished fringe measurements (1978).
19. Y. Ohmachi, *J. Opt. Soc. Am.* **63**, 630 (1973).
20. D. J. Treacy, P. C. Taylor, and P. B. Klein, *Solid State Commun.* **32**, 423 (1979).

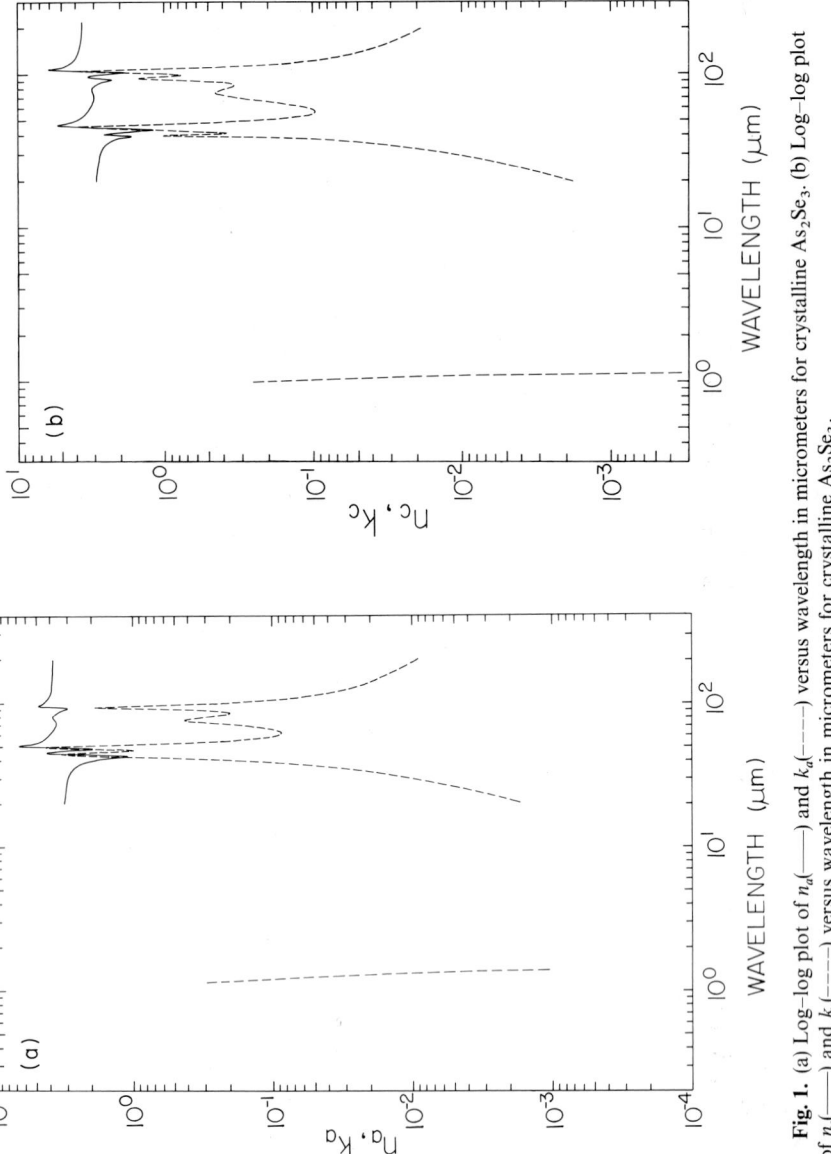

Fig. 1. (a) Log–log plot of n_a(——) and k_a(– – –) versus wavelength in micrometers for crystalline As_2Se_3. (b) Log–log plot of n_c(——) and k_c(– – –) versus wavelength in micrometers for crystalline As_2Se_3.

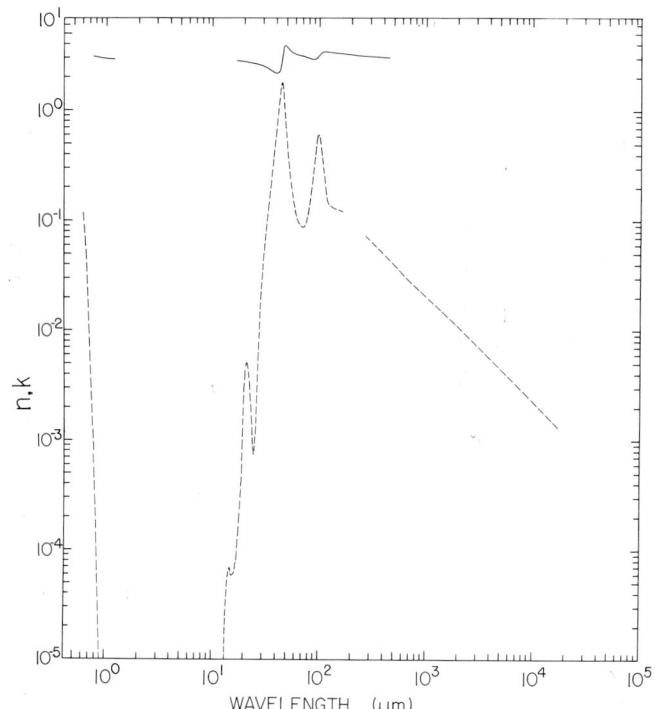

Fig. 2. Log–log plot of n (——) and k(– – –) versus wavelength in micrometers for vitreous As$_2$Se$_3$. Two significant figures are sometimes not enough to show a smooth variation in n and k, and so the curves have been drawn through these points to smooth out steps.

TABLE I

Values of n and k for Crystalline As$_2$Se$_3$ from Various References[a]

eV	cm^{-1}	μm	n_a	n_c	k_a	k_c
2.194	17,700	0.565			0.30 [2]	
2.168	17,480	0.572			0.25	
2.141	17,270	0.579			0.20	
2.123	17,120	0.584			0.17	
2.098	16,920	0.591			0.13	
2.094	16,890	0.592				0.26 [2]
2.091	16,860	0.593				0.26
2.073	16,720	0.598			0.10	0.23
2.060	16,610	0.602				0.20
2.049	16,530	0.605			0.079	0.17
2.036	16,420	0.609				0.15
2.023	16,310	0.613				0.12
2.013	16,230	0.616			0.050	
2.009	16,210	0.617				0.097

(continued)

TABLE I (*Continued*)

Crystalline As$_2$Se$_3$

eV	cm^{-1}	μm	n_a	n_c	k_a	k_c
2.000	16,130	0.620				0.082
1.987	16,030	0.624				0.063
1.977	15,940	0.627			0.031	
1.974	15,920	0.628				0.051
1.962	15,820	0.632				0.038
1.953	15,750	0.635				0.030
1.949	15,720	0.636			0.020	
1.937	15,630	0.640				0.022
1.925	15,530	0.644				0.017
1.922	15,500	0.645			0.012	
1.905	15,360	0.651			8.6×10^{-3}	
1.893	15,270	0.655			6.4	
1.881	15,170	0.659			5.2	
1.859	14,990	0.667			3.1	
1.848	14,900	0.671				1.7×10^{-3}
1.845	14,880	0.672			2.0	
1.842	14,860	0.673				1.2×10^{-3}
1.831	14,770	0.677			1.3×10^{-3}	9.0×10^{-4}
1.826	14,730	0.679				6.4
1.821	14,680	0.681				4.7
1.818	14,660	0.682			8.6×10^{-4}	
1.815	14,640	0.683				3.4
1.807	14,580	0.686			5.5	
1.802	14,530	0.688			4.1	
0.06199	500.0	20.0	3.2 [1]	2.9 [1]	1.7×10^{-3} [1]	1.8×10^{-3} [1]
0.05904	476.2	21.0	3.1	2.9	2.1	2.2
0.05636	454.5	22.0	3.1	2.9	2.5	2.6
0.05391	434.8	23.0	3.1	2.9	3.0	3.1
0.05166	416.7	24.0	3.1	2.8	3.6	3.7
0.04959	400.0	25.0	3.1	2.8	4.3	4.5
0.04769	384.6	26.0	3.1	2.8	5.2	5.3
0.04592	370.4	27.0	3.0	2.8	6.3	6.4
0.04428	357.1	28.0	3.0	2.8	7.6	7.7
0.04275	344.8	29.0	3.0	2.8	9.2×10^{-3}	9.3×10^{-3}
0.04133	333.3	30.0	3.0	2.7	0.011	0.011
0.04065	327.9	30.5	2.9	2.7	0.012	0.012
0.04000	322.6	31.0	2.9	2.7	0.014	0.014
0.03936	317.5	31.5	2.9	2.7	0.015	0.015
0.03875	312.5	32.0	2.9	2.7	0.017	0.017
0.03815	307.7	32.5	2.9	2.6	0.019	0.019
0.03757	303.0	33.0	2.8	2.6	0.021	0.021
0.03701	298.6	33.5	2.8	2.6	0.024	0.023
0.03647	294.1	34.0	2.8	2.6	0.027	0.027
0.03594	289.9	34.5	2.7	2.6	0.031	0.030
0.03542	285.7	35.0	2.7	2.5	0.037	0.034
0.03493	281.7	35.5	2.7	2.5	0.040	0.040
0.03444	277.8	36.0	2.6	2.4	0.046	0.046

TABLE I (*Continued*)

Crystalline As$_2$Se$_3$

eV	cm^{-1}	μm	n_a	n_c	k_a	k_c
0.03397	274.0	36.5	2.6	2.4	0.053	0.054
0.03351	270.3	37.0	2.5	2.3	0.063	0.066
0.03306	266.7	37.5	2.4	2.3	0.083	0.081
0.03263	263.2	38.0	2.4	2.2	0.089	0.10
0.03220	259.7	38.5	2.3	2.1	0.11	0.14
0.03179	256.4	39.0	2.2	2.0	0.14	0.21
0.03139	253.2	39.5	2.1	1.7	0.19	0.41
0.03100	250.0	40.0	1.9	1.7	0.38	1.0
0.03061	247.0	40.5	2.0	2.6	0.33	0.95
0.03024	244.0	41.0	1.7	2.4	0.41	0.46
0.02988	241.0	41.5	1.4	2.1	0.62	0.38
0.02952	238.1	42.0	1.2	1.9	1.0	0.44
0.02917	235.3	42.5	1.1	1.6	1.6	0.61
0.01883	232.6	43.0	1.2	1.3	2.2	0.94
0.02850	229.9	43.5	1.6	1.2	2.8	1.4
0.02818	227.3	44.0	2.3	1.2	3.3	2.0
0.02786	224.7	44.5	3.4	1.5	3.3	2.7
0.02755	222.2	45.0	4.2	2.0	2.5	3.3
0.02725	219.8	45.5	4.1	3.0	1.6	3.9
0.02695	217.4	46.0	3.7	4.4	1.2	3.7
0.02666	215.1	46.5	3.3	5.2	1.0	2.6
0.02638	212.8	47.0	2.8	5.3	1.1	1.6
0.02610	210.5	47.5	2.3	5.0	1.5	1.0
0.02583	208.3	48.0	2.0	4.7	2.2	0.71
0.02556	206.2	48.5	2.2	4.5	3.2	0.52
0.02530	204.1	49.0	3.0	4.3	4.3	0.40
0.02505	202.0	49.5	4.8	4.2	4.8	0.32
0.02480	200.0	50.0	6.5	4.0	3.6	0.26
0.02455	198.0	50.5	6.6	3.9	2.0	0.22
0.02431	196.1	51.0	6.1	3.9	1.2	0.19
0.02407	194.2	51.5	5.7	3.8	0.76	0.17
0.02384	192.3	52.0	5.4	3.7	0.54	0.15
0.02362	190.5	52.5	5.2	3.7	0.41	0.14
0.02339	188.7	53.0	5.0	3.6	0.33	0.13
0.02317	186.9	53.5	4.8	3.6	0.27	0.12
0.02296	185.2	54.0	4.7	3.5	0.23	0.11
0.02275	183.5	54.5	4.6	3.5	0.20	0.11
0.02254	181.8	55.0	4.5	3.5	0.17	0.10
0.02234	180.2	55.5	4.5	3.4	0.15	0.10
0.02214	178.6	56.0	4.4	3.4	0.14	0.099
0.02194	177.0	56.5	4.3	3.4	0.13	0.098
0.02175	175.4	57.0	4.3	3.4	0.12	0.097
0.02156	173.9	57.5	4.2	3.3	0.11	0.097
0.02138	172.4	58.0	4.2	3.3	0.10	0.097
0.02119	170.9	58.5	4.1	3.3	0.098	0.098
0.02101	169.5	59.0	4.1	3.3	0.094	0.10

(*continued*)

TABLE I (*Continued*)

Crystalline As$_2$Se$_3$

eV	cm^{-1}	μm	n_a	n_c	k_a	k_c
0.02084	168.1	59.5	4.1	3.3	0.091	0.10
0.02066	166.7	60.0	4.0	3.2	0.089	0.10
0.02049	165.3	60.5	4.0	3.2	0.087	0.11
0.02033	163.9	61.0	4.0	3.2	0.086	0.11
0.02016	162.6	61.5	4.0	3.2	0.085	0.12
0.02000	161.3	62.0	3.9	3.2	0.085	0.12
0.01984	160.0	62.5	3.9	3.2	0.086	0.13
0.01968	158.7	63.0	3.9	3.1	0.087	0.13
0.01953	157.5	63.5	3.9	3.1	0.088	0.14
0.01937	156.3	64.0	3.8	3.1	0.091	0.15
0.01922	155.0	64.5	3.8	3.1	0.093	0.15
0.01907	153.8	65.0	3.8	3.1	0.097	0.16
0.01893	152.7	65.5	3.8	3.1	0.10	0.17
0.01879	151.5	66.0	3.8	3.1	0.11	0.18
0.01864	150.4	66.5	3.7	3.0	0.11	0.20
0.01851	149.3	67.0	3.7	3.0	0.12	0.21
0.01837	148.1	67.5	3.7	3.0	0.13	0.22
0.01823	147.1	68.0	3.7	3.0	0.14	0.24
0.01810	146.0	68.5	3.7	3.0	0.15	0.25
0.01797	144.9	69.0	3.6	3.0	0.16	0.27
0.01784	143.9	69.5	3.6	3.0	0.18	0.28
0.01771	142.9	70.0	3.6	3.0	0.19	0.30
0.01759	141.8	70.5	3.6	3.0	0.21	0.32
0.01746	140.8	71.0	3.6	3.0	0.23	0.33
0.01734	139.9	71.5	3.6	3.0	0.26	0.35
0.01722	138.9	72.0	3.6	3.0	0.28	0.37
0.01710	137.9	72.5	3.6	3.0	0.31	0.38
0.01698	137.0	73.0	3.6	3.0	0.33	0.40
0.01687	136.1	73.5	3.6	3.0	0.36	0.41
0.01675	135.1	74.0	3.6	3.0	0.38	0.42
0.01664	134.2	74.5	3.6	3.0	0.40	0.43
0.01653	133.3	75.0	3.7	3.0	0.41	0.44
0.01642	132.4	75.5	3.7	3.0	0.42	0.44
0.01631	131.6	76.0	3.7	3.1	0.42	0.45
0.01621	130.7	76.5	3.7	3.1	0.41	0.45
0.01610	129.9	77.0	3.7	3.1	0.40	0.44
0.01600	129.0	77.5	3.8	3.1	0.38	0.44
0.01590	128.2	78.0	3.8	3.1	0.37	0.43
0.01579	127.4	78.5	3.8	3.1	0.35	0.43
0.01569	126.6	79.0	3.8	3.1	0.33	0.42
0.01560	125.8	79.5	3.8	3.1	0.31	0.41
0.01550	125.0	80.0	3.8	3.1	0.29	0.40
0.01540	124.2	80.5	3.8	3.1 [1]	0.27	0.39
0.01531	123.4	81.0	3.8 [1]	3.1	0.26	0.38
0.01521	122.7	81.5	3.7	3.1	0.24	0.37
0.01512	122.0	82.0	3.7	3.1	0.23	0.37
0.01503	121.2	82.5	3.7	3.0	0.22	0.36

TABLE I (*Continued*)

Crystalline As$_2$Se$_3$

eV	cm^{-1}	μm	n_a	n_c	k_a	k_c
0.01494	120.5	83.0	3.7	3.0	0.21	0.35
0.01485	119.8	83.5	3.7	3.0	0.21	0.35
0.01476	119.0	84.0	3.7	3.0	0.20	0.34
0.01467	118.3	84.5	3.6	3.0	0.20	0.34
0.01459	117.6	85.0	3.6	2.9	0.20	0.34
0.01450	117.0	85.5	3.6	2.9	0.20	0.34
0.01442	116.3	86.0	3.5	2.9	0.21	0.34
0.01433	115.6	86.5	3.5	2.9	0.21	0.34
0.01425	114.9	87.0	3.5	2.8	0.22	0.35
0.01417	114.3	87.5	3.4	2.8	0.24	0.36
0.01409	113.6	88.0	3.4	2.8	0.26	0.37
0.01401	113.0	88.5	3.4	2.7	0.28	0.39
0.01393	112.3	89.0	3.3	2.7	0.32	0.42
0.01384	111.7	89.5	3.2	2.6	0.37	0.45
0.01378	111.1	90.0	3.2	2.6	0.43	0.49
0.01370	110.5	90.5	3.1	2.5	0.52	0.55
0.01362	109.9	91.0	3.0	2.4	0.64	0.62
0.01355	109.3	91.5	3.0	2.4	0.81	0.71
0.01348	108.7	92.0	3.0	2.3	1.0	0.83
0.01340	108.1	92.5	3.0	2.3	1.3	0.97
0.01333	107.5	93.0	3.2	2.4	1.6	1.1
0.01326	107.0	93.5	3.5	2.4	1.8	1.3
0.01319	106.4	94.0	4.0	2.6	1.9	1.5
0.01312	105.8	94.5	4.4	2.8	1.8	1.5
0.01305	105.3	95.0	4.7	3.0	1.5	1.5
0.01298	104.7	95.5	4.8	3.2	1.2	1.5
0.01292	104.2	96.0	4.8	3.3	0.94	1.3
0.01284	103.6	96.5	4.8	3.3	0.74	1.2
0.01278	103.1	97.0	4.7	3.3	0.59	1.0
0.01272	102.6	97.5	4.7	3.3	0.48	0.92
0.01265	102.0	98.0	4.6	3.2	0.40	0.84
0.01259	101.5	98.5	4.5	3.1	0.34	0.79
0.01252	101.0	99.0	4.5	3.0	0.29	0.77
0.01246	100.5	99.5	4.5	2.8	0.25	0.78
0.01240	100.0	100.0	4.4	2.7	0.22	0.81
0.01228	99.01	101.0	4.3	2.4	0.18	1.0
0.01216	98.04	102.0	4.3	2.1	0.15	1.4
0.01204	97.09	103.0	4.2	1.9	0.12	2.1
0.01192	96.15	104.0	4.2	2.2	0.11	3.0
0.01181	95.24	105.0	4.2	3.0	0.094	3.9
0.01170	94.34	106.0	4.1	4.7	0.084	4.1
0.01159	93.46	107.0	4.1	6.0	0.076	3.0
0.01148	92.59	108.0	4.1	6.1	0.069	1.7
0.01137	91.74	109.0	4.1	5.7	0.064	1.0
0.01127	90.91	110.0	4.1	5.3	0.059	0.70
0.01117	90.09	111.0	4.0	5.1	0.055	0.51

(*continued*)

TABLE I (*Continued*)

Crystalline As$_2$Se$_3$

eV	cm^{-1}	μm	n_a	n_c	k_a	k_c
0.01107	89.29	112.0	4.0	4.9	0.051	0.39
0.01097	88.50	113.0	4.0	4.8	0.048	0.31
0.01088	87.72	114.0	4.0	4.7	0.045	0.26
0.01078	86.96	115.0	4.0	4.6	0.043	0.22
0.01069	86.21	116.0	4.0	4.5	0.041	0.19
0.01060	85.47	117.0	4.0	4.4	0.039	0.17
0.01051	84.75	118.0	4.0	4.3	0.037	0.15
0.01042	84.03	119.0	4.0	4.3	0.036	0.14
0.01033	83.33	120.0	3.9	4.2	0.034	0.13
0.01025	82.64	121.0	3.9	4.2	0.033	0.12
0.01016	81.97	122.0	3.9	4.2	0.032	0.11
0.01008	81.30	123.0	3.9	4.1	0.030	0.10
0.009999	80.65	124.0	3.9	4.1	0.029	0.094
0.009919	80.00	125.0	3.9	4.1	0.028	0.088
0.009840	79.37	126.0	3.9	4.0	0.028	0.083
0.009763	78.74	127.0	3.9	4.0	0.027	0.079
0.009686	78.12	128.0	3.9	4.0	0.026	0.075
0.009611	77.52	129.0	3.9	4.0	0.025	0.072
0.009537	76.92	130.0	3.9	4.0	0.024	0.069
0.009465	76.34	131.0	3.9	4.0	0.024	0.066
0.009393	75.76	132.0	3.9	3.9	0.023	0.063
0.009322	75.19	133.0	3.9	3.9	0.023	0.061
0.009253	74.63	134.0	3.9	3.9	0.022	0.059
0.009184	74.07	135.0	3.9	3.9	0.022	0.057
0.009117	73.53	136.0	3.9	3.9	0.021	0.055
0.009050	72.99	137.0	3.9	3.9	0.021	0.053
0.008984	72.46	138.0	3.9	3.9	0.020	0.051
0.008920	71.94	139.0	3.9	3.9	0.020	0.050
0.008856	71.43	140.0	3.9	3.8	0.019	0.048
0.008551	68.97	145.0	3.9	3.8	0.018	0.043
0.008266	66.67	150.0	3.8	3.8	0.016	0.038
0.007999	64.52	155.0	3.8	3.7	0.015	0.034
0.007749	62.50	160.0	3.8	3.7	0.014	0.032
0.007514	60.61	165.0	3.8	3.7	0.013	0.029
0.007293	58.82	170.0	3.8	3.7	0.012	0.027
0.007085	57.14	175.0	3.8	3.7	0.012	0.025
0.006888	55.55	180.0	3.8	3.7	0.011	0.024
0.006702	54.05	185.0	3.8	3.7	0.010	0.022
0.006526	52.63	190.0	3.8	3.6	0.010	0.021
0.006358	51.28	195.0	3.8	3.6	9.5×10^{-3}	0.020
0.006199	50.00	200.0	3.8	3.6	9.1	0.019

^{a} The radiation electric field is parallel to the a and c axes of the crystal. References are indicated in brackets and a reference applies down the column until a new bracket appears. Usually, two significant figures are given.

TABLE II

Values of n and k for Vitreous As$_2$Se$_3^a$

eV	cm^{-1}	μm	n	k
2.056	16,580	0.603		0.12 [2]
2.039	16,450	0.608		0.11
2.026	16,340	0.612		0.11
2.006	16,180	0.618		9.9×10^{-2}
1.990	16,050	0.623		9.0
1.965	15,850	0.631		7.6
1.946	15,700	0.637		6.8
1.925	15,530	0.644		5.6
1.905	15,360	0.651		4.5
1.876	15,130	0.661		3.1
1,848	14,900	0.671		2.1
1.826	14,730	0.679		1.4
1.810	14,600	0.685		1.2×10^{-2}
1.794	14,470	0.691		8.9×10^{-3}
1.771	14,290	0.700		6.2
1.749	14,100	0.709		4.5
1.729	13,950	0.717		3.3
1.715	13,830	0.723		2.6
1.701	13,720	0.729		2.2×10^{-3}
1.647	13,280	0.753		4.6×10^{-4}
1.629	13,140	0.761	3.07 [19]	4.0
1.596	12,870	0.777	3.06	2.7
1.579	12,740	0.785	3.05	1.9
1.562	12,590	0.794	3.05	1.3×10^{-4}
1.544	12,450	0.803	3.04	9.4×10^{-5}
1.529	12,330	0.811	3.03	6.3
1.512	12,200	0.820	3.03	4.2
1.494	12,050	0.830	3.02	2.8
1.476	11,910	0.840	3.01	1.8
1.442	11,630	0.86	3.00	
1.409	11,360	0.88	2.99	
1.378	11,110	0.90	2.98	
1.348	10,870	0.92	2.97	
1.319	10,640	0.94	2.96	
1.292	10,420	0.96	2.95	
1.265	10,200	0.98	2.94	
1.240	10,000	1.00	2.93	
1.216	9,804	1.02	2.93	
1.192	9,615	1.04	2.92	
1.170	9,434	1.06	2.92	
1.148	9,259	1.08	2.92	
1.127	9,091	1.10	2.90	
1.107	8,929	1.12	2.90	
1.088	8,772	1.14	2.89	
1.069	8,621	1.16	2.89	
1.051	8,475	1.18	2.89	

(*continued*)

TABLE II (*Continued*)

Vitreous As$_2$Se$_3$

eV	cm^{-1}	μm	n	k
1.033	8,333	1.20	2.88	
0.2555	1,980	5.05		1.6×10^{-7} [15]
0.2380	1,919	5.21		9.9×10^{-8}
0.2344	1,890	5.29		1.1×10^{-7}
0.1345	1,085	9.22		4.4
0.1339	1,080	9.26		3.7
0.1333	1,075	9.30		4.4
0.1308	1,055	9.48		4.5
0.1302	1,050	9.52		4.9
0.1289	1,040	9.62		6.1
0.1277	1,030	9.71		6.6
0.1227	990	10.10		8.4 [15]
0.1215	980	10.20		8.9
0.1203	970	10.31		9.9×10^{-7}
0.1196	965	10.36		1.0×10^{-6}
0.1178	950	10.53		1.1
0.1170	943.4	10.60		1.3
0.1165	940	10.64		1.1
0.1159	935	10.70		1.1
0.1147	925	10.81		1.2
0.1141	920	10.87		1.1
0.1116	900	11.11		1.8
0.1103	890	11.24		1.8
0.1091	880	11.36		1.8
0.1079	870	11.49		1.8
0.1066	860	11.63		1.9
0.1054	850	11.76		1.9
0.1041	840	11.90		2.3
0.1029	830	12.05		2.9
0.1017	820	12.20		3.9
0.1004	810	12.35		4.9
0.09919	800	12.50		7.0×10^{-6}
0.09795	790	12.66		1.0×10^{-5}
0.09671	780	12.82		1.5
0.09547	770	12.99		2.2
0.09423	760	13.16		2.9
0.09299	750	13.33		3.7
0.09175	740	13.51		4.3
0.09051	730	13.70		5.5
0.08927	720	13.89		6.1
0.08803	710	14.08		6.7
0.08679	700	14.29		6.8
0.08555	690	14.49		6.9
0.08431	680	14.71		5.9
0.08307	670	14.93		5.9
0.08183	660	15.15		6.0
0.08059	650	15.38		6.1

TABLE II (*Continued*)

Vitreous As$_2$Se$_3$

eV	cm^{-1}	μm	n	k
0.07935	640	15.63		6.2
0.07811	630	15.87		6.3
0.07687	620	16.13		7.7
0.07563	610	16.39		7.8
0.07439	600	16.67		9.3×10^{-5} [16]
0.07315	590	16.95	2.8 [16]	1.2×10^{-4}
0.07191	580	17.24	2.8	1.4
0.07067	570	17.54	2.8	1.8
0.06943	560	17.86	2.8	2.8
0.06819	550	18.18	2.8	3.3
0.06695	540	18.52	2.8	4.3
0.06633	535	18.69	2.8	5.2
0.06571	530	18.87	2.8	7.2×10^{-4}
0.06509	525	19.05	2.8	1.2×10^{-3}
0.06447	520	19.23	2.8	1.7
0.06385	515	19.42	2.8	2.2
0.06323	510	19.61	2.8	2.8
0.06261	505	19.80	2.8	3.3
0.06199	500	20.00	2.8	4.0
0.06137	495	20.20	2.7	4.3
0.06075	490	20.41	2.7	4.9
0.06024	485.9	20.58	2.7	5.2
0.06013	485	20.62	2.7	5.1
0.05951	480	20.83	2.7	5.1
0.05889	475	21.05	2.7	5.2
0.05827	470	21.28	2.7	4.9
0.05765	465	21.51	2.7	4.6
0.05703	460	21.74	2.7	4.2
0.05641	455	21.98	2.7	3.7
0.05579	450	22.22	2.7	3.2
0.05517	445	22.47	2.7	2.5
0.05455	440	22.73	2.7	2.2
0.05393	435	22.99	2.7	1.7
0.05331	430	23.26	2.7	1.4
0.05269	425	23.53	2.7	1.1×10^{-3}
0.05207	420	23.81	2.7	8.5×10^{-4}
0.05145	415	24.10	2.7	7.3
0.05083	410	24.39	2.7	8.3
0.05021	405	24.69	2.7	9.4×10^{-4}
0.04959	400	25.0	2.7 [11]	1.2×10^{-3} [11]
0.04862	392.2	25.5	2.6	1.6
0.04769	384.6	26.0	2.6	3.9
0.04679	377.4	26.5	2.6	5.0
0.04592	370.4	27.0	2.6	8.0×10^{-3}
0.04509	363.6	27.5	2.6	1.2×10^{-2}

(*continued*)

TABLE II (*Continued*)

Vitreous As_2Se_3

eV	cm^{-1}	μm	n	k
0.04428	357.1	28.0	2.6	1.7
0.04350	350.9	28.5	2.6	2.2
0.04275	344.8	29.0	2.6	2.8
0.04203	339.0	29.5	2.6	3.6
0.04133	333.3	30.0	2.5	4.4
0.04065	327.9	30.5	2.5	5.3
0.04000	322.6	31.0	2.5	6.3
0.03936	317.5	31.5	2.5	7.3
0.03875	312.5	32.0	2.5	8.2
0.03815	307.7	32.5	2.5	9.3×10^{-2}
0.03757	303.0	33.0	2.4	0.11
0.03701	298.5	33.5	2.4	0.12
0.03647	294.1	34.0	2.4	0.14
0.03594	289.9	34.5	2.4	0.16
0.03542	285.7	35.0	2.3	0.18
0.03493	281.7	35.5	2.3	0.21
0.03444	277.8	36.0	2.3	0.24
0.03397	274.0	36.5	2.3	0.28
0.03351	270.3	37.0	2.2	0.32
0.03306	266.7	37.5	2.2	0.37
0.03263	263.2	38.0	2.2	0.42
0.03220	259.7	38.5	2.2	0.48
0.03179	256.4	39.0	2.2	0.54
0.03139	253.2	39.5	2.2	0.60
0.03100	250.0	40.0	2.2	0.67
0.03061	246.9	40.5	2.2	0.73
0.03024	243.9	41.0	2.2	0.81
0.02988	241.0	41.5	2.2	0.89
0.02952	238.1	42.0	2.2	1.0
0.02917	235.3	42.5	2.2	1.1
0.02883	232.6	43.0	2.2	1.3
0.02850	229.9	43.5	2.3	1.4
0.02818	227.3	44.0	2.5	1.6
0.02786	224.7	44.5	2.7	1.7
0.02755	222.2	45.0	2.9	1.8
0.02725	219.8	45.5	3.2	1.8
0.02695	217.4	46.0	3.5	1.7
0.02666	215.1	46.5	3.7	1.5
0.02638	212.8	47.0	3.8	1.3
0.02610	210.5	47.5	3.8	1.1
0.02583	208.3	48.0	3.9	0.94
0.02556	206.2	48.5	3.8	0.80
0.02530	204.1	49.0	3.8	0.68
0.02505	202.0	49.5	3.8	0.59
0.02480	200.0	50.0	3.7	0.52
0.02455	198.0	50.5	3.7	0.45
0.02431	196.1	51.0	3.7	0.40

TABLE II (*Continued*)

Vitreous As$_2$Se$_3$

eV	cm^{-1}	μm	n	k
0.02407	194.2	51.5	3.6	0.36
0.02384	192.3	52.0	3.6	0.33
0.02362	190.5	52.5	3.6	0.30
0.02339	188.7	53.0	3.5	0.27
0.02317	186.9	53.5	3.5	0.25
0.02296	185.2	54.0	3.4	0.23
0.02275	183.5	54.5	3.5	0.22
0.02254	181.8	55.0	3.4	0.20
0.02234	180.2	55.5	3.4	0.19
0.02214	178.6	56.0	3.4	0.18
0.02194	177.0	56.5	3.4	0.17
0.02175	175.4	57.0	3.4	0.16
0.02156	173.9	57.5	3.4	0.15
0.02138	172.4	58.0	3.3	0.14
0.02119	170.9	58.5	3.3	0.14
0.02101	169.5	59.0	3.3	0.13
0.02184	168.1	59.5	3.3	0.13
0.02066	166.7	60.0	3.3	0.12
0.02049	165.3	60.5	3.3	0.12
0.02033	163.9	61.0	3.3	0.11
0.02016	162.6	61.5	3.3	0.11
0.02000	161.3	62.0	3.2	0.10
0.01984	160.0	62.5	3.2	0.10
0.01968	158.7	63.0	3.2	0.10
0.01953	157.5	63.5	3.2	0.10
0.01937	156.2	64.0	3.2	0.10
0.01922	155.0	64.5	3.2	9.6×10^{-2}
0.01907	153.8	65.0	3.2	9.4
0.01893	152.7	65.5	3.2	9.3
0.01879	151.5	66.0	3.2	9.1
0.01864	150.4	66.5	3.2	9.0
0.01851	149.2	67.0	3.2	8.9
0.01837	148.1	67.5	3.2	8.8
0.01823	147.1	68.0	3.2	8.8
0.01810	146.0	68.5	3.1	8.7
0.01797	144.9	69.0	3.1	8.7
0.01784	143.9	69.5	3.1	8.7
0.01771	142.9	70.0	3.1	8.7
0.01759	141.8	70.5	3.1	8.7
0.01746	140.8	71.0	3.1	8.7
0.01734	139.9	71.5	3.1	8.7
0.01722	138.9	72.0	3.1	8.8
0.01710	137.9	72.5	3.1	8.8
0.01698	137.0	73.0	3.1	8.9
0.01687	136.1	73.5	3.1	9.0

(*continued*)

TABLE II (*Continued*)

Vitreous As_2Se_3

eV	cm^{-1}	μm	n	k
0.01675	135.1	74.0	3.1	9.1
0.01664	134.2	74.5	3.1	9.2
0.01653	133.3	75.0	3.1	9.4
0.01642	132.5	75.5	3.1	9.6×10^{-2}
0.01631	131.6	76.0	3.0	0.10
0.01621	130.7	76.5	3.0	0.10
0.01610	129.9	77.0	3.0	0.10
0.01600	129.0	77.5	3.0	0.10
0.01590	128.2	78.0	3.0	0.11
0.01579	127.4	78.5	3.0	0.11
0.01569	126.6	79.0	3.0	0.11
0.01560	125.8	79.5	3.0	0.12
0.01550	125.0	80.0	3.0	0.12
0.01540	124.2	80.5	3.0	0.13
0.01531	123.5	81.0	3.0	0.13
0.01521	122.7	81.5	3.0	0.14
0.01512	121.9	82.0	3.0	0.14
0.01503	121.2	82.5	3.0	0.15
0.01494	120.5	83.0	3.0	0.15
0.01485	119.8	83.5	2.9	0.16
0.01476	119.0	84.0	2.9	0.17
0.01467	118.3	84.5	2.9	0.18
0.01459	117.6	85.0	2.9	0.19
0.01450	117.0	85.5	2.9	0.19
0.01442	116.3	86.0	2.9	0.21
0.01433	115.6	86.5	2.9	0.22
0.01425	114.9	87.0	2.9	0.23
0.01417	114.3	87.5	2.9	0.24
0.01409	113.6	88.0	2.9	0.25
0.01401	113.0	88.5	2.9	0.27
0.01393	112.4	89.0	2.9	0.29
0.01384	111.7	89.5	2.9	0.30
0.01378	111.1	90.0	2.9	0.32
0.01370	110.5	90.5	2.9	0.34
0.01362	109.9	91.0	2.9	0.36
0.01355	109.3	91.5	2.9	0.38
0.01348	108.7	92.0	2.9	0.40
0.01340	108.1	92.5	2.9	0.42
0.01333	107.5	93.0	2.9	0.44
0.01326	106.9	93.5	2.9	0.47
0.01319	106.4	94.0	2.9	0.49
0.01312	105.8	94.5	2.9	0.51
0.01305	105.3	95.0	3.0	0.53
0.01298	104.7	95.5	3.0	0.55
0.01292	104.2	96.0	3.0	0.56
0.01285	103.6	96.5	3.0	0.58
0.01278	103.1	97.0	3.1	0.59

TABLE II (*Continued*)

Vitreous As$_2$Se$_3$

eV	cm^{-1}	μm	n	k
0.01272	102.6	97.5	3.1	0.60
0.01265	102.0	98.0	3.1	0.60
0.01259	101.5	98.5	3.2	0.60
0.01252	101.0	99.0	3.2	0.60
0.01246	100.5	99.5	3.2	0.60
0.01240	100.0	100.0	3.2	0.59
0.01228	99.01	101.0	3.3	0.57
0.01216	98.04	102.0	3.3	0.54
0.01204	97.09	103.0	3.4	0.51
0.01192	96.15	104.0	3.4	0.47
0.01181	95.24	105.0	3.4	0.43
0.01170	94.34	106.0	3.4	0.40
0.01159	93.46.	107.0	3.4	0.37
0.01148	92.59	108.0	3.4	0.34
0.01137	91.74	109.0	3.4	0.31
0.01127	90.91	110.0	3.4	0.28
0.01117	90.09	111.0	3.4	0.26
0.01107	89.28	112.0	3.4	0.24
0.01097	88.49	113.0	3.4	0.23
0.01088	87.72	114.0	3.4	0.21
0.01078	86.96	115.0	3.4	0.20
0.01025	82.64	121.0	3.4	0.14
0.007606	61.35	163.0	3.3	0.12
0.006199	50.00	200.0	3.2	
0.004592	37.04	270.0	3.1	7.2×10^{-2} [7]
0.002799	22.57	443.0	3.0	4.5
0.001826	14.73	679.0	3.0	2.8
0.001273	10.27	974.0	3.0	2.1
0.0006491	5.236	1910.0	3.0	1.1×10^{-2}
0.0004376	3.530	2833.0	3.0	7.5×10^{-3}
0.0002903	2.341	4271.0	3.0	5.0
0.0001716	1.384	7224.0	3.0	3.1
0.00009047	0.7297	13704	3.0	1.6×10^{-3}
0.00005621	0.4534	22056	3.0	9.9×10^{-4}
0.00002774	0.2237	44699	3.0	5.2
0.00001439	0.1161	86153	3.0	2.6

^{a} The material is now isotropic with one set of n and k. At 400 cm^{-1}, data of Lucovsky [11] and Treacy and Taylor [16] were matched by averaging. At 163 and 200 μm, data of Strom and Taylor [7] and Lucovsky [11] were matched by averaging. Two significant figures are given in most cases.

Arsenic Sulfide (As₂S₃)

D. J. TREACY

Physics Department
U.S. Naval Academy
Annapolis, Maryland

The values of n and k listed in Tables III and IV are room-temperature values. The discussion will cover both the crystalline and vitreous phases.

Crystalline As₂S₃   The crystalline phase has been investigated, primarily, as a comparison with the vitreous phase, and there are substantial gaps in the data. The crystalline form of As₂S₃, orpiment, is a naturally occurring mineral, but it is difficult to obtain large crystals of good optical quality. The crystal exhibits micaceous cleavage, and this makes it difficult to obtain optical data perpendicular to its layers. The crystalline form is monoclinic [1]; hence the results will be tabulated as n_a, k_a ($E \parallel a$); n_b, k_b ($E \parallel b$); and n_c, k_c ($E \parallel c$) in Table III. The data are plotted in Fig. 3.

In the spectral region from 100 to 25 μm, the spectra are reflection spectra [2, 3] with some transmission measurements at the position of weaker bands. The spectra were fit employing a superposition of several oscillators. In Zallen et al. [2] the criterion used to determine the oscillator fit was to minimize the root mean square of the deviation of the theoretical fit from the observed reflectivity spectra. The two studies agree essentially but differ systematically with n from Zallen et al. [2] in general being greater than Treacy and Taylor [3]. The systematic difference in n is due to the choice of the high-frequency dielectric constant. In Zallen et al. [2] higher reflectivities were observed at short wavelength, leading to a higher dielectric constant. Both choices of the dielectric constant, however, are probably in error because one overestimates and the other underestimates the index of refraction as determined from accurate fringe measurements by Evans and Young [4] in the near infrared and the visible.

This is a fairly common problem in attempting to determine optical constants from reflectivity data, for the reflectivity can vary over a substantial range, particularly in crystals showing micaceous cleavage. For the spectral region from 100 to 25 μm, the data presented for $E \parallel a$ and $E \parallel c$ are from Zallen

641

et al. [2] and are probably accurate to about 10% in n and k. The data for $E \| b$ are from Treacy and Taylor [3] and are probably good within the same limits. A point was added to the table at 35.8 μm to account for a weak mode that was observed in transmission studies.

In the spectral range from $\lambda = 21.7$ to 9.35 μm (the multiphonon region) the data for $E \| a$, $E \| c$, are taken from Klein *et al.* [5]. These are transmission measurements and are probably good to about 10% with a somewhat better accuracy in the stronger absorption bonds. The values of k in the relatively transparent region were limited by the quality of the natural crystals and probably are representative of impurity absorption and scattering. The values of k observed in this region decrease with wavelength near 25 μm faster than the oscillator fits used from 100 to 25 μm.

In the spectral range from 2 to 0.42 μm, the data for n are taken from Evans and Young [4]. The measurements for $E \| a$ and $E \| c$ were very accurate fringe measurements that involved an absolute determination of the order of interference. The results for n_a and n_c are probably accurate to better than 1% and are in substantial agreement with an earlier determination [6]. The measurements for $E \| b$ were again fringe measurements, but an intermediate step involving the birefringence was necessary.

The values of k for $E \| a$ and $E \| c$ are a mixture from Evans and Young [4] and Zallen *et al.* [7]. The data at small values of k were taken from Evans and Young [4]; for the moderate values of k Evans and Young [4] and Zallen *et al.* [7] agree, and for large values of k the data are from Zallen *et al.* [7]. Values of k for low temperatures may be obtained from Evans and Young [4] ($T = 77$ K) and from Zallen *et al.* [7] ($T = 10$ K). It is difficult to evaluate what limits of confidence can be placed on these data and what effect, if any, impurities have at low absorption levels. Both measurements agree for moderate values of k and are probably good to about 10%. At high values of k the data from Zallen *et al.* were used, but these values of k are probably good only within about 20%.

Since the reflectivity data for $\lambda < 0.4$ μm [7] have not been analyzed but rather presented for their qualitative features, no further constants will be listed.

Vitreous As$_2$S$_3$ The glassy (amorphous) form of As$_2$S$_3$ has been investigated in substantially greater depth than the crystalline form due to its possible technological applications. The values for n and k are listed in Table IV and plotted in Fig. 4.

There is one very accurate low-frequency measurement of the dielectric constant and the loss factor [8]. The loss is sufficiently low that it may be considered negligible, and the index of refraction is determined directly from the real part of the dielectric constant.

At this point it should be emphasized that there are substantial differences between commercially available forms of a-As$_2$S$_3$. It should also be noted that the commercially available forms of a-As$_2$S$_3$ differ from a-As$_2$S$_3$ that was prepared specifically for its purity. It is worthwhile when using this set of tables to make allowance for the specific type of a-As$_2$S$_3$ in use.

The measurements reported by Fontanella [8] are accurate to substantially better than 1% within one batch of commercial material but are not better than 3% when different melts and suppliers are considered. This measurement [8] was taken at 300 K.

In the region from $\lambda = 1.2 \times 10^4$ to 116 μm, the data are taken from Strom and Taylor [9]. The first point in this study was taken from microwave perturbation techniques and is probably good within a factor of 2. The remainder of the points were taken in transmission by using a laser source. These points are probably good to within 20%.

From $\lambda = 100$ μm to $\lambda = 25$ μm, the data are taken from Lucovsky [10]. These data were taken as a reflectance spectrum and fit with oscillator parameters. The caveat given earlier concerning reflectance spectra applies here, but since the glass can be polished better than the crystal, the values of n and k are probably somewhat better than 15%.

The values of n from 24.69 to 12.5 μm were taken from Young [15]. When these values are compared with the extrapolation of the oscillator fit in Lucovsky [10], it is seen that the values given by the oscillator fit do not rise as quickly as the measured points as wavelength decreases. These measured values of n were determined from fringe measurements in transmission. The absolute value of the order of the fringes was obtained, and the measurements are probably good to about 1%.

The values of k for $\lambda = 25$ to 12.5 μm were taken from Klein et al. [11]. This was a sample in–sample out transmission spectrum, and the values of k listed are within 10%.

The values of n for 12.2 to 0.5600 μm were taken from Rodney et al. [12]. These are accurate prism measurements and the original values were listed to six significant figures. The data were fit with a five-term Sellmeier equation, and the fit was typically less than one part in 10^4. The authors note, however, that this is for a particular sample and that the difference between samples is of the order of 4% in the visible and of the order of 0.1% near 12 μm. (The values from Rodney et al. [12] and Young [15] agree to better than 1% at 12 μm.) This underscores the earlier comment concerning the difference between commercial suppliers and different melts of glass from any single commercial source.

The values of k for $\lambda = 12.2$ to 6.6 μm are taken from Maklad et al. [16]. At 12.5 μm Klein et al. [11] and Maklad et al. [16] agree. Most points were taken in transmission on a double-beam instrument and are good to within 20%. The points between 9.2 and 10.9 μm were taken by using laser calorimetry. For the points taken on the double-beam instrument, it is difficult to estimate the absolute magnitude of scattering losses, and the values of k listed can be interpreted as an upper bound.

The value of k at $\lambda = 1.0$ μm is taken from Wood and Tauc [13] and is probably good to 20%.

The values of k from $\lambda = 0.8$ to 0.56 μm were taken from Zallen et al. [7]. These were done in transmission with a foreprism monochrometer instrument and are probably good to within 20%. In the region of overlap between Zallen et al. and Wood and Tauc [13], the values quoted in Wood and Jauc [13] are slightly higher (about 25% higher at 0.5 μm).

The values of n and k from $\lambda = 0.5585$ to 0.22 μm were taken from Young [15]. For the values of n, 0.05 was added to the data presented by Young [15]. This was done because (a) this is well within the uncertainty between melts quoted in Rodney et al. [12], (b) there is greater uncertainty in the values of thickness for the thin samples quoted in Young [15], and (c) the curves run parallel to each other in the region of overlap. It must be emphasized that the differences in the index of refraction are representative differences between melts. The values of k listed here were taken on a prism-grating double monochrometer and should be accurate to better than 20%. In the region in which Zallen et al. [7] was used, the values from Young [15] agree quite well, but Zallen et al. [7] is consistently quite a bit lower for values of λ less than 0.56 μm.

REFERENCES

1. N. Morimoto, *Mineral J. Sapporo* **1**, 160 (1954).
2. R. Zallen, M. L. Slade, and A. T. Ward, *Phys. Rev. B* **3**, 4257 (1971).
3. D. J. Treacy and P. C. Taylor, *Phys. Rev. B* **11**, 2941 (1975).
4. B. L. Evans and P. A. Young, *Proc. R. Soc. Ser. A* **297**, 230 (1967).
5. P. G. Klein, P. C. Taylor, and D. J. Treacy, *Phys. Rev. B* **16**, 4501 (1977).
6. E. S. Larsen and H. Berman, *U. S. Geol. Surv. Bull.* **848**, 213 (1934).
7. R. Zallen, R. E. Drew, R. L. Emerald, and M. L. Slade, *Phys. Rev. Lett.* **26**, 1564 (1971).
8. J. J. Fontanella, private communication (1975).
9. U. Strom and P. C. Taylor, *in* "Amorphous and Liquid Semiconductors" (J. Stuke and W. Brenig, eds.), p. 375, Taylor and Francis, London, 1974; U. Strom and P. C. Taylor, *Phys. Rev. B.* **16**, 5512 (1977).
10. G. Lucovsky, *Phys. Rev. B* **6**, 1480 (1972).
11. P. B. Klein, P. C. Taylor, and D. J. Treacy, *Phys. Rev. B* **16**, 4511 (1977).
12. W. S. Rodney, I. H. Malitson, and T. A. King, *J. Opt. Soc. Am.* **48**, 633 (1958).
13. D. L. Wood and J. Tauc, *Phys. Rev. B* **5**, 3144 (1972).
14. E. L. Zorina, *Sov. Phys. Solid State* **7**, 269 (1965).
15. P. A. Young, *J. Phys. C* **4**, 93 (1971).
16. M. S. Maklad, R. K. Mohr, R. E. Howard, P. B. Macedo, and C. T. Moynihan, *Solid State Commun.* **15**, 855 (1974).

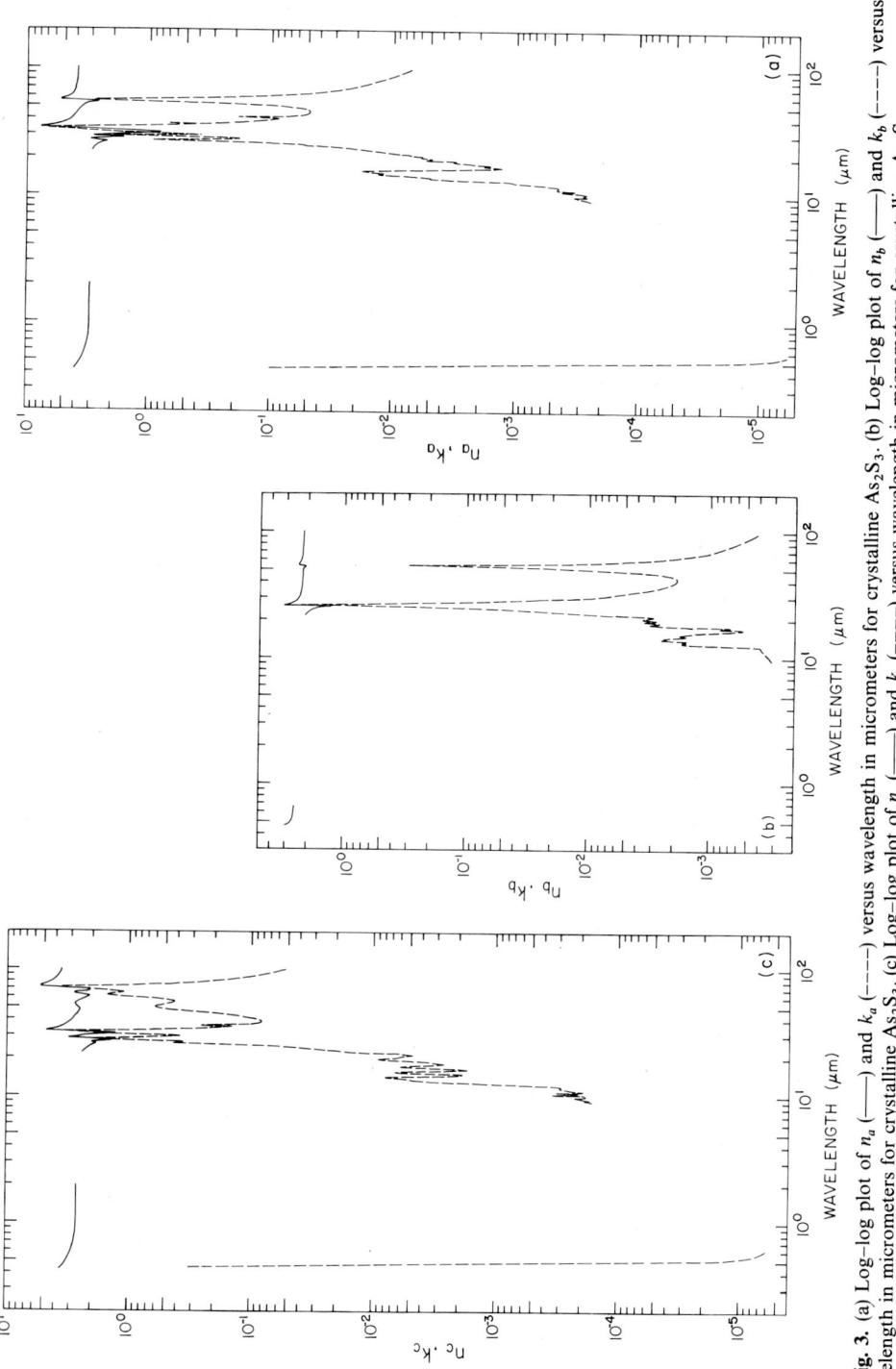

Fig. 3. (a) Log–log plot of n_a (——) and k_a (– – –) versus wavelength in micrometers for crystalline As$_2$S$_3$. (b) Log–log plot of n_b (——) and k_b (– – –) versus wavelength in micrometers for crystalline As$_2$S$_3$. (c) Log–log plot of n_c (——) and k_c (– – –) versus wavelength in micrometers for crystalline As$_2$S$_3$.

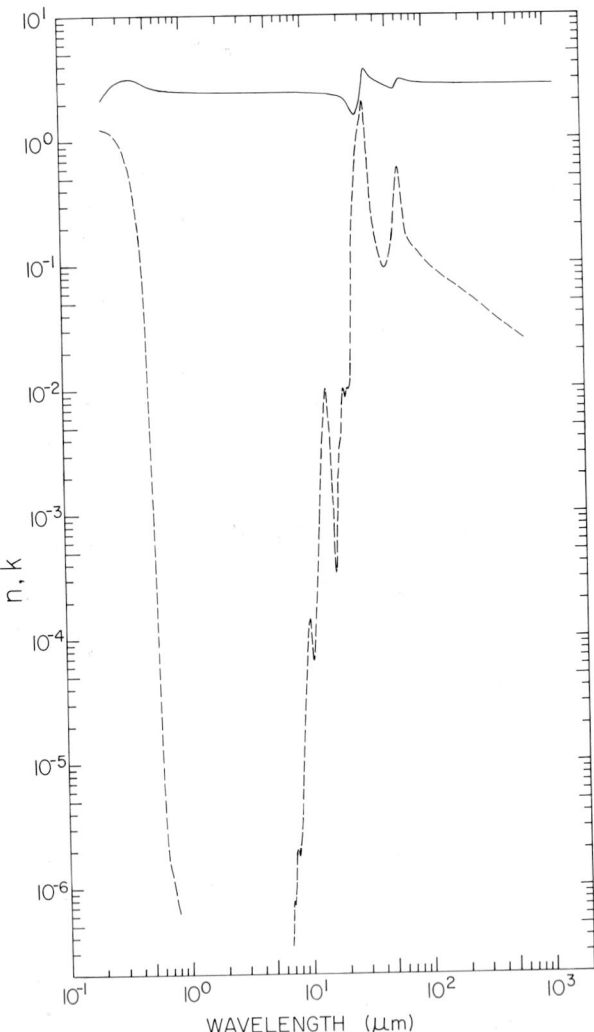

Fig. 4. Log–log plot of n (———) and (–––) k versus wavelength in micrometers for vitreous As_2S_3. Two significant figures are sometimes not enough to show a smooth variation in n and k, so the curves have been drawn through these points to smooth out steps.

TABLE III

Values of n and k for Crystalline As$_2$S$_3$ from Various References[a]

eV	cm^{-1}	μm	n_a	n_b	n_c	k_a	k_b	k_c
2.914	23,500	0.4255	3.92 [4]		3.60 [4]			
2.852	23,000	0.4348	3.84		3.51	9.8 × 10^{-2} [7]		
2.790	22,500	0.4444	3.77		3.43	3.5		0.32 [7]
2.728	22,000	0.4545	3.70		3.36			9.4 × 10^{-2}
2.666	21,500	0.4651	3.64	3.02 [4]	3.29	1.2 × 10^{-2}		2.6 × 10^{-2}
2.604	21,000	0.4762	3.58	2.79	3.24	3.7 × 10^{-3}		6.1 × 10^{-3} [4]
2.542	20,500	0.4878	3.53	2.69	3.18	1.0 × 10^{-3} [4]		8.5 × 10^{-4}
2.480	20,000	0.5000	3.48	2.64	3.13	2.5 × 10^{-4}		1.4 × 10^{-4}
2.418	19,500	0.5128	3.44	2.62	3.08	4.8 × 10^{-5}		1.9 × 10^{-5}
2.356	19,000	0.5263	3.39	2.60	3.04	2.4		1.3 × 10^{-5}
2.294	18,500	0.5405	3.35	2.58	3.00	1.2 × 10^{-5}		9.4 × 10^{-6}
2.232	18,000	0.5556	3.32	2.57	2.96	7.3 × 10^{-6}		8.2
2.170	17,500	0.5714	3.28	2.56	2.93	6.8		7.3
2.108	17,000	0.5882	3.25	2.55	2.90	5.9		6.7
2.046	16,500	0.6061	3.22	2.54	2.87			6.4
1.984	16,000	0.6250	3.19	2.53	2.84			6.2
1.922	15,500	0.6452	3.16	2.52	2.81			
1.860	15,000	0.6667	3.14	2.51	2.79			
1.798	14,500	0.6897	3.11		2.76			
1.736	14,000	0.7143	3.09		2.74			
1.674	13,500	0.7407	3.07		2.72			
1.612	13,000	0.7692	3.05		2.70			
1.550	12,500	0.8000	3.03		2.68			
1.488	12,000	0.8333	3.01		2.67			
1.426	11,500	0.8696	2.99		2.65			
1.364	11,000	0.9091	2.98		2.64			

(continued)

TABLE III (*Continued*)

Crystalline As$_2$S$_3$

eV	cm^{-1}	μm	n_a	n_b	n_c	k_a	k_b	k_c
1.302	10,500	0.9524	2.98		2.64			
1.240	10,000	1.000	2.98		2.64			
1.178	9,500	1.053	2.98		2.64			
1.116	9,000	1.111	2.98		2.64			
1.054	8,500	1.177	2.98		2.64			
0.9919	8,000	1.250	2.98		2.64			
0.9299	7,500	1.333	2.98		2.64			
0.8679	7,000	1.429	2.98		2.64			
0.8059	6,500	1.539	2.98		2.64			
0.7439	6,000	1.667	2.98		2.64			
0.6819	5,500	1.818	2.98		2.64			
0.6199	5,000	2.000	2.98		2.64			
0.1327	1,070	9.346				2.4×10^{-4} [5]		1.6×10^{-4} [5]
0.1314	1,060	9.434				2.6		1.9
0.1302	1,050	9.524				2.6		1.7
0.1289	1,040	9.615				2.9		1.9
0.1277	1,030	9.709				3.0		1.9
0.1265	1,020	9.804				3.2	3.1×10^{-4} [5]	2.0
0.1252	1,010	9.901				3.2	3.2	2.0
0.1240	1,000	10.00				3.4	3.2	2.3
0.1227	990	10.10				2.6	3.2	1.8
0.1215	980	10.20				2.8	3.2	1.8
0.1203	970	10.31				2.7	3.3	1.9
0.1190	960	10.42				2.6	3.3	3.4
0.1178	950	10.53				2.7	3.3	1.9
0.1165	940	10.64				2.8	3.4	1.9
0.1153	930	10.75				2.9	3.4	2.1
0.1141	920	10.87				3.5	3.5	2.6

0.1128	910	10.99	4.2	3.5	3.2
0.1116	900	11.11	3.3	3.5	1.9
0.1103	890	11.24	4.4	3.6	2.5
0.1091	880	11.36	4.7	3.6	2.5
0.1079	870	11.49	4.6	3.7	2.5
0.1066	860	11.63	4.5	3.7	2.6
0.1054	850	11.76	4.6	3.8	2.9
0.1041	840	11.90	4.6	3.8	3.0
0.1029	830	12.05	4.7	3.8	2.9
0.1017	820	12.20	5.1	3.8	3.1
0.1004	810	12.35	7.3	3.9	3.6
0.09919	800	12.50	8.0	4.0	6.8
0.09869	796	12.56	9.0×10^{-4}	5.0	6.0
0.09795	790	12.66	1.0×10^{-3}	6.1	8.1×10^{-4}
0.09720	784	12.76	1.1	7.1	1.0×10^{-3}
0.09646	778	12.85	1.4	9.2	1.3
0.09571	772	12.95	1.6	9.3×10^{-4}	2.1
0.09497	766	13.05	1.8	1.3×10^{-3}	2.6
0.09423	760	13.16	2.1	1.4	3.2
0.09348	754	13.26	3.5	1.7	3.6
0.09274	748	13.37	4.7	1.7	4.1
0.09199	742	13.48	4.8	1.6	4.6
0.09125	736	13.59	5.0	1.7	4.9
0.09051	730	13.70	4.9	1.7	5.2
0.08976	724	13.81	5.2	1.6	5.5
0.08902	718	13.93	5.8	1.8	6.2
0.08828	712	14.04	5.9	1.8	7.0
0.08753	706	14.16	7.1	2.3	7.5
0.08679	700	14.29	8.3	2.5	7.7
0.08617	695	14.39	8.7×10^{-3}	2.4	5.5
0.08555	690	14.49	0.013	2.3	4.5

(continued)

TABLE III (*Continued*)

Crystalline As$_2$S$_3$

eV	cm^{-1}	μm	n_a	n_b	n_c	k_a	k_b	k_c
0.08493	685	14.60				0.013	2.4	2.9
0.08431	680	14.71				0.013	2.3	2.9
0.08369	675	14.81				0.013	2.1	1.9
0.08307	670	14.93				0.012	1.9	2.5
0.08245	665	15.04				0.013	1.7	2.6
0.08183	660	15.15				0.015	1.6	3.6
0.08121	655	15.27				0.016	1.8	4.4
0.08059	650	15.38				0.017	1.8	5.6
0.07997	645	15.50				0.018	1.7	6.7
0.07935	640	15.63				0.017	1.6	5.8
0.07873	635	15.75				0.012	1.5	4.4
0.07811	630	15.87				8.2×10^{-3}	1.3	3.3
0.07749	625	16.00				6.7	1.0×10^{-3}	2.9
0.07687	620	16.13				4.5	9.0×10^{-4}	2.3
0.07625	615	16.26				1.9	9.0	1.7
0.07563	610	16.39				1.7	6.5	2.3
0.07501	605	16.53				1.5	5.3	2.9
0.07439	600	16.67				1.3	5.3	3.5
0.07377	595	16.81				1.5	6.7	3.9
0.07315	590	16.95				1.7	8.1	4.8
0.07253	585	17.09				1.8	8.1	5.7
0.07191	580	17.24				1.6	6.8	5.9
0.07129	575	17.39				1.7	8.3	5.4
0.07067	570	17.54				1.8	9.7×10^{-4}	4.5
0.07005	565	17.70				2.1	1.4×10^{-3}	3.7
0.06943	560	17.86				2.4	1.7	2.8
0.06881	555	18.02				2.7	2.1	2.6
0.06819	550	18.18				3.0	2.6	2.9

0.06757	545	18.35				3.2	2.9	3.1
0.06695	540	18.52				3.1	3.1	3.5
0.06633	535	18.69				3.3	2.7	4.0
0.06571	530	18.87				4.8	2.6	5.1
0.06509	525	19.05				5.3	2.7	6.5
0.06447	520	19.23				5.2	2.9	7.6
0.06385	515	19.42				4.8	3.4	8.0
0.06323	510	19.61				5.3	3.3	8.9
0.06261	505	19.80				5.7	2.8	8.5
0.06199	500	20.00				5.9	2.9	7.2
0.06150	496	20.16				5.8	3.2	6.3
0.06075	490	20.41				6.5	3.4	6.2
0.06001	484	20.66				7.1	3.5	5.9
0.05926	478	20.92				7.5	3.2	5.5
0.05852	472	21.19				8.1	3.0	4.7
0.05778	466	21.46		2.1 [3]		9.9×10^{-3}	3.1	5.5
0.05703	460	21.74		2.1		0.011	3.3	6.0×10^{-3} [2]
0.05636	454.6	22.0	2.9 [2]	2.0	2.4 [2]	0.013 [2]	7.7×10^{-3} [3]	0.011 [2]
0.05510	444.4	22.5	2.9	2.0	2.4	0.016	0.011	0.017
0.05391	434.8	23.0	2.9	2.0	2.3	0.019	0.016	0.023
0.05276	425.5	23.5	2.8	1.9	2.3	0.023	0.024	0.031
0.05166	416.7	24.0	2.8	1.9	2.2	0.029	0.042	0.041
0.05060	408.2	24.5	2.7	1.7	2.1	0.050	0.080	0.057
0.04959	400.0	25.0	2.6	1.5	2.0	0.053	0.24	0.089
0.04862	392.2	25.5	2.5	1.3	1.9	0.087	1.1	0.18
0.04769	384.6	26.0	2.3	3.1	1.9	0.20	1.4	0.44
0.04679	377.4	26.5	2.3	2.9	2.0	0.84	0.30	0.37
0.04592	370.4	27.0	2.9	2.7	1.6	0.41	0.11	0.42
0.04508	363.6	27.5	2.5	2.5	1.4	0.19	0.057	1.0
0.04428	357.1	28.0	2.2	2.5	2.0	0.20	0.036	2.0
0.04350	350.9	28.5	1.6	2.4	3.1	0.50	0.024	1.3

(continued)

TABLE III (*Continued*)

Crystalline As$_2$S$_3$

eV	cm^{-1}	μm	n_a	n_b	n_c	k_a	k_b	k_c
0.04275	344.8	29.0	2.9	2.4	2.8	1.8	0.018	0.54
0.04203	339.0	29.5	2.4	2.4	2.4	0.39	0.014	0.39
0.04133	333.3	30.0	1.8	2.3	2.1	0.45	0.011	0.43
0.04065	327.9	30.5	1.2	2.3	1.6	1.4	9.4×10^{-3}	0.75
0.03999	322.6	31.0	0.83	2.3	1.3	1.7	8.0	1.6
0.03936	317.5	31.5	0.86	2.3	1.8	2.6	6.9	2.7
0.03874	312.5	32.0	1.1	2.3	3.5	3.6	6.0	3.4
0.03815	307.7	32.5	1.8	2.3	4.8	4.9	5.3	1.9
0.03757	303.0	33.0	3.5	2.3	4.5	6.3	4.7	0.79
0.03701	298.5	33.5	7.0	2.3	4.1	5.5	4.3	0.43
0.03647	294.1	34.0	7.6	2.3	3.8	2.4	3.9	0.28
0.03594	289.9	34.5	6.7	2.3	3.6	1.2	3.6	0.20
0.03542	285.7	35.0	6.0	2.3	3.5	0.69	3.3	0.16
0.03492	281.7	35.5	5.6	2.3	3.3	0.48	3.1	0.15
0.03459	279.0	35.9				0.55		
0.03444	277.8	36.0	5.3	2.3	3.3	0.38	2.9	0.26
0.03397	274.0	36.5	5.1	2.3	3.3	0.26	2.7	0.11
0.03351	270.2	37.0	4.9	2.3	3.2	0.20	2.5	0.094
0.03306	266.7	37.5	4.7	2.2	3.1	0.16	2.4	0.087
0.03263	263.2	38.0	4.6	2.2	3.1	0.14	2.3	0.084
0.03220	259.7	38.5	4.5	2.2	3.0	0.12	2.2	0.082
0.03179	256.4	39.0	4.4	2.2	3.0	0.10	2.1	0.083
0.03139	253.2	39.5	4.4	2.2	3.0	0.094	2.1	0.084
0.03100	250.0	40.0	4.3	2.2	2.9	0.087	2.0	0.087
0.03061	246.9	40.5	4.2	2.2	2.9	0.094	1.9	0.092
0.03024	243.9	41.0	4.2	2.2	2.9	0.19	1.9	0.097
0.02988	241.0	41.5	4.2	2.2	2.8	0.080	1.9×10^{-3}	0.10
0.02952	238.1	42.0	4.1	2.2	2.8	0.063	1.9	0.11

0.02917	235.3	42.5	4.1	2.2	2.8	0.058	1.9	0.12
0.02883	232.6	43.0	4.1	2.2	2.7	0.054	1.9	0.14
0.02850	229.9	43.5	4.0	2.2	2.7	0.052	1.9	0.15
0.02818	227.3	44.0	4.0	2.2	2.7	0.050	1.9	0.17
0.02786	224.7	44.5	4.0	2.2	2.6	0.048	2.0	0.19
0.02755	222.2	45.0	3.9	2.2	2.6	0.048	2.0	0.21
0.02725	219.8	45.5	3.9	2.2	2.6	0.048	2.1	0.24
0.02695	217.4	46.0	3.9	2.2	2.6	0.048	2.3	0.28
0.02666	215.1	46.5	3.8	2.2	2.5	0.049	2.5	0.31
0.02638	212.8	47.0	3.8	2.2	2.5	0.051	2.6	0.36
0.02610	210.5	47.5	3.8	2.2	2.5	0.053	2.9	0.40
0.02583	208.3	48.0	3.7	2.2	2.5	0.056	3.3	0.46
0.02556	206.2	48.5	3.7	2.2	2.5	0.061	3.8	0.50
0.02530	204.1	49.0	3.6	2.2	2.6	0.068	4.5	0.55
0.02505	202.0	49.5	3.5	2.2	2.6	0.076	5.4	0.58
0.02480	200.0	50.0	3.5	2.2	2.6	0.088	6.8	0.60
0.02455	198.0	50.5	3.5	2.2	2.7	0.11	8.9 × 10^{-3}	0.61
0.02431	196.1	51.0	3.4	2.2	2.7	0.13	0.012	0.60
0.02407	194.2	51.5	3.3	2.2	2.7	0.16	0.018	0.59
0.02384	192.3	52.0	3.2	2.1	2.8	0.21	0.029	0.56
0.02362	190.5	52.5	3.1	2.1	2.8	0.31	0.054	0.53
0.02339	188.7	53.0	2.9	2.1	2.8	0.47	0.12	0.51
0.02317	186.9	53.5	2.7	2.4	2.8	0.79	0.28	0.48
0.02296	185.2	54.0	2.6	2.4	2.7	1.4	0.27	0.46
0.02275	183.5	54.5	2.9	2.4	2.7	2.2	0.11	0.44
0.02254	181.8	55.0	4.0	2.3	2.7	2.7	0.054	0.43
0.02234	180.2	55.5	5.1	2.3	2.6	2.1	0.031	0.42
0.02214	178.6	56.0	5.3	2.3	2.6	1.2	0.020	0.42
0.02194	177.0	56.5	5.2	2.3	2.5	0.73	0.014	0.43
0.02175	175.4	57.0	5.0	2.3	2.5	0.47	0.011	0.45
0.02156	173.9	57.5	4.8	2.3	2.4	0.32	8.3 × 10^{-3}	0.48

(continued)

TABLE III (*Continued*)

Crystalline As$_2$S$_3$

eV	cm^{-1}	μm	n_a	n_b	n_c	k_a	k_b	k_c
0.02138	172.4	58.0	4.7	2.3	2.3	0.24	6.7	0.52
0.02119	170.9	58.5	4.6	2.3	2.3	0.18	5.6	0.57
0.02101	169.5	59.0	4.5	2.3	2.2	0.15	4.7	0.65
0.02084	168.1	59.5	4.4	2.3	2.2	0.12	4.1	0.75
0.02066	166.7	60.0	4.4	2.3	2.1	0.10	3.6	0.87
0.02049	165.3	60.5	4.3	2.3	2.1	0.088	3.2	1.0
0.02033	163.9	61.0	4.3	2.3	2.1	0.077	2.9	1.2
0.02016	162.6	61.5	4.3	2.3	2.3	0.068	2.6	1.3
0.02000	161.3	62.0	4.2	2.3	2.4	0.061	2.4	1.4
0.01984	160.0	62.5	4.2	2.2	2.6	0.055	2.2	1.5
0.01968	158.7	63.0	4.2	2.2	2.7	0.050	2.1	1.4
0.01953	157.5	63.5	4.2	2.2	2.8	0.046	1.9	1.4
0.01937	156.3	64.0	4.2	2.2	2.8	0.042	1.8	1.3
0.01922	155.0	64.5	4.1	2.2	2.8	0.039	1.7	1.2
0.01907	153.9	65.0	4.1	2.2	2.7	0.036	1.6	1.1
0.01893	152.7	65.5	4.1	2.2	2.6	0.034	1.5	1.1
0.01879	151.5	66.0	4.1	2.2	2.5	0.032	1.5	1.1
0.01864	150.4	66.5	4.1	2.2	2.4	0.030	1.4	1.2
0.01851	149.3	67.0	4.1	2.2	2.2	0.029	1.3	1.3
0.01837	148.2	67.5	4.1	2.2	2.1	0.027	1.3	1.5
0.01823	147.1	68.0	4.1	2.2	2.1	0.026	1.2	1.8
0.01810	146.0	68.5	4.1	2.2	2.1	0.025	1.2	2.1
0.01797	144.9	69.0	4.0	2.2	2.2	0.024	1.1	2.5
0.01784	143.9	69.5	4.0	2.2	2.5	0.023	1.1	2.9
0.01771	142.9	70.0	4.0	2.2	2.9	0.022	1.1	3.3
0.01759	141.8	70.5	4.0	2.2	3.5	0.021	1.0×10^{-3}	3.5
0.01746	140.9	71.0	4.0	2.2	4.2	0.020	9.9×10^{-4}	3.5
0.01734	139.9	71.5	4.0	2.2	4.8	0.019	9.6	3.2

0.01722	138.9	72.0	4.0	2.2	5.3	0.019	9.4	2.7
0.01710	137.9	72.5	4.0	2.2	5.4	0.018	9.1	2.1
0.01698	137.0	73.0	4.0	2.2	5.4	0.018	8.9	1.7
0.01687	136.1	73.5	4.0	2.2	5.3	0.017	8.6	1.3
0.01675	135.1	74.0	4.0	2.2	5.2	0.017	8.4	1.1
0.01664	134.2	74.5	4.0	2.2	5.1	0.016	8.2	0.86
0.01653	133.3	75.0	4.0	2.2	4.9	0.016	8.1	0.72
0.01642	132.5	75.5	4.0	2.2	4.8	0.015	7.9	0.61
0.01631	131.6	76.0	4.0	2.2	4.7	0.015	7.7	0.52
0.01621	130.7	76.5	4.0	2.2	4.6	0.015	7.6	0.46
0.01610	129.9	77.0	4.0	2.2	4.6	0.014	7.4	0.40
0.01600	129.0	77.5	4.0	2.2	4.5	0.014	7.3	0.36
0.01590	128.2	78.0	3.9	2.2	4.4	0.014	7.1	0.32
0.01579	127.4	78.5	3.9	2.2	4.4	0.013	7.0	0.29
0.01569	126.6	79.0	3.9	2.2	4.3	0.013	6.9	0.27
0.01560	125.8	79.5	3.9	2.2	4.3	0.013	6.7	0.25
0.01550	125.0	80.0	3.9	2.2	4.2	0.013	6.6	0.23
0.01540	124.2	80.5	3.9	2.2	4.2	0.012	6.5	0.21
0.01531	123.5	81.0	3.9	2.2	4.1	0.012	6.4	0.20
0.01521	122.7	81.5	3.9	2.2	4.1	0.012	6.3	0.18
0.01512	122.0	82.0	3.9	2.2	4.1	0.012	6.2	0.17
0.01503	121.2	82.5	3.9	2.2	4.1	0.011	6.1	0.16
0.01494	120.5	83.0	3.9	2.2	4.0	0.011	6.0	0.15
0.01485	119.8	83.5	3.9	2.2	4.0	0.011	5.9	0.15
0.01476	119.1	84.0	3.9	2.2	4.0	0.0108	5.8	0.14
0.01467	118.3	84.5	3.9	2.2	4.0	0.0106	5.8	0.13
0.01459	117.7	85.0	3.9	2.2	3.9	0.0105	5.7	0.13
0.01450	117.0	85.5	3.9	2.2	3.9	0.0103	5.6	0.12
0.01442	116.3	86.0	3.9	2.2	3.9	0.0101	5.5	0.12
0.01433	115.6	86.5	3.9	2.2	3.9	0.0100	5.5	0.11
0.01425	114.9	87.0	3.9	2.2	3.9	9.8×10^{-3}	5.4	0.11

(continued)

TABLE III (*Continued*)

Crystalline As$_2$S$_3$

eV	cm^{-1}	μm	n_a	n_b	n_c	k_a	k_b	k_c
0.01417	114.3	87.5	3.9	2.2	3.8	9.7	5.3	0.10
0.01409	113.6	88.0	3.9	2.2	3.8	9.6	5.3	0.099
0.01401	113.0	88.5	3.9	2.2	3.8	9.4	5.2	0.095
0.01393	112.4	89.0	3.9	2.2	3.8	9.3	5.1	0.092
0.01385	111.7	89.5	3.9	2.2	3.8	9.2	5.1	0.089
0.01378	111.1	90.0	3.9	2.2	3.8	9.0	5.0	0.086
0.01370	110.5	90.5	3.9	2.2	3.8	9.0	5.0	0.083
0.01362	109.9	91.0	3.9	2.2	3.8	8.8	4.9	0.081
0.01355	109.3	91.5	3.9	2.2	3.7	8.7	4.8	0.078
0.01348	108.7	92.0	3.9	2.2	3.7	8.6	4.8	0.076
0.01340	108.1	92.5	3.9	2.2	3.7	8.5	4.7	0.074
0.01333	107.5	93.0	3.9	2.2	3.7	8.4	4.7	0.072
0.01326	107.0	93.5	3.9	2.2	3.7	8.3	4.7	0.070
0.01319	106.4	94.0	3.9	2.2	3.7	8.2	4.6	0.068
0.01310	105.8	94.5	3.9	2.2	3.7	8.1	4.6	0.067
0.01305	105.3	95.0	3.9	2.2	3.7	8.0	4.5	0.065
0.01298	104.7	95.5	3.9	2.2	3.7	7.9	4.5	0.064
0.01292	104.2	96.0	3.9	2.2	3.7	7.8	4.4	0.062
0.01285	103.6	96.5	3.9	2.2	3.7	7.7	4.4	0.061
0.01278	103.1	97.0	3.9	2.2	3.7	7.7	4.3	0.059
0.01272	102.6	97.5	3.9	2.2	3.6	7.6	4.3	0.058
0.01265	102.0	98.0	3.9	2.2	3.6	7.5	4.3	0.057
0.01259	101.5	98.5	3.9	2.2	3.6	7.4	4.2	0.056
0.01252	101.0	99.0	3.9	2.2	3.6	7.3	4.2	0.055
0.01246	100.5	99.5	3.9	2.2	3.6	7.3	4.1	0.053
0.01240	100.0	100	3.9	2.2	3.6	7.2	4.1	0.052

[a] The radiation electric field is parallel to the a, b, and c axes of the crystal.

TABLE IV
Values of n and k for Vitreous As$_2$S$_3$ from Various References[a]

eV	cm^{-1}	μm	n	k
5.636	45,460	0.2200	2.10 [15]	1.24 [15]
5.390	43,480	0.2300	2.20	1.23
5.166	41,670	0.2400	2.34	1.22
4.959	40,000	0.2500	2.48	1.21
4.769	38,460	0.2600	2.60	1.19
4.592	37,040	0.27	2.73	1.15
4.428	35,710	0.28	2.79	1.11
4.275	34,480	0.29	2.85	1.05
4.133	33,330	0.30	2.90	1.01
3.999	32,260	0.31	2.98	0.95
3.874	31,250	0.32	3.02	0.89
3.757	30,300	0.33	3.05	0.82
3.647	29,410	0.34	3.07	0.73
3.542	28,570	0.35	3.09	0.66
3.444	27,780	0.36	3.10	0.61
3.351	27,030	0.37	3.11	0.55
3.263	26,320	0.38	3.11	0.49
3.179	25,610	0.39	3.10	0.42
3.100	25,000	0.40	3.09	0.34
3.024	24,390	0.41	3.06	0.28
2.952	23,810	0.42	3.04	0.24
2.883	23,260	0.43	3.02	0.20
2.818	22,730	0.44	3.00	0.16
2.80	22,520	0.4428	3.00	0.15
2.79	22,500	0.4444	2.99	0.15
2.78	22,420	0.4460	2.99	0.14
2.77	22,340	0.4476	2.99	0.14
2.76	22,220	0.4492	2.99	0.13
2.75	22,180	0.4508	2.98	0.126
2.74	22,100	0.4525	2.98	0.122
2.73	22,020	0.4542	2.98	0.117
2.72	21,940	0.4558	2.97	0.112
2.71	21,860	0.4575	2.97	0.105
2.70	21,780	0.4592	2.97	0.099
2.69	21,700	0.4609	2.96	0.096
2.68	21,620	0.4626	2.96	0.091
2.67	21,540	0.4644	2.95	0.084
2.66	21,460	0.4661	2.95	0.078
2.65	21,370	0.4679	2.94	0.076
2.64	21,290	0.4696	2.94	0.071
2.63	21,210	0.4714	2.93	0.067
2.62	21,130	0.4732	2.93	0.060
2.61	21,050	0.4750	2.92	0.055
2.60	20,970	0.4769	2.92	0.052
2.59	20,890	0.4787	2.91	0.047
2.58	20,810	0.4806	2.90	0.044

(*continued*)

TABLE IV (*Continued*)

Vitreous As_2S_3

eV	cm^{-1}	μm	n	k
2.57	20,730	0.4824	2.89	0.040
2.56	20,650	0.4843	2.89	0.035
2.55	20,570	0.4862	2.88	0.030
2.54	20,490	0.4881	2.88	0.026
2.53	20,410	0.4901	2.87	0.022
2.52	20,260	0.4920	2.87	0.019
2.51	20,250	0.4940	2.86	0.017
2.50	20,160	0.4959	2.85	0.016
2.49	20,080	0.4979	2.84	0.014
2.48	20,000	0.4999	2.83	0.013
2.47	19,920	0.5020	2.82	0.011
2.46	19,840	0.5040	2.82	9.8×10^{-3}
2.45	19,760	0.5061	2.81	8.7
2.44	19,680	0.5081	2.80	8.0
2.43	19,600	0.5102	2.80	7.0
2.42	19,520	0.5123	2.79	6.4
2.41	19,440	0.5145	2.78	5.5
2.40	19,360	0.5166	2.78	4.7
2.39	19,280	0.5188	2.77	4.1
2.38	19,200	0.5209	2.76	3.6
2.37	19,120	0.5231	2.76	3.2
2.36	19,040	0.5254	2.76	2.7
2.35	18,950	0.5276	2.75	2.4
2.34	18,870	0.5298	2.74	2.1
2.33	18,790	0.5321	2.73	1.8
2.32	18,710	0.5344	2.72	1.6
2.31	18,630	0.5367	2.72	1.3
2.30	18,550	0.5391	2.72	1.1×10^{-3}
2.29	18,470	0.5414	2.71	9.2×10^{-4}
2.28	18,390	0.5438	2.71	8.1
2.27	18,310	0.5462	2.70	7.0
2.26	18,230	0.5486	2.70	6.0
2.25	18,150	0.5510	2.70	5.1
2.24	18,070	0.5535	2.69	4.3
2.23	17,990	0.5560	2.69	3.8
2.22	17,910	0.5585	2.69	3.3
2.214	17,860	0.56	2.69 [12]	2.4×10^{-4} [7]
2.138	17,240	0.58	2.66	5.6×10^{-5}
2.066	16,670	0.60	2.64	$1.5 \times 10^{-|5}$
2.000	16,130	0.62	2.62	5.1×10^{-6}
1.937	15,630	0.64	2.60	2.6
1.879	15,150	0.66	2.59	1.7
1.823	14,710	0.68	2.57	1.5
1.771	14,290	0.70	2.56	1.2
1.722	13,890	0.72	2.55	1.03×10^{-6}
1.675	13,510	0.74	2.54	9.1×10^{-7}
1.631	13,160	0.76	2.53	7.4

TABLE IV (*Continued*)

Vitreous As$_2$S$_3$

eV	cm^{-1}	μm	n	k
1.590	12,820	0.78	2.53	6.5
1.550	12,500	0.80	2.52	6.2
1.512	12,200	0.82	2.51	
1.476	11,900	0.84	2.51	
1.442	11,630	0.86	2.50	
1.409	11,360	0.88	2.50	
1.378	11,110	0.90	2.50	
1.348	10,870	0.92	2.49	
1.319	10,640	0.94	2.49	
1.291	10,420	0.96	2.48	
1.265	10,200	0.98	2.48	
1.240	10,000	1.0	2.48	2.4×10^{-7} [13]
1.033	8,333	1.2	2.46	
0.8856	7,143	1.4	2.44	
0.7749	6,250	1.6	2.44	
0.6888	5,556	1.8	2.43	
0.6199	5,000	2.0	2.43	
0.5636	4,546	2.2	2.42	
0.5166	4,167	2.4	2.42	
0.4769	3,846	2.6	2.42	
0.4428	3,571	2.8	2.42	
0.4133	3,333	3.0	2.42	
0.3874	3,125	3.2	2.41	
0.3647	2,941	3.4	2.41	
0.3444	2,778	3.6	2.41	
0.3263	2,632	3.8	2.41	
0.3100	2,500	4.0	2.41	
0.2952	2,381	4.2	2.41	
0.2818	2,273	4.4	2.41	
0.2695	2,174	4.6	2.41	
0.2583	2,083	4.8	2.41	
0.2480	2,000	5.0	2.41	
0.2384	1,923	5.2	2.41	
0.2296	1,852	5.4	2.41	
0.2214	1,786	5.6	2.40	
0.2138	1,724	5.8	2.40	
0.2066	1,667	6.0	2.40	
0.2000	1,613	6.2	2.40	
0.1937	1,563	6.4	2.40	
0.1879	1,515	6.6	2.40	
0.1860	1,500	6.667	2.40	3.4 [16]
0.1829	1,475	6.780	2.40	5.4
0.1798	1,450	6.897	2.40	6.5
0.1767	1,425	7.018	2.40	6.4
0.1736	1,400	7.143	2.40	7.4
0.1705	1,375	7.273	2.40	8.7×10^{-7}

(*continued*)

TABLE IV (*Continued*)

Vitreous As₂S₃

eV	cm^{-1}	μm	n	k
0.1674	1,350	7.407	2.40	1.2×10^{-6}
0.1643	1,325	7.547	2.40	1.4
0.1612	1,300	7.692	2.40	1.4
0.1581	1,275	7.843	2.39	1.2
0.1550	1,250	8.000	2.39	1.3
0.1519	1,225	8.163	2.39	1.8
0.1488	1,200	8.333	2.39	3.1
0.1457	1,175	8.511	2.39	5.5×10^{-6}
0.1426	1,150	8.696	2.39	1.04×10^{-5}
0.1395	1,125	8.889	2.39	1.6
0.1364	1,100	9.091	2.38	2.9
0.1339	1,080	9.259	2.38	3.5
0.1314	1,060	9.434	2.38	4.5
0.1289	1,040	9.615	2.38	6.4
0.1265	1,020	9.804	2.38	9.9×10^{-5}
0.1240	1,000	10.00	2.38	1.3×10^{-4}
0.1215	980	10.20	2.38	1.4
0.1190	960	10.42	2.38	1.03×10^{-4}
0.1165	940	10.64	2.38	9.3×10^{-5}
0.1141	920	10.87	2.37	6.5
0.1116	900	11.11	2.37	9.3×10^{-5}
0.1011	880	11.36	2.37	1.3×10^{-4}
0.1066	860	11.63	2.37	1.9
0.1041	840	11.91	2.36	2.8
0.1017	820	12.20	2.36	3.6
0.09919	800	12.50	2.36 [15]	5.0 [11]
0.09795	790	12.66	2.36	6.5
0.09671	780	12.82	2.36	9.0×10^{-4}
0.09547	770	12.99	2.36	1.8×10^{-3}
0.09423	760	13.16	2.35	2.4
0.09299	750	13.33	2.35	3.0
0.09175	740	13.51	2.35	3.8
0.09051	730	13.70	2.35	4.6
0.08989	725	13.79	2.35	5.3
0.08927	720	13.89	2.34	6.1
0.08865	715	13.99	2.34	6.7
0.08803	710	14.09	2.34	7.7
0.08679	700	14.29	2.34	8.6
0.08555	690	14.49	2.34	9.7×10^{-3}
0.08493	685	14.60	2.34	1.0×10^{-2}
0.08431	680	14.71	2.34	9.6×10^{-3}
0.08307	670	14.93	2.33	8.3
0.08183	660	15.15	2.33	6.0
0.08059	650	15.39	2.33	5.3
0.07935	640	15.63	2.33	4.4
0.07811	630	15.87	2.32	3.0
0.07687	620	16.13	2.32	1.8×10^{-3}

TABLE IV (*Continued*)

Vitreous As$_2$S$_3$

eV	cm^{-1}	μm	n	k
0.07563	610	16.39	2.31	9.1×10^{-4}
0.07439	600	16.67	2.31	4.6
0.07315	590	16.95	2.31	3.4
0.07191	580	17.24	2.30	4.1
0.07067	570	17.54	2.30	8.4×10^{-4}
0.06943	560	17.86	2.29	1.1×10^{-3}
0.06819	550	18.18	2.28	1.9
0.06695	540	18.52	2.27	2.4
0.06571	530	18.87	2.26	3.5
0.06447	520	19.23	2.25	4.0
0.06323	510	19.61	2.24	5.5
0.06199	500	20.00	2.23	9.4×10^{-3}
0.06100	492	20.33	2.22	0.010
0.06050	488	20.49	2.22	0.011
0.05951	480	20.83	2.21	9.6×10^{-3}
0.05852	472	21.19	2.20	8.4
0.05802	468	21.37	2.19	8.6×10^{-3}
0.05703	460	21.74	2.17	0.010
0.05579	450	22.22	2.15	9.9×10^{-3}
0.05269	425	23.53	2.07	0.014
0.05083	410	24.39	2.01	0.050
0.05021	405	24.69	1.94	0.098
0.04959	400.0	25.0	1.79 [10]	0.20 [10]
0.04862	392.1	25.5	1.73	0.25
0.04769	384.6	26.0	1.67	0.33
0.04679	377.4	26.5	1.61	0.42
0.04592	370.4	27.0	1.57	0.55
0.04508	363.6	27.5	1.55	0.70
0.04428	357.1	28.0	1.56	0.86
0.04350	350.9	28.5	1.62	1.0
0.04275	344.8	29.0	1.70	1.1
0.04203	339.0	29.5	1.79	1.3
0.04133	333.3	30.0	1.86	1.4
0.04065	327.9	30.5	1.96	1.5
0.03999	322.6	31.0	2.12	1.8
0.03936	317.5	31.5	2.43	2.0
0.03874	312.5	32.0	2.88	2.0
0.03815	307.7	32.5	3.33	1.8
0.03757	303.0	33.0	3.59	1.4
0.03701	298.5	33.5	3.65	1.1
0.03647	294.1	34.0	3.61	0.83
0.03594	289.9	34.5	3.53	0.64
0.03542	285.7	35.0	3.45	0.51
0.03492	281.7	35.5	3.37	0.42
0.03444	277.8	36.0	3.30	0.35
0.03397	274.0	36.5	3.24	0.30

(*continued*)

TABLE IV (*Continued*)

Vitreous As_2S_3

eV	cm^{-1}	μm	n	k
0.03351	270.3	37.0	3.19	0.26
0.03306	266.7	37.5	3.14	0.23
0.03263	263.2	38.0	3.10	0.21
0.03220	259.7	38.5	3.07	0.19
0.03179	256.4	39.0	3.03	0.17
0.03139	253.2	39.5	3.00	0.16
0.03100	250.0	40.0	2.98	0.15
0.03024	243.9	41	2.93	0.13
0.02952	238.1	42	2.89	0.12
0.02883	232.6	43	2.85	0.11
0.02818	227.3	44	2.82	0.10
0.02755	222.2	45	2.79	0.096
0.02695	217.4	46	2.76	0.094
0.02638	212.8	47	2.74	0.094
0.02583	208.3	48	2.71	0.096
0.02530	204.1	49	2.68	0.10
0.02480	200.0	50	2.66	0.11
0.02431	196.1	51	2.63	0.12
0.02384	192.3	52	2.61	0.13
0.02339	188.7	53	2.58	0.15
0.02296	185.2	54	2.55	0.18
0.02254	181.8	55	2.53	0.22
0.02214	178.6	56	2.50	0.27
0.02175	175.4	57	2.50	0.34
0.02138	172.4	58	2.51	0.42
0.02101	169.5	59	2.56	0.51
0.02066	166.7	60	2.64	0.57
0.02032	163.9	61	2.76	0.60
0.02000	161.3	62	2.87	0.58
0.01968	158.7	63	2.96	0.52
0.01937	156.3	64	3.01	0.44
0.01908	153.9	65	3.03	0.37
0.01879	151.5	66	3.04	0.31
0.01850	149.3	67	3.03	0.26
0.01823	147.1	68	3.02	0.22
0.01797	144.9	69	3.00	0.19
0.01771	142.9	70	2.99	0.17
0.01722	138.9	72	2.96	0.15
0.01676	135.1	74	2.94	0.15
0.01631	131.6	76	2.92	0.15
0.01590	128.2	78	2.90	0.14
0.01550	125.0	80	2.89	0.14
0.01512	122.0	82	2.87	0.14
0.01476	119.1	84	2.86	0.14
0.01442	116.3	86	2.85	0.14
0.01409	113.6	88	2.84	0.13
0.01378	111.1	90	2.84	0.12

TABLE IV (*Continued*)

Vitreous As$_2$S$_3$

eV	cm^{-1}	μm	n	k
0.01348	108.7	92	2.83	0.12
0.01319	106.4	94	2.82	0.11
0.01292	104.2	96	2.82	0.11
0.01265	102.0	98	2.81	0.10
0.01240	100	100	2.81	0.10
0.01066	86	116	2.79 [9]	0.095 [9]
0.008183	66	152	2.76	0.072
0.006447	52	192	2.75	0.054
0.005393	43.5	230	2.75	0.055
0.004587	37	270	2.74	0.047
0.004029	32.5	308	2.74	0.044
0.003162	25.5	392	2.74	0.033
0.002418	19.5	513	2.74	0.031
0.002170	17.5	571	2.74	0.027
0.001984	16	625	2.74	0.025
0.001048	8.45	1,180	2.73	8.8×10^{-3}
0.0001033	0.833	12,000	2.73	1.3
4.129×10^{-12}	3.33×10^{-8}	3×10^{11}	2.73 [8]	

[a] Reference are indicated in brackets.

Cubic Carbon (Diamond)

DAVID F. EDWARDS*,†

University of California
Los Alamos National Laboratory
Los Alamos, New Mexico

H. R. PHILIPP

General Electric Research and Development Center
Schenectady, New York

Diamond is classified into four types depending on its optical and electrical properties. Common to each type is an absorption band in the 2- to 6-μm infrared region, which, as explained later, is due to a multiphonon absorption [1, 2]. Most natural diamonds are of type Ia and, in addition to the intrinsic 2- to 6-μm band, contain nitrogen [3] as an impurity that absorbs in the 6- to 12-μm region. The ultraviolet absorption edge of type Ia diamonds is at about 291 nm. Synthetic diamonds are type Ib and contain nitrogen as an impurity in a dispersed form. This, too, absorbs in the 6- to 12-μm region. Type IIa diamonds are effectively free of nitrogen and exhibit only the intrinsic 2- to 6-μm absorption. The ultraviolet absorption edge is at about 222 nm. Type IIb diamonds exhibit semiconducting properties [4] at ordinary temperatures, have a narrow absorption band from 7 to 10 μm, and are generally blue in color. Irradiation with neutrons or electrons will produce absorption in the 6- to 12-μm band (characteristic of type I diamonds) in previously transparent type IIa diamonds [1].

Type IIa diamonds have the optical properties best suited for infrared optical components. They absorb only in the intrinsic 2- to 6-μm band and their ultraviolet transmittance extends to 222 nm. Optical components with flat surfaces (e.g., windows, prisms, wedge plates, interferometer plates, infrared diffusers) can be produced to specification from diamond. It has not been cost-effective to produce polished curved surfaces (e.g., lenses) of optical quality on diamond. However, it should be possible to produce diamond Fresnel lenses, diffraction gratings, and other components with flat surface patterns [5].

* Work performed under the auspices of the U.S. Department of Energy.
† Present address: Lawrence Livermore National Laboratory, Livermore, California.

665

Seal [6], Vishnerskiy *et al.* [7], and others have shown that the index of refraction of diamond sensitively depends on the properties of the crystal. The crystal internal structure, defects, impurities, and mechanical inclusions can alter the index. This is true of natural as well as synthetically grown diamonds. As stated earlier, the presence of nitrogen as an impurity can shift the ultraviolet absorption edge (222 nm for nitrogen free to 291 nm for a few parts in 10^4) and produce infrared absorption bands (6 to 12 μm). The effect of impurities on the index of refraction is greater in the infrared than the visible regions. For example, Vishnevskiy *et al.* [7] measured a 0.4% index variation for 26 diamonds (22 synthetic and 4 natural) at 580 nm, and Seal [6], working at 10 μm, saw a 4.7% variation for 11 samples (types Ia and IIa natural stones).

Tabulated in Table V and plotted in Fig. 5 are the room-temperature indices of refraction and extinction coefficients for natural types Ia and IIa diamonds; no values are included for synthetic diamond. For the visible region (226 to 722 nm) three results [7–9] were for diamonds either identified as or assumed to be type Ia. Two samples were IIa diamonds [10]. For common wavelengths the visible indices of types Ia and IIa diamonds as measured by different investigators [7–11] agree within three parts or less in the fourth decimal place. Because of this small difference, no distinction is made as to type for the visible indices in Table I. For the infrared, the sample-to-sample index variation is much greater for type Ia diamonds than for type IIa diamonds. Therefore, the listed infrared indices ($\lambda \geq 2.5$ μm) are for type IIa diamond exclusively.

The visible refractive indices were determined from the dispersion of a small diamond prism. Peter [10], using type IIa diamond samples, measured the index from 226 to 643 nm. His measurements are perhaps the most complete for the visible. Other investigations [7–9, 11] have also reported refractive indices for the visible region but for limited wavelengths. Each used some variation of a prism measuring technique with the exception of Bartoshinskii *et al.* [9], who developed a technique for determining the index using total internal reflections from {111} crystallographic planes in diamond.

The infrared index values from 2.5 to 25 μm are the measurements of Edwards and Ochoa [12] and are for a type IIa diamond. They used the channel spectrum technique [13], and it is the only known report for the infrared. The details of the channel technique can be found elsewhere in this publication and will not be repeated here. (See the critique on silicon in this handbook.)

The visible and infrared indices of refraction given in Table V as n_c are obtained by a fit of data with the Herzberger-type dispersion formula [14]

$$n = A + BL + CL^2 + D\lambda^2 + E\lambda^4,$$

where

$$L = 1/(\lambda^2 - 0.028)$$

with λ in micrometers. As explained by Herzberger et al. [14], the 0.028 is the square of the mean asymptote for the short-wavelength abrupt rise in the index for 14 materials (diamond not included). The coefficients, evaluated by least squares, are $A = 2.37837$, $B = 1.18897 \times 10^{-2}$ $C = -1.0083 \times 10^{-4}$, $D = -2.3676 \times 10^{-5}$, and $E = 3.24263 \times 10^{-8}$. The agreement between measured and calculated values is 0.001 in the IR from 20 to 2.5 μm but then deteriorates to 0.003 from 2.5 to 0.30 μm and to 0.01 at 0.23 μm. Peter [10], on the other hand, was able to fit his data from 0.60 to 0.22 μm with a Sellmeier-type formula to 0.0003. This suggests that further work is needed to fit the total spectrum from IR to UV because of apparent systematic variations over this long spectral range among various investigators.

The index data of Table V can also be fit to the Sellmeier-type formula

$$n^2 = \varepsilon + A\lambda_1^2 + B\lambda_1^2/(\lambda^2 - \lambda_1^2).$$

In view of the difficulties of fitting the Herzberger-type formula to the wide spectral range 20 to 0.22 μm, we have not done a similar fit for the Sellmeier-type formula.

For the infrared values in the region 2.5 to 25 μm, the relative uncertainty in the measured index is set by the relative uncertainty in the sample thickness. Edwards and Ochoa [12] report an index uncertainty of about 10^{-3}.

Results for the extinction coefficient in the UV–visible transparent region have been given by Peter [10] and Clark et al. [15].

Lax and Burstein [2] and Smith et al. [1] have verified that the 2- to 6-μm band in all diamond types is the result of two- and three-phonon absorption processes. Hardy and Smith [1] measured the absorption coefficient in this band for two diamonds with good optical surfaces. One type Ia and one type IIa were measured giving identical results within experimental error. Their absorption coefficient values are given in Table V for selected multiphonon frequencies in the 2 to 6 μm band.

The index of refraction n of cubic carbon (diamond) has been precisely documented in the region of optical transparency [10, 11]. However, in the region of intrinsic absorption $h\nu > 5.2$ eV, few attempts have been made to obtain a self-consistent set of optical constants n and k for this material. In the region of relatively low absorption, difficulties occur in preparing thin, intrinsic specimens for transmission measurements [16–19]. In the region of strong absorption, this information must be obtained by analysis of reflectance data. In some of these studies, problems in the evaluation of the reflectance and in its analysis were apparent [15, 20–23]. However, two studies were subsequently published [24, 25] that indicated reasonable but far from exact agreement in the reflectance values and their analysis by the Kramers–

Kronig technique. In these two studies the reflectance data differ by as much as $\pm 10\%$ in their relative values for $h\nu > 13$ eV, and the need for additional work is indicated. The results of Philipp and Taft [24], rather than those of Roberts and Walker [25], are given in Table V for two reasons. First, it is not clear whether the results of [25] are for room temperature (~ 300 K) or for 77 K. The data they analyze show two peaks in reflectance near 7 eV that only show up clearly at low temperature [26]. Second, their results give negative k value for $h\nu \geq 31.25$ eV [27], indicating a problem with their reflectance values or extrapolation procedures.

In most of the table entries for the ultraviolet, the values for n are given to four places and those for k to three places. The accuracy of these values does not in general warrant this precision. This is done to show the possible presence of weak structure in the optical properties that can often be discerned by comparison of precise but not necessarily accurate values over narrow energy ranges. In the region $4.5 \leq h\nu \leq 5.8$ eV, the k values are taken from the transmission data of Clark et al. [15] and Peter [10]. The table entries were obtained from a smooth curve drawn through these data and do not show up all of the fine details present in their measurements. The values of n for $h\nu \geq 6.0$ eV are taken from Philipp and Taft [24].

It is recommended that anyone who uses Table V carefully examine the material and references cited by the preceding authors.

REFERENCES

1. J. R. Hardy and S. D. Smith, *Philos. Mag.* **6**, 1163 (1961); S. D. Smith, J. R. Hardy, and E. W. J. Mitchell, *Rep. Int. Conf. Phys. Semicond., 6th, Exeter*, p. 529. Institute of Physics and the Physical Society, London, 1962.
2. M. Lax and E. Burstein, *Phys. Rev.* **97**, 39 (1955).
3. W. Kaiser and W. L. Bond, *Phys. Rev.* **115**, 857 (1959).
4. J. F. H. Custers, Physica **18**, 489 (1952).
5. D. Drukker and Zn. N. V., Amsterdam, The Netherlands (1979).
6. M. Seal, D. Drukker, and Zn. N. V., private communication (1979).
7. A. S. Vishnevskiy and V. G. Malogolovets, *Dopov. Akad. Nauk. Ukr. RSR Ser. A* **35**, 892 (1973).
8. S. von Rosch, *Opt. Acta* **12**, 253 (1965).
9. Z. V. Bartoshinskii, N. S. Pidzyraelo, I. V. Stefanski, and N. I. Tretyak, *Mineral. Sb. Lvov.* **29**, 24 (1975).
10. F. Peter, *Z. Phys.* **15**, 358 (1923).
11. Martens, *Ann. Physik* **8**, 463 (1902).
12. D. F. Edwards and E. Ochoa, *J. Opt. Soc. Am.* **71**, 607 (1981).
13. E. V. Loewenstein and D. R. Smith, *Appl. Opt.* **10**, 577 (1971).
14. M. Herzberger and C. D. Salzberg, *J. Opt. Soc. Am.* **52**, 420 (1962); M. Herzberger, *Opt. Acta* **6**, 197 (1959).
15. C. D. Clark, P. J. Dean, and P. V. Harris, *Proc. R. Soc. London* **A 277**, 312 (1964).
16. C. D. Clark, *J. Phys. Chem. Solids* **8**, 481 (1959).
17. J. F. H. Custers and F. A. Raal, *Nature* **179**, 268 (1957).
18. E. W. J. Mitchell, *J. Phys Chem. Solids* **8**, 444 (1959).
19. W. Kaiser and W. L. Bond, *Phys. Rev.* **115**, 857 (1959).

20. S. Robin, *Rev. Opt.* **33**, 377 (1954).
21. J. R. Nelson and W. C. Crocker, *Bull, Am. Phys. Soc.* **5**, 431 (1960).
22. H. R. Philipp and E. A. Taft, *Phys. Rev.* **127**, 159 (1962).
23. W. C. Walker and J. Osantowski, *Phys. Rev.* **134**, A153 (1964).
24. H. H. Philipp and E. A. Taft, *Phys. Rev.* **136**, A1445 (1964).
25. R. A. Roberts and W. C. Walker, *Phys. Rev.* **161**, 730 (1967).
26. R. A. Roberts, D. M. Roessler, and W. C. Walker, *Phys. Rev. Lett.* **17**, 302 (1966).
27. R. A. Roberts, Technical Report, NASA Research grant No. NSG91-60, Supplement 4, January 1967.

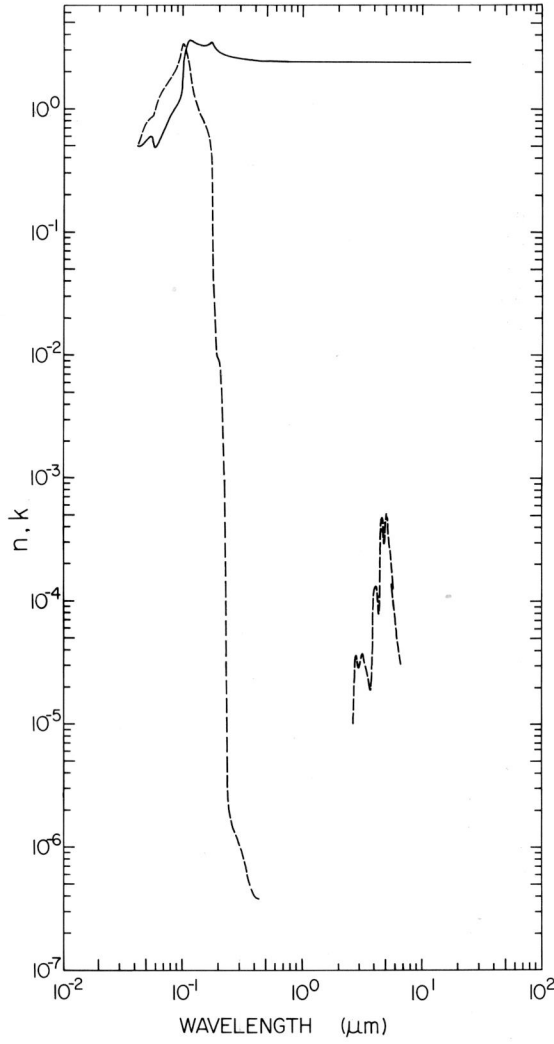

Fig. 5. Log–log plot of n(———) and k(-----) versus wavelength in micrometers for cubic carbon.

TABLE V

Values of n and k for Cubic Carbon Obtained from Various References[a]

eV	μm	cm^{-1}	n	n_c	k
30.0	0.04133		0.493 [24]		0.490 [24]
29.5	0.04203		0.493		0.521
29.0	0.04275		0.494		0.552
28.5	0.04350		0.496		0.584
28.0	0.04428		0.501		0.616
27.5	0.04508		0.508		0.648
27.0	0.04592		0.516		0.679
26.5	0.04678		0.527		0.708
26.0	0.04769		0.538		0.736
25.5	0.04862		0.550		0.763
25.0	0.04959		0.562		0.787
24.5	0.05060		0.574		0.810
24.0	0.05166		0.586		0.829
23.5	0.05276		0.595		0.843
23.0	0.05391		0.597		0.850
22.5	0.05510		0.579		0.846
22.0	0.05635		0.518		0.888
21.5	0.05767		0.483		0.969
21.0	0.05904		0.487		1.052
20.5	0.06048		0.503		1.130
20.0	0.06199		0.527		1.203
19.5	0.06358		0.557		1.273
19.0	0.06525		0.589		1.341
18.5	0.06702		0.626		1.408
18.0	0.06888		0.665		1.476
17.5	0.07085		0.707		1.546
17.0	0.07293		0.753		1.619
16.5	0.07514		0.805		1.692
16.0	0.07749		0.861		1.767
15.5	0.07999		0.917		1.845
15.0	0.08266		0.972		1.929
14.75	0.08405		0.994		1.978
14.5	0.08551		1.018		2.034
14.25	0.08700		1.041		2.102
14.0	0.08856		1.070		2.178
13.75	0.09017		1.098		2.268
13.5	0.09184		1.129		2.379
13.25	0.09357		1.166		2.522
13.0	0.09537		1.223		2.722
12.8	0.09686		1.312		2.953
12.6	0.09839		1.532		3.265
12.4	0.09998		1.983		3.382
12.2	0.1016		2.383		3.354
12.0	0.1033		2.736		3.228
11.75	0.1055		3.090		2.986
11.5	0.1078		3.346		2.693
11.25	0.1102		3.507		2.380

TABLE V (*Continued*)

Cubic Carbon

eV	μm	cm^{-1}	n	n_c	k
11.0	0.1127		3.582		2.078
10.75	0.1153		3.600		1.813
10.5	0.1181		3.565		1.581
10.25	0.1210		3.514		1.403
10.0	0.1240		3.453		1.258
9.75	0.1272		3.398		1.147
9.5	0.1305		3.348		1.055
9.25	0.1340		3.308		0.978
9.0	0.1378		3.272		0.910
8.75	0.1417		3.247		0.855
8.5	0.1459		3.228		0.806
8.25	0.1503		3.232		0.765
8.0	0.1550		3.251		0.712
7.8	0.1590		3.276		0.659
7.6	0.1631		3.306		0.592
7.5	0.1653		3.321		0.553
7.4	0.1675		3.335		0.515
7.3	0.1698		3.376		0.473
7.2	0.1722		3.437		0.388
7.15	0.1734		3.464		0.307
7.1	0.1746		3.444		0.210
7.0	0.1771		3.322		9.35×10^{-2}
6.9	0.1797		3.220		5.24
6.8	0.1823		3.146		3.44
6.7	0.1851		3.085		2.20
6.6	0.1879		3.031		1.47
6.5	0.1907		2.985		1.10×10^{-2}
6.4	0.1937		2.944		9.87×10^{-3}
6.3	0.1968		2.910		9.74
6.2	0.2000		2.879		9.30
6.1	0.2033	49,200	2.852		8.62
6.0	0.2066	48,390	2.826		7.99
5.8	0.2138	46,780	2.780		5.02
5.6	0.2214	45,170	2.740		1.48×10^{-3} [15]
5.55	0.2234	44,760			5.48×10^{-4}
5.5	0.2254	44,360			3.41×10^{-5}
5.47386	0.226504	44,149.3	2.7151 [10]	2.7029	2.662 [10]
5.45	0.2275	43,960			1.48 [15]
5.4	0.2296	43,550			1.04×10^{-5}
5.36064	0.231288	43,236.1	2.6950		$\left.\begin{array}{l} \\ \end{array}\right\} 7.10 \times 10^{-6}$ [10]
5.34131	0.232125	43,080.2	2.6917		
5.35	0.2317	43,150			6.45 [15]
5.32267	0.232938	42,929.9	2.6881	2.6849	
5.3	0.2339	42,750			2.98
5.0	0.25	40,000		2.6383	
4.8187	0.25730	38,865	2.6145	2.6205	1.47 [10]

(*continued*)

TABLE V (*Continued*)

Cubic Carbon

eV	μm	cm^{-1}	n	n_c	k
4.5112	0.27484	36,385	2.5786		1.29×10^{-6}
4.1596	0.29807	33,549	2.5429	2.5465	
4.133	0.30	33,330		2.5439	
3.961	0.313	31,950			8.97×10^{-7}
3.576	0.3467	28,840	2.4954	2.4955	
3.542	0.35	28,570		2.4929	
3.434	0.3611	27,690	2.4854	2.4849	
3.3924	0.36548	27,361			4.65
3.100	0.40	25,000		2.4627	
3.064	0.4047	24,710	2.4626	2.4605	3.86
2.8447	0.43584	22,944			3.82×10^{-7}
2.755	0.45	22,220		2.4432	
2.6503	0.46782	21,376	2.4408	2.4380	
2.5504	0.48614	20,570	2.4354 [8]	2.4332	
2.480	0.50	20,000		2.4299	
2.4379	0.50858	19,663	2.4306 [10]	2.4280	
2.2705	0.54608	18,312	2.4235 [8]	2.4210	
2.254	0.55	18,180		2.4203	
2.105	0.589	16,980	2.4175	2.4147	
2.066	0.60	16,670		2.4133	
1.9257	0.64385	15,531	2.4111 [10]	2.4084	
1.907	0.65	15,380		2.4079	
1.8892	0.65629	15,237	2.4104 [8]	2.4073	
1.771	0.70	14,290		2.4036	
1.653	0.75	13,330		2.4003	
1.550	0.80	12,500		2.3975	
1.459	0.85	11,760		2.3953	
1.378	0.90	11,110		2.3934	
1.305	0.95	11,760		2.3918	
1.240	1.0	10,000		2.3905	
0.8266	1.5	6,667		2.3837	
0.6199	2.0	5,000		2.3813	
0.4959	2.5	4,000		2.3801	
0.48	2.583	3,872			6.17×10^{-6} [1]
0.47	2.638	3,791			1.57×10^{-5}
0.46	2.695	3,710			3.22
0.45	2.755	3,630			3.62
0.44	2.818	3,549			3.14
0.43	2.883	3,468			2.87
0.42	2.952	3,388			2.94
0.4133	3.0	3,333		2.3795	
0.41	3.024	3,307			3.25
0.40	3.099	3,226			3.58
0.39	3.179	3,146			3.67
0.38	3.263	3,065			3.11
0.37	3.351	2,984			2.80
0.36	3.444	2,904			2.47

TABLE V (*Continued*)
Cubic Carbon

eV	μm	cm^{-1}	n	n_c	k
0.3542	3.5	2,857			
0.35	3.542	2,823			2.11
0.34	3.646	2,742			1.89
0.33	3.757	2,662			2.99×10^{-5}
0.32	3.874	2,581			1.11×10^{-4}
0.31	3.999	2,500		2.3787	1.30
0.30	4.133	2,420			1.32
0.2955	4.203	2,379			1.07×10^{-4}
0.29	4.275	2,339			7.82×10^{-5}
0.285	4.350	2,299			1.45×10^{-4}
0.28	4.428	2,258			2.75
0.275	4.508	2,218			3.52
0.27	4.592	2,178			4.39
0.2675	4.635	2,158			4.79
0.265	4.679	2,137			4.06
0.2625	4.723	2,117			3.38
0.26	4.768	2,097			2.96
0.255	4.862	2,057			3.87
0.2525	4.910	2,037			4.69
0.25	4.959	2,016			5.21
0.2480	5.0	2,000		2.3783	
0.2475	5.009	1,996			4.82
0.245	5.060	1,976			4.87
0.2425	5.113	1,956			4.68
0.24	5.166	1,936			3.82
0.235	5.276	1,895			2.90
0.23	5.391	1,855			2.36
0.2254	5.5	1,818			1.4 [2]
0.225	5.510	1,815			1.93 [1]
0.22	5.636	1,774			1.21
0.2066	6.0	1,667		2.3779	5.7×10^{-5} [2]
0.1907	6.5	1,538			3.1
0.1771	7.0	1,429			
0.1550	8.0	1,250		2.3772	
0.1378	9.0	1,111			
0.1240	10.0	1,000		2.3765	
0.1033	12.0	833.3		2.3757	
0.08856	14.0	714.3		2.3750	
0.07749	16.0	625.0		2.3745	
0.06888	18.0	555.6		2.3741	
0.06199	20.0	500.0		2.3741	

[a] The column headed n_c contains refractive index values calculated with the Herzberger-type formula. References are indicated in brackets.

Lithium Fluoride (LiF)

EDWARD D. PALIK and W. R. HUNTER*

Naval Research Laboratory

Washington, D.C.

The data are tabulated in Table VI and plotted in Fig. 6. These data are for room-temperature samples unless otherwise specified. When investigators make measurements at other than room temperature, they usually say so; consequently, we assume no statement of temperature implies room temperature, $\sim 20°C$.

The Henke et al. model [1] has been used to calculate n and k from approximately 6 to 123 Å. Lukirskii et al. [2] have measured n and k for vacuum-deposited films about 1000 Å thick (as determined by multiple-beam interferometry) on glass substrates at discrete wavelengths from 23.6 to 113 Å. They used a version of the reflectance method to find n and k. Brown et al. [3] measured the transmittance of thin vacuum-deposited films of different thicknesses on Formvar substrates from 53.9 Å (230 eV) to 206.6 Å (60 eV) and calculated the absorption coefficient from their measurements. The film thicknesses were determined by a quartz-crystal monitor that had been calibrated by the Tolansky technique (multiple-beam, interferometry). The spectral features did not sharpen much at 77 K. Milgram and Givens [4] also used thin vacuum-deposited films on celluloid substrates. They evaporated LiF from Pt boats at about 900°C at a pressure of $\sim 5 \times 10^{-5}$ torr, and by comparing the transmission of different thicknesses using a photographic method [5], they obtained values for the absorption coefficient. Film thicknesses were determined by using multiple-beam interferometry. Their data between 206 and 496 Å are included. Data of Rao et al. [6] are also included between 413 Å (30 eV) and 496 Å (25 eV). They used freshly cleaved crystals kept in vacuum continually while reflectance was measured with synchrotron radiation over the energy range 6–35 eV. Sample temperature was 100 K. Kramers–Kronig analysis was used. Roessler and Walker [7] measured the near-normal incidence reflectance of freshly cleaved crystal from 7 to 25 eV (1770–496 Å) at 300 K and used a Kramers–Kronig

---

* Present address: Sachs/Freeman Associates, Bowie, Maryland.

HANDBOOK OF OPTICAL CONSTANTS OF SOLIDS

analysis. Data of Rao *et al.* [6] and Roessler and Walker [7] are in reasonable agreement from 25 to 7 eV.

Tomiki and Miyata [8] have measured the absorption coefficient on thin cleaved slabs of UV-quality LiF at 298.5 K in the band-edge region to study the Urbach tail. The band-edge location is very temperature dependent. At room temperature there was some evidence for extrinsic absorption in the edge. At photon energies less than the edge, the absorption bottomed out to values near $k = 8 \times 10^{-7}$, which varied by a factor of 2 in different samples. We tabulate results for an intermediate sample. The steeply changing values of k in the band edge are much smaller than those of Roessler and Walker [7] and are to be preferred.

A number of investigators have measured the refractive index of LiF at wavelengths longer than the absorption edge. Early measurements were made by Gyulai [9] and Schneider [10], who used prisms of LiF. The most recent measurements that use prisms are those of Laporte and Subtil [11] at 25°C, which are reported here. They obtained index values from 1043.5 to 2000 Å. Li [12] has analyzed much of the existing n data and fitted it with a dispersion relation at 293 K:

$$n^2 = 1 + \frac{0.92549\lambda^2}{\lambda^2 - (0.07376)^2} + \frac{6.96747\lambda^2}{\lambda^2 - (32.79)^2}.$$

We list these values from 0.18 to 11 μm. Estimated uncertainty is given as ± 0.0002 in the 0.35–3-μm region. This increases to ± 0.005 for 0.25–0.35 and 3.0–5.0 μm, to ± 0.001 for 0.15–0.25 and 5.0–7.0 μm. The results of Laporte and Subtil [11] at 0.2 μm are 0.002 larger than Li's [12] value and are preferable to shorter wavelength.

Calorimetry measurements in the UV–visible region have been done by Harrington *et al.* [13] Values of k listed in Table VI are somewhat smaller than the results of Tomiki and Miyata [8] obtained nearer to the band edge. The origin of this extrinsic absorption is not known.

Deutsch [14] has measured the absorption coefficient from 1500 to 2300 cm^{-1} with slabs of different thicknesses. He noted that his α data and those of Hohls [15] from 900 to 1800 cm^{-1} plotted on a log scale fall on a straight line. We read the data from this graph. This is absorption due to two- and three-phonon processes and contains little structure at room temperature [16, 17]. This is consistent with theoretical calculations. Klier [18] has also measured the absorption at 300 K from 5.5 to 14 μm. His data nicely bridge the data of Deutsh [14] and the oscillator-model results of Kachare *et al.* [19].

Jasperse *et al.* [20] measured the reflectance of LiF in the 200–800-cm^{-1} region at many temperatures, including 295 K. The average angle of incidence was 11°. High-purity, single-crystal LiF was polished flat to $\frac{1}{2}$ wavelength at the sodium d line. The sample was then annealed in a vacuum furnance for two days at a temperature of about $\frac{3}{4}$ of the melting temperature (842°C). Presumably, the surface mechanical damage was annealed out. The reflection standard was an aluminum mirror assumed to have 98.5% reflectivity over the entire spectral range. This will introduce some uncertainty in R since the reflectivity of Al[†] does vary steady through this spectral region from 0.994 to 0.988. A two-oscillator model was used to fit the data. The pertinent equations are

$$\varepsilon' = n^2 - k^2 = \varepsilon_\infty + \sum_i \frac{4\pi\rho_i v_i^2(v_i^2 - v^2)}{(v_i^2 - v^2)^2 + (\gamma_i v)^2},$$

$$\varepsilon'' = 2nk = \sum_i \frac{4\pi\rho_i v \gamma_i v^2}{(v_i^2 - v^2)^2 + (\gamma_i v)^2}.$$

Kachare et al. [19] reanalyzed the data forcing a fit to R by comparing the entire reflectance curve with a calculated one and minimizing a certain error function. While their values of the oscillator parameters were very near those of Japerse et al. for 295 K, they noted a large deviation in the damping constants at low temperature, suggesting that the classical oscillator model was not adequate for obtaining accurate values of the damping constant for strongly anharmonic lattice modes of ionic crystals. The values of Kachare et al. [19] were $\varepsilon_\infty = 1.96$, $v_1 = 307.5\,\text{cm}^{-1}$, $v_2 = 501.4\,\text{cm}^{-1}$, $4\pi\rho_1 = 6.67$, $4\pi\rho_2 = 0.116$, $\gamma_1/v_1 = 0.057$, and $\gamma_2/v_2 = 0.173$. We have used these to calculate n and k in the 200–1000-cm^{-1} region. The results for k are in reasonable agreement with the composite results of Seger and Genzel [21]. The refractive index of Li is comparable near 10 μm. The oscillator model gives $n = 2.96$ at 400 μm which is much smaller than the value of $n = 3.70 \pm 0.03$ measured by Genzel and Klier [22]: This suggests that the oscillator strength $(4\pi\rho_i)$ is too small in the fit of Kachare et al. [19].

Genzel and Klier [22] and Seger and Genzel [21] have measured the absorption coefficient from 70 to 1000 μm at 20°C and in some cases the refractive index, also. Transmittances of slabs of various thicknesses were measured along with some reflectance. Results of Stolen and Dransfeld [23] for k at 320 μm at 300 K are similar. Eldridge [17] discusses the two-phonon difference-band region below v_1.

The data of Seger and Genzel appear to extrapolate nicely to the lower-frequency results of Owens [24] at 116, 35.4, and 9.8 GHz (2.584, 8.469, 30.590 mm) at 25°C, Owens used a filled-resonant-cavity method and transmission through a dielectric plug in a section of waveguide to determine ε' and ε'', the real and imaginary parts of the dielectric function, from a measurement of the loss tangent $\tan \delta = \varepsilon'/\varepsilon''$ and ε'. His value of $n = 3.02$ at $\lambda = 2.584$ mm is the same as that of Dianov and Irisova [25] at $\lambda = 2$ mm at 300 K.

Breckenridge [26] used a capacitor technique to measure ε' and $\tan \delta$ over a wide range of frequencies from 4.5×10^7 to 10^2 Hz at 80°C. We list values of n and k obtained at 4.5×10^7 and 1×10^7 Hz (666.2 and 3335 cm).

[†] See the discussion of the optical properties of metallic aluminum by Smith et al. in this handbook.

It appears that k is leveling off in this region if extrapolation from higher frequency is made in Fig. 6, suggesting some extrinsic absorption mechanism.

Theoretical calculations supports a decreasing exponential dependence of that absorption as a function of frequency in the multiphonon region (above v_1) with essentially no structure. In the difference multiphonon, region, an increasing exponential dependence can be observed for NaCl, (below v_1), but some structure appears to be present for other alkali halides. The exponential dependence with no structure for NaCl is probably fortuitous, and some structure for the other materials is consistent with theoretical expectations. While there is general agreement about the sum multiphonon region, the theoretical situation in the difference multiphonon region is still unsettled.

The magnitude and wavelength dependence of the millimeter-wave absorption coefficients of the alkali halides is consistent with recent theoretical calculations of Hardy and Karo [27] and of Sparks et al. [28] However, these two theoretical attempts to calculate the absorption as due to multiphonon difference processes are based on somewhat different models. The shape of the experimental data does suggest that the absorption, at least above 35 GHz, is essentially intrinsic in character.

ACKNOWLEDGMENT

We thank M. Hass of the Naval Research Laboratory for discussions of the far-infrared, submillimeter properties of LiF.

REFERENCES

1. B. L. Henke, P. Lee, T. J. Tanaka, R. L. Shimabukuro, and B. K. Fujikawa, "Low Energy X-ray Diagnostics—1981" (D. T. Attwood and B. L. Henke eds.). AIP Conf. Proc. No. 75, American Institute of Physics, New York, 1981.
2. A. P. Lukirskii, E. P. Savinov, O. A. Ershov, and Y. F. Shepelev, Opt. Spektrosk. **16**, 310 (1964); A. P. Lukirskii, E. P. Savinov, O. A. Ershov, and Y. F. Shepelev, Opt. Spectrosc. **16**, 168 (1964).
3. F. C. Brown, C. Gahwiller, A. B. Kunz, and N. O. Lipari, Phys. Rev. Lett. **25**, 927 (1970).
4. A. Milgram and M. P. Givens, Phys. Rev. **125**, 1506 (1962).
5. W. M. Cady and D. H. Tomboulian, Phys. Rev. **59**, 381 (1941).
6. K. K. Rao, T. J. Moravec, J. C. Rife, and R. N. Dexter, Phys. Rev. B **12**, 5937 (1975).
7. D. M. Roessler and W. C. Walker, J. Opt. Soc. Am. **57**, 835 (1967).
8. T. Tomiki and T. Miyata, J. Phys. Soc. Jpn. **27**, 658 (1969).
9. Z. Gyulai, Z. Physik **46**, 80 (1927).
10. E. G. Schneider, Phys. Rev. **49**, 341 (1936).
11. P. Laporte and J. L. Subtil, J. Opt. Soc. Am. **72**, 1558 (1982).
12. H. H. Li, J. Phys. Chem. Ref. Data **9**, 561 (1980).
13. J. A. Harrington, B. L. Bobbs, M. Braunstein, R. K. Kim, R. Stearns, and R. Braunstein, Appl. Opt. **17**, 1541 (1978).
14. T. F. Deutsch, J. Phys. Chem. Solids **34**, 2091 (1973).
15. H. W. Hohls, Ann. Physik **29**, 433 (1937).
16. V. V. Mitskevich, Soc. Phys.-Solid State **4**, 2224 (1963); V. V. Mitskerich, Fiz. Tverd. Tela **4**, 3035 (1962).
17. J. E. Eldridge, Phys. Rev. B **6**, 1510 (1972).

18. M. Klier, *Z. Physik* **150**, 49 (1958).

19. A. Kachare, G. Andermann, and L. R. Brantley, *J. Phys. Chem. Solids* **33**, 467 (1972).

20. J. R. Jasperse, A. Kahan, J. N. Plendl, and S. S. Mitra, *Phys. Rev.* **146**, 526 (1966).

21. G. Seger and L. Genzel, *Z. Physik* **169**, 66 (1962).

22. L. Genzel and M. Klier, *Z. Physik* **144**, 25 (1956).

23. R. Stolen and K. Dransfeld, *Phys. Rev.* **139**, A1295 (1965).

24. J. C. Owens, *Phys. Rev.* **181**, 1228 (1969).

25. E. M. Dianov and N. A. Irisova, *Sov. Phys.-Solid State* **8**, 1807 (1967); E. M. Dianov and
    N. A. Irisova, *Fiz. Tverd. Tela* **8**, 2265 (1966).

26. R. G. Breckenridge, *J. Chem. Phys.* **16**, 959 (1948).

27. J. R. Hardy and A. M. Karo, *Phys. Rev. B* **26**, 3327 (1982).

28. M. Sparks, D. F. King, and D. L. Mills *Phys. Rev. B* **26**, 6987 (1982).

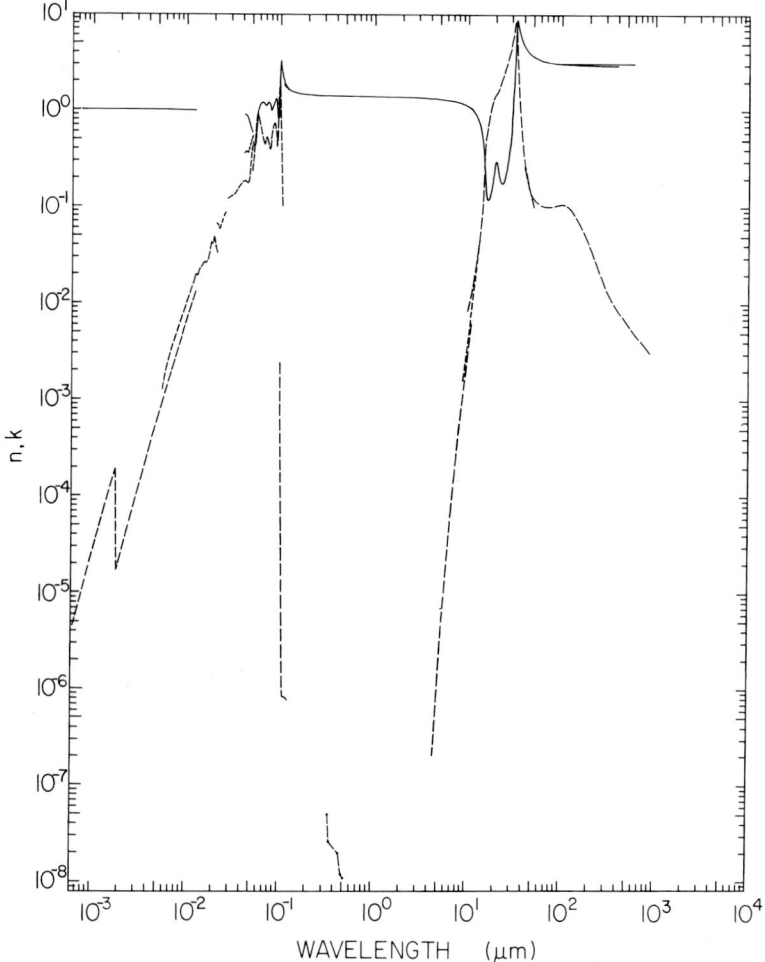

Fig. 6. Log–log plot of n (——) and k (----) versus wavelength in micrometers for lithium fluoride. Note the incredibly small values of k in the transparent region centered near 1 μm.

TABLE VI

Values of n and k for Lithium Fluoride Obtained from Various References[a]

eV	Å	n	k
2000	6.199	0.9999347 [1]	4.33×10^{-6} [1]
1952	6.351	0.9999314	4.74
1905	6.508	0.9999279	5.19
1860	6.665	0.9999244	5.68
1815	6.831	0.9999206	6.23
1772	6.997	0.9999167	6.82
1730	7.166	0.9999126	7.46
1688	7.345	0.9999082	8.18
1648	7.523	0.9999037	8.94
1609	7.705	0.999899	9.78×10^{-6}
1570	7.897	0.999894	1.07×10^{-5}
1533	8.087	0.999889	1.17
1496	8.287	0.999883	1.28
1460	8.492	0.999878	1.40
1426	8.694	0.999872	1.53
1392	8.906	0.999866	1.67
1358	9.129	0.999859	1.83
1326	9.350	0.999852	2.00
1294	9.581	0.999845	2.18
1263	9.816	0.999838	2.38
1233	10.05	0.999830	2.60
1204	10.30	0.999822	2.83
1175	10.55	0.999814	3.09
1147	10.81	0.999805	3.36
1119	11.08	0.999796	3.67
1093	11.34	0.999787	3.99
1067	11.62	0.999778	4.35
1041	11.91	0.999768	4.75
1016	12.20	0.999757	5.18
992	12.50	0.999747	5.64
968	12.81	0.999736	6.14
945	13.12	0.999726	6.67
923	13.43	0.999715	7.23
901	13.76	0.999704	7.85
879	14.10	0.999692	8.53
858	14.45	0.999681	9.28×10^{-5}
838	14.79	0.999670	1.0×10^{-4}
818	15.16	0.999660	1.10
798	15.54	0.999650	1.20
779	15.92	0.999643	1.31
760	16.31	0.999638	1.42
742	16.71	0.999637	1.53
725	17.10	0.999643	1.62
707	17.54	0.999658	1.76
690	17.97	0.999758	1.90×10^{-4}
674	18.39	0.999663	1.70×10^{-5}
658	18.84	0.999606	1.86

TABLE VI (*Continued*)

Lithium Fluoride

eV	Å	n	k
642	19.31	0.999556	2.04
627	19.77	0.999517	2.23
612	20.26	0.999481	2.43
597	20.77	0.999443	2.67
583	21.27	0.999407	2.91
569	21.79	0.999369	3.18
555	22.34	0.999330	3.49
542	22.87	0.999291	3.80
529	23.44	0.999250	4.16×10^{-5}
525.4	23.6	0.99886 [2]	1.1×10^{-4} [2]
516	24.03	0.999206 [1]	4.55×10^{-5} [1]
504	24.60	0.999162	4.96
492	25.20	0.999116	5.42
480	25.83	0.999066	5.93
469	26.43	0.999017	6.46
458	27.07	0.99896	7.04
447	27.74	0.99891	7.69
436	28.44	0.99885	8.42
426	29.10	0.99879	9.16×10^{-5}
415	29.87	0.99872	1.01×10^{-4}
406	30.54	0.99866	1.09
396	31.31	0.99859	1.19
394.9	31.4	0.99814 [2]	2.0 [2]
386	32.12	0.99851 [1]	1.31 [1]
377	32.89	0.99843	1.43
368	33.69	0.99835	1.55
359	34.53	0.99826	1.70
351	35.32	0.99818	1.84
342	36.25	0.99807	2.02
334	37.12	0.99798	2.20
326	38.03	0.99787	2.40
318	38.99	0.99776	2.63
311	39.86	0.99765	2.84
303	40.92	0.99752	3.12
296	41.88	0.99740	3.39
289	42.90	0.99727	3.70
282	43.96	0.99713	4.03
281.8	44.0	0.99592 [2]	8.9 [2]
275	45.08	0.99697 [1]	4.41 [1]
269	46.09	0.99683	4.77
262	47.32	0.99666	5.23
256	48.43	0.99650	5.68
250	49.59	0.99632	6.17
244	50.81	0.99613	6.72
238	52.09	0.99593	7.34
232	53.44	0.99576	8.02

(*continued*)

TABLE VI (*Continued*)

Lithium Fluoride

eV	Å	n	k
227	54.62	0.99552	8.66
221	56.10	0.99528	9.51×10^{-4}
220	56.36		1.26×10^{-3} [3]
216	57.40	0.99505	1.03 [1]
211	58.76	0.99481	1.12
210	59.04		1.83 [3]
206	60.18	0.99456	1.21 [1]
201	61.68	0.99429	1.32
200	61.99		2.12 [3]
196	63.25	0.99399	1.44 [1]
191	64.91	0.99368	1.57
190	65.26		2.70 [3]
187	66.30	0.99341	1.69 [1]
185	67.0	0.99180 [2]	3.12 [2]
182	68.12	0.99305 [1]	1.85 [1]
180	68.88		3.29 [3]
178	69.65	0.99274	2.00 [1]
174	71.25	0.99241	2.16
170	72.93	0.99206	2.33
			4.06 [3]
166	74.69	0.99168	2.53 [1]
162	76.53	0.99128	2.74
160	77.49		4.75 [3]
158	78.47	0.99086	2.98 [1]
154	80.51	0.99040	3.25
150	82.65	0.9899	3.54
			5.79 [3]
147	84.34	0.9895	3.78 [1]
143	86.70	0.9890	4.14
140	88.56	0.9885	4.45
			7.05 [3]
136	91.16	0.9879	4.90 [1]
133	93.22	0.9874	5.28
130	95.37	0.9869	5.69
			8.56 [3]
127	97.62	0.9864	6.13 [1]
124	99.98	0.9858	6.61
121	102.5	0.9852	7.15×10^{-3}
120	103.3		1.16×10^{-2} [3]
118	105.1	0.9845	7.75×10^{-3} [1]
115	107.8	0.9839	8.42
112	110.7	0.9832	9.13
110	112.7	0.9827	9.65×10^{-3}
			1.44×10^{-2} [3]
109.7	113	0.98330 [2]	1.17 [2]
107	115.9	0.9819 [1]	1.05 [1]
104	119.2	0.9811	1.16
102	121.5	0.9806	1.24

TABLE VI (*Continued*)

Lithium Fluoride

eV	Å	*n*	*k*
100	124.0	0.9801	1.32
			1.94 [3]
97.5	127.2		2.00
95	130.5		1.96
92.5	134.0		2.06
90	137.8		2.28
87.5	141.7		2.40
85	145.9		2.48
82.5	150.3		2.62
80	155.0		2.64
77.5	160.0		2.61
75	165.3		2.63
72.5	171.0		2.86
70	177.1		3.81
67.5	183.7		4.22
65	190.7		4.16
62.5	198.4		7.74
60	206.6		3.78
			6.74 [4]
57.5	215.6		6.52
			3.26 [3]
55	225.4		5.92 [4]
52.5	236.2		6.77
50	248.0		7.89
47.5	261.0		8.72×10^{-2}
45	275.5		0.123
42.5	291.7		0.130
40	310.0		0.133
37.5	330.6		0.142
35	354.2		0.158
32.5	381.5		0.173
30	413.3	0.869 [7]	0.184
			0.367 [6]
29	427.5	0.861	0.376
			0.180 [4]
28	442.8	0.844	0.176
			0.363 [6]
27	459.2	0.749	0.401
			0.194 [4]
26	476.9	0.662	0.296
			0.467 [6]
25	495.9	0.558	0.521
			0.434 [4]
		0.47 [7]	0.23 [7]
24.5	506.1	0.44	0.31
24	516.6	0.43	0.42

(*continued*)

TABLE VI (*Continued*)

Lithium Fluoride

eV	Å	n	k
23.7	523.1	0.42	0.50
23.6	525.4	0.43	0.56
23.4	529.9	0.43	0.59
23.3	532.1	0.44	0.62
23.2	534.4	0.45	0.66
23.1	536.7	0.46	0.70
23.05	537.9	0.47	0.72
23	539.1	0.48	0.74
22.95	540.2	0.50	0.76
22.9	541.4	0.51	0.78
22.8	543.8	0.55	0.81
22.6	548.6	0.62	0.87
22.4	553.5	0.70	0.91
22.2	558.5	0.80	0.94
22	563.6	0.91	0.93
21.8	568.7	0.99	0.88
21.6	574.0	1.04	0.84
21.4	579.4	1.08	0.80
21	590.4	1.14	0.73
20.5	604.8	1.18	0.65
20	619.9	1.20	0.58
19.5	635.8	1.21	0.53
19	652.6	1.19	0.47
18.6	666.6	1.17	0.44
18.5	670.2	1.15	0.44
18.4	673.8	1.14	0.44
18.3	677.5	1.13	0.45
18	688.8	1.11	0.48
17.7	700.5	1.13	0.52
17.6	704.5	1.14	0.52
17.5	708.5	1.16	0.53
17.4	712.6	1.18	0.52
17.3	716.7	1.19	0.51
16.9	733.6	1.20	0.47
16.4	756.0	1.16	0.40
16.2	765.3	1.12	0.39
16.1	770.1	1.09	0.39
16	774.9	1.06	0.41
15.9	779.8	1.02	0.42
15.8	784.7	0.99	0.47
15.7	789.7	0.98	0.52
15.5	799.9	1.00	0.59
15.3	810.4	1.04	0.64
15.1	821.1	1.08	0.68
14.9	832.1	1.12	0.71
14.7	843.4	1.17	0.73
14.6	849.2	1.19	0.74

TABLE VI (*Continued*)

Lithium Fluoride

eV	Å	n	k
14.5	855.1	1.22	0.74
14.4	861.0	1.25	0.73
14.3	867.0	1.27	0.72
14.2	873.1	1.30	0.71
14.1	879.3	1.32	0.68
14	885.6	1.32	0.64
13.95	888.9	1.32	0.62
13.9	892.0	1.32	0.59
13.85	895.2	1.31	0.56
13.8	898.4	1.29	0.52
13.75	901.7	1.24	0.46
13.7	905.0	1.15	0.42
13.65	908.3	1.05	0.43
13.6	911.7	0.95	0.50
13.55	915.0	0.88	0.58
13.5	918.4	0.84	0.67
13.45	921.8	0.82	0.76
13.4	925.3	0.81	0.84
13.35	928.7	0.80	0.92
13.3	932.2	0.80	1.01
13.25	935.7	0.80	1.10
13.2	939.3	0.81	1.20
13.15	942.9	0.84	1.33
13.1	946.5	0.91	1.43
13.05	950.1	0.97	1.53
13	953.7	1.04	1.64
12.95	957.4	1.13	1.77
12.9	961.1	1.26	1.90
12.85	964.9	1.42	2.02
12.8	968.6	1.64	2.13
12.75	972.4	1.92	2.20
12.7	976.3	2.24	2.19
12.65	980.1	2.57	2.08
12.6	984.0	2.89	1.90
12.5	991.9	3.34	1.17
12.4	999.9	2.93	0.48

eV	cm^{-1}	μm	n	k
12.3		0.1008	2.68	0.33
12.2		0.1016	2.51	0.23
12.1		0.1025		2.42×10^{-3} [8] 4.89×10^{-4}
12.0		0.1033	2.28	0.11 [7]
11.9		0.1042		9.86×10^{-5} [8] 2.32×10^{-5}

(*continued*)

TABLE VI (*Continued*)

Lithium Fluoride

eV	cm^{-1}	μm	n	k
11.88		0.10435	2.003 [11]	
11.85		0.10465	1.985	
11.8		0.1051		4.18×10^{-6}
11.78		0.10525	1.951	
11.75		0.1055	2.08 [7]	4×10^{-2} [7]
11.75		0.10552	1.937 [11]	
11.70		0.10600	1.914	1.60×10^{-6} [8]
11.6		0.1069		1.10×10^{-6}
11.59		0.10695	1.875	
11.50		0.10781	1.844	
11.5		0.1078	1.94 [7]	9.44×10^{-7}
11.46		0.10815	1.833 [11]	
11.4		0.1088		8.40
11.33		0.10940	1.798	
11.3		0.1097		8.29
11.25		0.1102	1.84 [7]	
11.24		9.11025	1.777 [11]	
11.2		0.1107		8.37
11.13		0.11142	1.751	
11.1		0.1117		8.62
11.01		0.11265	1.729	
11.0		0.1127	1.77 [7]	8.07
11.00		0.11271	1.728 [11]	
10.82		0.11455	1.699	
10.8		0.1148		8.22
10.66		0.11625	1.677	
10.6		0.1170		8.19
10.52		0.11785	1.659	
10.50		0.11808	1.656	
10.5		0.1181	1.67 [7]	
10.4		0.1192		8.06
10.35		0.11980	1.639 [11]	
10.20		0.12153	1.624	7.74
10.09		0.12285	1.614	
10.00		0.12398	1.606	7.70×10^{-7}
			1.60 [7]	
9.927		0.12490	1.599 [11]	
9.717		0.12760	1.583	
9.526		0.13015	1.569	
9.357		0.13250	1.558	
9.184		0.13500	1.548	
9		0.1375	1.53 [7]	
9.004		0.13770	1.538 [11]	
8.856		0.14000	1.530	
8.551		0.14500	1.515	
8.266		0.15000	1.503	

TABLE VI (*Continued*)

Lithium Fluoride

eV	cm^{-1}	μm	n	k
8		0.15500	1.493	
			1.49 [7]	
7.749		0.16000	1.484 [11]	
7.293		0.17000	1.469	
7		0.1771	1.46 [7]	
7.000		0.17712	1.461 [11]	
6.888		0.18000	1.458	
			1.4533 [12]	
6.812		0.182	1.4516	
6.738		0.184	1.4500	
6.666		0.186	1.4484	
6.595		0.188	1.4469	
6.526		0.190	1.4455	
			1.449 [11]	
6.458		0.192	1.4441 [12]	
6.391		0.194	1.4428	
6.326		0.196	1.4415	
6.262		0.198	1.4403	
6.199	50,000	0.200	1.4391	
			1.441 [11]	
6.048	48,780	0.205	1.4363 [12]	
5.904	47,620	0.210	1.4337	
5.767	46,510	0.215	1.4313	
5.636	45,450	0.220	1.4291	
5.510	44,440	0.225	1.4271	
5.391	43,480	0.230	1.4252	
5.276	42,550	0.235	1.4235	
5.166	41,670	0.240	1.4219	
5.061	40,820	0.245	1.4203	
4.959	40,000	0.250	1.4189	
4.862	39,220	0.255	1.4176	
4.769	38,460	0.260	1.4164	
4.679	37,740	0.265	1.4152	
4.592	37,040	0.270	1.4141	
4.509	36,360	0.275	1.4131	
4.428	35,710	0.280	1.4121	
4.350	35,090	0.285	1.4112	
4.275	34,480	0.290	1.4103	
4.203	33,900	0.295	1.4095	
4.133	33,330	0.300	1.4087	
4.000	32,260	0.31	1.4073	
3.875	31,250	0.32	1.4060	
3.757	30,300	0.33	1.4048	
3.647	29,410	0.34	1.4037	
3.542	28,570	0.35	1.4028	

(*continued*)

TABLE VI (*Continued*)

Lithium Fluoride

eV	cm^{-1}	μm	n	k
3.531	28,480	0.3511		5.0 × 10^{-8} [13]
3.444	27,780	0.36	1.4019	
3.408	27,490	0.3638		2.6
3.351	27,030	0.37	1.4010	
3.263	26,320	0.38	1.4003	
3.179	25,640	0.39	1.3996	
3.100	25,000	0.40	1.3989	
2.952	23,810	0.42	1.3978	
2.818	22,730	0.44	1.3968	
2.708	21,840	0.4579		1.5
2.695	21,740	0.46	1.3959	
2.583	20,830	0.48	1.3951	
2.541	20,490	0.4880		1.2
2.480	20,000	0.50	1.3944	
2.410	19,440	0.5145		1.1 × 10^{-8}
2.384	19,230	0.52	1.3938	
2.296	18,520	0.54	1.3933	
2.214	17,860	0.56	1.3928	
2.138	17,240	0.58	1.3923	
2.066	16,670	0.60	1.3919	
2.000	16,130	0.62	1.3915	
1.937	15,630	0.64	1.3912	
1.879	15,150	0.66	1.3908	
1.823	14,710	0.68	1.3905	
1.771	14,290	0.70	1.3902	
1.722	13,890	0.72	1.3900	
1.675	13,510	0.74	1.3897	
1.631	13,160	0.76	1.3894	
1.590	12,820	0.78	1.3892	
1.550	12,500	0.80	1.3890	
1.512	12,200	0.82	1.3888	
1.476	11,900	0.84	1.3886	
1.442	11,630	0.86	1.3884	
1.409	11,360	0.88	1.3882	
1.378	11,110	0.90	1.3880	
1.348	10,870	0.92	1.3878	
1.319	10,640	0.94	1.3876	
1.292	10,420	0.96	1.3875	
1.265	10,200	0.98	1.3873	
1.240	10,000	1.00	1.3871	
1.181	9,524	1.05	1.3867	
1.127	9,091	1.10	1.3863	
1.087	8,696	1.15	1.3859	
1.033	8,333	1.20	1.3855	
0.9919	8,000	1.25	1.3851	
0.9537	7,692	1.30	1.3847	
0.9184	7,407	1.35	1.3844	

TABLE VI (*Continued*)

Lithium Fluoride

eV	cm^{-1}	μm	n	k
0.8856	7,143	1.40	1.3840	
0.8551	6,897	1.45	1.3836	
0.8266	6,667	1.50	1.3832	
0.7999	6,452	1.55	1.3828	
0.7749	6,250	1.60	1.3823	
0.7514	6,061	1.65	1.3819	
0.7293	5,882	1.70	1.3815	
0.7085	5,714	1.75	1.3810	
0.6888	5,556	1.80	1.3806	
0.6702	5,405	1.85	1.3801	
0.6526	5,263	1.90	1.3796	
0.6358	5,128	1.95	1.3791	
0.6199	5,000	2.0	1.3787	
0.5904	4,762	2.1	1.3777	
0.5636	4,545	2.2	1.3766	
0.5391	4,348	2.3	1.3755	
0.5166	4,167	2.4	1.3744	
0.4959	4,000	2.5	1.3731	
0.4769	3,846	2.6	1.3719	
0.4592	3,704	2.7	1.3706	
0.4428	3,571	2.8	1.3693	
0.4275	3,448	2.9	1.3679	
0.4133	3,333	3.0	1.3665	
0.4000	3,226	3.1	1.3650	
0.3875	3,125	3.2	1.3635	
0.3757	3,030	3.3	1.3619	
0.3647	2,941	3.4	1.3602	
0.3542	2,857	3.5	1.3585	
0.3444	2,778	3.6	1.3568	
0.3351	2,703	3.7	1.3550	
0.3263	2,632	3.8	1.3531	
0.3179	2,564	3.9	1.3512	
0.3100	2,500	4.0	1.3493	
0.3024	2,439	4.1	1.3473	
0.2952	2,381	4.2	1.3452	
0.2883	2,326	4.3	1.3431	
0.2852	2,300	4.348		2.1 × 10^{-7} [14]
0.2818	2,273	4.4	1.3409	
0.2755	2,222	4.5	1.3387	
0.2728	2,200	4.545		4.3
0.2695	2,174	4.6	1.3364	
0.2638	2,128	4.7	1.3340	
0.2604	2,100	4.762		8.0 × 10^{-7}
0.2583	2,083	4.8	1.3316	
0.2530	2,041	4.9	1.3291	
0.2480	2,000	5.0	1.3266	

(*continued*)

TABLE VI (*Continued*)

Lithium Fluoride

eV	cm^{-1}	μm	n	k
0.2480	2,000	5.000		1.8×10^{-6}
0.2431	1,961	5.1	1.3240	
0.2384	1,923	5.2	1.3213	
0.2356	1,900	5.263		3.6
0.2339	1,887	5.3	1.3186	
0.2296	1,852	5.4	1.3158	
0.2254	1,818	5.5	1.3130	
0.2232	1,800	5.556		7.5×10^{-6}
0.2214	1,786	5.6	1.3101	
0.2175	1,754	5.7	1.3071	
0.2138	1,724	5.8	1.3041	
0.2108	1,700	5.882		1.5×10^{-5}
0.2101	1,695	5.9	1.3010	
0.2066	1,667	6.0	1.2978	
0.2033	1,639	6.1	1.2946	
0.2000	1,613	6.2	1.2912	
0.1984	1,600	6.250		3.1
0.1968	1,587	6.3	1.2879	
0.1937	1,563	6.4	1.2844	
0.1907	1,538	6.5	1.2809	
0.1879	1,515	6.6	1.2773	
0.1860	1,500	6.667		6.9×10^{-5}
0.1851	1,493	6.7	1.2736	
0.1823	1,471	6.8	1.2698	
0.1797	1,449	6.9	1.2660	
0.1771	1,429	7.0	1.2621	
0.1746	1,408	7.1	1.2581	
0.1736	1,400	7.143		1.4×10^{-4}
0.1722	1,389	7.2	1.2541	
0.1698	1,370	7.3	1.2499	
0.1675	1,351	7.4	1.2457	
0.1653	1,333	7.5	1.2414	
0.1631	1,316	7.6	1.2369	
0.1612	1,300	7.692		2.9
0.1610	1,299	7.7	1.2325	
0.1590	1,282	7.8	1.2279	
0.1569	1,266	7.9	1.2232	
0.1550	1,250	8.0	1.2184	6.0 [18]
0.1531	1,235	8.1	1.2136	
0.1512	1,220	8.2	1.2086	
0.1494	1,205	8.3	1.2036	
0.1488	1,200	8.333		6.0×10^{-4} [14]
0.1476	1,190	8.4	1.1984	
0.1459	1,176	8.5	1.1931	
0.1442	1,163	8.6	1.1878	
0.1425	1,149	8.7	1.1823	
0.1409	1,136	8.8	1.1768	

TABLE VI (*Continued*)

Lithium Fluoride

eV	cm^{-1}	µm	n	k
0.1393	1,124	8.9	1.1711	
0.1378	1,111	9.0	1.1653	1.5 × 10^{-3} [18]
0.1364	1,100	9.091		1.2 [14]
0.1362	1,099	9.1	1.1594	
0.1348	1,087	9.2	1.1533	
0.1333	1,075	9.3	1.1472	
0.1319	1,064	9.4	1.1409	
0.1305	1,053	9.5	1.1345	
0.1292	1,042	9.6	1.1280	
0.1278	1,031	9.7	1.1213	
0.1265	1,021	9.8	1.1145	
0.1252	1,010	9.9	1.1076	
0.1240	1,000	10.0	1.1005	2.6
			1.11 [19]	8.10 [19]
				3.3 × 10^{-3} [18]
0.1216	980.4	10.2	1.0859 [12]	
0.1192	961.5	10.4	1.0706	
0.1178	950	10.53	1.07 [19]	1.02 × 10^{-2} [19]
0.1170	943.4	10.6	1.0547 [12]	
0.1148	925.9	10.8	1.0381	
0.1127	909.1	11.0	1.0208	8.0 × 10^{-3} [18]
0.1116	900	11.11	1.01 [19]	1.31 × 10^{-2} [19]
				6.2 × 10^{-3} [14]
0.1054	850	11.76	0.947	1.76 × 10^{-2} [19]
0.1033	833.3	12.0		1.9 [18]
0.09919	800	12.50	0.856	2.49 [19]
0.09671	780	12.82	0.810	2.84
0.09537	769.2	13		3.7 [18]
0.09423	760	13.16	0.755	3.53 [19]
0.09184	740.7	13.5		5.0 [18]
0.09175	740	13.51	0.691	4.36 [19]
0.08927	720	13.89	0.611	5.61
0.08679	700	14.29	0.508	7.74
0.08555	690	14.49	0.443	9.54 × 10^{-2}
0.08431	680	14.71	0.367	0.124
0.08307	670	14.93	0.279	0.177
0.08183	660	15.15	0.201	0.268
0.08059	650	15.38	0.157	0.373
0.07935	640	15.63	0.137	0.471
0.07811	630	15.87	0.127	0.561
0.07687	620	16.13	0.122	0.645
0.07563	610	16.39	0.121	0.725
0.07439	600	16.67	0.124	0.804
0.07315	590	16.95	0.128	0.881
0.07191	580	17.24	0.136	0.957
0.07067	570	17.54	0.146	1.03

(*continued*)

TABLE VI (*Continued*)

Lithium Fluoride

eV	cm⁻¹	μm	n	k
0.06943	560	17.86	0.161	1.11
0.06819	550	18.18	0.179	1.18
0.06695	540	18.52	0.203	1.26
0.06571	530	18.87	0.233	1.33
0.06447	520	19.23	0.265	1.39
0.06323	510	19.61	0.293	1.43
0.06199	500	20.00	0.306	1.47
0.06075	490	20.41	0.297	1.51
0.05951	480	20.83	0.270	1.57
0.05827	470	21.28	0.237	1.65
0.05703	460	21.74	0.210	1.76
0.05579	450	22.22	0.191	1.88
0.05455	440	22.73	0.180	2.01
0.05331	430	23.26	0.177	2.16
0.05207	420	23.81	0.181	2.32
0.05083	410	24.39	0.191	2.50
0.04959	400	25.00	0.208	2.71
0.04835	390	25.64	0.234	2.94
0.04711	380	26.32	0.269	3.21
0.04587	370	27.03	0.325	3.53
0.04463	360	27.78	0.407	3.93
0.04401	355	28.17	0.465	4.16
0.04339	350	28.57	0.539	4.43
0.04277	345	28.99	0.637	4.75
0.04215	340	29.41	0.772	5.12
0.04154	335	29.85	0.964	5.57
0.04092	330	30.30	1.25	6.10
0.04030	325	30.77	1.71	6.77
0.03968	320	31.25	2.51	7.57
0.03906	315	31.75	3.98	8.37
0.03844	310	32.26	6.43	8.35
0.03782	305	32.79	8.58	6.35
0.03720	300	33.33	8.76	3.91
0.03658	295	33.90	8.05	2.43
0.03596	290	34.48	7.32	1.62
0.03534	285	35.09	6.70	1.16
0.03472	280	35.71	6.22	0.875
0.03410	275	36.36	5.82	0.687
0.03348	270	37.04	5.50	0.554
0.03286	265	37.74	5.23	0.458
0.03224	260	38.46	5.00	0.386
0.03162	255	39.22	4.81	0.331
0.03100	250	40.00	4.64	0.287
				0.26 [21]
0.02976	240	41.67	4.36	0.222 [19]
0.02852	230	43.48	4.14	0.177
0.02728	220	45.45	3.96	0.145

TABLE VI (*Continued*)

Lithium Fluoride

eV	cm⁻¹	μm	n	k
0.02604	210	47.62	3.81	0.120
0.02480	200	50.00	3.69	0.102
				0.12 [21]
0.02066	166.7	60	3.39	0.105
0.01771	142.9	70		0.101
0.01550	125.0	80	3.17	0.101
0.01378	111.1	90		0.103
0.01240	100.0	100	3.067 [21]	0.106
			3.09 [19]	
0.01033	83.33	120		9.6×10^{-2}
0.08266	66.67	150	3.067 [21]	6.5
0.06199	50.00	200	3.067	4.0
			2.99 [19]	
0.04959	40.00	250	3.067 [21]	2.2
0.04133	33.33	300	3.067	1.4×10^{-2}
0.03875	31.25	320		8.66×10^{-3} [23]
0.03100	25.00	400	3.067	8.5 [21]
			2.96 [19]	
0.02480	20.00	500	3.067 [21]	6.3
0.02066	16.67	600	3.067	5.0
0.01771	14.29	700		4.1
0.01550	12.50	800		3.5
0.01378	11.11	900		3.1
4.798×10^{-4}	3.870	2,584	3.023 [24]	1.19×10^{-3} [24]
1.464×10^{-4}	1.181	8,469	3.023	6.20×10^{-4}
4.053×10^{-5}	0.3269	30,590	3.023	2.63×10^{-4}
1.861×10^{-7}	1.501×10^{-3}	6.662×10^{6}	3.018 [26]	1.6×10^{-5} [26]
3.718×10^{-8}	2.999×10^{-4}	3.335×10^{7}	3.018	1.6

[a] References are given in brackets.

Lithium Niobate (LiNbO₃)

EDWARD D. PALIK

Naval Research Laboratory
Washington, D.C.

Lithium niobate is a uniaxial, trigonal crystal useful for second-harmonic-generation applications of laser radiation. Index matching at ω and 2ω in the near IR–visible region is accomplished by temperature tuning of the ordinary (o) refractive index $n_o(\omega)$ $(E \perp c$ axis) and the extraordinary (e) refractive index $n_e(2\omega)$ $(E \parallel c$ axis). Two types of material are considered stoichiometric and congruent. A mole ratio of $Li_2O/(LiO_2 + Nb_2O_5) = 0.500$ corresponds to stoichiometric material. This is grown from a melt containing excess Li. A congruently melting composition crystal grows from a melt of the *same* composition, corresponding to a mole ration of $Li_2O/(LiO_2 + Nb_2O_5) = 0.486$. This material is used in most applications.

The UV data are taken from the reflection data of Wiesendanger and Güntherodt [1] obtained at 293 K on cleaved surfaces; stoichiometry was not specified. Kramers–Kronig (KK) analyses for both polarizations $(E\parallel$ and $\perp$ to the c axis) were performed. Above ~ 7 eV, the difference in $R_{\parallel}$ and $R_{\perp}$ is negligible. An analysis by Barner *et al.* [2] from 4 to 14.5 eV with unpolarized light and their subsequent KK analysis is not used here because of the large differences with Wiesendanger and Güntherodt [1]. It is not obvious that n, k (unpolarized) are indeed averages of n_o, n_e and k_o, k_e. Since $n_{o,e}$ and $k_{o,e}$ were calculated from values of ε_1 and ε_2 as read from a graph, the numbers are probably good to two figures, although three figures are listed in Table VII. Table VII is plotted in Fig. 7 for both o and e rays.

The absorption edge of congruent material has been studied by Redfield and Burke [3] over a wide temperature range. A series of samples were mechanically polished. This produced spurious absorption in the small-absorption region when compared to the transmission of the next thicker sample. The c axis was in the plane of the sample, so that dichroism might be measured. However, very little difference was observed, so that only $E \parallel c$ data are given. The tabulated data are for 300 K.

The refractive index in the transparent region has been measured for various materials, MgO doped [4], stoichiometric [5], and congruent [6]. Generally, the minimum-deviation technique has been used with the optic axis

695

ISBN 0-12-544420-6

parallel to both faces of the prism. We tabulate the data from Boyd *et al.* [5] and Nelson and Mikulyak [6]. Boyd *et al.* [5] do not state room temperature explicitly. Nelson and Mikulyak [6] used 24.5°C and state an accuracy of ± 0.0002. Differences in n in the two sets of data are due to the slight differences in stoichiometry in the two kinds of materials.

In the infrared, the lattice vibration bands have been studied in reflection by Barker and Loudon [7], Axe and O'Kane [8], and Poplavko *et al.* [9]. The refractive indices at 300 K have been provided by Barker and Loudon [7] in the transparent region by analysis of reflectivity directly as $R = (\sqrt{\varepsilon_1} - 1)^2/(\sqrt{\varepsilon_1} + 1)^2$. Barker and Loudon [7] fit oscillator models through the data of Boyd *et al.* [5] and through their own data by using one band-gap oscillator and eight lattice oscillators for $E \parallel c$ and one band-gap oscillator and five lattice oscillators for $E \perp c$. Since the given numbers did not reproduce the long-wavelength values for $n_o(\perp)$ and $n_e(\parallel)$ given in a graph by Barker and Loudon [7], we chose to read the given graph of experimental data directly. The results are tabulated from 5 to 11 μm.

In Barker and Loudon [7], Axe and O'Kane [8], and Poplavko [9], the reflectivity was measured and KK analyzed. As an example of the differences, the vibration mode near 150 cm^{-1} has an imaginary peak $\varepsilon_{02} = 230$, 158, and 200, respectively.

We tabulate the data of Axe and O'Kane [8] at longer wavelength because it is more complete. The samples were probably stoichiometric material polished with 0.05-μm alumina and etched with HF–HNO$_3$ until certain reflection features present in mechanically polished surfaces cleared up or became sharper. Reflectivity was measured with an aluminized reference surface. Temperature was not stated but was no doubt near 300 K. Dispersion-relation analysis yielded graphs of ε_1 and ε_2 that we read and then converted to n and k. While two decimal places are given, the numbers are probably good to the first significant figure.

In the far infrared, Bosomworth [10] and Sakai [11] have measured n_o and n_e by utilizing channel spectra, and the former has also measured the absorption coefficient, all at 300 K. We tabulate the refractive indices of Sakai [11] because they cover a somewhat wider spectral range and are virtually identical to the value of Bosomworth [10] (both read off graphs). We read the values from a smooth line put through the data points that scatter by ± 0.05. The extinction coefficients of Bosomworth [10] are tabulated for a Czochralski-grown sample polished with diamond grit. In the 60-cm^{-1} region, k_o and k_e from the KK analysis [8] are much larger than measured in transmission [10], suggesting that dispersion-relation analysis is poor when $n \gg k$.

Poplavko *et al.* [9] found that $n_o = 7.19$, $k_o = 0.8$ and $n_e = 5.48$, $k_e = 0.2$ at 300 μm. These n values are significantly larger than those of Barker and Loudon [7] and Axe and O'Kane [8]; the k values are one to two orders of magnitude larger, which is troublesome. Permittivity measurements at

3000 μm using a waveguide-resonance reflection method yielded ε_1 and tan $\delta = \varepsilon_2/\varepsilon_1$; then $n_o = 7.21$, $k_o = 0.04$ and $n_e = 5.66$, $k_e = 0.01$, but the variation in ε_{01} from sample to sample was typically 52–48 due to differences in stoichiometry and structure defects in the crystals. Measurements of ε_1 and ε_2 down to ~ 50 Hz with standard bridge and resonance measurements methods showed a piezoelectric resonance near 10^6–10^7 Hz. In the frequency range 126–132 GHz (2380–2270 μm), Vinogradov et al. [12] determined $n_o = 7.2 \pm 0.2$ and tan $\delta = (2.5 \pm 0.5) \times 10^{-3}$. This gave $\varepsilon_{01} = 51.8$, $\varepsilon_{02} = 0.13$, so that $n_o = 7.20$ and $k_o = 9 \times 10^{-3}$. Again we see the large variation in the data of Barker and Loudon [7], Axe and O'Kane [8], Poplavko et al. [9], Bosomworth [10], and Sakai [11], which is probably sample dependent.

REFERENCES

1. E. Wiesendanger and G. Güntherodt, *Solid State Commun.* **14**, 303 (1974).
2. K. Barner, R. Braunstein, and H. A. Weakliem, *Phys. Stat. Solidi B* **68**, 525 (1975).
3. D. Redfield and W. J. Burke, *J. Appl. Phys.* **45**, 4566 (1974).
4. G. D. Boyd, W. L. Bond, and H. L. Carter, *J. Appl. Phys.* **38**, 1941 (1967).
5. G. D. Boyd, R. C. Miller, K. Nassau, W. L. Bond, and A. Savage, *Appl. Phys. Lett.* **5**, 934 (1964).
6. D. F. Nelson and R. M. Mikulyak, *J. Appl. Phys.* **45**, 3688 (1974).
7. A. S. Barker and R. Loudon, *Phys. Rev.* **158**, 433 (1967).
8. J. D. Axe and D. F. O'Kane, *Appl. Phys. Lett.* **9**, 58 (1966).
9. Y. M. Poplavko, V. V. Meriakri, V. N. Aleshechkin, V. G. Tsykalov, E. F. Ushatkin, and A. S. Knyazev, *Sov. Phys. Solid State* **15**, 991 (1973); Y. M. Poplavko, V. V. Meriakri, V. N. Aleshechkin, V. G. Tsykalov, E. F. Ushatkin, and A. S. Knyazev, *Fiz. Tverd. Tela.* **15**, 1473 (1973).
10. D. R. Bosomworth, *Appl. Phys. Lett.* **9**, 330 (1966).
11. K. Sakai, *Appl. Opt.* **11**, 2894 (1972).
12. P. N. Vinogradov, N. A. Prisova, and G. V. Kozlov, *Sov. Phys. Solid State* **12**, 605 (1970); P. N. Vinogradov, N. A. Prisova, and G. V. Kozlov, *Fiz. Tverd. Tela.* **12**, 781 (1970).

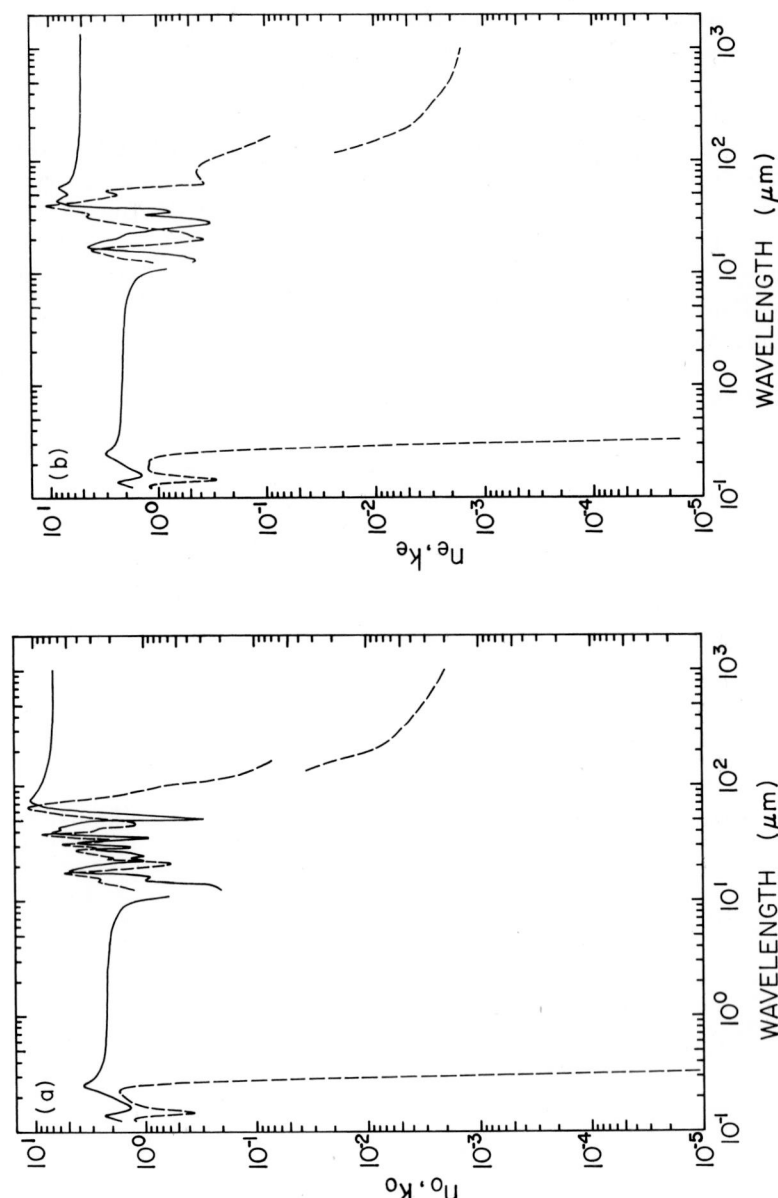

Fig. 7. (a) Log–log plot of n_o (———) and k_o (– – –) versus wavelength in micrometers for lithium niobate. (b) Log–log plot of n_e (———) and k_e (– – –) versus wavelength in micrometers for lithium niobate.

TABLE VII
Values of n and k for Lithium Niobate Obtained from Various References[a]

eV	cm^{-1}	μm	$n_o(\perp)$	$k_o(\perp)$	$n_e(\parallel)$	$k_e(\parallel)$
10.0	80,650	0.1240	1.67 [1]	1.22 [1]	1.73 [1]	1.18 [1]
9.75	78,640	0.1272	1.88	1.28	1.97	1.22
9.5	76,620	0.1305	2.04	1.17	2.14	1.12
9.25	74,610	0.1340	2.19	1.09	2.32	1.03
9.0	72,590	0.1378	2.32	0.75	2.42	0.72
8.75	70,570	0.1417	2.12	0.45	2.17	0.34
8.5	68,560	0.1459	1.82	0.36	1.89	0.29
8.25	66,540	0.1503	1.58	0.46	1.63	0.40
8.0	64,520	0.1550	1.43	0.59	1.48	0.57
7.5	60,490	0.1653	1.38	0.98	1.48	0.94
7.0	56,460	0.1771	1.54	1.30	1.75	1.24
6.5	52,430	0.1907	1.82	1.48	2.07	1.26
6.0	48,390	0.2066	2.16	1.60	2.30	1.22
5.75	46,380	0.2156	2.36	1.67	2.46	1.22
5.5	44,360	0.2254	2.68	1.75	2.76	1.16
5.25	42,340	0.2362	3.22	1.60	2.98	1.00
5.0	40,330	0.2480	3.62	1.27	3.10	0.64
4.75	38,310	0.2610	3.52	0.43	2.99	0.25
4.5	36,290	0.2755	3.19	0.11	2.79	0.06
4.25	34,280	0.2917	2.91	0.00	2.59	0.00
						1.02×10^{-2} [3]
4.2	33,880	0.2952				5.87×10^{-3}
4.15	33,470	0.2988			3.09	
4.1	33,070	0.3024				1.71×10^{-3}
4.05	32,670	0.3061				9.01×10^{-4}
4.0	32,260	0.3100	2.76		2.49	4.19
3.95	31,860	0.3139				2.12×10^{-4}
3.9	31,460	0.3179				7.84×10^{-5}
3.85	31,050	0.3220			3.33	
3.8	30,650	0.3263				1.69×10^{-5}
3.75	30,250	0.3306				8.16×10^{-6}
3.7	29,840	0.3351			5.07	

eV	cm^{-1}	μm	$n_o(\perp)$	$n_e(\parallel)$	$n_o(\perp)$	$n_e(\parallel)$
3.0642	24,714	0.40463	2.4317 [6]	2.3260 [6]		
2.952	23,810	0.42			2.4089 [5]	2.3025 [5]
2.8447	22,944	0.43584	2.3928	2.2932		
2.755	22,220	0.45			2.3780	2.2772
2.6503	21,376	0.46782	2.3634	2.2683		
2.5831	20,834	0.47999	2.3541	2.2605		
2.480	20,000	0.50			2.3410	2.2457
2.4379	19,663	0.50858	2.3356	2.2448		
2.2705	18,313	0.54607	2.3165	2.2285		
2.254	18,180	0.55			2.3132	2.2237
2.1489	17,332	0.57696	2.3040	2.2178		
2.1415	17,272	0.57897	2.3032	2.2171		
2.1102	17,019	0.58756	2.3002	2.2147		

(*continued*)

TABLE VII (*Continued*)

Lithium Niobate

eV	cm^{-1}	μm	$n_o(\perp)$	$n_e(\parallel)$	$n_o(\perp)$	$n_e(\parallel)$
2.066	16,670	0.60			2.2967	2.2082
1.9257	15,532	0.64385	2.2835	2.2002		
1.8566	14,974	0.66782	2.2778	2.1953		
1.771	14,290	0.70			2.2716	2.1874
1.7549	14,154	0.70652	2.2699	2.1886		
1.550	12,500	0.80			2.2571	2.1745
1.5321	12,357	0.80926	2.2541	2.1749		
1.4224	11,472	0.87168	2.2471	2.1688		
1.378	11,110	0.90			2.2448	2.1641
1.3251	10,688	0.93564	2.2412	2.1639		
1.2915	10,417	0.95998	2.2393	2.1622		
1.240	10,000	1.00			2.2370	2.1567
1.2227	9,961.9	1.0140	2.2351	2.1584		
1.1352	9,156.34	1.09214	2.2304	2.1545		
1.0745	8,666.11	1.15392	2.2271	2.1517		
1.0707	8,636.03	1.15794	2.2269	2.1515		
1.033	8,333	1.20			2.2269	2.1478
0.96284	7,765.78	1.28770	2.2211	2.1464		
0.8856	7,143	1.40			2.2184	2.1417
0.86103	6,944.59	1.43997	2.2151	2.1413		
0.7749	6,250	1.60			2.2113	2.1361
0.75683	6,104.22	1.63821	2.2083	2.1356		
0.6888	5,556	1.80			2.2049	2.1306
0.64871	5,232.18	1.91125	2.1994	2.1280		
0.6199	5,000	2.00			2.1974	2.1250
0.56762	4,578.17	2.18428	2.1912	2.1211		
0.5636	4,545	2.20			2.1909	2.1183
0.51662	4,166.75	2.39995	2.1840	2.1151		
0.5166	4,167	2.40			2.1850	2.1129
0.4769	3,846	2.60			2.1778	2.1071
0.47412	3,824.03	2.61504	2.1765	2.1087		
0.45410	3,662.53	2.73035	2.1724	2.1053		
0.4428	3,571	2.80			2.1703	2.1009
0.42793	3,451.45	2.89733	2.1657	2.0999		
0.4133	3,333	3.00			2.1625	2.0945
0.40631	3,277.10	3.05148	2.1594	2.0946		
0.3875	3,125	3.20			2.1543	2.0871
0.3647	2,941	3.40			2.1456	2.0804
0.3444	2,778	3.60			2.1363	2.0725
0.3263	2,632	3.80			2.1263	2.0642
0.3100	2,500	4.00			2.1155	2.0553

eV	cm^{-1}	μm	$n_o(\perp)$	$k_o(\perp)$	$n_e(\parallel)$	$k_e(\parallel)$
0.2480	2,000	5.0	2.05 [7]		2.00 [7]	
0.2066	1,667	6.0	1.97		1.93	
0.1771	1,429	7.0	1.84		1.83	
0.1550	1,250	8.0	1.71		1.72	
0.1378	1,111	9.0	1.52		1.55	
0.1240	1,000	10	1.20		1.30	

TABLE VII (*Continued*)

Lithium Niobate

eV	cm^{-1}	μm	$n_o(\perp)$	$k_o(\perp)$	$n_e(\parallel)$	$k_e(\parallel)$
0.1127	909.1	11	0.60		0.81	
0.09919	800	12.50	0.20 [8]	1.24 [8]	0.46 [8]	1.10 [8]
0.09671	780	12.82	0.21	1.43	0.44	1.48
0.09423	760	13.16	0.22	1.56	0.45	1.78
0.09175	740	13.51	0.23	1.74	0.46	2.17
0.08927	720	13.89	0.26	1.94	0.53	2.34
0.08679	700	14.29	0.54	2.30	0.56	2.67
0.08431	680	14.71	0.90	2.51	0.75	3.01
0.08307	670	14.93	0.90	2.60		
0.08183	660	15.15	0.99	2.52	1.05	3.33
0.07935	640	15.63	0.95	2.52	1.44	3.83
0.07811	630	15.87			1.85	4.05
0.07687	620	16.13	0.90	2.88	2.52	4.16
0.07563	610	16.39			3.42	4.09
0.07439	600	16.67	1.10	3.63	4.02	3.48
0.07315	590	16.95			4.24	2.83
0.07191	580	17.24	1.83	4.62	4.47	2.12
0.07067	570	17.54	3.01	5.30		
0.06943	560	17.86	4.77	4.55	4.17	1.20
0.06819	550	18.18	4.88	2.97		
0.06695	540	18.52	4.51	2.10	3.67	0.68
0.06447	520	19.23	4.00	1.24	3.20	0.47
0.06199	500	20.00	3.27	0.84	2.67	0.37
0.05951	480	20.83	2.52	0.59	2.28	0.44
0.05703	460	21.74	1.78	0.62	2.06	0.48
0.05455	440	22.73	1.03	1.44	1.82	0.55
0.05331	430	23.26	1.20	2.11		
0.05207	420	23.81	1.30	1.92	1.27	0.79
0.04959	400	25.00	1.01	2.45	0.83	1.09
0.04711	380	26.32	1.33	3.36	0.48	1.65
0.04587	370	27.03	1.73	3.74		
0.04463	360	27.78	2.61	4.10	0.32	2.47
0.04339	350	28.57	2.64	2.83		
0.04215	340	29.41	1.34	2.96	0.37	3.34
0.04092	330	30.30	1.89	4.75		
0.03968	320	31.25	3.39	5.52	0.70	4.30
0.03844	310	32.26	4.09	3.04	1.16	4.51
0.03720	300	33.33	2.80	1.96	1.20	4.35
0.03596	290	34.48			0.75	4.64
0.03472	280	35.71	0.92	4.34	0.80	5.62
0.03348	270	37.04	2.19	6.84	0.94	6.91
0.03286	265	37.74	4.19	8.34		
0.03224	260	38.46	5.71	7.91	3.04	9.86
0.03100	250	40.00	6.79	2.94	6.58	10.4
0.02976	240	41.67	5.76	2.68	7.48	7.34
0.02852	230	43.48	5.82	2.31	8.55	4.91
0.02728	220	45.45	4.94	1.21	8.32	3.36

(*continued*)

TABLE VII (*Continued*)

Lithium Niobate

eV	cm^{-1}	μm	$n_o(\perp)$	$k_o(\perp)$	$n_e(\parallel)$	$k_e(\parallel)$
0.02604	210	47.62			7.66	2.60
0.02480	200	50.00	2.54	1.57	6.78	2.35
0.02356	190	52.63			7.08	2.85
0.02232	180	55.56	0.98	5.09	7.86	2.41
0.02108	170	58.82	2.04	7.36	8.03	0.75
0.01984	160	62.50	5.43	10.1	6.93	0.36
0.01860	150	66.67	7.96	9.92	6.33	0.38
0.01736	140	71.43	10.1	5.68	6.17	0.40
0.01612	130	76.92	10.6	3.06	6.01	0.43
0.01488	120	83.33	9.83	1.67	5.84	0.43
0.01240	100	100	8.57	0.70	5.66	0.35
0.01054	85	117.6				2.3×10^{-2} [10]
0.009919	80	125	7.87	0.13	5.52	0.18 [8]
						1.8×10^{-2} [10]
0.009299	75	133.3		3.5×10^{-2} [10]		
0.008679	70	142.9		3.0		1.1
0.007439	60	166.7	7.34	7 [8]	5.38	9×10^{-2} [8]
			7.06 [11]	1.9×10^{-2} [10]	5.22 [11]	8.0×10^{-3} [10]
0.006199	50	200	6.90	9.5×10^{-3}	5.165	4.8
0.004959	40	250	6.85 [8]	6.0		3.6
			6.80 [11]		5.125	
0.003720	30	333.3	6.70	4.5	5.10	2.9
0.002480	20	500	6.64	3.2	5.07	2.0
0.001240	10	1000	6.61	2.0	5.06	1.6

^{a} The reference is given in brackets at the beginning of the tabulation of n and k and is understood to refer to all n's or all k's below it until a new reference appears. When an exponent of 10 is given, all numbers below it have the same exponent until the power of 10 changes or the next number has 10^{-1}, which is usually written as a decimal. Boyd *et al.* [5] n values are for stoichiometric material and Nelson and Mikulyak [6] n values are for congruent material. We specify $n_o(\perp)$ and $n_e(\parallel)$ for redundancy.

Potassium Chloride (KCl)

EDWARD D. PALIK
Naval Research Laboratory
Washington, D.C.

The optical constants n and k are collected in Table VIII and plotted in Fig. 8.

The spectral region 3000–300 μm primarily contains absorption due to two- and three-phonon absorption processes. Here, k has been determined by transmission measurements of slabs. The data of Genzel et al. [1] are tabulated in Table VIII. Measurement was with a grating monochromator utilizing an Hg lamp and the harmonics of a 12-mm Klystron as sources. The tabulated numbers were read off a wavelength graph with a smooth line drawn through the data points, which scattered by $\lesssim 10\%$. The more-limited results of Stolen and Dransfeld [2] at 300 K were in good agreement, with the temperature dependence studied in detail.

The spectral region 200–30 μm has beeen studied by Johnson and Bell [3] by asymmetric Fourier-transform spectroscopy. This technique yields the phase and amplitude of the reflectivity experimentally, thus circumventing Kramers–Kronig (KK) analysis of R or oscillator-model fits. At the reststrahlen minimum, a power reflectance of 0.0004 was measured, which suggests little polishing damage. Resolution was 2 cm^{-1}. An accuracy of 0.005 in amplitude reflectance was stated. Shoulders on the high-frequency side of ν_T are due to multiple-phonon absorption processes. We have read graphs to obtain n and k. Bravo to workers who put fiduciary marks on all four sides of a graph and include all the digits from 2 to 9 on the log scales! We typically give three figures, but our reading ability is probably somewhat less than this.

For the region 50–12.5 μm, Deutsch [4] has collected values of absorption coefficient α from several sources all more or less at room temperature. Transmission of slab samples was used. On semilog paper these form a straight line to within the scatter, and we have read the values on the line. Boyer et al. [5] treat the temperature dependence of such absorption in numerous materials including KCl.

HANDBOOK OF OPTICAL CONSTANTS OF SOLIDS

ISBN 0-12-544420-6

Calorimetry measurements at single laser wavelengths have been carried out in the transparent region [5–7] and the k values are listed. These represent bulk values with uncertainties of factors of 2 from sample to sample and lab to lab. Surface preparation included chemical polishing.

In the transparent region 35–0.18 μm, Li [8] has assembled and analyzed a great deal of n data from many references. Minimum deviation was the experimental method in most cases. He has fitted a Sellmeier-type equation with nine adjustable parameters to represent the data to the fifth decimal place. The dispersion formula has the form

$$n^2 = 1.26486 + \frac{0.30523\lambda^2}{\lambda^2 - (0.100)^2} + \frac{0.41620\lambda^2}{\lambda^2 - (0.131)^2}$$

$$+ \frac{0.18870\lambda^2}{\lambda^2 - (0.162)^2} + \frac{2.6200\lambda^2}{\lambda^2 - (70.42)^2} \tag{1}$$

at 293 K. This indicates that $\lambda_T = 70.42$ μm ($\nu_T = 142.0$ cm^{-1}). Li [8] has provided a temperature-dependence formula of the form

$$2n\frac{dn}{dT} = -11.13(n^2 - 1) + 0.19 + \frac{3.393\lambda^4}{(\lambda^2 - 0.02624)^2} + \frac{142.56\lambda^4}{(\lambda^2 - 4958.98)^2}. \tag{2}$$

Thus, a 3-K difference in room temperature between labs begins to affect the fourth decimal place. For the spectral region 0.24–21 μm, Li [8] states that the fourth decimal place is meaningful, with estimated uncertainties in the range 0.0001–0.0005. To either side, the uncertainty moves to the third decimal place.

At and above the fundamental band gap, a variety of n and k data have been assembled. Some overlap is provided to show discrepancies and variations. For 295 K, Tomika [9] has measured transmission of thin slabs to determine k in the region 6.8–7.9 eV, including the Urbach tail and exciton effects. Two discharge lamps (hydrogen and CO_2) were utilized. Much attention was given to the growth and preparation of pure samples to minimize impurity absorption effects.

For the region 6.4–11.6 eV, reflectivity of cleaved samples has been fitted by KK analysis to yield n and k as listed [10]. A great deal of temperature data was obtained by Tomika et al. [11]. In this region absorption is due primarily to a number of interband transitions between lower-lying valence and higher-lying conduction bands.

In the spectral region 5–26 eV, R was measured by Roessler and Walker [12] with KK analysis. More recent measurements have been performed [13] that we choose to emphasize. As usual, differences of 20% in n and k appear in overlap regions, although the features seem the same. Antinori et al. [13] used a continuum He discharge as a source. Reflectivity was reproducible among various cleaved samples to within 5%. As Table II in Antinori et al. [13] emphasizes, the reflectivity at 18.4 eV varied by a factor of 2 among seven different references. As Fig. 5 in Antinori et al. [13] indicates, spectra in this region are also affected by resolution; two examples of spectra of 1.5 and 3 Å resolution are shown to differ significantly. More structure was resolved in room-temperature spectra than previously had been seen. Absorption in this region is due primarily to plasma oscillations of the 6p electrons of Cl^- filling the valence-band states (~ 14 eV) and to transitions of 3p electrons of K^+ to the bottom of the conduction band (~ 20 eV).

Balzarotti et al. [14] have measured reflectivity in the region from 20 to 44 eV by utilizing synchrotron radiation. Kramers–Kronig analysis values of n and k are listed. Peaks in this spectral region are primarily due to electron transitions from 3p levels of K^+ and 3s levels of Cl^- into the conduction band. No doubt, cleaved samples were used.

Above this energy, absolute values of k become sparse, although a great deal of work has been done at the various L and K absorption edges. The work is collected by Haelbich et al. [15], and it proved very useful as a starting point for the present work. They give a figure summarizing absorption coefficient α in inverse centimeters for the range 50–280 eV. However, some of the original references contain only relative values. Therefore, we have found some other references to supplement these. For the spectral region 50–200 eV, the data of Lukirskii and Zimkina [16] as given in Haelbich et al. [15] indicated a featureless decrease in k to high energy. Lukirskii et al. [17] report measurements of n and k at two x-ray wavelengths. The k values are about a factor of 2 smaller than those given in Haelbich et al. [15], but we list them along with the n values since they seem to join the high-energy data better.

For the spectral region 200–240 eV, the data of Brown et al. [18], obtained with synchrotron radiation, are only relative in the original paper. We use the work of Aita et al. [19], who used an x-ray tube as a continuous source to study the $L_{2,3}$ edge of Cl^- in evaporated films. Iguchi et al. [20] used synchrotron radiation to measure relative absorption in the 200–280 eV region.

Finally, the K absorption edge of Cl^- near 2823 eV has been studied by Parratt et al. [21] with an x-ray source spectrometer, and k is tabulated. They pointed out a resolution (spectral window) problem that distorted the measurement of α as a function of sample thickness near the absorption edge. Kiyono and Sugiura [22] also measured relative absorption in this region.

The K absorption edge of potassium near 3610 eV has been studied by Sugiura [23]. It is not clear that the L edges of potassium have been measured in the 290–300 eV region.

REFERENCES

1. L. Genzel, H. Happ, and R. Weber, *Z. Physik* **154**, 13 (1959).
2. R. Stolen and K. Dransfeld, *Phys. Rev.* **139A**, 1295 (1965).
3. K. W. Johnson and E. E. Bell, *Phys. Rev.* **187**, 1044 (1969).
4. T. F. Deutsch, *J. Phys. Chem. Solids* **34**, 209 (1973).
5. L. L. Boyer, J. A. Harrington, M. Hass, and H. R. Rosenstock, *Phys. Rev.* **11B**, 1665 (1975).
6. S. D. Allen and J. A. Harrington, *Appl. Opt.* **17**, 1679 (1978).
7. J. A. Harrington, B. L. Bobbs, M. Braunstein, R. K. Kim, R. Stearns, and R. Braunstein, *Appl. Opt.* **171**, 541 (1978).
8. H. H. Li, *J. Phys. Chem. Ref. Data* **5**, 329 (1976).
9. T. Tomika, *J. Phys. Soc. Jpn.* **23**, 1280 (1967).
10. T. Tomika, *J. Phys. Soc. Jpn.* **22**, 463 (1967).
11. T. Tomika, T. Miyata, and H. Tsukamoto, *J. Phys. Soc. Jpn.* **35**, 495 (1973).
12. D. M. Roessler and W. C. Walker, *J. Opt. Soc. Am.* **58**, 279 (1968).
13. M. Antinori, A. Balzarotti, and M. Piacentini, *Phys. Rev. B* **7**, 1541 (1973); M. Antinori, A. Balzarotti, and M. Piacentini, private communication (1982).
14. A Balzarotti, A. Bianconi, E. Burattini, and G. Strinati, *Solid State Commun.* **15**, 1431 (1974).
15. R. P. Haelbich, M. Iwan, and E. E. Koch, "Physik Daten, Physics Data." Fach-information Zentrum, Karlsruhe, 1977.
16. A. P. Lukirskii and T. M. Zimkina, *in* "Rontgen Spektren und Chemische Bindung," p. 187. Karl-Marx Universitat, Leipzig, 1966.
17. A. P. Lukirskii, E. P. Savinov, O. A. Ershov, and Y. F. Shepelev, *Opt. Spectrosc.* **16**, 168 (1964); A. P. Lukirskii, E. P. Savinov, O. A. Erskov, and Y. F. Shepelev, *Opt. Specktrosk.* **16**, 310 (1964).
18. F. C. Brown, C. Gahwiller, H. Fujita, A. B. Kunz, W. Scheifley, and N. Carrera, *Phys. Rev. B* **2**, 2126 (1970).
19. O. Aita, I. Nagakura, and T. Sagawa, *J. Phys. Soc. Jpn.* **30**, 1414 (1971).
20. Y. Iguchi, T. Sagawa, S. Sato, M. Watanabe, H. Yamashita, A. Ejiri, M. Sasanuma, S. Nakai, M. Nakamura, S. Yamaguchi, Y. Nakai, and Y. Oshio, *Solid State Commun.* **6**, 575 (1968).
21. L. G. Parratt, C. F. Hempstead, and E. L. Jossem, *Phys. Rev.* **105**, 1228 (1957).
22. S. Kiyono and C. Sugiura, *Tech. Repts. Tohoku Univ.* **30**, 9 (1965).
23. C. Sugiura, *Sci. Repo. Tohoku Univ.* **45**, 248 (1961).

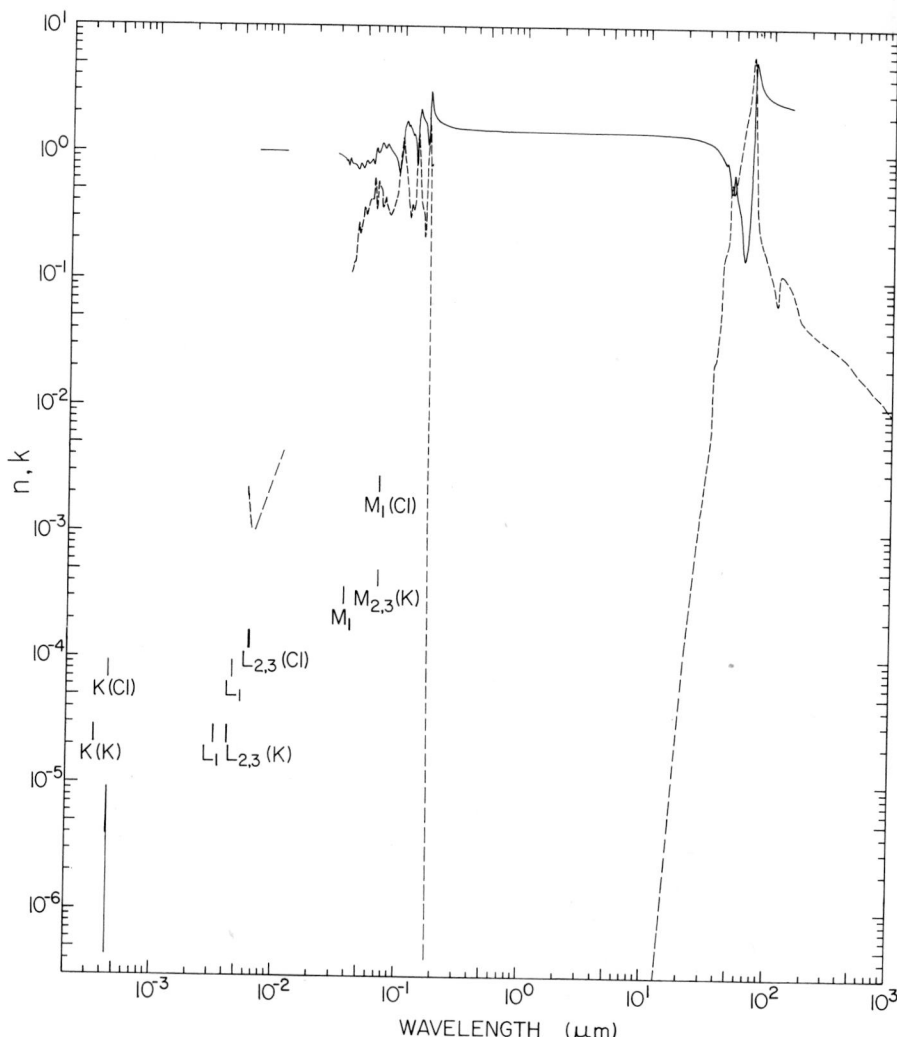

Fig. 8. Log–log plot of n (——) and k (-----) versus wavelength in micrometers for potassium chloride.

TABLE VIII

Values of n and k for Potassium Chloride Obtained from Various References[a]

eV	Å	n	k
2860.3	4.3347		3.93×10^{-6} [21]
2859.3	4.3362		3.83
2858.3	4.3377		3.73
2857.3	4.3392		3.63
2856.3	4.3408		3.52
2855.3	4.3423		3.39

(continued)

TABLE VIII (*Continued*)

Potassium Chloride

eV	Å	n	k
2854.3	4.3438		3.32
2853.3	4.3453		3.60
2852.3	4.3468		3.91
2851.3	4.3484		4.36
2850.3	4.3499		4.74
2849.3	4.3514		4.61
2848.3	4.3530		4.16
2847.3	4.3545		4.09
2846.3	4.3560		4.13
2845.3	4.3575		4.20
2844.3	4.3590		4.06
2843.3	4.3606		3.82
2842.8	4.3614		3.71
2842.3	4.3621		3.71
2841.8	4.3629		3.82
2841.3	4.3637		3.89
2840.8	4.3644		3.54
2840.3	4.3652		2.54
2839.8	4.3660		2.08
2839.3	4.3668		1.95
2838.8	4.3675		2.09
2838.3	4.3683		2.57
2837.8	4.3691		3.65
2837.3	4.3698		4.80
2836.8	4.3706		4.94
2836.3	4.3714		4.87
2835.8	4.3721		5.85
2835.3	4.3729		5.95
2834.8	4.3737		5.85
2834.3	4.3745		4.53
2833.8	4.3752		3.55
2833.3	4.3760		3.66
2832.8	4.3768		4.53
2832.3	4.3775		5.85
2831.8	4.3783		7.49
2831.3	4.3791		9.34
2830.8	4.3799		9.06
2830.3	4.3806		6.45
2829.8	4.3814		1.57
2829.3	4.3822		1.05×10^{-6}
2828.8	4.3830		6.98×10^{-7}
2828.3	4.3837		4.19×10^{-7}
219	56.61		1.82×10^{-3} [19]
218.5	56.74		1.82
218	56.87		1.85
217.5	57.00		1.88
217	57.13		1.87
216.5	57.27		1.89
216	57.40		1.92
215.5	57.53		1.91

TABLE VIII (*Continued*)

Potassium Chloride

eV	Å	n	k
215	57.67		1.84
214.5	57.80		1.84
214	57.94		1.84
213.5	58.07		1.85
213	58.21		2.02
212.5	58.34		2.19
212	58.48		2.14
211.5	58.62		1.87
211	58.76		1.82
210.5	58.90		1.91
210	59.04		1.71
209.5	59.18		1.48
209	59.32		1.45
208.5	59.46		1.42
208	59.61		1.46
207.5	59.75		1.55
207	59.89		1.56
206.5	60.04		1.51
206	60.19		1.55
205.5	60.33		1.56
205	60.48		1.52
204.5	60.63		1.40
204	60.78		1.23
203.5	60.92		1.31
203	61.07		1.46
202.5	61.23		1.15
202	61.38		1.13
201.5	61.53		1.22
201	61.68		1.04
200.5	61.84		1.03
200	61.99		1.03
185.1	67	0.99874 [17]	1.01 [17]
109.7	113	0.99578	4.22×10^{-3}

eV	cm^{-1}	μm	n	k
43		0.02883	0.96 [13, 14]	3.0×10^{-2} [13, 14]
42		0.02952	0.955	2.5
41		0.03024	0.945	2.0
40		0.03099	0.925	1.8
39		0.03179	0.91	2.0
38		0.03263	0.89	3.0
37		0.03351	0.87	5.0
36.5		0.03397	0.845	7.04
36		0.03444	0.86	8.0
35.5		0.03493	0.830	9.16×10^{-2}
35		0.03542	0.87	0.115
34.5		0.03594	0.851	0.105
34		0.03646	0.89	0.12

(*continued*)

TABLE VIII (*Continued*)

Potassium Chloride

eV	cm^{-1}	μm	n	k
33.5		0.03701	0.854	7.83×10^{-2}
33.1		0.03746	0.821	6.77
32.7		0.03792	0.800	9.24×10^{-2}
32.3		0.03839	0.797	0.111
31.9		0.03887	0.795	0.121
31.5		0.03936	0.797	0.131
31.1		0.03987	0.799	0.134
30.7		0.04039	0.795	0.134
30.3		0.04092	0.779	0.134
29.9		0.04147	0.756	0.145
29.5		0.04203	0.743	0.175
29.1		0.04261	0.730	0.207
28.9		0.04290	0.729	0.234
28.7		0.04320	0.743	0.265
28.5		0.04350	0.785	0.275
28.3		0.04381	0.805	0.261
28.1		0.04412	0.816	0.245
27.9		0.04444	0.814	0.228
27.5		0.04509	0.789	0.230
27.1		0.04575	0.775	0.241
26.7		0.04644	0.755	0.263
26.3		0.04714	0.745	0.302
26.1		0.04750	0.750	0.335
25.9		0.04787	0.779	0.359
25.7		0.04824	0.805	0.362
25.5		0.04862	0.824	0.358
25.3		0.04901	0.836	0.357
25.1		0.04940	0.851	0.346
24.7		0.05020	0.849	0.317
24.3		0.05102	0.813	0.328
23.9		0.05188	0.806	0.363
23.5		0.05276	0.819	0.388
23.1		0.05367	0.834	0.413
22.9		0.05414	0.849	0.418
22.7		0.05462	0.860	0.420
22.5		0.05510	0.868	0.420
22.3		0.05560	0.872	0.418
21.9		0.05661	0.814	0.416
21.7		0.05714	0.790	0.488
21.5		0.05767	0.820	0.569
21.3		0.05821	0.918	0.612
21.1		0.05876	1.01	0.587
20.9		0.05932	1.06	0.530
20.7		0.05990	1.07	0.449
20.5		0.06048	1.02	0.355
20.3		0.06108	0.926	0.411
20.1		0.06168	0.910	0.495

TABLE VIII (*Continued*)

Potassium Chloride

eV	cm^{-1}	μm	n	k
19.9		0.06230	0.982	0.584
19.7		0.06294	1.00	0.497
19.5		0.06358	1.01	0.533
19.1		0.06491	1.09	0.525
18.7		0.06630	1.17	0.489
18.3		0.06775	1.15	0.387
17.9		0.06927	1.11	0.367
17.7		0.07005	1.07	0.382
17.5		0.07085	1.07	0.411
17.3		0.07167	1.09	0.428
17.1		0.07251	1.13	0.420
16.9		0.07336	1.14	0.376
16.7		0.07424	1.13	0.357
16.3		0.07606	1.09	0.322
15.9		0.07798	1.05	0.317
15.5		0.07999	1.01	0.328
15.1		0.08211	0.965	0.344
14.7		0.08434	0.931	0.367
14.3		0.08670	0.883	0.415
13.9		0.08920	0.791	0.441
13.5		0.09184	0.690	0.731
13.1		0.09465	0.841	0.950
12.9		0.09611	0.914	1.04
12.7		0.09763	1.04	1.22
12.5		0.09919	1.34	1.29
12.3		0.1008	1.57	1.12
12.1		0.1025	1.67	0.959
11.9		0.1042	1.74	0.767
11.7		0.1060	1.73	0.640
11.6		0.1069	1.67 [10]	0.58 [10]
11.5		0.1078	1.66	0.53
			1.69	0.520 [13]
11.4		0.1088	1.65	0.49 [10]
11.3		0.1097	1.63	0.45
			1.62	0.425 [13]
11.2		0.1107	1.60	0.41 [10]
11.1		0.1117	1.57	0.38
			1.55	0.374 [13]
11.0		0.1127	1.54	0.35 [10]
10.9		0.1137	1.50	0.325
			1.50	0.363 [13]
10.8		0.1148	1.47	0.31 [10]
10.7		0.1159	1.42	0.315
			1.40	0.336 [13]
10.6		0.1170	1.38	0.36 [10]
10.55		0.1175	1.36	0.38

(*continued*)

TABLE VIII (*Continued*)

Potassium Chloride

eV	cm^{-1}	μm	n	k
10.5		0.1181	1.37	0.385
10.4		0.1192	1.395	0.38
10.3		0.1204	1.35	0.37
10.2		0.1216	1.30	0.35
10.1		0.1228	1.24	0.355
10.0		0.1240	1.16	0.38
9.9		0.1252	1.05	0.41
9.8		0.1265	0.87	0.58
9.75		0.1272		0.71
9.7		0.1278	0.80	0.90
9.65		0.1285		1.1
9.6		0.1291	0.94	1.27
9.55		0.1298		1.41
9.5		0.1305	1.38	1.495
9.45		0.1312		1.48
9.4		0.1319	1.81	1.38
9.35		0.1326		1.28
9.3		0.1333	2.07	1.13
9.2		0.1348	2.18	0.85
9.1		0.1362	2.12	0.64
9.0		0.1378	1.99	0.50
8.9		0.1393	1.90	0.43
8.8		0.1409	1.85	0.40
8.7		0.1425	1.81	0.39
8.6		0.1442	1.79	0.37
8.5		0.1459	1.75	0.345
8.4		0.1476	1.73	0.34
8.3		0.1494	1.68	0.265
8.2		0.1512	1.54	0.215
8.1		0.1531	1.36	0.25
8.0		0.1550	1.15	0.46
7.95		0.1560		0.67
7.9		0.1569	1.13	0.90
7.85		0.1579		1.12
7.8		0.1589	1.30	1.36
7.75		0.1600		1.54
7.7		0.1610	1.92	1.64
				1.67 [9]
7.65		0.1621		1.6 [10]
7.6		0.1631	2.63	1.34
				1.43 [9]
7.55		0.1642	2.92	1.1 [10]
7.5		0.1653	3.0	0.51
				0.494 [9]
7.45		0.1664	2.8	0.2 [10]
7.4		0.1675	2.61	6×10^{-2}
				5.60×10^{-2} [9]

TABLE VIII (*Continued*)

Potassium Chloride

eV	cm^{-1}	μm	n	k
7.3		0.1698	2.34	3.11×10^{-3}
7.2		0.1722	2.19	1.78×10^{-4}
7.1		0.1746	2.08	7.96×10^{-6}
7.0		0.1771	2.0	8.46×10^{-7}
6.9		0.1797	1.94	2.29×10^{-7}
6.888		0.180	1.89324 [8]	
6.812		0.182	1.86392	
6.8		0.1823	1.89 [10]	1.45×10^{-7}
6.738		0.184	1.83883 [8]	
6.7		0.1850	1.85 [10]	
6.666		0.186	1.81705 [8]	
6.6		0.1879	1.82 [10]	
6.595		0.1888	1.79789 [8]	
6.525		0.190	1.78087	
6.5		0.1907	1.80 [10]	
6.4		0.1937	1.78	
6.199	50,000	0.20	1.71739 [8]	
5.904	47,620	0.21	1.67539	
5.636	45,450	0.22	1.64517	
5.391	43,480	0.23	1.62226	
5.166	41,670	0.24	1.60426	
4.959	40,000	0.25	1.58972	
4.769	38,460	0.26	1.57775	
4.592	37,040	0.27	1.56772	
4.428	35,710	0.28	1.55921	
4.275	34,480	0.29	1.55191	
4.133	33,330	0.30	1.54558	
3.999	32,260	0.31	1.54005	
3.874	31,250	0.32	1.53518	
3.757	30,300	0.33	1.53087	
3.647	29,410	0.34	1.52703	
3.542	28,570	0.35	1.52358	
3.473	28,010	0.357		5.1×10^{-10} [7]
3.444	27,780	0.36	1.52049	
3.351	27,030	0.37	1.51769	
3.263	26,320	0.38	1.51515	
3.179	25,640	0.39	1.51283	
3.100	25,000	0.40	1.51072	
2.952	23,810	0.42	1.50701	
2.818	22,730	0.44	1.50386	
2.695	21,740	0.46	1.50115	
2.616	21,100	0.474		7.6×10^{-11}
2.583	20,830	0.48	1.49882	
2.480	20,000	0.50	1.49679	
2.384	19,230	0.52	1.49501	
2.296	18,520	0.54	1.49344	

(*continued*)

TABLE VIII (*Continued*)

Potassium Chloride

eV	cm^{-1}	μm	n	k
2.214	17,860	0.56	1.49205	
2.138	17,240	0.58	1.49081	
2.066	16,670	0.60	1.48969	
2.000	16,130	0.62	1.48869	
1.937	15,630	0.64	1.48779	
1.879	15,150	0.66	1.48697	
1.823	14,710	0.68	1.48623	
1.771	14,290	0.70	1.48555	
1.722	13,890	0.72	1.48493	
1.675	13,510	0.74	1.48436	
1.631	13,160	0.76	1.48384	
1.590	12,820	0.78	1.48336	
1.550	12,500	0.80	1.48291	
1.512	12,200	0.82	1.48250	
1.476	11,900	0.84	1.48212	
1.442	11,630	0.86	1.48176	
1.409	11,360	0.88	1.48143	
1.378	11,110	0.90	1.48111	
1.348	10,870	0.92	1.48082	
1.319	10,640	0.94	1.48055	
1.291	10,420	0.96	1.48030	
1.265	10,200	0.98	1.48006	
1.240	10,000	1.00	1.47983	
1.127	9,091	1.1	1.47887	
1.033	8,333	1.2	1.47813	
0.9537	7,692	1.3	1.47755	
0.8856	7,143	1.4	1.47708	
0.8265	6,667	1.5	1.47668	
0.7749	6,250	1.6	1.47634	
0.7293	5,882	1.7	1.47605	
0.6888	5,556	1.8	1.47580	
0.6525	5,263	1.9	1.47557	
0.6199	5,000	2.0	1.47536	
0.5636	4,545	2.2	1.47498	
0.5166	4,167	2.4	1.47464	
0.4769	3,046	2.6	1.47433	
0.4428	3,571	2.8	1.47403	1.27×10^{-10} [6]
0.4133	3,333	3.0	1.47374	
0.3874	3,125	3.2	1.47345	
0.3647	2,941	3.4	1.47315	
0.3444	2,778	3.6	1.47285	
0.3263	2,632	3.8	1.47254	1.97×10^{-10}
0.3100	2,500	4.0	1.47223	
0.2952	2,381	4.2	1.47190	
0.2818	2,273	4.4	1.47156	
0.2695	2,174	4.6	1.47122	
0.2583	2,083	4.8	1.47085	

TABLE VIII (*Continued*)

Potassium Chloride

eV	cm^{-1}	μm	n	k
0.2480	2,000	5.0	1.47048	
0.2384	1,923	5.2	1.47010	
0.2339	1,887	5.3		2.11×10^{-11}
0.2296	1,852	5.4	1.46970	
0.2214	1,786	5.6	1.46928	
0.2138	1,724	5.8	1.46886	
0.2066	1,667	6.0	1.46842	
0.2000	1,613	6.2	1.46796	
0.1937	1,563	6.4	1.46749	
0.1879	1,515	6.6	1.46701	
0.1823	1,471	6.8	1.46651	
0.1771	1,429	7.0	1.46600	
0.1722	1,389	7.2	1.46547	
0.1675	1,351	7.4	1.46493	
0.1631	1,316	7.6	1.46437	
0.1590	1,282	7.8	1.46379	
0.1550	1,250	8.0	1.46290	
0.1512	1,220	8.2	1.46260	
0.1476	1,190	8.4	1.46198	
0.1442	1,163	8.6	1.46134	
0.1409	1,136	8.8	1.46069	
0.1378	1,111	9.0	1.46002	
0.1348	1,087	9.2	1.45934	
0.1319	1,064	9.4	1.45864	
0.1291	1,042	9.6	1.45792	
0.1265	1,020	9.8	1.45719	
0.1240	1,000	10.0	1.45644	
0.1216	980.4	10.2	1.45567	
0.1192	961.5	10.4	1.45489	
0.1170	943.4	10.6	1.45409	2.53×10^{-8} [5]
0.1148	925.9	10.8	1.45327	
0.1127	909.1	11.0	1.45244	
0.1107	892.9	11.2	1.45159	
0.1088	877.2	11.4	1.45072	
0.1069	862.1	11.6	1.44984	
0.1051	847.5	11.8	1.44893	
0.1033	833.3	12.0	1.44801	
0.1016	819.7	12.2	1.44707	
0.09999	806.5	12.4	1.44611	
0.09919	800	12.50		1.49×10^{-7} [4]
0.09840	793.7	12.6	1.44514	
0.09686	781.3	12.8	1.44415	
0.09537	769.2	13.0	1.44313	
0.09393	757.6	13.2	1.44210	
0.09252	746.3	13.4	1.44105	
0.09116	735.3	13.6	1.43999	

(*continued*)

TABLE VIII (*Continued*)

Potassium Chloride

eV	cm^{-1}	μm	n	k
0.08984	724.6	13.8	1.43890	
0.08856	714.3	14.0	1.43779	
0.08731	704.2	14.2	1.43667	
0.08679	700	14.29		1.14×10^{-6}
0.08610	694.4	14.4	1.43552	
0.08492	684.9	14.6	1.43343	
0.08377	675.7	14.8	1.43317	
0.08265	666.7	15.0	1.43197	
0.08157	657.9	15.2	1.43074	
0.08051	649.4	15.4	1.42950	
0.07948	641.0	15.6	1.42823	
0.07847	632.9	15.8	1.42694	
0.07749	625.0	16.0	1.42563	
0.07653	617.3	16.2	1.42	
0.07560	609.8	16.4	1.42295	
0.07469	602.4	16.6	1.42158	
0.07439	600	16.67		8.76×10^{-6}
0.07380	595.2	16.8	1.42018	
0.07293	588.2	17.0	1.41877	
0.07208	581.4	17.2	1.41733	
0.07125	574.7	17.4	1.41587	
0.07044	568.2	17.6	1.41438	
0.06965	561.8	17.8	1.41287	
0.06888	555.6	18.0	1.41134	
0.06812	549.4	18.2	1.40979	
0.06738	543.5	18.4	1.40821	
0.06666	537.6	18.6	1.40661	
0.06595	531.9	18.8	1.40498	
0.06525	526.3	19.0	1.40333	
0.06457	520.8	19.2	1.40165	
0.06391	515.5	19.4	1.39995	
0.06326	510.2	19.6	1.39822	
0.06262	505.1	19.8	1.39647	
0.06199	500.0	20.0	1.39469	7.64×10^{-5}
0.06048	487.8	20.5	1.39012	
0.05904	476.2	21.0	1.38538	
0.05767	465.1	21.5	1.38047	
0.05636	454.5	22.0	1.37537	
0.05510	444.4	22.5	1.37009	
0.05391	434.8	23.0	1.36461	
0.05276	425.5	23.5	1.35892	
0.05166	416.7	24.0	1.35303	
0.05060	408.2	24.5	1.34692	
0.04959	400.0	25.0	1.34059	6.57×10^{-4}
0.04862	392.2	25.5	1.33402	
0.04769	384.6	26.0	1.32721	
0.04679	377.4	26.5	1.32014	

TABLE VIII (*Continued*)

Potassium Chloride

eV	cm^{-1}	μm	n	k
0.04592	370.4	27.0	1.31281	
0.04508	363.6	27.5	1.30520	
0.04428	357.1	28.0	1.29731	
0.04350	350.9	28.5	1.28911	
0.04275	344.8	29.0	1.28060	
0.04215	340	29.41	1.30 [3]	
0.04203	339.0	29.5	1.27175 [8]	
0.04133	333.3	30.0	1.2626	
0.04065	327.9	30.5	1.2530	
0.03999	322.6	31.0	1.2431	
0.03967	320	31.25	1.28 [3]	
0.03936	317.5	31.5	1.2327 [8]	
0.03874	312.5	32.0	1.2220	
0.03843	310	32.26	1.25 [3]	
0.03815	307.7	32.5	1.2108 [8]	
0.03757	303.0	33.0	1.1991	
0.03719	300	33.33	1.20 [3]	6.63×10^{-3}
0.03701	298.5	33.5	1.8669 [8]	
0.03647	294.1	34.0	1.1741	
0.03595	290	34.48	1.15 [3]	
0.03594	289.9	34.5	1.1608 [8]	
0.03542	285.7	35.0	1.1469	
0.03471	280	35.71	1.10 [3]	2.3×10^{-2} [3]
0.03409	275	36.36		2.5
0.03348	270	37.04	1.05	2.8
0.03268	265	37.74		3.2
0.03224	260	38.46	0.99	4.2
0.03100	250	40.00	0.93	8.3×10^{-2}
0.02976	240	41.67	0.85	0.16
0.02914	235	42.55	0.86	0.17
0.02852	230	43.48	0.80	0.18
0.02790	225	44.44	0.70	0.20
0.02728	220	45.45	0.53	0.35
0.02666	215	46.51	0.61	0.58
0.02604	210	47.62	0.70	0.52
0.02542	205	48.78		0.50
0.02480	200	50.00	0.57	0.53
0.02356	190	52.63	0.41	0.77
0.02232	180	55.56	0.31	1.05
0.02170	175	57.14	0.23	
0.02108	170	58.82	0.17	1.5
0.02046	165	60.61	0.15	
0.01984	160	62.50	0.17	2.2
0.01922	155	64.52	0.24	2.9
0.01860	150	66.67	0.44	4.0
0.01798	145	68.97	1.1	5.5

(*continued*)

TABLE VIII (*Continued*)

Potassium Chloride

eV	cm^{-1}	μm	n	k
0.01736	140	71.43	5.3	4.0
0.01674	135	74.07	5.0	0.80
0.01612	130	76.92	4.1	0.32
0.01550	125	80.00	3.5	0.23
0.01488	120	83.33	3.2	0.20
0.01364	110	90.91	2.9	0.15
0.01240	100	100.0	2.7	0.11
0.01178	95	105.3		9.0×10^{-2}
0.01116	90	111.1	2.6	6.6×10^{-2}
0.01054	85	117.6		0.11
0.009919	80	125.0	2.5	0.11
0.009299	75	133.3		
0.008679	70	142.9	2.4	9.2×10^{-2}
0.007439	60	166.7	2.3	5.3
0.006199	50	200.0	2.2	
0.004133	33.33	300		3.1×10^{-2} [1]
0.003874	31.25	320		2.65 [2]
0.003100	25.00	400		2.5 [1]
0.002480	20.00	500		1.9
				1.83 [2]
0.002066	16.67	600		1.6 [1]
0.001771	14.29	700		1.3
0.001550	12.50	800		1.2
0.001378	11.11	900		1.0
				1.07 [2]
0.001240	10.00	1,000		9.0×10^{-3} [1]
0.0008265	6.667	1,500		5.5
0.0006199	5.000	2,000		3.7
0.0004133	3.333	3,000		2.0

^{a} The original references for the n and k data are indicated in brackets. A given reference applies down the column until a new one appears. The smaller values of k are given in powers of 10 until 10^{-2}, whereupon we revert to decimals only.

Silicon Dioxide (SiO$_2$), Type α (Crystalline)

H. R. PHILIPP

General Electric Research and Development Center
Schenectady, New York

The room-temperature optical properties of type α, crystalline SiO$_2$ have been the subject of numerous studies. In the infrared spectral region, this information is reasonably detailed. In the vacuum ultraviolet, however, few attempts have been made to obtain a self-consistent set of optical constants n and k for this material. These parameters are generally obtained by Kramers–Kronig (KK) analysis of reflectance data, which are difficult to measure with high accuracy. In addition, the analysis in the case of SiO$_2$ requires extrapolation of the reflectance into spectral regions in which no data exist. This introduces considerable uncertainty in the derived n and k values. In the region of low absorption between the infrared and vacuum ultraviolet bands, the index of refraction can be evaluated from prism data, and this has been accomplished with great precision. The measured k values, on the other hand, can be strongly influenced by the presence of impurity and defect absorption, and for this reason no k values are given in the wavelength range 6.0 μm $\geq \lambda \geq$ 0.1476 μm, although some data are available for the users' consideration and will be referenced.

A set of optical parameters n and k are given in Table IX and plotted in Fig. 9 for the wavelength region 333.3 μm $\geq \lambda \geq$ 0.04768 μm. They were obtained from a variety of sources. The wavelength λ index of refraction n and extinction coefficient k are given to four places in most of the table entries. The accuracy of these values does not, in general, warrant this precision. This is done to show the possible presence of weak structure in the optical properties that can often be discerned by comparison of precise, but not necessarily accurate, values over small wavelength ranges. The n_o and k_o values refer to the ordinary (o) ray (electric field perpendicular to the optic axis), and n_e and k_e refer to the extraoridinary (e) ray (electric field parallel to the optic axis).

HANDBOOK OF OPTICAL CONSTANTS OF SOLIDS

In the sub-millimeter-wave region from 1000 to 200 μm the n_o results of several workers have been combined and fitted by Simonis[†] with a frequency-squared dependence. This fit gives values in agreement with all these experiments to ± 0.0015, except for the results of Loewenstein et al. [1], which are consistently 0.005 higher. Most of these results were obtained by analysis of channel spectra. In the far infrared $333 \geq \lambda \geq 50$ μm the optical constants determined by Loewenstein et al. [1] are used. The estimated experimental error in n is ± 0.001, but at least in the region 333–200 μm it is considerably more. The error in k is considerably larger than the error in n. Their data extend to shorter wavelength ($\lambda = 30.3$ μm) for temperature $T = 1.5$ K but were not evaluated at room temperature. Because of the strong temperature dependence of the absorption in this region and the presence of a resonance absorption at $\lambda \approx 37.88$ μm, considerable caution should be taken in using the low-temperature data to estimate room-temperature values.

In the infrared, 32.00 μm $\geq \lambda \geq 2.000$ μm, the values were obtained from the analysis of Spitzer and Kleinman [2]. They carefully measured the reflectance and fitted the results to classical dispersion theory. Their n and k values reproduce the measured reflectance and independently obtained transmission curves in detail over most of the spectral range. They also compare their results with values derived by KK analysis of the same reflectance data. They find that the agreement in k is quite good for $k > 0.1$ but is poor for $k \lesssim 0.1$. We believe that this inconsistency is due, at least in part, to the lack of adequate precision in the reflectance data that is necessary to evaluate accurately small k values by KK analysis. We note in this connection that in the regions of very low reflectance, the measured values are generally appreciably higher (on a relative basis) than those calculated from the n and k values. This could result from a small scattered-light component in the measuring beam.

The n and k values given in the table were computed from the dispersion parameters of Spitzer and Kleinman [2] that do not consider absorption processes (resonances) above $\lambda \approx 35$ μm. The results for the e ray join, more or less smoothly, in the region 32 μm $\leq \lambda \leq 50$ μm with the results of Loewenstein et al. [1], while those for the o ray, which show the presence of a resonance near 38 μm, do not. At the shorter wavelengths, the results of Spitzer and Kleinman [2] agree very well with index of refraction data for SiO_2 obtained from prism measurements [3]. For the o ray, the precise prism measurements extend as high as 4 μm, while for the e ray, they extend only to 2 μm. It should be noted here that less precise index of refraction values given in the American Institute of Physics Handbook [3] for the range 4.2 μm $\leq \lambda \leq 7.0$ μm do not agree very closely with the dispersion analysis results [2] that are given in the table.

No k values are given in the table for infrared wavelengths below 6.00 μm. Both Saksena [4] and Drummond [5] have examined the absorption spectra

[†] See Chapter 8 of this handbook.

of α-SiO$_2$ for $\lambda \leq 6.00\ \mu$m, and the results show a variety of features. It is difficult to determine which of these features are intrinsic to SiO$_2$, and, hence, these results are not considered further here, although they do contain a wealth of information.

In the wavelength region $2.00\ \mu\text{m} \geq \lambda \geq 0.185\ \mu$m, the precise index of refraction values of the American Institute of Physics Handbook [3] are given in table. Some time ago Sosman [6] prepared a list of most probable values of the refraction indices of SiO$_2$, and these values are also given in the table, in which they are denoted by brackets. These latter values are generally more precise than those of the American Institute of Physic Handbook [3]; however, I cannot comment on their accuracy.

For $\lambda < 0.185\ \mu$m, the optical constants given in the table were obtained from the reflectance measurements of Philipp [7] augmented by the transmission measurements of Appleton et al. [8] and Philipp [9] in the tail of the fundamental absorption edge. Both Lamy [10] and Philipp [7] have measured the reflectance and determined n and k by KK analysis. The analysis of Lamy [10] uses the reflectance values of Philipp [7] for $\lambda \leq 0.10\ \mu$m. The n values from these two sources agree very closely for wavelengths down to $\sim 0.11\ \mu$m. The k values show very good agreement for $\lambda \lesssim 0.08\ \mu$m. The differences found below these wavelengths of $\sim 10\%$ must be attributed to the extrapolation procedures used in the KK analysis. The choice of using Philipp's [7] values over those of Lamy [10] was an arbitrary one. Care should be exercised in using the table results, especially for $\lambda < 0.08\ \mu$m, until data are available for $\lambda \lesssim 0.05\ \mu$m, which more specifically details the reflectance extrapolation.

The k values determined from the KK analysis of Phillipp [7] are tabulated for wavelengths of $0.1278\ \mu$m and below. Above this wavelength, where they become small, they are unreliable as determined by this technique. In the range $0.1476\ \mu\text{m} \geq \lambda \geq 0.1409\ \mu$m, they are obtained from the absorption measurements of Appleton et al. [8] and Philipp [9], which agree reasonably closely. A smooth curve was drawn between the absorption data and the KK results, and the values so determined are given in the table, in which they are delineated by brackets; they are not reliable. No k values are given for $\lambda > 0.1476\ \mu$m because they may have a component associated with a defect or impurity absorption.

It is recommended that anyone who uses the table carefully examine the material and references cited in the above-named references [1–10] as well as other references [11–15].

REFERENCES

1. E. V. Loewenstein, D. R. Smith, and R. L. Morgan, *Appl. Opt.* **12**, 398 (1973).
2. W. G. Spitzer and D. A. Kleinman, *Phys. Rev.* **121**, 1324 (1961).
3. "American Institute of Physics Handbook," 3rd ed., Chapter 6, p. 27. McGraw-Hill, New York, 1972.

4. B. D. Saksena, *Proc. Phys. Soc. London* **72**, 9 (1958).

5. D. G. Drummond, *Proc. R. Soc. London* **153**, 328 (1935).

6. R. B. Sosman, "The Properties of Silica," p. 591. Chem. Catalog Co. (Tudor), New York, 1927.

7. H. R. Philipp, *Solid State Commun.* **4**, 73 (1966).

8. A. Appleton, T. Chiranjivi, and M. Jafaripour-Ghazvini, *in* "The Physics of SiO_2 and Its Interfaces" (S. T. Pantelides, ed.), p. 94. Pergamon, New York, 1978.

9. H. R. Philipp, *J. Phys. Chem. Solids* **32**, 1935 (1971).

10. P. L. Lamy, *Appl. Opt.* **16**, 2212 (1977).

11. I. Simon and H. O. McMahon, *J. Chem. Phys.* **21**, 23 (1953).

12. J. Reitzel, *J. Chem. Phys.* **23**, 2407 (1955).

13. E. Loh, *Solid State Commun.* **2**, 269 (1964).

14. S. Roberts and E. Coon, *J. Opt. Soc. Am.* **53**, 1023 (1962).

15. D. E. McCarthy, *Appl. Opt.* **2**, 591 (1963).

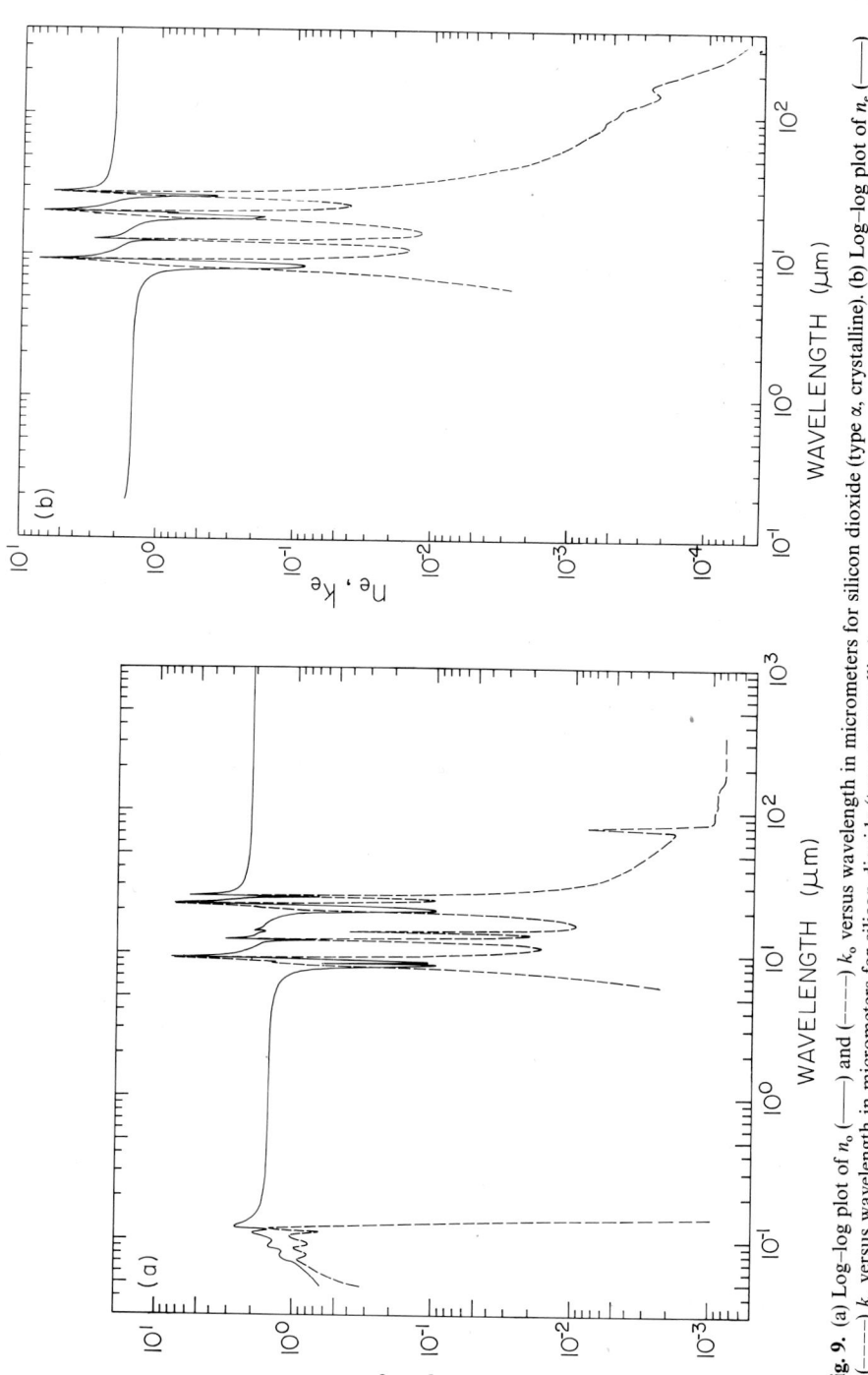

Fig. 9. (a) Log–log plot of n_o (——) and (———) k_o versus wavelength in micrometers for silicon dioxide (type α, crystalline). (b) Log–log plot of n_e (——) and (———) k_e versus wavelength in micrometers for silicon dioxide (type α, crystalline).

TABLE IX

Values of n and k for Silicon Dioxide (Type α, Crystalline) from Various References[a]

eV	cm^{-1}	μm	n_o	k_o	n_e	k_e
26		0.04769	0.6285 [7]	0.3243 [7]		
25.5		0.04862	0.6432	0.3609		
25		0.04959	0.6517	0.3907		
24.5		0.05061	0.6591	0.4231		
24		0.05166	0.6723	0.4540		
23.5		0.05276	0.6862	0.4822		
23		0.05391	0.6998	0.5097		
22.5		0.05510	0.7138	0.5387		
22		0.5636	0.7304	0.5667		
21.5		0.05767	0.7470	0.5959		
21		0.05904	0.7664	0.6272		
20.5		0.06048	0.7900	0.6559		
20		0.06199	0.8131	0.6844		
19.5		0.06358	0.8371	0.7154		
19		0.06526	0.8642	0.7488		
18.5		0.06702	0.8949	0.7847		
18		0.06888	0.9289	0.8273		
17.75		0.06985	0.9499	0.8619		
17.5		0.07085	0.9894	0.8987		
17.4		0.07125	1.013	0.9108		
17.25		0.07187	1.052	0.9236		
17		0.07293	1.118	0.9153		
16.75		0.07402	1.169	0.8917		
16.5		0.07514	1.205	0.8599		
16.25		0.07629	1.224	0.8223		
16		0.07749	1.221	0.7896		
15.75		0.07872	1.210	0.7848		

15.5	0.07999	1.202	0.7841
15.25	0.08130	1.191	0.7968
15	0.08266	1.184	0.8265
14.8	0.08377	1.187	0.8694
14.6	0.08492	1.217	0.9290
14.4	0.08610	1.285	0.9724
14.3	0.08670	1.335	0.9829
14.2	0.08731	1.386	0.9709
14	0.08856	1.457	9187
13.8	0.08984	1.484	0.8462
13.6	0.09116	1.469	0.7910
13.4	0.09252	1.447	0.7761
13.2	0.09392	1.430	0.7715
13	0.09537	1.410	0.7753
12.8	0.09686	1.394	0.8017
12.6	0.09839	1.390	0.8406
12.4	0.09998	1.400	0.8908
12.2	0.1016	1.435	0.9573
12	0.1033	1.504	1.015
11.9	0.1042	1.551	1.035
11.8	0.1051	1.602	1.048
11.7	0.1060	1.661	1.056
11.6	0.1069	1.722	1.046
11.5	0.1078	1.782	1.027
11.4	0.1088	1.839	0.9939
11.3	0.1097	1.892	0.9452
11.2	0.1107	1.924	0.8730
11.1	0.1117	1.925	0.7958
11	0.1127	1.891	0.7210
10.9	0.1137	1.805	0.6597
10.85	0.1143	1.738	0.6418

(continued)

TABLE IX (Continued)

Silicon Dioxide (Type α, Crystalline)

eV	cm⁻¹	μm	n_o	k_o	n_e	k_e
10.83		0.1145	1.688	0.6356		
10.8		0.1148	1.626	0.6732		
10.75		0.1153	1.569	0.7594		
10.7		0.1159	1.528	0.8463		
10.65		0.1164	1.506	0.9666		
10.6		0.1170	1.521	1.114		
10.55		0.1175	1.607	1.289		
10.5		0.1181	1.769	1.400		
10.45		0.1186	1.959	1.454		
10.4		0.1192	2.159	1.452		
10.37		0.1196	2.289	1.423		
10.35		0.1198	2.373	1.378		
10.3		0.1204	2.515	1.229		
10.25		0.1210	2.590	1.072		
10.2		0.1215	2.619	0.9326		
10.1		0.1227	2.621	0.7123		
10		0.1240	2.581	0.5441		
9.9		0.1252	2.529	0.4161		
9.8		0.1265	2.464	0.3148		
9.7		0.1278	2.397	0.2402		
9.6		0.1291	2.335	(0.17)		
9.5		0.1305	2.275	(0.12)		
9.4		0.1319	2.222	(0.082)		
9.3		0.1333	2.174	(0.056)		
9.2		0.1348	2.133	(0.036)		
9.1		0.1362	2.095	(0.023)		
9		0.1378	2.060	(0.013)		

8.9		0.1393	2.028	(7.0 × 10⁻³)	
8.8		0.1409	1.999	3.1 × 10⁻³ [8, 9]	
8.7		0.1425	1.971	1.0 × 10⁻³	
8.6		0.1442	1.946	2.3 × 10⁻⁴	
8.5		0.1459	1.922	3.5 × 10⁻⁵	
8.4		0.1476	1.899	2.7 × 10⁻⁶	
8.3		0.1494	1.877		
8.2		0.1512	1.856		
8.1		0.1531	1.838		
8		0.1550	1.820		
7.8		0.1590	1.789		
7.6		0.1631	1.761		
7.4		0.1675	1.737		
7.2		0.1722	1.716		
7		0.1771	1.698		
6.75		0.1837	1.679		
6.7		0.1851	1.6775 [3]		1.68988 [3]
6.68484	53,917.9	0.185467	(1.67578) [6]		(1.68997) [6]
6.40458	51,657.4	0.193583	(1.65999)		(1.67343)
6.262	50,510	0.1980	1.65087 [3]	1.66394 [3]	(1.66227) [6]
6.1972	49,985	0.20006	(1.64927) [6]		(1.65842)
6.1210	49,371	0.20255	(1.64557)		(1.65562)
6.06422	48,912.2	0.204448	(1.64288)		(1.64671)
5.8740	47,378	0.21107	(1.63432)		(1.64262)
5.78168	46,633.3	0.214439	(1.63039)		(1.63698)
5.64935	45,566.0	0.219462	(1.62497)		(1.62992)
5.47374	44,149.5	0.226503	(1.61818)		1.62555 [3]
5.367	43,290	0.2310	1.61395 [3]		(1.62559)
5.36049	43,236.1	0.231288	(1.61401)		(1.61650)
5.10642	41,186.8	0.242796	(1.60525)		(1.61139)
4.95275	39,947.4	0.250329	(1.60032)		

(continued)

TABLE IX (*Continued*)

Silicon Dioxide (Type α, Crystalline)

eV	cm⁻¹	μm	n_o	k_o	n_e	k_e
4.81849	38,864.5	0.257304	(1.59622)		(1.60714)	
4.71136	38,000.4	0.263155	(1.59309)		(1.60389)	
4.51061	36,381.2	0.274867	(1.58752)		(1.59813)	
4.25531	34,322.0	0.291358	(1.58098)		(1.59136)	
4.08625	32,958.5	0.303412	(1.57955)		(1.58720)	
3.97023	32,022.6	0.312279	(1.57433)		(1.584485)	
3.81186	30,745.3	0.325253	(1.570915)		(1.58095)	
3.647	29,410	0.3400	1.56747 [3]		1.57737 [3]	
3.64261	29,380.2	0.340365	(1.56747) [6]		(1.577385) [6]	
3.4566	27,880	0.35868	(1.563915)		(1.573705)	
3.147	25,380	0.3940	1.55846 [3]		1.56805 [3]	
3.12416	25,198.6	0.396848	(1.55813) [6]		(1.56772) [6]	
3.06388	24,712.3	0.404656	(1.557156)		(1.56671)	
3.02266	24,379.9	0.410174	(1.556502)		(1.566031)	
2.857	23,040	0.4340	1.55396 [3]		1.56339 [3]	
2.85641	23,093.0	0.434047	(1.553963) [6]		(1.563405) [6]	
2.84470	22,944.5	0.435834	(1.553790)		(1.563225)	
2.65023	21,376.0	0.467815	(1.551027)		(1.560368)	
2.58300	20,833.7	0.479991	(1.550118)		(1.559428)	
2.55037	20,570.5	0.486133	(1.549683)		(1.558979)	
2.138	17,240	0.5080	1.54822 [3]		1.55746 [3]	
2.43779	19,662.5	0.508582	(1.548229) [6]		(1.557475) [6]	
2.39180	19,291.5	0.518362	(1.547651)		(1.556877)	
2.32241	18,731.9	0.53385	(1.546799)		(1.555996)	
2.27043	18,312.6	0.546072	(1.546174)		(1.555350)	
2.14107	17,269.2	0.579066	(1.544667)		(1.553791)	
2.11010	17,019.5	0.587563	(1.544316)		(1.553428)	

2.1039	16,970	0.58929	(1.544246)	(1.553355)
2.104	16,970	0.5893	1.54424 [3]	1.55335 [3]
1.9748	15,928	0.62782	(1.542819) [6]	(1.551880) [6]
1.92564	15,531.6	0.643847	(1.542288)	(1.551332)
1.88917	15,237.4	0.656278	(1.541899)	(1.550929)
1.85653	14,974.2	0.667815	(1.541553)	(1.550573)
1.84831	14,907.9	0.670786	(1.541466)	(1.550483)
1.75482	14,153.9	0.706520	(1.540488)	(1.549472)
1.70273	13,733.7	0.728135	(1.539948)	(1.548913)
1.61752	13,046.4	0.766494	(1.539071)	(1.548005)
1.614	13,020	0.7680	1.53903 [3]	1.54794 [3]
1.55998	12,582.4	0.794763	(1.538478) [6]	(1.547392) [6]
1.489	12,010	0.8325	1.53773 [3]	1.54661 [3]
1.4678	11,839	0.84467	(1.537525) [6]	(1.54640) [6]
1.251	10,090	0.9914	1.53514 [3]	1.54392 [3]
1.23982	10,000.0	1.00000	(1.53503) [6]	(1.54381) [6]
1.22263	9,861.35	1.01406	(1.53483)	(1.54360)
1.14477	9,233.35	1.08303	(1.53387)	(1.54260)
1.0695	8,626.6	1.1592	1.53283 [3]	1.54152 [3]
1.03318	8,333.33	1.20000	(1.53232) [6]	(1.54098) [6]
0.953706	7,692.31	1.30000	(1.53102)	(1.53962)
0.94860	7,651.1	1.3070	1.53090 [3]	1.53951 [3]
0.88825	7,164.4	1.3958	1.52977	1.53832
0.885584	7,142.86	1.40000	(1.52972) [6]	(1.53826) [6]
0.83817	6,760.4	1.4792	1.52865 [3]	1.53716 [3]
0.810545	6,537.61	1.52961	(1.52800) [6]	(1.53646) [6]
0.80435	6,487.6	1.5414	1.52781 [3]	1.53630 [3]
0.774886	6,250.00	1.60000	(1.52703) [6]	(1.53545) [6]
0.73733	5,947.1	1.6815	1.52583 [3]	1.53422 [3]
0.70388	5,677.3	1.7614	1.52468	1.53301
0.688788	5,555.55	1.80000	(1.52413) [6]	(1.53242) [6]

(continued)

TABLE IX (Continued)

Silicon Dioxide (Type α, Crystalline)

eV	cm⁻¹	μm	n_o	k_o	n_e	k_e
0.63721	5,139.5	1.9457	1.52184 [3]		1.53004 [3]	
0.60388	4,870.7	2.0531	1.52005		1.52823	
0.602362	4,858.47	2.05820	(1.51998) [6]		(1.52814) [6]	
0.5391	4,348	2.300	1.51561 [3]			
0.495927	4,000.00	2.50000	(1.51156) [6]		(1.5195)	
0.4769	3,846	2.600	1.50986 [3]			
0.4133	3,333	3.000	1.49953		1.507 [2]	
0.3542	2,857	3.500	1.48451		1.492	
0.3100	2,500	4.000	1.46617		1.473	
0.2952	2,381	4.200	1.457 [2]		1.465	
0.2818	2,273	4.400	1.448		1.455	
0.2695	2,174	4.600	1.437		1.444	
0.2583	2,083	4.800	1.425		1.432	
0.2480	2,000	5.000	1.412		1.419	
0.2384	1,923	5.200	1.398		1.404	
0.2296	1,852	5.400	1.382		1.388	
0.2214	1,786	5.600	1.364		1.369	
0.2138	1,724	5.800	1.343		1.348	
0.2066	1,667	6.000	1.320	2.270 × 10⁻³ [2]	1.325	2.672 × 10⁻³ [2]
0.2033	1,639	6.100	1.307	2.545	1.311	3.000
0.2000	1,613	6.200	1.294	2.862	1.297	3.378
0.1968	1,587	6.300	1.279	3.229	1.282	3.816
0.1937	1,563	6.400	1.263	3.657	1.266	4.328
0.1907	1,538	6.500	1.245	4.160	1.248	4.928
0.1879	1,515	6.600	1.226	4.755	1.229	5.639
0.1851	1,493	6.700	1.206	5.467	1.207	6.486
0.1823	1,471	6.800	1.183	6.328	1.184	7.507

0.1797	1,449	6.900	1.158	7.380	1.159	8.747×10^{-3}
0.1771	1,429	7.000	1.130	8.683	1.130	1.027×10^{-2}
0.1759	1,418	7.050	1.115	9.455×10^{-3}	1.115	1.117
0.1746	1,408	7.100	1.099	1.032×10^{-2}	1.099	1.217
0.1734	1,399	7.150	1.083	1.131	1.081	1.330
0.1722	1,389	7.200	1.065	1.243	1.063	1.457
0.1710	1,379	7.250	1.046	1.370	1.043	1.601
0.1698	1,370	7.300	1.025	1.517	1.023	1.764
0.1687	1,361	7.350	1.004	1.687	1.000	1.951
0.1675	1,351	7.400	0.9801	1.885	0.9764	2.166
0.1664	1,342	7.450	0.9549	2.117	0.9505	2.413
0.1653	1,333	7.500	0.9277	2.391	0.9226	2.700
0.1642	1,325	7.550	0.8982	2.717	0.8923	3.035
0.1631	1,316	7.600	0.8660	3.108	0.8592	3.430
0.1621	1,307	7.650	0.8308	3.583	0.8230	3.899
0.1610	1,299	7.700	0.7921	4.163	0.7829	4.462
0.1600	1,290	7.750	0.7492	4.883	0.73383	5.150
0.1590	1,282	7.800	0.7013	5.788	0.6881	6.007
0.1579	1,274	7.850	0.6472	949	0.6308	7.112
0.1569	1,266	7.900	0.5953	8.487×10^{-2}	0.5642	8.609×10^{-2}
0.1560	1,258	7.950	0.5132	0.1063	0.4850	0.1081
0.1550	1,250	8.000	0.4276	0.1391	0.3885	0.1450
0.1540	1,242	8.050	0.3274	0.1962	0.2767	0.2179
0.1531	1,235	8.100	0.2318	0.2958	0.1890	0.3399
0.1521	1,227	8.150	0.1728	0.4190	0.1454	0.4685
0.1512	1,220	8.200	0.1410	0.5380	0.1230	0.5857
0.1503	1,212	8.250	0.1216	0.6504	0.1094	0.6947
0.1494	1,205	8.300	0.1088	0.7599	0.1003	0.7992
0.1485	1,198	8.350	0.1005	0.8699	0.09390	0.9019
0.1476	1,190	8.400	0.0965	0.9842	0.08950	1.005

(continued)

TABLE IX (*Continued*)

Silicon Dioxide (Type α, Crystalline)

eV	cm⁻¹	μm	n_o	k_o	n_e	k_e
0.1467	1,183	8.450	0.09826	1.109	0.08671	1.110
0.1459	1,176	8.500	0.1136	1.255	0.08538	1.218
0.1450	1,170	8.550	0.1913	1.459	0.08549	1.331
0.1442	1,163	8.600	0.6590	1.421	0.08711	1.450
0.1433	1,156	8.650	0.2427	1.178	0.09042	1.578
0.1425	1,149	8.700	0.1309	1.408	0.09570	1.717
0.1417	1,143	8.750	0.1115	1.591	0.1034	1.868
0.1409	1,136	8.800	0.1102	1.763	0.1143	2.038
0.1401	1,130	8.850	0.1167	1.941	0.1296	2.230
0.1393	1,124	8.900	0.1290	2.134	0.1513	2.453
0.1385	1,117	8.950	0.1478	2.352	0.1828	2.719
0.1378	1,111	9.000	0.1757	2.606	0.2308	3.047
0.1370	1,105	9.050	0.2178	2.911	0.3092	3.470
0.1362	1,099	9.100	0.2846	3.296	0.4518	4.050
0.1355	1,093	9.150	0.4006	3.808	0.7618	4.924
0.1348	1,087	9.200	0.6330	4.546	1.689	6.406
0.1340	1,081	9.250	1.230	5.733	5.862	7.634
0.1333	1,075	9.300	3.583	7.566	7.460	2.497
0.1326	1,070	9.350	7.929	4.184	5.808	0.9409
0.1319	1,064	9.400	6.336	1.331	4.868	0.5056
0.1312	1,058	9.450	5.180	0.6460	4.291	0.3239
0.1305	1,053	9.500	4.497	0.3920	3.899	0.2293
0.1298	1,047	9.550	4.048	0.2685	3.614	0.1732
0.1292	1,042	9.600	3.727	0.1982	3.395	0.1367
0.1285	1,036	9.650	3.485	0.1539	3.221	0.1116
0.1278	1,031	9.700	3.295	0.1240	3.079	9.335×10^{-2}
0.1272	1,026	9.750	3.142	0.1027	2.961	7.968

0.1265	1,020	9.800	3.014	8.691 × 10^{-2}	2.860	6.913
0.1259	1,015	9.850	2.907	7.485	2.773	6.078
0.1252	1,010	9.900	2.814	6.540	2.697	5.405
0.1246	1,005	9.950	2.734	5.785	2.631	4.854
0.1240	1,000	10.00	2.663	5.170	2.571	4.396
0.1234	995.0	10.05	2.600	4.662	2.518	4.011
0.1228	990.1	10.10	2.544	4.238	2.469	3.684
0.1221	985.2	10.15	2.493	3.879	2.425	3.403
0.1216	980.4	10.20	2.447	3.573	2.385	3.161
0.1210	975.6	10.25	2.405	3.310	2.348	2.950
0.1204	970.9	10.30	2.366	3.083	2.314	2.766
0.1198	966.2	10.35	2.330	2.885	2.282	2.603
0.1192	961.5	10.40	2.297	2.712	2.253	2.460
0.1186	956.9	10.45	2.266	2.560	2.225	2.333
0.1181	952.4	10.50	2.237	2.427	2.199	2.220
0.1175	947.9	10.55	2.210	2.309	2.175	2.120
0.1170	943.4	10.60	2.184	2.205	2.151	2.030
0.1164	939.0	10.65	2.159	2.114	2.129	1.949
0.1159	934.6	10.70	2.136	2.033	2.108	1.878
0.1153	930.2	10.75	2.113	1.962	2.088	1.814
0.1148	925.9	10.80	2.092	1.901	2.069	1.757
0.1143	921.7	10.85	2.071	1.847	2.051	1.707
0.1137	917.4	10.90	2.051	1.802	2.033	1.662
0.1132	913.2	10.95	2.032	1.764	2.016	1.623
0.1127	909.1	11.00	2.013	1.734	1.999	1.590
0.1122	905.0	11.05	1.995	1.710	1.983	1.561
0.1117	900.9	11.10	1.977	1.694	1.967	1.538
0.1112	896.9	11.15	1.959	1.685	1.952	1.519
0.1107	892.9	11.20	1.941	1.683	1.937	1.505
0.1102	888.9	11.25	1.924	1.690	1.922	1.496
0.1097	885.0	11.30	1.906	1.706	1.907	1.492

(continued)

TABLE IX (*Continued*)

Silicon Dioxide (Type α, Crystalline)

eV	cm⁻¹	μm	n_o	k_o	n_e	k_e
0.1092	881.1	11.35	1.888	1.730	1.892	1.493
0.1088	877.2	11.40	1.870	1.766	1.878	1.499
0.1083	873.4	11.45	1.852	1.814	1.863	1.511
0.1078	869.6	11.50	1.833	1.877	1.849	1.530
0.1073	865.8	11.55	1.814	1.956	1.834	1.556
0.1069	862.1	11.60	1.794	2.055	1.819	1.590
0.1064	858.4	11.65	1.773	2.180	1.804	1.632
0.1060	854.7	11.70	1.751	2.335	1.789	1.685
0.1055	851.1	11.75	1.727	2.530	1.773	1.751
0.1051	847.5	11.80	1.701	2.775	1.756	1.830
0.1046	843.9	11.85	1.672	3.088	1.739	1.927
0.1042	840.3	11.90	1.641	3.491	1.721	2.044
0.1038	836.8	11.95	1.605	4.021	1.702	2.187
0.1033	833.3	12.00	1.564	4.732	1.682	2.363
0.1029	829.9	12.05	1.516	5.715	1.660	2.579
0.1025	826.4	12.10	1.457	7.123	1.637	2.848
0.1020	823.0	12.15	1.384	9.241×10^{-2}	1.611	3.188
0.1016	819.7	12.20	1.289	0.1265	1.581	3.622
0.1012	816.3	12.25	1.162	0.1868	1.548	4.187
0.1008	813.0	12.30	0.9858	0.3097	1.510	4.940
0.1004	809.7	12.35	0.7844	0.5904	1.466	5.971
0.09999	806.5	12.40	0.7289	1.072	1.412	7.434
0.09958	803.2	12.45	0.9288	1.655	1.345	9.608×10^{-2}
0.09919	800.0	12.50	1.587	2.269	1.260	0.1304
0.09879	796.8	12.55	2.900	2.106	1.148	0.1896
0.09840	793.7	12.60	3.236	1.105	0.9971	0.3041

0.09801	790.5	12.65	0.4893	2.956	0.8281	0.5464
0.09762	787.4	12.70	0.2805	2.718	0.7711	0.9621
0.09724	784.3	12.75	0.1823	2.552	0.9404	1.473
0.09686	781.3	12.80	0.1291	2.433	1.482	1.990
0.09648	778.2	12.85	9.703×10^{-2}	2.344	2.539	1.968
0.09611	775.2	12.90	7.628	2.275	3.011	1.099
0.09574	772.2	12.95	6.208	2.220	2.834	0.5529
0.09537	769.2	13.00	5.194	2.174	2.625	0.3183
0.09501	766.3	13.05	4.446	2.136	2.469	0.2061
0.09464	763.4	13.10	3.879	2.102	2.355	0.1451
0.09429	760.5	13.15	3.441	2.073	2.269	0.1085
0.09393	757.6	13.20	3.098	2.047	2.202	8.480×10^{-2}
0.09357	754.7	13.25	2.825	2.024	2.148	6.861
0.09322	751.9	13.30	2.607	2.002	2.104	5.705
0.09287	749.1	13.35	2.433	1.983	2.066	4.851
0.09253	746.3	13.40	2.296	1.964	2.034	4.201
0.09218	743.5	13.45	2.189	1.947	2.006	3.696
0.09184	740.7	13.50	2.109	1.930	1.981	3.295
0.09150	738.0	13.55	2.055	1.915	1.959	2.972
0.09117	735.3	13.60	2.026	1.899	1.939	2.708
0.09083	732.6	13.65	2.022	1.884	1.920	2.489
0.09050	729.9	13.70	2.047	1.869	1.903	2.305
0.09017	727.3	13.75	2.107	1.854	1.887	2.151
0.08984	724.6	13.80	2.208	1.839	1.873	2.019
0.08952	722.0	13.85	2.366	1.823	1.859	1.906
0.08920	719.4	13.90	2.601	1.807	1.846	1.809
0.08888	716.8	13.95	2.953	1.789	1.833	1.725
0.08856	714.3	14.00	3.484	1.769	1.821	1.651
0.08825	711.7	14.05	4.311	1.746	1.810	1.587
0.08793	709.2	14.10	5.665	1.719	1.799	1.531

(continued)

TABLE IX (Continued)
Silicon Dioxide (Type α, Crystalline)

eV	cm^{-1}	μm	n_o	k_o	n_e	k_e
0.08762	706.7	14.15	1.686	8.025×10^{-2}	1.788	1.482
0.08731	704.2	14.20	1.649	0.1246	1.778	1.439
0.08701	701.8	14.25	1.621	0.2119	1.768	1.401
0.08670	699.3	14.30	1.670	0.3539	1.758	1.368
0.08640	696.9	14.35	1.867	0.4101	1.749	1.339
0.08610	694.4	14.40	1.999	0.2820	1.740	1.314
0.08580	692.0	14.45	1.998	0.1644	1.730	1.292
0.08551	689.7	14.50	1.964	0.1016	1.721	1.273
0.08521	687.3	14.55	1.931	6.871×10^{-2}	1.713	1.257
0.08492	684.9	14.60	1.903	5.016	1.704	1.243
0.08463	682.6	14.65	1.881	3.885	1.695	1.231
0.08434	680.3	14.70	1.861	3.150	1.687	1.222
0.08406	678.0	14.75	1.845	2.646	1.678	1.215
0.08377	675.7	14.80	1.831	2.287	1.670	1.210
0.08349	673.4	14.85	1.818	2.022	1.661	1.206
0.08321	671.1	14.90	1.806	1.821	1.653	1.204
0.08293	668.9	14.95	1.795	1.664	1.644	1.204
0.08266	666.7	15.0	1.785	1.540	1.636	1.205
0.08238	664.5	15.05	1.776	1.439	1.628	1.208
0.08211	662.3	15.10	1.767	1.358	1.619	1.213
0.08184	660.1	15.15	1.758	1.290	1.611	1.218
0.08157	657.9	15.20	1.750	1.234	1.602	1.226
0.08130	655.7	15.25	1.742	1.186	1.594	1.234
0.08103	653.6	15.30	1.734	1.146	1.585	1.245
0.08077	651.5	15.35	1.726	1.112	1.576	1.256
0.08051	649.4	15.40	1.719	1.083	1.568	1.269
0.08025	647.2	15.45	1.711	1.059	1.559	1.284

0.07999	645.2	15.50	1.704	1.037	1.550	1.300
0.07973	643.1	15.55	1.697	1.019	1.541	1.317
0.07948	641.0	15.60	1.690	1.004×10^{-2}	1.532	1.336
0.07922	639.0	15.65	1.683	9.912×10^{-3}	1.522	1.357
0.07897	636.9	15.70	1.676	9.805	1.513	1.379
0.07872	634.9	15.75	1.669	9.716	1.504	1.404
0.07847	632.9	15.80	1.663	9.644	1.494	1.429
0.07822	630.9	15.85	1.656	9.588	1.484	1.457
0.07798	628.9	15.90	1.649	9.546	1.474	1.487
0.07773	627.0	15.95	1.642	9.518	1.464	1.519
0.07749	625.0	16.00	1.635	9.501	1.453	1.553
0.07725	623.1	16.05	1.629	9.497	1.443	1.589
0.07701	621.1	16.10	1.622	9.503	1.432	1.628
0.07677	619.2	16.15	1.615	9.520	1.421	1.670
0.07653	617.3	16.20	1.608	9.547	1.410	1.714
0.07630	615.4	16.25	1.601	9.583	1.398	1.761
0.07606	613.5	16.30	1.594	9.629	1.386	1.811
0.07583	611.6	16.35	1.587	9.683	1.374	1.865
0.07560	609.8	16.40	1.580	9.747	1.361	1.923
0.07537	607.9	16.45	1.573	9.820	1.349	1.984
0.07514	606.1	16.50	1.566	9.901	1.335	2.050
0.07491	604.2	16.55	1.559	9.991×10^{-3}	1.322	2.121
0.07469	602.4	16.60	1.551	1.009×10^{-2}	1.308	2.196
0.07446	600.6	16.65	1.544	1.020	1.293	2.278
0.07424	598.8	16.70	1.537	1.031	1.278	2.365
0.07402	597.0	16.75	1.529	1.044	1.263	2.459
0.07380	595.2	16.80	1.521	1.058	1.247	2.561
0.07358	593.5	16.85	1.514	1.072	1.230	2.671
0.07336	591.7	16.90	1.506	1.087	1.213	2.790
0.07315	590.0	16.95	1.498	1.104	1.195	2.919
0.07293	588.2	17.00	1.490	1.121	1.175	3.060

(continued)

TABLE IX (Continued)

Silicon Dioxide (Type α, Crystalline)

eV	cm^{-1}	μm	n_o	k_o	n_e	k_e
0.07272	586.5	17.05	1.481	1.140	1.157	3.213
0.07250	584.8	17.10	1.473	1.159	1.136	3.382
0.07229	583.1	17.15	1.464	1.180	1.115	3.567
0.07208	581.4	17.20	1.456	1.202	1.093	3.771
0.07187	579.7	17.25	1.447	1.225	1.069	3.997
0.07167	578.0	17.30	1.438	1.249	1.045	4.249
0.07146	576.4	17.35	1.429	1.275	1.019	4.532
0.07126	574.7	17.40	1.420	1.302	0.9913	4.850
0.07105	573.1	17.45	1.410	1.331	0.9620	5.211
0.07085	571.4	17.50	1.400	1.361	0.9309	5.623
0.07065	569.8	17.55	1.390	1.393	0.8976	6.100
0.07045	568.2	17.60	1.380	1.427	0.8619	6.655
0.07025	566.6	17.65	1.370	1.462	0.8234	7.313
0.07005	565.0	17.70	1.359	1.500	0.7817	8.102
0.06985	563.4	17.75	1.348	1.540	0.7361	9.068 × 10^{-2}
0.06965	561.8	17.80	1.337	1.581	0.6860	0.1028
0.06946	560.2	17.85	1.326	1.626	0.6307	0.1184
0.06927	558.7	17.90	1.314	1.672	0.5689	0.1394
0.06907	557.1	17.95	1.302	1.722	0.5001	0.1689
0.06888	555.6	18.00	1.289	1.774	0.4251	0.2123
0.06869	554.0	18.05	1.277	1.830	0.3501	0.2764
0.06850	552.5	18.10	1.263	1.889	0.2886	0.3607
0.06831	551.0	18.15	1.250	1.951	0.2481	0.4527
0.06812	549.5	18.20	1.236	2.018	0.2243	0.5420
0.06794	547.9	18.25	1.221	2.089	0.2109	0.6256
0.06775	546.4	18.30	1.207	2.165	0.2037	0.7034
0.06757	545.0	18.35	1.191	2.245	0.2003	0.7763

0.06738	543.5	18.40	1.175	2.332	0.1990	0.8449
0.06720	542.0	18.45	1.159	2.424	0.1983	0.9101
0.06702	540.5	18.50	1.142	2.524	0.1972	0.9728
0.06684	539.1	18.55	1.124	2.631	0.1950	1.034
0.06666	537.6	18.60	1.105	2.746	0.1913	1.096
0.06648	536.2	18.65	1.086	2.871	0.1865	1.160
0.06630	534.8	18.70	1.066	3.006	0.1814	1.226
0.06612	533.3	18.75	1.045	3.153	0.1768	1.295
0.06595	531.9	18.80	1.023	3.313	0.1733	1.367
0.06577	530.5	18.85	1.000	3.489	0.1713	1.442
0.06560	529.1	18.90	0.9761	3.682	0.1712	1.520
0.06543	527.7	18.95	0.9507	3.896	0.1730	1.601
0.06526	526.3	19.00	0.9239	4.134	0.1768	1.685
0.06508	524.9	19.05	0.8955	4.401	0.1827	1.773
0.06491	523.6	19.10	0.8654	4.701	0.1911	1.866
0.06474	522.2	19.15	0.8333	5.044	0.2023	1.963
0.06457	520.8	19.20	0.7989	5.438	0.2169	2.065
0.06441	519.5	19.25	0.7620	5.897	0.2359	2.174
0.06424	518.1	19.30	0.7221	6.441	0.2608	2.291
0.06407	516.8	19.35	0.6786	7.099	0.2941	2.417
0.06391	515.5	19.40	0.6308	7.915	0.3398	2.552
0.06374	514.1	19.45	0.5777	8.963 × 10^{-2}	0.4042	2.695
0.06358	512.8	19.50	0.5180	0.1037	0.4971	2.841
0.06342	511.5	19.55	0.4501	0.1240	0.6278	2.971
0.06326	510.2	19.60	0.3723	0.1558	0.7847	3.044
0.06310	508.9	19.65	0.2879	0.2097	0.8957	3.037
0.06293	507.6	19.70	0.2149	0.2924	0.8839	3.033
0.06278	506.3	19.75	0.1696	0.3862	0.8042	3.144
0.06262	505.1	19.80	0.1439	0.4746	0.7526	3.371
0.06246	503.8	19.85	0.1284	0.5554	0.7629	3.673
0.06230	502.5	19.90	0.1183	0.6300	0.8405	4.035

(continued)

TABLE IX *(Continued)*

Silicon Dioxide (Type α, Crystalline)

eV	cm^{-1}	μm	n_o	k_o	n_e	k_e
0.06215	501.3	19.95	0.1113	0.69998	0.9974	4.464
0.06199	500.0	20.00	0.1065	0.7665	1.270	4.976
0.06184	498.8	20.05	0.1030	0.8304	1.743	5.589
0.06168	497.5	20.10	0.1006	0.8924	2.597	6.267
0.06153	496.3	20.15	9.894×10^{-2}	0.9530	4.127	6.707
0.06138	495.0	20.20	9.794	1.013	6.142	5.987
0.06123	493.8	20.25	9.747	1.072	7.094	4.070
0.06107	492.6	20.30	9.746	1.130	6.818	2.521
0.06092	491.4	20.35	9.786	1.189	6.247	1.643
0.06078	490.2	20.40	9.863	1.247	5.724	1.149
0.06063	489.0	20.45	9.976×10^{-2}	1.306	5.296	0.8527
0.06048	487.8	20.50	0.1012	1.365	4.951	0.6618
0.06033	486.6	20.55	0.1031	1.425	4.669	0.5316
0.06019	485.4	20.60	0.1052	1.486	4.435	0.4386
0.06004	484.3	20.65	0.1078	1.548	4.238	0.3697
0.05989	483.1	20.70	0.1107	1.611	4.069	0.3171
0.05975	481.9	20.75	0.1140	1.676	3.923	0.2759
0.05961	480.8	20.80	0.1177	1.742	3.795	0.2430
0.05946	479.6	20.85	0.1219	1.811	3.682	0.2162
0.05932	478.5	20.90	0.1266	1.881	3.581	0.1941
0.05918	477.3	20.95	0.1319	1.954	3.490	0.1756
0.05904	476.2	21.00	0.1379	2.030	3.408	0.1600
0.05890	475.1	21.05	0.1446	2.108	3.334	0.1466
0.05876	473.9	21.10	0.1521	2.191	3.265	0.1351
0.05862	472.8	21.15	0.1606	2.277	3.203	0.1251
0.05848	471.7	21.20	0.1703	2.368	3.145	0.1164
0.05834	470.6	21.25	0.1812	2.464	3.091	0.1087

0.05821	469.5	21.30	0.1937	2.565	3.041	0.1020
0.05807	468.4	21.35	0.2081	2.673	2.994	9.592 × 10^{-2}
0.05794	467.3	21.40	0.2248	2.789	2.950	9.054
0.05780	466.2	21.45	0.2442	2.914	2.909	8.571
0.05767	465.1	21.50	0.2670	3.048	2.870	8.137
0.05753	464.0	21.55	0.2941	3.195	2.834	7.745
0.05740	463.0	21.60	0.3267	3.356	2.799	7.390
0.05727	461.9	21.65	0.3664	3.533	2.766	7.068
0.05713	460.8	21.70	0.4154	3.731	2.734	6.776
0.05700	459.8	21.75	0.4773	3.955	2.704	6.509
0.05687	458.7	21.80	0.5569	4.209	2.676	6.265
0.05674	457.7	21.85	0.6622	4.502	2.648	6.043
0.05661	456.6	21.90	0.8059	4.846	2.622	5.839
0.05648	455.6	21.95	1.010	5.253	2.596	5.651
0.05636	454.5	22.00	1.314	5.742	2.572	5.480
0.05623	453.5	22.05	1.793	6.326	2.548	5.323
0.05610	452.5	22.10	2.595	6.983	2.526	5.178
0.05597	451.5	22.15	3.966	7.519	2.504	5.046
0.05585	450.5	22.20	5.976	7.259	2.482	4.924
0.05572	449.4	22.25	7.562	5.603	2.462	4.812
0.05560	448.4	22.30	7.753	3.682	2.441	4.710
0.05547	447.5	22.35	7.250	2.412	2.422	4.617
0.05535	446.4	22.40	6.658	1.671	2.403	4.532
0.05523	445.4	22.45	6.139	1.225	2.384	4.454
0.05510	444.4	22.50	5.709	0.9407	2.366	4.384
0.05498	443.5	22.55	5.354	0.7490	2.348	4.320
0.05486	442.5	22.60	5.058	0.6136	2.330	4.264
0.05474	441.5	22.65	4.807	0.5142	2.313	4.213
0.05462	440.5	22.70	4.593	0.4390	2.296	4.168
0.05450	439.6	22.75	4.406	0.3806	2.279	4.129
0.05438	438.6	22.80	4.243	0.3343	2.263	4.095

(continued)

TABLE IX (*Continued*)

Silicon Dioxide (Type α, Crystalline)

eV	cm^{-1}	μm	n_o	k_o	n_e	k_e
0.05426	437.6	22.85	4.098	0.2969	2.247	4.067
0.05414	436.7	22.90	3.968	0.2663	2.231	4.044
0.05402	435.7	22.95	3.851	0.2408	2.215	4.025
0.05391	434.8	23.00	3.745	0.2195	2.199	4.012
0.05379	433.8	23.05	3.648	0.2015	2.184	4.004
0.05367	432.9	23.10	3.558	0.1861	2.169	4.000
0.05356	432.0	23.15	3.475	0.1728	2.153	4.001
0.05344	431.0	23.20	3.398	0.1614	2.138	4.007
0.05333	430.1	23.25	3.325	0.1515	2.123	4.018
0.05321	429.2	23.30	3.257	0.1429	2.108	4.034
0.05310	428.3	23.35	3.192	0.1354	2.093	4.054
0.05298	427.4	23.40	3.131	0.1289	2.077	4.079
0.05287	426.4	23.45	3.072	0.1233	2.062	4.110
0.05276	425.5	23.50	3.015	0.1184	2.047	4.145
0.05265	424.6	23.55	2.961	0.1142	2.032	4.186
0.05253	423.7	23.60	2.908	0.1106	2.017	4.232
0.05242	422.8	23.65	2.857	0.1076	2.001	4.284
0.05231	421.9	23.70	2.806	0.1052	1.986	4.341
0.05220	421.1	23.75	2.757	0.1033	1.970	4.405
0.05209	420.2	23.80	2.708	0.1020	1.954	4.475
0.05198	419.3	23.85	2.659	0.1012	1.938	4.552
0.05188	418.4	23.90	2.610	0.1009	1.922	4.635
0.05177	417.5	23.95	2.651	0.1012	1.906	4.727
0.05166	416.7	24.00	2.512	0.1020	1.889	4.826
0.05155	415.8	24.05	2.461	0.1036	1.872	4.933
0.05144	414.9	24.10	2.409	0.1059	1.855	5.049
0.05134	414.1	24.15	2.355	0.1090	1.838	5.175

0.05123	413.2	24.20	2.300	0.1131	1.820	5.311
0.05113	412.4	24.25	2.241	0.1184	1.802	5.458
0.05102	411.5	24.30	2.179	0.1252	1.783	5.617
0.05092	410.7	24.35	2.112	0.1336	1.764	5.789
0.05081	409.8	24.40	2.041	0.1444	1.744	5.975
0.05070	409.0	24.45	1.962	0.1581	1.724	6.176
0.05061	408.2	24.50	1.876	0.1757	1.703	6.393
0.05050	407.3	24.55	1.779	0.1988	1.682	6.629
0.05040	406.5	24.60	1.668	0.2298	1.660	6.885
0.05030	405.7	24.65	1.540	0.2730	1.637	7.163
0.05020	404.9	24.70	1.390	0.3360	1.614	7.465
0.05009	404.0	24.75	1.214	0.4336	1.589	7.795
0.04999	403.2	24.80	1.015	0.5932	1.564	8.155
0.04989	402.4	24.85	0.8305	0.8431	1.537	8.550
0.04979	401.6	24.90	0.7156	1.161	1.510	8.984
0.04969	400.8	24.95	0.6717	1.502	1.481	9.461
0.04959	400.0	25.00	0.6787	1.853	1.451	9.989×10^{-2}
0.04949	399.2	25.05	0.7284	2.224	1.419	0.1058
0.04940	398.4	25.10	0.8260	2.630	1.386	0.1123
0.04930	397.6	25.15	0.9915	3.087	1.350	0.1196
0.04920	396.8	25.20	1.270	3.618	1.313	0.1279
0.04910	396.0	25.25	1.759	4.233	1.273	0.1372
0.04900	395.3	25.30	2.661	4.854	1.231	0.1479
0.04891	394.5	25.35	4.194	5.028	1.186	0.1602
0.04881	393.7	25.40	5.689	3.971	1.137	0.1745
0.04872	392.9	25.45	5.982	2.469	1.085	0.1914
0.04862	392.2	25.50	5.633	1.515	1.029	0.2116
0.04853	391.4	25.55	5.220	0.9975	0.9678	0.2362
0.04843	390.6	25.60	4.872	0.7043	0.9017	0.2665
0.04834	389.9	25.65	4.595	0.5253	0.8305	0.3047
0.04824	389.1	25.70	4.373	0.4086	0.7547	0.3537

(continued)

TABLE IX (*Continued*)

Silicon Dioxide (Type α, Crystalline)

eV	cm⁻¹	μm	n_o	k_o	n_e	k_e
0.04815	388.3	25.75	4.193	0.3284	0.6769	0.4168
0.04805	387.6	25.80	4.044	0.2708	0.6020	0.4962
0.04796	386.8	25.85	3.919	0.2280	0.5364	0.5907
0.04787	386.1	25.90	3.812	0.1954	0.4845	0.6952
0.04778	385.4	25.95	3.720	0.1698	0.4465	0.8037
0.04769	384.6	26.00	3.639	0.1494	0.4198	0.9127
0.04759	383.9	26.05	3.568	0.1328	0.4018	1.020
0.04750	383.1	26.10	3.505	0.1191	0.3905	1.127
0.04741	382.4	26.15	3.448	0.1077	0.3842	1.232
0.04732	381.7	26.20	3.397	9.810×10^{-2}	0.3820	1.337
0.04723	381.0	26.25	3.350	8.988	0.3833	1.442
0.04714	380.2	26.30	3.308	8.280	0.3878	1.548
0.04705	379.5	26.35	3.269	7.666	0.3952	1.656
0.04696	378.8	26.40	3.233	7.129	0.4055	1.765
0.04687	378.1	26.45	3.199	6.656	0.4189	1.877
0.04679	377.4	26.50	3.168	6.237	0.4345	1.992
0.04670	376.6	26.55	3.140	5.865	0.4554	2.112
0.04661	375.9	26.60	3.113	5.531	0.4793	2.236
0.04652	375.2	26.65	3.087	5.230	0.5078	2.367
0.04644	374.5	26.70	3.063	4.959	0.5416	2.504
0.04635	373.8	26.75	3.041	4.713	0.5818	2.649
0.04626	373.1	26.80	3.020	4.489	0.6297	2.803
0.04618	372.4	26.85	2.999	4.284	0.6873	2.968
0.04609	371.7	26.90	2.980	4.096	0.7569	3.145
0.04600	371.1	26.95	2.962	3.924	0.8419	3.336
0.04592	370.4	27.00	2.945	3.764	0.9469	3.544
0.04583	369.7	27.05	2.928	3.617	1.078	3.770

0.04575	369.0	27.10	2.913	3.480	1.246	4.015
0.04567	368.3	27.15	2.898	3.353	1.461	4.279
0.04558	367.6	27.20	2.883	3.235	1.744	4.559
0.04550	367.0	27.25	2.869	3.125	2.119	4.843
0.04541	366.3	27.30	2.856	3.021	2.618	5.099
0.04533	365.6	27.35	2.843	2.924	3.263	5.267
0.04525	365.0	27.40	2.831	2.833	4.043	5.246
0.04517	364.3	27.45	2.819	2.748	4.854	4.934
0.04509	363.6	27.50	2.808	2.667	5.509	4.327
0.04500	363.0	27.55	2.797	2.591	5.870	3.574
0.04492	362.3	27.60	2.786	2.519	5.951	2.855
0.04484	361.7	27.65	2.776	2.451	5.855	2.263
0.04476	361.0	27.70	2.766	2.386	5.677	1.809
0.04468	360.4	27.75	2.756	2.325	5.473	1.467
0.04460	359.7	27.80	2.747	2.267	5.270	1.209
0.04452	359.1	27.85	2.738	2.212	5.079	1.012
0.04444	358.4	27.90	2.729	2.159	4.905	0.8598
0.04436	357.8	27.95	2.720	2.108	4.748	0.7397
0.04428	357.1	28.00	2.712	2.060	4.607	0.6437
0.04420	356.5	28.05	2.704	2.014	4.479	0.5657
0.04412	355.9	28.10	2.696	1.971	4.364	0.5016
0.04404	355.2	28.15	2.689	1.929	4.260	0.4482
0.04397	354.6	28.20	2.681	1.888	4.166	0.4033
0.04389	354.0	28.25	2.674	1.850	4.080	0.3651
0.04381	353.4	28.30	2.667	1.813	4.002	0.3323
0.04373	352.7	28.35	2.660	1.777	3.930	0.3040
0.04366	352.1	28.40	2.653	1.743	3.863	0.2794
0.04358	351.5	28.45	2.647	1.710	3.802	0.2578
0.04350	350.9	28.50	2.641	1.678	3.745	0.2388
0.04343	350.3	28.55	2.634	1.647	3.692	0.2220
0.04335	349.7	28.60	2.628	1.618	3.643	0.2069

(continued)

TABLE IX (*Continued*)
Silicon Dioxide (Type α, Crystalline)

eV	cm^{-1}	μm	n_o	k_o	n_e	k_e
0.04327	349.0	28.65	2.622	1.589	3.597	0.1935
0.04320	348.4	28.70	2.617	1.562	3.554	0.1814
0.04312	347.8	28.75	2.611	1.535	3.514	0.1705
0.04305	347.2	28.80	2.605	1.510	3.476	0.1607
0.04297	346.6	28.85	2.600	1.485	3.440	0.1517
0.04290	346.0	28.90	2.595	1.461	3.407	0.1435
0.04283	345.4	28.95	2.589	1.438	3.375	0.1360
0.04275	344.8	29.00	2.584	1.415	3.345	0.1291
0.04246	342.5	29.20	2.565	1.332	3.238	0.1066
0.04217	340.1	29.40	2.547	1.258	3.151	8.996×10^{-2}
0.04189	337.8	29.60	2.530	1.192	3.077	7.722
0.04160	335.6	29.80	2.515	1.132	3.013	6.724
0.04133	333.3	30.00	2.500	1.079×10^{-2}	2.959	5.926
0.04065	327.9	30.50	2.468	9.643×10^{-3}	2.849	4.508
0.04000	322.6	31.00	2.440	8.721	2.766	3.590
0.03936	317.5	31.50	2.416	7.962	2.701	2.957
0.03875	312.5	32.00	2.395	7.326	2.648	2.500
0.02480	200.0	50.00	2.214 [1]	2.86 [1]	2.262 [1]	1.71×10^{-3} [1]
0.02356	190.0	52.63	2.200	2.68	2.248	1.47
0.02232	180.0	55.56	2.188	2.48	2.237	1.28
0.02108	170.0	58.82	2.178	2.29	2.226	1.12×10^{-3}
0.01984	160.0	62.50	2.169	2.14	2.217	9.95×10^{-4}
0.01860	150.0	66.67	2.161	1.91	2.208	9.02
0.01736	140.0	71.43	2.154	1.82	2.200	7.96
0.01705	137.5	72.73	2.152	1.85		
0.01674	135.0	74.07	2.150	2.36		
0.01643	132.5	75.47	2.148	3.60		

0.01612	130.0	76.92	2.147	6.18	2.193	6.73
0.01581	127.5	78.43	2.147	6.74		
0.01550	125.0	80.00	2.148	3.69		
0.01519	122.5	81.63	2.146	1.56		
0.01488	120.0	83.33	2.144	1.06×10^{-3}	2.187	5.97
0.01364	110.0	90.91	2.136	9.40×10^{-4}	2.182	5.79
0.01240	100.0	100.0	2.132	9.55	2.176	4.77
0.01116	90.0	111.1	2.128	8.84	2.172	4.42
0.009919	80	125.0	2.124	8.95	2.168	2.98
0.008679	70	142.9	2.121	9.10	2.164	2.27
0.007439	60	166.7	2.119[b]	7.96	2.162	2.65
0.006199	50	200.0	2.1114	7.96	2.159	1.59×10^{-4}
0.004959	40	250.0	2.1096	7.96	2.157	7.96×10^{-5}
0.003720	30	333.3	2.1081	7.96	2.156	5.30
0.002480	20	500.0	2.1070			
0.001240	10	1000	2.1063			

[a] Optical constants n and k for both the ordinary and extraordinary rays are listed. References from which the data were obtained are listed in brackets. Values of k enclosed by brackets in the wavelength region 0.1291 μm $\leq \lambda \leq$ 0.1378 μm were obtained from a smooth curve drawn between absorption data and KK results. Therefore, no reference is applicable to these results. Values of n enclosed by brackets in the wavelength region 0.185467 μm $\leq \lambda \leq$ 2.50000 μm were taken from Sosman [6].

[b] Data from George J. Simonis in Chapter 8 of this handbook.

Silicon Dioxide (SiO$_2$) (Glass)

H. R. PHILIPP

General Electric Research and Development Center
Schenectady, New York

The room-temperature optical properties of SiO$_2$ glass have been the subject of numerous studies; however, few attempts have been made to analyze this information to obtain a self-consistent set of optical constants, n and k, for this material, especially in the regions of strong absorption in the infrared and vacuum ultraviolet. Considering the technological importance of this material, it is somewhat surprising that this information is not presently available in precise detail. There are, of course, reasons for this. When the absorption is high, the optical constants are usually obtained by Kramers–Kronig (KK) analysis of reflectance data that are difficult to measure with high accuracy. The KK analysis itself may require extrapolations into spectral regions for which no data exist, thereby introducing additional uncertainties in the derived n and k values. In the region of low absorption, the index of refraction can be evaluated from prism data, and this has been accomplished with great precision for SiO$_2$. However the measured k values can be strongly affected by impurity and defect absorption. For SiO$_2$, the presence of water or OH absorption in the samples makes the determination of the intrinsic k values extremely difficult (if not impossible) in certain parts of the infrared and vacuum ultraviolet spectral regions. The importance of SiO$_2$ in fiber-optical applications may give people a strong incentive to clean up this material as much as possible.

A set of optical parameters n and k are given in Table X and plotted in Fig. 10 for the wavelength region 500 μm $\geq \lambda \geq$ 0.05 μm. They were obtained from a variety of sources. The wavelength λ and index of refraction n values are given to four places, and the extinction coefficient k values are given to three places in most of the table entries. The accuracy of these values does not, in general, warrant this precision. This is done to show the possible presence of weak structure in the optical properties that can often be discerned by comparison of precise, but not necessarily accurate, values over small wavelength ranges.

HANDBOOK OF OPTICAL CONSTANTS OF SOLIDS

In the far infrared, $500 \, \mu m \geq \lambda \geq 100 \, \mu m$, the values given in the table were obtained from Randall and Rawcliffe [1], who used an interferometric technique. The n values appear to be reasonably precise, while the k values may be in error by as much as $\pm 30\%$.

In the infrared $40.00 \, \mu m \geq \lambda \geq 3.846 \, \mu m$, the values given in the table were taken from Philipp [2]. They were obtained by KK analysis of reflectance data augmented by absorption measurements. His reflectance values are similar to those of Miller [3] but generally slightly lower in absolute magnitude. While Miller also performed a KK analysis on his data, his computing technique leads to nonphysical results in certain parts of the spectrum and hence are not used in this documentation. Miller correctly points out that on the short-wavelength side of the 9-μm absorption peak, the reflectance shows a deep minimum whose absolute value is difficult to evaluate. Philipp [2] used absorption measurements and a KK analysis on n and α (absorption coefficient $\alpha = 4\pi k/\lambda$) parameters in addition to that on reflectance R and phase ϕ to evaluate R in this region.

In the regions of strong infrared absorption ($k \gtrsim 0.5$), the values for n and k given in the table should be reasonably good ($\pm 15\%$), although there is certainly room for improvement, and further optical studies should be undertaken to improve their accuracy. In the regions of weak infrared absorption, the k values as given by KK analysis are not reliable. Hence, a second set of k values determined from absorption data [2] are given in the table for certain wavelength ranges. They are listed under the heading k_{ABS}. While they are perhaps better than the KK derived values, they should also be treated as unreliable. This is mainly due to the possible presence of impurities and defects in the samples used. In fact, k values must be carefully examined for wavelengths below $7.634 \, \mu m$, where impurity absorption, especially that associated with OH bonding, can mask intrinsic SiO_2 absorption. Drummond [4] has examined the absorption spectrum of SiO_2 for $\lambda \lesssim 7.5 \, \mu m$, and his results show a variety of features (some multiphonon bands) that have been duplicated for the most part by Galeener and Lucovsky [5] in the 3.7–7.5-μm region, with α roughly 20% larger. Therefore, we include some of these k values.

For wavelengths in the range $3.7067 \, \mu m \geq \lambda \geq 0.213856 \, \mu m$, the index of refraction of SiO_2 has been determined with great precision by Malitson [6], who used prism data. His results have been fitted to three- and four-term Sellmeier equations, and the values generated by Brixner [7] are given in the table. It should be pointed out that the very precise Sellmeier-equation fits to index of refraction data are not very reliable for wavelengths outside the fitted range. This is probably related to the fact that these fits tend to be purely mathematical and do not necessarily relate to the actual physical absorption processes that determine the spectral dependence of the index of refraction.

No k values are given in this wavelength range. For $\lambda \gtrsim 0.6$ μm, the measured absorption is dominated by the presence of OH impurity bands. For $\lambda \lesssim 0.6$ μm, the ultimate lower limit to the intrinsic absorption has been estimated by Keck et al. [8]; however, the reliability of their values are questionable, especially at the shorter wavelengths.

For $\lambda < 0.2139$ μm, the optical constants given in the table were obtained from a variety of sources. Both Philipp [9] and Lamy [10] have evaluated n and k down to $\lambda = 0.05$ μm from KK analysis of reflectance data. Lamy's [10] results make use of reflectance data by Platzoder and Steinmann [11] for $\lambda \leq 0.1$ μm. For $\lambda > 0.11$ μm, the n and k values of Philipp [9] and Lamy [10] differ somewhat, while for $\lambda < 0.11$ μm they are in reasonably good agreement. The decision to use the results of Philipp [9] (n values) rather than those of Lamy [10] for $\lambda > 0.11$ μm was based on the close agreement of Philipp's k values in the region from 0.11 to 0.12 μm with the results of Weinberg et al. [12], who made very careful transmission measurements on thin layers of SiO$_2$ prepared by thermal oxidation of silicon. This agreement was accomplished with a slight modification of Philipp's results as described in Weinberg et al. [12]. The k values given in the table for 0.11 μm $\leq \lambda \leq$ 0.13 μm are taken from this work [12]. For $\lambda > 0.13$ μm, where absorption in SiO$_2$ is weak, the k values were obtained from a smooth curved drawn through the transmission results of Weinberg et al. [12], their thickest sample, those of Appleton et al. [13] and those of Kaminow et al. [14], sample B$_2$. Care should be taken in using the k values for $\lambda \gtrsim 0.13$ μm because they may have a component of impurity absorption (especially OH) and defect absorption in them. For this reason, no k values are given for $\lambda \geq 0.163$ μm.

For $\lambda \leq 0.11$ μm, where the results of Philipp [9] and Lamy [10] are in reasonably good agreement, the n and k values given in the table were obtained from a smooth curve drawn through both sets of data. The deviations of each set of data points from the smoothed curve is generally less than 5%. This does not attest to their absolute accuracy, however, because in both cases the KK evaluations are based on somewhat arbitrary extrapolations of the reflectance for $\lambda < 00.5$ μm as required by the analysis. In addition, the reflectance values themselves may be in error.

Hunter [15] has determined n and k for fused silica in the region from 0.3 to 0.03 μm by measuring reflectivity versus angle of incidence. We use these data from 600 to 300 Å, which show good overlap with the data of Philipp [9] and Lamy [10]. For the region 300–82 Å, data of Rife and Osantowski [16] for a high-silica glass with 7.5% TiO$_2$ were used. They measured reflectivity as a function of angle of incidence utilizing synchrotron radiation.

Since there are no experimental data at shorter wavelengths, we have calculated n and k based on the results of Henke et al. [17] in the region 80.5–6.1 Å by assuming the atomic scattering factors of Si and O in appropriate mixture. (See the section by Lynch and Hunter for metals in this handbook.) The agreement in the overlap region near 82 Å with data of Rife and Osantowski [16] is remarkable. The same calculated n and k would be obtained for α-quartz, assuming that the density of fused silica and α-quartz are the same. These calculated values are at least a starting point for the optical properties of glass in the far-UV and x-ray regions.

It is recommended that anyone who uses the table carefully examine the material and references cited in the above-named references as well as other references [18–25] listed.

REFERENCES

1. C. M. Randall and R. D. Rawcliffe, *Appl. Opt.* **6**, 1889 (1967).
2. H. R. Philipp, *J. Appl. Phys.* **50**, 1053 (1979).
3. M. Miller, *Czech. J. Phys. B* **18**, 354 (1968).
4. D. G. Drummond, *Proc. R. Soc. London* **153**, 328 (1935).
5. F. L. Galeener and G. Lucovsky, *Proc. Int. Conf. Light Scat. Solids, 3rd, Campenas,* 1975 (M. Balkanski, R. C. C. Leite, and S. P. S. Porto, eds.), p. 641. Flammarion Sciences, Paris, 1976.
6. I. H. Malitson, *J. Opt. Soc. Am.* **55**, 1205 (1965).
7. B. Brixner, *J. Opt. Soc. Am.* **57**, 674 (1967).
8. D. B. Keck, R. D. Maurer, and P. C. Schultz, *Appl. Phys. Lett.* **22**, 307 (1973).
9. H. R. Philipp, *Solid State Commun.* **4**, 73 (1966); H. R. Philipp, *J. Phys. Chem. Solids* **32**, 1935 (1971).
10. P. L. Lamy, *Appl. Opt.* **16**, 2212 (1977).
11. K. Platzoder and W. Steinmann, *J. Opt. Soc. Am.* **58**, 588 (1968).
12. Z. A. Weinberg, G. W. Rubloff, and E. Bassous, *Phys. Rev. B* **19**, 3107 (1979).
13. A. Appleton, T. Chiranjivi, and M. Jafaripour-Ghazvini, *in* "The Physics of SiO$_2$ and Its Interfaces" (S. T. Pantelides, ed.), p. 94. Pergamon, New York, 1978.
14. I. P. Kaminow, B. G. Bagley, and C. G. Olson, *Appl. Phys. Lett.* **32**, 98 (1978).
15. W. R. Hunter, private communication of unpublished data (1983).
16. J. Rife and J. Osantowski, *J. Opt. Soc. Am.* **70**, 1513 (1980).
17. B. L. Henke, P. Lee, T. J. Tanaka, R. L. Shimabukuro, and B. K. Fujikawa, "Low Energy X-Ray Diagnostics–1981" (D. T. Attwood and B. L. Henke, eds.). AIP Conference Proc., No. 75, American Institute of Physcis, New York, 1981.
18. I. Simon and H. O. McMahon, *J. Chem. Phys.* **21**, 23 (1953).
19. J. Reitzel, *J. Chem. Phys.* **23**, 2407 (1955).
20. P. E. Jellyman and J. P. Procter, *J. Soc. Glass Technol.* **39**, 173 (1955).
21. Gouq-Jen Su, N. F. Borrelli, and A. R. Miller, *Phys. Chem. Glasses* **3**, 167 (1962).
22. D. Crozier and R. W. Douglas, *Phys. Chem. Glasses* **6**, 240 (1965).
23. J. E. Diai, R. E. Gong, and J. N. Fordemwalt, *J. Electrohem. Soc.* **115**, 327 (1968).
24. J. F. Osantowski, *J. Opt. Soc. Am.* **64**, 834 (1974).
25. R. J. Powell and M. Morad, *J. Appl. Phys.* **49**, 2499 (1978).

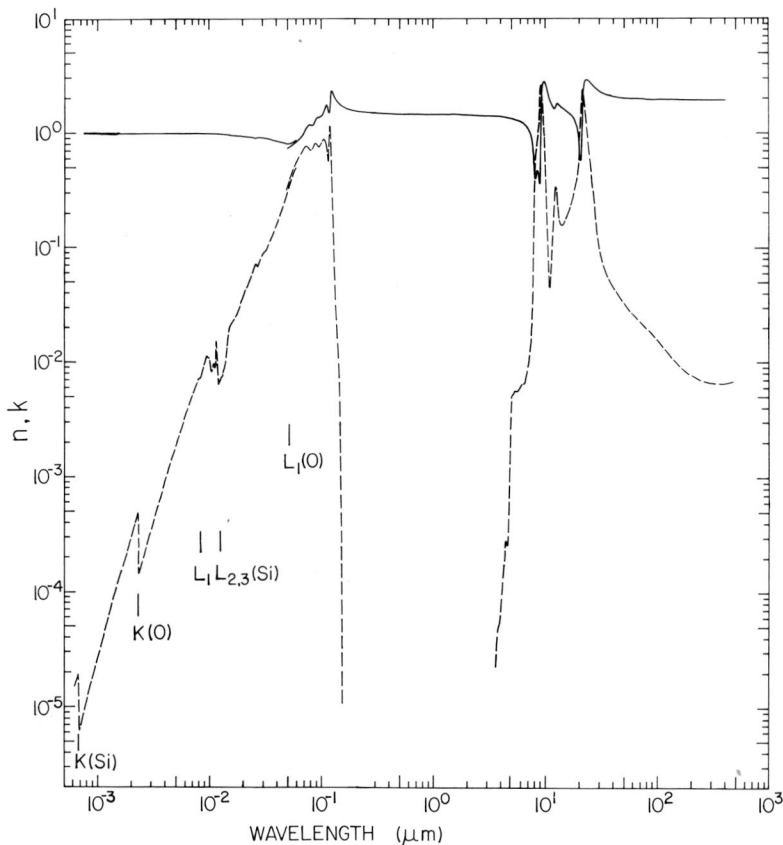

Fig. 10. Log–log plot of n (——) and k (----) versus wavelength in micrometers for silicon dioxide (glass).

TABLE X

Values of n and k for Silicon Dioxide Glass Obtained from Various References[a]

eV	cm^{-1}	Å	n	k
2000		6.199	0.99993 [17]	1.503×10^{-5} [17]
1952		6.351	0.99992	1.636
1905		6.508	0.99992	1.781
1860		6.665	0.99991	1.936×10^{-5}
1815		6.831	0.99991	$6.300 \times 10^{-6\cdot}$
1772		6.997	0.99991	6.904
1730		7.166	0.99990	7.564
1688		7.345	0.99990	8.298
1648		7.523	0.99989	9.083
1609		7.705	0.99989	9.941×10^{-6}

(*continued*)

TABLE X (*Continued*)

Silicon Dioxide (Glass)

eV	cm^{-1}	Å	n	k
1570		7.897	0.99988	1.090×10^{-5}
1533		8.087	0.99987	1.193
1496		8.287	0.99987	1.308
1460		8.492	0.99986	1.432
1426		8.694	0.99985	1.562
1392		8.906	0.99985	1.708
1358		9.129	0.99984	1.872
1326		9.350	0.99983	2.044
1294		9.581	0.99982	2.238
1263		9.816	0.99981	2.447
1233		10.05	0.99981	2.673
1204		10.30	0.99980	2.916
1175		10.55	0.99979	3.189
1147		10.81	0.99978	3.483
1119		11.08	0.99976	3.812
1093		11.34	0.99975	4.155
1067		11.62	0.99974	4.537
1041		11.91	0.99973	4.965
1016		12.20	0.99971	5.423
992		12.50	0.99970	5.915
968		12.81	0.99968	6.468
945		13.12	0.99967	7.061
923		13.43	0.99965	7.686
901		13.76	0.99964	8.364
879		14.10	0.99962	9.121
858		14.45	0.99960	9.928×10^{-5}
838		14.79	0.99958	1.080×10^{-4}
818		15.16	0.99956	1.179
798		15.54	0.99954	1.289
779		15.92	0.99951	1.407
760		16.31	0.99949	1.535
742		16.71	0.99946	1.671
725		17.10	0.99944	1.813
707		17.54	0.99941	1.981
690		17.97	0.99938	2.149
674		18.39	0.99935	2.326
658		18.84	0.99932	2.536
642		19.31	0.99928	2.771
627		19.77	0.99925	3.009
612		20.26	0.99921	3.270
597		20.77	0.99917	3.560
583		21.27	0.99913	3.862
569		21.79	0.99909	4.199
555		22.34	0.99904	4.578
542		22.87	0.99899	4.971
529		23.44	0.99894	1.486
516		24.03	0.99889	1.621

TABLE X (*Continued*)

Silicon Dioxide (Glass)

eV	cm^{-1}	Å	n	k
504		24.60	0.99884	1.76
492		25.20	0.99878	1.91
480		25.83	0.99872	2.08
469		26.43	0.99866	2.26
458		27.07	0.99859	2.45
447		27.74	0.99852	2.67
436		28.44	0.99845	2.90
426		29.10	0.99837	3.15
415		29.87	0.99828	3.44
406		30.54	0.99821	3.71
396		31.31	0.99812	4.04
386		32.12	0.99802	4.40
377		32.89	0.99792	4.77
368		33.69	0.99782	5.18
359		34.53	0.99771	5.63
351		35.32	0.99760	6.07
342		36.25	0.99747	6.62
334		37.12	0.99735	7.17
326		38.03	0.99722	7.77
318		38.99	0.99708	8.44
311		39.86	0.99695	9.09
303		40.92	0.99678	9.91×10^{-4}
296		41.88	0.99663	1.07×10^{-3}
289		42.90	0.99646	1.16
282		43.96	0.99629	1.25
275		45.08	0.99609	1.36
269		46.09	0.99592	1.46
262		47.32	0.99570	1.59
256		48.43	0.99549	1.71
250		49.59	0.99527	1.85
244		50.81	0.99504	2.00
238		52.09	0.99478	2.16
232		53.44	0.99451	2.34
227		54.62	0.99427	2.50
221		56.10	0.99395	2.72
216		57.40	0.99367	2.92
211		58.76	0.99336	3.14
206		60.18	0.99304	3.37
201		61.68	0.99269	3.63
196		63.25	0.99231	3.90
191		64.91	0.99190	4.21
187		66.30	0.99155	4.47
182		68.12	0.99108	4.82
178		69.65	0.99068	5.11
174		71.25	0.99024	5.44
170		72.93	0.9898	5.80

(*continued*)

TABLE X (*Continued*)

Silicon Dioxide (Glass)

eV	cm^{-1}	Å	n	k
166		74.69	0.9893	6.19
162		76.53	0.9887	6.62
158		78.47	0.9882	6.99
154		80.51	0.9875	7.30
151.2		82	0.9871 [16]	7.3 [16]
147.6		84	0.9865	7.6
144.2		86	0.986	8.1
140.9		88	0.9855	8.9
137.8		90	0.9853	9.9×10^{-3}
134.8		92	0.9854	1.07×10^{-2}
131.9		94	0.9858	1.14
129.2		96	0.9868	1.11
126.5		98	0.9872	1.11×10^{-2}
124.0		100	0.9874	9.7×10^{-3}
121.6		102	0.9865	8.5
119.2		104	0.9851	8.3
117.0		106	0.9841	9.6
114.8		108	0.9848	9.9
112.7		110	0.9844	9.0
110.7		112	0.9822	9.3×10^{-3}
108.8		114	0.9828	1.59×10^{-2}
106.9		116	0.9867	1.06×10^{-2}
105.1		118	0.9858	7.5×10^{-3}
103.3		120	0.9839	6.5
101.6		122	0.9823	6.8
99.99		124	0.9813	7.0
98.40		126	0.9803	7.3
96.86		128	0.9794	7.6
95.37		130	0.9789	7.6
93.93		132	0.9778	8.3
92.53		134	0.977	8.7
91.17		136	0.9761	9.3×10^{-3}
89.84		138	0.9747	1.03×10^{-2}
82.66		150	0.9634	1.99
77.49		160	0.9608	2.26
72.93		170	0.9562	2.44
68.88		180	0.9509	2.81
65.26		190	0.9458	3.35
61.99		200	0.9416	3.74
59.04		210	0.9386	4.37
56.36		220	0.9328	4.62
53.91		230	0.9271	5.2
51.66		240	0.9222	5.78
49.59		250	0.9164	6.5
47.69		260	0.9105	7.26
45.92		270	0.9207	6.8
44.28		280	0.9175	7.5

TABLE X (*Continued*)

Silicon Dioxide (Glass)

eV	cm^{-1}	Å	n	k
42.75		290	0.9137	8.2
41.33		300	0.913 [15]	9.0 [15]
40.00		310	0.907	9.2
38.75		320	0.901	9.4
37.57		330	0.895	9.8 × 10^{-2}
36.47		340	0.888	0.107
35.42		350	0.882	0.113
34.44		360	0.877	0.120
33.51		370	0.870	0.128
32.63		380	0.866	0.137
31.79		390	0.858	0.144
31.00		400	0.851	0.156
30.24		410	0.845	0.169
29.52		420	0.839	0.180
28.83		430	0.833	0.190
28.18		440	0.827	0.202
27.55		450	0.822	0.218
26.95		460	0.817	0.233
26.38		470	0.813	0.250
25.83		480	0.808	0.270
25.30		490	0.804	0.282

eV	cm^{-1}	μm	n	k
25		0.04959	0.733 [9, 10]	0.325 [9, 10]
24.80		0.0500	0.803 [15]	0.300 [15]
24.31		0.0510	0.804	0.322
24		0.05166	0.753 [9, 10]	0.375 [9, 10]
23.84		0.0520	0.806 [15]	0.343 [15]
23.39		0.0530	0.811	0.366
23		0.05391	0.774 [9, 10]	0.434 [9, 10]
22.96		0.540	0.817 [15]	0.385 [15]
22.54		0.0550	0.822	0.408
22.14		0.0560	0.829	0.430
22		0.05636	0.797 [9, 10]	0.480 [9, 10]
21.75		0.0570	0.833 [15]	0.450 [15]
21.38		0.0580	0.843	0.470
21.01		0.0590	0.851	0.482
21		0.05904	0.827 [9, 10]	0.530 [9, 10]
20.66		0.0600	0.862 [15]	0.497 [15]
20		0.06199	0.859 [9, 10]	0.585 [9, 10]
19.5		0.06358	0.879	0.613
19		0.06526	0.902	0.645
18.5		0.06701	0.927	0.677
18		0.06888	0.957	0.712
17.75		0.06985	0.975	0.731

(*continued*)

TABLE X (*Continued*)

Silicon Dioxide (Glass)

eV	cm^{-1}	μm	n	k
17.5		0.07085	0.999	0.750
17.25		0.07187	1.030	0.763
17		0.07293	1.072	0.768
16.75		0.07402	1.124	0.765
16.5		0.07514	1.137	0.755
16.25		0.07630	1.156	0.737
16		0.07749	1.172	0.717
15.75		0.07872	1.178	0.703
15.5		0.07999	1.172	0.696
15.25		0.08130	1.167	0.699
15		0.08266	1.168	0.711
14.75		0.08405	1.175	0.739
14.5		0.08551	1.195	0.771
14.25		0.08700	1.225	0.799
14		0.08856	1.265	0.808
13.75		0.09017	1.320	0.795
13.5		0.09184	1.363	0.775
13.25		0.09357	1.371	0.755
13		0.09537	1.368	0.747
12.75		0.09724	1.372	0.766
12.5		0.09919	1.383	0.793
12.25		0.1012	1.410	0.824
12		0.1033	1.475	0.861
11.8		0.1051	1.554	0.874
11.6		0.1069	1.635	0.859
11.4		0.1088	1.716	0.810
11.2		0.1107	1.766 [9]	0.718 [12]
11		0.1127	1.739	0.569
10.9		0.1137	1.687	0.565
10.8		0.1148	1.587	0.618
10.7		0.1159	1.513	0.725
10.6		0.1170	1.492	0.914
10.5		0.1181	1.567	1.11
10.45		0.1187	1.645	1.136
10.4		0.1192	1.772	1.13
10.35		0.1198	1.919	1.045
10.3		0.1204	2.048	0.925
10.25		0.1210	2.152	0.810
10.2		0.1215	2.240	0.715
10.1		0.1228	2.332	0.460
10		0.1240	2.330	0.323
9.9		0.1252	2.292	0.236
9.8		0.1265	2.243	0.168
9.7		0.1278	2.190	0.119
9.6		0.1291	2.140	0.077
9.5		0.1305	2.092	0.0561 [12–14]
9.4		0.1319	2.047	0.0430

TABLE X (*Continued*)

Silicon Dioxide (Glass)

eV	cm^{-1}	μm	n	k
9.3		0.1333	2.006	0.0339
9.2		0.1348	1.969	0.0271
9.1		0.1362	1.935	0.0228
9.0		0.1378	1.904	1.89×10^{-2}
8.9		0.1393	1.876	1.56
8.8		0.1409	1.850	1.32
8.7		0.1425	1.825	1.09×10^{-2}
8.6		0.1442	1.803	8.38×10^{-3}
8.5		0.1459	1.783	5.57
8.4		0.1476	1.764	3.17
8.3		0.1494	1.747	1.40×10^{-3}
8.2		0.1512	1.730	4.63×10^{-4}
8.1		0.1531	1.716	1.22×10^{-4}
8.0		0.1550	1.702	3.2×10^{-5}
7.8		0.1590	1.676	4.7×10^{-6}
7.6		0.1631	1.653	
7.4		0.1675	1.633	
7.2		0.1722	1.616	
7.0		0.1771	1.600	
6.75		0.1837	1.582	
6.5		0.1907	1.567	
6.25		0.1984	1.554	
6.0	48,390	0.2066	1.543	
5.7976	46,760.4	0.213856	1.53429 [7]	
5.7819	46,633.5	0.214438	1.53371	
5.4680	44,102.0	0.226747	1.52276	
5.3858	43,438.8	0.230209	1.52009	
5.2131	42,046.3	0.237833	1.51474	
5.1674	41,677.4	0.239938	1.51338	
4.9939	40,278.4	0.248272	1.50841	
4.6751	37,706.8	0.265204	1.50004	
4.5940	37,052.8	0.269885	1.49805	
4.5040	36,326.9	0.275278	1.49592	
4.4226	35,670.1	0.280347	1.49404	
4.2848	34,559.0	0.289360	1.49099	
4.1784	33,700.9	0.296728	1.48873	
4.1034	33,096.1	0.302150	1.48719	
3.7542	30,279.3	0.330259	1.48053	
3.7105	29,926.9	0.334148	1.47976	
3.6427	29,380.2	0.340365	1.47858	
3.5770	28,850.0	0.346620	1.47746	
3.4340	27,696.9	0.361051	1.47512	
3.3967	27,396.1	0.365015	1.47453	
3.0640	24,712.3	0.404656	1.46961	
2.8448	22,944.5	0.435835	1.46669	
2.6503	21,375.9	0.467816	1.46429	

(*continued*)

TABLE X (*Continued*)

Silicon Dioxide (Glass)

eV	cm^{-1}	μm	n	k
2.5504	20,570.5	0.486133	1.46313	
2.4379	19,662.5	0.508582	1.46187	
2.2705	18,312.5	0.546074	1.46008	
2.1489	17,332.3	0.576959	1.45885	
2.1411	17,269.2	0.579065	1.45877	
2.1102	17,019.5	0.587561	1.45847	
2.1041	16,970.4	0.589262	1.45841	
1.9257	15,531.6	0.643847	1.45671	
1.8892	15,237.6	0.656272	1.45637	
1.8566	14,974.2	0.667815	1.45608	
1.7549	14,153.9	0.706519	1.45515	

eV	cm^{-1}	μm	n	k	k_{ABS}
1.4550	11,735.6	0.852111	1.45248		
1.3863	11,181.3	0.894350	1.45185		
1.2228	9,862.13	1.01398	1.45025		
1.1449	9,233.87	1.08297	1.44941		
1.0985	8,860.06	1.12866	1.44888		
0.91018	7,341.1	1.3622	1.44621		
0.88874	7,168.15	1.39506	1.44584		
0.84372	6,805.0	1.4695	1.44497		
0.81061	6,538.00	1.52952	1.44427		
0.74663	6,021.9	1.6606	1.44267		
0.7376	5,949	1.681	1.44241		
0.73225	5,906.0	1.6932	1.44226		
0.72543	5,850.93	1.70913	1.44205		
0.68384	5,515.51	1.81307	1.44069		
0.62934	5,075.91	1.97009	1.43851		
0.60243	4,858.9	2.0581	1.43722		
0.57598	4,645.5	2.1526	1.43576		
0.53317	4,300.30	2.32542	1.43292		
0.50868	4,102.7	2.4374	1.43095		
0.38221	3,082.7	3.2439	1.41314		
0.37953	3,061.1	3.2668	1.41253		
0.37542	3,027.9	3.3026	1.41155		
0.3623	2,922	3.422	1.40819		
0.35354	2,851.4	3.5070	1.40568		
0.34863	2,811.8	3.5564	1.40418		
0.3410	2,750	3.636			2.25×10^{-5} [4]
0.3348	2,700	3.704			3.39
0.33449	2,697.8	3.7067	1.39936		
0.3286	2,650	3.774			3.93
0.3224	2,600	3.846	1.395 [2]		4.96
0.3162	2,550	3.922			5.18
0.3100	2,500	4.000			5.79
0.3038	2,450	4.082			7.99×10^{-5}
0.2976	2,400	4.167	1.383		1.07×10^{-4}

TABLE X (*Continued*)

Silicon Dioxide (Glass)

eV	cm^{-1}	μm	n	k	k_{ABS}
0.2914	2,350	4.255			1.32
0.2852	2,300	4.348			2.13
0.2821	2,275	4.396			2.65
0.2790	2,250	4.444			2.84
0.2759	2,225	4.494			2.84
0.2728	2,200	4.545	1.365		2.56
0.2666	2,150	4.651			2.62
0.2604	2,100	4.762			4.85×10^{-4}
0.2542	2,050	4.878			1.82×10^{-3}
0.2480	2,000	5.000	1.342		3.98
0.2449	1,975	5.063			5.12
0.2418	1,950	5.128			5.18
0.2356	1,900	5.263			5.49
0.2325	1,875	5.333			5.69
0.2294	1,850	5.405			5.72
0.2232	1,800	5.556	1.306		5.63
0.2170	1,750	5.714			
0.2108	1,700	5.882	1.278		5.94
0.2046	1,650	6.061			6.32
0.2015	1,625	6.154			6.46
0.1984	1,600	6.250	1.239		6.52
0.1922	1,550	6.452	1.212		6.57
0.1860	1,500	6.667	1.175		7.16
0.1835	1,480	6.757	1.158		
0.1810	1,460	6.849	1.135		
0.1798	1,450	6.897			8.51×10^{-3}
0.1785	1,440	6.944	1.107		
0.1761	1,420	7.042	1.084		
0.1736	1,400	7.143	1.053		1.06×10^{-2}
0.1711	1,380	7.246	1.014		
0.1686	1,360	7.353	0.9702		
0.1674	1,350	7.407	0.9488		1.48
0.1661	1,340	7.463	0.9175		
0.1649	1,330	7.519	0.8897		
0.1637	1,320	7.576	0.8600		
0.1624	1,310	7.634	0.8213		
0.1612	1,300	7.692	0.7719 [2]	3.72×10^{-2} [2]	4.0×10^{-2} [2]
0.1599	1,290	7.752	0.7037	4.74	5.2
0.1587	1,280	7.813	0.6232	7.68×10^{-2}	7.3
0.1575	1,270	7.874	0.5456	0.132	
0.1562	1,260	7.937	0.4677	0.216	
0.1550	1,250	8.000	0.4113	0.323	
0.1537	1,240	8.065	0.3931	0.446	
0.1525	1,230	8.130	0.4020	0.553	
0.1513	1,220	8.197	0.4329	0.635	
0.1500	1,210	8.265	0.4530	0.704	

(*continued*)

TABLE X (*Continued*)

Silicon Dioxide (Glass)

eV	cm^{-1}	μm	n	k	k_{ABS}
0.1488	1,200	8.333	0.4600	0.771	
0.1475	1,190	8.403	0.4730	0.840	
0.1463	1,180	8.475	0.4746	0.903	
0.1451	1,170	8.547	0.4656	0.978	
0.1438	1,160	8.621	0.4563	1.07	
0.1426	1,150	8.696	0.4309	1.17	
0.1413	1,140	8.772	0.3915	1.32	
0.1401	1,130	8.850	0.3563	1.53	
0.1389	1,120	8.929	0.3705	1.85	
0.1376	1,110	9.009	0.5846	2.27	
0.1364	1,100	9.091	1.043	2.55	
0.1351	1,090	9.174	1.616	2.63	
0.1333	1,075	9.302	2.250	2.26	
0.1302	1,050	9.524	2.760	1.65	
0.1271	1,025	9.756	2.839	0.962	
0.1240	1,000	10.00	2.694	0.509	
0.1209	975	10.26	2.448	0.231	6.8×10^{-2}
0.1178	950	10.53	2.224	0.102	3.1
0.1147	925	10.81	2.038	4.60×10^{-2}	2.3
0.1116	900	11.11	1.869	5.06	2.2
0.1091	880	11.36	1.784	7.75×10^{-2}	2.6
0.1066	860	11.63	1.690	0.116	7.7×10^{-2}
0.1054	850	11.76	1.652	0.152	0.12
0.1041	840	11.90	1.619	0.204	0.17
0.1029	830	12.05	1.615	0.267	
0.1017	820	12.20	1.658	0.323	
0.1004	810	12.35	1.701	0.341	
0.09919	800	12.50	1.753	0.343	
0.09795	790	12.66	1.789	0.314	
0.09671	780	12.82	1.811	0.275	0.16
0.09547	770	12.99	1.810	0.227	0.11
0.09423	760	13.16	1.779	0.192	8.5×10^{-2}
0.09299	750	13.33	1.756	0.177	8.0
0.08989	725	13.79	1.698	0.157	7.5
0.08679	700	14.29	1.643	0.157	7.2
0.08369	675	14.81	1.598	0.168	7.7
0.08059	650	15.38	1.555	0.182	8.8×10^{-2}
0.07749	625	16.00	1.502	0.202	0.11
0.07439	600	16.67	1.450	0.235	0.14
0.07191	580	17.24	1.401	0.264	0.16
0.06943	560	17.86	1.337	0.298	0.18
0.06695	540	18.52	1.235	0.341	0.21
0.06534	527	18.97	1.161	0.377	0.25
0.06447	520	19.23	1.050	0.415	0.32
0.06323	510	19.61	0.8857	0.524	
0.06199	500	20.00	0.6616	0.822	

TABLE X (*Continued*)

Silicon Dioxide (Glass)

eV	cm^{-1}	μm	n	k	k_{ABS}
0.06075	490	20.41	0.5777	1.28	
0.05951	480	20.83	0.7517	1.86	
0.05889	475	21.05	1.002	2.22	
0.05827	470	21.28	1.484	2.39	
0.05703	460	21.74	2.308	2.29	
0.05455	440	22.73	2.936	1.29	
0.05207	420	23.81	2.912	0.738	
0.04959	400	25.00	2.739	0.397	
0.04649	375	26.67	2.537	0.199	0.20
0.04339	350	28.57	2.388		0.13
0.04030	325	30.77	2.284		9.2×10^{-2}
0.03720	300	33.33	2.210		6.7
0.03410	275	36.36	2.147		5.6
0.03100	250	40.00	2.100		4.6
0.01240	100	100.0	1.967 [1]	1.59×10^{-2} [1]	
0.009919	80	125.0	1.962	1.19×10^{-2}	
0.007439	60	166.7	1,959	8.62×10^{-3}	
0.004959	40	250.0	1.957	6.96	
0.002480	20	500.0	1.955	7.96	

[a] References are indicated in brackets. Data marked k_{ABS} were obtained from transmission measurements.

Silicon Monoxide (SiO)
(Noncrystalline)

H. R. PHILIPP

General Electric Research and Development Center

Schenectady, New York

Silicon monoxide (SiO) is used as a protective layer and antireflecting coating in optical applications and as the dielectric material in certain microelectronic devices. These layers are generally prepared by the rapid evaporation of silicon monoxide under high-vacuum conditions. If the deposition is carried out slowly in the presence of a partial pressure of oxygen, the condensate acquires excess oxygen and the O-to-Si-atom ratio rises above unity. These materials are often labeled SiO_x ($x > 1.0$). The optical properties of SiO and SiO_x are different, and the values given here should only be applied to materials of stoichiometry SiO. For completeness, references will be given that also describe the optical behavior of SiO_x materials.

Although SiO has some important applications, it is a rather dull material from an optical point of view. For this and other reasons, there is not very much information available on its optical properties. However, in the visible and near-ultraviolet spectral regions in which data from several sources can be compared, the results agree with one another quite well. This would indicate that SiO is a reasonably well defined material, when prepared in the absence of oxygen, with definitive optical properties.

A set of optical parameters n and k for noncrystalline SiO are given in Table XI and plotted in Fig. 11 for the wavelength region $14.0\ \mu m \geq \lambda \geq 0.050\ \mu m$. The wavelength λ, index of refraction n, and extinction coefficient k are given to four places in most of the table entries. The accuracy of the values does not warrant this precision. This is done to show the possible presence of weak structure in the optical properties that can often be discerned by comparison of precise, but not necessarily accurate, values over small wavelength ranges.

In the infrared, $14.0\ \mu m \geq \lambda \geq 1.00\ \mu m$, the values given in the table were obtained from Hass and Salzberg [1]. The values were derived through analysis of transmittance and reflectance measurements on a series of SiO layers

765

of various thicknesses. Although the overall accuracy of these values cannot be simply specified, they fit the measured transmittance and reflectance to within $\pm 2\%$.

In the region 1.00 μm $\geq \lambda \geq$ 0.049 μm, the values are taken from Philipp [2]. They were derived from Kramers–Kronig (KK) analysis of reflectance data augmented by transmission measurements in the region of low absorption. These results agree reasonably well with those of Hass and Salzberg [1], whose measurements extend to 0.24 μm, as well as with the absorption data of Cremer et al. [3]. The results of Philipp are given in the table because they cover a wider wavelength range and more explicitly evaluate the extinction coefficient in the region of low absorption compared to the other measurements. They are not necessarily more accurate than those of Hass and Salzberg [1] and Cremer [3]. The differences that are found in these measurements [1–3], $\pm 5\%$ or less in n and slightly larger in k, may be due in part to sample preparation (evapoaration rate, vacuum conditions, substrate temperature, etc.), and as stated earlier, it is encouraging that they agree as well as they do. It should also be pointed out that the table values at the shorter wavelengths, especially those below 0.100 μm, depend somewhat on the extrapolation used in the KK analysis and should be treated with caution until more definitive measurements are made for $\lambda \leq 0.05$ μm, which more clearly delineate the proper reflectance extrapolation.

It is recommended that anyone who uses the table carefully examine the measurements and references cited in the references discussed [1–3] as well as other references [4–7] that also describe the optical behavior of materials of stoichiometry SiO_x ($x > 1$).

REFERENCES

1. G. Hass and C. D. Salzberg, *J. Opt. Soc. Am.* **44**, 181 (1954).
2. H. R. Philipp, *J. Phys. Chem. Solids* **32**, 1935 (1971).
3. E. Cremer, T. Kraus, and E. Ritter, *Z. Elektrochem.* **62**, 939 (1958).
4. E. Ritter, *Opt. Acta* **9**, 197 (1962).
5. A. P. Bradford and G. Hass, *J. Opt. Soc. Am.* **53**, 1096 (1963).
6. G. Hass, *J. Am. Ceramic Soc.* **33**, 353 (1950).
7. A. P. Bradford, G. Hass, M. McFarland, and E. Ritter, *Appl. Opt.* **4**, 971 (1965).

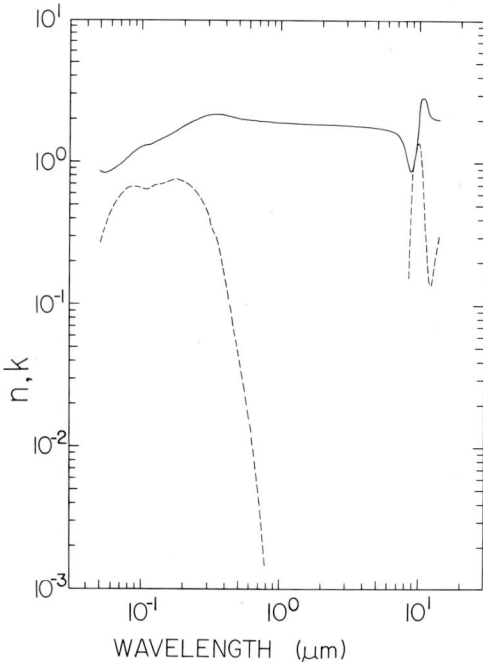

Fig. 11. Log–log plot n (——) and k (– – –) versus wavelength in micrometers for silicon monoxide (noncrystalline).

TABLE XI

Values of n and k for Silicon Monoxide (Noncrystalline) from Various References [a]

eV	cm^{-1}	μm	n	k
25		0.04959	0.8690 [2]	0.2717 [2]
24		0.05166	0.8444	0.3060
23		0.05391	0.8371	0.3505
22.5		0.05510	0.8391	0.3761
22		0.05636	0.8454	0.3987
21.5		0.05767	0.8519	0.4222
21		0.05904	0.8610	0.4456
20.5		0.06048	0.8721	0.4688
20		0.06199	0.8853	0.4919
19.5		0.06358	0.9007	0.5140
19		0.06526	0.9178	0.5362
18.5		0.06702	0.9376	0.5578
18		0.06888	0.9596	0.5771
17.5		0.07085	0.9825	0.5961
17		0.07293	1.008	0.6147
16.5		0.07514	1.036	0.6309
16		0.07749	1.066	0.6453
15.5		0.07999	1.098	0.6566

(continued)

TABLE XI (*Continued*)

Silicon Monoxide (Noncrystalline)

eV	cm^{-1}	μm	n	k
15		0.08266	1.132	0.6651
14.5		0.08551	1.166	0.6692
14		0.08856	1.199	0.6698
13.5		0.09184	1.231	0.6666
13		0.09537	1.259	0.6602
12.5		0.09919	1.283	0.6523
12		0.1033	1.307	0.6464
11.5		0.1078	1.311	0.6293
11		0.1127	1.320	0.6529
10.5		0.1181	1.345	0.6701
10		0.1240	1.378	0.6843
9.5		0.1305	1.412	0.6920
9		0.1378	1.445	0.7002
8.5		0.1459	1.482	0.7153
8		0.1550	1.530	0.7333
7.5		0.1653	1.593	0.7473
7		0.1771	1.667	0.7479
6.5		0.1907	1.746	0.7348
6	48,390	0.2066	1.829	0.7084
5.75	46,380	0.2156	1.871	0.6890
5.5	44,360	0.2254	1.914	0.6663
5.25	42,340	0.2362	1.957	0.6383
5	40,330	0.2480	2.001	0.6052
4.8	38,710	0.2583	2.034	0.5723
4.6	37,100	0.2695	2.066	0.5364
4.4	35,490	0.2818	2.094	0.4948
4.2	33,880	0.2952	2.119	0.4499
4	32,260	0.3100	2.141	0.4006
3.8	30,650	0.3263	2.157	0.3453
3.6	29,040	0.3444	2.162	0.2872
3.4	27,420	0.3647	2.160	0.2287
3.2	25,810	0.3875	2.144	0.1706
3	24,200	0.4133	2.116	0.1211
2.8	22,580	0.4428	2.085	0.08374
2.6	20,970	0.4769	2.053	0.05544
2.4	19,360	0.5166	2.021	0.03533
2.2	17,740	0.5636	1.994	0.02153
2	16,130	0.6199	1.969	0.01175
1.8	14,520	0.6888	1.948	0.00523
1.6	12,900	0.7749	1.929	0.00151
1.4	11,290	0.8856	1.913	
1.240	10,000	1.000	1.87 [1]	
0.6199	5,000	2.000	1.84	
0.4133	3,333	3.000	1.82	
0.3100	2,500	4.000	1.80	
0.2480	2,000	5.000	1.75	

TABLE XI (*Continued*)

Silicon Monoxide (Noncrystalline)

eV	cm^{-1}	μm	n	k
0.2066	1,667	6.000	1.70	
0.1771	1,492	7.000	1.60	
0.1653	1,333	7.500	1.42	
0.1550	1,250	8.000	1.15	
0.1459	1,176	8.500	0.90	0.18 [1]
0.1378	1,111	9.000	0.91	0.75
0.1305	1,053	9.500	1.20	1.20
0.1240	1,000	10.00	2.00	1.38
0.1181	952.4	10.50	2.85	0.90
0.1153	930.2	10.75	2.86	0.58
0.1127	909.1	11.00	2.82	0.40
0.1078	869.6	11.50	2.50	0.20
0.1033	833.3	12.00	2.13	0.14
0.09537	769.2	13.00	2.04	0.20
0.08856	714.3	14.00	2.01	0.30

[a] References are indicated in brackets.

Silicon Nitride (Si$_3$N$_4$) (Noncrystalline)

H. R. PHILIPP

General Electric Research and Development Center
Schenectady, New York

Noncrystalline silicon nitride (Si$_3$N$_4$) is an important material in integrated-circuit technology. In this application it is used in thin-film form and can be prepared by a variety of deposition techniques, including (a) pyrolytic decomposition of a mixture of gases containing silicon and nitrogen (for example, SiH$_4$ and NH$_3$), (b) sputtering, and (c) rf glow-discharge techniques. Silicon oxynitride (SiO$_x$N$_y$) films can also be readily formed by adding small amounts of NO or O$_2$ to the reactive gases. The optical properties of these films are strongly dependent on the deposition temperature and the Si-to-N- and Si-to-O-atom ratios. Other studies also indicate the presence of chemically bound hydrogen in the film. Thus, the stoichiometry of an arbitrary film can be more accurately indicated by SiO$_x$N$_y$H$_z$, and the refractive index and other optical properties will be a function of x, y, and z. Samples prepared by high-temperature pyrolysis are generally considered to have, or come close to, the ideal stoichiometry, Si$_3$N$_4$ (exclusive of hydrogen content).

While there have been a number of studies on the optical properties of silicon nitride materials, the data have generally been limited to the evaluation of the index of refraction in the visible region of the spectrum [1–14] and to infrared measurements of the lattice absorption bands near 11 μm and other bands, particularly those which show up N–H and Si–H bonding [1–7, 9, 12, 15–20]. The infrared measurements have been used mainly for structural and chemical evaluation purposes. While Taft [12] measured the absorption coefficient of silicon nitride films prepared by several different techniques in the region of 11-μm wavelength, no attempt has been made to determine a self-consistent set of optical parameters for this material in the infrared.

In the region of strong electronic absorption, both Philipp [21] and Bauer [22] have evaluated n and k from the visible into the vacuum ultraviolet region. The results of Philipp [21], which extend to 24 eV, were determined by Kramers–Kronig analysis of reflectance and absorption data. The films

771

he used were prepared by pyrolytic decomposition at 1000°C of a mixture of SiH_4 and NH_3 gases in the ratio 1 to 40,000. Bauer's [22] measurements extend to only 7.5 eV but were obtained on samples prepared by a variety of deposition techniques including pyrolysis of a mixture of SiO_4 and NH_3 gases at substrate temperatures in the range 720 to 1000°C. His n and k values were evaluated by error-function analysis of reflectance and transmission data. For pyrolytic silicon nitride films, the results of Philipp [21] and Bauer [22] show reasonable but not precise agreement.

A set of optical parameters n and k for noncrystalline silicon nitride are given in Table XII and are plotted in Fig. 12 for the energy range 1–24 eV (1.24–0.0517-μm wavelength). They are taken entirely from the work of Philipp [21]. These results are not necessarily more accurate than other data but were primarily chosen because they cover the widest energy range. In addition, Theeten et al. [23] in a study of Si_3N_4 in the region from 1.5 to 5.8 eV using spectroscopic ellipsometry found good agreement with the values of Philipp [21] (and it is in this energy range that the largest differences occur between the results of Philipp [21] and Bauer [22]). In most of the table entries, the index of refraction n values are given to four places and the extinction coefficient k values are given to three places. The accuracy of these values does not warrant this precision. This is done to show the possible presence of weak structure in the optical properties that can often be discerned by comparison of precise, but not necessarily accurate, values over small energy ranges. It is also recommended that anyone who uses the table carefully examine the material and references cited in the references. Since the optical properties of silicon nitride depend on the method and temperature of deposition, care must be taken in using the table values for materials prepared by using techniques other than high-temperature pyrolysis of gases with a high N-to-Si ratio.

REFERENCES

1. V. Doo, D. Nichols, and G. Silvey, *J. Electrochem. Soc.* **133**, 1279 (1966).
2. K. Bean, P. Gleim, R. Yeakley, and W. Runyan, *J. Electrochem. Soc.* **114**, 733 (1967).
3. T. Chu, C. Lee, and G. Gruber, *J. Electrochem. Soc.* **114**, 717 (1967).
4. V. Doo, D. Kerr, and D. Nichols, *J. Electrochem. Soc.* **115**, 61 (1968).
5. D. Brown, P. Gray, F. Heumann, H. Philipp, and E. Taft, *J. Electrochem. Soc.* **115**, 311 (1968).
6. M. Grieco, F. Worthing, and B. Schwartz, *J. Electrochem. Soc.* **115**, 525 (1968).
7. R. Levitt and W. Zwicker, *J. Electrochem. Soc.* **114**, 1192 (1967).
8. B. Deal, P. Fleming, and P. Castro, *J. Electrochem. Soc.* **115**, 300 (1968).
9. G. Brown, W. Robinette, Jr., and H. Carlson, *J. Electrochem. Soc.* **115**, 948 (1968).
10. E. V. Shitova, I. A. Yasneva, and N. A. Genkina, *Opt. Spectrosc.* **43**, 140 (1977); E. V. Shitova, I. A. Yasneva, and N. A. Genkina, *Opt. Spektrosk.* **43**, 244 (1977).
11. Y. N. Volgin, O. P. Borisov, Y. I. Ukhanov, and N. I. Sukhanova, *J. Appl. Spectrosc.* **24**, 115 (1976); Y. N. Volgin, O. P. Borisov, Y. I. Ukhanov and N. I. Sukhanova, *Zh. Priklad. Spektrosk.* **24**, 164 (1976).
12. E. A. Taft, *J. Electrochem. Soc.* **118**, 1341 (1971).
13. M. J. Rand and D. R. Wonsidler, *J. Electrochem. Soc.* **125**, 99 (1978).

14. T. Wittberg, J. Hoenigman, W. Moddeman, C. Cothern, and M. Gulett, *J. Vac. Sci. Technol.* **15**, 348 (1978).
15. S. Yoshioka and S. Takayanagi, *J. Electrochem. Soc.* **114**, 962 (1967).
16. H. J. Stein, *Appl. Phys. Lett.* **32**, 379 (1978).
17. S. M. Hu, *J. Electrochem. Soc.* **113**, 693 (1966).
18. H. J. Stein and H. A. R. Wagener, *J. Electrochem. Soc.* **124**, 908 (1977).
19. P. S. Peercy, H. J. Stein, B. L. Doyle, and S. T. Picraux, *J. Electron. Mater.* **8**, 11 (1979).
20. L. F. Cordes, *Appl. Phys. Lett.* **11**, 383 (1967).
21. H. R. Philipp, *J. Electrochem. Soc.* **120**, 295 (1973).
22. J. Bauer, *Phys. Status. Solidi A* **39**, 411 (1977).
23. J. B. Theeten, D. E. Aspnes, F. Simondet, M. Errman, and P. C. Mürau, *J. Appl. Phys.* **52**, 6788 (1981).

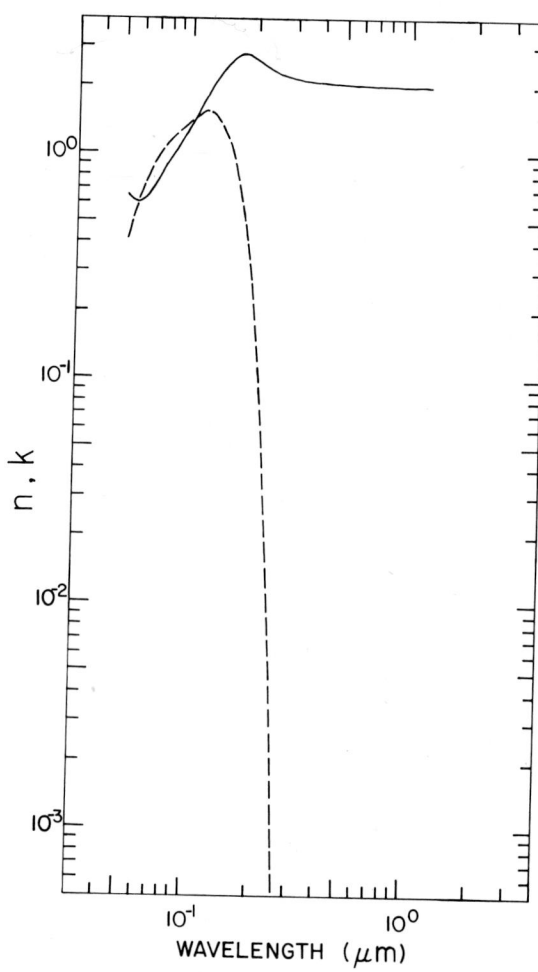

Fig. 12. Log–log plot of n (——) and k (----) versus wavelength in micrometers for silicon nitride (noncrystalline).

TABLE XII

**Values of *n* and *k* for Silicon Nitride (Noncrystalline)
from Various References[a]**

eV	cm^{-1}	μm	n	k
24		0.05166	0.655 [21]	0.420 [21]
23		0.05391	0.625	0.481
22		0.05636	0.611	0.560
21		0.05904	0.617	0.647
20		0.06199	0.635	0.743
19		0.06526	0.676	0.841
18		0.06888	0.735	0.936
17		0.07293	0.810	1.03
16		0.07749	0.902	1.11
15		0.08266	1.001	1.18
14		0.08856	1.111	1.26
13		0.09537	1.247	1.35
12	96,790	0.1033	1.417	1.43
11	88,720	0.1127	1.657	1.52
10.5	84,690	0.1181	1.827	1.53
10	80,650	0.1240	2.000	1.49
9.5	76,620	0.1305	2.162	1.44
9	72,590	0.1378	2.326	1.32
8.5	68,560	0.1459	2.492	1.16
8	64,520	0.1550	2.651	0.962
7.75	62,510	0.1600	2.711	0.866
7.5	60,490	0.1653	2.753	0.750
7.25	58,470	0.1710	2.766	0.612
7	56,460	0.1771	2.752	0.493
6.75	54,440	0.1837	2.724	0.380
6.5	52,430	0.1907	2.682	0.273
6.25	50,410	0.1984	2.620	0.174
6	48,390	0.2066	2.541	0.102
5.75	46,380	0.2156	2.464	5.7×10^{-2}
5.5	44,360	0.2254	2.393	2.9
5.25	42,340	0.2362	2.331	1.1×10^{-2}
5	40,330	0.2480	2.278	4.9×10^{-3}
4.75	38,310	0.2610	2.234	1.2×10^{-3}
4.5	36,290	0.2755	2.198	2.2×10^{-4}
4.25	34,280	0.2917	2.167	
4	32,260	0.3100	2.141	
3.5	28,230	0.3542	2.099	
3	24,200	0.4133	2.066	
2.5	20,160	0.4959	2.041	
2	16,130	0.6199	2.022	
1.5	12,100	0.8266	2.008	
1	8,065	1.240	1.998	

[a] Data obtained from Philipp [21].

Sodium Chloride (NaCl)

J. E. ELDRIDGE

Department of Physics
University of British Columbia
Vancouver, British Columbia, Canada

EDWARD D. PALIK

Naval Research Laboratory
Washington, D.C.

The room-temperature values of n and k tabulated here were obtained from the following works and references therein: The centimeter–millimeter-wave-region results, 0.33 cm^{-1} (30 mm) to 5 cm^{-1} (2 mm), are from Owens [1]; the 3.1-cm^{-1} (3200-μm) to 10-cm^{-1} (1000-μm) results are from Stolen and Dransfeld [2], Genzel et al. [3], and Dianov and Irisova [4]; the far-infrared reststrahlen results, 10 cm^{-1} (1000 μm) to 500 cm^{-1} (20 μm), are from Eldridge and Staal [5]; the multiphonon-tail data, 500 cm^{-1} (20 μm) to 900 cm^{-1} (11.1 μm), are from Harrington et al. [6]; values of k in the transparent region, 943 cm^{-1} (10.6 μm) to 3571 cm^{-1} (2.8 μm), are from Allen and Harrington [7]; while those in the visible, 19,430 cm^{-1} (0.5145 μm) to 28,489 cm^{-1} (0.3511 μm), are from Harrington et al. [8]; values of n from 500 cm^{-1} (20 μm) to 50,000 cm^{-1} (0.2 μm), also in the transparent region, are from Li [9], who consolidated various data [10–13]; the visible and ultraviolet values are from Miyata and Tomiki [14] and Roessler and Walker [15]; L$_{2,3}$ absorption-edge data for Cl^{-} are from Aita et al. [16]. A composite smooth-curve plot of log n and log k versus log wavelength is given in Fig. 13. Numerical values are given in Table XIII.

In the centimeter–millimeter-wave region a filled resonant-cavity method was used [1] to measure the real part of the dielectric function ε' and the loss tangent, tan $\delta = \varepsilon''/\varepsilon'$. From these, n and k could be extracted. Other results in the millimeter-wave region were obtained with transmission measurements made on slab samples of different thicknesses to determine n and k [2–4].

While not listed in the table, Breckenridge [17] gives 85°C data over a wide range of frequencies, 10^2–10^{10} Hz by measuring the capacitance and susceptance of blocks of NaCl. Generally, k is less than the last value given in the table, as was noted by Owens [1].

Above 10 cm^{-1} the far-infrared results were obtained mainly by the technique of dispersive-reflection spectroscopy [5]. In this technique a flat-polished (to within 0.3 μm) crystal of the material under investigation replaces the mirror in one of the arms of a Michelson interferometer. The temperature was 290 K. The resulting interferogram is asymmetric and contains information on both the reflectance amplitude r and phase change θ where $\hat{r} = re^{i\theta}$ for the crystal. When ratioed with a background spectrum, obtained when the usual mirror is in place, both r and θ, and consequently *all* of the real and imaginary optical properties, can be obtained simultaneously and directly.

This obviates the need for a Kramers–Kronig (KK) analysis of power reflectivity measurements, which has serious and well-known limitations [18]. For instance, the KK analysis will give poor values of n or k if either one is much smaller than the other. It works best when the values are comparable, since they both then have an effect of the power reflectivity. Of course, one also has to measure this power reflectivity over a large range for the integral to be at all valid. Furthermore, the power reflectivity becomes extremely small following a strong resonance such as the transverse-optic lattice mode in alkali halides. This occurs in the region of the longitudinal-optic frequency. All of the fine structure in the optical properties at this frequency is therefore usually lost. Another difficult region is the reststrahlen peak when the reflectivity can be very close to unity and an accurate value for k difficult to obtain. One has then to resort to measurements of transmission through thin films, with their concomitant dependence on interference effects and the quality and properties of the surfaces.

The dispersive-reflection technique is therefore obviously superior, but it is not without its difficulties. The main one of these is the physical interchange of reference mirror and sample in exactly the same position (either that or the exact relative position must be known). In Eldridge and Staal [5] this problem was overcome by aluminizing a portion of the large NaCl crystal, which would then act as the reference surface. (A correction must be made to the reflectivity and phase change of aluminum, depending on the film thickness.) In order to avoid the other main problem of ratioing spectra taken subsequent to one another, with the associated errors due to micrometer mirror-drive backlash, thermal drift, and component drift, a mask was placed in front of the aluminized sample and switched from background to sample at each step of the moving mirror. Dual interferograms were therefore simultaneously obtained. The eight-sector design of the mask and aluminized sample also relaxed the condition of flatness required of the polished sample, since both sample and reference were equally affected.

The results obtained cover the reststrahlen region between 100 cm^{-1} (100 μm) and 340 cm^{-1} (29.4 μm). Below this, and up to 500 cm^{-1} (20 μm), the results were supplemented with transmission measurements through thin, wedge-shaped, polished single crystals.

One can see in Fig. 13 that there is structure after the resonance, between 200 and 300 cm^{-1} (50 to 33 μm), and a distinct shoulder at 20 cm^{-1} (500 μm). In the past a multioscillator model has been used to fit reflection and transmission data [3]. However, there is only one oscillator, the TO mode, but it has an extremely frequency-dependent damping. The primary component of this damping is the coupling of the TO mode, through cubic anharmonicity, to two phonons. Energy and crystal momentum are conserved. This accounts for most of the structure in the damping. Three-phonon and higher-order processes also contribute strongly, depending on the temperature and the frequency, but they contain little structure, especially at room temperature. There is also some damping around the resonance due to the isotopic disorder of the chlorine ion, although this is extremely small at room temperature. All of the damping contributions mentioned can now be calculated with high accuracy, by use of the theories that have been developed, in conjunction with good lattice-dynamical data, which are obtained by inelastic neutron scattering (see Eldridge and Staal [5]). The agreement between experiment and theory is excellent. The similarities in the optical-phonon spectra of many alkali halides are discussed by Bilz [19].

Beyond the reststrahlen peak, in the "transparent" region (20–0.2 μm), the refractive index n has been determined by Li [9], who has fitted a multitude of experimental data [10–13] with a dispersion formula of the Sellmeier type, assuming $k = 0$. Temperature no doubt varied from 293 to 298 K for various laboratories. The formula for 293 K is

$$n^2 = 1.00055 + \frac{0.198\lambda^2}{\lambda^2 - (0.05)^2} + \frac{0.48398\lambda^2}{\lambda^2 - (0.1)^2} + \frac{0.38693\lambda^2}{\lambda^2 - (0.128)^2}$$

$$+ \frac{0.25998\lambda^2}{\lambda^2 - (0.158)^2} + \frac{0.08796\lambda^2}{\lambda^2 - (40.5)^2} + \frac{3.17064\lambda^2}{\lambda^2 - (60.98)^2} + \frac{0.30028\lambda^2}{\lambda^2 - (120.34)^2}. \qquad (1)$$

The 15 adjustable parameters were first chosen as input parameters for least-square fitting based on previous fits of more restricted data. The first 5 terms constitute the dispersion due to all interband oscillators above the band gap. Note that Li [9] used three far-IR classical oscillators to account for dispersion well above the TO phonon frequency (60.98 μm). This choice is based on earlier work of Genzel et al. [3] and Geick [18], who measured and collected n and k data, interpreting them as indicating oscillators at these frequencies.

Li's [9] fit agrees with the experiment generally to a few units in the fourth decimal place ($\pm 0.03\%$). We have therefore used this fit in the table for the region from 0.2 to 20 μm as representative of experimental data. Typical measurements of n were done with a prism utilizing the angle of minimum deviation [10–13, 20, 21] with four to five decimal places quoted.

Interestingly, even present-day determinations of n for laser-window materials use this technique and achieve an accuracy of several parts in 10^6, quoting five decimal places [22]. Since the temperature dependence of n is -3.27×10^{-5} K^{-1} at 2000 cm^{-1}, one sees that control of room temperature (293 K) is important, since a variation of 1 K from laboratory to laboratory leads to changes of three units in the fifth decimal place [9].

The values of k between 550 cm^{-1} (18.2 μm) and 900 cm^{-1} (11.1 μm) in Table XIII were taken from Harrington et al. [6] and are the average transmission results through single crystals, taken at two laboratories with different spectrometers. Corrections for the low-temperature emittance of the sample were also employed. Sample temperature was 300 K. These agree at the 500-cm^{-1} end with the infrared results of Eldridge and Staal [5] and at the high-energy end with the laser-calorimetry results of Allen and Harrington [7]. These authors obtained values at five distinct laser wavelengths between 10.6 and 2.8 μm. Here the values of k are so small that the calorimetry method is essential. Up to 1000 cm^{-1} or so (i.e., above 10 μm), the absorption is multiphonon damping of the TO resonance as previously mentioned, and agreement between experiment and theory is good. Beyond 1000 cm^{-1}, the absorption is greater than predicted by the theory, and one must assume that much of it is extrinsic and due to impurity levels in the gap [7, 8]. Certainly, as materials are purified, this absorption decreases. One also has a sizable contribution from surface absorption, which can, however, be dynamically separated from the bulk absorption in the laser-calorimetric technique. Surface polish is still very important. The bulk values are quoted in Table XIII.

In the UV region near the band gap, there are two detailed measurements [14, 15]. Thin transmission samples at 298 K have been used to determine n and k in the Urbach tail [14], while in the exciton region (0.158 μm) and above the band gap, near-normal-incidence reflectance measurements of cleaved samples have been made followed by KK analysis. In Miyaka and Tomiki [14] the reflectance at wavelengths shorter than 0.092 μm has been approximated by extrapolation to zero, while the contribution of the reststrahlen region at long wavelength has been added in performing the KK integrals. The experimentally determined values of k in the Urbach tail are fitted smoothly to the values of k above the band gap, since KK values determined below the gap are erratic. There is no reliable oscillator model for interband absorption to give the detailed values and structure shown in Fig. 13, although in Eq. (1) lossless oscillators are assumed above the fundamental band gap to account for the dispersion below the band gap.

The second set of data [15] was measured over a wider range from 0.048 to 0.247 μm (at a temperature of 300 K) and a method of KK analysis used that utilized a restricted data range (as justified in Roessler [23]). Far-infrared contributions were neglected. The two sets of data differ by as much as 20% at peaks and valleys in n and k. Perusal of the reflectance data indicates that the reflectance of Miyata and Tomiki [14] was higher at peaks than the Roessler and Walker [15] data, although not always at the valleys. Each work quotes experimental uncertainties in R to $\pm 3\%$ in the range $0.2479–0.1033 \ \mu m$ with Roessler and Walker [15] giving $\pm 10\%$ for $0.1033–0.0476 \ \mu m$. Such variations are enough to account for the differences in n and k. At low temperature adsorbed gases were a serious problem affecting R; this did not seem to be a problem at room temperature, although pressures of about 3×10^{-7} torr were reported in Miyata and Tomiki [14]. This suggests that monolayers of adsorbed molecules such as water and hydrocarbons form within minutes after cleaving. Each work is presumably an absolute reflectance measurement involving comparison of the direct intensity with the reflected intensity with no additional mirror reflection involved. The use of refractive-index data in the transparent region by Miyata and Tomiki [14] as a secondary standard seems to be reasonable.

In the interband spectral region from 0.04 to 0.16 μm, the structure in k is intrinsic due to a series of excitions associated with various interband transitions [15, 24]. The Urbach tail at wavelengths slightly longer than 0.16 μm is due to intrinsic rather than impurity excitons, as temperature dependence shows [14]. Extrapolation of this tail to longer wavelength yields values of k much smaller than observed experimentally [8], which suggests absorption due to color centers or metal-ion impurities.

Since we have a classic case of two reflectance measurements differing slightly with the subsequent differences in n and k upon KK analysis, we elect to display both sets of data for this one material.

Above 25 eV there are few data giving any sort of absolute values for n and k. The available data have been reviewed by Haelbich et al. [25].

Relative absorption in the 25–72 eV region has been measured by Le Comte et al. [26] for samples at 80 K.

The room-temperature, $L_{2,3}$ absorption edge of Cl^- in NaCl has been measured on an absolute scale by Aita et al. [16] in the 200–210-eV region at 300 K and is tabulated. The samples were evaporated polycrystalline films on transparent substrate films. The source was an x-ray tube emitting continuous radiation. Brown et al. [27], using synchrotron radiation, have made measurements in a somewhat wider spectral region but only in arbitrary units.

The absorption in arbitrary units has been measured in the region 200–280 eV by Iguchi et al. [28]. The K absorption edge of Na^+ near 1076 eV has been measured by Rule [29], and the K absorption edge of Cl^- near 2825 eV has been measured by Sugiura [30].

REFERENCES

1. J. C. Owens, *Phys. Rev.* **181**, 1228 (1969).
2. R. Stolen and K. Dransfeld, *Phys. Rev.* **139A**, 1295 (1965).
3. L. Genzel, H. Happ, and R. Weber, *Z. Physik* **154**, 13 (1959).

4. E. M. Dianov and N. A. Irisova, *J. Appl. Spectros.* **5**, 187 (1966); E. M. Dianov and N. A. Irisova, *Zhur. Prikl. Spektrosk.* **5**, 251 (1966).
5. J. E. Eldridge and P. R. Staal, *Phys. Rev. B* **16**, 4608 (1977).
6. J. A. Harrington, C. J. Duthler, F. W. Patten, and M. Hass, *Solid State Commun.* **18**, 1043 (1976).
7. S. Allen and J. A. Harrington, *Appl. Opt.* **17**, 1679 (1978).
8. J. A. Harrington, B. L. Bobbs, M. Braunstein, R. K. Kim, R. Stearns, and R. Braunstein, *Appl. Opt.* **17**, 1541 (1978).
9. H. H. Li, *J. Phys. Chem. Refer. Data* **5**, 329 (1976).
10. H. Rubens and E. F. Nichols, *Ann. Phys. Chem.* **60**, 418 (1897).
11. H. Rubens and A. Trowbridge, *Ann. Phys. Chem.* **60**, 724 (1897).
12. F. Paschen, *Ann. Physik* **26**, 120 (1908).
13. F. F. Martens, *Ann. Physik* **6**, 603 (1901); F. F. Martens, *Ann. Physik* **8**, 459 (1902).
14. T. Miyata and T. Tomiki, *J. Phys. Soc. Jpn.* **24**, 1286 (1968); T. Miyata and T. Tomiki, *J. Phys. Soc. Jpn.* **22**, 209 (1967); T. Miyata and T. Tomiki, private communication (1981).
15. D. M. Roessler and W. C. Walker, *J. Opt. Soc. Am.* **58**, 279 (1968); D. M. Roessler and W. C. Walker, *Phys. Rev.* **166**, 599 (1968).
16. O. Aita, I. Nagakura, and T. Sagawa, *J. Phys. Soc. Jpn.* **30**, 1414 (1971).
17. R. G. Breckenridge, *J. Chem. Phys.* **16**, 959 (1948).
18. R. Geick, *Z. Physik* **166**, 122 (1962).
19. H. Bilz, *in* "Correlation Functions and Quasiparticle Interactions in Condensed Matter" (J. W. Halley ed.), p. 531. Plenum, New York, 1978.
20. S. P. Langley, *Ann. Astrophy. Obs. Smithsonian Inst.* **1**, 219 (1902).
21. W. W. Coblentz, *J. Opt. Soc. Am.* **4**, 443 (1920).
22. A. Feldman, D. Horowitz, R. M. Waxler, and M. J. Dodge, NBA Technical Note 933, Optical Materials Characterization. U.S. Department of Commerce, National Bureau of Standards, Washington, D.C., February 1979.
23. D. M. Roessler, *Brit. J. Appl. Phys.* **16**, 1119 (1965); D. M. Roessler, *Brit. J. Appl. Phys.* **17**, 1313 (1966).
24. T. Tomiki, T. Miyata, and H. Tsukamoto, *Z. Naturforsch.* **29**, 145 (1974).
25. R. P. Haelbich, M. Iwan, and E. E. Koch, "Physik Daten-Physics Data", p. 57. Fach-information Zentrum, Karlsruhe, 1977.
26. A. Le Comte, A. Savary, M. Morlais, and S. Robin, *Opt. Commun.* **4**, 296 (1971).
27. F. C. Brown, C. Gähwiller, H. Fujita, A. B. Kunz, W. Scheifley, and N. Carrera, *Phys. Rev. B* **2**, 2126 (1970).
28. Y. Iguchi, T. Sagawa, S. Sato, M. Watanabe, H. Yamashita, A. Ejiri, M. Sasanuma, S. Nakai, M. Nakamura, S. Yamaguchi, Y. Nakai, and T. Oshio, *Solid State Commun.* **6**, 575 (1968).
29. K. C. Rule, *Phys. Rev.* **66**, 199 (1944).
30. C. Sugiura, *Phys. Rev. B* **6**, 170 (1972).

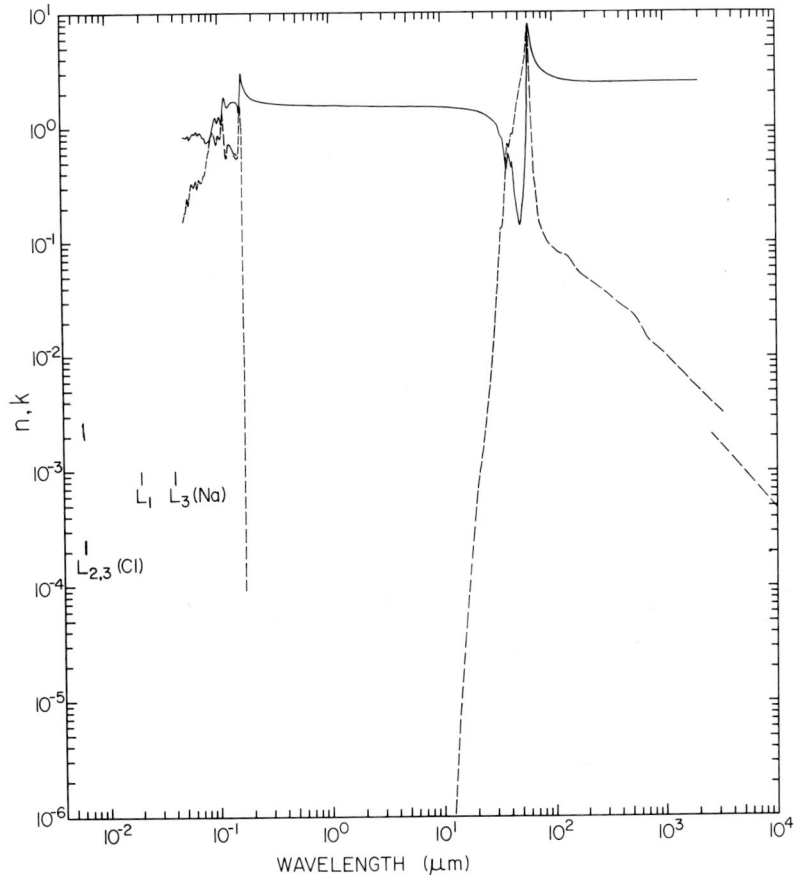

Fig. 13. Log–log plot of n (——) and k (- - - -) versus wavelength in micrometers for sodium chloride.

TABLE XIII

Values of n and k for Sodium Chloride from Various References[a]

eV	cm^{-1}	Å	n	k	n	k
209.5		59.18		2.54×10^{-3} [16]		
209		59.32		2.56		
208.5		59.46		2.58		
208		59.61		2.55		
207.5		59.75		2.56		
207		59.89		2.59		
206.5		60.04		2.64		
206		60.19		2.62		

(continued)

TABLE XIII (*Continued*)

Sodium Chloride

eV	cm⁻¹	Å	n	k	n	k
205.75		60.26		2.57		
205.5		60.33		2.45		
205.25		60.41		2.48		
205		60.48		2.53		
204.75		60.55		2.60		
204.5		60.63		2.68		
204.25		60.70		2.65		
204		60.78		2.64		
203.75		60.85		2.52		
203.5		60.92		2.38		
203.25		61.00		2.14		
203		61.07		2.08		
202.75		61.15		2.04		
202.5		61.23		2.06		
202.25		61.30		2.24		
202		61.38		2.22		
201.5		61.53		1.96		
201		61.68		1.91		
200.5		61.84		1.91		
200		61.99		1.92		

eV	cm⁻¹	μm	n	k	n	k
26.0		0.04769	0.83 [15]	0.15 [15]		
25.6		0.04842	0.82	0.17		
25.4		0.04880	0.83	0.17		
25.2		0.04919	0.83	0.18		
25.0		0.04959	0.83	0.18		
24.9		0.04979	0.82	0.18		
24.7		0.05019	0.82	0.18		
24.3		0.05101	0.81	0.21		
24.1		0.05144	0.81	0.22		
24.0		0.05166	0.82	0.23		
23.9		0.05187	0.82	0.23		
23.8		0.05209	0.83	0.24		
23.4		0.05298	0.85	0.23		
23.3		0.05320	0.85	0.22		
23.2		0.05343	0.85	0.21		
23.1		0.05366	0.83	0.21		
23.0		0.05391	0.82	0.21		
22.8		0.05437	0.80	0.23		
22.6		0.05485	0.79	0.26		
22.5		0.05510	0.80	0.27		
22.4		0.05534	0.80	0.28		
22.3		0.05559	0.81	0.29		
22.2		0.05584	0.82	0.30		
22.0		0.05636	0.83	0.31		

TABLE XIII (*Continued*)

Sodium Chloride

eV	cm⁻¹	μm	n	k	n	k
21.9		0.05661	0.84	0.32		
21.8		0.05687	0.85	0.32		
21.5		0.05766	0.87	0.32		
21.1		0.05875	0.88	0.31		
20.9		0.05931	0.88	0.30		
20.8		0.05960	0.88	0.30		
20.6		0.06018	0.88	0.30		
20.5		0.06048	0.87	0.31		
20.2		0.06137	0.87	0.33		
20.1		0.06168	0.88	0.33		
20.0		0.06199	0.88	0.34		
19.9		0.06230	0.89	0.34		
19.8		0.06261	0.91	0.34		
19.6		0.06325	0.92	0.32		
19.5		0.06358	0.92	0.31		
19.4		0.06390	0.91	0.30		
19.3		0.06423	0.90	0.29		
19.2		0.06457	0.88	0.29		
19.1		0.06490	0.87	0.30		
18.9		0.06559	0.86	0.33		
18.8		0.06594	0.86	0.34		
18.7		0.06629	0.87	0.35		
18.6		0.06666	0.88	0.35		
18.5		0.06702	0.89	0.35		
18.4		0.06737	0.90	0.35		
18.3		0.06774	0.90	0.34		
18.2		0.06811	0.90	0.33		
18.0		0.06888	0.89	0.33		
17.5		0.07085	0.85	0.33		
17.3		0.07166	0.84	0.34		
17.1		0.07250	0.82	0.36		
17.0		0.07293	0.82	0.37		
16.9		0.07335	0.82	0.38		
16.8		0.07379	0.82	0.38		
16.7		0.07423	0.82	0.38		
16.5		0.07514	0.80	0.38		
16.4		0.07559	0.79	0.39		
16.3		0.07601	0.77	0.40		
16.2		0.07652	0.74	0.43		
16.1		0.07700	0.74	0.45		
15.9		0.07797	0.74	0.49		
15.7		0.07896	0.74	0.53		
15.5		0.07999	0.75	0.57		
15.4		0.08050	0.76	0.59		
15.3		0.08103	0.76	0.60		
15.2		0.08156	0.77	0.62		

(*continued*)

TABLE XIII (*Continued*)

Sodium Chloride

eV	cm⁻¹	μm	n	k	n	k
14.9		0.08320	0.77	0.67		
14.8		0.08376	0.76	0.71		
14.7		0.08433	0.78	0.75		
14.65		0.08460	0.79	0.77		
14.6		0.08491	0.81	0.78		
14.55		0.08520	0.82	0.79		
14.5		0.08551	0.84	0.80		
14.45		0.08570	0.85	0.81		
14.4		0.08609	0.86	0.82		
14.3		0.08669	0.88	0.84		
14.25		0.08700	0.89	0.85		
14.2		0.08730	0.91	0.86		
14.1		0.08792	0.94	0.88		
14.0		0.08856	0.98	0.89		
13.9		0.08919	1.02	0.90		
13.8		0.08983	1.06	0.90		
13.7		0.09049	1.10	0.89		
13.5		0.09184	1.16	0.85		
13.40		0.09255			1.096 [14]	0.916 [14]
13.30		0.09322	1.20 [15]	0.81 [15]		
13.20		0.09390			1.197	0.868
13.1		0.09464	1.21	0.74		
13.05		0.09500			1.169	0.759
13.0		0.09537	1.19	0.71		
12.95		0.09574	1.17	0.70		
12.93		0.09587			1.104	0.765
12.90		0.09611	1.16	0.70		
12.88		0.09626			1.088	0.794
12.85		0.09648	1.14	0.71		
12.8		0.09686	1.12	0.71		
12.76		0.09712			1.083	0.834
12.7		0.09762	1.10	0.75		
12.65		0.09802			1.082	0.896
12.60		0.09840	1.10	0.81		
12.55		0.09879	1.14	0.83		
12.51		0.09907			1.119	0.967
12.45		0.09958	1.15	0.84		
12.40		0.09998	1.17	0.85		
12.36		0.1003			1.197	1.02
12.35		0.1004	1.19	0.85		
12.31		0.1007			1.230	1.02
12.30		0.1008	1.21	0.85		
12.20		0.1016	1.24	0.83	1.304	0.999
12.10		0.1025	1.24	0.79		
12.09		0.1025			1.320	0.940
12.0		0.1033	1.22	0.79		
11.99		0.1034			1.290	0.921

TABLE XIII (*Continued*)

Sodium Chloride

eV	cm^{-1}	μm	n	k	n	k
11.92		0.1040			1.272	0.931
11.90		0.1042	1.20	0.79		
11.84		0.1047			1.234	0.937
11.80		0.1051	1.16	0.80		
11.78		0.1052			1.209	0.987
11.71		0.1058			1.204	1.05
11.70		0.1060	1.12	0.86		
11.65		0.1064	1.10	0.89		
11.64		0.1065			1.219	1.14
11.60		0.1069	1.07	0.96		
11.58		0.1071			1.255	1.205
11.55		0.1073	1.09	1.06		
11.51		0.1077			1.338	1.28
11.50		0.1078	1.15	1.14		
11.46		0.1082			1.422	1.34
11.45		0.1083	1.24	1.19		
11.43		0.1085			1.491	1.36
11.40		0.1088	1.32	1.21		
11.38		0.1090			1.588	1.35
11.35		0.1092	1.40	1.23		
11.32		0.1095			1.682	1.34
11.30		0.1097	1.49	1.22		
11.25		0.1102	1.57	1.19	1.809	1.29
11.23		0.1104			1.864	1.27
11.20		0.1107	1.64	1.15	1.918	1.21
11.10		0.1117	1.75	1.04		
11.08		0.1119			2.002	1.02
11.03		0.1124			2.003	0.938
11.00		0.1127	1.82	0.93		
10.94		0.1133			1.990	0.834
10.90		0.1137	1.82	0.78		
10.82		0.1146			1.975	0.706
10.77		0.1151			1.947	0.650
10.68		0.1161			1.873	0.569
10.63		0.1166			1.821	0.543
10.60		0.1170	1.67	0.56		
10.55		0.1175			1.730	0.546
10.50		0.1181	1.58	0.53	1.707	0.563
10.45		0.1186	1.54	0.54		
10.42		0.1189			1.654	0.575
10.40		0.1192	1.51	0.56		
10.35		0.1198	1.49	0.59	1.637	0.623
10.29		0.1205			1.637	0.645
10.20		0.1216			1.640	0.673
10.15		0.1221	1.50	0.70		
10.13		0.1224			1.664	0.694

(*continued*)

TABLE XIII (*Continued*)

Sodium Chloride

eV	cm⁻¹	μm	n	k	n	k
10.10		0.1227	1.52	0.71		
10.08		0.1230			1.673	0.687
10.06		0.1232	1.54	0.71		
10.00		0.1240	1.55	0.71	1.678	0.697
9.908		0.1251			1.711	0.715
9.825		0.1262			1.742	0.700
9.80		0.1265	1.59	0.70		
9.757		0.1271			1.751	0.681
9.662		0.1283			1.764	0.670
9.60		0.1291	1.63	0.68		
9.586		0.1293			1.770	0.658
9.557		0.1297			1.779	0.655
9.539		0.1300			1.776	0.647
9.444		0.1313			1.787	0.623
9.40		0.1319	1.65	0.65		
9.356		0.1325			1.780	0.617
9.214		0.1345			1.792	0.597
9.20		0.1348	1.66	0.63		
9.093		0.1363			1.791	0.585
9.00		0.1378	1.67	0.60		
8.947		0.1386			1.804	0.564
8.90		0.1393	1.66	0.59		
8.894		0.1394			1.792	0.549
8.840		0.1402			1.789	0.552
8.80		0.1409	1.66	0.60		
8.775		0.1413			1.789	0.546
8.75		0.1417	1.66	0.59		
8.70		0.1425	1.66	0.59		
8.652		0.1433			1.776	0.529
8.60		0.1442	1.65	0.58	1.765	0.521
8.550		0.1450			1.749	0.534
8.50		0.1459	1.64	0.58		
8.490		0.1460			1.745	0.539
8.45		0.1467	1.64	0.58	1.746	0.539
8.409		0.1474			1.735	0.531
8.40		0.1476	1.63	0.58		
8.367		0.1482			1.719	0.531
8.35		0.1485	1.61	0.57		
8.325		0.1489			1.695	0.534
8.30		0.1494	1.59	0.58		
8.290		0.1496			1.682	0.542
8.267		0.1500			1.665	0.542
8.25		0.1503	1.55	0.58		
8.238		0.1505			1.638	0.550
8.20		0.1512	1.50	0.60	1.604	0.567
8.169		0.1517			1.547	0.598
8.15		0.1521	1.44	0.65		

TABLE XIII (*Continued*)

Sodium Chloride

eV	cm^{-1}	μm	n	k	n	k
8.127		0.1525			1.483	0.691
8.10		0.1531	1.38	0.74		
8.091		0.1532			1.461	0.784
8.08		0.1534	1.36	0.80	1.451	0.829
8.062		0.1538			1.446	0.878
8.05		0.1540	1.33	0.89		
8.040		0.1542			1.470	0.974
8.03		0.1544	1.34	0.99		
8.024		0.1545			1.490	1.01
8.00		0.1550	1.38	1.10		
7.98		0.1554	1.42	1.18	1.550	1.20
7.95		0.1560	1.51	1.29		
7.93		0.1563	1.60	1.36	1.727	1.40
7.918		0.1566			1.809	1.45
7.90		0.1569	1.74	1.43	1.929	1.51
7.885		0.1572			2.025	1.53
7.88		0.1573	1.84	1.46		
7.859		0.1577			2.206	1.52
7.85		0.1579	2.00	1.49		
7.83		0.1583	2.12	1.49		
7.802		0.1589			2.563	1.45
7.78		0.1594	2.42	1.41	2.695	1.39
7.767		0.1596			2.778	1.37
7.75		0.1600	2.59	1.33	2.911	1.31
7.738		0.1602			3.019	1.21
7.73		0.1604	2.72	1.26	3.078	1.12
7.710		0.1608			3.161	0.991
7.70		0.1610	2.91	1.05		
7.684		0.1613			3.252	0.737
7.65		0.1621	2.90	0.59	3.180	0.383
7.613		0.1628			3.019	0.151
7.60		0.1631	2.69		2.974	0.126
7.588		0.1634			2.902	8.67×10^{-2}
7.567		0.1638			2.827	2.76×10^{-2}
7.55		0.1642	2.55			
7.50		0.1653	2.46		2.636	
7.45		0.1664	2.40			
7.419		0.1671				5.67×10^{-4}
7.40		0.1675	2.34			
7.353		0.1686				1.11×10^{-7}
7.30		0.1698	2.25			
7.276		0.1704				1.37×10^{-5}
7.27		0.1704				1.37×10^{-5}
7.205		0.1721				2.02×10^{-6}
7.20		0.1722	2.17			2.02×10^{-6}
7.112		0.1743			2.081	4.43×10^{-7}

(*continued*)

TABLE XIII (*Continued*)

Sodium Chloride

eV	cm^{-1}	μm	n	k	n	k
7.00		0.1771	2.03			
6.996		0.1772				1.83
6.75		0.1837	1.91			
6.50		0.1907	1.84			
6.23		0.1990			1.792	
6.199	50,000	0.20	1.7899 [9]			
6.00	48,390	0.2066	1.75 [15]			
5.904	47,620	0.21	1.74676 [9]			
5.636	45,450	0.22	1.71499			
5.50	44,360	0.2254	1.69 [15]			
5.391	43,480	0.23	1.69056 [9]			
5.166	41,670	0.24	1.67118			
5.00	40,330	0.2480	1.65 [15]			
4.959	40,000	0.25	1.65537 [9]			
4.769	38,460	0.26	1.64227			
4.592	37,040	0.27	1.63124			
4.428	35,710	0.28	1.62184			
4.275	34,480	0.29	1.61374			
4.133	33,330	0.30	1.60669			
3.875	31,250	0.32	1.59506			
3.647	29,410	0.34	1.58590			
3.531	28,480	0.3511		6.4×10^{-10} [8]		
3.444	27,780	0.36	1.57852			
3.408	27,490	0.3638		4.3		
3.263	26,320	0.38	1.57248			
3.100	25,000	0.40	1.56746			
2.952	23,810	0.42	1.56324			
2.818	22,730	0.44	1.55965			
2.708	21,840	0.4579		2.9		
2.695	21,740	0.46	1.55657			
2.583	20,830	0.48	1.55390			
2.541	20,490	0.4880		1.1×10^{-10}		
2.480	20,000	0.50	1.55157			
2.410	19,440	0.5145		4.9×10^{-11}		
2.384	19,230	0.52	1.54953			
2.296	18,520	0.54	1.54773			
2.214	17,860	0.56	1.54613			
2.138	17,240	0.58	1.54471			
2.066	16,670	0.60	1..54343			
2.000	16,130	0.62	1.54228			
1.937	15,630	0.64	1.54124			
1.879	15,150	0.66	1.54030			
1.823	14,710	0.68	1.53944			
1.771	14,290	0.70	1.53865			
1.722	13,890	0.72	1.53794			
1.675	13,510	0.74	1.53728			
1.631	13,160	0.76	1.53667			

TABLE XIII (*Continued*)

Sodium Chloride

eV	cm^{-1}	μm	n	k	n	k
1.590	12,820	0.78	1.53611			
1.550	12,500	0.80	1.53560			
1.512	12,200	0.82	1.53512			
1.476	11,900	0.84	1.53467			
1.442	11,630	0.86	1.53426			
1.409	11,360	0.88	1.53387			
1.378	11,110	0.90	1.53351			
1.348	10,870	0.92	1.53317			
1.319	10,640	0.94	1.53285			
1.292	10,420	0.96	1.53255			
1.265	10,200	0.98	1.53227			
1.240	10,000	1.0	1.53200			
1.127	9,091	1.1	1.53088			
1.033	8,333	1.2	1.53000			
0.9537	7,692	1.3	1.52930			
0.8856	7,143	1.4	1.52873			
0.8266	6,667	1.5	1.52824			
0.7749	6,250	1.6	1.52782			
0.7293	5,882	1.7	1.52745			
0.6888	5,556	1.8	1.52712			
0.6526	5,263	1.9	1.52682			
0.6199	5,000	2.0	1.52654			
0.5904	4,762	2.1	1.52627			
0.5636	4,545	2.2	1.52602			
0.5391	4,348	2.3	1.52578			
0.5166	4,167	2.4	1.52554			
0.4959	4,000	2.5	1.52531			
0.4769	3,846	2.6	1.52509			
0.4592	3,704	2.7	1.52486			
0.4428	3,571	2.8	1.52463	$(1.3 \pm 0.1) \times 10^{-9}$ [7]		
0.4275	3,448	2.9	1.52441			
0.4133	3,333	3.0	1.52418			
0.4000	3,226	3.1	1.52395			
0.3875	3,125	3.2	1.52372			
0.3757	3,030	3.3	1.52349			
0.3647	2,941	3.4	1.52325			
0.3542	2,857	3.5	1.52301			
0.3444	2,778	3.6	1.52277			
0.3351	2,703	3.7	1.52252			
0.3263	2,632	3.8	1.52226	$(1.8 \pm 0.2) \times 10^{-9}$		
0.3179	2,564	3.9	1.52201			
0.3100	2,500	4.0	1.52174			
0.3024	2,439	4.1	1.52148			
0.2952	2,381	4.2	1.52121			
0.2883	2,326	4.3	1.52093			
0.2818	2,273	4.4	1.52065			

(*continued*)

TABLE XIII (*Continued*)

Sodium Chloride

eV	cm^{-1}	μm	n	k	n	k
0.2755	2,222	4.5	1.52036			
0.2695	2,174	4.6	1.52006			
0.2638	2,128	4.7	1.51977			
0.2583	2,083	4.8	1.51946			
0.2530	2,041	4.9	1.51915			
0.2480	2,000	5.0	1.51883			
0.2431	1,961	5.1	1.51851			
0.2384	1,923	5.2	1.51818			
0.2339	1,887	5.3	1.51785	$(1.7 \pm 0.5) \times 10^{-9}$		
0.2296	1,852	5.4	1.51751			
0.2254	1,818	5.5	1.51716			
0.2214	1,786	5.6	1.51681			
0.2175	1,754	5.7	1.51645			
0.2138	1,724	5.8	1.51609			
0.2101	1,695	5.9	1.51572			
0.2066	1,667	6.0	1.51534			
0.2000	1,613	6.2	1.51457			
0.1937	1,563	6.4	1.51377			
0.1879	1,515	6.6	1.51294			
0.1823	1,471	6.8	1.51209			
0.1771	1,429	7.0	1.51122			
0.1722	1,389	7.2	1.51031			
0.1675	1,351	7.4	1.50938			
0.1631	1,316	7.6	1.50842			
0.1590	1,282	7.8	1.50744			
0.1550	1,250	8.0	1.50643			
0.1512	1,220	8.2	1.50539			
0.1476	1,190	8.4	1.50432			
0.1442	1,163	8.6	1.50322			
0.1409	1,136	8.8	1.50210			
0.1378	1,111	9.0	1.50094			
0.1348	1,087	9.2	1.49996			
0.1337	1,079	9.27		$(2 \pm 1.5) \times 10^{-8}$		
0.1319	1,064	9.4	1.49855			
0.1292	1,042	9.6	1.49731			
0.1265	1,020	9.8	1.49604			
0.1240	1,000	10.0	1.49473			
0.1216	980.4	10.2	1.49340			
0.1192	961.5	10.4	1.49204			
0.1170	943.4	10.6	1.49065	$(8 \pm 0.8) \times 10^{-8}$		
0.1148	925.9	10.8	1.48922			
0.1127	909.1	11.0	1.48776			
0.1116	900.0	11.11		1.5×10^{-7} [6]		
0.1107	892.9	11.2	1.48628			
0.1088	877.2	11.4	1.48476			
0.1069	862.1	11.6	1.48320			
0.1054	850.0	11.76		4.5×10^{-7}		

TABLE XIII (*Continued*)

Sodium Chloride

eV	cm^{-1}	μm	n	k	n	k
0.1051	847.5	11.8	1.48162			
0.1033	833.3	12.0	1.48000			
0.1016	819.7	12.2	1.47834			
0.09999	806.5	12.4	1.47665			
0.09919	800.0	12.50		1.4×10^{-6}		
0.09840	793.7	12.6	1.47493			
0.09686	781.3	12.8	1.47318			
0.09537	769.2	13.0	1.47138			
0.09393	757.6	13.2	1.46956			
0.09299	750.0	13.33		3.8×10^{-6}		
0.09252	746.3	13.4	1.46769			
0.09116	735.3	13.6	1.46579			
0.08984	724.6	13.8	1.46385			
0.08856	714.3	14.0	1.46188			
0.08731	704.2	14.2	1.45986			
0.08679	700.0	14.29		1.1×10^{-5}		
0.08670	699.3	14.3				
0.08610	694.4	14.4	1.45781			
0.08492	684.9	14.6	1.45572			
0.08377	675.7	14.8	1.45359			
0.08266	666.7	15.0	1.4514			
0.08157	657.9	15.2	1.4492			
0.08059	650.0	15.38		3.6×10^{-5}		
0.08051	649.4	15.4	1.4469			
0.07948	641.0	15.6	1.4446			
0.07847	632.9	15.8	1.4423			
0.07749	625.0	16.0	1.4399			
0.07653	617.3	16.2	1.4375			
0.07560	609.8	16.4	1.4351			
0.07469	602.4	16.6	1.4325			
0.07439	600.0	16.67		8.6×10^{-5}		
0.07380	595.2	16.8	1.4300			
0.07293	588.2	17.0	1.4274			
0.07208	581.4	17.2	1.4247			
0.07125	574.7	17.4	1.4220			
0.07044	568.2	17.6	1.4193			
0.06965	561.8	17.8	1.41649			
0.06888	555.5	18.0	1.41364			
0.06819	550.0	18.18		2.2×10^{-4}		
0.06812	549.4	18.2	1.41074			
0.06738	543.5	18.4	1.40779			
0.06666	537.6	18.6	1.40478			
0.06595	531.9	18.8	1.40173			
0.06526	526.3	19.0	1.39861			
0.06457	520.8	19.2	1.39544			
0.06391	515.5	19.4	1.39222			

(*continued*)

TABLE XIII (*Continued*)

Sodium Chloride

eV	cm^{-1}	μm	n	k	n	k
0.06326	510.2	19.6	1.38893			
0.06262	505.0	19.8	1.38559			
0.06199	500.0	20.0	1.3822			
			1.41 [5]	6.2×10^{-4} [5]		
0.05951	480	20.83	1.38	8.7×10^{-4}		
0.05703	460	21.74	1.35	1.1×10^{-3}		
0.05455	440	22.73	1.33	1.5		
0.05207	420	23.81	1.32	2.3		
0.04959	400	25.00	1.27	3.5		
0.04711	380	26.32	1.23	5.3		
0.04463	360	27.78	1.17	9.8		
0.04215	340	29.41	1.12	1.7×10^{-2}		
0.03967	320	31.25	1.01	4.5×10^{-2}		
0.03720	300	33.33	0.85	0.11		
0.03658	295	33.90	0.83	0.13		
0.03596	290	34.48	0.80	0.13		
0.03534	285	35.09	0.78	0.14		
0.03472	280	35.71	0.69	0.17		
0.03410	275	36.36	0.59	0.22		
0.03348	270	37.04	0.49	0.31		
0.03286	265	37.74	0.42	0.50		
0.03224	260	38.46	0.45	0.66		
0.03162	255	39.22	0.57	0.71		
0.03124	252	39.68	0.59	0.67		
0.03100	250	40.00	0.58	0.66		
0.03038	245	40.82	0.51	0.69		
0.02976	240	41.67	0.46	0.81		
0.02938	237	42.19	0.44	0.87		
0.02914	235	42.55	0.46	0.87		
0.02889	233	42.92	0.49	0.87		
0.02852	230	43.48	0.43	0.87		
0.02790	225	44.44	0.30	1.00		
0.02728	220	45.45	0.25	1.15		
0.02666	215	46.51	0.20	1.36		
0.02604	210	47.62	0.18	1.56		
0.02542	205	48.78	0.16	1.74		
0.02480	200	50.00	0.14	1.99		
0.02418	195	51.28	0.15	2.29		
0.02356	190	52.63	0.19	2.63		
0.02294	185	54.05	0.24	3.02		
0.02232	180	55.56	0.35	3.72		
0.02170	175	57.14	0.59	4.57		
0.02108	170	58.82	1.35	6.03		
0.02046	165	60.61	6.17	7.41		
0.02021	163	61.35	7.76	6.03		
0.01984	160	62.50	6.92	2.14		
0.01922	155	64.52	5.50	0.87		

TABLE XIII *(Continued)*

Sodium Chloride

eV	cm^{-1}	μm	n	k	n	k
0.01860	150	66.67	4.52	0.380		
0.01798	145	68.97	4.07	0.309		
0.01736	140	71.43	3.72	0.219		
0.01674	135	74.07	3.47	0.145		
0.01612	130	76.92	3.31	0.135		
0.01488	120	83.33	3.02	0.110		
0.01364	110	90.91	2.85	0.095		
0.01240	100	100.0	2.74	0.087		
0.01116	90	111.1	2.63	0.078		
0.009919	80	125.0	2.57	0.077		
0.008679	70	142.9	2.52	0.068		
0.007439	60	166.7	2.48	0.055		
0.006199	50	200.0	2.45	0.047		
0.005579	45	222.2	2.44	0.045		
0.004959	40	250.0	2.44	0.041		
0.004339	35	285.7	2.43	0.036		
0.004132	33.3	300		0.030 [2]		
0.003720	30	333.3	2.43	0.032 [5]		
0.003100	25	400.0	2.43	0.027		
0.002480	20	500.0	2.43	0.024		
				0.024 [2]		
0.001860	15	666.7	2.43	0.014 [5]		
0.001240	10	1000	2.43	0.006 [5]		
				0.01 [3]		
0.001215	9.8	1020		8.1 × 10^{-3} [2]		
0.001127	9.091	1100		9.5 [3]		
0.001033	8.333	1200		8.8		
0.0008856	7.143	1400		6.5		
0.0007872	6.349	1575		5.76		
0.0006888	5.556	1800		5.4		
0.0006199	5.000	2000	2.43 [4]			
0.0005904	4.762	2100		4.3		
0.0004959	4.000	2500		4.4		
0.0004797	3.869	2584	2.43 [1]	2.1 [1]		
0.0003875	3.125	3200		3.3 × 10^{-3} [3]		
0.0001464	1.181	8469	2.43	5.8 × 10^{-4} [1]		
0.00004053	0.3269	30590	2.43	2.5		

[a] Tabulation of n and k as a function of photon energy (electron volts), wave number (inverse centimeters), and wavelength (micrometers). References for the data are given in brackets and apply to all data below the brackets until a new reference appears; k is often very small and exponents often are given for numbers smaller than 10^{-2}. The power of 10 applies for all data below the exponent until a new exponent appears or the exponent becomes 10^{-1}, whereupon the number is given in decimal form.

Titanium Dioxide
(TiO$_2$) (Rutile)

M. W. RIBARSKY
School of Physics
Georgia Institute of Technology
Atlanta, Georgia

Titanium dioxide exists in three crystal structures: rutile, brookite, and anatase. The work reported here is all on rutile (tetragonal) single crystals. Rutile is easily reduced, and the crystal will have some oxygen vacancies or titanium interstitials. Since the structure cannot be treated as a slightly perturbed cubic crystal, there are pronounced polarization effects. It is therefore necessary to measure the optical constants for $E \parallel c$ axis and $E \perp c$ axis.

Compiled in Table XIV and plotted in Fig. 14 are room-temperature values of n and k for wavelengths between 36 and 0.11 μm.

There are evidently very few reports of optical constants for single-crystal TiO$_2$ in the ultraviolet above 20 eV. In the far infrared, Smith and Loewenstein [1] measured n_o ($\perp$) at 1.5 K from channeled spectra in the range 400–100 μm. Meriakri and Ushatkin [2] measured n at 1030 μm at an unspecified room temperature. The technique involved the use of a millimeter-wave source whose frequency could be varied periodically over a short range and a Michelson interferometer in which the path length could be varied. A beating fringe pattern was obtained from which an average n could be obtained. Parker [3] measured the susceptability of oriented rutile in a capacitor at low frequency and determined dc values of $n_{\parallel}^2 = 170$ and $n_{\perp}^2 = 86$ at 300 K.

The infrared results of 36.4 to 11.1 μm are from Gervais and Piriou [4]; the results in the near infrared and visible 1.5 to 0.425 μm are from Devore [5], and overlapping sets of results for the visible to ultraviolet regions are from Cardona and Harbeke [6] (1.2–0.11 μm) and from Vos and Krusemeyer [7] (0.413–0.207 μm). Cardona and Harbeke [6] have also taken the reflectivity from 0.11 to 0.056 μm with unpolarized light.

All the results were obtained from power reflectivity measurements. In the case of Cardona and Harbeke [6] and Vos and Krusemeyer [7] a Kramers–

HANDBOOK OF OPTICAL CONSTANTS OF SOLIDS

Kronig (KK) analysis was used to obtain the optical constants. The uncertainty due to truncation of the reflectivity to a finite frequency range was somewhat relieved by extrapolation to previous measurements at frequencies below the measured region. Above the measured region, a standard power-law fit to the reflectivity [7] was used, or a fit of the expansion coefficients in an approximate expression for the phase of the complex reflectivity was made [6]. Neither procedure took account of the higher interband transitions in any direct way. Such uncertainties introduce errors in the magnitudes of the optical constants and in the positions of peaks and other features [8]. These errors are more pronounced near the boundaries of the spectrum.

Most of the spectra were derived from reflectivity measurements. All the experiments used single crystals with the orientation of the c axis determined by x-ray diffraction. The infrared reflectivity spectra were taken with a single-beam monochromator on optically polished rutile crystals [4]. Errors were of the order $\Delta R/R = 2\%$ and $\Delta E = 2$–5 cm^{-1}. The very near-infrared refractive-index results [5] were taken directly from the measured amount of refraction of light from a mercury arc by a single-crystal rutile prism. Cardona and Harbeke [6] measured the reflectivity spectra below 6 eV with a grating spectrometer and used a vacuum-UV spectrometer at higher energies. Vos and Krusemeyer [7] took the absolute specular reflectance with a relative uncertainty of less than 0.5%. They cut their samples from flame-fusion-grown boules of rutile that were polished and etched in NaOH.

The optical constants in the infrared were tabulated by using the following four-parameter semiquantum model [4] for the dielectric function

$$\tilde{\varepsilon} = \varepsilon_1 \pm i\varepsilon_2 = \varepsilon_\infty \prod_j \frac{\Omega_{j\text{LO}}^2 - \omega^2 \pm i\gamma_{j\text{LO}}\omega}{\Omega_{j\text{TO}}^2 - \omega^2 \mp i\gamma_{j\text{TO}}\omega}. \tag{1}$$

Each TO (LO) phonon mode is assumed to be an oscillator with frequency Ω_{TO} (Ω_{LO}) and damping γ_{TO} (γ_{LO}). In general, the parameters are frequency and time dependent, but tests of Eq. (1) for realistic cases have shown that the parameters vary slowly with frequency even for wide reflectivity bands [9]. The temperature dependences of γ_{TO} and γ_{LO} for $E \| c$ indicate large anharmonicities due perhaps to sixth-order coupling at higher temperatures. The model dielectric function was used in the expression for the power reflectivity, and the resulting fit to the reflectivity data was within the experimental error. The results tabulated here use this model with the parameters given in Tables I and II of Gervais and Piriou [4].

The data show that there are only one TO mode and one LO mode in the A_{2u} ($E \| c$) spectrum but that there is also a secondary mode at 592 cm^{-1}. This could be a two-phonon peak since high phonon densities exist at energies about half the secondary oscillator energy [10]. However, it could also be a forbidden mode, such as the A_{1g} mode, activated by substitutional impurities

[4]. The E_u ($E \perp c$) spectrum has three TO and LO modes and a secondary peak near the same energy as the A_{2u} peak.

There are apparently no recent measurements of the optical constants in the range 1.5–11.1 μm. However, in 1921 Liebisch and Rubens [11] measured the index of refraction of a rutile prism in this range. For the region in which their measurements overlap those of Devore [5] (0.7–1.5 μm), Liebisch and Rubens's [11] results are in agreement with Devore's [5] for $E \perp c$. However, for $E \| c$ the refractive-index curve of Liebisch and Rubens [11] has the same shape as Devore's [5] but is offset by approximately -0.1. A similar offset for $E \| c$ exists between the curves of Devore [5] and Cronemeyer [12] in the range 0.4–0.6 μm. However, Devore's [5] values are quite close to the original measurements of Baerwald [13]. There are no measurements of the extinction coefficient in the range 0.45–1.5 μm, but it is presumably quite small and structureless.

Devore has fit his refractive-index results with the equations

$$n_\perp^2 = 5.913 + 2.441 \times 10^7/(\lambda^2 - 0.803 \times 10^7),$$
$$n_\|^2 = 7.197 + 3.322 \times 10^7/(\lambda^2 - 0.843 \times 10^7),$$

(2)

where λ is in angstroms. These equations give results within 0.001–0.003 of the experimental results, and they are used here to obtain the tabulated results from 0.425 to 1.5 μm. At 0.405 μm Cronemeyer [12] gives $k = 0.086$.

The data of Cardona and Harbeke [6] and of Vos and Krusemeyer [7] are significantly different in the region of overlap (0.21–0.41 μm). Vos and Krusemeyer [7] ascertain that bulk pretreatment of their samples did not affect the reflectivity and that there were no errors in their measuring procedure. Gupta and Ravindra [14] point out that the values of ε_∞ obtained by integrating ε_2 over the spectral range are in better agreement with experiment for the results of Cardona and Harbeke [6] than for those of Vos and Krusemeyer [7]. However, the Vos and Krusemeyer [7] results are over a more limited energy range and the truncated KK analysis can introduce shifts in the optical-constant spectra [8] that would affect the evaluation of ε_∞. We have therefore elected to present both the results of Cardona and Harbeke [6] and of Vos and Krusemeyer [7].

There is a rapid rise in all the spectra above the fundamental edge at 3.05 eV (0.405 μm). Comparison with a semiempirical band structure [15] indicates that the rise is due to nearly parallel conduction and valence bands. The calculations also show that the structures in ε_2 for both directions of polarization are due to transitions to two groupings of conduction bands, one from 3 to 6 eV and the other from 6 to 9 eV above the valence bands. The calculated ε_2 for $E \perp c$ is not as close to experiment as for $E \| c$ [15]. The shift to lower energy of the main peak in the experimental spectrum may indicate the presence of excitonic effects.

REFERENCES

1. D. R. Smith and E. V. Loewenstein, *Appl. Opt.* **15**, 859 (1976).
2. V. V. Meriakri and E. F. Ushatkin, *Prib. Tekh. Eksp.* **16**, 143 (1973); V. V. Meriakri and E. F. Ushatkin, *Instrum. Exp. Tech.* **16**, 498 (1973).
3. R. A. Parker, *Phys. Rev.* **124**, 1719 (1961).
4. F. Gervais and B. Piriou, *Phys. Rev. B* **10**, 1642 (1974).
5. J. R. Devore, *J. Opt. Soc. Am.* **41**, 416 (1951).
6. M. Cardona and G. Harbeke, *Phys. Rev.* **137A**, 1467 (1965).
7. K. Vos and H. J. Krusemeyer, *J. Phys. C* **10**, 3893 (1977).
8. W. E. Wall, M. W. Ribarsky, and J. R. Stevenson, *J. Appl. Phys.* **51**, 661 (1980).
9. F. Gervais and B. Piriou, *J. Phys. C* **7**, 2374 (1973).
10. G. A. Samara and P. S. Peercy, *Phys. Rev. B* **7**, 1131 (1973).
11. T. Liebisch and H. Rubens, *Sitzungsber. Preuss. Akad. Wiss. Phys.-Math. Kl.* 211 (1921).
12. D. C. Cronemeyer, *Phys. Rev.* **87**, 876 (1952).
13. C. Baerwald, *Z. Krist.* **7**, 167 (1883).
14. V. P. Gupta and N. M. Ravindra, *Phys. Chem. Solids* **41**, 591 (1980).
15. K. Vos, *J. Phys. C* **10**, 3917 (1977).

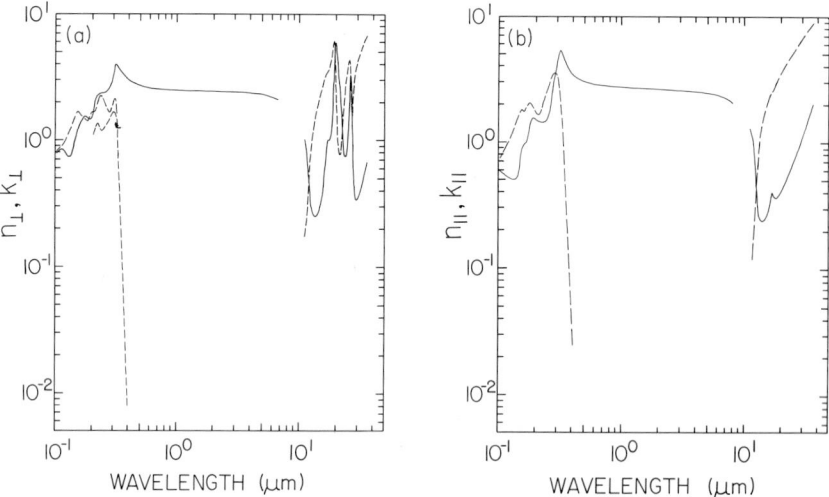

Fig. 14. (a) Log–log plot of $n_\perp$ (——), $k_\perp$ (– – – –) versus wavelength in micrometers for titanium dioxide. (b) Log–log plot of $n_{\parallel}$ (——), $k_{\parallel}$ (– – – –) versus wavelength in micrometers for titanium dioxide. The $\perp$ and $\parallel$ symbols refer to radiation electric field E perpendicular and parallel to the c axis of the crystal. The values of n at 1030 μm given in Meriakri and Ushatkin [2] are not shown.

TABLE XIV

Values of n and k for Titanium Dioxide Obtained from Various References[a]

eV	cm^{-1}	μm	$n(\|)$	$n(\perp)$	$k(\|)$	$k(\perp)$
12.04		0.103	0.590 [6]	0.790 [6]	0.760 [6]	0.820 [6]
11.07		0.112	0.510	0.854	0.908	0.893
10.25		0.121	0.560	0.816	1.05	0.947
9.116		0.136	0.490	0.740	1.42	1.31
8.377		0.148	0.670	1.07	1.74	1.67
8.051		0.154			1.81	
7.798		0.159	0.950	1.32	1.77	1.57
7.293		0.170	1.04	1.51	1.94	1.50
7.167		0.173			2.01	
6.965		0.178	1.31	1.54	2.03	1.38
6.391		0.194	1.55	1.46	1.77	1.65
6.048	48,780	0.205		1.98 [7]		1.10 [7]
5.904	47,620	0.210	1.46	1.46 [6]	1.65	1.68 [6]
				2.02 [7]		1.20 [7]
5.713	46,080	0.217	1.82 [7]		0.971 [7]	
5.610	45,250	0.221		2.26		1.35
5.510	44,440	0.225				2.21 [6]
5.438	43,860	0.228	1.81		1.17	
5.344	43,100	0.232	1.45 [6]	2.31 [6]	2.15 [6]	2.29
5.188	41,840	0.239		2.36 [7]		1.19 [7]
4.940	39,840	0.251	2.01 [7]		1.71 [7]	
4.862	39,220	0.255		2.37		1.29
4.661	37,590	0.266	1.97 [6]	2.44 [6]	3.36 [6]	1.67 [6]
4.592	37,040	0.270	2.53 [7]		2.09 [7]	
4.558	36,760	0.272	2.59 [6]	2.49 [6]	3.47 [6]	1.65 [6]
4.541	36,630	0.273	2.92	2.50	3.49	1.65
4.492	36,230	0.276		2.59 [7]		1.41 [7]
4.350	35,090	0.285	3.03 [7]		2.21 [7]	
					3.56 [6]	
4.305	34,720	0.288	3.56 [6]	2.83 [6]	3.55	1.82 [6]
4.217	34,010	0.294	4.23	3.02	3.48	1.97
4.174	33,670	0.297		3.00 [7]		1.68 [7]
4.105	33,110	0.302				2.13 [6]
4.052	32,680	0.306	3.84 [7]		1.95 [7]	
4.038	32,570	0.307		3.42		1.61 [7]
3.899	31,450	0.318	5.38 [6]	4.00 [6]	2.18 [6]	1.79 [5]
3.827	30,870	0.324			1.44	1.44
3.791	30,580	0.327		3.87 [7]		0.810 [7]
3.701	29,850	0.335	4.22 [7]		0.788 [7]	
3.604	29,070	0.344	4.36 [6]	4.30 [6]		
3.463	27,930	0.358		3.38 [7]		0.117 [7]
3.444	27,780	0.360	3.87 [7]		0.251	
3.195	25,770	0.388	3.49	2.88		
3.100	25,000	0.400	3.40 [5]	3.00 [5]		
3.084	24,870	0.402		3.00 [7]		0.008

(*continued*)

TABLE XIV (Continued)

Titanium Dioxide

eV	cm^{-1}	μm	$n(\|)$	$n(\perp)$	$k(\|)$	$k(\perp)$
3.009	24,270	0.412	3.24 [7]		0.022	
2.952	23,810	0.420	3.29 [5]	2.91 [5]		
2.818	22,730	0.440	3.20	2.84		
2.817	22,670	0.441			0.080 [6]	0.040 [6]
2.695	21,740	0.460	3.13	2.79		
2.583	20,830	0.480	3.08	2.75		
2.480	20,000	0.500	3.03	2.71		
2.384	19,230	0.520	3.00	2.68		
2.300	18,550	0.539	2.95 [6]			
2.296	18,520	0.540	2.97 [5]	2.66 [5]		
2.214	17,860	0.560	2.94	2.64		
2.138	17,240	0.580	2.92	2.62		
2.066	16,670	0.600	2.90	2.60		
2.000	16,130	0.620	2.88	2.59		
1.937	15,630	0.640	2.87	2.58		
1.879	15,150	0.660	2.85	2.57		
1.823	14,710	0.680	2.84	2.56		
1.771	14,290	0.700	2.83	2.55		
1.722	13,890	0.720	2.82	2.54		
1.675	13,510	0.740	2.81	2.54		
1.631	13,160	0.760	2.81	2.53		
1.590	12,820	0.780	2.80	2.52		
1.550	12,500	0.800	2.79	2.52		
1.512	12,200	0.820	2.79	2.52		
1.476	11,900	0.840	2.78	2.51		
1.442	11,630	0.860	2.78	2.51		
1.409	11,360	0.880	2.77	2.50		
1.378	11,110	0.900	2.77	2.50		
1.348	10,870	0.920	2.76	2.50		
1.319	10,640	0.940	2.76	2.49		
1.291	10,420	0.960	2.76	2.49		
1.265	10,200	0.980	2.75	2.49		
1.240	10,000	1.00	2.75	2.49		
1.216	9,804	1.02	2.75	2.48		
1.192	9,615	1.04	2.74	2.48		
1.170	9,434	1.06	2.74	2.48		
1.148	9,259	1.08	2.74	2.48		
1.127	9,091	1.10	2.74	2.48		
1.107	8,929	1.12	2.74	2.47		
1.088	8,772	1.14	2.73	2.47		
1.069	8,621	1.16	2.73	2.47		
1.051	8,475	1.18	2.73	2.47		
1.033	8,333	1.20	2.73	2.47		
1.016	8,197	1.22	2.73	2.47		
0.9999	8,065	1.24	2.72	2.47		
0.9840	7,937	1.26	2.72	2.46		

TABLE XIV (*Continued*)

Titanium Dioxide

eV	cm^{-1}	μm	$n(\parallel)$	$n(\perp)$	$k(\parallel)$	$k(\perp)$
0.9686	7,813	1.28	2.72	2.46		
0.9537	7,692	1.30	2.72	2.46		
0.9393	7,576	1.32	2.72	2.46		
0.9252	7,463	1.34	2.72	2.46		
0.9116	7,353	1.36	2.72	2.46		
0.8984	7,246	1.38	2.72	2.46		
0.8856	7,143	1.40	2.72	2.46		
0.8731	7,042	1.42	2.71	2.46		
0.8610	6,944	1.44	2.71	2.46		
0.8492	6,849	1.46	2.71	2.46		
0.8377	6,757	1.48	2.71	2.46		
0.8265	6,667	1.50	2.71	2.45		
0.5123	4,132	2.42	2.59 [12]	2.40 [12]		
0.3668	2,959	3.38	2.58	2.41		
0.3271	2,639	3.79	2.57	2.39		
0.2897	2,336	4.28	2.51	2.34		
0.2535	2,045	4.89	2.49	2.32		
0.2164	1,745	5.73	2.43	2.24		
0.1839	1,484	6.74		2.11		
0.1729	1,395	7.17	2.27			
0.1542	1,244	8.04	2.06			
0.1107	892.9	11.2	1.30 [4]	1.01 [4]	0.075 [4]	0.176 [4]
0.1088	877.2	11.4	1.21	0.905	0.091	0.209
0.1069	862.1	11.6	1.11	0.784	0.111	0.256
0.1051	847.5	11.8	1.01	0.649	0.136	0.328
0.1033	833.3	12.0	0.882	0.513	0.171	0.440
0.1016	819.7	12.2	0.743	0.406	0.224	0.591
0.09999	806.5	12.4	0.587	0.340	0.310	0.751
0.09840	793.7	12.6	0.441	0.301	0.450	0.902
0.09686	781.2	12.8	0.345	0.278	0.625	1.04
0.09537	769.2	13.0	0.296	0.264	0.791	1.17
0.09393	757.6	13.2	0.269	0.255	0.940	1.30
0.09252	746.3	13.4	0.254	0.250	1.07	1.42
0.09116	735.3	13.6	0.245	0.249	1.20	1.53
0.08984	724.6	13.8	0.241	0.250	1.31	1.65
0.08856	714.3	14.0	0.239	0.253	1.42	1.76
0.08731	704.2	14.2	0.239	0.258	1.52	1.87
0.08610	694.4	14.4	0.240	0.265	1.62	1.98
0.08492	684.9	14.6	0.243	0.275	1.72	2.09
0.08377	675.7	14.8	0.247	0.287	1.81	2.21
0.08265	666.7	15.0	0.253	0.303	1.90	2.32
0.08157	657.9	15.2	0.259	0.322	1.98	2.44
0.08051	649.4	15.4	0.268	0.347	2.07	2.56
0.07948	641.0	15.6	0.278	0.378	2.15	2.69
0.07847	632.9	15.8	0.290	0.418	2.24	2.81

(*continued*)

TABLE XIV (*Continued*)

Titanium Dioxide

eV	cm^{-1}	μm	$n(\parallel)$	$n(\perp)$	$k(\parallel)$	$k(\perp)$
0.07749	625.0	16.0	0.306	0.468	2.32	2.94
0.07653	617.3	16.2	0.326	0.533	2.39	3.07
0.07560	609.8	16.4	0.351	0.614	2.46	3.19
0.07469	602.4	16.6	0.377	0.710	2.52	3.30
0.07380	595.2	16.8	0.398	0.814	2.57	3.39
0.07293	588.2	17.0	0.406	0.908	2.61	3.45
0.07208	581.4	17.2	0.401	0.972	2.66	3.49
0.07125	574.7	17.4	0.389	1.00	2.71	3.54
0.07044	568.2	17.6	0.378	1.00	2.77	3.62
0.06965	561.8	17.8	0.370	1.00	2.84	3.76
0.06888	555.6	18.0	0.366	1.02	2.91	3.94
0.06812	549.5	18.2	0.364	1.06	2.98	4.16
0.06738	543.5	18.4	0.365	1.15	3.05	4.43
0.06666	537.6	18.6	0.368	1.31	3.12	4.74
0.06595	531.9	18.8	0.372	1.56	3.19	5.10
0.06525	526.3	19.0	0.377	1.96	3.25	5.51
0.06457	520.8	19.2	0.382	2.62	3.32	5.91
0.06391	515.5	19.4	0.389	3.64	3.39	6.12
0.06326	510.2	19.6	0.395	4.96	3.45	5.72
0.06262	505.1	19.8	0.403	5.88	3.52	4.51
0.06199	500.0	20.0	0.410	5.95	3.59	3.17
0.06138	495.0	20.2	0.418	5.53	3.65	2.21
0.06078	490.2	20.4	0.427	5.01	3.72	1.63
0.06019	485.4	20.6	0.435	4.51	3.78	1.27
0.05961	480.8	20.8	0.444	4.06	3.84	1.05
0.05904	476.2	21.0	0.453	3.66	3.91	0.909
0.05848	471.7	21.2	0.463	3.29	3.97	0.826
0.05794	467.3	21.4	0.472	2.96	4.04	0.783
0.05740	463.0	21.6	0.482	2.64	4.10	0.772
0.05687	458.7	21.8	0.492	2.33	4.16	0.792
0.05636	454.5	22.0	0.502	2.03	4.23	0.846
0.05585	450.5	22.2	0.513	1.73	4.29	0.941
0.05535	446.4	22.4	0.524	1.45	4.35	1.09
0.05486	442.5	22.6	0.535	1.21	4.42	1.28
0.05438	438.6	22.8	0.546	1.03	4.48	1.51
0.05391	434.8	23.0	0.557	0.904	4.54	1.75
0.05344	431.0	23.2	0.569	0.822	4.60	1.98
0.05298	427.4	23.4	0.581	0.772	4.67	2.19
0.05253	423.7	23.6	0.593	0.744	4.73	2.40
0.05209	420.2	23.8	0.605	0.733	4.79	2.59
0.05166	416.7	24.0	0.617	0.737	4.85	2.78
0.05123	413.2	24.2	0.630	0.756	4.92	2.96
0.05081	409.8	24.4	0.643	0.790	4.98	3.14
0.05040	406.5	24.6	0.656	0.841	5.04	3.32
0.04999	403.2	24.8	0.670	0.916	5.11	3.51
0.04959	400.0	25.0	0.683	1.02	5.17	3.70

TABLE XIV (*Continued*)

Titanium Dioxide

eV	cm^{-1}	μm	$n(\parallel)$	$n(\perp)$	$k(\parallel)$	$k(\perp)$
0.04920	396.8	25.2	0.697	1.17	5.23	3.90
0.04881	393.7	25.4	0.711	1.39	5.29	4.09
0.04843	390.6	25.6	0.726	1.70	5.36	4.26
0.04805	387.6	25.8	0.740	2.15	5.42	4.33
0.04769	384.6	26.0	0.755	2.70	5.48	4.18
0.04732	381.7	26.2	0.771	3.18	5.55	3.67
0.04696	378.8	26.4	0.786	3.30	5.61	2.93
0.04661	375.9	26.6	0.802	3.02	5.67	2.26
0.04626	373.1	26.8	0.818	2.52	5.74	1.84
0.04592	370.4	27.0	0.834	1.95	5.80	1.66
0.04558	367.6	27.2	0.851	1.40	5.87	1.72
0.04525	365.0	27.4	0.867	0.968	5.93	1.95
0.04492	362.3	27.6	0.885	0.704	5.99	2.24
0.04460	359.7	27.8	0.902	0.553	6.06	2.51
0.04428	357.1	28.0	0.920	0.465	6.12	2.75
0.04397	354.6	28.2	0.938	0.411	6.19	2.95
0.04366	352.1	28.4	0.956	0.377	6.25	3.13
0.04335	349.7	28.6	0.975	0.356	6.32	3.30
0.04305	347.2	28.8	0.994	0.344	6.38	3.44
0.04275	344.8	29.0	1.01	0.336	6.45	3.58
0.04246	342.5	29.2	1.03	0.333	6.51	3.71
0.04217	340.1	29.4	1.05	0.332	6.58	3.83
0.04189	337.8	29.6	1.07	0.334	6.64	3.94
0.04160	335.6	29.8	1.09	0.336	6.71	4.05
0.04133	333.3	30.0	1.12	0.341	6.78	4.15
0.04105	331.1	30.2	1.14	0.346	6.84	4.25
0.04078	328.9	30.4	1.16	0.351	6.91	4.35
0.04052	326.8	30.6	1.18	0.358	6.98	4.45
0.04025	324.7	30.8	1.20	0.365	7.04	4.54
0.03999	322.6	31.0	1.23	0.372	7.11	4.63
0.03974	320.5	31.2	1.25	0.380	7.18	4.72
0.03948	318.5	31.4	1.28	0.389	7.25	4.81
0.03923	316.5	31.6	1.30	0.398	7.31	4.90
0.03899	314.5	31.8	1.33	0.407	7.38	4.98
0.03874	312.5	32.0	1.35	0.416	7.45	5.07
0.03850	310.6	32.2	1.38	0.426	7.52	5.16
0.03827	308.6	32.4	1.40	0.436	7.59	5.24
0.03803	396.7	32.6	1.43	0.446	7.66	5.33
0.03780	304.9	32.8	1.46	0.457	7.73	5.41
0.03757	303.0	33.0	1.48	0.468	7.79	5.49
0.03734	301.2	33.2	1.51	0.480	7.86	5.58
0.03712	299.4	33.4	1.54	0.491	7.93	5.66
0.03690	297.6	33.6	1.57	0.503	8.00	5.75
0.03668	295.9	33.8	1.60	0.516	8.08	5.83
0.03647	294.1	34.0	1.63	0.528	8.15	5.92

(*continued*)

TABLE XIV (*Continued*)

Titanium Dioxide

eV	cm^{-1}	μm	$n(\|)$	$n(\perp)$	$k(\|)$	$k(\perp)$
0.03625	292.4	34.2	1.66	0.541	8.22	6.00
0.03604	290.7	34.4	1.69	0.555	8.29	6.09
0.03583	289.0	34.6	1.73	0.569	8.36	6.17
0.03563	287.4	34.8	1.76	0.583	8.43	6.26
0.03542	285.7	35.0	1.79	0.598	8.50	6.34
0.03522	284.1	35.2	1.83	0.613	8.58	6.43
0.03502	282.5	35.4	1.86	0.628	8.65	6.52
0.03483	280.9	35.6	1.90	0.644	8.72	6.60
0.03463	279.3	35.8	1.93	0.661	8.79	6.69
0.03444	277.8	36.0	1.97	0.678	8.87	6.78
0.03425	276.2	36.2	2.01	0.695	8.94	6.87
0.03406	274.7	36.4	2.05	0.713	9.01	6.96
0.00120	9.71	1,030	12.85	9.15 [2]		

[a] References are indicated in brackets. While we use the notation $\|$, $\perp$, these are the same as e, o and c, a used by others.